THE
ELECTROMAGNETIC
FIELD

McGRAW-HILL BOOK COMPANY
New York
St. Louis
San Francisco
Düsseldorf
Johannesburg
Kuala Lumpur
London
Mexico
Montreal
New Delhi
Panama
Paris
São Paulo
Singapore
Sydney
Tokyo
Toronto

ALBERT SHADOWITZ
Professor of Physics
Fairleigh Dickinson University

The Electromagnetic Field

This book was set in Press Roman by Scripta Graphica.
The editors were Jack L. Farnsworth and Joan Stern;
the cover was designed by Barbara Ellwood;
the production supervisor was Judi Frey.
The drawings were done by Vantage Art, Inc.
The Maple Press Company was printer and binder.

Library of Congress Cataloging in Publication Data

Shadowitz, A.
The electromagnetic field.

1. Electromagnetic fields. I. Title.
QC665.E4S5 1975 537 74-5105
ISBN 0-07-056368-3

THE
ELECTROMAGNETIC
FIELD

1 2 3 4 5 6 7 8 9 0 M A M M 7 9 8 7 6 5 4

For EDITH

CONTENTS

PREFACE

This textbook in electricity and magnetism is intended for junior or senior students of physics or electrical engineering. In each chapter of the book every section is followed by several worked-out examples, all illustrating the practical application of the preceding theory. Each section concludes with a collection of 15 to 25 problems, pertinent only to the subject matter of that section. Answers are provided for the odd-numbered problems in the back of the book. There are more than 1200 problems in the text, strictly grouped by subject; most of them have been tested by my students.

It is not really possible to study the electromagnetic field without employing considerable mathematics. The examples and problems have a large number of equations; they are the tools of the trade and the student must learn to use them. But in the text I try to emphasize the physical significance. Long derivations are indented and marked by a bracket at the left to call attention to the fact that these sections can be skipped over lightly. Some derivations are presented as appendixes at the end of the chapter. Sometimes even the simplest approach requires more sophistication than the student possesses; one cannot really avoid the use of elliptic integrals in some parts of magnetostatics. In such cases I include the material but, again, with a bracket at the left; the instructor may then choose to skip it. Some entire sections, so indicated, may be omitted without destroying the continuity.

In Chapter 1, which serves as the mathematical foundation for the rest of the text and contains very little physics, I have indicated where a jump may be made to and from these sections in succeeding chapters which are more physical. The grouping of the mathematics in one chapter, however, is a great convenience for reference.

I present electrostatics and magnetostatics by chapters in an interwoven manner rather than in two parallel strands. This permits me to compare and contrast the two disciplines while they are both still fresh in the student's mind. Corresponding equations are given similar numbers, preceded by the letter E or M, for easy reference. If the instructor prefers the customary sequence, there is no difficulty in making the order of chapters as follows: 2, 4, 6 (Electrostatics); 3, 5, 7 (Magnetostatics).

The mksa system is used throughout. This is strictly a matter of taste. Any choice, regardless of the reasons, is bound to displease someone.

There are seven appendixes placed at the ends of the first seven chapters. Some of these are derivations pertaining to the text of that particular chapter. Others are reference materials (e.g., the derivation of the Helmholtz theorem) which I prefer to present separately. One appendix is a lengthy discussion of the failure of Newton's third law for differential current elements. This material, in separate places, has long been known in the research literature but seems not to have filtered down to the textbooks. In addition to the appendixes at the ends of Chapters 1 through 7, there are four Reference Tables grouped together at the end of the book. These are simply collections of facts–physical constants, units, dimensions, factors for conversion to the cgs system, etc.

A recent trend in texts on electricity and magnetism is the inclusion of a chapter on special relativity. Indeed, so many years after Einstein's 1905 paper, On the Electrodynamics of Moving Bodies, it is time that relativity be included in books on classical electricity and magnetism. However, I feel that this material should be presented to the student only after he has been exposed to the description of the magnetic phenomena themselves and to the ordinary explanations. Accordingly, relativity appears here only in Chapters 13 and 14, long after the chapters on magnetostatics (3, 5, and 7). The chapters between 7 and 13 are concerned with other matters. Chapter 8 is devoted to three special methods for solving various problems in electrostatics. Chapter 9 deals with metallic conduction, and Chapter 10 is concerned with ferromagnetism–both in a phenomenological fashion. Chapter 11 discusses the basic phenomena associated with variations in time. Then Chapter 12 treats electric circuits.

The last five chapters in the book deal with waves of one kind or another. They are, respectively: plane waves, transmission lines, reflection and refraction, guided waves, and radiation. The importance of electromagnetic waves easily merits so much attention.

A classical subject is, by definition, old hat and far behind the frontiers of present-day research. Yet scores of articles appear each year on the properties of the electromagnetic field, and no less than two books have been written recently on this subject by Nobel Prize winners. Evidently there are still topics that deserve a careful, sophisticated approach. I have included a number of such recent discussions and have added a few of my own. Some examples follow.

Is it possible for a vector to be proportional to its curl? (Yes. But so far it has not been possible to make use of this.) Was Maxwell correct in elevating displacement current to an equal level with conventional current as a cause of the magnetic field? (No. We follow Feynman's lead in refusing even to give **D** a name.) Does the energy-flow interpretation of the Poynting vector apply to static fields? (Yes. Pugh and Pugh have shown it is absolutely essential for the conservation of angular momentum of the entire system.) If the longitudinal **E** within a conductor is caused by the charges in the battery producing the current then why does not **E** fall off with distance instead of remaining constant? (**E** is only indirectly caused by the battery charges. A more immediate cause is the charge on the surface of the conductors. This charge may be looked at as due either to capacitance or to the transient waves.) Is a superconductor current a conduction current (there is a conductor) or a convection current (there is no battery)? (Either. Or neither. We give a mathematical criterion.) How do betatron electrons know whether the Kerst condition for **B** is satisfied in a region of space they do not traverse? (They do not know; neither do they care. They are more concerned with **A** than with **B**.) If **E** and **B** are the fundamental vectors, while **D** and **H** have only secondary importance, how can such a basic concept as energy flux involve **H** rather than **B**? (The entry of momentum density here, also not an insignificant concept, clarifies the picture.) Can two fields be the same everywhere except at one point? (Consider the point electric dipole and the point magnetic dipole.) Butler has shown that the expression for magnetic energy density in a wave must contain a minus in front of the usual formula. Can this be so? (Are the currents conduction or convection currents?)

I hope that this list is sufficient to show that the study of the electromagnetic field can be of interest not only to the student but to the teacher as well.

My own background is divided almost equally between physics and electronic engineering. Engineers and physicists are actually two different kinds of people, with quite different approaches not only to textbooks, but to life. In writing this book I have tried to appeal to the practicality of the engineer, but I have also attempted to stress the significance of the various physical concepts. Now, the attempt to reconcile two somewhat conflicting points of view can be a risky business: instead of making sense to both sides, one can wind up appealing to neither. It is a risk I have had to take, because that is the way I think.

I am indebted to the American Journal of Physics, which, over the past decade or so, has published several hundred articles devoted to the pedagogic side of electricity and magnetism. Many of these have been invaluable in showing me aspects of the subject that were of pressing interest to others. I would like to acknowledge an informative discussion with my colleague, Ralph Hautau, on the physical significance of the vector potential. I also thank an unknown reviewer, kindly furnished by the publishers, for his very careful perusal of the manuscript and for the many, many excellent suggestions he made, most of which I adopted.

ALBERT SHADOWITZ

THE ELECTROMAGNETIC FIELD

1

THE DEL OPERATOR

This chapter does not really deal with electricity and magnetism: it is needed as the mathematical foundation for the vector treatment of this subject. If this chapter is going to be taught as part of the course it is probably wise to split the treatment of this chapter into two parts, alternating them with parts of the following chapters. In this way the physics of the situation is brought to the attention of the student a bit sooner. The following is one suggested order of procedure.

1	Chap. 1, Sec. 1-4:	the gradient and the divergence
2	Chap. 2:	the electric field and its divergence
3	Chap. 1, Sec. 5-8:	the curl and the laplacian
4	Chap. 4:	potential and the curl of the electric field

The laplacian, dealt with in the last section of Chap. 1, is not really utilized until Chap. 8, which is largely devoted to several special methods for the solution of Laplace's equation. But it is not inappropriate to include this last section with the others. This entire chapter than becomes useful for future reference.

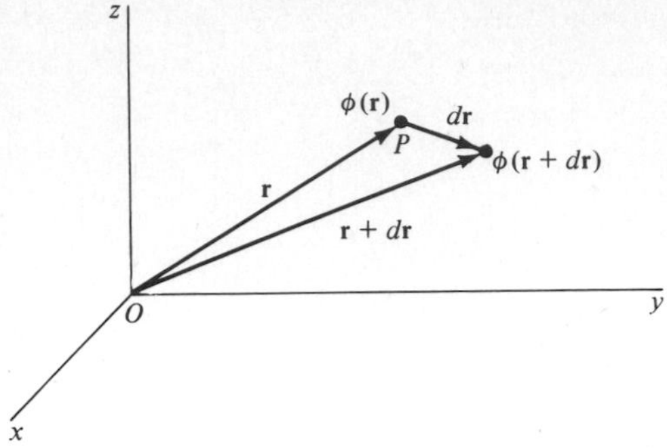

FIGURE 1-1

1-1 THE GRADIENT

Suppose a scalar quantity ϕ, such as the relative humidity of the air, is given as a function of each point in some region: $\phi(\mathbf{r}) = \phi(x,y,z)$. At one point, P, this function has the value $\phi(x,y,z)$; at another point $P + dP$, removed from P by the differential distance $d\mathbf{r}$ (see Fig. 1-1), the value of the function is given by

$$\phi(x + dx, y + dy, z + dz) = \phi(x,y,z) + \left[\left(\frac{\partial\phi}{\partial x}\right)_P dx + \left(\frac{\partial\phi}{\partial y}\right)_P dy + \left(\frac{\partial\phi}{\partial z}\right)_P dz\right] + \cdots$$

Neglecting higher order differentials, the increment in the scalar function between the two points P and $P + dP$ is given by

$$\begin{aligned} d\phi &= \phi(x + dx, y + dy, z + dz) - \phi(x,y,z) \\ &= \left(\frac{\partial\phi}{\partial x}\right) dx + \left(\frac{\partial\phi}{\partial y}\right) dy + \left(\frac{\partial\phi}{\partial z}\right) dz \\ &= \left(\hat{\mathbf{x}}\frac{\partial\phi}{\partial x} + \hat{\mathbf{y}}\frac{\partial\phi}{\partial y} + \hat{\mathbf{z}}\frac{\partial\phi}{\partial z}\right) \cdot \left(\hat{\mathbf{x}}\, dx + \hat{\mathbf{y}}\, dy + \hat{\mathbf{z}}\, dz\right) \end{aligned}$$

Here $\hat{\mathbf{x}}$ is a unit vector in the direction of the vector x, i.e., $\mathbf{x} = \hat{\mathbf{x}}x$; similarly for $\hat{\mathbf{y}}$ and $\hat{\mathbf{z}}$. The second bracket on the right in the last form is $d\mathbf{r}$. In the first bracket the subscript P is implicit on the derivatives: the derivatives are to be evaluated as numbers, representing the values of the derivatives at P. If the first bracket is written as $\nabla\phi$ (del phi), this equation becomes

$$\boxed{d\phi = (\nabla\phi) \cdot d\mathbf{r}}$$

$\nabla\phi$, defined by this equation, is called the gradient of ϕ and is applicable in any coordinate system, cartesian or not; it is sometimes written grad ϕ. Note that $\nabla\phi$ is completely different from $\Delta\phi$ (delta phi), which is a finite increment in ϕ between two points which are separated by a finite distance.

As seen from the equation for $d\phi$ above, the value of $\nabla\phi$ in the cartesian coordinate system is

$$\nabla\phi = \hat{\mathbf{x}}\frac{\partial\phi}{\partial x} + \hat{\mathbf{y}}\frac{\partial\phi}{\partial y} + \hat{\mathbf{z}}\frac{\partial\phi}{\partial z}$$

Appendix 1 gives the expressions for $\nabla\phi$ in the spherical and cylindrical coordinate systems. In any system $\nabla\phi$ is a vector function obtained when del operates on the scalar function ϕ. The del operator is an operator, not a function; it has no meaning by itself.

So $d\phi = (\nabla\phi) \cdot d\mathbf{r} = |\nabla\phi|\, dr \cos\theta$, where θ is the angle between $\nabla\phi$ and $d\mathbf{r}$. When $d\mathbf{r}$ is a vector which lies along the direction of $\nabla\phi$ the resultant change in ϕ will have its maximum value: $|\nabla\phi|\, dr$. $\nabla\phi$ therefore acts like a vector which points in the direction of the maximum rate of change of ϕ; and the sense of the arrow representing $\nabla\phi$ is such that the arrow points in the direction of increasing ϕ. The magnitude of this vector is

$$|\nabla\phi| = \sqrt{\left(\frac{\partial\phi}{\partial x}\right)^2 + \left(\frac{\partial\phi}{\partial y}\right)^2 + \left(\frac{\partial\phi}{\partial z}\right)^2}$$

The vector $-\nabla\phi$ has the same magnitude as $\nabla\phi$ but points in the direction directly opposite to that of the maximum rate of increase.

As an example of a gradient, we may consider a portion of a hill. The height above sea level h at any point on the hill is some function of position: $h = h(x,y)$. A two-dimensional simplified version of the gradient,

$$\nabla h = \hat{\mathbf{x}}\frac{\partial h}{\partial x} + \hat{\mathbf{y}}\frac{\partial h}{\partial y}$$

gives the magnitude and direction of the maximum rate of change of height at a given point, as in Fig. 1-2. This is a topographical two-dimensional map using the height as a parameter. The magnitude of ∇h, here

$$|\nabla h| = \sqrt{\left(\frac{\partial h}{\partial x}\right)^2 + \left(\frac{\partial h}{\partial y}\right)^2}$$

gives the rate of change of the height with respect to a horizontal displacement. The direction of ∇h is that in which this rate of change of height is a maximum, going toward higher altitude. For a horizontal displacement $d\mathbf{r} = \hat{\mathbf{x}}\, dx + \hat{\mathbf{y}}\, dy$ whose direction is not along the direction of ∇h (here ∇h is in the xy plane) the change of height will then be $dh = |\nabla h|\, dr \cos\theta$, where θ is the angle between ∇h and $d\mathbf{r}$.

The directional derivative of the height in a particular direction is defined as

$$\frac{dh}{dr}$$

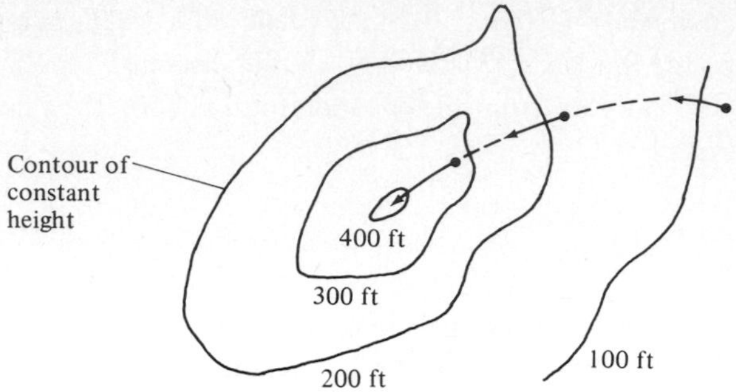

Arrows show magnitude and direction of the two-dimensional gradient of the height. The curve to which the arrows are tangent is orthogonal to the constant-height contours. The magnitude of the arrows along such a curve does not remain constant from one point to another.

FIGURE 1-2

and is equal to $|\nabla h| \cos\theta$. This is less than the maximum rate of change in height, obtained for the same dr when $d\mathbf{r}$ is along the direction of ∇h. For $\theta = 180°$ the height will decrease at a maximum rate, the horizontal displacement being opposite in direction to that of the gradient. The height h in this two-dimensional analog of the gradient is to be distinguished from the dimension z of the ordinary three-dimensional gradient.

The directional derivative of a function is thus the rate of change of that function in some particular direction. If the directional derivative is positive in some particular direction, then the function is increasing in that direction, at that point. The gradient of a function at a point is the directional derivative in that direction for which the rate of change is maximum.

A scalar function whose gradient is zero at a point is a function which has either a maximum, a minimum, a max-min (saddle point), or an inflection there. If the gradient is zero everywhere then the scalar function is a constant; when the value of this constant is arbitrary it is often convenient to set it equal to zero. For instance, we may take $\nabla h = 0$ on the surface of the oceans. Then sea level can be taken to be at zero height, the datum level from which other heights are measured.

Examples

1. The gradient of ϕ, where

$$\phi = 17x - \frac{2xy}{z} + y^2 z^3$$

$$\nabla\phi = \left(\hat{\mathbf{x}}\frac{\partial}{\partial x} + \hat{\mathbf{y}}\frac{\partial}{\partial y} + \hat{\mathbf{z}}\frac{\partial}{\partial z}\right)\left(17x - \frac{2xy}{z} + y^2z^3\right)$$

$$= \hat{\mathbf{x}}\frac{\partial}{\partial x}\left(17x - \frac{2xy}{z} + y^2z^3\right) + \hat{\mathbf{y}}\frac{\partial}{\partial y}\left(17x - \frac{2xy}{z} + y^2z^3\right)$$

$$+ \hat{\mathbf{z}}\frac{\partial}{\partial z}\left(17x - \frac{2xy}{z} + y^2z^3\right)$$

$$= \hat{\mathbf{x}}\left(17 - \frac{2y}{z}\right) + \hat{\mathbf{y}}\left(\frac{-2x}{z} + 2yz^3\right) + \hat{\mathbf{z}}\left(\frac{2xy}{z^2} + 3y^2z^2\right)$$

2. The value of $\nabla\phi$ in Example 1 at the point (2, 0, -1).

$$\nabla\phi = \hat{\mathbf{x}}\left(17 - \frac{0}{-1}\right) + \hat{\mathbf{y}}\left(\frac{-4}{-1} + 0\right) + \hat{\mathbf{z}}\left(\frac{0}{1} + 0\right) = \hat{\mathbf{x}}17 + \hat{\mathbf{y}}4$$

The gradient of ϕ at this point lies in the xy plane. Several contours in the xy plane near (2, 0, -1) are shown in Fig. 1-3.

3. $\nabla\phi$ at (1,2,3) in the case above

$$\nabla\phi = \hat{\mathbf{x}}\left(17 - \frac{2 \times 2}{3}\right) + \hat{\mathbf{y}}\left(\frac{-2 \times 1}{3} + 2 \times 2 \times 3^3\right)$$

$$+ \hat{\mathbf{z}}\left(\frac{2 \times 1 \times 2}{3^2} + 3 \times 2^2 \times 3^2\right)$$

$$= \hat{\mathbf{x}}(17 - 1.33) + \hat{\mathbf{y}}(-0.67 + 108) + \hat{\mathbf{z}}(0.44 + 108)$$

$$= \hat{\mathbf{x}}15.67 + \hat{\mathbf{y}}107.33 + \hat{\mathbf{z}}108.44$$

To show this on a graph we would need a three-dimensional map with equipotential surfaces.

4. ∇r i.e., the gradient of a function $\nabla\phi$ when ϕ varies directly with r, the constant of proportionality being unity.

$$\nabla r = \left(\hat{\mathbf{x}}\frac{\partial}{\partial x} + \hat{\mathbf{y}}\frac{\partial}{\partial y} + \hat{\mathbf{z}}\frac{\partial}{\partial z}\right)\sqrt{x^2 + y^2 + z^2}$$

$$= \hat{\mathbf{x}}\frac{\partial}{\partial x}\sqrt{x^2 + y^2 + z^2} + \hat{\mathbf{y}}\frac{\partial}{\partial y}\sqrt{x^2 + y^2 + z^2} + \hat{\mathbf{z}}\frac{\partial}{\partial z}\sqrt{x^2 + y^2 + z^2}$$

$$= \hat{\mathbf{x}}\left(\frac{1}{2}\frac{1}{\sqrt{x^2 + y^2 + z^2}}2x\right) + \hat{\mathbf{y}}\left(\frac{1}{2}\frac{1}{\sqrt{x^2 + x^2 + z^2}}2y\right)$$

$$+ \hat{\mathbf{z}}\left(\frac{1}{2}\frac{1}{\sqrt{x^2 + y^2 + z^2}}2z\right)$$

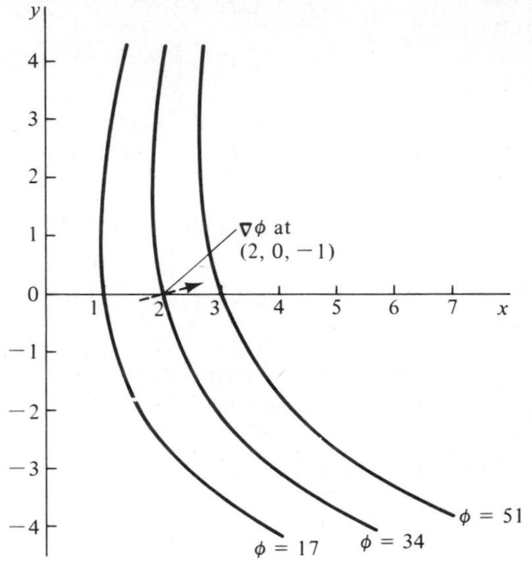

FIGURE 1-3 $\phi = 17x - \frac{2xy}{z} + y^2z^3$ in the $z = -1$ plane near $x = 2, y = 0$. At any point $\nabla\phi$ is orthogonal to ϕ = constant

$$= \frac{\hat{\mathbf{x}}x + \hat{\mathbf{y}}y + \hat{\mathbf{z}}z}{\sqrt{x^2 + y^2 + z^2}}$$

$$= \frac{\mathbf{r}}{r}$$

$$= \hat{\mathbf{r}}$$

The gradient of the function $\phi = r$ is a vector of unit magnitude, everywhere directed radially outward.

5. $\nabla(1/r)$.

$$\nabla\left(\frac{1}{r}\right) = \left(\hat{\mathbf{x}}\frac{\partial}{\partial x} + \hat{\mathbf{y}}\frac{\partial}{\partial y} + \hat{\mathbf{z}}\frac{\partial}{\partial z}\right)(x^2 + y^2 + z^2)^{-1/2}$$

$$= \hat{\mathbf{x}}\left(-\frac{1}{2}\right)(x^2 + y^2 + z^2)^{-3/2}(2x)$$

$$+ \hat{\mathbf{y}}\left(-\frac{1}{2}\right)(x^2 + y^2 + z^2)^{-3/2}(2y)$$

$$+ \hat{\mathbf{z}}\left(-\frac{1}{2}\right)(x^2 + y^2 + z^2)^{-3/2}(2z)$$

$$= -\frac{\hat{\mathbf{x}}x + \hat{\mathbf{y}}y + \hat{\mathbf{z}}z}{(x^2 + y^2 + z^2)^{3/2}}$$

$$= -\frac{\mathbf{r}}{r^3}$$

$$= -\frac{\hat{\mathbf{r}}}{r^2}$$

This gradient is directed radially inward and decreases in magnitude with increase of the distance from the origin. It is shown in Fig. 1-4. This is a very useful, widely used, result.

6. Given a surface $\phi(x,y,z) = \text{const}$. Then $\nabla\phi$ is a vector which is everywhere orthogonal to this surface.

Consider two points an infinitesimal distance apart. Then

$$d\phi = \frac{\partial\phi}{\partial x}dx + \frac{\partial\phi}{\partial y}dy + \frac{\partial\phi}{\partial z}dz$$

If both points are on the surface, $\phi = \text{const}$ and $d\phi = (\nabla\phi) \cdot d\mathbf{r} = 0$. So: either $\nabla\phi = 0$, or $d\mathbf{r} = 0$, or $\nabla\phi \perp d\mathbf{r}$. Unless ϕ is a constant *everywhere* the first possibility is excluded; the second case is excluded by definition; this leaves $\nabla\phi \perp d\mathbf{r}$. But $d\mathbf{r}$ is in the surface if both points are shown on the surface. So $\nabla\phi$ is everywhere perpendicular to the plane which is tangent to the surface at any point.

7. The unit normal to a surface $\phi(x,y,z) = \text{const}$.

$\nabla\phi$ is a vector normal to the surface, with magnitude $|\nabla\phi|$. So $\nabla\phi/|\nabla\phi|$ is normal to the surface and has unit magnitude.

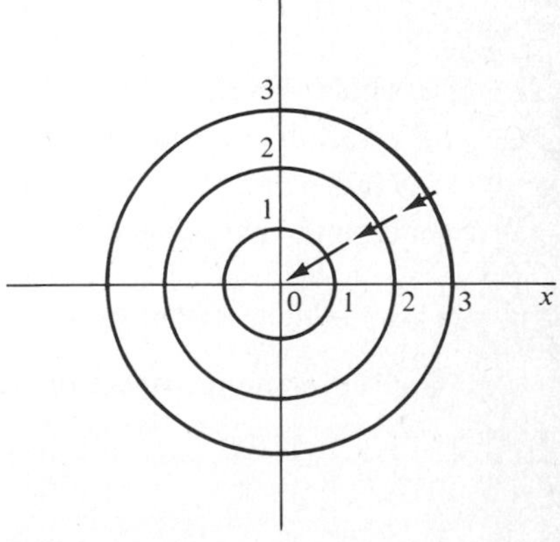

FIGURE 1-4

8. The unit normal to the surface $\phi = 3x + 2xyz - z^2 - 2$ at (0,1,2).

$$\nabla\phi = \hat{\mathbf{x}}(3 + 2yz) + \hat{\mathbf{y}}(2xz) + \hat{\mathbf{z}}(2xy - 2z)$$

At (0,1,2) we have

$$\nabla\phi = \hat{\mathbf{x}}7 - \hat{\mathbf{z}}4$$

So

$$|\nabla\phi| = \sqrt{(7)^2 + (-4)^2} = \sqrt{65}$$

Then

$$\hat{\mathbf{n}} = \frac{\nabla\phi}{|\nabla\phi|} = \hat{\mathbf{x}}\frac{7}{\sqrt{65}} - \hat{\mathbf{z}}\frac{4}{\sqrt{65}}$$

9. $\nabla(1/R)$, considering the point (x_s,y_s,z_s) in Fig. 1-5 as fixed while the point (x,y,z) is variable.

$$\mathbf{R} = \hat{\mathbf{x}}(x - x_s) + \hat{\mathbf{y}}(y - y_s) + \hat{\mathbf{z}}(z - z_s)$$

$$R = [(x - x_s)^2 + (y - y_s)^2 + (z - z_s)^2]^{1/2}$$

$$\begin{aligned}\nabla\left(\frac{1}{R}\right) &= \hat{\mathbf{x}}\frac{\partial}{\partial x}[(x - x_s)^2 + (y - y_s)^2 + (z - z_s)^2]^{-1/2} \\ &\quad + \hat{\mathbf{y}}\frac{\partial}{\partial y}[(x - x_s)^2 + (y - y_s)^2 + (z - z_s)^2]^{-1/2} \\ &\quad + \hat{\mathbf{z}}\frac{\partial}{\partial z}[(x - x_s)^2 + (y - y_s)^2 + (z - z_s)^2]^{-1/2} \\ &= \hat{\mathbf{x}}\left(-\frac{1}{2}\right)[(x - x_s)^2 + (y - y_s)^2 + (z - z_s)^2]^{-3/2}2(x - x_s)(1) \\ &\quad + \hat{\mathbf{y}}\left(-\frac{1}{2}\right)[(x - x_s)^2 + (y - y_s)^2 + (z - z_s)^2]^{-3/2}2(y - y_s)(1) \\ &\quad + \hat{\mathbf{z}}\left(-\frac{1}{2}\right)[(x - x_s)^2 + (y - y_s)^2 + (z - z_s)^2]^{-3/2}2(z - z_s)(1) \\ &= -\frac{\hat{\mathbf{x}}(x - x_s) + \hat{\mathbf{y}}(y - y_s) + \hat{\mathbf{z}}(z - z_s)}{R^3} = -\frac{\mathbf{R}}{R^3} = -\frac{\hat{\mathbf{R}}}{R^2}\end{aligned}$$

This is a similar result to that obtained in Example 5. That is, let $(x_s,y_s,z_s) \rightarrow (0,0,0)$; then the result here is identically the same as the previous one. This shows that the gradient is independent of the particular origin chosen. The slope at a point on a hill does not depend on the particular origin that a surveyor selects for his measurements.

10. $\nabla_s(1/R)$ in the previous example.

Here ∇_s is the del operator with $\mathbf{r}_s$ considered as the variable and $\mathbf{r}$ considered as fixed:

$$\nabla_s = \hat{\mathbf{x}}\frac{\partial}{\partial x_s} + \hat{\mathbf{y}}\frac{\partial}{\partial y_s} + \hat{\mathbf{z}}\frac{\partial}{\partial z_s}$$

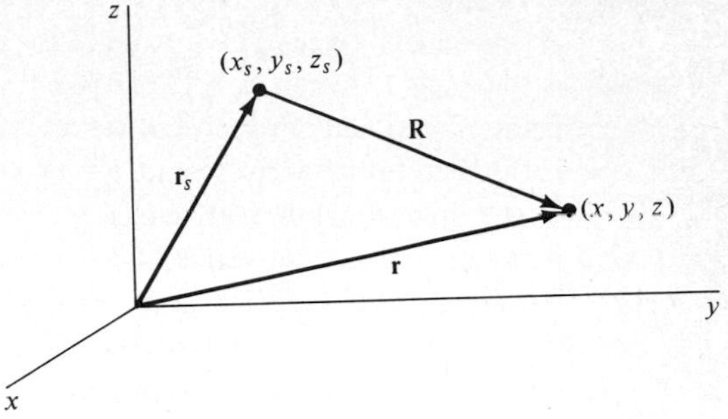

FIGURE 1-5

The unit vectors are the same as the previous ones.

$$\nabla_s\left(\frac{1}{R}\right) = \hat{\mathbf{x}}\frac{\partial}{\partial x_s}\,[(x - x_s)^2 + (y - y_s)^2 + (z - z_s)^2]^{-1/2}$$
$$+ \hat{\mathbf{y}}\frac{\partial}{\partial y_s}\,[(x - x_s)^2 + (y - y_s)^2 + (z - z_s)^2]^{1/2}$$
$$+ \hat{\mathbf{z}}\frac{\partial}{\partial z_s}\,[(x - x_s)^2 + (y - y_s)^2 + (z - z_s)^2]^{-1/2}$$
$$= \hat{\mathbf{x}}\left(-\frac{1}{2}\right)[(x - x_s)^2 + (y - y_s)^2 + (z - z_s)^2]^{-3/2}\,2(x - x_s)(-1)$$
$$+ \hat{\mathbf{y}}\left(-\frac{1}{2}\right)[(x - x_s)^2 + (y - y_s)^2 + (z - z_s)^2]^{-3/2}\,2(y - y_s)(-1)$$
$$+ \hat{\mathbf{z}}\left(-\frac{1}{2}\right)[(x - x_s)^2 + (y - y_s)^2 + (z - z_s)^2]^{-3/2}\,2(z - z_s)(-1)$$
$$= \frac{\hat{\mathbf{x}}(x - x_s) + \hat{\mathbf{y}}(y - y_s) + \hat{\mathbf{z}}(z - z_s)}{R^3} = \frac{\mathbf{R}}{R^3} = \frac{\hat{\mathbf{R}}}{R^2}$$

This result is the negative of the previous case; this result and the previous one are widely used.

PROBLEMS

1 If $\phi = x/y$, find $\nabla\phi$.
2 Plot $\phi = x/y$ for $\phi = \frac{1}{2}$, 1, and 2. On the $\phi = 1$ curve show $\nabla\phi$, by arrows, at $x = -2$, -1, 1, and 2.
3 If $\phi = xy$, find $\nabla\phi$.
4 Plot $\phi = xy$ for $\phi = 1$ and $\phi = 2$. On the $\phi = 1$ curve show $\nabla\phi$, by arrows, at $x = \frac{1}{2}$, 1, and 2.
5 If $\phi = e^{-x} \sin y$, find $\nabla\phi$.

6 If $\phi = \sinh z$, find $\nabla\phi$.

7 What is the angle between $(\hat{\mathbf{x}} - \hat{\mathbf{y}})$ and $(2\hat{\mathbf{x}} + \hat{\mathbf{y}})$?

8 What is the angle between $(\hat{\mathbf{x}} + \hat{\mathbf{y}})$ and $(2\hat{\mathbf{x}} - \hat{\mathbf{y}})$?

9 What is the directional derivative of $\phi = xy$ at (1,1) in the direction of $\hat{\mathbf{x}}$?

10 What is the directional derivative of $\phi = xy$ at (1,1) in the direction of $(\hat{\mathbf{x}} - \hat{\mathbf{y}})$?

11 What is the directional derivative of $\phi = x^2 + y^2$ at (2,0) in the direction of $(\hat{\mathbf{x}} + \hat{\mathbf{y}})$?

12 What is the directional derivative of $\phi = x^2 + y^2$ at (2,0) in the direction of $\hat{\mathbf{y}}$?

13 Let $x \to X = x \cos\theta + y \sin\theta$ and $y \to Y = -x \sin\theta + y \cos\theta$ while $z \to Z = z$. This represents a counterclockwise rotation of the axes through θ in the xy plane to give a new set of coordinates. Using the chain rule of calculus, for $\phi|X,Y,Z|x,y,z$, find the expression to which $\hat{\mathbf{x}}(\partial\phi/\partial x) + \hat{\mathbf{y}}(\partial\phi/\partial y) + \hat{\mathbf{z}}(\partial\phi/\partial z)$ transforms under the rotation. Is the gradient form-invariant under rotation?

14 $d\phi = (\nabla\phi) \cdot d\mathbf{r}$. Can this be written $d\mathbf{r} \cdot (\nabla\phi)$? Can it be written $(d\mathbf{r} \cdot \nabla)\phi$? Can it be written $(\nabla \cdot d\mathbf{r})\phi$?

15 (*a*) Repeat Prob. 13 with the expression $\hat{\mathbf{x}}3(\partial\phi/\partial x) + \hat{\mathbf{y}}(\partial\phi/\partial y) + \hat{\mathbf{z}}(\partial\phi/\partial z)$. To what expression does this transform?

(*b*) Is this form-invariant under rotation?

(*c*) Under translation of the origin: $x \to X + a$, $y \to Y + b$, $z \to Z + c$?

(*d*) Does the magnitude remain invariant after the rotation, even if the form is not invariant?

(*e*) What significance is there to a function (as distinguished from a quantity consisting of scalar components) which looks like a vector (i.e., it has components) but which is not form-invariant under either rotation or translation?

16 The rotation of a rigid body with angular velocity ω about the z axis is given by $V_x = -\omega y$, $V_y = \omega x$, $V_z = 0$. Is there a ϕ such that $V = \nabla\phi$? See Fig. 1-6*a*.

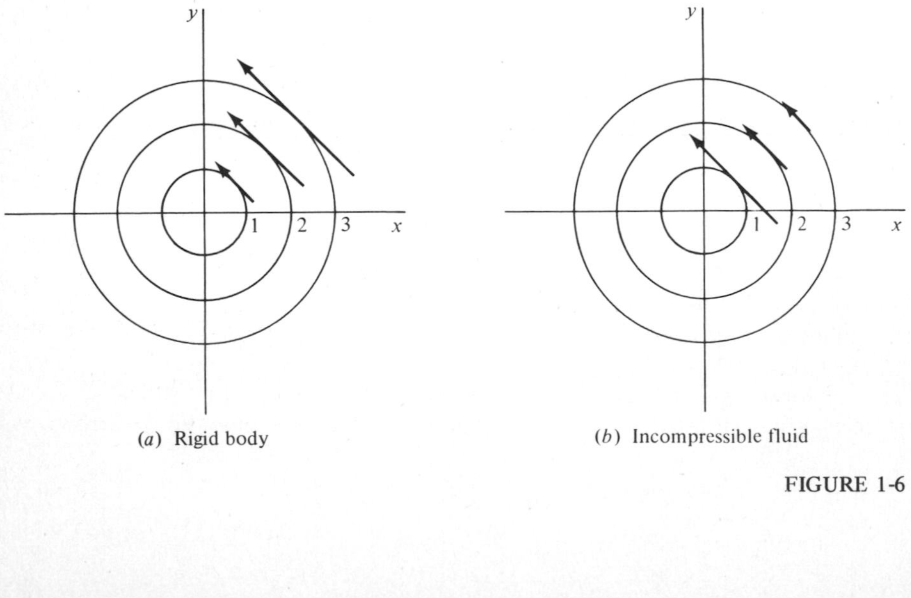

(*a*) Rigid body (*b*) Incompressible fluid

FIGURE 1-6

17 What are the necessary and sufficient conditions of integrability such that for a given **F**, it is possible to find a ϕ that satisfies $\mathbf{F} = \nabla\phi$?

18 The rotation of an incompressible fluid produced by the rotation of a cylinder of radius a about the z, axis with angular velocity ω is given by $V_x = -\omega a^2 y/(x^2 + y^2)$, $V_y = \omega a^2 x/(x^2 + y^2)$, $V_z = 0$. Is there a ϕ such that $\mathbf{V} = \nabla\phi$? See Fig. 1-6*b*.

1-2 FLUX

A closed surface is a boundary surface which divides a volume into two parts: an inside and an outside. The surface itself is unbounded–no curve in the surface acts as an edge. An elemental area of the closed surface is represented by $d\mathbf{S}$, a vector of magnitude dS which points in the direction from the inside volume toward the outside volume; the actual area of magnitude dS is orthogonal to the vector $d\mathbf{S}$ which represents it. The surface of a sphere is a closed surface; there, at all points, $d\mathbf{S}$ points radially outward. Of the two possible directions for $d\mathbf{S}$, the universal convention is to take the direction going from inside to outside as the positive one.

An open surface is one in which the surface is bounded by a curve. The page of a book is an open surface, and the edge of the page is the bounding curve. For an open surface $d\mathbf{S}$ again has the magnitude dS; and the vector $d\mathbf{S}$ is orthogonal to the actual area which it represents. Here, also, there are two possibilities for the direction of $d\mathbf{S}$ but this time no definition of the positive sense is possible: there are two directions, one the negative of the other. The direction which is chosen as positive is, however, related to the positive sense of traversing the perimeter (the bounding curve C) by the following convention: if a right-hand screw is turned in such a way as to follow in general the positive sense of the perimeter, then the screw will advance in the direction of the positive normal to the surface. If the positive sense of traversal of the perimeter of a page is taken as counterclockwise (CCW) then the normal to the page is up; if the positive sense of traversal of the edge is clockwise (CW) then the normal of the page is down, as is the vector representing its area. See Fig. 1-7.

It is interesting to note in passing that it is not possible to give rigorous definitions–without the use of examples–of the words right, left, clockwise, or counterclockwise. One cannot, e.g., make any of these concepts perfectly clear to, say, an intelligent Martian except by an actual comparison with a right shoe, or with the face of a clock, or with a view of the earth revolving about the sun *against the background* constellations, or with a cobalt 60 nucleus–in which beta decay occurs in a preferential manner, such that the south end emits more electrons than does the north end. The reader is referred to Martin Gardner's book, *The Ambidextrous Universe*,[1] for further details.

[1] Published by the New American Library, New York, 1969.

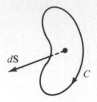

FIGURE 1-7

The flux of a vector field **F** is defined for an open surface Σ by

$$\int_{\Sigma} \mathbf{F} \cdot d\mathbf{S}$$

One integral sign is used for simplicity: it actually represents a double integration when performed, using coordinate variables. The flux for this case has two values, one the negative of the other, depending on which normal is taken as positive. For a closed surface Σ the flux of a vector **F** is uniquely defined:

$$\oint_{\Sigma} \mathbf{F} \cdot d\mathbf{S}$$

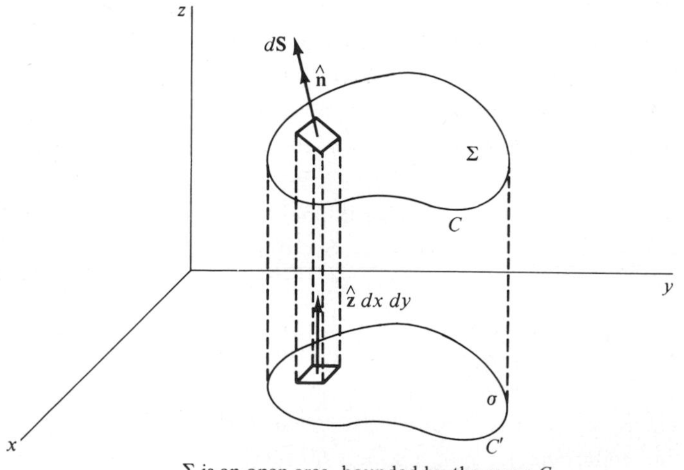

Σ is an open area, bounded by the curve C
σ is an open area, bounded by the curve C', and is the projection of Σ on the xy plane
C' is the projection of C on the xy plane

FIGURE 1-8

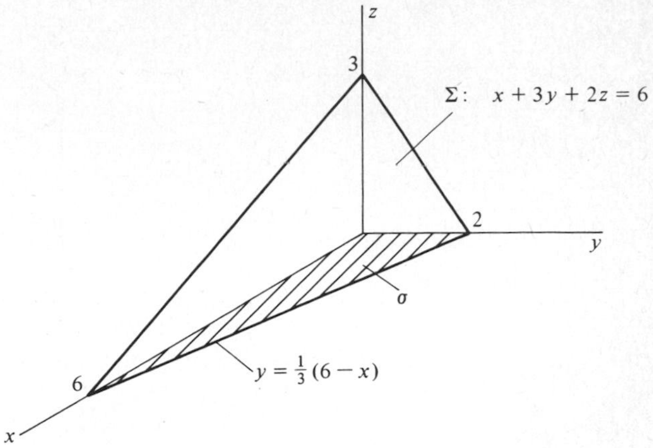

FIGURE 1-9

Again the single integral sign represents a double integration: x and y, or r and θ, or any two appropriate variables. The small circle on the integral sign indicates that Σ is a closed surface and that the integration is to be performed over the whole surface.

To evaluate the flux of a vector through a curved surface (or even through a plane surface, when this is not parallel to a coordinate plane) one may use a theorem which relates a surface integral over an open surface Σ to an integration over the area σ which is projected by Σ on a coordinate plane. For example, let the projected area be on the xy plane, as in Fig. 1-8. Let $(\hat{\mathbf{n}},\hat{\mathbf{z}})$ represent the angle between $\hat{\mathbf{n}}$, the unit vector which is normal to the plane, and $\hat{\mathbf{z}}$, the unit vector along the z axis. Cos $(\hat{\mathbf{n}},\hat{\mathbf{z}})$ is the cosine of this angle. There $dS \cos(\hat{\mathbf{n}},\hat{\mathbf{z}}) = dx\,dy$. The theorem states that

$$\int_\Sigma \mathbf{F}\cdot d\mathbf{S} = \int_\Sigma \mathbf{F}\cdot\hat{\mathbf{n}}\,dS = \iint_\sigma \mathbf{F}\cdot\hat{\mathbf{n}}\,\frac{dx\,dy}{|\cos(\hat{\mathbf{n}},\hat{\mathbf{z}})|}$$

If the projection had been taken on the yz plane the result would have been $\iint_\sigma \mathbf{F}\cdot\hat{\mathbf{n}}(dy\,dz/|\cos(\hat{\mathbf{n}},\hat{\mathbf{x}})|)$.

Examples

1. The flux of $\mathbf{F} = \hat{\mathbf{x}}2 - \hat{\mathbf{y}}3x + \hat{\mathbf{z}}y$ through Σ where Σ is the plane surface $x + 3y + 2z = 6$ in the first octant. The surface is shown in Fig. 1-9.

Let $f = x + 3y + 2z - 6$. Then $\nabla f = \hat{\mathbf{x}} + \hat{\mathbf{y}}3 + \hat{\mathbf{z}}2$ is a vector normal to this plane (as is its negative). $|\nabla f| = \sqrt{1+9+4} = \sqrt{14}$. One unit normal is given by

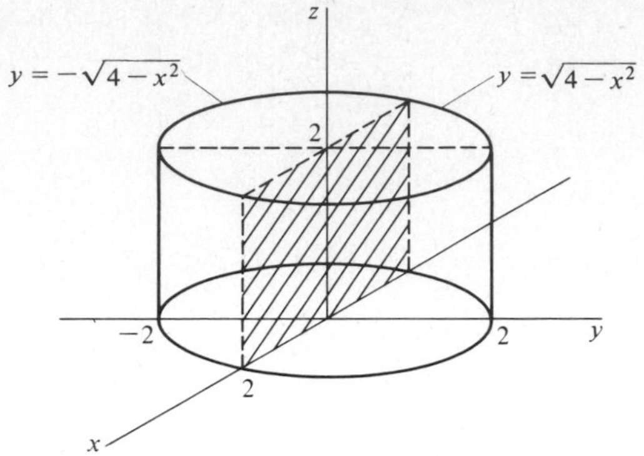

FIGURE 1-10

$$\hat{\mathbf{n}} = \hat{\mathbf{x}}\,\frac{1}{\sqrt{14}} + \hat{\mathbf{y}}\,\frac{3}{\sqrt{14}} + \hat{\mathbf{z}}\,\frac{2}{\sqrt{14}}$$

Then $\hat{\mathbf{n}} \cdot \hat{\mathbf{z}} = 2/\sqrt{14}$. But $\hat{\mathbf{n}} \cdot \hat{\mathbf{z}} = (1)(1)\cos(\hat{\mathbf{n}} \cdot \hat{\mathbf{z}})$. So $\cos(\hat{\mathbf{n}},\hat{\mathbf{z}}) = 2/\sqrt{14}$. Also, $\mathbf{F} \cdot \hat{\mathbf{n}} = 2/\sqrt{14} - 9x/\sqrt{14} + 2y/\sqrt{14}$. Then the flux is

$$\Phi = \iint_\sigma \left(\frac{2}{\sqrt{14}} - \frac{9x}{\sqrt{14}} + \frac{2y}{\sqrt{14}}\right)\left(\frac{\sqrt{14}}{2}\,dx\,dy\right)$$

The plane $x + 3y + 2z = 6$ has the intercept $x + 3y = 6$ in the xy plane (put $z = 0$ in the equation of the plane). Then the integration over σ may be taken with y varying from 0 to $(6 - x)/3$ and with x varying from 0 to 6:

$$\Phi = \int_0^6 dx \int_0^{(6-x)/3} \left(1 - \frac{9}{2}\,x + y\right) dy = \int_0^6 dx\left(4 - 10x + \frac{14}{9}\,x^2\right) = -44$$

The normal selected as positive was the one up and out, toward the reader. The negative sign for the flux shows that the net component of $\mathbf{F}$ through Σ is actually down and left through the plane.

2. The flux of $\mathbf{F} = \hat{\mathbf{x}}z + \hat{\mathbf{y}}x - \hat{\mathbf{z}}3y^2z$ out of Σ, where Σ is a closed surface consisting of (1) the cylinder $x^2 + y^2 = 4$ between $z = 0$ and $z = +2$; (2) the circular area with radius $r = 2$ which bounds the cylinder at $z = 0$; (3) the corresponding circular area at $z = +2$. The surface is shown in Fig. 1-10.

The surface consists of three parts as follows: top, bottom, and side. Over the top surface $d\mathbf{S} = \hat{\mathbf{z}}\,dS$ so $\mathbf{F} \cdot d\mathbf{S} = -3y^2z\,dx\,dy$ and

$$\int_{\text{top}} \mathbf{F}\cdot d\mathbf{S} = \int_{-2}^{2} -3z\,dx \int_{-\sqrt{4-x^2}}^{\sqrt{4-x^2}} y^2\,dy$$

$$= -3z \int_{-2}^{2} dx \left[\frac{2}{3}(4-x^2)^{3/2}\right]$$

$$= -2(2) \int_{-2}^{2} (4-x^2)^{3/2}\,dx = -4(6\pi) = -24\pi$$

Over the bottom surface $d\mathbf{S} = -\hat{\mathbf{z}}\,dS$ so $\mathbf{F}\cdot d\mathbf{S} = 3y^2 z\,dx\,dy$ and

$$\int_{\text{bottom}} \mathbf{F}\cdot d\mathbf{S} = \int_{-2}^{2} 3z\,dx \int_{-\sqrt{4-x^2}}^{\sqrt{4-x^2}} y^2\,dy = 3z \int_{-2}^{2} dx\left[\frac{2}{3}(4-x^2)^{3/2}\right]$$

But here $z = 0$ so the value of this contribution to the flux is zero. Over the curved side one can use a projection on a coordinate plane; here it is not possible to employ the xy plane. Take the xz plane, as shown in Fig. 1-10.

$$\int_{\text{side}} \mathbf{F}\cdot d\mathbf{S} = \int_{\text{side}} \mathbf{F}\cdot\hat{\mathbf{n}}\,dS = \int_{\sigma} \mathbf{F}\cdot\hat{\mathbf{n}}\,\frac{dz\,dx}{\cos(\hat{\mathbf{n}},\hat{\mathbf{y}})}$$

The gradient to the cylindrical surface, $\nabla(x^2+y^2-4) = \hat{\mathbf{x}}2x + \hat{\mathbf{y}}2y$, is normal to this surface. Its magnitude is $\sqrt{(2x)^2+(2y)^2} = 2\sqrt{x^2+y^2} = 2\sqrt{4} = 4$, so the unit normal to the side surface is $\hat{\mathbf{x}}x/2 + \hat{\mathbf{y}}y/2$. Then $\mathbf{F}\cdot\hat{\mathbf{n}} = (x/2)(y+z)$. From $\hat{\mathbf{n}}\cdot\hat{\mathbf{y}} = y/2$ and $\hat{\mathbf{n}}\cdot\hat{\mathbf{y}} = (1)(1)\cos(\hat{\mathbf{n}},\hat{\mathbf{y}})$ one obtains $\cos(\hat{\mathbf{n}},\hat{\mathbf{y}}) = y/2 = \pm\frac{1}{2}\sqrt{4-x^2}$. Therefore

$$\int_{\text{side}} \mathbf{F}\cdot d\mathbf{S} = \iint_{\sigma} \frac{x}{2}(y+z)\,\frac{dz\,dx}{y/2}$$

$$= \int_{0}^{2} dz \int_{-2}^{2} x\,dx + \int_{0}^{2} z\,dz \int_{-2}^{2} \frac{x\,dx}{\pm\sqrt{4-x^2}}$$

Instead of evaluating the flux through the side of the entire cylinder it is convenient to consider the four quadrants individually.

Out of the first quadrant the flux is

$$\int_0^2 dz \int_0^2 x\,dx + \int_0^2 z\,dz \int_0^2 \frac{x\,dx}{+\sqrt{4-x^2}} = 4 + 4 = 8$$

From the second quadrant the flux is

$$\int_0^2 dz \int_{-2}^0 x\,dx + \int_0^2 z\,dz \int_{-2}^0 \frac{x\,dx}{+\sqrt{4-x^2}} = -4 - 4 = -8$$

The third quadrant yields

$$\int_0^2 dz \int_{-2}^0 x\,dx + \int_0^2 z\,dz \int_{-2}^0 \frac{x\,dx}{-\sqrt{4-x^2}} = -4 + 4 = 0$$

The fourth quadrant gives, similarly,

$$\int_0^2 dz \int_0^2 x\,dx + \int_0^2 z\,dz \int_0^2 \frac{x\,dx}{-\sqrt{4-x^2}} = 4 - 4 = 0$$

The net flux through the entire curved side is zero. The total flux out of the entire volume is the same as that out of the top circular area: -24π. The minus sign shows the flux is actually inward.

PROBLEMS

1 Find the flux of $\mathbf{F} = \hat{\mathbf{x}}xy + \hat{\mathbf{y}}yz + \hat{\mathbf{z}}zx$ out of the unit cube in the first octant with one corner at the origin and edges parallel to the axes.

2 Repeat Prob. 1 for the unit cube $-1 \leqslant x \leqslant 0$, $-1 \leqslant y \leqslant 0$, $-1 \leqslant z \leqslant 0$.

3 In Example 1 take the unit normal of the opposite sign. Calculate the flux.

4 Calculate the flux in Example 1 using a projection on the zx plane.

5 Calculate the flux in Example 1 using a projection on the yz plane.

6 Suppose that, instead of evaluating $\oint_\Sigma \mathbf{F} \cdot \hat{\mathbf{n}}\, dS$, one wishes to evaluate the integral $\int_\Sigma f\hat{\mathbf{n}}\, dS$. This is also a surface integral. What does this integral become if the region of integration is σ, the area projected by Σ on a coordinate plane?

7 Find the flux of $\mathbf{F} = \hat{\mathbf{x}}x + \hat{\mathbf{y}}y + \hat{\mathbf{z}}z$ out of the unit cube in the first octant with one corner at the origin and edges parallel to the axes.

8 Find the flux of $\mathbf{F} = \hat{\mathbf{x}}x + \hat{\mathbf{y}}y + \hat{\mathbf{z}}z$ out of a sphere of unit radius with center at the origin.

9 Find the flux of $\mathbf{F} = \hat{\mathbf{x}}(x + y) + \hat{\mathbf{y}}(-x + y) + \hat{\mathbf{z}}(-2z)$ through the upper hemisphere of unit radius centered at the origin.

10 Repeat Prob. 9 for the lower hemisphere.

1-3 THE DIVERGENCE

The divergence of the vector function **F** at a point P may now be defined. This is given in terms of (1) the total outward flux of **F** through a closed surface Σ which surrounds P; and (2) the volume $\mathcal{V}$ inside Σ. Take a surface Σ_1 surrounding P and having an inside volume $\mathcal{V}_1$. Form the ratio

$$\frac{1}{\mathcal{V}_1} \oint_{\Sigma_1} \mathbf{F} \cdot d\mathbf{S}$$

which gives the flux per unit volume enclosed. Then take another surface Σ_2 which is everywhere closer than Σ_1 to the point P, and form the ratio

$$\frac{1}{\mathcal{V}_2} \oint_{\Sigma_2} \mathbf{F} \cdot d\mathbf{S}$$

When this process is continued indefinitely, then if the ratio approaches a unique limit this is called the divergence of **F** at P. (There are, however, exceptional cases, when, e.g., $\mathbf{F} = \hat{\mathbf{r}}/r^2$ no limit exists at $r = 0$.) Mathematically the expression for the divergence, which holds in any coordinate system, is

$$\boxed{\operatorname{div} \mathbf{F} = \lim_{\mathcal{V} \to 0} \left(\frac{1}{\mathcal{V}} \oint_{\Sigma} \mathbf{F} \cdot d\mathbf{S} \right)}$$

The divergence of **F** is the source strength of the flux of **F**, defined at a point. We will now derive a specific formula for the divergence of **F** when **F** is given in the cartesian coordinate system.

Consider a small but finite rectangular box with edges Δx, Δy, Δz, which has the point $P(x,y,z)$ at its center, as in Fig. 1-11. (Eventually the box will be made infinitesimal, with all sides approaching zero length.) Here $\oint_{\text{box}} \mathbf{F} \cdot d\mathbf{S}$ becomes the sum of six integrals: over the front, back, right, left, top, and bottom, respectively. Consider the front integral. $d\mathbf{S}$ for the front face is $\hat{\mathbf{x}}\, dy\, dz$. The value of **F** on the front face is obtained from its value, $\mathbf{F}(x,y,z)$, at the center of the box by means of a Taylor expansion. The only component of **F** which contributes to the front integral is the x component. Let $(F_x)_P$ stand for the x component of **F** evaluated at the center P. The word *small* applied to the box means that the zero- and first-order terms in the expansion are sufficient:

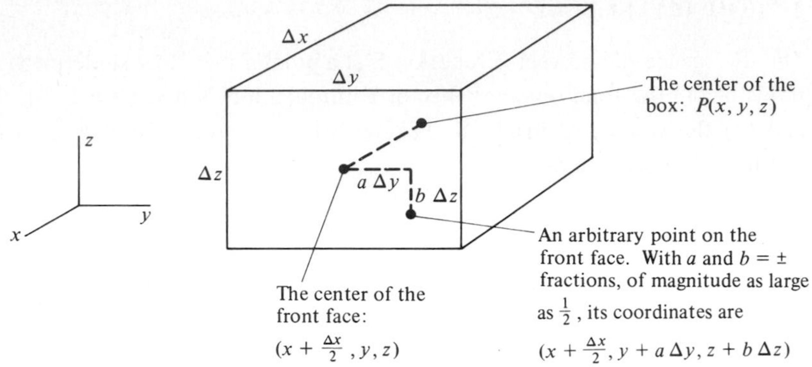

FIGURE 1-11

$$\int_{\text{front}} \mathbf{F}\cdot d\mathbf{S} = \int_{\text{front}} \left\{ (F_x)_P + \left[\left(\frac{\partial F_x}{\partial x}\right)_P \left(\frac{\Delta x}{2}\right) + \left(\frac{\partial F_x}{\partial y}\right)_P (a\,\Delta y) + \left(\frac{\partial F_x}{\partial z}\right)_P (b\,\Delta z) \right] + \cdots \right\} dy\,dz$$

The factors a and b arise from considering a point on the front face at an arbitrary position: $x + (\Delta x/2)$, $y + a\,\Delta y$, $z + b\,\Delta z$ with $-\frac{1}{2} \leqslant a \leqslant \frac{1}{2}$, $-\frac{1}{2} \leqslant b \leqslant \frac{1}{2}$. The derivatives, evaluated at P, are numerics and can be moved to the left of the integral sign. So the terms involving a and b are, when the integration is performed,

$$\left(\frac{\partial F_x}{\partial y}\right)_P a(\Delta y)^2\,\Delta z + \left(\frac{\partial F_x}{\partial z}\right)_P b(\Delta z)^2\,\Delta y$$

These terms, and all higher order terms, contain higher powers of Δx, Δy, Δz than does the first term. Therefore their contribution to the sum becomes vanishingly small compared to that of the first-order term when Δx, Δy, and $\Delta z \to 0$, and they may be neglected. For a small enough box, all points on the front face may thus be considered to be at the center of the front face. Since $(F_x)_P$ is also a number, not a function, it may also be moved to the left of the integral sign, and the front integral becomes

$$(F_x)\Delta y\,\Delta z + \frac{1}{2}\left(\frac{\partial F_x}{\partial x}\right)\Delta x\,\Delta y\,\Delta z$$

where it is implicit that the brackets be evaluated at P.

In the case of the second integral, the one over the rear face, $d\mathbf{S}_{\text{back}} = -\hat{\mathbf{x}}\, dy\, dz$; also,

$$\hat{\mathbf{x}}(F_{\text{back}})_x = \hat{\mathbf{x}} \left\{ (F_x) + \left[\left(\frac{\partial F_x}{\partial x}\right)_P \left(-\frac{\Delta x}{2}\right) + \left(\frac{\partial F_x}{\partial y}\right)_P (a\,\Delta y) + \left(\frac{\partial F_x}{\partial z}\right)_P (b\,\Delta z) \right] + \cdots \right\}$$

So the second integral becomes

$$-(F_x)\Delta y\,\Delta z + \frac{1}{2}\left(\frac{\partial F_x}{\partial x}\right)\Delta x\,\Delta y\,\Delta z$$

Accordingly, the sum of the first two integrals, over the front and back faces of the box, is

$$\left(\frac{\partial F_x}{\partial x}\right)\Delta x\,\Delta y\,\Delta z$$

By symmetry, the third and fourth integrals give

$$\left(\frac{\partial F_y}{\partial y}\right)\Delta x\,\Delta y\,\Delta z$$

Similarly, the last two integrals yield

$$\left(\frac{\partial F_z}{\partial z}\right)\Delta x\,\Delta y\,\Delta z$$

The total flux of $\mathbf{F}$ for the entire surface of the infinitesimal box is

$$\oint_{\text{small box}} \mathbf{F}\cdot d\mathbf{S} = \left[\left(\frac{\partial F_x}{\partial x}\right) + \left(\frac{\partial F_y}{\partial y}\right) + \left(\frac{\partial F_z}{\partial z}\right)\right]\Delta x\,\Delta y\,\Delta z$$

Since the volume of the finite box is $\mathcal{V} = \Delta x\,\Delta y\,\Delta z$,

$$\text{div}\,\mathbf{F} = \lim_{\mathcal{V}\to 0}\left[\frac{1}{\mathcal{V}}\oint_{\text{small box}} \mathbf{F}\cdot d\mathbf{S}\right] = \lim_{\mathcal{V}\to 0}\left[\frac{1}{\Delta x\,\Delta y\,\Delta z}\oint_{\text{small box}} \mathbf{F}\cdot d\mathbf{S}\right]$$

$$= \left(\frac{\partial F_x}{\partial x}\right) + \left(\frac{\partial F_y}{\partial y}\right) + \left(\frac{\partial F_z}{\partial z}\right)$$

$$= \left(\hat{\mathbf{x}}\frac{\partial}{\partial x} + \hat{\mathbf{y}}\frac{\partial}{\partial y} + \hat{\mathbf{z}}\frac{\partial}{\partial z}\right) \cdot \left(\hat{\mathbf{x}}F_x + \hat{\mathbf{y}}F_y + \hat{\mathbf{z}}F_z\right)$$

An explicit expression for div **F** in the cartesian system is, then, given by

$$\boxed{\text{div } \mathbf{F} = \nabla \cdot \mathbf{F} = \frac{\partial F_x}{\partial x} + \frac{\partial F_y}{\partial y} + \frac{\partial F_z}{\partial z}}$$

The subscripts on the derivatives are no longer needed since the box had degenerated to a point. In cartesian coordinates the del symbol has the same form, when used as a divergence operator, that it has when employed as a gradient operator. In noncartesian coordinates ∇ also has the same form, when used as a divergence operator, that it has when used as a gradient operator *provided* ∇ operates on the unit vectors themselves as well as on the different components of a vector. In the final expressions which are obtained, however, the appearance differs from one system to another. Even in a given system the appearance is different for the gradient, divergence, curl, and laplacian. Appendix 1, at the end of this chapter, gives the necessary details.

A vector whose divergence is everywhere zero is called a solenoidal vector. This nomenclature arises from the fact, shown later, that every magnetic field, including that created by a solenoid, has zero divergence. Used in the divergence, del operates on a vector function to give a scalar function; for the gradient, del operates on a scalar function to give a vector function.

Examples

1. The divergence of $\mathbf{F} = \hat{\mathbf{x}} + \hat{\mathbf{y}}y^2 + \hat{\mathbf{z}}xy$ at (1,2,3).

$$\nabla \cdot (\hat{\mathbf{x}} + \hat{\mathbf{y}}y^2 + \hat{\mathbf{z}}xy) = \frac{\partial}{\partial x}(1) + \frac{\partial}{\partial y}(y^2) + \frac{\partial}{\partial z}(xy) = 0 + 2y + 0 = 2y$$

At (1,2,3) this has the value 2(2) = 4. Since this is positive, the net flux through a small sphere surrounding (1,2,3) would be outward.

2. $(1/v)\oint_\Sigma \mathbf{F} \cdot d\mathbf{S}$, when $\mathbf{F} = \hat{\mathbf{x}}x^2 + \hat{\mathbf{y}}y^2 + \hat{\mathbf{z}}z^2$, for a cube of unit size with sides parallel to the axes and center at (1,1,1).

$$\int_{\text{front}} \mathbf{F} \cdot d\mathbf{S} = \int_{\text{front}} (F_{\text{front}})_x dy\, dz = \int_{\text{front}} \left(\frac{3}{2}\right)^2 dy\, dz = \frac{9}{4}\int_{1/2}^{3/2} dy \int_{1/2}^{3/2} dz = \frac{9}{4}$$

$$\int_{\text{back}} \mathbf{F}\cdot d\mathbf{S} = -\int_{\text{back}} (F_{\text{back}})_x\, dy\, dz = -\int_{\text{back}} \left(\frac{1}{2}\right)^2 dy\, dz = -\frac{1}{4}\int_{1/2}^{3/2} dy \int_{1/2}^{3/2} dz = -\frac{1}{4}$$

The situation for the front and back faces is shown in Fig. 1-12. These two sides together give 2. By symmetry the left and right sides also give 2, as do the top and bottom. The sum for all six faces is 6.

The values for the individual faces here are different from those that would be obtained using only the first two terms of a Taylor expansion from the center, as done in

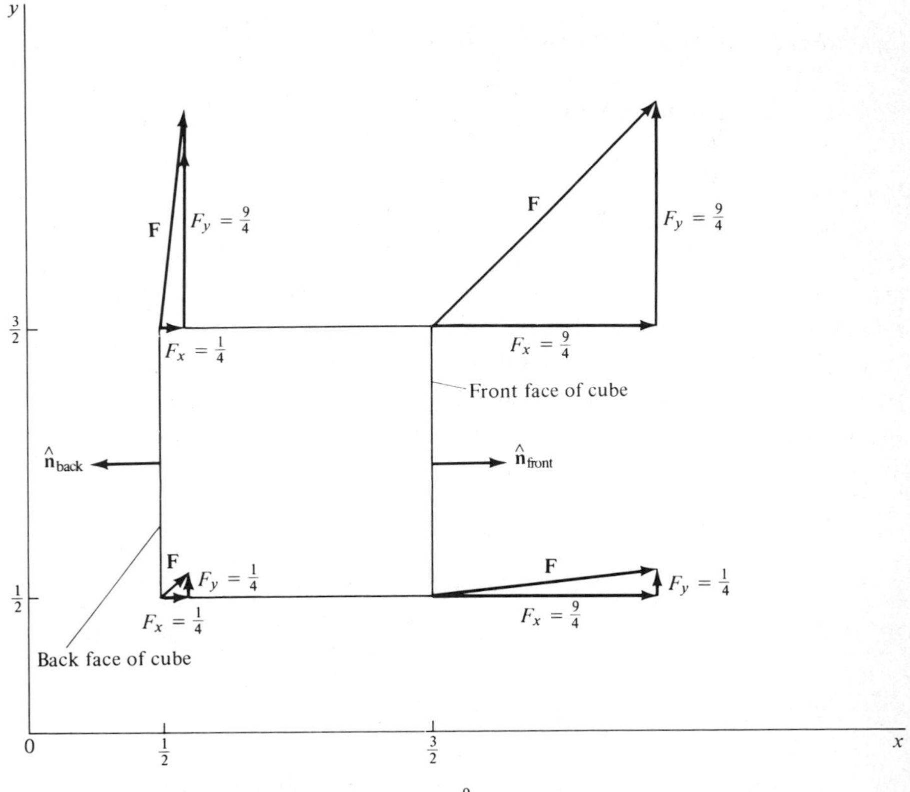

F_x has the constant value $\frac{9}{4}$ over the front face. The flux is $\frac{9}{4}$.

F_x has the constant value $\frac{1}{4}$ over the back face. The flux is $-\frac{1}{4}$ because here $\hat{\mathbf{n}}$ points toward $-x$.

FIGURE 1-12

the derivation for the divergence formula in cartesian coordinates. Instead of $\frac{9}{4}$ and $-\frac{1}{4}$ one would then obtain 2 and 0. (It is accidental that the sum of the two is 2 in both cases.) Cutting the expansion off after two terms is only an approximation for a finite cube, an approximation which gets worse as the size of the cube increases.

The divergence of **F** is $2x + 2y + 2z$. At (1,1,1) this equals 6. By accident $(1/\upsilon)\oint_\Sigma \mathbf{F} \cdot d\mathbf{S}$ for this finite box gives the value of $\nabla \cdot \mathbf{F}$ at its center.

3. $\nabla \cdot \mathbf{r}$ in cartesian, cylindrical, and spherical coordinates. Use Appendix 1 for the latter two.

Cartesian $[\mathbf{r} = (x,y,z)]$: $\quad \nabla \cdot \mathbf{r} = \dfrac{\partial x}{\partial x} + \dfrac{\partial y}{\partial y} + \dfrac{\partial z}{\partial z} = 1 + 1 + 1 = 3$

Cylindrical $[\mathbf{r} = (r,0,0)]$: $\quad \nabla \cdot \mathbf{r} = \dfrac{1}{r}\dfrac{\partial}{\partial r}(r^2) + \dfrac{1}{r}\dfrac{\partial(0)}{\partial \phi} + \dfrac{\partial(0)}{\partial z} = 2 + 0 + 0 = 2$

Spherical $[\mathbf{r} = (r,0,0)]$: $\quad \nabla \cdot \mathbf{r} = \dfrac{1}{r^2}\dfrac{\partial}{\partial r}(r^3) + \dfrac{1}{r \sin\theta}\dfrac{\partial(0)}{\partial \theta} + \dfrac{1}{r \sin\theta}\dfrac{\partial(0)}{\partial \phi}$

$$= 3 + 0 + 0 = 3$$

The reason the second answer differs from the other two is that the **r** of the cylindrical case is the distance to the z axis while the **r** of the other two cases is the distance to the origin. It is unfortunate that they both have the same symbol—they are actually different quantities.

The fact that $\nabla \cdot \mathbf{r} > 0$ means that a function which increases linearly with distance from the origin (or from the z axis) will be one which has more flux leaving a closed surface than entering it.

4. If $\mathbf{B} = \nabla\phi$, the value of $\nabla \cdot \mathbf{B}$ in terms of ϕ.

$$\mathbf{B} = \hat{\mathbf{x}}\frac{\partial \phi}{\partial x} + \hat{\mathbf{y}}\frac{\partial \phi}{\partial y} + \hat{\mathbf{z}}\frac{\partial \phi}{\partial z}$$

so

$$\nabla \cdot \mathbf{B} = \nabla \cdot (\nabla \phi) = \frac{\partial B_x}{\partial x} + \frac{\partial B_y}{\partial y} + \frac{\partial B_z}{\partial z}$$

Then $\nabla \cdot \mathbf{B} = \dfrac{\partial}{\partial x}\dfrac{\partial \phi}{\partial x} + \dfrac{\partial}{\partial y}\dfrac{\partial \phi}{\partial y} + \dfrac{\partial}{\partial z}\dfrac{\partial \phi}{\partial z} = \dfrac{\partial^2 \phi}{\partial x^2} + \dfrac{\partial^2 \phi}{\partial y^2} + \dfrac{\partial^2 \phi}{\partial z^2}$

$$= \left[\left(\hat{\mathbf{x}}\frac{\partial}{\partial x} + \hat{\mathbf{y}}\frac{\partial}{\partial y} + \hat{\mathbf{z}}\frac{\partial}{\partial z}\right) \cdot \left(\hat{\mathbf{x}}\frac{\partial}{\partial x} + \hat{\mathbf{y}}\frac{\partial}{\partial y} + \hat{\mathbf{z}}\frac{\partial}{\partial z}\right)\right]\phi = (\nabla \cdot \nabla)\phi = \nabla^2 \phi$$

This del operator is called the laplacian and

$$\frac{\partial^2 \phi}{\partial x^2} + \frac{\partial^2 \phi}{\partial y^2} + \frac{\partial^2 \phi}{\partial z^2} = 0$$

is called Laplace's equation.

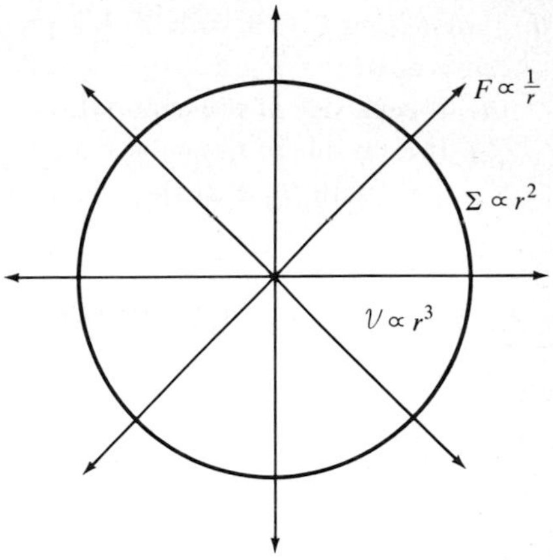

FIGURE 1-13

5. $\nabla \cdot (\hat{\mathbf{r}}/r)$.

(Except where otherwise specified, r is the distance to the origin; i.e., it is the r in spherical coordinates. See Fig. 1-13. This convention will be employed throughout this text.)

$$\nabla \cdot \mathbf{F} = \frac{1}{r^2}\frac{\partial}{\partial r}(r^2 F_r) + \frac{1}{r\sin\theta}\frac{\partial}{\partial\theta}(F_\theta \sin\theta) + \frac{1}{r\sin\theta}\frac{\partial F_\phi}{\partial\phi}$$

$$= \frac{1}{r^2}\frac{\partial}{\partial r}\left(r^2\frac{1}{r}\right) + \frac{1}{r\sin\theta}\frac{\partial(0)}{\partial\theta} + \frac{1}{r\sin\theta}\frac{\partial(0)}{\partial\phi}$$

$$= \frac{1}{r^2}$$

PROBLEMS

1 Find $\nabla \cdot (\hat{\mathbf{r}}/r^2)$.
2 Find $\nabla \cdot (\hat{\mathbf{r}}/r^3)$.
3 Find $\nabla \cdot (\hat{\mathbf{x}}10x)$.
4 Find $\nabla \cdot (\hat{\mathbf{y}}10y)$.
5 Find $\nabla \cdot (\hat{\mathbf{x}}10y)$.
6 Find $\nabla \cdot (\hat{\mathbf{x}}10x + \hat{\mathbf{y}}10y + \hat{\mathbf{z}}z)$.
7 Find $\nabla^2(xyz)$.
8 Find $(1/\mathcal{V})\oint_\Sigma \mathbf{F} \cdot d\mathbf{S}$ if $\mathbf{F} = \hat{\mathbf{z}}z$ and $\mathcal{V}$ is a cube of unit size with sides parallel to the axes, centered at the origin.
9 Find $\nabla \cdot (\hat{\mathbf{x}}xyz + \hat{\mathbf{y}}xyz + \hat{\mathbf{z}}xyz)$.

10 Find $(1/\mathcal{V})\oint \mathbf{F} \cdot d\mathbf{S}$ with $\mathbf{F} = \hat{\mathbf{x}}x^2y$ for a cube with sides parallel to the axes and centered at the origin:
(*a*) If each side of the cube = 1.
(*b*) If each side of the cube = 0.1.
Compare with $\nabla \cdot \mathbf{F}$ at the origin.

1-4 GAUSS' DIVERGENCE THEOREM

A very useful theorem may be obtained directly from the definition of div **F** applied to the small box above:

$$\operatorname{div} \mathbf{F} = \lim_{\Delta\tau \to 0} \frac{1}{\Delta\tau} \oint_{\text{box}} \mathbf{F} \cdot d\mathbf{S}$$

or

$$\oint_{\text{box}} \mathbf{F} \cdot d\mathbf{S} = \lim_{\Delta\tau \to 0} [(\nabla \cdot \mathbf{F})\Delta\tau]$$

Given a finite volume of any shape, not necessarily rectangular, this volume can be divided into a large number of small boxes, and the equation above for the flux out of a box can be applied to each of them. At the interface between two such boxes the flux contribution from one box is equal in magnitude but opposite in direction to that from the second box: these cancel if the function **F** is continuous. As seen from Fig. 1-14, the net flux for the entire volume, when each box is made infinitesimal and the number of boxes becomes infinite, is then only the flux out of the finite, bounding, closed surface. This is Gauss' divergence theorem.

Mathematically, let Σ_i be the surface of the ith box. Then

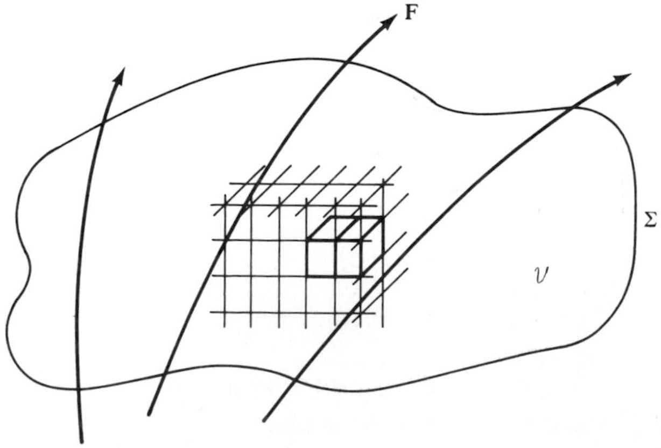

FIGURE 1-14

$$\sum_{i=1}^{N} \oint_{\Sigma_i} \mathbf{F} \cdot d\mathbf{S}_i = \sum_{i=1}^{N} \lim_{\Delta\tau_i \to 0} (\nabla \cdot \mathbf{F}) \Delta\tau_i$$

$$\lim_{N\to\infty} \left[\sum_{i=1}^{N} \oint_{\Sigma_i} \mathbf{F} \cdot d\mathbf{S}_i \right] = \lim_{N\to\infty} \left[\sum_{i=1}^{N} \lim_{\Delta\tau_i \to 0} (\nabla \cdot \mathbf{F}) \Delta\tau_i \right]$$

In the limit, the left side becomes the surface integral for only *the closed outer surface* Σ, enclosing the entire volume $\mathcal{V}$, while the right side becomes the integral for its volume $\mathcal{V}$:

$$\boxed{\oint_{\Sigma} \mathbf{F} \cdot d\mathbf{S} = \int_{\mathcal{V}} \nabla \cdot \mathbf{F} \, d\tau}$$

Note that in actual evaluation the integral on the left side is usually a double integral over two independent variables, while the integral on the right is a triple integral. Note also that the surface in this theorem is a closed surface, as shown by the circle on the surface integral sign. Beside being very useful in various derivations, this theorem is sometimes applied to specific calculations when one side or the other is much easier to evaluate than the other.

Examples

1. $\int_{\mathcal{V}} \nabla \cdot \mathbf{F} \, d\tau$ for $\mathbf{F} = \hat{\mathbf{x}}x^2 + \hat{\mathbf{y}}y^2 + \hat{\mathbf{z}}z^2$, if $\mathcal{V}$ is a unit cube centered at (1,1,1) with sides parallel to the axes.

$$\nabla \cdot \mathbf{F} = 2x + 2y + 2z$$

$$\int_{\mathcal{V}} \nabla \cdot \mathbf{F} \, d\tau = \int_{1/2}^{3/2} dx \int_{1/2}^{3/2} dy \int_{1/2}^{3/2} (2x + 2y + 2z) dz$$

$$= \int_{1/2}^{3/2} dx \int_{1/2}^{3/2} dy \left[2xz + 2yz + z^2\right]_{z=1/2}^{3/2} = \int_{1/2}^{3/2} dx \int_{1/2}^{3/2} dy (2x + 2y + 2)$$

$$= \int_{1/2}^{3/2} dx \left[2xy + y^2 + 2y\right]_{y=1/2}^{3/2} = \int_{1/2}^{3/2} dx (2x + 2 + 2) = 2 \int_{1/2}^{3/2} (x + 2) dx$$

$$= 2 \left[\frac{x^2}{2} + 2x\right]_{1/2}^{3/2} = 2 + 4 = 6$$

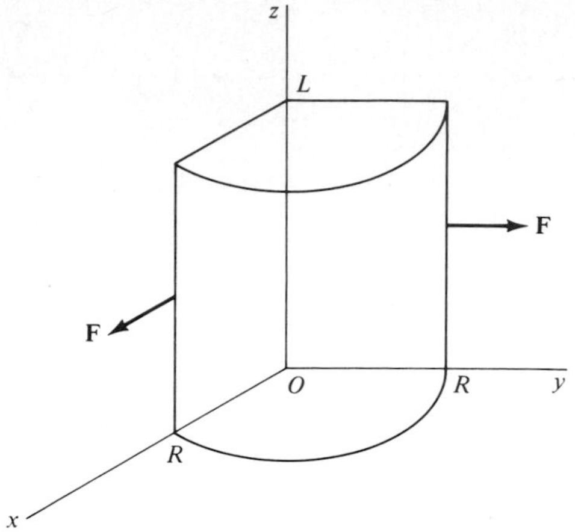

FIGURE 1-15 One–quarter of the cylinder is shown.

This is the same answer obtained, in Example 2 of the previous section, for $(1/\mathcal{V})\oint_\Sigma \mathbf{F} \cdot d\mathbf{S}$. This illustrates the applicability of Gauss' divergence theorem.

2. To show that Gauss' divergence theorem applies to $\mathbf{F} = ra\hat{\mathbf{r}}$ (here r is cylindrical). $\mathcal{V}$ is a cylinder between $z = 0$ and $z = L$, having radius R, and centered on the z axis. See Fig. 1-15.

$$\oint \mathbf{F}\cdot d\mathbf{S} = \int_{\text{bottom}} \mathbf{F}\cdot d\mathbf{S} + \int_{\text{side}} \mathbf{F}\cdot d\mathbf{S} + \int_{\text{top}} \mathbf{F}\cdot d\mathbf{S}$$

$$\int_{\text{bottom}} \mathbf{F}\cdot d\mathbf{S} = \int_{\text{top}} \mathbf{F}\cdot d\mathbf{S} = 0$$

since for these $d\mathbf{S} = \pm\hat{\mathbf{z}}\, dS$ while $\mathbf{F} = \hat{\mathbf{r}}ar$ and $\hat{\mathbf{r}} \perp \hat{\mathbf{z}}$.

$$\int_{\text{side}} \mathbf{F}\cdot d\mathbf{S} = \int_{\text{side}} (\hat{\mathbf{r}}aR)\cdot(\hat{\mathbf{r}}\, dS) = \int_{\text{side}} aR\, dS$$

$$= \int_{z=0}^{L} dz \int_{\phi=0}^{2\pi} aR(R)\, d\phi = 2\pi aR^2 L$$

Then
$$\oint_\Sigma \mathbf{F}\cdot d\mathbf{S} = 2\pi a R^2 L$$

$$\nabla\cdot\mathbf{F} = \frac{1}{r}\frac{\partial}{\partial r}(rar) + \frac{1}{r}\frac{\partial}{\partial\phi}(0) + \frac{\partial}{\partial z}(0) = 2a$$

$$\int_{\mathcal{V}} \nabla\cdot\mathbf{F}\, d\tau = \int_{\mathcal{V}} 2a\, d\tau = 2a(\pi R^2 L) \qquad \text{Q.E.D.}$$

PROBLEMS

1 (*a*) Express the unit vectors of a cylindrical coordinate system $(\hat{\mathbf{r}},\hat{\boldsymbol{\phi}},\hat{\mathbf{z}})$ in terms of the unit vectors $(\hat{\mathbf{x}},\hat{\mathbf{y}},\hat{\mathbf{z}})$ of a cartesian system. Illustrate this by a diagram.
(*b*) Repeat (*a*), but in the reverse direction.
(*c*) Express the unit vectors of a spherical coordinate system $(\hat{\mathbf{r}},\hat{\boldsymbol{\theta}},\hat{\boldsymbol{\phi}})$ in terms of the unit vectors $(\hat{\mathbf{x}},\hat{\mathbf{y}},\hat{\mathbf{z}})$ of a cartesian system.
(*d*) Repeat (*c*), but in the reverse direction.

2 (*a*) Express the unit vectors of a spherical coordinate system $(\hat{\mathbf{r}},\hat{\boldsymbol{\theta}},\hat{\boldsymbol{\phi}})$ in terms of the unit vectors $(\hat{\mathbf{r}},\hat{\boldsymbol{\phi}},\hat{\mathbf{z}})$ of a cylindrical system.
(*b*) Repeat (*a*) in the reverse direction.

3 Given $\mathbf{F} = \hat{\mathbf{r}}ar$ in spherical coordinates,
(*a*) Convert $\mathbf{F}$ to cylindrical coordinates.
(*b*) Show that Gauss' divergence theorem holds, using the $\mathbf{F}$ of (*a*), for a right cylinder of radius R extending from $z = 0$ to $z = L$ symmetric about the z axis.

4 Try to apply Gauss' divergence theorem to $\mathbf{F} = \hat{\mathbf{r}}a/r$ (cylindrical coordinates) with the volume of Prob. 3. What difficulty is encountered with $\nabla\cdot\mathbf{F}$ at $r = 0$? If this one point is ignored Gauss' divergence theorem does not apply. Can this difficulty be resolved by taking $\mathbf{F} = \hat{\mathbf{r}}a/r^n$ and setting $n = 1$ at the very end? (This problem is actually an example of the Dirac delta function.)

5 Apply both sides of Gauss' divergence theorem to a unit cube with center at the origin if (*a*) $\mathbf{F} = \hat{\mathbf{x}}x$; (*b*) $\mathbf{F} = \hat{\mathbf{x}}|x|$.

6 Repeat Prob. 5 with the cube changed to a sphere of unit radius.

7 Given $\mathbf{F} = \hat{\mathbf{r}}r$ (spherical coordinates), find $\oint_\Sigma \mathbf{F}\cdot d\mathbf{S}$ if Σ is a unit cube with one corner at the origin having edges parallel to the axes in the first octant. This is a case where the triple integration is easier to perform than the double integration.

8 Repeat Prob. 7 for the case where the edge of the cube which originally went from (0,0,0) to (1,0,0) is tilted up in the zx plane by $\theta°$; i.e., the cube is rotated $\theta°$ about the y axis.

9 Let $\mathbf{F} = \hat{\mathbf{x}}a + \hat{\mathbf{y}}b + \hat{\mathbf{z}}c$, where a, b, c are constants. Suppose Σ is the surface of a unit sphere centered at (x_0, y_0, z_0). Find $\oint_\Sigma \mathbf{F}\cdot d\mathbf{S}$. See Fig. 1-16.

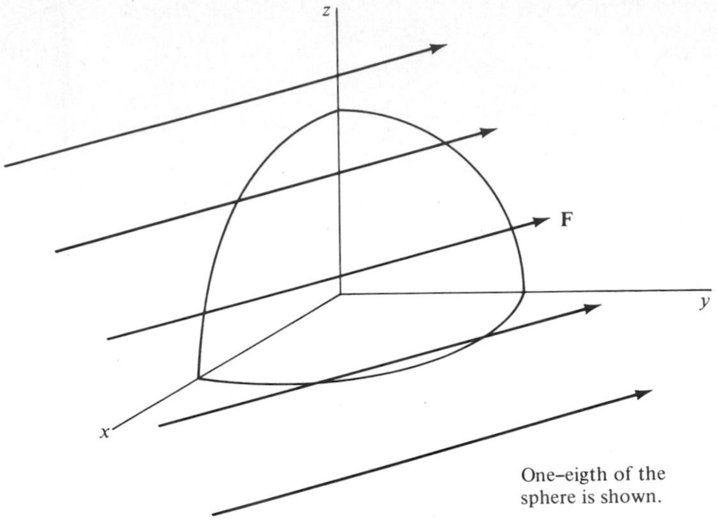

FIGURE 1-16

10 Apply Gauss' divergence theorem to $\mathbf{F} = \hat{\mathbf{x}}x^2z + \hat{\mathbf{y}}y - \hat{\mathbf{z}}2xz^2$ if $\mathcal{V}$ is the portion of the cylinder $x^2 + y^2 = 4$ in the first octant between $z = 0$ and $z = 3$.

11 Given $\mathbf{F} = \hat{\mathbf{r}}ar$ (spherical) and the volume of Prob. 3. Find $\int \mathbf{F} \cdot d\mathbf{S}$ for the bottom, the top, and the sides, using spherical coordinates.

12 Apply Gauss' divergence theorem to the product $\mathbf{C} \times \mathbf{F}$, where $\mathbf{F}$ is a vector function of position and $\mathbf{C}$ is a constant vector, to prove that

$$\int_{\mathcal{V}} (\nabla \times \mathbf{F}) d\tau = -\oint_{\Sigma} \mathbf{F} \times d\mathbf{S}$$

1-5 LINE INTEGRALS

A closed line is one on which it is possible to begin at any point, traverse the entire curve in a given sense, and return to the starting point. It is not necessary that this path lie in a plane. The closed curve may be considered to be the boundary of an open surface, but this surface is not unique—there is an infinite number of such surfaces for a given curve. On the other hand, every two-sided open surface has a unique curve for its boundary. (There are some one-sided surfaces, such as the Möbius surface, which we will ignore.)

An open line has a beginning and an end; it is not possible to return to the starting point if one traverses the open curve in a given sense.

The integral $\int_a^b f\,ds$ is one type of line integral of a function f along a curve, either open or closed, between the two points a and b. The independent variable here is s, the

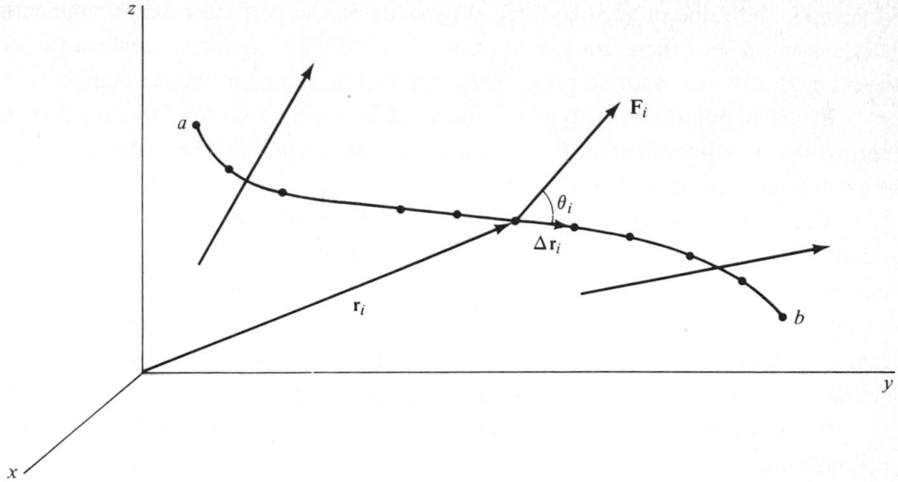

FIGURE 1-17

distance along the curve. (In the ordinary definite integral, $\int_a^b f\,dx$, the independent variable is x, the distance along some straight-line axis.) The function f in this line integral is a scalar function of the position.

A simple extension of this idea defines a line integral of a vector function:

$$\int_a^b \mathbf{F} \cdot d\mathbf{r} = \int_a^b F \cos \theta \, dr$$

where θ is the angle between $\mathbf{F}$ and $d\mathbf{r}$ at each point along the curve. A visualization of such a line integral, as the limit of the infinite series

$$\lim_{\substack{\Delta r_i \to 0 \\ N \to \infty}} \sum_{i=1}^{N} \mathbf{F}_i \cdot \Delta \mathbf{r}_i$$

is shown in Fig. 1-17.

A simple but important example of this line integral is provided by the case of the work done when a force $\mathbf{F}$ moves a body along some path C between a and b:

$$W = \int_a^b \mathbf{F} \cdot d\mathbf{r}$$

In general, the value of such an integral depends on the particular path C connecting the points a and b, but there are special cases of considerable importance where the value of the integral is the same for all paths. These will be considered in the next chapter.

$\mathbf{F} \cdot d\mathbf{r}$ is not the only type of integrand involving a vector function that may be used to define a line integral. $\mathbf{F} \times d\mathbf{r}$ is also such an integrand. The first type, however, is by far the most common one.

The vector line integral given for the work, above, becomes a scalar line integral which is much simpler to evaluate when the angle between $\mathbf{F}$ and $d\mathbf{r}$ is a constant, for then $\cos\theta$ may be taken to the left of the integral sign. If, in addition, $\mathbf{F}$ is constant in magnitude along the path C then the work becomes simply $FL\cos\theta$, where L is the length of the path C between a and b. But in general, neither θ nor $\mathbf{F}$ is constant along C, and neither factor may be moved to the left of the integral.

When the curve C is a closed one it is possible to make the two end points of the integration coincide. Then a change is made in the symbol for the line integral over the closed path:

$$\int_{\text{path } C\ a}^{b} \mathbf{F}\cdot d\mathbf{r} \rightarrow \oint_{\text{path } C} \mathbf{F}\cdot d\mathbf{r}$$

This integral is called the circulation of $\mathbf{F}$ around the path C. There are two senses, one the opposite of the other, in which this circulation may be taken. There is nothing to favor one sense over the other but, whichever one is picked, the positive direction of the normal to any open surface which is bounded by C is then fixed by the right-hand screw convention mentioned in the previous discussion of the divergence. See Fig. 1-7. A rule often followed is this: the sense of the circulation is such that the open area bounded by C is to be kept on the left-hand side in traversing C.

The small circle on the integral sign above indicates a closed path about an open area; the small circle used on an integral sign in connection with the flux of a vector refers to a closed area. When $\mathbf{F} \cdot d\mathbf{r}$ is written in terms of its coordinate components it appears that three integrations are required: $\oint F_x\, dx + \oint F_y\, dy + \oint F_z\, dz$, but, parameterizing $d\mathbf{r}$ by using a single scalar variable, we can actually require only the single integration $\oint \mathbf{F} \cdot d\mathbf{r}$. In the case of the flux, the integral sign usually stands for a double integration and there are two independent variables.

Examples

1. $\int_{a\ \text{path } C_1}^{b} \mathbf{F} \cdot d\mathbf{r}$ when $a = (0,0,0)$, $b = (1,1,1)$, $\mathbf{F} = \hat{\mathbf{x}}(2x + y^2) - \hat{\mathbf{y}}(3yz) + \hat{\mathbf{z}}$, and C_1 is the path consisting of the three segments (0,0,0) to (1,0,0), (1,0,0) to (1,1,0), and (1,1,0) to (1,1,1). This is shown in Fig. 1-18.

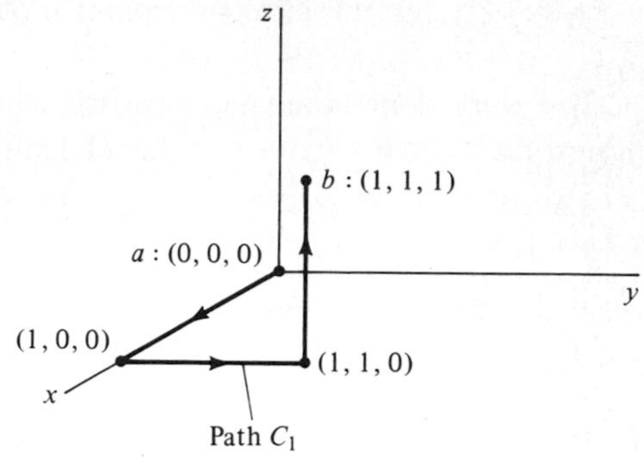

FIGURE 1-18

$\mathbf{F} \cdot d\mathbf{r} = (2x + y^2)dx - 3yz\,dy + dz$. Along the first segment $dy = dz = 0$, so $\int_1 \mathbf{F} \cdot d\mathbf{r}$ is $\int_0^1 (2x + y^2)dx = \int_0^1 2x\,dx = 1$. On the second segment $dz = dx = 0$ and $\int_2 \mathbf{F} \cdot d\mathbf{r}$ is $-\int_0^1 3yz\,dy$. Since $z = 0$ along this segment, this contribution vanishes. For the third segment $dx = dy = 0$ and $\int_3 \mathbf{F} \cdot d\mathbf{r} = \int_0^1 dz = 1$. So

$$\int_{\text{path } C_1\ (0,0,0)}^{(1,1,1)} \mathbf{F} \cdot d\mathbf{r} = 1 + 0 + 1 = 2$$

2. A repeat of Example 1, but along a different path. C_2 is taken as the straight line from (0,0,0) to (1,1,1), as shown in Fig. 1-19.

In parametric representation the path C_2 is given by $x = t,\ y = t,\ z = t$ so $\mathbf{F} \cdot d\mathbf{r} = (2t + t^2)\,dt - 3t^2\,dt + dt$ or $\mathbf{F} \cdot d\mathbf{r} = (-2t^2 + 2t + 1)\,dt$. At (0,0,0) $t = 0$, while at (1,1,1) $t = 1$. Then

$$\int_{\text{path } C_2\ (0,0,0)}^{(1,1,1)} \mathbf{F} \cdot d\mathbf{r} = \int_0^1 (-2t^2 + 2t + 1)dt = \left[-2\left(\frac{t^3}{3}\right) + 2\left(\frac{t^2}{2}\right) + t\right]_0^1 = \frac{4}{3}$$

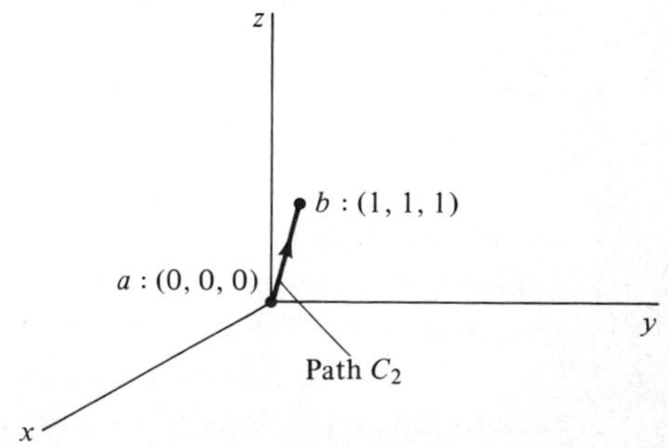

FIGURE 1-19

Note that this result is different from that obtained along C_1.

3. The work done in moving a particle along the curve $y = x^2$ from the origin to (1,1) against the force $\mathbf{F} = \hat{\mathbf{x}}xy + \hat{\mathbf{y}}y^2$. Let (1,1) be the point A.

$$W = \int_0^A \mathbf{F}\cdot d\mathbf{r} = \int_0^A (\hat{\mathbf{x}}xy + \hat{\mathbf{y}}y^2)\cdot(\hat{\mathbf{x}}dx + \hat{\mathbf{y}}dy) = \int_0^A (xy\,dx + y^2 dy)$$

$$= \int_0^1 [x(x^2)dx + (x^2)^2(2x\,dx)] = \int_0^1 (x^3 + 2x^5)dx = \left[\frac{x^4}{4} + 2\frac{x^6}{6}\right]_0^1 = \frac{7}{12}$$

If the force is in newtons and the distance in meters, the work is in joules.

4. A repeat of Example 3, but employing parametric representation.

Let $x = t$. Then $y = t^2$. So $xy\,dx = t(t^2)dt = t^3\,dt$ and $y^2\,dy = (t^2)^2(2t\,dt) = 2t^5\,dt$. Point O is always given by $t = 0$, point A by $t = 1$. Then

$$W = \int_0^1 (t^3 + 2t^5)dt = \tfrac{7}{12}$$

PROBLEMS

1 Evaluate $\int_C f\,dr$ where $f = xy^2$ and C is the straight line $y = 3x$ from (−2, −6) to (1,3). In parametric representation, if $x = t$ then $y = 3t$; also $dr = (dr/dt)\,dt$ and $(d\mathbf{r}/dt) = \sqrt{1+9} = \sqrt{10}$.

2 Evaluate the case above when C consists of two straight lines: (a) from (−2, −6) to (4,0); (b) from (4,0) to (1,3).

3 Find $\int_C(xyz\,dx + x^2y\,dy)$ when C is given by $y = x^2$ from (0,0,1) to (2,4,1). Here z is constant.

4 (a) Change C of Prob. 3 to $y = 2x$ with the same $f(x,y,z)$ and the same end points. (b) Repeat with $y = 3x$.

5 Find the work in moving a particle along the curve $y = x^2$ from the origin to (1,1) against the force $\mathbf{F} = \hat{\mathbf{x}}xy - \hat{\mathbf{y}}y^2$.

6 Find the work in moving a particle from the origin to (1,1) along $y = x^2$ and then back to the origin along $y^2 = x$; the force to be overcome is $\mathbf{F} = \hat{\mathbf{x}}(x + y) + \hat{\mathbf{y}}xy$.

1-6 THE CURL

The curl of **F**, at a point P, may now be defined. Consider a vector whose magnitude is the ratio of the circulation of **F**, along a path C around P, to an area ΔS bounded by C. Assume C lies in a plane, so ΔS can be uniquely defined; then take this ratio in the limit as ΔS and C shrink to zero. It will be assumed that in the limit this ratio exists and is unique.

The direction of the vector defined above is that of the normal to ΔS, the positive sense being related to the sense of circulation in the way the linear motion of a right-hand screw is related to the rotary motion. The magnitude and direction of this vector will depend on the path C, which is quite arbitrary up to this point. Now assume that the same procedure is followed, to give a similar vector at this point, in some other direction. Then let this be done for all possible directions. It will be assumed that for some direction the magnitude of the vector will be a maximum. The vector determined by this procedure, in the direction that maximizes the magnitude, is defined as the curl of **F** at P. The operations above gave the component of curl **F** in the direction normal to ΔS.

If $\hat{\mathbf{n}}$ is a unit vector in any particular direction and C_n is a closed path in a plane perpendicular to $\hat{\mathbf{n}}$, then, for any coordinate system, if ΔS_n is the area enclosed by C_n,

$$(\text{curl } \mathbf{F})_n = (\text{curl } \mathbf{F})\cdot\hat{\mathbf{n}} = \lim_{\Delta S_n \to 0}\left[\frac{1}{\Delta S_n}\oint_{C_n} \mathbf{F}\cdot d\mathbf{r}\right]$$

Figure 1-20 shows this procedure geometrically, it being assumed that C_n is made infinitesimal. The result obtained is the component of curl **F** along the direction of the

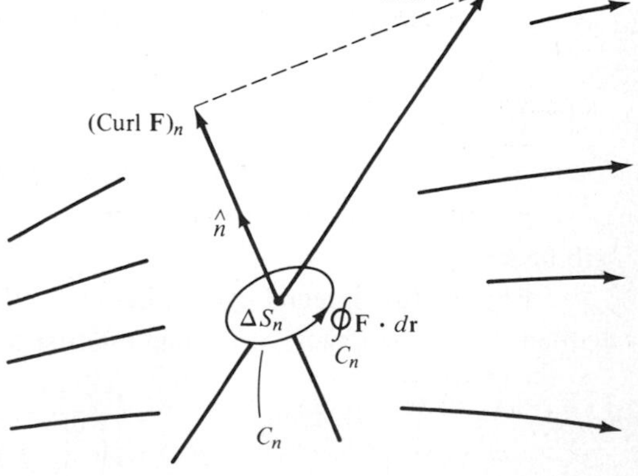

FIGURE 1-20

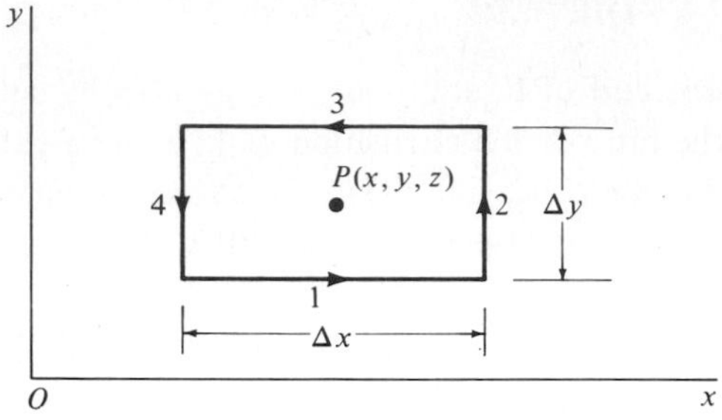

FIGURE 1-21

unit vector $\hat{\mathbf{n}}$. If $\hat{\mathbf{n}}$ is chosen in the direction of curl **F** then the component of curl **F** along this direction is a maximum: the component equals the magnitude of the vector.

We will now find an expression for curl **F** in cartesian coordinates.

Consider the $\hat{\mathbf{z}}$ component of the curl of **F**:

$$(\text{curl } \mathbf{F})_z = (\text{curl } \mathbf{F}) \cdot \hat{\mathbf{z}} = \lim_{\Delta S_z \to 0} \left[\frac{1}{\Delta S_z} \oint_{C_z} \mathbf{F} \cdot d\mathbf{r} \right]$$

Let ΔS_z be the small rectangular area in a plane z = constant, as shown in Fig. 1-21, with edges Δx and Δy and the point $P(x,y,z)$ at its center. Then $\hat{\mathbf{n}} = \hat{\mathbf{z}}$ and ΔS_z is $\Delta x \, \Delta y$. The circulation of **F** is $\oint_{C_z} \mathbf{F} \cdot d\mathbf{r} = \oint_{C_z} (F_x \, dx + F_y \, dy)$ and the contour of integration C_z may be divided into four parts, all with z = const:

1 From $[x - (\Delta x/2), y - (\Delta y/2)]$ to $[x + (\Delta x/2), y - (\Delta y/2)]$
2 From $[x + (\Delta x/2), y - (\Delta y/2)]$ to $[x + (\Delta x/2), y + (\Delta y/2)]$
3 From $[x + (\Delta x/2), y + (\Delta y/2)]$ to $[x - (\Delta x/2), y + (\Delta y/2)]$
4 From $[x - (\Delta x/2), y + (\Delta y/2)]$ to $[x - (\Delta x/2), y - (\Delta y/2)]$

Then $\oint_{C_z} = \int_1 + \int_2 + \int_3 + \int_4$, where each integrand is $(F_x \, dx + F_y \, dy)$.

The values of the components of **F** along the four parts of the path of integration are obtained by Taylor expansions. The small rectangular area will subsequently become a differential area, so in the expansion only the first two terms will be kept.

For the first integral $dy = 0$. Let $(F_x) = (F_x)_P$ represent the x component of **F** at P; then the x component of **F** along the first segment of the contour is

$$(F_x) + \left[\left(\frac{\partial F_x}{\partial y} \right)_P \left(-\frac{1}{2} \Delta y \right) + \left(\frac{\partial F_x}{\partial x} \right)_P (a \, \Delta x) \right]$$

where a is a factor (similar to that used in the derivation of the cartesian form of the divergence) whose magnitude determines the position of a point along the first segment: $-\frac{1}{2} \leqslant a \leqslant \frac{1}{2}$. The term containing a will have positive and negative contributions which cancel out. Letting the subscript on the derivative be implicit,

$$\int_{\text{path 1}} (F_x\,dx + F_y\,dy) = \int_{x-(\Delta x/2)}^{x+(\Delta x/2)} \left[(F_x) - \frac{1}{2}\left(\frac{\partial F_x}{\partial y}\right)\Delta y \right] dx$$

$$= (F_x)\Delta x - \frac{1}{2}\left(\frac{\partial F_x}{\partial y}\right)\Delta y\,\Delta x$$

For the second integral $dx = 0$ so only the y component of $\mathbf{F}$ need be evaluated:

$$(F_y) + \left[\left(\frac{\partial F_y}{\partial x}\right)\left(\frac{1}{2}\Delta x\right) + \left(\frac{\partial F_y}{\partial y}\right)(a\,\Delta y) \right]$$

The term in a gives zero as before; so, with F_y and its derivatives evaluated at P,

$$\int_{\text{path 2}} (F_x\,dx + F_y\,dy) = \int_{y-(\Delta y/2)}^{y+(\Delta y/2)} \left[(F_y) + \frac{1}{2}\left(\frac{\partial F_y}{\partial x}\right)\Delta x \right] dy$$

$$= (F_y)\Delta y + \frac{1}{2}\left(\frac{\partial F_y}{\partial x}\right)\Delta x\,\Delta y$$

In the case of the third integral $dy = 0$ again, so F_y need not be considered and the integrand becomes

$$(F_x) + \left(\frac{\partial F_x}{\partial y}\right)\left(\frac{1}{2}\Delta y\right)$$

Then

$$\int_{\text{path 3}} (F_x\,dx + F_y\,dy) = \int_{x+(\Delta x/2)}^{x-(\Delta x/2)} \left[(F_x) + \frac{1}{2}\left(\frac{\partial F_x}{\partial y}\right)\Delta y \right] dx$$

(Note that the expression for the differential here is $+dx$ just as for the first integral; but the limits are reversed. It would also be correct to keep the limits the same as before if, instead of using $+dx$, $-dx$ were employed; but it would be incorrect to reverse both the limits and the sign of the independent variable.) So

$$\int_3 (F_x\,dx + F_y\,dy) = -(F_x)\Delta x - \frac{1}{2}\left(\frac{\partial F_y}{\partial y}\right)\Delta y\,\Delta x$$

Similarly, for the fourth integral $dx = 0$ and the integrand is

$$(F_y) + \left(\frac{\partial F_y}{\partial x}\right)\left(-\frac{1}{2}\,\Delta x\right)$$

so

$$\int_4 (F_x dx + F_y dy) = \int_{y+(\Delta y/2)}^{y-(\Delta y/2)} \left[(F_y) - \frac{1}{2}\left(\frac{\partial F_y}{\partial x}\right)\Delta x\right] dy$$

$$= -(F_y)\Delta y + \frac{1}{2}\left(\frac{\partial F_y}{\partial x}\right)\Delta x\,\Delta y$$

Adding the four contributions,

$$\int_{C_z} \mathbf{F}\cdot d\mathbf{r} = \left(\frac{\partial F_y}{\partial x} - \frac{\partial F_x}{\partial y}\right)\Delta x\,\Delta y$$

Consequently

$$(\text{curl}\,\mathbf{F})_z = \lim_{\substack{\Delta y\to 0\\ \Delta x\to 0}} \left[\frac{1}{\Delta x\,\Delta y}\left(\frac{\partial F_y}{\partial x} - \frac{\partial F_x}{\partial y}\right)\Delta x\,\Delta y\right] = \left(\frac{\partial F_y}{\partial x} - \frac{\partial F_x}{\partial y}\right)$$

The y and x components of curl **F** may be obtained by cyclical permutation of x, y, and z. Letting $x \to y$, $y \to z$, $z \to x$ gives

$$(\text{curl}\,\mathbf{F})_x = \frac{\partial F_z}{\partial y} - \frac{\partial F_y}{\partial z}$$

Another cyclical permutation of x, y, and z gives

$$(\text{curl}\,\mathbf{F})_y = \frac{\partial F_x}{\partial z} - \frac{\partial F_z}{\partial x}$$

In cartesian coordinates the expression for curl **F** is, therefore,

$$\boxed{\text{curl}\,\mathbf{F} = \hat{\mathbf{x}}\left(\frac{\partial F_z}{\partial y} - \frac{\partial F_y}{\partial z}\right) + \hat{\mathbf{y}}\left(\frac{\partial F_x}{\partial z} - \frac{\partial F_z}{\partial x}\right) + \hat{\mathbf{z}}\left(\frac{\partial F_y}{\partial x} - \frac{\partial F_x}{\partial y}\right)}$$

Using the rules for expansion of a determinant this result may also be written

$$\text{curl } \mathbf{F} = \begin{vmatrix} \hat{\mathbf{x}} & \hat{\mathbf{y}} & \hat{\mathbf{z}} \\ \dfrac{\partial}{\partial x} & \dfrac{\partial}{\partial y} & \dfrac{\partial}{\partial z} \\ F_x & F_y & F_z \end{vmatrix}$$

Care must be taken, in expanding this determinant, to have a factor from the second row always precede the factor from the third row. Since

$$\mathbf{A} \times \mathbf{F} = \begin{vmatrix} \hat{\mathbf{x}} & \hat{\mathbf{y}} & \hat{\mathbf{z}} \\ A_x & A_y & A_z \\ F_x & F_y & F_z \end{vmatrix}$$

the expression for curl **F** can also be written

$$\text{curl } \mathbf{F} = \nabla \times \mathbf{F}$$

where the del operator in cartesian coordinates has the same value that it has for the gradient and for the divergence.

When the del symbol is used as a curl operator it transforms one vector field into another vector field. Example 5, below, applies this concept to the easily visualized case of fluid flow, where **F** is a velocity field.

In the case of the gradient at a given point, there is a maximum rate of change of a scalar function in some direction; along any other direction the rate of change of the function equals the maximum rate of change multiplied by the cosine of the angle between the two directions. The maximum rate of change and the particular direction specify the gradient vector. Similarly, in the case of the curl at a given point, there is a maximum circulation of the vector function per unit of enclosed infinitesimal area in some direction; along any other direction the circulation per unit area equals the maximum value multiplied by the cosine of the angle between the two directions. The maximum circulation per unit area (in the limit as the area becomes infinitesimal) and the particular direction specify the curl vector.

A vector whose curl is *everywhere* zero in some singly connected region is called irrotational, or lamellar. The first name means there is no rotation; i.e., no circulation. The second stems from a type of flow of fluids, in thin parallel layers or lamellae, without vortices. In the special case when an irrotational vector field is independent of time the field is called conservative. (In a static gravitational or electric field the energy of a test particle is a function of the position only, not of the path by which it reached the position. When a test particle is moved from (1) a starting position, via some path, to (2) a final position which coincides with the starting point, its energy is conserved. Hence the name conservative.)

The significance of the divergence and curl of a vector are brought out by Helmholtz's theorem, which will simply be stated here. (The derivation of this important theorem is given in Appendix 4 at the end of Chap. 4.) Take a finite region R which contains all the sources of a vector field **F**; or, stated mathematically, let $\nabla \cdot \mathbf{F}$ and $\nabla \times \mathbf{F}$ vanish everywhere outside R. If $\nabla \cdot \mathbf{F} = b$ and $\nabla \times \mathbf{F} = \mathbf{c}$ are given everywhere within R, then **F** is uniquely defined. (The theorem is not applicable to the infinite domain.) The theorem further states the formula by which **F** may be found:

$$\mathbf{F}(\mathbf{r}_t) = -\nabla\left[\frac{1}{4\pi}\int \frac{b(\mathbf{r}_s)}{R}\,d\tau_s\right] + \nabla \times \left[\frac{1}{4\pi}\int \frac{\mathbf{c}(\mathbf{r}_s)}{R}\,d\tau_s\right]$$

where $R = |\mathbf{r}_s - \mathbf{r}_t|$ is the distance between source point and test point. In principle then, a knowledge of the div and the curl of a vector field is sufficient to learn everything about the vector field itself. There is a trivial added constant to **F** which may be conveniently taken to be zero.

Examples

1. The $\nabla \times \mathbf{F}$, if $\mathbf{F} = \hat{\mathbf{x}}2xy + \hat{\mathbf{y}}y^2z + \hat{\mathbf{z}}x^2y^2$.

$$\nabla \times \mathbf{F} = \left(\hat{\mathbf{x}}\frac{\partial}{\partial x} + \hat{\mathbf{y}}\frac{\partial}{\partial y} + \hat{\mathbf{z}}\frac{\partial}{\partial z}\right) \times (\hat{\mathbf{x}}2xy + \hat{\mathbf{y}}y^2z + \hat{\mathbf{z}}x^2y^2)$$

$$= \begin{vmatrix} \hat{\mathbf{x}} & \hat{\mathbf{y}} & \hat{\mathbf{z}} \\ \partial_x & \partial_y & \partial_z \\ 2xy & y^2z & x^2y^2 \end{vmatrix} = \hat{\mathbf{x}}(2xy - y^2) + \hat{\mathbf{y}}(0 - 2xy^2) + \hat{\mathbf{z}}(0 - 2x)$$

$$= \hat{\mathbf{x}}(2xy - y^2) - \hat{\mathbf{y}}2xy^2 - \hat{\mathbf{z}}2x$$

2. The conditions on **F** imposed by $\nabla \times \mathbf{F} = 0$.

$$\hat{\mathbf{x}}\left(\frac{\partial F_y}{\partial z} - \frac{\partial F_z}{\partial y}\right) + \hat{\mathbf{y}}\left(\frac{\partial F_z}{\partial x} - \frac{\partial F_x}{\partial z}\right) + \hat{\mathbf{z}}\left(\frac{\partial F_x}{\partial y} - \frac{\partial F_y}{\partial x}\right) = 0$$

Then

$$\frac{\partial F_y}{\partial z} = \frac{\partial F_z}{\partial y} \qquad \frac{\partial F_z}{\partial x} = \frac{\partial F_x}{\partial z} \qquad \frac{\partial F_x}{\partial y} = \frac{\partial F_y}{\partial x}$$

If $\mathbf{F} = \nabla\phi$, i.e., if

$$F_x = \frac{\partial \phi}{\partial x} \qquad F_y = \frac{\partial \phi}{\partial y} \qquad F_z = \frac{\partial \phi}{\partial z}$$

then these three equations are satisfied. So $\mathbf{F} = \nabla\phi$ is a sufficient condition that $\nabla \times \mathbf{F}$ vanishes; it can also be shown that $\mathbf{F} = \nabla\phi$ is a necessary condition that $\nabla \times \mathbf{F} = 0$.

3. $\nabla \times (\hat{\mathbf{r}}/r)$, using cartesian coordinates.

$$\nabla \times \frac{\hat{\mathbf{r}}}{r} = \nabla \times \frac{\mathbf{r}}{r^2} = (\hat{\mathbf{x}}\,\partial_x + \hat{\mathbf{y}}\,\partial_y + \hat{\mathbf{z}}\,\partial_z) \times \left(\frac{\hat{\mathbf{x}}x + \hat{\mathbf{y}}y + \hat{\mathbf{z}}z}{x^2 + y^2 + z^2}\right)$$

$$= \begin{vmatrix} \hat{\mathbf{x}} & \hat{\mathbf{y}} & \hat{\mathbf{z}} \\ \partial_x & \partial_y & \partial_z \\ \dfrac{x}{(x^2+y^2+z^2)} & \dfrac{y}{(x^2+y^2+z^2)} & \dfrac{z}{(x^2+y^2+z^2)} \end{vmatrix}$$

$$= \hat{\mathbf{x}}\left[\partial_y \frac{z}{(x^2+y^2+z^2)} - \partial_z \frac{y}{(x^2+y^2+z^2)}\right]$$

$$+ \hat{\mathbf{y}}\left[\partial_z \frac{x}{(x^2+y^2+z^2)} - \partial_x \frac{z}{(x^2+y^2+z^2)}\right]$$

$$+ \hat{\mathbf{z}}\left[\partial_x \frac{y}{(x^2+y^2+z^2)} - \partial_y \frac{x}{(x^2+y^2+z^2)}\right]$$

$$= \hat{\mathbf{x}}\left[\frac{0 - z(2y)}{(x^2+y^2+z^2)^2} - \frac{0 - y(2z)}{(x^2+y^2+z^2)^2}\right]$$

$$+ \hat{\mathbf{y}}\left[\frac{0 - x(2z)}{(x^2+y^2+z^2)^2} - \frac{0 - z(2x)}{(x^2+y^2+z^2)^2}\right]$$

$$+ \hat{\mathbf{z}}\left[\frac{0 - y(2z)}{(x^2+y^2+z^2)^2} - \frac{0 - x(2y)}{(x^2+y^2+z^2)^2}\right]$$

$$= \hat{\mathbf{x}}\,\frac{-2zy + 2yz}{(x^2+y^2+z^2)^2} + \hat{\mathbf{y}}\,\frac{-2xz + 2xz}{(x^2+y^2+z^2)^2} + \hat{\mathbf{z}}\,\frac{-2yx + 2yx}{(x^2+y^2+z^2)^2} = 0$$

4. $\nabla \times (\hat{\mathbf{r}}/r)$ in spherical coordinates.
From Appendix 1,

$$\nabla \times \mathbf{F} = \hat{\mathbf{r}}\left\{\frac{1}{r\sin\theta}\left[\frac{\partial}{\partial\theta}(\sin\theta\, F_\phi) - \frac{\partial F_\theta}{\partial\phi}\right]\right\} + \hat{\boldsymbol{\theta}}\left\{\frac{1}{r\sin\theta}\frac{\partial F_r}{\partial\phi} - \frac{1}{r}\frac{\partial}{\partial r}(rF_\phi)\right\}$$

$$+ \hat{\boldsymbol{\phi}}\left\{\frac{1}{r}\frac{\partial}{\partial r}(rF_\theta) - \frac{1}{r}\frac{\partial F_r}{\partial\theta}\right\}$$

(Note that the del operator for the curl in spherical coordinates is different from the del operator for the divergence or gradient in spherical coordinates.)

If $\mathbf{F} = \hat{\mathbf{r}}/r$, then $F_r = 1/r$, $F_\theta = 0$, $F_\phi = 0$.

$$\nabla \times \frac{\hat{\mathbf{r}}}{r} = \hat{\mathbf{r}}\left[\frac{1}{r \sin\theta}(0-0)\right] + \hat{\boldsymbol{\theta}}\left[\frac{1}{r\sin\theta}\frac{\partial}{\partial\phi}\frac{1}{r} - 0\right] + \hat{\boldsymbol{\phi}}\left[\frac{1}{r}(0) - \frac{1}{r}\frac{\partial}{\partial\theta}\frac{1}{r}\right]$$
$$= \hat{\mathbf{r}}(0) + \hat{\boldsymbol{\theta}}(0) + \hat{\boldsymbol{\phi}}(0) = 0$$

This is true, also, at the origin. Even though $|\mathbf{F}|$ is infinite at $r = 0$, $\nabla \times \mathbf{F} = 0$ there.

5. The construction of a curl meter for measuring the curl of the velocity field of water in a river.

The curl can be approximated by the circulation around the perimeter of a small, fixed circle. Figure 1-22 shows a device for measuring the circulation of a velocity field. The smaller the width of the vanes, the better the approximation to the curl but the less sensitive the devices. The torque of the blades about the axis will be proportional to the component of the velocity field normal to the blades, integrated around the axis. Then the angular velocity of the blades would give a measure of the curl component along the direction of the axis; or, if the rotation were to be opposed by a spring, the equilibrium angular deflection of the blades from their zero position would serve this purpose.

Looking in the direction of the arrow, a tendency to rotate clockwise would correspond to a positive curl component in the direction of the arrow. To find the curl itself, it would be necessary to change the direction of the arrow until a maximum rate of rotation was found, or else to employ three such mutually orthogonal devices and to combine the results to give the magnitude.

For measuring the curl of an electric field one could substitute, for the blades, equal charges pasted to the ends of insulating spokes.

PROBLEMS

1 Find $\nabla \times \mathbf{r}$.

2 Find $\nabla \times \hat{\mathbf{r}}/r^2$. This would apply, e.g., to the earth's gravitational field.

3 Suppose the water in a stream has a velocity which is zero at the river bed and increases linearly to a value v_m at the surface. If the velocity is in the x direction, the river bed is $z = 0$ while the surface is $z = h$, what is the curl of the velocity?

4 Let the water velocity in Prob. 3 increase from zero at the river bed to v_m at $z = h/2$ and then decrease linearly from v_m at $z = h/2$ to zero at $z = h$. What is the curl of the velocity?

5 A rigid rod of radius R turns about its axis with constant angular velocity $\boldsymbol{\omega}$. The velocity of an arbitrary point in the rod is then $\mathbf{v} = \boldsymbol{\omega} \times \mathbf{r}$. Find $\nabla \times \mathbf{v}$.

6 Prove that if $\nabla \times \mathbf{F} = 0$ then $\mathbf{F} = \nabla\phi$ is a necessary condition.

7 Suppose $\mathbf{F} = -\hat{\mathbf{x}}y/2 + \hat{\mathbf{y}}x/2$. What is $\nabla \times \mathbf{F}$? Draw curves of $\mathbf{F}$ for $|\mathbf{F}| = 1$ and $|\mathbf{F}| = 2$. What does $\mathbf{F}$ represent?

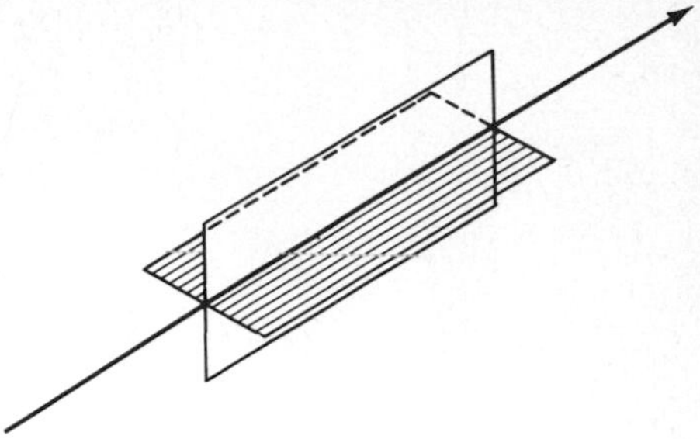

FIGURE 1-22

8 Suppose $\mathbf{F} = -\hat{\mathbf{x}}y$. Find $\nabla \times \mathbf{F}$. Draw curves for $|\mathbf{F}| = 1$ and $|\mathbf{F}| = 2$. What does **F** represent?

9 If $\mathbf{F} = \hat{\mathbf{x}}(-y/2 + x/2) + \hat{\mathbf{y}}x/2$, what is $\nabla \times \mathbf{F}$? Draw curves for $|\mathbf{F}| = 1$ and $|\mathbf{F}| = 2$. Compare this with Prob. 7.

10 If $\mathbf{F} = (\hat{\mathbf{x}}x + \hat{\mathbf{y}}y)/(x^2 + y^2)$, what is $\nabla \times \mathbf{F}$? Draw a curve for $|\mathbf{F}| = 1$ and indicate **F** on it at four points. This function represents the electric field produced by an infinitely long charged wire. What is $\nabla \times \mathbf{F}$ at $x = y = 0$? **F** is also $\hat{\mathbf{r}}(1/r)$ (cylindrical). Find $\nabla \times \mathbf{F}$ at $r = 0$ in cylindrical coordinates.

11 If $\mathbf{F} = \hat{\boldsymbol{\phi}}(1/r)$ (cylindrical) $= (-\hat{\mathbf{x}}y + \hat{\mathbf{y}}x)/(x^2 + y^2)$ find $\nabla \times \mathbf{F}$ everywhere, including the z axis. Do this both in cartesian and cylindrical coordinates. This function represents the magnetic field produced by a current flowing in an infinitely long wire on the z axis.

12 Draw a curve for $|\mathbf{F}| = 1$ in Prob. 11 and indicate **F** on it at four points. Note that, in comparing Prob. 10 and Prob. 11, $\nabla \times \mathbf{F} = 0$ in both cases except at $r = 0$. These two cases are very different physically. That is why a conservative field must have $\nabla \times \mathbf{F} = 0$ *everywhere.*

1-7 STOKES' THEOREM

Stokes' theorem is one which relates an integral over an open surface to another integral over the closed curve which bounds that surface. It is obtained by applying the definition of the curl to an open surface bounded by a closed curve, assuming the function we are dealing with is continuous. This is quite analogous to the way Gauss' divergence theorem was obtained by applying the definition of the divergence to a volume bounded by a closed surface. A finite open area of any shape, not necessarily flat, can be divided into an infinite number of infinitesimal rectangles each having a bounding closed curve. The equations

$$(\nabla \times \mathbf{F})\cdot\hat{\mathbf{n}} = \lim_{\Delta S \to 0} \left[\frac{1}{\Delta S} \oint_{\text{small rect}} \mathbf{F}\cdot d\mathbf{r} \right]$$

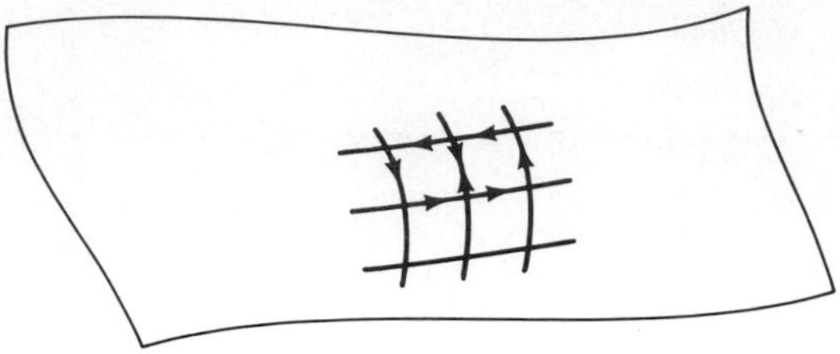

FIGURE 1-23

or

$$\oint_{\text{small rect}} \mathbf{F}\cdot d\mathbf{r} = \lim_{\Delta S\to 0} [(\nabla \times \mathbf{F})\cdot\Delta\mathbf{S}]$$

may be applied to each of these rectangles, as shown in Fig. 1-23. At the edge common to two such rectangles the contribution to the circulation around one rectangle is equal in magnitude but opposite in direction to the contribution, along that edge, to the circulation around the second rectangle; this is assuming the function is continuous. In the sum of the circulations about the two rectangles these two contributions cancel each other. The net circulation for all the rectangles is only the circulation around the boundary curve of the entire, finite, open area. This is Stokes' theorem. Mathematically,

$$\sum_{i=1}^{N} \oint_{C_i} \mathbf{F}\cdot d\mathbf{r}_i = \sum_{i=1}^{N} \lim_{\Delta S_i\to 0} [(\nabla \times \mathbf{F})\cdot\Delta\mathbf{S}_i]$$

$$\lim_{N\to\infty}\left[\sum_{i=1}^{N} \oint_{C_i} \mathbf{F}\cdot d\mathbf{r}_i\right] = \lim_{N\to\infty}\left\{\sum_{i=1}^{N} \lim_{\Delta S_i\to 0} [(\nabla \times \mathbf{F})\cdot\Delta\mathbf{S}_i]\right\}$$

So, for the curve C bounding the open area Σ,

$$\boxed{\oint_C \mathbf{F}\cdot d\mathbf{r} = \int_\Sigma (\nabla \times \mathbf{F})\cdot d\mathbf{S}}$$

Just as Gauss' divergence theorem gives a connection between a triple integral and a double integral, so Stokes' theorem gives a connection between a double integral and a single integral. It is not possible to combine these, however, to obtain a relation between a triple integral and a single integral, for one theorem applies only to a closed surface while the other applies only to an open surface.

Examples

1. A check of Stokes' theorem for $\mathbf{F} = \hat{\mathbf{x}}(x+y) - \hat{\mathbf{y}}2x^2 + \hat{\mathbf{z}}xy$ and the upper hemisphere of $x^2 + y^2 + z^2 = 1$.

$$\nabla \times \mathbf{F} = \begin{vmatrix} \hat{\mathbf{x}} & \hat{\mathbf{y}} & \hat{\mathbf{z}} \\ \partial_x & \partial_y & \partial_z \\ x+y & -2x^2 & xy \end{vmatrix} = \hat{\mathbf{x}}(x-0) + \hat{\mathbf{y}}(0-y) + \hat{\mathbf{z}}(-4x-1)$$

$$= \hat{\mathbf{x}}x - \hat{\mathbf{y}}y - \hat{\mathbf{z}}(4x+1)$$

$$\int_\Sigma (\nabla \times \mathbf{F})\cdot d\mathbf{S} = \int_\Sigma [\hat{\mathbf{x}}x - \hat{\mathbf{y}}y - \hat{\mathbf{z}}(4x+1)]\cdot\hat{\mathbf{n}}\, dS$$

$$= \int_\sigma [\hat{\mathbf{x}}x - \hat{\mathbf{y}}y - \hat{\mathbf{z}}(4x+1)]\cdot\hat{\mathbf{n}} \frac{dx\,dy}{\cos(\hat{\mathbf{n}},\hat{\mathbf{z}})}$$

Here σ is the projection of Σ on the xy plane as shown in Fig. 1-24. The gradient to the surface $f = x^2 + y^2 + z^2 - 1$ is $\nabla f = \hat{\mathbf{x}}2x + \hat{\mathbf{y}}2y + \hat{\mathbf{z}}2z$. Then $|\nabla f| = \sqrt{4x^2 + 4y^2 + 4z^2} = 2\sqrt{x^2 + y^2 + z^2} = 2$, and $\hat{\mathbf{n}} = \nabla f/|\nabla f| = \hat{\mathbf{x}}x + \hat{\mathbf{y}}y + \hat{\mathbf{z}}z$. So

$$[\hat{\mathbf{x}}x - \hat{\mathbf{y}}y - \hat{\mathbf{z}}(4x+1)]\cdot\hat{\mathbf{n}} = [\hat{\mathbf{x}}x - \hat{\mathbf{y}}y - \hat{\mathbf{z}}(4x+1)]\cdot[\hat{\mathbf{x}}x + \hat{\mathbf{y}}y + \hat{\mathbf{z}}z]$$

$$= x^2 - y^2 - (4x+1)z$$

Also, $\hat{\mathbf{n}}\cdot\hat{\mathbf{z}} = (1)(1)\cos(\hat{\mathbf{n}},\hat{\mathbf{z}}) = (x)(0) + (y)(0) + (z)(1) = z$ so $\cos(\hat{\mathbf{n}},\hat{\mathbf{z}}) = z$. The integral is, therefore,

$$\int_\Sigma (\nabla \times \mathbf{F})\cdot d\mathbf{S} = \int_{x=-1}^{1}\int_{y=-\sqrt{1-x^2}}^{\sqrt{1-x^2}} [x^2 - y^2 - (4x+1)z]\frac{dx\,dy}{z}$$

$$= \int_{x=-1}^{1} dx \int_{y=-\sqrt{1-x^2}}^{\sqrt{1-x^2}} \frac{x^2 - y^2}{\sqrt{1-x^2-y^2}}\,dy$$

$$- \int_{x=-1}^{1} dx \int_{y=-\sqrt{1-x^2}}^{\sqrt{1-x^2}} (4x+1)\,dy$$

$$= \int_{x=-1}^{1} x^2\,dx \int_{y=-\sqrt{1-x^2}}^{\sqrt{1-x^2}} \frac{dy}{\sqrt{(1-x^2)-y^2}}$$

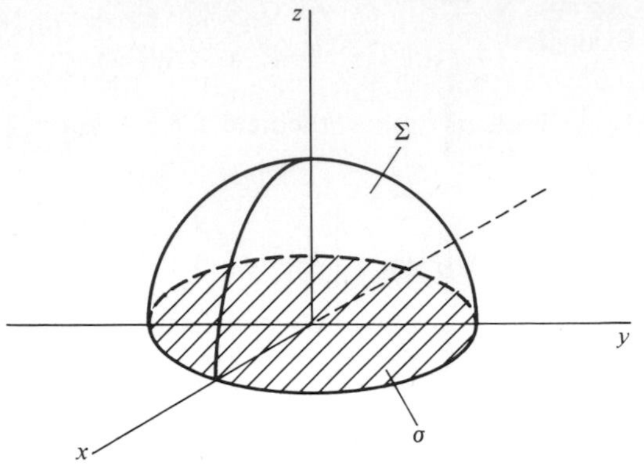

FIGURE 1-24

$$- \int_{x=-1}^{1} dx \int_{y=-\sqrt{1-x^2}}^{\sqrt{1-x^2}} \frac{y^2 dy}{\sqrt{(1-x^2)-y^2}}$$

$$- \int_{x=-1}^{1} (4x+1) dx \int_{y=-\sqrt{1-x^2}}^{\sqrt{1-x^2}} dy$$

$$= \int_{x=-1}^{1} x^2 \, dx \left[\sin^{-1} \frac{y}{\sqrt{1-x^2}} \right]_{y=-\sqrt{1-x^2}}^{+\sqrt{1-x^2}}$$

$$- \int_{x=-1}^{1} dx \left[-\frac{y}{2} \sqrt{(1-x^2)-y^2} \right.$$

$$\left. + \frac{1-x^2}{2} \sin^{-1} \frac{y}{\sqrt{1-x^2}} \right]_{y=-\sqrt{1-x^2}}^{+\sqrt{1-x^2}}$$

$$- \int_{x=-1}^{1} (4x+1) \, dx \left(2\sqrt{1-x^2} \right)$$

$$= \int_{-1}^{1} x^2 \, dx \left[\frac{\pi}{2} - \left(-\frac{\pi}{2} \right) \right]$$

$$- \int_{-1}^{1} dx \left[0 + \frac{(1-x^2)}{2} \frac{\pi}{2} - 0 - \frac{(1-x^2)}{2} \left(-\frac{\pi}{2} \right) \right]$$

$$- \int_{-1}^{1} 8x\sqrt{1-x^2}\,dx - 2\int_{-1}^{1} \sqrt{1-x^2}\,dx$$

$$= \pi \int_{-1}^{1} x^2\,dx - \frac{\pi}{2}\int_{-1}^{1} (1-x^2)\,dx$$

$$-8\int_{-1}^{1} x\sqrt{1-x^2}\,dx - 2\int_{-1}^{1} \sqrt{1-x^2}\,dx$$

$$= \pi\left[\frac{x^3}{3}\right]_{-1}^{1} - \frac{\pi}{2}\left[x - \frac{x^3}{3}\right]_{-1}^{1} - 8\left(-\frac{1}{3}\right)[(1-x^2)^{3/2}]_{-1}^{1}$$

$$-2\left(\frac{1}{2}\right)[x\sqrt{1-x^2} - \sin^{-1} x]_{-1}^{1}$$

$$= \frac{\pi}{3}[1-(-1)] - \frac{\pi}{2}\left[1 - \frac{1}{3} - (-1) + \frac{-1}{3}\right]$$

$$+ \frac{8}{3}(0) - \left[0 + \frac{\pi}{2} - 0 - \left(-\frac{\pi}{2}\right)\right]$$

$$= \frac{\pi}{3}(2) - \frac{\pi}{2}\left(\frac{4}{3}\right) + 0 - \pi$$

$$= -\pi$$

So much for the surface integral. Now the line integral.

$$\oint_C \mathbf{F}\cdot d\mathbf{r} = \oint_C [\hat{\mathbf{x}}(x+y) - \hat{\mathbf{y}}(2x^2) + \hat{\mathbf{z}}xy]\cdot[\hat{\mathbf{x}}\,dx + \hat{\mathbf{y}}\,dy + \hat{\mathbf{z}}\,dz]$$

$$= \oint_C [(x+y)dx - 2x^2\,dy]$$

Let $x = \cos\phi$, $y = \sin\phi$. Then $dx = -\sin\phi\,d\phi$, $dy = \cos\phi\,d\phi$

$$\oint_C \mathbf{F}\cdot d\mathbf{r} = \int_0^{2\pi} [(\cos\phi + \sin\phi)(-\sin\phi\,d\phi) - 2\cos^3\phi\,d\phi]$$

$$= 0 - \pi - \frac{2}{3} [\sin \phi(\cos^2 \phi) + 2 \sin \phi]_0^{2\pi}$$

$$= -\pi$$

This checks Stokes' theorem for this case. The line integral here is much easier to evaluate than the surface integral.

2. A check of Stokes' theorem for $\mathbf{F} = \hat{\mathbf{y}}x + \hat{\mathbf{z}}y$ and the unit square in the xy plane with center at the origin and sides parallel to the axes.

$$\nabla \times \mathbf{F} = \begin{vmatrix} \hat{\mathbf{x}} & \hat{\mathbf{y}} & \hat{\mathbf{z}} \\ \partial_x & \partial_y & \partial_z \\ 0 & x & y \end{vmatrix} = \hat{\mathbf{x}} + \hat{\mathbf{z}}$$

$$\int_\Sigma (\nabla \times \mathbf{F}) \cdot d\mathbf{S} = \int_\Sigma (\hat{\mathbf{x}} + \hat{\mathbf{z}}) \cdot \hat{\mathbf{z}}\, dS = \int_\Sigma dS = 1$$

$$\oint_C \mathbf{F} \cdot d\mathbf{r} = \oint_C (\hat{\mathbf{y}}x + \hat{\mathbf{z}}y) \cdot (\hat{\mathbf{x}}\, dx + \hat{\mathbf{y}}\, dy) = \oint_C x\, dy$$

$$= 0 + \frac{1}{2} \int_{-1/2}^{1/2} dy + 0 + \left(-\frac{1}{2}\right) \int_{1/2}^{-1/2} dy = \frac{1}{2} + \frac{1}{2} = 1 \qquad \text{Check.}$$

PROBLEMS

1 Check Stokes' theorem for $\mathbf{F} = \hat{\mathbf{y}}x + \hat{\mathbf{z}}y$ and the unit square with corners at (0,0), (0,1), (1,1), (1,0); i.e., calculate both the double integral and the single integral.

2 It will be shown in Chap. 3 that the magnetic field **B** is solenoidal ($\nabla \cdot \mathbf{B} = 0$) and that it can be obtained from a vector potential ($\mathbf{B} = \nabla \times \mathbf{A}$). Assuming this to be so, find the fallacy in the following argument: $\Phi = \int \mathbf{B} \cdot d\mathbf{S} = \int \nabla \cdot \mathbf{B}\, d\tau = 0$. Then $\int (\nabla \times \mathbf{A}) \cdot d\mathbf{S} = 0 = \oint \mathbf{A} \cdot d\mathbf{r}$. So $\mathbf{A} = \nabla \phi$. But then $\mathbf{B} = \nabla \times \nabla \phi \equiv 0$.

3 Find $\int (\nabla \times \mathbf{F}) \cdot d\mathbf{S}$ with $\mathbf{F} = \hat{\mathbf{x}}(x + y) - \hat{\mathbf{y}}2x^2 + \hat{\mathbf{z}}xy$ by actual integration over the closed surface of a right circular cylinder extending between $z = 0$ and $z = L$, having unit radius, and symmetric with respect to the z axis.

4 If in Example 1 the same vector $\mathbf{F} = \hat{\mathbf{x}}(x + y) - \hat{\mathbf{y}}2x^2 + \hat{\mathbf{z}}xy$ is used but the area is taken to be the bottom hemisphere of $x^2 + y^2 + z^2 = 1$, what will be the new value of the surface integral?

5 Without actually evaluating the surface integral, find $\oint_\Sigma (\nabla \times \mathbf{F}) \cdot d\mathbf{S}$ for the **F** of Example 1 but with the Σ of a pyramid consisting of four equilateral triangles where one—the base—is in the xy plane with its center at the origin. One side, parallel to the x axis, is in the third and fourth quadrants.

6 Check Stokes' theorem for $\mathbf{F} = -\hat{\mathbf{x}}y/2 + \hat{\mathbf{y}}x/2$ and the circle $x^2 + y^2 = 1$ using cartesian coordinates.

7 Prove that the area enclosed by a curve C in the xy plane equals

$$\oint_C [-(y/2)\,dx + (x/2)\,dy]$$

8 Do Prob. 6 using polar coordinates.

9 How many answers are there to the question: what is $\int_\Sigma (\nabla \times \mathbf{F}) \cdot d\mathbf{S}$ for a given **F** and an open Σ? How are they related?

10 Show $\oint d\mathbf{S} = 0$.

11 Apply Stokes' theorem to the product $f\mathbf{C}$, where f is a scalar function of position and **C** is a vector with constant components. Find $\int_\Sigma (\nabla f) \times d\mathbf{S}$ in terms of a contour integral.

1-8 THE LAPLACIAN

Suppose one takes the divergence of a vector function which is, itself, the gradient of a scalar function f: $\nabla \cdot \mathbf{F} = \nabla \cdot (\nabla f)$. In cartesian coordinates,

$$\nabla \cdot \mathbf{F} = \left(\hat{\mathbf{x}}\frac{\partial}{\partial x} + \hat{\mathbf{y}}\frac{\partial}{\partial y} + \hat{\mathbf{z}}\frac{\partial}{\partial z}\right) \cdot \left(\hat{\mathbf{x}}\frac{\partial f}{\partial x} + \hat{\mathbf{y}}\frac{\partial f}{\partial y} + \hat{\mathbf{z}}\frac{\partial f}{\partial z}\right)$$

$$= \frac{\partial^2 f}{\partial x^2} + \frac{\partial^2 f}{\partial y^2} + \frac{\partial^2 f}{\partial z^2}$$

This is defined as the laplacian of f. It turns out that $\nabla \cdot (\nabla f) = (\nabla \cdot \nabla) f$ only in cartesian coordinates, but the symbol ∇^2 is employed for the laplacian in all coordinate systems. Laplacian $f \equiv \nabla^2 f$ means, always, div(grad f).

The laplacian operator may also be applied to a vector function $\mathbf{F}(\mathbf{r})$: $\nabla^2 \mathbf{F}$. This means that the div grad of each of the components of **F**, multiplied by the appropriate unit vector. In cartesian coordinates

$$\nabla \cdot \nabla\, \mathbf{F} = \nabla^2 \mathbf{F} = \nabla^2 (\hat{\mathbf{x}}F_x + \hat{\mathbf{y}}F_y + \hat{\mathbf{z}}F_z) = \hat{\mathbf{x}}\,\nabla^2 F_x + \hat{\mathbf{y}}\nabla^2 F_y + \hat{\mathbf{z}}\,\nabla^2 F_z$$

$$= \hat{\mathbf{x}}\left(\frac{\partial^2 F_x}{\partial x^2} + \frac{\partial^2 F_x}{\partial y^2} + \frac{\partial^2 F_x}{\partial z^2}\right) + \hat{\mathbf{y}}\left(\frac{\partial^2 F_y}{\partial x^2} + \frac{\partial^2 F_y}{\partial y^2} + \frac{\partial^2 F_y}{\partial z^2}\right)$$

$$+ \hat{\mathbf{z}}\left(\frac{\partial^2 F_z}{\partial x^2} + \frac{\partial^2 F_z}{\partial y^2} + \frac{\partial^2 F_z}{\partial z^2}\right)$$

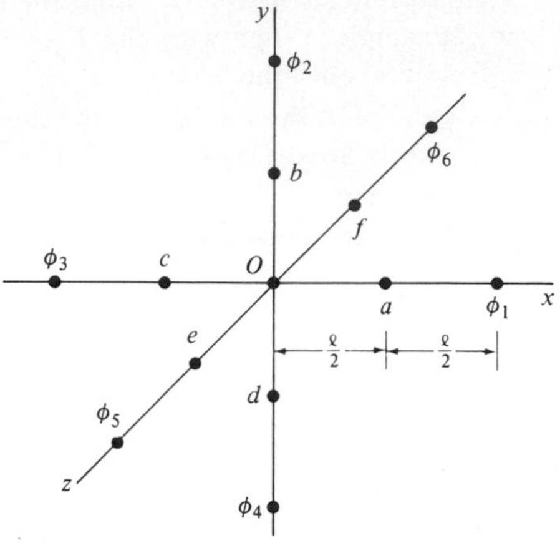

FIGURE 1-25

In other coordinate systems this quantity is considerably more complicated.

The physical significance of the laplacian may be brought out by reference to Fig. 1-25. ϕ_0 is the value of a function at the point O; $\phi_1, \ldots, \phi_6$ are the values of the function at six points equally spaced about O and separated from O by the distance l. Points a, b, c, d, e, and f are situated midway between these points and O. Then

$$\left(\frac{\partial \phi}{\partial x}\right)_a \simeq \frac{\phi_1 - \phi_0}{l}$$

The approximation is better the closer a is to O. Similarly,

$$\left(\frac{\partial \phi}{\partial x}\right)_c \simeq \frac{\phi_0 - \phi_3}{l}$$

Consequently

$$\left(\frac{\partial^2 \phi}{\partial x^2}\right)_0 \simeq \frac{(\phi_1 - \phi_0)/l - (\phi_0 - \phi_3)/l}{l} \simeq \frac{1}{l^2}(\phi_1 + \phi_3 - 2\phi_0)$$

For the same reason,

$$\left(\frac{\partial^2 \phi}{\partial y^2}\right)_0 \simeq \frac{1}{l^2}(\phi_2 + \phi_4 - 2\phi_0) \qquad \text{and} \qquad \left(\frac{\partial^2 \phi}{\partial z^2}\right)_0 \simeq \frac{1}{l^2}(\phi_5 + \phi_6 - 2\phi_0)$$

Then $(\nabla^2 \phi)_2 \simeq (6/l^2)(\bar{\phi} - \phi_0)$ where $\bar{\phi} = \frac{1}{6}(\phi_1 + \phi_2 + \phi_3 + \phi_4 + \phi_5 + \phi_6)$ is the average value of ϕ at the six points equally spaced from O.

By an extension of this argument to an infinite number of very near points, equally spaced from O (i.e., the surface of a small sphere, of radius r, about O) the laplacian of ϕ at O is found to satisfy a similar equation:

$$\nabla^2 \phi \simeq -\frac{6}{r^2}(\phi_0 - \bar{\phi})$$

Maxwell called $-\nabla^2\phi$ the concentration of ϕ. When $-\nabla^2\phi$ is zero at a point then the average value of the function on a small sphere about that point is the same as the value of the function at the point; when $-\nabla^2\phi$ is positive at some point then the value of ϕ at that point exceeds the average value near that point; if $-\nabla^2\phi$ is negative the value of the function at that point is less than the average value on the small sphere. The "$\simeq$" instead of an "=" signifies that the above relations are only approximately true; other terms, of order r^{-4}, are also present.

Examples

1. In the next chapter it will be shown that the electrostatic potential in free space ϕ satisfies Poisson's equation

$$\nabla^2 \phi = -\frac{\rho}{\epsilon_0}$$

where ρ is the electric charge density at the point where $\nabla^2\phi$ is being evaluated and ϵ_0 is a constant. We discuss the relation between ρ and the average value of ϕ on the surface of a small sphere about the point in question.

(*a*) If $\rho = 0$ then $\nabla^2\phi = 0$. Here $\bar{\phi}$ on the sphere equals ϕ_0 and Poisson's equation becomes Laplace's equation $\nabla^2\phi = 0$. This happens at any point which is free of charge.

(*b*) If $\rho > 0$ then $\nabla^2\phi < 0$. Here there is a deficiency of ϕ on the sphere: $\bar{\phi}$ is less than ϕ_0. At a point which has positive charge, the potential has a higher value than the value of the average potential at nearby points.

(*c*) If $\rho < 0$ then $\nabla^2\phi > 0$. Here $\bar{\phi}$ on the sphere is greater than ϕ_0 at the center. A point with negative charge has a lower potential than the average potential of nearby points.

2. (This example requires some knowledge of functions of a complex variable and may be omitted without any loss of continuity.) To show that the real and imaginary components of an analytic function of a complex variable (*a*) satisfy Laplace's equation and (*b*) give families of curves which are orthogonal to each other.

(*a*) Let $z = x + iy$ be a point in the complex plane and $w = f(z) = u(x,y) + iv(x,y)$ be a complex function of the position in this plane. Then

$$\frac{\Delta w}{\Delta z} = \frac{\left(\frac{\partial u}{\partial x} + i\frac{\partial v}{\partial x}\right)\Delta x + \left(\frac{\partial u}{\partial y} + i\frac{\partial v}{\partial y}\right)\Delta y}{\Delta x + i\Delta y}$$

$$= \frac{\left(\frac{\partial u}{\partial x} + i\frac{\partial v}{\partial x}\right) + \frac{\Delta y}{\Delta x}\left(\frac{\partial u}{\partial y} + i\frac{\partial v}{\partial y}\right)}{1 + i\frac{\Delta y}{\Delta x}}$$

$$= \left(\frac{\partial u}{\partial x} + i\frac{\partial v}{\partial x}\right)\left[\frac{1 + \dfrac{i\frac{\Delta y}{\Delta x}\left(\frac{\partial v}{\partial y} - i\frac{\partial u}{\partial y}\right)}{\left(\frac{\partial u}{\partial x} + i\frac{\partial v}{\partial x}\right)}}{1 + i\frac{\Delta y}{\Delta x}}\right]$$

If $dw/dz = \lim_{\Delta z \to 0} (\Delta w/\Delta z)$ is to be independent of the direction of Δz, then

$$\frac{\frac{\partial v}{\partial y} - i\frac{\partial u}{\partial y}}{\frac{\partial u}{\partial x} + i\frac{\partial v}{\partial x}}$$

must equal unity. This gives the Cauchy-Riemann equations:

$$\frac{\partial u}{\partial x} = \frac{\partial v}{\partial y} \qquad \frac{\partial u}{\partial y} = -\frac{\partial v}{\partial x}$$

These are necessary conditions for w to be analytic at the point considered. They can also be shown to be sufficient conditions.

Differentiating the first with respect to x, the second with respect to y, and comparing gives

$$\frac{\partial^2 u}{\partial x^2} + \frac{\partial^2 u}{\partial y^2} = 0$$

If the first is differentiated with respect to y and the second with respect to x there results

$$\frac{\partial^2 v}{\partial x^2} + \frac{\partial^2 v}{\partial y^2} = 0$$

Thus, u and v individually satisfy Laplace's equation in two dimensions.

(*b*) ∇u is orthogonal to u = const; ∇v is orthogonal to v = const. If ∇u is orthogonal to ∇v then the family of curves u = const is orthogonal to the family v = const. Taking the two-dimensional gradients,

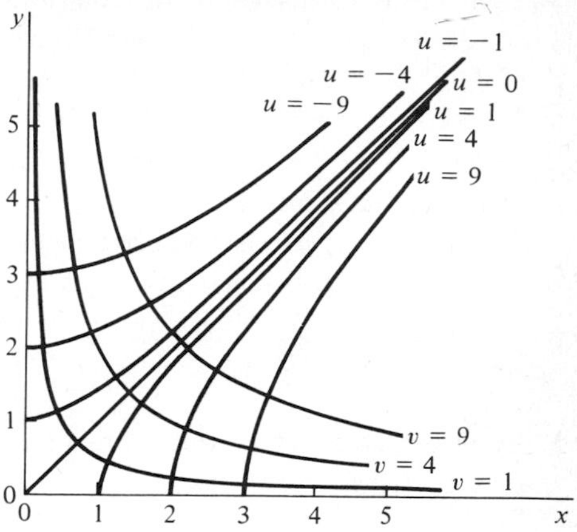

FIGURE 1-26

$$u = x^2 - y^2$$
$$v = 2xy$$

One family of curves gives the equipotentials; the other, orthogonal to the first, gives the lines of flux

$$\nabla u \cdot \nabla v = \left(\hat{\mathbf{x}}\frac{\partial u}{\partial x} + \hat{\mathbf{y}}\frac{\partial u}{\partial y}\right) \cdot \left(\hat{\mathbf{x}}\frac{\partial v}{\partial x} + \hat{\mathbf{y}}\frac{\partial v}{\partial y}\right)$$

$$= \frac{\partial u}{\partial x}\frac{\partial v}{\partial x} + \frac{\partial u}{\partial y}\frac{\partial v}{\partial y} = \frac{\partial u}{\partial x}\left(-\frac{\partial u}{\partial y}\right) + \frac{\partial u}{\partial y}\left(\frac{\partial u}{\partial x}\right) = 0$$

This method of taking the functions in the complex plane is useful in solving two-dimensional problems in electrostatics. The two orthogonal families correspond to the lines of flux and the equipotentials. Figure 1-26 shows the field lines, $u = x^2 - y^2$, and the equipotentials, $v = 2xy$, for the case of a right angle with charged conducting walls.

PROBLEMS

1 Find $\nabla^2\phi$ for (*a*) $\phi = \text{const}$, for (*b*) $\phi = ax + by$, for (*c*) $\phi = ax^2 + bxy + cy^2$, and for (*d*) $\phi = \Sigma_{n=0}^{N} a_n x^n$.

2 Laplace's equation is a partial, second-order, homogeneous differential equation. Explain what these four qualifying adjectives mean. How many arbitrary functions are there in the general solution? Change the equation to Poisson's equation; how is the general solution of the previous case modified?

3 Find $\nabla^2\phi$ for $\phi = \sqrt{x}$, $\phi = \sqrt{x^3}$, $\phi = \sqrt{x^5}$. Graph ϕ and $\nabla^2\phi$ for $0 \leqslant x \leqslant 1$ at invervals of $\Delta x = \frac{1}{4}$.

4 Compare the behavior of the solutions of Prob. 3 at $x = 0$. Explain this on the basis of (*a*) the value of n in $x^{n/2}$ and (*b*) continuity.

5 If $\phi e^{i(\mathbf{k}\cdot\mathbf{r}-\omega t)}$ find $\nabla^2\phi$. This function represents a wave traveling in the direction $\mathbf{k} = \hat{\mathbf{x}}k_x + \hat{\mathbf{y}}k_y + \hat{\mathbf{z}}k_z$ with velocity $u = \omega/k$, where $k = |\mathbf{k}|$.

6 In Prob. 5 find $\partial^2\phi/\partial t^2$. From this and the answer to Prob. 5 write a differential equation that ϕ satisfies.

7 Does $x/(x^2 + y^2)$ satisfy Laplace's equation? Does $\sin x$? Does $\sin x\, e^y$?

8 If f and g are two scalar functions of $\mathbf{r}$ prove the identity

$$\nabla^2(fg) = f\nabla^2 g + g\nabla^2 f + 2(\nabla f)\cdot(\nabla g)$$

9 By means of the chain rule, obtain the expression for $\nabla^2 f$ in polar coordinates from that in cartesian coordinates.

10 Fill in the boxes in the following chart with the appropriate choices from among grad, div, curl, laplacian:

Final function \ Original function	Scalar	Vector
Scalar		
Vector		

11 Find the value of n in $-\nabla^2\psi = n/r^2(\psi - \psi_{av})$ if ψ is given at the center of a small sphere of radius r and ψ_{av} is its average value within the sphere, rather than on the surface of the sphere.

12 Discuss the meaning of the heat conduction partial differential equation,

$$\frac{\partial\Theta}{\partial t} = k\nabla^2\Theta$$

from the viewpoint of Maxwell's concentration of the temperature: $-\nabla^2\Theta$.

13 Change Prob. 11 to a cube of side a. Find n in $-\nabla^2\psi = (n/a^2)(\psi - \psi_{av})$ if ψ_{av} is the average value of ψ within the cube.

Appendix 1

A. THE DEL OPERATOR IN CYLINDRICAL AND SPHERICAL COORDINATES

Cylindrical

The del operator here is

$$\nabla = \hat{\mathbf{r}}\frac{\partial}{\partial r} + \hat{\boldsymbol{\phi}}\frac{1}{r}\frac{\partial}{\partial \phi} + \hat{\mathbf{z}}\frac{\partial}{\partial z}$$

This follows from

$$df = \nabla f \cdot d\mathbf{r} = \frac{\partial f}{\partial r}\,dr + \frac{\partial f}{\partial \phi}\,d\phi + \frac{\partial f}{\partial z}\,dz$$

and

$$d\mathbf{r} = \hat{\mathbf{r}}\,dr + \hat{\boldsymbol{\phi}} r\,d\phi + \hat{\mathbf{z}}\,dz$$

Next the following are needed:

$$\frac{\partial \hat{\mathbf{r}}}{\partial r} = 0 \qquad \frac{\partial \hat{\boldsymbol{\phi}}}{\partial r} = 0 \qquad \frac{\partial \hat{\mathbf{z}}}{\partial r} = 0$$

$$\frac{\partial \hat{\mathbf{r}}}{\partial \phi} = \hat{\boldsymbol{\phi}} \qquad \frac{\partial \hat{\boldsymbol{\phi}}}{\partial \phi} = -\hat{\mathbf{r}} \qquad \frac{\partial \hat{\mathbf{z}}}{\partial \phi} = 0$$

$$\frac{\partial \hat{\mathbf{r}}}{\partial z} = 0 \qquad \frac{\partial \hat{\boldsymbol{\phi}}}{\partial z} = 0 \qquad \frac{\partial \hat{\mathbf{z}}}{\partial z} = 0$$

From these equations the values of $\nabla \cdot \mathbf{F}$, $\nabla \times \mathbf{F}$, and $\nabla^2 f$ in cylindrical coordinates may be found. The expression for the divergence of $\mathbf{F}$, e.g., is obtained as follows:

$$\nabla \cdot \mathbf{F} = \left(\hat{\mathbf{r}}\frac{\partial}{\partial r} + \hat{\boldsymbol{\phi}}\frac{1}{r}\frac{\partial}{\partial \phi} + \hat{\mathbf{z}}\frac{\partial}{\partial z}\right) \cdot (\hat{\mathbf{r}}F_r + \hat{\boldsymbol{\phi}}F_\phi + \hat{\mathbf{z}}F_z)$$

$$\nabla \cdot \mathbf{F} = \hat{\mathbf{r}}\frac{\partial}{\partial r} \cdot (\hat{\mathbf{r}}F_r) + \hat{\mathbf{r}}\frac{\partial}{\partial r} \cdot (\hat{\boldsymbol{\phi}}F_\phi) + \hat{\mathbf{r}}\frac{\partial}{\partial r} \cdot (\hat{\mathbf{z}}F_z)$$

$$+ \hat{\boldsymbol{\phi}}\frac{1}{r}\frac{\partial}{\partial \phi} \cdot (\hat{\mathbf{r}}F_r) + \hat{\boldsymbol{\phi}}\frac{1}{r}\frac{\partial}{\partial \phi} \cdot (\hat{\boldsymbol{\phi}}F_\phi) + \hat{\boldsymbol{\phi}}\frac{1}{r}\frac{\partial}{\partial \phi} \cdot (\hat{\mathbf{z}}F_z)$$

$$+ \hat{\mathbf{z}}\frac{\partial}{\partial z} \cdot (\hat{\mathbf{r}}F_r) + \hat{\mathbf{z}}\frac{\partial}{\partial z} \cdot (\hat{\boldsymbol{\phi}}F_\phi) + \hat{\mathbf{z}}\frac{\partial}{\partial z} \cdot (\hat{\mathbf{z}}F_z)$$

$$\nabla \cdot \mathbf{F} = \hat{\mathbf{r}} \cdot \left[\hat{\mathbf{r}}\frac{\partial F_r}{\partial r} + F_r\frac{\partial \hat{\mathbf{r}}}{\partial r}\right] + \hat{\mathbf{r}} \cdot \left[\hat{\boldsymbol{\phi}}\frac{\partial F_\phi}{\partial r} + F_\phi\frac{\partial \hat{\phi}}{\partial r}\right] + \hat{\mathbf{r}} \cdot \left[\hat{\mathbf{z}}\frac{\partial F_z}{\partial r} + F_z\frac{\partial \hat{\mathbf{z}}}{\partial r}\right]$$

$$+\hat{\boldsymbol{\phi}}\cdot\left[\frac{\hat{\mathbf{r}}}{r}\frac{\partial F_r}{\partial\phi}+\frac{F_r}{r}\frac{\partial\hat{\mathbf{r}}}{\partial\phi}\right]+\hat{\boldsymbol{\phi}}\cdot\left[\frac{\hat{\boldsymbol{\phi}}}{r}\frac{\partial F_\phi}{\partial\phi}+\frac{F_\phi}{r}\frac{\partial\hat{\boldsymbol{\phi}}}{\partial\phi}\right]$$

$$+\hat{\boldsymbol{\phi}}\cdot\left[\frac{\hat{\mathbf{z}}}{r}\frac{\partial F_z}{\partial\phi}+\frac{F_z}{r}\frac{\partial\hat{\mathbf{z}}}{\partial\phi}\right]$$

$$+\hat{\mathbf{z}}\cdot\left[\hat{\mathbf{r}}\frac{\partial F_r}{\partial z}+F_r\frac{\partial\hat{\mathbf{r}}}{\partial z}\right]$$

$$+\hat{\mathbf{z}}\cdot\left[\hat{\boldsymbol{\phi}}\frac{\partial F_\phi}{\partial z}+F_\phi\frac{\partial\hat{\boldsymbol{\phi}}}{\partial z}\right]+\hat{\mathbf{z}}\cdot\left[\hat{\mathbf{z}}\frac{\partial F_z}{\partial z}+F_z\frac{\partial\hat{\mathbf{z}}}{\partial z}\right]$$

$$\nabla\cdot\mathbf{F}=\hat{\mathbf{r}}\cdot\hat{\mathbf{r}}\frac{\partial F_r}{\partial r}+\hat{\mathbf{r}}\cdot\hat{\boldsymbol{\phi}}\frac{\partial F_\phi}{\partial r}+\hat{\mathbf{r}}\cdot\hat{\mathbf{z}}\frac{\partial F_z}{\partial r}$$

$$+\hat{\boldsymbol{\phi}}\cdot\frac{\hat{\mathbf{r}}}{r}\frac{\partial F_r}{\partial\phi}+\hat{\boldsymbol{\phi}}\cdot\hat{\boldsymbol{\phi}}\frac{F_r}{r}+\hat{\boldsymbol{\phi}}\cdot\frac{\hat{\boldsymbol{\phi}}}{r}\frac{\partial F_\phi}{\partial\phi}-\hat{\boldsymbol{\phi}}\cdot\hat{\mathbf{r}}\frac{F_\phi}{r}+\hat{\boldsymbol{\phi}}\cdot\frac{\hat{\mathbf{z}}}{r}\frac{\partial F_z}{\partial\phi}$$

$$+\hat{\mathbf{z}}\cdot\hat{\mathbf{r}}\frac{\partial F_r}{\partial z}+\hat{\mathbf{z}}\cdot\hat{\boldsymbol{\phi}}\frac{\partial F_\phi}{\partial z}+\hat{\mathbf{z}}\cdot\hat{\mathbf{z}}\frac{\partial F_z}{\partial z}$$

$$\nabla\cdot\mathbf{F}=\frac{\partial F_r}{\partial r}+\frac{F_r}{r}+\frac{1}{r}\frac{\partial F_\phi}{\partial\phi}+\frac{\partial F_z}{\partial z}$$

Proceeding in this fashion, the other results are similarly obtained. The relations are all summarized below.

Gradient:
$$\nabla f=\hat{\mathbf{r}}\frac{\partial f}{\partial r}+\hat{\boldsymbol{\phi}}\frac{1}{r}\frac{\partial f}{\partial\phi}+\hat{\mathbf{z}}\frac{\partial f}{\partial z}$$

Divergence:
$$\nabla\cdot\mathbf{F}=\frac{1}{r}\frac{\partial}{\partial r}(rF_r)+\frac{1}{r}\frac{\partial F_\phi}{\partial\phi}+\frac{\partial F_z}{\partial z}$$

Curl:
$$\nabla\times\mathbf{F}=\hat{\mathbf{r}}\left[\frac{1}{r}\frac{\partial F_z}{\partial\phi}-\frac{\partial F_\phi}{\partial z}\right]$$

$$+\hat{\boldsymbol{\phi}}\left[\frac{\partial F_r}{\partial z}-\frac{\partial F_z}{\partial r}\right]$$

$$+\hat{\mathbf{z}}\left[\frac{1}{r}\frac{\partial}{\partial r}(rF_\phi)-\frac{1}{r}\frac{\partial F_r}{\partial\phi}\right]$$

Laplacian:
$$\nabla^2 f=\frac{1}{r}\frac{\partial}{\partial r}\left(\frac{r\partial f}{\partial r}\right)+\frac{1}{r^2}\frac{\partial^2 f}{\partial\phi^2}+\frac{\partial^2 f}{\partial z^2}$$

Spherical

The del operator here is

$$\nabla = \hat{\mathbf{r}} \frac{\partial}{\partial r} + \hat{\boldsymbol{\theta}} \frac{1}{r} \frac{\partial}{\partial \theta} + \hat{\boldsymbol{\phi}} \frac{1}{r \sin \theta} \frac{\partial}{\partial \phi}$$

The derivatives of the unit vectors are:

$$\frac{\partial \hat{\mathbf{r}}}{\partial r} = 0 \qquad \frac{\partial \hat{\boldsymbol{\theta}}}{\partial r} = 0 \qquad \frac{\partial \hat{\boldsymbol{\phi}}}{\partial r} = 0$$

$$\frac{\partial \hat{\mathbf{r}}}{\partial \phi} = \hat{\boldsymbol{\theta}} \qquad \frac{\partial \hat{\boldsymbol{\theta}}}{\partial \theta} = -\hat{\mathbf{r}} r \qquad \frac{\partial \hat{\boldsymbol{\phi}}}{\partial \theta} = 0$$

$$\frac{\partial \hat{\mathbf{r}}}{\partial \phi} = \hat{\boldsymbol{\phi}} \sin \theta \qquad \frac{\partial \hat{\boldsymbol{\theta}}}{\partial \phi} = \hat{\boldsymbol{\phi}} \cos \theta \qquad \frac{\partial \hat{\boldsymbol{\phi}}}{\partial \phi} = -\hat{\mathbf{r}} \sin \theta - \hat{\boldsymbol{\theta}} \cos \theta$$

The following results hold:

Gradient:

$$\nabla f = \hat{\mathbf{r}} \frac{\partial f}{\partial r} + \hat{\boldsymbol{\theta}} \frac{1}{r} \frac{\partial f}{\partial \theta} + \hat{\boldsymbol{\phi}} \frac{1}{r \sin \theta} \frac{\partial f}{\partial \phi}$$

Divergence:

$$\nabla \cdot \mathbf{F} = \frac{1}{r^2} \frac{\partial}{\partial r} (r^2 F_r) + \frac{1}{r \sin \theta} \frac{\partial}{\partial \theta} (\sin \theta F_\theta) + \frac{1}{r \sin \theta} \frac{\partial F_\phi}{\partial \phi}$$

Curl:

$$\nabla \times \mathbf{F} = \hat{\mathbf{r}} \left[\frac{1}{r \sin \theta} \left\{ \frac{\partial (\sin \theta F_\phi)}{\partial \theta} - \frac{\partial F_\theta}{\partial \phi} \right\} \right] + \hat{\boldsymbol{\theta}} \left[\frac{1}{r \sin \theta} \frac{\partial F_r}{\partial \phi} - \frac{1}{r} \frac{\partial}{\partial r} r(F_\phi) \right] + \hat{\boldsymbol{\phi}} \left[\frac{1}{r} \frac{\partial}{\partial r} (r F_\theta) - \frac{1}{r} \frac{\partial F_r}{\partial \theta} \right]$$

Laplacian:

$$\nabla^2 f = \frac{1}{r^2} \frac{\partial}{\partial r} \left(r^2 \frac{\partial f}{\partial r} \right) + \frac{1}{r^2 \sin \theta} \frac{\partial}{\partial \theta} \left(\sin \theta \frac{\partial f}{\partial \theta} \right) + \frac{1}{r^2 \sin^2 \theta} \frac{\partial^2 f}{\partial \phi^2}$$

B. VECTOR IDENTITIES

$$(ab) = (a \nabla b) + (b \nabla a)$$

$$\nabla(\mathbf{A} \cdot \mathbf{B}) = [(\mathbf{A} \cdot \nabla)\mathbf{B}] + [(\mathbf{B} \cdot \nabla)\mathbf{A}] + [\mathbf{A} \times (\nabla \times \mathbf{B})] + [\mathbf{B} \times (\nabla \times \mathbf{A})]$$

$$\nabla \cdot (a\mathbf{A}) = (\mathbf{A} \cdot \nabla a) + (a \nabla \cdot \mathbf{A})$$

$$\nabla \cdot (\mathbf{A} \times \mathbf{B}) = [\mathbf{B} \cdot (\nabla \times \mathbf{A})] - [\mathbf{A} \cdot (\nabla \times \mathbf{B})]$$

$$\nabla \times (a\mathbf{A}) = (\nabla a \times \mathbf{A}) + (a \nabla \times \mathbf{A})$$

$$\nabla \times (\mathbf{A} \times \mathbf{B}) = [(\nabla \cdot B)\mathbf{A}] - [(\nabla \cdot \mathbf{A})\mathbf{B}] + [(\mathbf{B} \cdot \nabla)\mathbf{A}] - [(\mathbf{A} \cdot \nabla)\mathbf{B}]$$

$$\mathbf{A} \cdot (\mathbf{B} \times \mathbf{C}) = \mathbf{B} \cdot (\mathbf{C} \times \mathbf{A}) = \mathbf{C} \cdot (\mathbf{A} \times \mathbf{B})$$

$$\mathbf{A} \times (\mathbf{B} \times \mathbf{C}) = \mathbf{B}(\mathbf{A} \cdot \mathbf{C}) - \mathbf{C}(\mathbf{A} \cdot \mathbf{B})$$

$$\nabla \times (\nabla \times \mathbf{C}) = \nabla(\nabla \cdot \mathbf{C}) - \nabla^2 \mathbf{C}$$

2

THE ELECTROSTATIC DIVERGENCE IN VACUUM

2-1 ORIGINS

Electricity and magnetism is now a classical branch of physics and its development has followed a classical pattern. In the beginning the rate of progress in our understanding of the subject was unbelievably slow. Although by 600 B.C. the ancient Greeks knew that amber (Greek: *elektron*), when rubbed, would attract small quantities of straw, silk, or other light objects, nothing further was done with this knowledge, and nothing further was learned about electricity, for 2200 years. Twenty-two centuries of standing still!

In A.D. 1600 William Gilbert, an Englishman, published a book which showed that amber was not unique: he listed many other substances possessing the electrification property of amber. Amber was merely the first such substance discovered, a result of the fact that it existed in abundance in its natural state in many parts of Greece.

This book marked a turning point in the development of our knowledge of electricity; the subject was now leaving the stage of infancy and entering adolescence. A graph of knowledge versus time would here begin to show an exponential rise. It took only another 133 years before DuFay found that the electrification of substances by rubbing could result in repulsion as well as attraction. Then, 14 years later, Benjamin Franklin, recognizing the division of electric charge into two types, labeled them as positive and negative. He arbitrarily made the choice of which was positive and which negative. The designations chosen had no basic significance; the reverse choice could have been made with equal validity and would, indeed, be preferable in many cases. Franklin also surmised that the total charge was conserved in electrification.

The floodgates to knowledge were now opened by the industrial revolution, and between 1750 and 1900 there occurred a torrent of advances in this branch of science. A long list of important discoveries was made by many people from many countries. The experiments of Michael Faraday and the theoretical work of James Clerk Maxwell were probably the high levels in this tide. Physicists in England, France, and Germany (and even the United States) contributed to this progress, and the present book is predominantly a report on the work that was done during that period. Important advances in mathematics—by Gauss, Laplace, Lagrange, Euler, etc.—helped the physicists greatly.

The two great advances of general physics in the 20th century—relatively and quantum mechanics—have had comparatively little effect on this classical body of knowledge in electricity and magnetism. The role played by special relativity was simple: it explained the existence of the magnetic field in terms of a transformation of the electric field for a moving observer. This had broad implications, but it only slightly altered the perspective of the older work. The role played by quantum mechanics per se, excluding quantum electrodynamics, was even more narrow. It explained several effects for which classical theory proved to be inadequate. Superconductivity, the Aharanov-Pohm effect, ferromagnetism as a quantum *electric* effect, and the laser serve as examples. But most quantum effects are rather special cases, and most phenomena of classical theory may be treated without considering them.

Since the beginning of the 20th century the study of electricity and magnetism has been in its mature stage of development. A steady, but ever slower, accretion of knowledge has taken place, so that the graph is asymptotically approaching a plateau. When a body of knowledge is in this third stage it is called classical.

It should not be inferred from this that a classical subject is one that is in a state of senility and that a study of it is, at best, fruitless. Certainly in electricity and magnetism there is no end to the unsolved problems that remain, problems of fundamental importance. But it is clear today that the solutions to these problems cannot be found by the old methods. Possibly quantum electrodynamics will supply the answers to such questions as the following.

1 Why are there two kinds of charge? Only one kind of gravitational mass has been found so far. Is mass a simpler and more fundamental property than charge?

2 Must charge always be associated with mass? Mass is not always associated with charge.

3 Electric charge is quantized. Why does the minimum quantity of charge have the value it does have? If this minimum were either much larger or much smaller there would certainly be a great change in many phenomena. Some recent theories have postulated the existence of charge having one-third or two-thirds the charge of the present basic unit. Called quarks by Gell-Mann and dions by Schwinger, these speculations trace back to an article by Dirac* on magnetic monopoles in which he pointed out that the existence of a magnetic charge would lead to a minimum unit of electric charge, by the quantum laws of physics. But all experiments to date to find either magnetic monopoles or quarks have been unsuccessful.

4 Why are electrical forces so overwhelmingly larger than gravitational forces? It is interesting that, despite a huge disparity by a factor of 10^{43} or so, the presence of two kinds of charge causes the electrical forces to balance themselves out on the astronomical scale. Only the very weak gravitational forces are needed to account, e.g., for the motion of planets around a star.

2-2 COULOMB'S LAW

To commence the study of electricity, consider the effects which a number of charges, the source charges, produce on one particular charge, the test charge. The source charges are the source of the effect on the test charge; later, when we introduce the field concept, we will take the source charges to be the source of the field. For simplicity the charges are initially all considered to be in a vacuum and are taken to be at rest with respect to each other. What force is produced on the test charge by the stationary source charges?

The answer, supplied by experiment, is summed up in Coulomb's law. Consider, first, only two charges, q_s and q_t. Which charge is q_s, the source charge? That is arbitrary. Next, assume that the charges are mathematical points; this is merely the limiting case of two charges of finite size with the distance between the charges R very much larger than the dimensions of the charges. Protons are now known to have a spatial extent of $\sim 10^{-13}$ cm. The assumption that protons are point charges is reasonable if the separation between charges is always taken to be greater than 10^{-11} cm. With electrons the story is more complicated. They may literally be elementary point charges having no structure; they are surrounded, however, by a cloud of virtual quanta which, in effect, give the electron a finite size. By staying to separations between charges greater than 10^{-11} cm, the size of this cloud also becomes of negligible importance. Let the vector that extends *from* the position of q_s *to* that of q_t be $\mathbf{R}_{st}$, so $R = |\mathbf{R}_{st}|$ and the unit vector in this direction is $\hat{\mathbf{R}}_{st}$. If $\mathbf{F}_{st}$ is the force exerted *by* q_s *on* q_t, then Coulomb's law, as the written summary of the results of many experiments, is

$$\mathbf{F}_{st} = k\frac{q_s q_t}{R^2}\hat{\mathbf{R}}_{st}$$

*P.A.M., *Proc. Roy. Soc. London, Ser. A*, **133**:60 (1931); also *Phys. Rev.*, **74**:817 (1948).

Here k is a constant of proportionality that depends on the particular system of units employed.

The simplest choice for the system of units is obviously $k = 1$, and historically this is the first choice that was made. If F, i.e., $|\mathbf{F}_{st}|$, and R are in the cgs system, then Coulomb's law defines a unit of charge—the electrostatic unit of charge, or the esu. In the cgs system length, mass, and time are arbitrarily taken as primary quantities and all other quantities such as electric charge are, by definition, secondary quantities whose dimensions are to be expressed in terms of those of the primary quantities. The units of the latter are the centimeter, gram, and second, whence the name.

One main factor has led to a gradual abandonment of the cgs-esu system in favor of another system, the mksa system, based on four primary quantities: length, mass, time, and current. (The name of the system derives from the units—meter, kilogram, second, and ampere.) The size of the units that result from the choice $k = 1$ is not particularly suited to practical requirements, either on the atomic or on the macroscopic levels; those of the mksa system are very well adapted to the gross measurements of the everyday world, but are not any better suited for the atomic world. It is not surprising, therefore, that the switch from cgs to mksa is practically unanimous among engineers. Among research physicists, especially the older ones, the cgs system remains unchallenged. Their work deals either with the tiny elementary particles or the huge cosmos; if both cgs and mksa have units which are inconvenient for these worlds, why switch? The mksa system has become the favorite in physics textbooks, where both microscopic and macroscopic problems enter.

There is another, comparatively minor, factor. When secondary quantities are expressed dimensionally in terms of the primary quantities in the cgs system, the exponents of the primary dimensions are either integers or fractions. In the Coulomb law, for example, the unit of force on the left side would be the g cm s^{-2} (from $F = ma$). On the right side, $k = 1$ and is dimensionless; $\hat{\mathbf{R}}_{st}$ is dimensionless because it is the ratio of two lengths, so the units are $(q)^2$ cm^{-2}, where (q) is the unit of charge. The dimensions* on both sides of the equation must be the same, so that the two sides will be multiplied by the same factor for any change in the size of units. From [g cm s^{-2}] = [$(q)^2$ cm^{-2}] it follows that the esu unit of charge must have the dimensions [$cm^{3/2}$ $g^{1/2}$ s^{-1}]. In the mksa system, on the other hand, the exponents of the primary units are always integers, which are easier to deal with than fractions. Offsetting this, however, there can be four such exponents in the mksa system but only three in the cgs system.

One other reason besides established habit is offered by some proponents of the cgs system: the greater symmetry of Maxwell's equations in the cgs system. Since these are the basic differential equations of electricity and magnetism it is felt that this should be a

*Throughout this book *dimensions* will be indicated by brackets about the corresponding units.

primary consideration. We will show in a later chapter that this argument is not valid—the equations are just as symmetric in one system as in the other. Basically, the choice boils down to a question of personal taste. We will employ the mksa system. By the use of Reference Table IV at the end of the book, any equation in this book may be converted to its cgs equivalent; similarly, Reference Table II-B there lists the relative sizes of the units in different systems.

From the viewpoint of a logical development of the subject, it would now be desirable to define the dimensions and size of the unit of charge here. Unfortunately, measurements of Coulomb's law are not very accurate, so, because of practical considerations (which always outweigh logical necessity), the size and the dimensions of the unit of charge in the mksa system are determined from an equation in magnetostatics, for, it turns out, greater accuracy can be obtained from measurements of force there. The name of the unit of charge in the mksa system is the coulomb (C). The charges of the electron in the two systems are related by $q_e \sim 1.6 \times 10^{-19}$ C $\sim 4.8 \times 10^{-10}$ esu. Neither magnitude is of the order of unity so neither unit is appropriate for atomic calculations. The size of the coulomb, however, is quite convenient for practical affairs: 1 C/s passing a given plane constitutes a current of 1 A (ampere), by definition. The operational definition of the ampere in terms of a measured force will be given in Chap. 3.

Electrostatics in Vacuum

The unit of force in the mksa system is the m kg s^{-2} and the unit of charge (from $q = It$) is the s A; it follows from Coulomb's law that the unit of k must be m^3 kg s^{-4} A^{-2}. For reasons of convenience in subsequent equations it seemed desirable (1) to have the proportionality constant in the denominator, and (2) to introduce a factor of 4π here so that it did not appear in other equations. The latter process is called rationalization. Both these aims were accomplished by setting the constant of proportionality, k, equal to $(4\pi\epsilon_0)^{-1}$. The new constant ϵ_0 is called the permittivity of free space; its unit is the inverse, dimensionally, of that of k: m^{-3} kg^{-1} s^4 A^2.

In the cgs system Coulomb's law gives a method for determining the unit of charge: F is measured for various values of r so that q_s and q_t may be found. By contrast, in the mksa system Coulomb's law is an instrument for the determination of the value of ϵ_0 (or k): F is measured for various values of r, q_s, and q_t so that ϵ_0 may be calculated. The value of ϵ_0, determined by experiment, is found to be $\epsilon_0 = 8.85 \times 10^{-12}$ m^{-3} kg^{-1} s^4 A^2. In Sec. 2-6 (Capacitance) of this chapter a unit called the farad (F) will be introduced, with 1 F = 1 m^{-2} kg^{-1} s^4 A^2; consequently the dimensions of ϵ_0 are [F/m]. Therefore $\epsilon_0 = 8.85 \times 10^{-12}$ F m^{-1}. A convenient constant to remember is (very closely)

$$\frac{1}{4\pi\epsilon_0} = 9 \times 10^9 \text{ m F}^{-1}$$

This constant will subsequently be found to be (exactly) $10^{-7} c^2$, where c is the speed of light in vacuum. Coulomb's law in vacuum for two charges, in the mksa system, is now

$$\boxed{\mathbf{F}_{st} = \frac{1}{4\pi\epsilon_0} \frac{q_s q_t}{R^2} \hat{\mathbf{R}}_{st}} \tag{E-1}$$

As above, $\hat{\mathbf{R}}_{st}$ is the unit vector that points from q_s toward q_t.

As an example of the use of Appendix 5 to convert an equation from the mksa to the cgs form, we will find Coulomb's law (cgs).

$$q_{\text{cgs}} = \frac{q_{\text{mksa}}}{\sqrt{4\pi\epsilon_0}}$$

and, implicitly,

$$F_{\text{cgs}} = F_{\text{mksa}} \qquad \text{and} \qquad R_{\text{cgs}} = R_{\text{mksa}}$$

Then

$$F_{\text{mksa}} = \frac{1}{4\pi\epsilon_0} \frac{q_{s(\text{mksa})} q_{t(\text{mksa})}}{R^2_{\text{mksa}}} \rightarrow F_{\text{cgs}}$$

$$= \frac{1}{4\pi\epsilon_0} \frac{(\sqrt{4\pi\epsilon_0}\, q_{s(\text{cgs})})(\sqrt{4\pi\epsilon_0}\, q_{t(\text{cgs})})}{R^2_{\text{cgs}}}$$

so that in the cgs system Coulomb's law is, as mentioned previously,

$$\mathbf{F} = \frac{q_s q_t}{R^2} \hat{\mathbf{R}}_{st}$$

A word of caution is necessary with respect to the meaning of relations such as $q_{\text{cgs}} = q_{\text{mksa}}/\sqrt{4\pi\epsilon_0}$. This does not imply that a cgs unit of charge is $\sqrt{9 \times 10^9} = 3\sqrt{10} \times 10^4$ times the mksa unit of charge, nor does it imply that a given charge contains $3\sqrt{10} \times 10^4$ times as many cgs units as mksa units. q_{cgs} and q_{mksa} have different dimensions and cannot be compared directly. There are three factors that must be determined before comparing two quantities such as q_{cgs} and q_{mksa}. (1) The factor $\sqrt{\epsilon_0}$ gives them both the same dimensions. Let brackets designate dimensions, though for simplicity the dimensions will be written in terms of units. Then $[q_{\text{mksa}}] = [\text{s A}]$. Since $[\epsilon_0] = [\text{m}^{-3}\ \text{kg}^{-1}\ \text{s}^4\ \text{A}^2]$, $[\sqrt{\epsilon_0}] = [\text{m}^{3/2}\ \text{kg}^{1/2}\ \text{s}^2\ \text{A}]$ and $[q_{\text{mksa}}/\sqrt{\epsilon_0}] = [\text{m}^{3/2}\ \text{kg}^{1/2}\ \text{s}^{-1}]$.

On the left side, $[q_{\text{cgs}}] = [\text{cm}^{3/2}\ \text{g}^{1/2}\ \text{s}^{-1}]$. So the dimentions of $(q_{\text{mksa}}/\sqrt{\epsilon_0})$ and q_{cgs} are the same, though the units differ. (2) The factor $1/\sqrt{4\pi}$ enters because one system is "rationalized" while the other is not. The conversion factors in Reference Table IV are obtained in each case by taking the product of these two factors. "Rationalized" here simply means that the factor 4π can be removed from some equations by altering the sizes of some units: the system is rationalized, but the 4π then appears in other equations where it did not appear before rationalization. (3) A factor must enter to change the units of one to match the other. In the present example it is necessary to change, for example, 1 $m^{3/2}\ kg^{1/2}\ s^{-1}$ to $cm^{3/2}\ g^{1/2}\ s^{-1}$:

$$\left(m \times \frac{100\ cm}{m}\right)^{3/2}\left(kg \times \frac{1000\ g}{kg}\right)^{1/2} s^{-1} = (10^3\ cm^{3/2})(10\sqrt{10}\ g^{1/2})(s^{-1})$$

$$= 10^4\sqrt{10}\ cm^{3/2}\ g^{1/2}\ s^{-1}$$

Since we started with $3\sqrt{10} \times 10^4\ m^{3/2}\ kg^{1/2}\ s^{-1}$ for each mksa unit of charge, this is equivalent to $(3\sqrt{10} \times 10^4)(10^4\sqrt{10})$ or 3×10^9 cgs units of charge. This checks with the values for an electron: 4.8×10^{-10} esu/$(1.6 \times 10^{-19}$ C) = 3×10^9 esu/C.

Reference Table IV gives the conversion factors needed to change an equation from one system to the other. Reference Table II-B gives the equivalent number of units, having different dimensions, which measure the same quantity. It is obtained, as above, from the product of the three factors.

The above fundamental law of force between two charges, at rest with respect to each other and to the observer, was first verified experimentally by Charles Augustin de Coulomb in 1785. Actually, the inverse-square behavior had been surmised even earlier, in 1767, by Joseph Priestly, on the basis of experiments by Benjamin Franklin. Figure 2-1 shows the relation between the vectors in Coulomb's law when q_s and q_t have the same signs. If q_s and q_t have opposite signs then the vector $\mathbf{F}_{st}$ will point in the direction $-\hat{\mathbf{R}}_{st}$, opposite to that shown. Thus, like charges repel each other while unlike charges attract each other.

In formulating this law no hypothesis is made concerning the mechanism by which the force is transmitted over the intervening distance in the vacuum. Either the force is transmitted instantaneously, i.e., with infinite speed, or it may be postulated that the speed of transmission of the force is finite, but that all transient effects have disappeared, leaving the steady-state condition, the one of interest. Either way, the situation being considered here is a static one.

If the labeling of the two charges is reversed then the new $\hat{\mathbf{R}}$ will be opposite to that shown in Fig. 2-1; otherwise Coulomb's law remains unchanged. So the force on the new q_t will be equal and opposite to that on the old q_t: Newton's third law (action is equal and opposite to reaction) is obeyed. This is typical of a central force, one directed along the line connecting the two charges.

It is simple to make a comparison of the relative magnitude of the electrical and gravitational forces between two charged bodies since the law of gravitational attraction is similar to Coulomb's law. An electron has the smallest quantum of charge and also the smallest known finite mass: 1.6×10^{-19} C and 9.1×10^{-31} kg. For two electrons separated by a distance of 1 mm,

$$F_{\text{elec}} = \frac{1}{4\pi\epsilon_0}\frac{q_s q_t}{R^2} = 9 \times 10^9\ \frac{(1.6 \times 10^{-19})^2}{(10^{-3})^2} = 2.3 \times 10^{-22}\ \text{N}$$

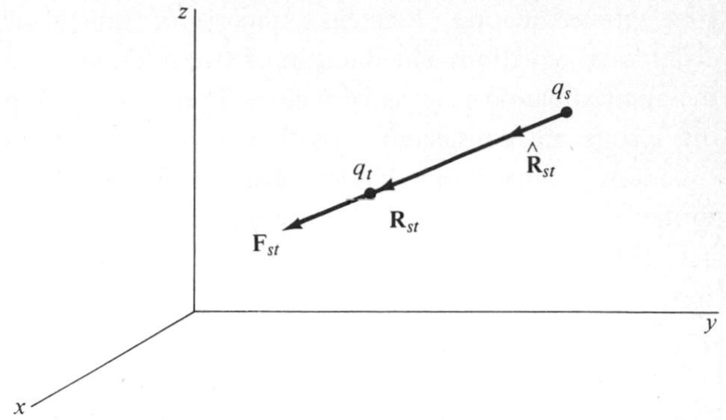

FIGURE 2-1

$$F_{\text{grav}} = G\frac{m_s m_t}{R^2} = 6.67 \times 10^{-11} \frac{(9.1 \times 10^{-31})^2}{(10^{-3})^2} = 5.5 \times 10^{-65}\ \text{N}$$

For electrons the electrical force is almost 10^{43} times as strong as the gravitational force; for other charged particles it is not very different. As a consequence, it is unnecessary to consider the gravitational force when electrical forces are present. In problems of astronomy, the fact that gravitational forces, alone, are employed to compute trajectories with very high accuracy is proof of the fact that very massive bodies are electrically neutral. If for some reason such a body should become electrically ionized, either positively or negatively, Coulomb's law operates on charges within the vicinity of the body to repel like charges and attract opposite charges until the body is neutral once more.

Now consider the force produced on a test charge by a number of source charges. Here the superposition principle enters: the force produced on q_t by $q_{s'}$ (at $r_{s'}$) and $q_{s''}$ (at $r_{s''}$) acting together is *assumed* to be the sum of the force produced on q_t when only $q_{s'}$ is present and the force produced on q_t when only $q_{s''}$ is present. The justification of this assumption is twofold. (1) Maxwell's differential equations, which govern the behavior of the electric field, are linear; i.e., the force or field strength and its derivatives appear only to the first power. Such linear differential equations possess the property that the sum of two solutions is also a solution. (2) The results of the assumption agree with experiment.

To show that the above common-sense assumption is not necessarily true, consider an analogy. Newton's law of universal gravitation also obeys a superposition principle: the resultant force when two gravitational masses attract a third, test, mass is the vector sum of the two forces. This is so because Poisson's equation, the basic differential equation here, is linear. But we know today that Newton's law of gravitation is only an extremely good approximation. Einstein's theory of general relativity is based on nonlinear differential equations and the sum of two solutions is then not an exact solution, though the approximation may be very close. There is no superposition principle in gravitation if the results are considered to sufficient accuracy. Since Newton's and Poisson's linear equations are much simpler to solve than Einstein's nonlinear equations, however, the former continue to find wide acceptance. This is similar to the situation with quantum mechanics: classical mechanics continues to be used where it is a good enough approximation because its equations are much more easily solved than are those of quantum mechanics.

Maxwell's equations have also been modified, by Born, to make them nonlinear. This is necessary, e.g., in order to consider the case of photon-photon scattering. Here the resultant waves would not be a simple superposition of the incident waves but would, in fact, have different frequencies. The collision cross section for this process is very small, so the effect has not yet been observed experimentally. A calculation made by quantum electrodynamics for this process also yields nonlinear results, thereby invalidating the superposition principle. The principle is widely employed, however, because it is the very simplest method and because the results, though only approximate, are accurate enough for all present requirements.

Assuming the superposition principle as valid, when a number of discrete source charges are present at the same time Coulomb's law becomes

$$\mathbf{F} = \frac{1}{4\pi\epsilon_0}\frac{q_1 q_t}{R_{1t}^2}\hat{\mathbf{R}}_{1t} + \frac{1}{4\pi\epsilon_0}\frac{q_2 q_t}{R_{2t}^2}\hat{\mathbf{R}}_{2t} + \cdots + \frac{1}{4\pi\epsilon_0}\frac{q_n q_t}{R_{nt}^2}\hat{\mathbf{R}}_{nt}$$

or

$$\mathbf{F} = \frac{q_t}{4\pi\epsilon_0}\sum_{i=1}^{n}\frac{q_i}{R_{it}^2}\hat{\mathbf{R}}_{it}$$

This gives the resultant force on the test charge q_t as the vector sum of n vectors, each one representing the force on q_t that is produced by the ith source charge q_i. The source charges also produce forces on each other but these are not of interest here: it is assumed that they are kept stationary at their separate positions by other (mechanical) forces. Figure 2-2 shows the modification made in Fig. 2-1 by the addition of a second source charge, of opposite sign to the first.

If the distribution of source charges is sufficiently dense, then a volume $d\tau_s$ may be selected which is macroscopically very small yet contains a very large number of discrete point charges. In that case a volume charge density ρ may be said to exist, as if the charge were a continuum instead of a discrete distribution, and the Coulomb force equation becomes

$$\mathbf{F}(\mathbf{r}_t) = \frac{q_t}{4\pi\epsilon_0}\int_{\mathcal{V}}\frac{\rho(\mathbf{r}_s)}{R^2}\hat{\mathbf{R}}_{st}\,d\tau_s$$

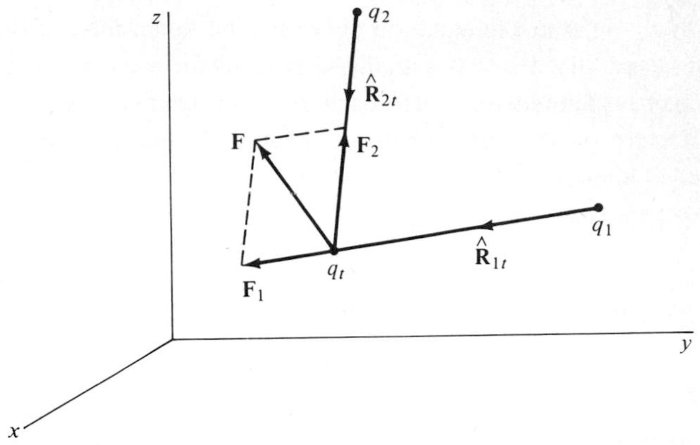

FIGURE 2-2

Figure 2-3 shows $\mathcal{V}$, the region which contains the source charge density. The position of the test charge here is shown as outside $\mathcal{V}$, but this is not necessary. $\mathbf{r}_t$ could also be inside $\mathcal{V}$, even if the test point is one at which a charge density ρ_s exists, provided, as will be seen, that ρ_s is finite.

One differential element of volume $d\tau_s$ is located at $\mathbf{r}_s$, the electric charge density there being $\rho(\mathbf{r}_s)$. $\hat{\mathbf{R}}_{st}$ is the unit vector from this differential volume element to the test charge, which is located at the point $\mathbf{r}_t$. The force on q_t that is produced by $\rho(\mathbf{r}_s)d\tau_s$ is a differential force; but when this is integrated over the infinite number of differential volumes $d\tau_s$ in $\mathcal{V}$, the resultant force $\mathbf{F}(\mathbf{r}_t)$ on the test charge is finite. The charge density $\rho(\mathbf{r}_s)$ will vary, in general, from point to point in $\mathcal{V}$; when it is a constant at all points in $\mathcal{V}$

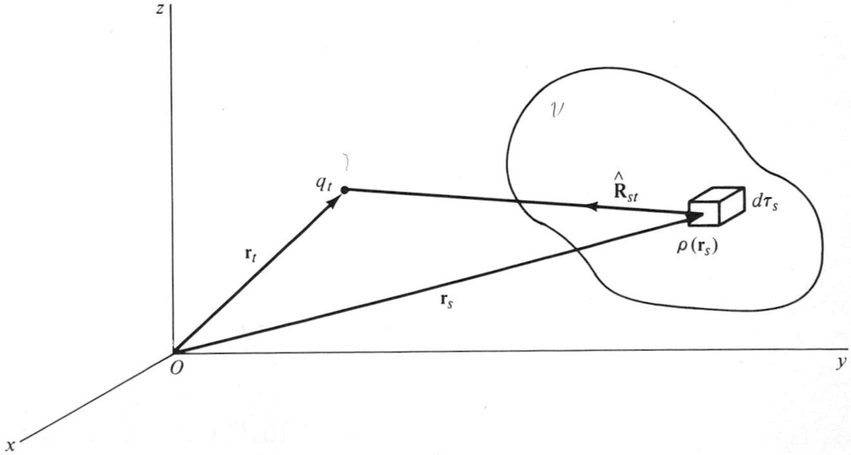

FIGURE 2-3

then the ρ in the equation above can be taken out, to the left of the integral sign. The integral sign itself is actually shorthand for a triple integration. The charge distribution may be limited to a surface; then the integral stands for a double integration, the charge density on the surface being designated by $\sigma(\mathbf{r}_s)$. Again, the distribution may be limited to a line; then the integral stands for a single integration, and the linear charge density is written $\lambda(\mathbf{r}_s)$.

Examples

1. Let $q_t = -1.6 \times 10^{-19}$ C be a test charge located at (1,2); and $q_s = +3.2 \times 10^{-19}$ C be a source charge at (2,1). To find $\mathbf{F} = \mathbf{F}_{st}$. This case is shown in Fig. 2-4.

$$\mathbf{R} = \mathbf{R}_{st} = \mathbf{r}_t - \mathbf{r}_s = (\hat{\mathbf{x}}x_t + \hat{\mathbf{y}}y_t) - (\hat{\mathbf{x}}x_s + \hat{\mathbf{y}}y_s) = \hat{\mathbf{x}}(x_t - x_s) + \hat{\mathbf{y}}(y_t - y_s)$$
$$= \hat{\mathbf{x}}(1-2) + \hat{\mathbf{y}}(2-1) = -\hat{\mathbf{x}} + \hat{\mathbf{y}}$$

So $R = |\mathbf{R}_{st}| = \sqrt{(-1)^2 + (1)^2} = \sqrt{2}$ and $R^2 = 2$. $\hat{\mathbf{R}} = \hat{\mathbf{R}}_{st} = \mathbf{R}/R = (-\hat{\mathbf{x}} + \hat{\mathbf{y}})/\sqrt{2}$.

$$\mathbf{F} = \frac{1}{4\pi\epsilon_0}\frac{q_s q_t}{R^2}\hat{\mathbf{R}}_{st}$$
$$= (9 \times 10^9)\frac{(3.2 \times 10^{-19})(-1.6 \times 10^{-19})}{2}\left(-\frac{\sqrt{2}}{2}\hat{\mathbf{x}} + \frac{\sqrt{2}}{2}\hat{\mathbf{y}}\right)$$
$$= 1.63 \times 10^{-28}(\hat{\mathbf{x}} - \hat{\mathbf{y}})\ \text{N}$$
$$F = |\mathbf{F}| = 1.63 \times 10^{-28}\sqrt{(1)^2 + (-1)^2} = 2.3 \times 10^{-28}\ \text{N}$$

The charge q_t feels a force opposite in direction to that of $\hat{\mathbf{R}}_{st}$, one directed toward q_s.

2. The force produced by q_t on q_s in the previous example.

$$\mathbf{F}_{ts} = \frac{1}{4\pi\epsilon_0}\frac{q_t q_s}{R^2}\hat{\mathbf{R}}_{ts}$$
$$\hat{\mathbf{R}}_{ts} = -\hat{\mathbf{R}}_{st} = \frac{\sqrt{2}}{2}\hat{\mathbf{x}} - \frac{\sqrt{2}}{2}\hat{\mathbf{y}}$$
$$\mathbf{F}_{ts} = (9 \times 10^9)\frac{(-1.6 \times 10^{-19})(+3.2 \times 10^{-19})}{2}\left(\frac{\sqrt{2}}{2}\hat{\mathbf{x}} - \frac{\sqrt{2}}{2}\hat{\mathbf{y}}\right)$$
$$= 1.63 \times 10^{-28}(-\hat{\mathbf{x}} + \hat{\mathbf{y}})\ \text{N}$$
$$= -\mathbf{F}_{st}$$

3. Let $q_t = -1.6 \times 10^{-19}$ C at (1,2); $q_s = -4.8 \times 10^{-19}$ C at (3,2). To find $\mathbf{F} = \mathbf{F}_{st}$. See Fig. 2-5.

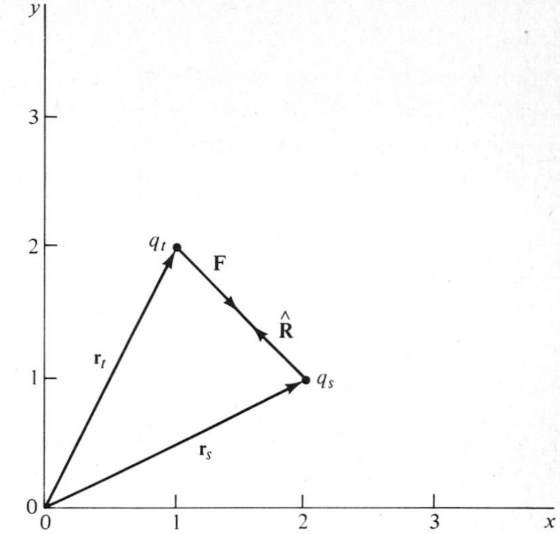

FIGURE 2-4

$$\mathbf{R} = \hat{\mathbf{x}}(1 - 3) + \hat{\mathbf{y}}(2 - 2) = -\hat{\mathbf{x}}\, 2$$

$$R = \sqrt{(-2)^2} = 2$$

$$\hat{\mathbf{R}} = -\hat{\mathbf{x}}$$

$$\mathbf{F} = (9 \times 10^9)\, \frac{(-4.8 \times 10^{-19})(-1.6 \times 10^{-19})}{4}\, (-\hat{\mathbf{x}})$$

$$= -1.73 \times 10^{-28}\, \hat{\mathbf{x}} \quad \text{newton}$$

4. Given $q_t = -1.6 \times 10^{-19}$ C at (1,2); $q_1 = +3.2 \times 10^{-19}$ C at (2,1); $q_2 = -4.8 \times 10^{-19}$ C at (3,2). To find $\mathbf{F}_t\,(\mathbf{r}_t)$.

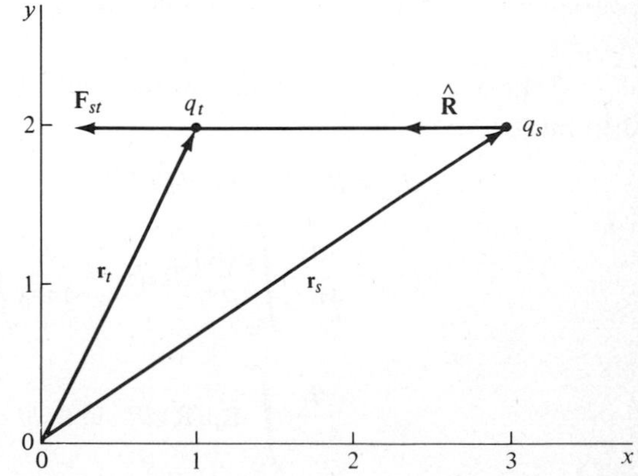

FIGURE 2-5

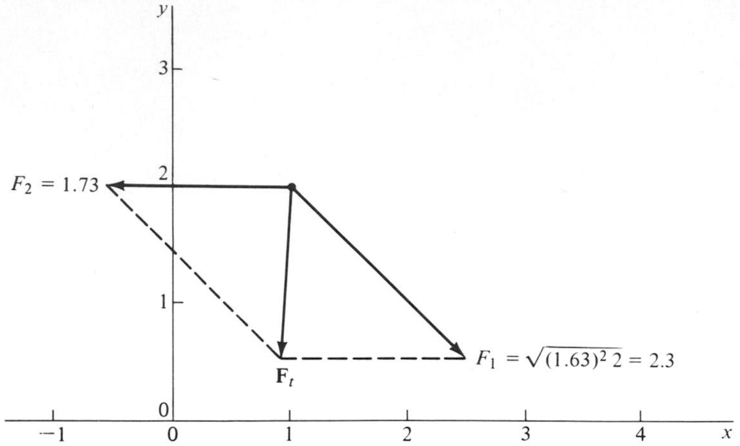

FIGURE 2-6

Apply the superposition theorem to the results of Examples 1 and 3, as in Fig. 2-6. To obtain the total force on the test charge it is necessary to add vectorially:

$$\begin{aligned}\mathbf{F}_t = \mathbf{F}_{1t} + \mathbf{F}_{2t} = \mathbf{F}_1 + \mathbf{F}_2 &= (\hat{\mathbf{x}}1.63 - \hat{\mathbf{y}}1.63)\,10^{-28} \\ &\quad + (-\hat{\mathbf{x}}1.73)\,10^{-28} \\ &= (-\hat{\mathbf{x}}0.1 - \hat{\mathbf{y}}1.63)\,10^{-28} \qquad \text{newton}\end{aligned}$$

$$F_t = |\mathbf{F}_t| = \sqrt{(-0.1)^2 + (-1.63)^2} \times 10^{-28} = 1.63 \times 10^{28}\ \text{N}$$

The direction of F_t is slightly west of south, the angular deviation from due south being $\theta = \tan^{-1}(0.1/1.63)$.

5. The integral for the force, in the case when the source charge distribution is a continuum, does not become infinite when the test point is taken inside the source volume.

Take a spherical coordinate system with the test point as origin within the charge distribution, as in Fig. 2-7.

$$d\tau_s = R^2 \sin\theta\, d\theta\, d\phi\, dR$$

$$\mathbf{F}(\mathbf{r}_t) = \frac{q_t}{4\pi\epsilon_0}\int_{\mathcal{V}} \frac{\rho(\mathbf{R})}{R^2}\,\hat{\mathbf{R}}\, d\tau_s = \frac{q_t}{4\pi\epsilon_0}\int_{\mathcal{V}} \frac{\rho(\mathbf{R})}{R^2}\,\hat{\mathbf{R}}R^2 \sin\theta\, d\theta\, d\phi\, dR$$

$$= \frac{q_t}{4\pi\epsilon_0}\int_{\mathcal{V}} \hat{\mathbf{R}}\rho(\mathbf{R})\,dR\, \sin\theta\, d\theta\, d\phi$$

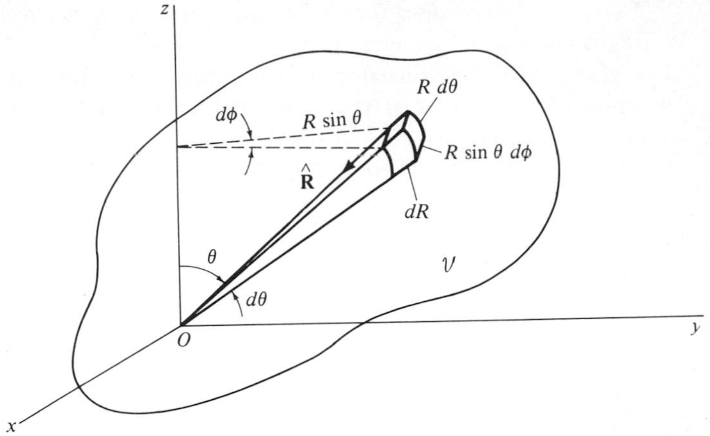

FIGURE 2-7

The integrand becomes infinite only if $\rho(\mathbf{R})$ is infinite at some point or points, regardless of whether this is at the origin or elsewhere. Any finite source charge density that exists at the test point does not make $\mathbf{F}$ infinite.

The Coulomb expression yields an infinite force when two point charges (finite charge, infinitesimal size, infinite charge density) are separated by an infinitesimal distance. But when one of the two charges is itself an infinitesimal, $\rho d\tau_s$, then the force it produces on a point test charge located there (at the same point) is finite.

PROBLEMS

1 Find the following ratio: the electrical force between two protons divided by the gravitational force between them. ($m_p = 1836\ m_e$, $q_p = -q_e$.)

2 If the earth and the sun each had the same charge q, what is the value needed to make $F_{\text{elec}} = 0.01\ F_{\text{grav}}$? Are these, with and without their neighborhoods, charged bodies? Consider, e.g., the Van Allen belt.

3 An electric charge $+Q$ is distributed with a $\rho(r)$ which does not depend on θ or ϕ throughout the volume of a sphere of radius R; i.e., it is spherically symmetric. Find the force it produces on an electron (charge $-e$) located at an outside point r meters from the center of the sphere. Let $\hat{\mathbf{r}}$ point from the origin, at the center of the sphere, to the electron.

4 If a charge q_s is at the origin, the force on q_t at (x,y,z) is

$$\mathbf{F} = \frac{1}{4\pi\epsilon_0} \frac{q_s q_t}{x^2 + y^2 + z^2} \cdot \frac{\hat{\mathbf{x}}x + \hat{\mathbf{y}}y + \hat{\mathbf{z}}z}{\sqrt{x^2 + y^2 + z^2}}$$

What is the expression for the vector force on q_t when q_s is removed from the origin and is relocated at (x_s,y_s,z_s)?

5 A charge $+10^{-16}$ C is located at each of the eight corners of a cube of side 1 m. The center of the cube is at the origin and its sides are parallel to the axes. Find the force on a charge $+10^{-16}$ C, at $x = 1$ m.

6 Move the test charge of Prob. 5 to the origin. Find the force on it.

7 If the eight source charges of Prob. 5 were placed at the origin instead of the cube corners, would the force on the test charge be larger or smaller than in Prob. 5? Why?

8 In Probs. 5 and 7, let the test charge be placed at $x = 10$ m. Compare the two resultant forces for this case.

9 A charge Q is uniformly distributed throughout the volume of a spherical shell of inner radius a and outer radius b. What is the force on a charge q located at $(c,0,0)$, where $c > b$? The center of the shell is at the origin.

10 In Prob. 9, change the location of q to the origin. Find the force on it.

11 In Prob. 9, change the location of q to a point with $c < a$. Find **F**, using Coulomb's law.

12 Suppose q in Prob. 9 is located at a point with $a < c < b$. Find **F**, using Coulomb's law.

13 The surface of a sphere of radius R has a uniform surface charge density, σ C/m^2. What is the force produced on a charge q at $x = r$, where $r > R$? Use double integration.

14 An electron is at $z = -0.1$ and a proton is at $z = +0.1$. Find the force on a charge q at (0,0,0); at (0.1,0,0); at (1,0,0); at (10,0,0).

15 (*a*) In Prob. 14, will the force on q be smaller when q is at (10,0,0) or at (0,0,10)? (*b*) Is the force at (10,0,0) equal to 0.01 that of the force at (1,0,0)? Why?

16 A finite charge Q is distributed throughout the volume of a sphere of radius R according to the density $\rho = \rho_0 r^{-n}$, where ρ_0 is a constant. For what values of n is the total charge finite? Repeat for a cylinder with $\rho = \rho_0 r^{-n}$ (r cylindrical).

2.3 THE ELECTRIC FIELD, E

Instead of dealing with the force produced by the charge q_s on the charge q_t, one may consider the force on q_t per unit charge of q_t. This is called the electric field intensity:

$$\boxed{\mathbf{E} = \frac{\mathbf{F}_{st}}{q_t}} \qquad \text{(E-2)}$$

For simplicity, this quantity is often simply called the electric intensity, but just as often it is termed the electric field. We will use the latter expression. The units in which **E** are expressed are N C^{-1}, or m kg s^{-1} A^{-1}.

To determine **E** operationally at some point, insert a test charge q_t there; measure the force on it, $\mathbf{F}_{st}$, produced by the source charges; and then divide this force by q_t. To insure that the act of measurement of **E** did not, itself, disturb the distribution of source charges, repeat this procedure with ever smaller test charges, taking the ratio only in the limit, as $q_t \to 0$. Since there is a minimum value of q_t this process is not always feasible; we may then assume the sources are kept in position by mechanical means.

The vector **E** is a quantity which may be assigned to every test point in space, $\mathbf{r}_t$, as soon as there is one source charge q_s at $\mathbf{r}_s$, say. If $R = |\mathbf{R}_{st}| = |\mathbf{r}_s - \mathbf{r}_t|$,

$$\boxed{\mathbf{E}(\mathbf{r}_t) = \frac{1}{4\pi\epsilon_0} \frac{q_s}{R^2} \hat{\mathbf{R}}_{st}} \tag{E-3}$$

The importance of this definition rests on the fact that **E** can be thought of as being produced by the source charge q_s quite independent of the presence or absence of the test charge q_t. **E** is said to define a vector field throughout all space. The short step from **F** to **E**, very simple mathematically, has very important physical consequences, for the field **E** assigns properties to a region of (even empty) space. The force that is produced on the test charge q_t, when it is placed at $\mathbf{r}_t$, is now said to be due to the field **E** which exists at $\mathbf{r}_t$, omitting the contribution of q_t to the total field there. We are now taking the view that charges do not produce forces on themselves. The force on q_t at $\mathbf{r}_t$ is produced by the *field* that would exist at $\mathbf{r}_t$ if q_t were not there.

In electrostatics the introduction of the two-step field process—source charges producing a field and the field producing a force on a test charge—is a matter of mathematical convenience only, and one could just as well stay with the original, one-step, action-at-a-distance process. But when time variations enter the picture one must allow for the finite time of propagation of the force, and the field concept becomes a necessity instead of a convenience. This distinction would not exist if the velocity of propagation of the force were infinite; action at a distance would then be a perfectly reasonable approach. In actual practice it is found that the velocity of propagation is so fast that for many applications, especially when the frequencies and the distances to be covered are small, the speed may be taken as infinite.

The inverse-square relation above for **E** is that which is produced by a single point charge. It will be shown below that for different configurations of source charges the resultant field will not vary as the inverse square of the distance, to the origin, say, or to any other point. Different variations with distance will obtain for different configurations. The inverse-square relation produced by a point source, however, is the primary one; the other cases may all be determined from this case, at least in principle. Just as with forces, when more than one source charge is present the superposition principle is found to operate. Here, also, this principle does not follow of necessity from any physical considerations. It may turn out, e.g., that for very large electric fields, much higher than can be obtained today, the superposition principle is simply not correct.

The direction of **E** is away from a positive source charge and toward a negative source charge: from $\mathbf{F} = q_t\mathbf{E}$, if $q_t > 0$, then **F** and **E** have the same direction. If an additional charge is added to some previously existing source charges then, if the added charge is to be considered the test charge, the value of **E** is to be considered as unaffected; but if the added charge is to be considered as an extra source charge (in order to find the **E** at some different, test, charge) then the added charge does change the value of **E**.

For a distribution of discrete source charges the electric field is given by

$$\mathbf{E}(\mathbf{r}_t) = \frac{1}{4\pi\epsilon_0}\sum_{i=1}^{N}\frac{q_i}{R_{it}^2}\hat{\mathbf{R}}_{it}$$

For a continuum distribution

$$\boxed{\mathbf{E}(\mathbf{r}_t) = \frac{1}{4\pi\epsilon_0}\int_{\mathcal{V}}\frac{\rho(\mathbf{r}_s)}{R^2}\hat{\mathbf{R}}_{st}\,d\tau_s} \qquad \text{(E-4)}$$

The subscript t here stands for test point or field point, even though there may actually not be a test charge at that point. Note the distinction between $\mathbf{r}_s$ and $\mathbf{r}_t$. The distance

$$R = |\mathbf{R}_{st}| = |\mathbf{r}_s - \mathbf{r}_t| \quad \text{and} \quad \hat{\mathbf{R}}_{st} = \frac{\mathbf{R}_{st}}{R} = \frac{\mathbf{r}_s - \mathbf{r}_t}{|\mathbf{r}_s - \mathbf{r}_t|}$$

Examples

1. The electric field produced by a uniform charge distribution along an infinitely long straight line.

This configuration is shown in Fig. 2-8, the linear charge density being λ C/m. Because of the symmetry, **E** will not have a $\hat{\mathbf{y}}$ component: Why should E_y at P be up rather than down? (The line extends to infinity in either direction.) Neither will it have a $\hat{\mathbf{z}}$ component: Should E_z be out of the paper or in? So only E_x need be calculated. An element of source charge, $dq_s = \lambda\, dy$, is situated at a distance y from the origin. The latter is obtained by dropping a perpendicular from P to the y axis. dq_s produces a contribution $d\mathbf{E}$ to the total field **E** at P. The x component of this contribution will be

$$dE_x = \frac{1}{4\pi\epsilon_0}\frac{\lambda dy}{R^2}\cos\theta$$

This equation is not suitable for integration as it stands, for there are three quantities which vary with the position of dq_s—y, R, and θ. (The distance D, however, is

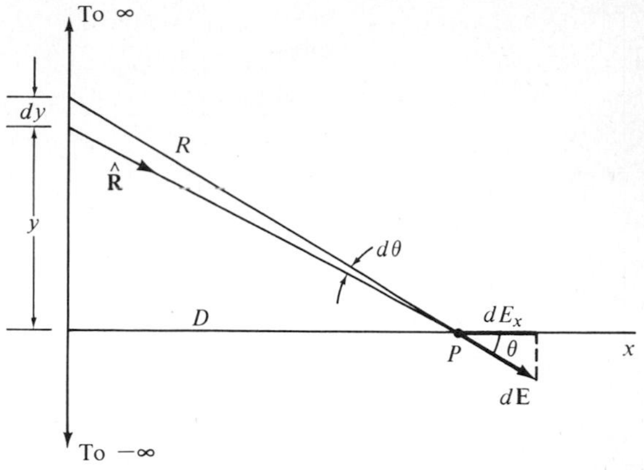

FIGURE 2-8

to be considered a fixed parameter in any given case.) It is necessary to express all three variables in terms of one, arbitrarily chosen, independent variable. We thus choose θ. Then $R = D/\cos\theta$. Also, $dy \cos\theta \simeq R\, d\theta$, so

$$dy = \frac{(D/\cos\theta)d\theta}{\cos\theta} = \frac{D\,d\theta}{\cos^2\theta}$$

or

$$dE_x = \left(\frac{\lambda}{4\pi\epsilon_0}\right)\frac{(D\,d\theta/\cos^2\theta)\cos\theta}{(D/\cos\theta)^2} = \frac{\lambda}{4\pi\epsilon_0}\,\frac{\cos\theta\,d\theta}{D}$$

Now there is only the one variable, θ. Integrating,

$$E_x = \frac{\lambda}{4\pi\epsilon_0 D}\int_{-\pi/2}^{\pi/2}\cos\theta\,d\theta = \frac{\lambda}{2\pi\epsilon_0 D}$$

Since $E_y = E_z = 0$, $\mathbf{E} = (\lambda/2\pi\epsilon_0 D)\,\hat{\mathbf{x}}$ for a point in the plane of the paper. For any point whatever this expression becomes, in cylindrical coordinates,

$$\mathbf{E} = \frac{\lambda}{2\pi\epsilon_0 D}\,\hat{\mathbf{r}}$$

For this configuration of source charges, **E** drops off inversely with the distance (instead of inversely with the square of the distance, as for a point charge).

2. The electric field produced by a uniformly charged plane circular disk at any point on the axis of symmetry perpendicular to the disk at its center.

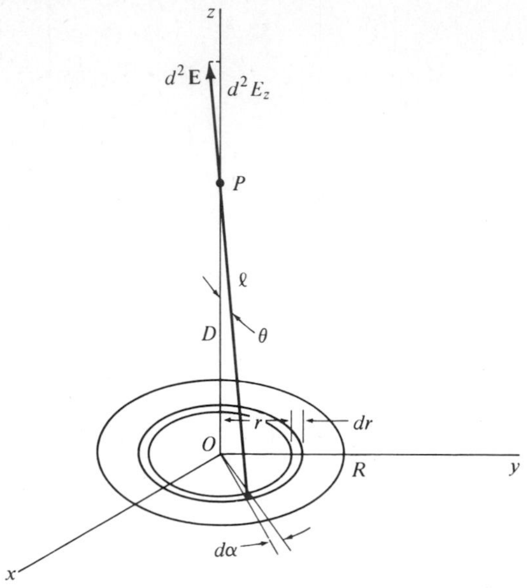

FIGURE 2-9

Figure 2-9 shows the disk of radius R. Let the surface charge density be σ C/m^2. The value of **E** is required at P, distant D from the disk center O. For a more general point (not on the z axis) the computation is considerably more involved, but here **E** will have only a z component, by symmetry.

Consider the annular ring shown between r and $r + dr$. Each element of charge within this ring will be at the same distance ℓ from P and will, thus, contribute the same differential magnitude d^2E_z to the electric field at P:

$$d^2E_z = \frac{1}{4\pi\epsilon_0}\frac{\sigma r\,d\alpha\,dr}{\ell^2}\cos\theta$$

This is written as a second-order differential; when it is integrated, to obtain the contribution of the entire annular ring, the result is the first-order differential

$$dE_z = \int_{\alpha=0}^{2\pi} d^2E_z = \frac{\sigma}{2\epsilon_0}\left(\frac{r\cos\theta}{\ell^2}\right)dr$$

To get the result for the entire disk it is necessary to express the three variables r, θ, ℓ in terms of one of them. Taking θ again,

$$dr = \frac{D\,d\theta}{\cos^2\theta} \qquad r = D\tan\theta \qquad \ell = \frac{D}{\cos\theta}$$

Then

$$E_z = \frac{\sigma}{2\epsilon_0} \int_0^{\tan^{-1}(R/D)} \frac{[D(\sin\theta/\cos\theta)]\,(\cos\theta)(D\,d\theta/\cos^2\theta)}{(D/\cos\theta)^2}$$

$$= \frac{\sigma}{2\epsilon_0} \int_0^{\tan^{-1}(R/D)} \sin\theta\, d\theta = \frac{\sigma}{2\epsilon_0}\left(1 - \frac{D}{\sqrt{R^2 + D^2}}\right)$$

The result above holds for the case when P is on the positive z axis. For a point on the other side of the disk

$$E_z = -\frac{\sigma}{2\epsilon_0}\left(1 - \frac{|D|}{\sqrt{R^2 + D^2}}\right) = -\frac{\sigma}{2\epsilon_0}\left(1 + \frac{D}{\sqrt{R^2 + D^2}}\right)$$

In general the result may be written

$$\mathbf{E} = \pm \frac{\sigma}{2\epsilon_0}\left(1 - \frac{|D|}{\sqrt{R^2 + D^2}}\right)\hat{\mathbf{z}} \quad \text{N/C}$$

where the plus applies for points on the $+z$ axis, the minus for points with $z < 0$.

Figure 2-10 is a graph of the electric field versus distance. At the plane of the disk there is a discontinuity, of amount σ/ϵ_0, in **E**. A positive test charge experiences a positive force (i.e., to the right) when it lies to the right of the disk; this becomes a negative force (i.e., to the left) when the test charge is to the left of the disk.

3. The **E** produced by a uniformly charged plane of infinite extent.

It is only necessary to let $R \to \infty$ in the previous example:

$$\mathbf{E} = \pm \frac{\sigma}{2\epsilon_0}\hat{\mathbf{z}} \quad \text{N/C}$$

Figure 2-11 gives a graph of this case. In the graph of the previous figure (Fig. 2-10), as $R \to \infty$ any finite value of D makes $D/R \to 0$, with $E_z = \pm\, \sigma/2\epsilon_0$.

E is a constant here, on any one side of the plane, independent of the distance from the plane. A test charge would experience a force of constant magnitude, independent of its distance from the plane. This is a most un-common-sense result; and the case may seem to be a rather impractical one, useful only for didactic purposes. This is not so: if the result of the previous example is examined it will be seen that the same answer is obtained whether $R \to \infty$ or $D \to 0$. This means that the value of **E** at a point very near the center of a uniformly charged plane disk of finite radius has the same value as that which obtains for the infinite plane, assuming $R \gg D$.

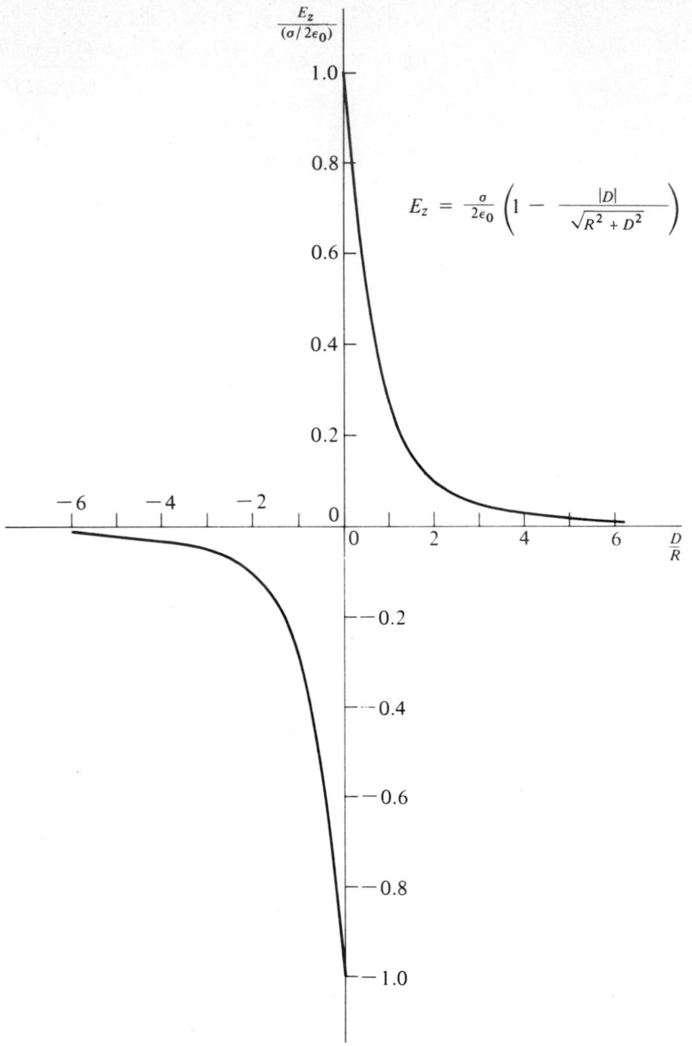

FIGURE 2-10

The result of the present example has even greater validity than this, for it can be shown that the normal component of **E** at *any* point an infinitesimal distance from a uniformly charged disk is $E_n = \pm\sigma/2\epsilon_0$, whether this point is on the axis of symmetry or not. Further, this result holds true even when the disk is not uniformly charged. Consequently the present result, instead of being academic, has very broad and useful application.

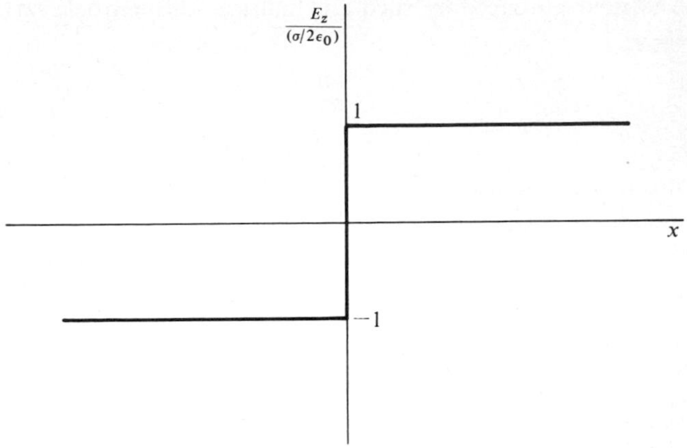

FIGURE 2-11

4. The **E** produced by a uniform charge spread on the surface of an infinitely long circular cylinder.

Figure 2-12 shows a cross section perpendicular to the axis of the cylinder for the case where P is outside the cylinder. The value of **E** may be calculated employing the result of Example 1, representing the cylinder in terms of line charges parallel to the axis. For an infinite uniformly charged line $E = \lambda/(2\pi\epsilon_0 D)$. In Fig. 2-12 let the charge on the cylindrical surface per unit length of the cylinder be Λ C/m. For a length ℓ the charge is then $\Lambda\ell$ C; the surface area for this length is $2\pi a\ell$ m^2; thus the surface charge density is $\sigma = \Lambda/(2\pi a)$ C/m^2.

On a surface which is ℓ meters long and which follows the arc length between ϕ and $\phi + d\phi$, the linear charge density is

$$\sigma(a\, d\phi)(1) = \frac{\Lambda\, d\phi}{2\pi} \qquad \text{C/m}$$

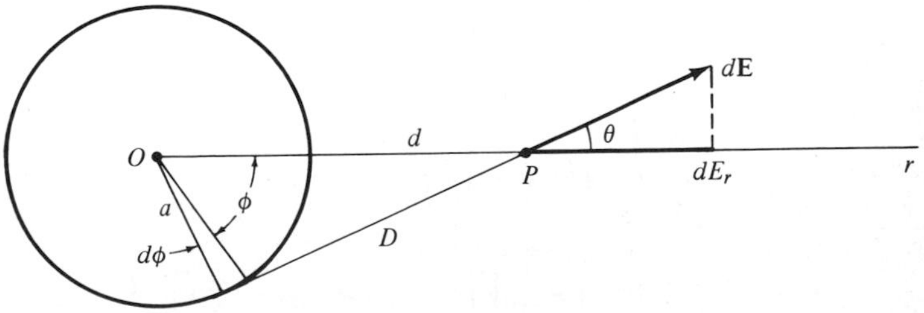

FIGURE 2-12

The field produced by such a cylindrical, differential, surface which extends from $-\infty$ to $+\infty$ is

$$d\mathbf{E} = \frac{(\Lambda\, d\phi/2\pi)}{2\pi\epsilon_0 D}\,\hat{\mathbf{D}}$$

From the symmetry of Fig. 2-12 the resultant electric field at P must lie along the r direction. Then

$$dE_r = \frac{\Lambda\, d\phi}{4\pi^2 \epsilon_0}\left(\frac{\cos\theta}{D}\right)$$

By the law of cosines

$$\cos\theta = \frac{d^2 + D^2 - a^2}{2Dd} \qquad \text{or} \qquad \left(\frac{\cos\theta}{D}\right) = \frac{d^2 + D^2 - a^2}{2D^2 d}$$

But $D^2 = a^2 + d^2 - 2ad\cos\phi$, so

$$\left(\frac{\cos\theta}{D}\right) = \frac{d - a\cos\phi}{a^2 + d^2 - 2ad\cos\phi}$$

and

$$dE_r = \frac{\Lambda\, d\phi}{4\pi^2 \epsilon_0}\left(\frac{d - a\cos\phi}{a^2 + d^2 - 2ad\cos\phi}\right)$$

Therefore

$$E_r = \frac{\Lambda}{4\pi^2 \epsilon_0}\int_0^{2\pi} \frac{d - a\cos\phi}{a^2 + d^2 - 2ad\cos\phi}\, d\phi = \frac{\Lambda}{2\pi^2 \epsilon_0}\int_0^{\pi} \frac{d - a\cos\phi}{a^2 + d^2 - 2ad\cos\phi}\, d\phi$$

The change in limits is permissible because the integrand behaves the same way going from π to 2π as it does going from π to 0. But it is not just a question of convenience–it is necessary to make this change, as we will see below. Using a table of integrals,

$$E_r = \frac{\Lambda}{2\pi^2 \epsilon_0}\left[\left(\frac{-a}{-2ad}\,\phi\right)_0^{\pi} + \frac{d(-2ad) - (a^2 + d^2)(-a)}{-2ad}\right.$$

$$\left.\times \int_0^{\pi} \frac{d\phi}{(a^2 + d^2) + (-2ad)\cos\phi}\right]$$

$$= \frac{\Lambda}{2\pi^2 \epsilon_0}\left\{\frac{\pi}{2d} + \left[\frac{1}{d}\tan^{-1}\left(\frac{d + a}{d - a}\tan\frac{\phi}{2}\right)\right]_0^{\pi}\right\}$$

$$= \frac{\Lambda}{2\pi^2 \epsilon_0} \left\{ \frac{\pi}{2d} + \frac{1}{d} \tan^{-1} \left[\frac{d+a}{d-a} \tan \frac{\pi}{2} \right] - \frac{1}{d} \tan^{-1} \left[\frac{d+a}{d-a} \tan 0 \right] \right\}$$

If we had not previously changed limits, the middle term would be

$$\frac{1}{d} \tan^{-1} \left[\frac{d+a}{d-a} \tan \pi \right]$$

But this lies on a different branch of the $\tan^{-1} \alpha$ function than the last term and it would give an incorrect answer.

When P is outside the cylinder $d > a$; so $(d+a)/(d-a) > 0$ and

$$E_r = \frac{\Lambda}{2\pi^2 \epsilon_0 d} \left(\frac{\pi}{2} + \frac{\pi}{2} - 0 \right) = \frac{\Lambda}{2\pi \epsilon_0 d}$$

This result is the same as that which would be obtained by removing the cylinder and letting its charge per unit length Λ reside completely on the axis. When P is an infinitesimal distance outside the cylinder we have $E = \Lambda/(2\pi\epsilon_0 a)$.

When P is inside the cylinder $d < a$; so $(d+a)/(d-a) < 0$ and

$$E_r = \frac{\Lambda}{2\pi^2 \epsilon_0 d} \left(\frac{\pi}{2} - \frac{\pi}{2} - 0 \right) = 0$$

$\mathbf{E} = 0$ inside the cylinder. This is true even when P is only an infinitesimal distance inside the cylindrical surface. There is a discontinuity in $\mathbf{E}$, of value $\Lambda/(2\pi\epsilon_0 a) = \sigma/\epsilon_0$ at the surface of the cylinder. Note that, as in the case of the plane disk, this is σ/ϵ_0 again.

PROBLEMS

1 A line of length ℓ is uniformly charged with λ C/m. Consider a point P whose distance from the line is D, as in Fig. 2-13. Find $\mathbf{E}$ at P in terms of θ_1 and θ_2. The positive sense of the angles is shown in the figure.

2 Suppose $D = 0$ in Prob. 1; i.e., P is actually on the charged line. Find $\mathbf{E}$.

3 Let $D = 0$ in Prob. 1 but suppose P is not on the charged line segment but is on an extension of the line of charge, distant d from the closest end of the charged segment. Find $\mathbf{E}$.

4 If you were given Prob. 3 and its answer, by what limiting process could you obtain the $\mathbf{E}$ at P which would be produced by a point charge?

5 Find E a distance of 10^{-15} m from a point electron. If another electron were placed there, what would be the repulsive force in newtons? In pounds?

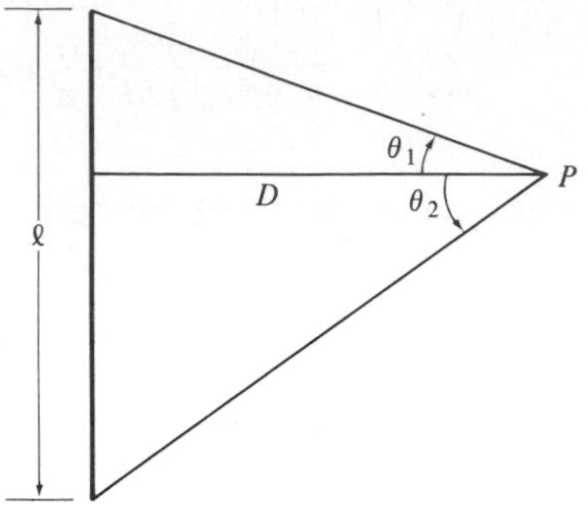

FIGURE 2-13

6 Find the **E** produced, at a point outside the sphere, by a spherical charge distribution with uniform charge density ρ. Use triple integration.

7 Find the **E** produced by a spherical charge distribution with uniform charge density at a point inside the sphere, using triple integration.

8 Find the **E** produced by a uniformly charged spherical shell at a point in the hollow interior, using triple integration.

9 One electron is placed at each of the following places: x = 1, 2, 3, 4, 5 (a total of five electrons). Find **E** at the origin.

10 One electron is placed at every positive integral value of x: 1, 2, 3, ... Find **E** at the origin.

11 (*a*) What is the Coulomb force between the proton and the electron in a hydrogen atom?

(*b*) How many times as large as g (the acceleration of gravity on the surface of the earth) is the centripetal acceleration of the electron?

12 Set up the integral for (do not attempt to solve) the following: Find **E** $(r,\phi,0)$ in the plane of a circle of radius a, having λ C/m on its periphery.

13 A circle of radius a has a uniform linear charge density, λ C/m, on its periphery. Find **E** at a point on the axis of symmetry, $(0,0,z)$, perpendicular to the plane of the circle at its center.

14 One electron is located at x = 1, another at x = −1. Give the location of all the points where **E** = 0.

15 A charge $+q$ is at (1,0,0), a charge $-q$ is at (−1,0,0). Give the expression for E_x at an arbitrary point in the xy plane.

16 Find the E_y in Prob. 15. (Problems 15 and 16 are almost the very simplest problem that can be formulated for finding an |**E**|. The complexity of this answer shows that in most cases this procedure is too difficult to follow.)

17 In Prob. 15 find the value of **E** at $(0,y)$.

18 Draw a curve for which the bracketed expression in the answer to Prob. 17 equals unity; and a second curve for which it equals 10.

2.4 GAUSS' LAW AND $\nabla \cdot \mathbf{E}$

A knowledge of both $\nabla \cdot \mathbf{E}$ and $\nabla \times \mathbf{E}$ everywhere within some finite domain is sufficient, by Helmholtz's theorem (see Appendix 4 at the end of Chap. 4) to determine **E** uniquely (to within an additive constant), provided there are no sources outside the domain. This provides an alternative method to that of Coulomb's law for finding **E**. A difficulty with Coulomb's law is that it requires a knowledge of the charge distribution. This is rarely known; it is also difficult to determine. So it is of great interest to investigate $\nabla \cdot \mathbf{E}$ and $\nabla \times \mathbf{E}$. We start with the former, but first–on the way to the determination of $\nabla \cdot \mathbf{E}$–we will find an equation which is extremely important in its own right. Called Gauss' law, this gives a relation between the flux of **E** coming out of a closed surface and the total charge within the volume bounded by that surface.

Consider the flux of **E** through a closed surface Σ about a region containing a point charge q at O, as shown in Fig. 2-14. The value of **E** produced by q at any point P on Σ is

$$\mathbf{E} = \hat{\mathbf{r}} \frac{1}{4\pi\epsilon_0} \frac{q}{r^2}$$

The flux of **E** through a differential element of the area Σ, one that contains P, is

$$\mathbf{E} \cdot d\mathbf{S} = \frac{q}{4\pi\epsilon_0} \frac{\hat{\mathbf{r}} \cdot d\mathbf{S}}{r^2}$$

But $\hat{\mathbf{r}} \cdot d\mathbf{S}$ is the projection of the vector $d\mathbf{S}$ on $\hat{\mathbf{r}}$; i.e., the projection of the area $d\mathbf{S}$ on a sphere of radius r about O. This area projected on the imagined sphere of radius r (call it dS') when divided by r^2 provides a measure of the solid angle contained within the elements of the cone which extend from O to both dS and dS'.

For, consider a different sphere, of radius R, drawn about O and completely within Σ. The elements of the cone will intersect an area dS'' on this sphere such that

$$\frac{dS'}{dS''} = \frac{r^2}{R^2}$$

This is true regardless of the value of r, so it is true everywhere on the surface Σ. Consequently, if $d\Omega' \equiv dS'/r^2$ and $d\Omega'' \equiv dS''/R^2$, then $d\Omega' = d\Omega''$. $d\Omega'$ is the differential solid angle subtended by the area dS' (or dS) at O. It is a dimensionless quantity (as is its equivalent in the plane, the plane angle). The unit solid angle is called the steradian (sr), by analogy with the radian. Integrating completely around O,

$$\oint d\Omega = \oint_{\Sigma'} \frac{dS}{r^2} = \frac{1}{r^2} \oint_{\Sigma'} r^2 \sin\theta \, d\theta \, d\phi = \int_0^{\pi} \sin\theta \, d\theta \int_0^{2\pi} d\phi = 4\pi$$

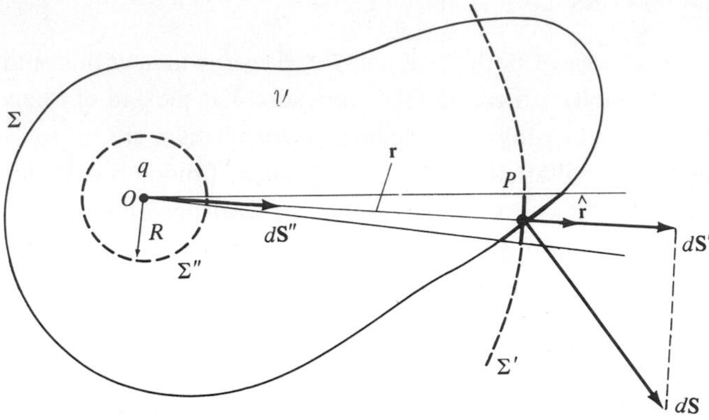

FIGURE 2-14

There are 4π sr about a point in three-dimensional space, and the area of a sphere is $4\pi r^2$; just as there are 2π radians about a point in a two-dimensional plane, and the circumference of a circle is $2\pi r$.

The flux of **E** through the entire sphere of radius R is the same as the flux of **E** through the entire sphere of radius r; and this is the same as the flux of **E** through the entire surface Σ. For

$$\oint_{\Sigma'} \mathbf{E} \cdot d\mathbf{S}' = \oint_{\Sigma'} \frac{q\hat{\mathbf{r}}}{4\pi\epsilon_0 r^2} \cdot d\mathbf{S}' = \frac{q}{4\pi\epsilon_0} \oint_{\Sigma'} \frac{\hat{\mathbf{r}} \cdot d\mathbf{S}'}{r^2} = \frac{q}{4\pi\epsilon_0} \oint_{\Sigma'} d\Omega' = \frac{q}{\epsilon_0}$$

and

$$\oint_{\Sigma''} \mathbf{E} \cdot d\mathbf{S}'' = \oint_{\Sigma''} \frac{q\hat{\mathbf{R}}}{4\pi\epsilon_0 R^2} \cdot d\mathbf{S}'' = \frac{q}{4\pi\epsilon_0} \oint_{\Sigma''} \frac{\hat{\mathbf{R}} \cdot d\mathbf{S}''}{R^2} = \frac{q}{4\pi\epsilon_0} \oint_{\Sigma''} d\Omega'' = \frac{q}{\epsilon_0}$$

Similarly,

$$\oint_{\Sigma} \mathbf{E} \cdot d\mathbf{S} = \frac{q}{4\pi\epsilon_0} \oint_{\Sigma} \frac{\hat{\mathbf{r}} \cdot d\mathbf{S}}{r^2} = \frac{q}{4\pi\epsilon_0} \int d\Omega = \frac{q}{\epsilon_0}$$

Note that this result does not depend on the position of q. If there are a number of charges inside Σ, then this procedure can be applied to each of them in turn. The flux from each of the individual charges would add to that of the others to give the total flux

$$\oint_{\Sigma} \mathbf{E} \cdot d\mathbf{S} = \frac{1}{\epsilon_0} Q \tag{E-5}$$

where Q is the total charge inside Σ. This extremely useful relation is known as Gauss' law. It is related to, but different from, Gauss' divergence theorem.

Gauss' law may now be used immediately to find $\nabla \cdot \mathbf{E}$. For the left-hand side of Gauss' law we have, by Gauss' divergence theorem,

$$\oint_{\Sigma} \mathbf{E} \cdot d\mathbf{S} = \int_{\mathcal{V}} \nabla \cdot \mathbf{E}\, d\tau$$

For the right-hand side we have, by the definition of charge density,

$$Q = \int_{\mathcal{V}} \rho\, d\tau$$

Then

$$\int_{\mathcal{V}} \nabla \cdot \mathbf{E}\, d\tau = \frac{1}{\epsilon_0} \int_{\mathcal{V}} \rho\, d\tau$$

Since this holds true for an arbitrary volume $\mathcal{V}$, the two integrands must be equal:

$$\nabla \cdot \mathbf{E} = \frac{\rho}{\epsilon_0} \tag{E-6}$$

The charge density in this equation is *any* charge density, whether free or bound. By a bound charge we mean, e.g., the charge of an ion in a solid lattice, where the average position of the ion is fixed in space because the ion is bound to the crystal lattice. The physical significance of this differential equation is the same as that of Gauss' law:

Positive charges are sources for lines of **E**, while negative charges are sinks for lines of **E**. It is not necessary, however, that a line of **E** begin or terminate on charge. Lines may terminate at infinity, e.g., even though there are no charges there. It is only necessary that the flux of lines entering any region about an arbitrary point equals the flux of lines of **E** leaving that region, provided only that there are no charges within the region. This is not only true at infinity but at any point where the charge density is zero. The lines may circulate indefinitely, without closing. (They may even close; but not in electrostatics, where $\nabla \times \mathbf{E} = 0$.)

Gauss' law is a macroscopic expression that relates the flux of the lines of **E** over any closed surface to the net total charge inside this closed surface. An important use is in the determination of **E** in cases with some degree of symmetry.

The expression for $\nabla \cdot \mathbf{E}$ is the differential or microscopic equivalent of Gauss' law. It is very important to all the subsequent theory because, together with the expression for $\nabla \times \mathbf{E}$, it determines **E** uniquely. This relation is one of the four differential equations which fix the values of the electric field and magnetic field vectors everywhere: two give the divergence and curl of the former, two give the divergence and curl of the latter vector. These equations, known collectively as Maxwell's equations, constitute the four cornerstones of the foundation upon which the theory of the electromagnetic field is generally erected. The first of these equations has just been derived from the experimentally obtained Coulomb's law. Although the $\nabla \cdot \mathbf{E}$ expression was obtained only for the case of electrostatics, it turns out that it is true in general, even when there are variations with time.

Examples

1. The electric field produced by a uniform charge distributed throughout the volume of a sphere.

One octant of the sphere, shown in Fig. 2-15, has a radius R. To find **E** outside the sphere, construct an imaginary sphere Σ_0, of radius r_0 larger than R, the actual radius. Apply Gauss' law to it:

$$\oint_{\Sigma_0} \mathbf{E}_0 \cdot d\mathbf{S}_0 = \frac{1}{\epsilon_0} Q_0$$

The charge Q_0 within Σ_0 equals Q, the total charge.

$$\oint_{\Sigma_0} \mathbf{E}_0 \cdot d\mathbf{S}_0 = \oint_{\Sigma_0} (E_0 \hat{\mathbf{r}}) \cdot (dS_0 \hat{\mathbf{r}}) = \oint_{\Sigma_0} E_0 \, dS_0$$

By symmetry, the value of E_0 must be constant on Σ_0. So

$$\oint_{\Sigma_0} \mathbf{E}_0 \cdot d\mathbf{S}_0 = E_0 \oint_{\Sigma_0} dS_0 = E_0 (4\pi r_o{}^2)$$

Then $E_0(4\pi r_o{}^2) = (1/\epsilon_0)Q$,

$$\mathbf{E} = \hat{\mathbf{r}} \left(\frac{Q}{4\pi\epsilon_0} \right) \frac{1}{r^2} \qquad (r > R)$$

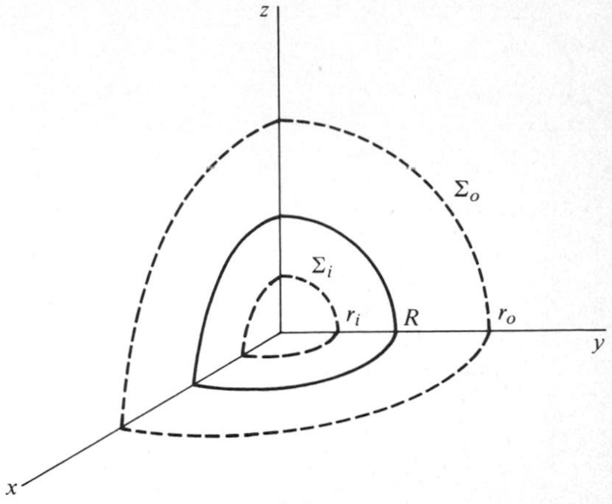

FIGURE 2-15

Next, to find $\mathbf{E}$ inside the sphere, construct a similar imaginary sphere of radius $r_i < R$ and apply Gauss' law to it:

$$\oint_{\Sigma_i} \mathbf{E}_i \cdot d\mathbf{S}_i = \frac{1}{\epsilon_0} Q_i$$

The total charge within Σ_i is

$$Q_i = \left(\frac{\frac{4}{3}\pi r_i^{\,3}}{\frac{4}{3}\pi R^3} \right) Q = \left(\frac{r_i}{R} \right)^3 Q$$

As before, $\oint_{\Sigma_i} \mathbf{E}_i \cdot d\mathbf{S}_i = E_i(4\pi r_i^{\,2})$. So $E_i(4\pi r_i^{\,2}) = (1/\epsilon_0)(r_i/R)^3\, Q$,

$$\mathbf{E} = \hat{\mathbf{r}} \left(\frac{Q}{4\pi\epsilon_0 R^3} \right) r \qquad (r < R)$$

At the surface of the sphere both equations give the same result:

$$\mathbf{E} = \hat{\mathbf{r}} \left(\frac{Q}{4\pi\epsilon_0 R^2} \right) \qquad (r = R)$$

Figure 2-16 shows how $\mathbf{E}$ varies with r.

A solution to this problem may also be obtained by a direct application of Coulomb's law. This would be considerably more complicated than the method above

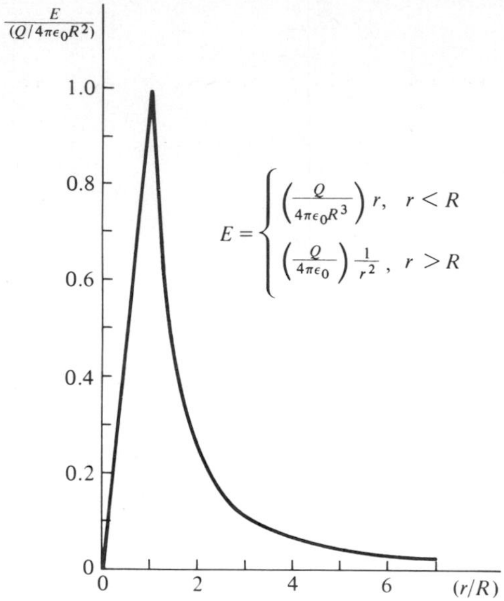

FIGURE 2-16

using Gauss' law, since a triple integration is involved. It is possible to use Gauss' law here, however, only because symmetry permits the transformation of $\oint \mathbf{E} \cdot d\mathbf{S}$ into $E \oint dS$.

Since $\mathbf{E}$ is force per unit test charge, the graph of Fig. 2-16 represents the force on a given charge at different distances from the center of the sphere. The curve also gives the gravitational force on a given mass at different distances from the center of a sphere of uniform mass density, since Newton's law of gravitation has the same form as Coulomb's law. If, as is the case with the earth, the mass density is not constant but varies in some manner that is independent of any angle and that is finite everywhere, then the portion of the curve for $r < R$ changes its shape but retains the same value at the end points, $r = 0$ and $r = R$. The rest of the curve, $r > R$, does not depend on the particular distribution of the charge density within the sphere, provided only that the total charge remains fixed and the distribution is spherically symmetric.

2. The $\mathbf{E}$ produced by a volume charge density distributed uniformly throughout an infinitely long cylinder of radius R.

Here, also, there is a high degree of symmetry and again Gauss' law may be applied. By symmetry, $\mathbf{E}$ can have only a radial component. (Note that this is not true if the length of the cylinder is finite; but it is approximately true near the center of a very long cylinder.) Figure 2-17 illustrates the case of infinite length, taking a finite portion which is typical.

First construct an imaginary cylinder of length ℓ which is coaxial with the actual cylinder but has a larger radius, $r_o > R$. Then $\oint_{\Sigma_0} \mathbf{E} \cdot d\mathbf{S}$ can be broken up into the sum

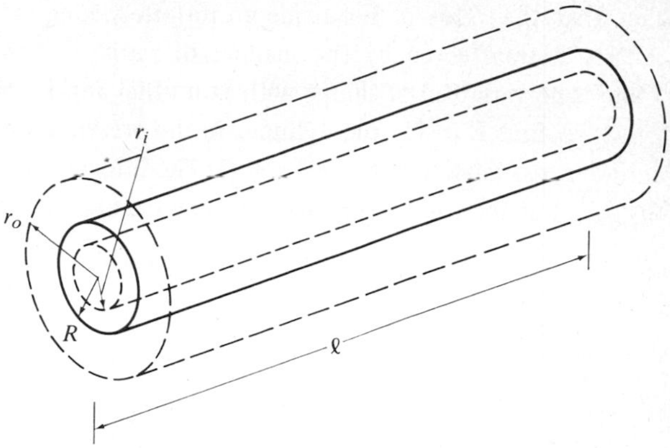

FIGURE 2-17

of three integrals: one over the left flat end, one over the curved central section, and one over the right flat end. For the two flat portions, however,

$$\mathbf{E} \cdot d\mathbf{S} = (\hat{\mathbf{r}}E) \cdot (\hat{\mathbf{n}}dS) = E\,dS(\hat{\mathbf{r}} \cdot \hat{\mathbf{n}}) = 0$$

For the middle portion $\int \mathbf{E}_0 \cdot d\mathbf{S}_0 = \int (E_0\hat{\mathbf{r}}) \cdot (dS_0\hat{\mathbf{r}}) = \int E_0\, dS_0 = E_0 \int dS_0 = E_0\,(2\pi r_o \ell)$. So

$$\oint_{\Sigma_0} \mathbf{E}_0 \cdot d\mathbf{S}_0 = 2\pi r_o \ell E_0$$

The total charge Q_0 within this outside imagined cylinder is the charge Q_ℓ within a length ℓ of the actual cylinder: $Q_0 = \rho\,(\pi R^2 \ell)$. So

$$2\pi r_o \ell E_0 = \frac{1}{\epsilon_0}\,[\rho \pi R^2 \ell]$$

and
$$\mathbf{E}_0 = \hat{\mathbf{r}} \left(\frac{\rho R^2}{2\epsilon_0}\right) \frac{1}{r_0} \qquad r_0 > R \qquad \text{(cylindrical coordinates)}$$

This equation may be put into a slightly different form by setting

$$\lambda = \frac{Q_\ell}{\ell} = \rho \pi R^2.$$

Then
$$E_0 = \hat{\mathbf{r}}\,\frac{\lambda}{2\pi\epsilon_0 r_o} \qquad \text{(cylindrical } r\text{)}$$

This is the same result obtained in Example 4 of Sec. 2-3 by direct application of Coulomb's law to an infinitely long cylindrical *surface* charge. This is an example of the fact that the value of **E** outside an infinitely long cylinder, uniformly charged along its length, is unaffected by the manner of radial distribution of the charge, provided that there is no variation of charge with azimuthal angle.

To find **E** inside the cylinder in the present case, change the radius of the imagined cylinder to a value less than R: $r_i < R$. The integral for the flux of **E** through the surface is $2\pi r_i \ell E_i$, but the total charge within the imagined cylinder is now $\rho\pi r_i^2 \ell$. So

$$2\pi r_i \ell E_i = \frac{1}{\epsilon_0}\rho\pi r_i^2 \ell$$

or

$$E_i = \hat{\mathbf{r}}\left(\frac{\rho}{2\epsilon_0}\right) r_i \qquad r_i < R \qquad \text{(cylindrical } r\text{)}$$

The results are graphed in Fig. 2-18.

Note that for $r < R$ the linear result here is different from that ($E = 0$) which holds for a cylindrical surface charge. (The volume charge density of the present example can be obtained only with a dielectric; inside an electrostatic metal (i.e., one in a field which is not changing with time) the charge density must be zero. In the case of surface charge density, either a dielectric or a metal may be used. Note, also, that the value of E falls off much more slowly ($1/r$) with increasing r than it does in the case of the sphere ($1/r^2$).

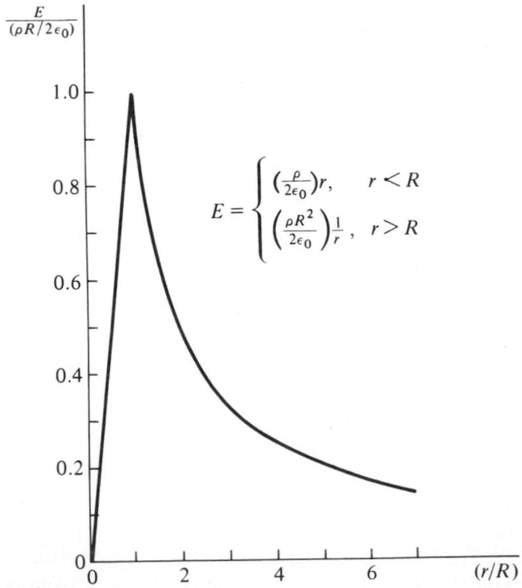

FIGURE 2-18

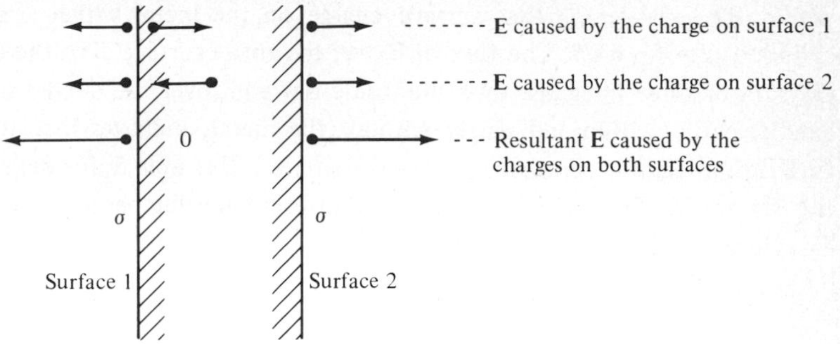

FIGURE 2-19

3. The value of **E** just outside the surface of a plane metal plate having a charge σ C/m^2 on each of the two parallel surfaces of the plate.

The value $E = \sigma/(2\epsilon_0)$ was obtained in connection with Fig. 2-11, the case of a uniformly charged infinite plane. Here the plate has two such surfaces, as shown in Fig. 2-19. The simplest way to obtain the answer for this case is by application of the superposition principle. In the region to the right of surface 1 in Fig. 2-19, the surface charge density σ on surface 1 produces a uniform $E = \sigma/2\epsilon_0$ to the right. This is true whether the point considered is inside the metal plate or in the region to the right of the metal. This is not the total **E** there, but only one constituent: that produced by the charge on surface 1. To the left of surface 1, similarly, the charge density σ on surface 1 produces a field $\sigma/2\epsilon_0$ to the left.

Similarly, the charge density σ on surface 2 will produce an **E** in the different regions as shown in Fig. 2-19. Combining both values gives

$$\mathbf{E} = \begin{cases} -\hat{\mathbf{x}}\,\dfrac{\sigma}{\epsilon_0} & \text{to the left of the metal plate} \\ 0 & \text{within the metal} \\ +\hat{\mathbf{x}}\,\dfrac{\sigma}{\epsilon_0} & \text{to the right of the metal plate} \end{cases}$$

In the region outside the plate, the two fields add and the answer is twice that of either field alone. In the region inside the plate the two fields buck and the answer is zero. The result would be unchanged if the plate were made of dielectric with a surface charge sprayed on it.

The answer may also be obtained by application of Gauss' law to a differential cylinder partially within and partially outside the plate. One flat surface is buried in the metal while the other is outside. There is no charge within the metal so Q, the charge within the cylinder, is that surface charge on the metal which is intercepted by the cylinder, $Q = \int_A \sigma\, dS$. The flux of **E** over the entire surface Σ of the cylinder consists of the sum of three integrals: over the inside flat end, over the curved sides (of differential length) partly within and partly without the metal, and over the outside flat end. The first flux integral is zero because $E = 0$ there; see Example 5, for detailed justification of this statement. The second integral has a differential value because the area is differential. The third integral has a finite value:

$$\int_A \mathbf{E} \cdot d\mathbf{S} = \int_A E\, dS$$

since **E** can have only a normal, but not a tangential, component close to the surface, outside the metal. So, for arbitrary A,

$$\int_A E\, dS = \frac{1}{\epsilon_0} \int_A \sigma\, dS$$

Consequently $E = \sigma/\epsilon_0$.

Although the calculation was made for a uniformly charged flat metal plate, the result has much broader validity, as was true for the infinite plane. For, the result here is applicable to a point just outside the surface of any conductor–flat or curved, uniformly charged or not, and whether the conductor boundary surfaces are parallel to each other or not. It is only necessary to let the area A become a differential to see this. Gauss' law can still be applied in the same way, to yield the same result. The σ that is pertinent is that which exists at the point of interest.

4. Is a line of **E**, the electric field, necessarily continuous in a real, physical situation? Must it begin on positive charge, proceed in a one-directional sense only, and end on negative charge?

The drawing of the lines of **E** is only a visualization. The only requirement that must be satisfied is

$$\nabla \cdot \mathbf{E} = \frac{1}{\epsilon_0} \rho$$

Fig. 2-20 shows the lines of **E** for a configuration of two charges, $+2q$ and $-q$. Within the dotted line there is a region where the lines of **E** that start on $+2q$ terminate on $-q$; outside the dotted line is a region where the lines of **E** that start on $+2q$ terminate, either on negative charge or on no charge, at infinity; either is permissible. The dotted line, itself, is also a line of **E**. It begins on $+2q$ and goes in a given sense with $|\mathbf{E}| \to 0$ at the

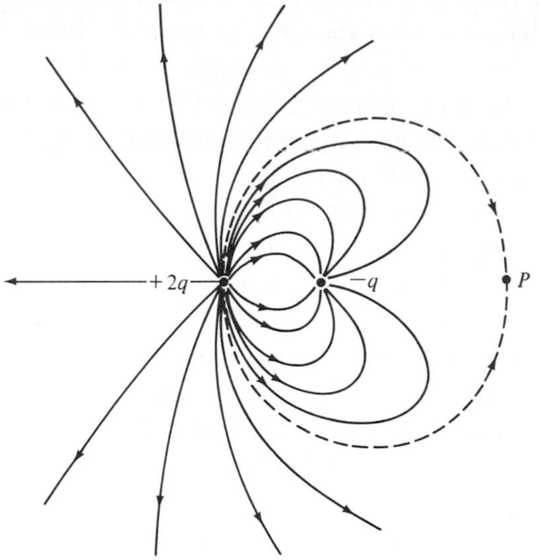

P is a point of unstable equilibrium The value of the electric field there is zero. but slightly to the left of *P* the field points left and slightly to the right of *P* it points right—in both cases away from *P*.
The up and down line through *P* goes through zero at *P* while reversing its direction there. In this direction *P* is a point of stable equilibrium.

FIGURE 2-20
(Modified slightly from Fig. 1.6d, p. 27, in William T. Scott, "The Physics of Electricity and Magnetism," 2d ed., Wiley, New York, 1966.)

point where the curve crosses the x axis to the right of $-q$; proceeding along the line beyond that point, the sense of **E** reverses. The line does not terminate, in one sense; in another, it terminates at P, where the value of **E** is zero. Near P, horizontally, the direction of **E** is away from P; near P, vertically, the direction of **F** is toward P.

5. Conclusions that can be drawn about the electric field within and on the surface of a metal in electrostatics.

The earlier result of zero electric field inside a charged cylindrical surface, Example 4 of Sec. 2-3, is applicable to a very great number of highly important cases, not just to that one. It applies both to dielectrics (insulators) on which a surface charge exists and to all electrostatic metals (conductors). In *electrostatic cases*, i.e., in situations where there is no flow of current or change of field with time, *the electric charge density within the interior of a metal must be zero.* For, the term metal means a substance which is a reservoir of a huge number of free electric charges, charges which are free to move in response to the slightest electric force exerted on them. (A typical value of this number is 10^{29} free electrons per cubic meter.) If there were *any* finite **E** *inside* a metal then the free charges would move in response to it; while this is occurring there is a current flow; the situation is then not electrostatic. Further charge motion will cease only if the free charges are permitted to redistribute themselves on the surface in such a way that the initial field is nullified; if this is not possible then the flow of charge continues and electrostatic conditions do not prevail. So in electrostatics the **E** everywhere inside a metal is zero.

On the surface of a conductor the situation is different. For the same reason that there can be no **E** in the interior, there can here be no tangential component of **E** in electrostatic cases. However, a normal component of **E**, outside the surface, is possible, as shown in Fig. 2-21. A component of **E** perpendicular to the surface would not ordinarily be able to remove the free charges from the metal. (If **E** is very large, of the order of millions of newtons per coulomb, then the free charges can literally be torn out of the surface. This is known as field emission. The phenomenon has been applied, in the field emission microscope, to a method for viewing the individual atoms in a crystal lattice.) In electrostatics, in summary, the **E** in the interior of a metal is zero, while the **E** at the surface exists only outside the metal and is normal to the surface. Problems 9 to 12 make use of this to prove some very broad conclusions, obtained by the mere application of Gauss' law.

In the case of the usual metals the free charges are exclusively the conduction electrons–electrons which have separated from the neutral metal atoms, leaving positive charged ions bound together in a lattice. The positive charges in metals are all bound charged and are not free to flow. If there is a superfluous free (i.e., mobile) surface charge it must be negative; the normal component of **E** at the surface, created by these free charges, will be directed toward the surface from the outside. Even though there are only free negative charges in the case of a metal, however, it is possible–and, indeed, quite common–to have a net positive surface charge density, with a concomitant normal component of **E** extending away from the surface to the outside. This occurs when there is a deficiency of free surface charge, below the usual amount that is present to neutralize the positive charge of the bound ions on the surface.

In the case of liquid metals the positive charges can also move with respect to the container. The mass of the positive ion, however, is thousands of times as great as that of the electron, the negative free charge; this greatly reduces the mobility (the average drift velocity divided by the electric field) of the ion compared to that of the electron; a typical ratio is 1:100. In semiconductors, also, the free carriers are both positive and negative in sign.

A monomolecular layer of a metal approximates a mathematical surface. For it the charge density, either negative (free) or positive (bound), is the same on both sides of the surface. For a metal plate which is thicker than one molecular layer there are two surfaces, and two possibilities may occur. If the plate is isolated then, in electrostatics, the charge density on one surface must be exactly equal to the charge density on the other surface, and the sign of both must be the same. Otherwise an **E** would be established within the plate, leading to the flow of current, contrary to the static assumption. If the plate is not isolated then the charge densities on the two surfaces can

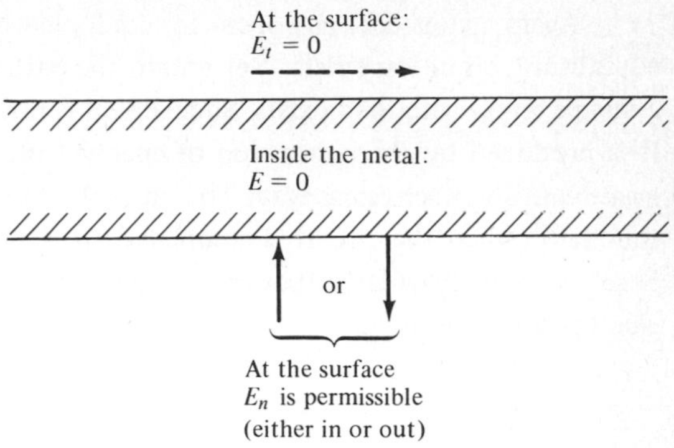

FIGURE 2-21

differ. Consider the case of two parallel plates, far removed from any other body. Such an assembly constitutes a parallel-plate capacitor, as shown in Fig. 2-22. The capacitor as a whole is isolated; its component plates are not. Here the charge on the inner surface of one plate is positive and exactly equal to the magnitude of the negative charge on the inner surface of the second plate. Both plates have zero net charge on their outer plane surfaces. $\mathbf{E} = 0$ within the plates.

Several qualifications are necessary with respect to the statement that $\mathbf{E} = 0$ inside an electrostatic conductor. The $\mathbf{E}$ referred to here is an average value—averaged over some extremely short time and averaged over some extremely small distance. Instantaneous electric fields do exist between the charges within an atom, as well as between the lattice ions and free electrons of the metal; furthermore, these instantaneous fields are very intense. However, they fluctuate both in space and time in such a way that their average value is zero—average, that is, over some macroscopically small region which nevertheless contains a large number of atoms.

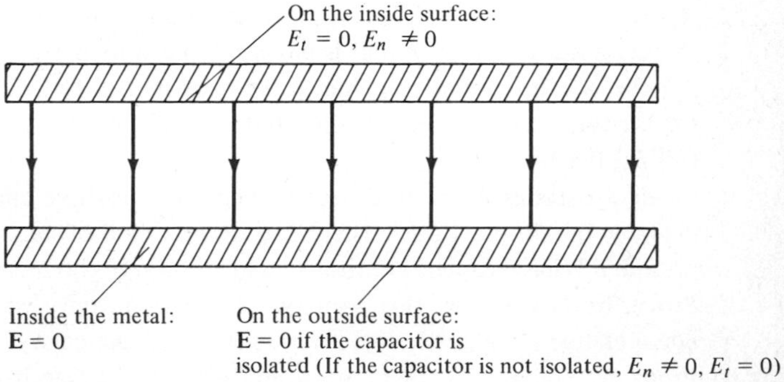

FIGURE 2-22

Again, consider an open-circuited electric battery. No current flows and an equilibrium situation exists. Yet within the battery electrolyte (a conductor, though, not a metal) an electric field exists. Such a field is not considered a classical electrostatic field. It is produced by the conversion of energy from the chemical state to the electrical state in a quantum-mechanical way. The current inside the battery which flows as a result of this field (when the circuit is completed) is, therefore, not produced by an electrostatic field. We will show later that the current outside the battery cannot be produced by an electrostatic field either.

PROBLEMS

1 Consider two infinitely long, coaxial, cylindrical conductors, one of radius r_i and one of larger radius r_o. The inner cylinder may be solid or hollow—it does not matter; r_o is the inner radius of the outer conductor. Assume a surface charge density σ on the outer surface of the inner conductor and a surface charge density $-\sigma$ on the inner surface of the outer conductor. This configuration constitutes a cylindrical capacitor and the result obtained here is pertinent to the subsequent section on capacitance, later in this chapter. Find **E** everywhere, using Gauss' law in cylindrical coordinates.

2 Find E, using Gauss' law in spherical coordinates, for two concentric spherical surfaces, one having radius r_i and surface charge density σ; and the other having radius r_o and surface charge density $-\sigma$. This is a spherical capacitor.

3 Let $\rho = a/\sqrt{r}$ from $r = 0$ to $r = R$ in spherical coordinates. Find **E**.

4 If $\rho = a/r^n$, what is the maximum value of n that gives (*a*) a finite value of the total charge Q? (*b*) A finite value of **E** at $r = 0$? (*c*) In Prob. 3 the charge density is infinite at $r = 0$, but the electric field is finite there. Is this case physically realizable?

5 Let $\rho = a/\sqrt{r}$ from $r = 0$ to $r = R$ in cylindrical coordinates. Find **E**.

6 Let $\rho = a/r^n$ in cylindrical coordinates. Find the expression for the total charge Q in a cylinder of length L and radius R about the z axis. Draw curves of $Q/2\pi La$ versus R for $n = 0, 0.5, 1.0, 1.5, 1.9$, plotting points at $R = 0, 0.1, 0.5, 1.0, 1.5, 2$. What is the maximum value of n that is physically permissible?

7 A cube is a body with a high degree of symmetry. Can Gauss' law be used to find **E** for a cubic, uniform, charge distribution? Write the integral for the $\hat{\mathbf{x}}$ face to show this explicitly.

8 In electrostatics lines of **E** usually begin on positive charges and end on negative charges, but in gravitation there is only one kind of mass (at least, that we know at present). What happens at infinity, where lines of gravitational field end?

9 Prove, by Gauss' law, that a charge placed in a cavity within a metal will induce an equal charge of opposite sign on the surface of the cavity.

10 Prove by Gauss' law that, when any external charge is put on a metal, or on the surfaces of cavities within the metal, or inside the metal itself, and when equilibrium (static) conditions are established, the charge will reside only on the outer surface.

11 Prove by Gauss' law that a charge outside a conductor does not produce any **E** within a cavity inside the conductor. The metal is said to shield the region inside the cavity from external fields outside the metal. Assume there are no charges in the cavity.

12 Prove by Gauss' law that an isolated hollow box of metal does not shield the outside world from the charges and fields within, or on the walls of, a cavity inside the metal. The results of Probs. 11 and 12 are not symmetric. Show that if the metal is grounded then the outside world will also be shielded from the charges within the cavity.

13 An experimental test of Gauss' law is also a test of the inverse-square law—the former was derived assuming the latter. The Cavendish sphere experiment is such a test. It consists of two concentric insulated metal spheres, the outer sphere having a small metal flap which may be opened for access to the inside. The flap carries a contact that touches the inner sphere only when the flap is closed. With the port closed and the two spheres in contact and uncharged, a charge is put on the outer surface of the outer sphere. It can be shown that for the field inside a charged spherical conductor to be zero, the necessary and sufficient condition is that the field from a charge element is directly proportional to $1/r^2$. So, if the field inside the charged outer sphere is not zero, there will be some flow of charge to the inner sphere. Opening the flap breaks the contact between the two spheres and permits measurement of the charge on the inner conductor. Such an experiment yielded the inverse-square law to within 2 parts in 10^9 (S. J. Plimpton and W. E. Lawton, *Phys. Rev.*, **50**:1066, 1936).

Prove that $dE \propto \sigma\, dS/r^2$ is a necessary condition that the electric field inside a uniformly charged spherical shell shall equal zero.

14 Starting from the equation $\int_{\mathcal{V}} \nabla \cdot \mathbf{E}\, d\tau = (1/\epsilon_0) \int_{\mathcal{V}} \rho\, d\tau$ we may conclude only that $\nabla \cdot \mathbf{E} = \rho/\epsilon_0 + \mu$, where $\int_{\mathcal{V}} \mu\, d\tau = 0$. What is the justification for taking $\mu = 0$?

15 Assume the electronic charge density of an atom of atomic number Z is $\rho_0 e^{-\alpha r}$. Find the **E**, using Gauss' law, that is produced by the electrons.

16 A metal can shield an inside region from the outside electromagnetic fields. Is there a substance which can perform a corresponding task for gravitational fields? Explain.

17 Prove, without brute force integration, that $dE \propto \sigma\, dS/r^2$ is a sufficient condition that the electric field inside a uniformly charged spherical surface shall equal zero.

18 Assume that $\nabla \cdot \mathbf{E} = 0$ and $\nabla \times \mathbf{E} = 0$ everywhere. Then (see Appendix 4) the Helmholtz theorem gives the following result for **E**:

$$\mathbf{E} = -\nabla\left[\frac{1}{4\pi}\int \frac{0}{R}\, d\tau_s\right] + \nabla \times \left[\frac{1}{4\pi}\int \frac{0}{R}\, d\tau_s\right] = 0 + 0 = 0$$

But the electric field $\mathbf{E} = \hat{\mathbf{z}}E_0$ has $\nabla \cdot \mathbf{E} = 0$ and $\nabla \times \mathbf{E} = 0$ everywhere. Reconcile this case with the Helmholtz theorem.

Appendix 2

THE DIRAC DELTA FUNCTION

Figure 2-23*a* gives a graph of a function which we will call the step function with a sloping riser $S(x)$. The derivative of $S(x)$, shown in Fig. 2-23*b*, is a rectangle of unit area—call it $\Delta(x)$. Suppose we let $\epsilon \to 0$. Then $S(x) \to s(x)$, the unit step function of Fig. 2-23*c*; and $\Delta(x) \to \delta(x)$, the delta function. It is not possible to draw a graph of this pathologic function: Figure 2-23*d* is merely suggestive, for the function $\delta(x)$ is zero everywhere except at $x = 0$, where there is a singularity, $\delta(0) = \infty$, such that the area under the curve is unity: $\int_{-\infty}^{\infty} \delta(x)\, dx = 1$. This equation really means

$$\int_{-\infty}^{\infty} [\lim_{\epsilon\to 0} \Delta(x)]\, dx = \lim_{\epsilon\to 0} \int_{-\infty}^{\infty} \Delta(x)\, dx = \lim_{\epsilon\to 0} \int_{0}^{\epsilon} \frac{1}{\epsilon}\, dx$$

$$= \lim_{\epsilon\to 0} \frac{1}{\epsilon} [x]_0^{\epsilon} = \lim_{\epsilon\to 0} \frac{1}{\epsilon} \cdot \epsilon = \lim_{\epsilon\to 0} 1 = 1$$

There is an additional defining property necessary for $\delta(x)$: when $\delta(x)$ is multiplied by any function $f(x)$ and the result is integrated from $-\infty$ to $+\infty$ the result is simply the value of $f(x)$ at $x = 0$ (where $\delta(x)$ has its singularity). Mathematically,

$$\int_{-\infty}^{\infty} f(x)\, \delta(x)\, dx = f(0)$$

To show this for the above function, note that the left side, L, really means

$$\lim_{\epsilon\to 0} \int_{-\infty}^{\infty} f(x)\, \Delta(x)\, dx$$

$$L = \lim_{\epsilon\to 0} \int_{0}^{\epsilon} f(x) \frac{1}{\epsilon}\, dx = \lim_{\epsilon\to 0} \left\{ \frac{1}{\epsilon} \int_{0}^{\epsilon} f(x)\, dx \right\} = \lim_{\epsilon\to 0} \left\{ \frac{1}{\epsilon} [I(\epsilon) - I(0)] \right\}$$

where the function $I(x)$ is the integral of $f(x)$: $I(x) = \int f(x) dx$. $I(\epsilon)$ may be obtained from $I(0)$ by a Taylor series expansion,

$$I(\epsilon) = I(0) + \left(\frac{dI}{dx}\right)_{x=0} \epsilon + \frac{1}{2!}\left(\frac{d^2 I}{dx^2}\right)_{x=0} \epsilon^2 + \cdots$$

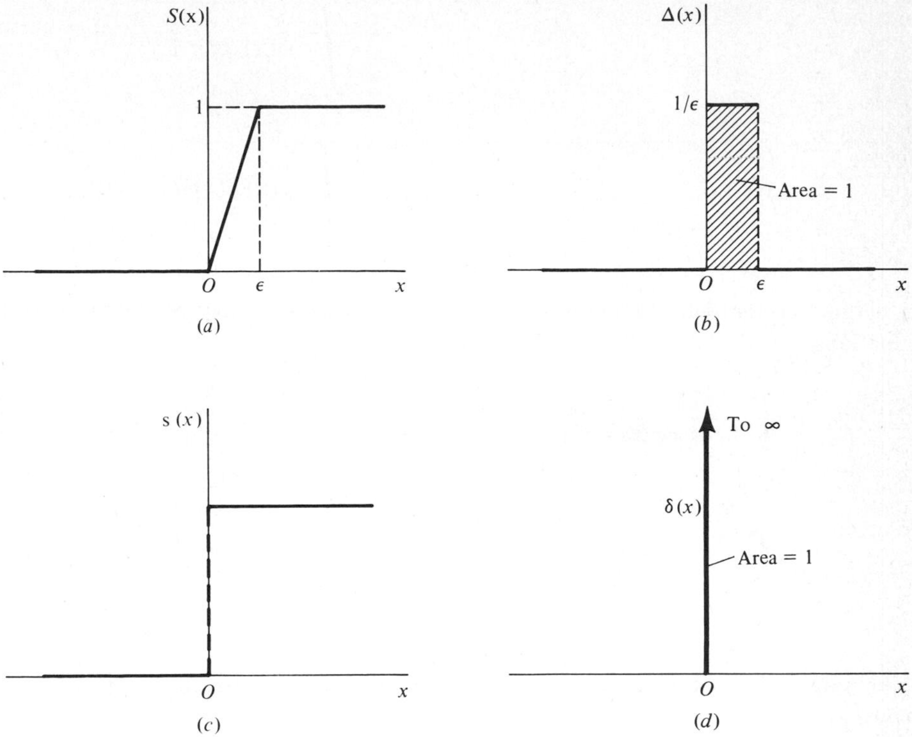

FIGURE 2-23

so

$$L = \lim_{\epsilon \to 0} \left\{ \frac{1}{\epsilon} \left[\left(\frac{dI}{dx} \right)_{x=0} \epsilon + \frac{1}{2!} \left(\frac{d^2 I}{dx^2} \right)_{x=0} \epsilon^2 + \cdots \right] \right\} = \left(\frac{dI}{dx} \right)_{x=0}$$

$$= \left[\frac{d}{dx} \left(\int f(x) dx \right) \right]_{x=0} = [f(x)]_{x=0}$$

The delta function need not be limited to one dimension. In general

$$\delta(r) = \delta(x)\, \delta(y)\, \delta(z)$$

Also, it is not limited to the derivative of the step function. Any function which has the properties

$$\delta(x) = \begin{cases} 0, & x \neq 0 \\ \infty, & x = 0 \end{cases}$$

$$\int_{-\infty}^{\infty} \delta(x)\, dx = 1$$

and

$$\int_{-\infty}^{\infty} f(x)\, \delta(x)\, dx = f(0)$$

is defined as the delta function. For example, the following may be shown to be delta functions:

1 $$\frac{1}{2\pi}\int_{-\infty}^{\infty} \cos kx\, dk = \delta(x)$$

2 $$\frac{1}{2\pi}\int_{-\infty}^{\infty} e^{ik(x-y)} dk = \delta\,(x-y)$$

(The singularity occurs at $x = y$, where the argument of the delta function, $x - y$, vanishes.)

3 $$\nabla^2\left(\frac{1}{R}\right) = -4\pi\,\delta\,(\mathbf{r}_s - \mathbf{r}_t) = -4\pi\,\delta\,(\mathbf{R})$$

$$= -4\pi\,\delta\,(x_s - x_t)\cdot\delta\,(y_s - y_t)\cdot\delta\,(z_s - z_t)$$

Here we are interested in (3). It is not too difficult to show that $\nabla^2\ (1/R) = 0$ when $R \neq 0$, simply by taking the derivatives. At $R = 0$ the resulting expression, $(3/R^3 - 3/R^3)$, becomes $\infty - \infty$, which is indeterminate. We can evaluate this at $R = 0$, and at the same time show that $\int \nabla^2\ (1/R)\, d\tau_s = -4\pi$, by considering a small sphere of radius a about $\mathbf{r}_t$. For simplicity take the center as the origin, so $\mathbf{R}$ becomes $\mathbf{r}_s$. If the sphere has volume $\mathcal{V}$ and area Σ,

$$\int_{\mathcal{V}} \nabla^2\ \frac{1}{R}\ d\tau_s = \int_{\mathcal{V}} \nabla^2 \frac{1}{r}\ d\tau = \int_{\mathcal{V}} \nabla\cdot\nabla\frac{1}{r}\ d\tau$$

$$= \oint_{\Sigma} \nabla\frac{1}{r}\cdot d\mathbf{S} \qquad \text{(by Gauss' law)}$$

$$= \oint_{\Sigma} \left[\hat{\mathbf{r}} \frac{\partial}{\partial r} \frac{1}{r} + \hat{\boldsymbol{\theta}} 0 + \hat{\boldsymbol{\phi}} 0 \right] \cdot \hat{\mathbf{r}}\, dS$$

$$= \oint_{\Sigma} \frac{1}{r^2}\, dS = - \oint_{\Sigma} d\Omega = -4\pi$$

Since the integral is finite but the integrand vanishes at all points other than $R = 0$, the value of the integrand at $R = 0$ must be infinite. Thus $\nabla^2 (1/R) = -4\pi\, \delta(R)$ satisfies the first two conditions defining the delta function. It may be shown, similarly, that the third condition is also satisfied,

$$\int f(\mathbf{r}_s)\, \nabla^2 \frac{1}{R}\, d\tau_s = -4\pi f(\mathbf{r}_t)$$

Then we may write

$$\nabla^2 \frac{1}{R} = -4\pi\, \delta(\mathbf{R})$$

3

THE MAGNETOSTATIC CURL IN VACUUM

3-1 A BACKWARD GLANCE

The word magnetism is derived from Magnesia, a region in Asia Minor whose inhabitants, historically, were familiar with stones (actually pieces of iron ore) that had the property of attracting similar small pieces of stone. The study of magnetism, consequently, began as the study of the mechanical attraction of some bodies for certain other bodies—not too much different from the study of electrification with which the investigation of electricity began. Despite this remarkable coincidence in beginnings, essentially no connection was made between the two disciplines and they remained completely isolated phenomena for millennia. In the 16th century many properties of magnetism were studied by William Gilbert (1540 to 1603), the same English physician who had investigated the connection between friction and electrostatic forces. After this there was no further investigation for almost 250 years—not until 1819 when H. C. Oersted found that a wire carrying electric current produced a force on a magnetized compass needle. The spirit of scientific investigation is not very old.

Electrostatic forces produced by rubbing various materials together never became of very great importance, either theoretically or practically. Eventually it was learned that all substances are capable of giving up or taking on some electrical charge. Rubbing (by some mechanism which is still not understood) simply made the transfer more efficient; and there the interest ended. Magnetostatic forces, on the other hand, proved to be extremely important in a practical way and had a decisive effect on the development of mankind—something which could scarcely have been predicted a priori. For it was learned that the earth, itself, acted like one of the pieces of lodestone from Magnesia, and the force that this natural magnet, the earth, exerted on small needles of iron soon led to the development of the compass. Without the compass ancient sailors could not have gone far beyond the sight of land, and America, Africa, and Asia would have remained terra incognita to the Western world for many more centuries. So while the magnetic compass does not rank in importance with the wheel, or with metal tools, it is not too far behind in the long-lasting nature of the effects it produced—not all of them beneficial.

A dozen years after Oersted's work Michael Faraday discovered electromagnetic induction, a further connection between electricity and magnetism. Actually an American physicist, Joseph Henry, learned about this first, but the brilliant Britisher published his results sooner. Maxwell's paper on the electromagnetic nature of light, as well as his "Treatise on Electricity and Magnetism," appeared in 1865; by 1888 Hertz had transmitted and detected electromagnetic waves; and from then on the two separate studies, electricity and magnetism, were treated more and more as different aspects of the same subject.

Albert Einstein's revolutionary 1905 paper, in which he introduced special relativity to a disinterested world, was written to clear up a question in electricity and magnetism: it was titled "On the Electrodynamics of Moving Bodies." Here it was made clear that a magnetic field was merely a mode of description required by an observer relative to whom source charges *and* test charges were moving with constant velocities. An electrostatic field, essentially, sufficed for an observer relative to whom either the source charges, alone, or the test charges, alone, were at rest. The magnetic field arose directly from the transformation equations between the two observers. An electric beam—a current of charges flowing in a stream—produced a magnetic field; but for an observer moving along with the charges there was only an electric field.

It has been recognized for more than 150 years that electric currents created magnetic field; and for two-thirds of a century it has been understood that the magnetic field was a convenience that was invented to help describe the effects of moving charges on moving charges. But, too, it was known that stationary electrons produced a small magnetic field; and the presence of magnetic fields created by permanent magnets was also well known. Here, apparently (but not really), no electric currents were present. It therefore seemed that magnetism had a certain life of its own, independent of the presence of electric charge. Two things remained to be demonstrated before it could be seen that this was illusory.

First, P. A. M. Dirac showed in 1927 that the magnetic field produced by a stationary electron was a relativistic effect intimately connected with the electron's charge. Dirac modified the Schrödinger equation of quantum mechanics to make it comply with the demands of special relativity and obtained the requirement that a stationary electron must have an intrinsic angular momentum, or spin. From Dirac's work it was evident that the permanent magnetism of electrons was connected with their charge and spin.

The second missing link connecting the magnetic fields of permanent magnets with electric currents and charges was provided shortly thereafter by Werner Heisenberg, who showed that ferromagnets–permanent magnets made of iron, of which the ancient lodestones of the Middle East were one type–owed their properties to cooperative *electrical* action of the atoms. There was an electric force, quite apart from the Coulomb force, which aligned the magnetic moments of the individual atoms parallel to each other. (Magnetic alignment forces, also present, were far too weak to account for ferromagnetism.) The permanent magnetism of macroscopic bodies was now a manifestation of the application of relativity and quantum mechanics to a collection of atoms and their electronic charges. All of the study of magnetism thereby becomes, really, a part of the study of electricity.

In this text electrostatics and magnetostatics are pursued in parallel but independent fashion; more accurately, they are interwined. Only when statics is completed, and time enters the picture, does it become either necessary or desirable to treat them in a combined, electromagnetic fashion. This first happens here in Chap. 9, where the two time-dependent Maxwell equations are discussed. Thereafter the two disciplines are wed to each other.

3-2 FORCES BETWEEN CONDUCTION CURRENTS

Electrostatics is the study of electricity when all charges, the sources as well as the test charge, are stationary with respect to the observer. In magnetostatics all charges are assumed to move with a constant speed, or at least with a constant average speed, relative to the observer. Strictly speaking, only the case of constant velocity should be considered–constant direction as well as constant speed; otherwise the charges are accelerating, and it becomes necessary to consider inertial forces and radiation. But many cases of interest occur where the accelerations are so small that their effects are negligible. Our only requirements for magnetostatics will be that the forces, fields, and currents are all constant with time and that there is negligible radiation. The study of the force produced by one electron on a second, relative to which it is moving, does not (strictly speaking) belong in magnetostatics, for such a force depends on the distance between the two electrons, and this varies with time. Since this is directly connected with the force between differential current elements, however, it is dealt with in some detail in Appendix 3, at the end of this chapter.

For simplicity, those fields created by the intrinsic magnetic moments of stationary sources will be ignored in most of this book. Electrons are considered, contrary to fact, to be particles without spin angular momentum or magnetic moment; permanent magnets, bodies displaying the magnetic moments of their atoms on a macroscopic scale, are considered only briefly. The addition of these magnetic fields would not alter the underlying outlook but would vastly complicate the mathematics.

An electric field I is defined as the net rate of transfer of a charge q through some specified region, just as a current of water is the rate of movement of mass across a given plane. Sometimes the current is filamentary, as when it flows through a long, very thin, conducting wire; sometimes it may have a volume distribution; sometimes the distribution will be limited to a surface. In the case of filamentary current, let there be a net linear charge density λ C/m, the charges having the constant average speed v m/s past a given point P, as in Fig. 3-1. Both λ and v are measured by an observer who is stationary with respect to the filamentary conductor in which the current is flowing. Then $I = \lambda v$. The unit of current, the ampere, is one of the four basic units (in the mksa system) in terms of which all other units are expressed. In this system the unit of charge, the coulomb, is defined in terms of amperes and seconds by the equation $q = It$: one coulomb is that charge which is transported in one second past the point P on a filamentary conductor when the steady current I is one ampere. The definition of the ampere, itself, will be given in the next section in terms of the force exerted by one current on another. This method of definition is roundabout but has practical advantages: the force between currents can be measured much more accurately than the force between charges.

Suppose a charge is permitted to move on a surface, instead of being forced to move along a filamentary curve. Such would be the case, for example, with a superconductor wire of finite cross section. Let the surface charge density be σ C/m^2; assume the average speed of the charges to be v across a length LL' which is in the surface and perpendicular to the velocity, as in Fig. 3-1*b*. Then $\mathbf{J}_l = \sigma \mathbf{v}$ A/m gives the surface current density—the current per unit length of LL'. Similarly, if there is a volume charge density ρ C/m^3, with a velocity v m/s as in Fig. 3-1*c*, then $\mathbf{J} = \rho \mathbf{v}$ A/m^2 is the volume current density.

It is convenient to divide all currents into two braod classes: conduction currents and convection currents. Much vagueness exists concerning the exact meaning of these terms. If all the charges present are of one sign only, then the current caused by their motion relative to an observer is a convection current. For another observer, moving along with the same velocity as the charges, there is no motion of the charges and they appear stationary to him; so there is no current for him and there would only be an electrostatic field. Two such parallel currents, moving in the same sense with the same velocity, would appear, to the moving observer, as two stationary line charges: they would repel each other. For the original observer, on the other hand, there would be two parallel beam currents producing both electric and magnetic forces on each other. The net

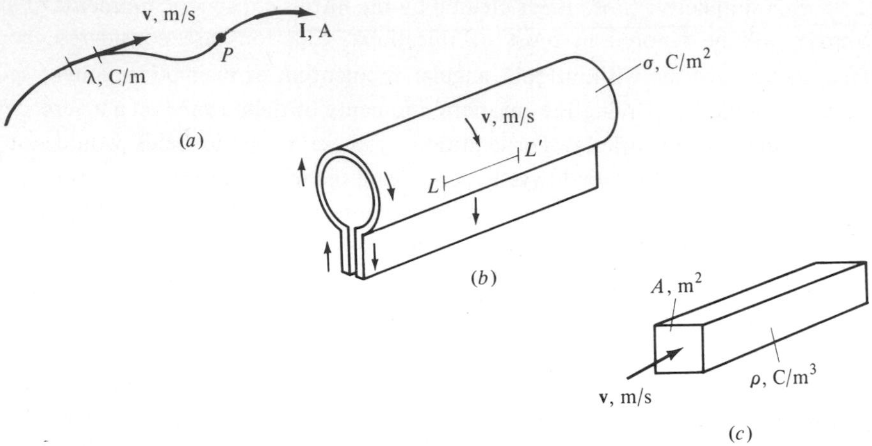

FIGURE 3-1

force for him would also be one of repulsion, for all observers agree on the question of attraction or repulsion. Such currents are called convection currents; for convection currents the electric forces dominate the magnetic forces.

In Chap. 14, Sec. 14-1, it will be shown that the criterion for a filamentary current I to be a convection current is that $(c\lambda)^2 - I^2 > 0$, where λ, the linear charge density, may consist of only one sign of charge (as above) or of both kinds; and c is the speed of light in vacuum. It will be shown that $(c\lambda)^2 - I^2$ is an invariant; that is, if λ' and I' are the values of the linear charge density and current in the filament for another observer, moving relative to the filament and the first observer, then

$$(c\lambda')^2 - (I')^2 = (c\lambda)^2 - (I)^2$$

The sign of this invariant quantity, therefore, has absolute significance: it is the same for all observers. If the primed observer is the one for whom $I' = 0$ the left side is positive; so, then, is the right side. A positive sign for this invariant is the criterion that the current be a convection current, even if both signs of charge are present and moving with different speeds. λ is the *net* charge density and I is the *net* current, for a given observer.

In ordinary conductors both kinds of charge are present but only one kind has a nonvanishing average drift velocity when current flows. The negative linear charge density λ_n of the moving electrons for a stationary observer is the negative of the positive linear charge density of the stationary ions λ_p for this observer. $\lambda = \lambda_n + \lambda_p$, the net linear charge density for this observer, is zero: $(c\lambda)^2 = 0$. So $(c\lambda)^2 - I^2$ is negative for him. Since it is an invariant, this quantity will be negative also for moving observers, for whom $\lambda \neq 0$. The criterion that a current be a conduction current is, simply, $(c\lambda)^2 - I^2 < 0$. For conduction currents an observer can be found for whom $\lambda = 0$; for convection currents an observer can be found for whom $I = 0$. It will be seen later in this chapter that two parallel conduction currents moving in the same direction attract each other, just the opposite of the result for two parallel convection currents. So this fact distinguishes conduction currents from convection currents: in the former the magnetic forces are greater than the electric forces. It is convenient to refer to conduction and convection currents, when it is not necessary to distinguish between them, as conventional current.

A third type of current is possible, one intermediate between conduction currents and convection currents: $(c\lambda)^2 - I^2 = 0$. This is not a common case; but it can occur. An infinitely long transmission line consisting of two parallel superconducting wires, or even such a line of finite length with a proper load (i.e., terminal) resistance R is one such case. For the line of finite length it is necessary that $R = R_0$, where R_0 is the characteristic impedance of the line, determined by its geometry. Then this condition is met, or when the line is of infinite length, the conductors exert no force on each other, regardless of the diameter of the wires or their spacing. *We will usually be concerned with conduction currents*, for this is the type commonly encountered in ordinary circuits. The same principles that apply here may also be adopted for the other types of current; though sometimes, viz., when dealing with energy, these give different answers.

We take the equation

$$\boxed{\mathbf{F}_{ST} = -\frac{\mu_0}{4\pi} \oint_T \oint_S \frac{(I_S\, d\mathbf{r}_s) \cdot (I_T\, d\mathbf{r}_t)}{R^2} \hat{\mathbf{R}}_{st}} \qquad \text{(M-1)}$$

as expressing the fundamental magnetic force between two filamentary currents in vacuum. See Fig. 3-2. Here the fact that the subscripts on $\mathbf{F}_{ST}$ are capital letters designates the fact that a complete circuit S, and a complete test circuit T, are being considered. (If the currents are conduction currents flowing in ordinary metal wires as in Fig. 3-2 then, for a stationary observer, the wires have a zero net charge even when current flows; but for convection currents this magnetic force would be one in addition to the electrical forces between the circuits.) This equation, which has the same basic importance for magnetostatics that the Coulomb force expression has for electrostatics, is based on many experiments; for example, the force between two long, straight, parallel

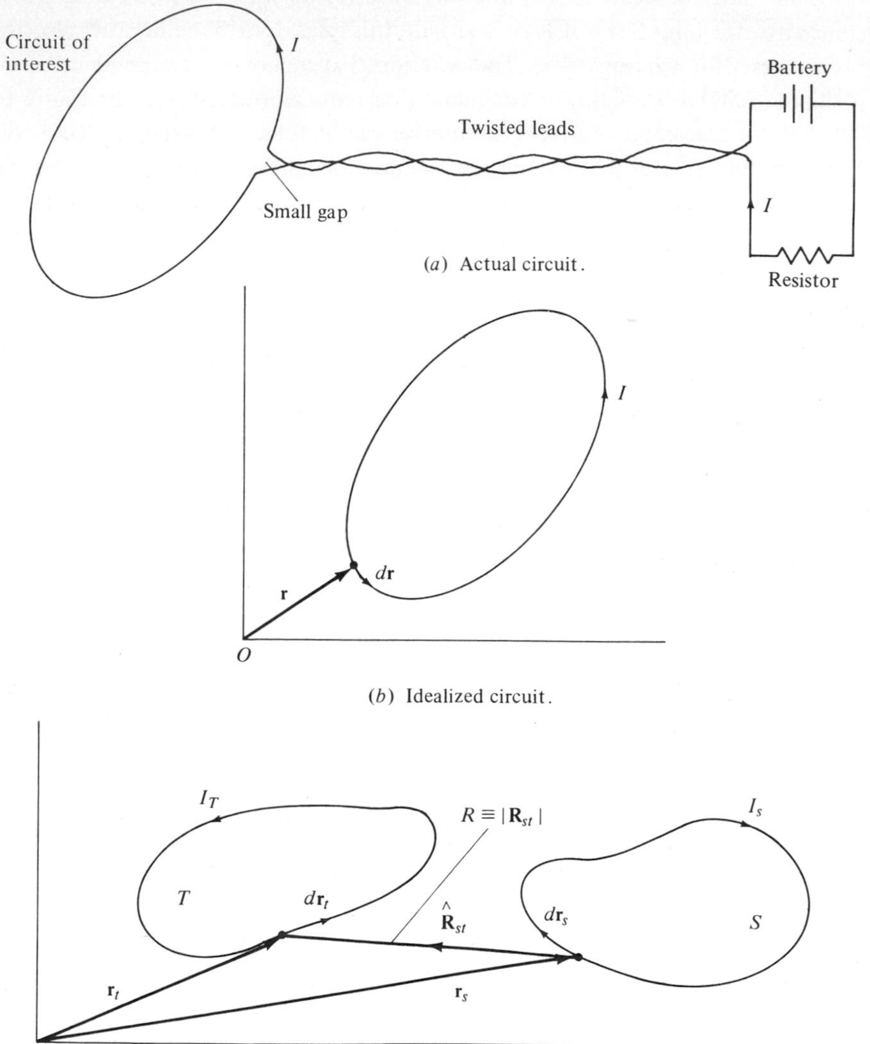

FIGURE 3-2

conductors. Since the first of these experiments was performed by Ampere, the equation is sometimes referred to as Ampere's force equation; but there are several other laws bearing Ampere's name, so we will merely refer to it as the basic magnetostatic force equation.

The quantities $\mathbf{F}_{ST}$, $\hat{\mathbf{R}}_{st}$, and R have meanings here similar to those in the case of the Coulomb law. μ_0 is a new constant, first introduced here, which is called the permeability of free space and is arbitrarily given the exact numerical value $4\pi \times 10^{-7}$ in the mksa system. From the equation above the unit of μ_0 must be the N/A^2, with dimensions $[\text{m kg s}^{-2}\ \text{a}^{-2}]$. In this equation μ_0 is in the numerator while in Coulomb's law ϵ_0 is in the denominator. In the continual comparison that will be made between the equations of electrostatics and their mates in magnetostatics this condition will continue to hold true: the quantity analogous to ϵ_0 is $1/\mu_0$, not μ_0. By assigning an a priori numerical value to μ_0 we are, in effect, fixing the size of the unit of current. Here the ampere is forced to be that unit which makes μ_0 have the value $4\pi \times 10^{-7}$. Fixing the size of the ampere also fixes the size of the coulomb. Then in Coulomb's law the numerical value of ϵ_0 is forced to be that which makes that equation hold true. Coulomb's law, in the mksa system, is really an expression for determining the value of the proportionality factor ϵ_0.

Figure 3-2*a* shows a battery supplying a constant current I to a closed circuit consisting of an idealized wire (one having no resistance) and a resistor. The battery, twisted leads, and resistor are far removed from the circuit proper, the latter being that part of interest here. It will be assumed that the battery and resistor are not of interest, that the effects of the two wires of the twisted pair neutralize each other, and that the gap (where the twisted pair connects to the circuit of interest) may be closed with negligible effect. The actual circuit is, then, idealized to that shown in Fig. 3-2*b*. The position vector of a point in the circuit with respect to an arbitrary origin is $\mathbf{r}$. An element of this circuit is $d\mathbf{r}$, the sense of this vector being taken to coincide with that of $\mathbf{I}$. $\mathbf{r}$ depends on the location of O but $d\mathbf{r}$ is independent of the origin.

In Figure 3-2*c* two idealized circuits are shown, one labeled S (for source) and one labeled T (for test). The directions of the two currents are arbitrary. $\mathbf{F}_{ST}$ is the force of the entire source circuit on the entire test circuit. The symbol $\oint_S$ in the basic magnetostatic force equation means that one given test element $d\mathbf{r}_t$ is selected in Fig. 3-2*c*; then an integration is carried out between it and each source element $d\mathbf{r}_s$ in succession, until the force of the entire source circuit S on this one test element has been evaluated. The test element is a differential, so the force on it that is due to the entire source circuit is also a differential. Both the differential current element and the differential force in this case are differentials of the first order. [There is a more basic force: that produced by a differential source element on a differential test element. This is a differential of the second order. Although this is *the* fundamental force, it is treated here only in Appendix 3 (at the end of this chapter) rather than in the main text. The reasons for this will be discussed shortly.]

The symbol $\oint_T$, similarly, means that the test element $d\mathbf{r}_t$ is changed, in succession, from one place to another until the entire idealized test circuit has been covered. At each new position of $\mathbf{r}_t$ the force of the entire source circuit is calculated on the $d\mathbf{r}_t$ there. When all the test elements have been considered, the vector sum of all the differential forces on all the individual test elements is the finite force of the entire source circuit on the entire test circuit.

The equation for the magnetic force between currents bears a certain resemblance to the corresponding Coulomb equation for electric force, but there are important distinctions.

1 With currents the force is dependent on the relative orientation of $d\mathbf{r}_s$ and $d\mathbf{r}_t$. This is shown by the dot product. In the case of the Coulomb force there is only one direction–that of the line joining the charges–while here there are three inherent directions.

2 The minus sign shows an attractive force for similarly directed current elements, unlike the repulsive force between two similar charges.

3 The basic force equation in magnetostatics appears in integral form.

It is incorrect to conclude from the basic magnetostatic force equation that the force exerted by one differential current element in one circuit on a differential current element of the second circuit is directed along $\hat{\mathbf{R}}_{st}$. The appearance of the expression is deceiving: the force exerted by $I_s\,d\mathbf{r}_s$ on $I_t\,d\mathbf{r}_t$ is not along $\hat{\mathbf{R}}_{st}$. Consider the force exerted by a distant source *circuit* on a differential *element* of the test circuit. This is obtained by taking a differential component of the basic magnetomotive force equation and adding an arbitrary function whose integral is zero. $\hat{\mathbf{R}}_{st}$ is approximately constant in this case and can be taken out of the integrand. So

$$d\mathbf{F}_{St} = -\hat{\mathbf{R}}_{St}\frac{\mu_0}{4\pi}I_S I_T \oint_S \frac{(d\mathbf{r}_s \cdot d\mathbf{r}_t)}{R^2} + d\mathbf{u}_t$$

Here $\mathbf{u}_t = \hat{\mathbf{x}}u_x + \hat{\mathbf{y}}u_y + \hat{\mathbf{z}}u_y$ is an arbitrary function of $\mathbf{r}_t$. (Calling $x_t = x$, $y_t = y$, $z_t = z$ for simplicity,

$$d\mathbf{u}_t = \frac{\partial \mathbf{u}_t}{\partial x}dx + \frac{\partial \mathbf{u}_t}{\partial y}dy + \frac{\partial \mathbf{u}_t}{\partial z}dz = (d\mathbf{r} \cdot \nabla)\mathbf{u}_t \Big)$$

Since $\mathbf{u}_t$ is arbitrary, $d\mathbf{u}_t$ need not be directed toward the test element; then neither must $d\mathbf{F}_{St}$. It will be shown below that the force is *not* a central one; i.e., it is not directed along the line of centers.

$d\mathbf{F}_{St}$ is obtained from a complete source circuit and a differential test element. A similar result is obtained for the differential force $d\mathbf{F}_{sT}$ produced by a differential source element on a distant test circuit:

$$d\mathbf{F}_{sT} = -\hat{\mathbf{R}}_{sT}\frac{\mu_0}{4\pi}I_S I_T \oint_T \frac{(d\mathbf{r}_s \cdot d\mathbf{r}_t)}{R^2} + d_s\mathbf{v}_s$$

(Here

$$\mathbf{v}_s = \hat{\mathbf{x}}v_x + \hat{\mathbf{y}}v_y + \hat{\mathbf{z}}v_z$$

is an arbitrary function of $\mathbf{r}_s$ and

$$d_s\mathbf{v}_s = \frac{\partial \mathbf{v}_s}{\partial x_s}dx_s + \frac{\partial \mathbf{v}_s}{\partial y_s}dy_s + \frac{\partial \mathbf{v}_s}{\partial z_s}dz_s = (d\mathbf{r}_s \cdot \nabla_s)\mathbf{v}_s\Big)$$

If both $d_s\mathbf{v}_s$ and $d\mathbf{u}_t$ differ from zero, then neither $d\mathbf{F}_{St}$ nor $d\mathbf{F}_{sT}$ are central forces.

$d^2\mathbf{F}_{st}$ is the second-order differential force which is exerted by one differential element of the source circuit on one differential element of the test circuit. Logically the expression for $d^2\mathbf{F}_{st}$ should serve as the basic magnetostatic force equation, the counterpart of the Coulomb equation in electrostatics. There are several reasons why this is not the course actually followed. (1) When one attempts to measure $d^2\mathbf{F}_{st}$ experimentally, using two steady, finite, current elements, $I_t\,\Delta\mathbf{r}_t$ and $I_s\,\Delta\mathbf{r}_s$, which are small enough to give accurate results, the resultant force is so small relative to the contribution made by the many other current elements which are also present that the measurement cannot be made. (2) The need for $d^2\mathbf{F}_{st}$, without integration, arises only rarely. (3) The expression for $d^2\mathbf{F}_{st}$ is rather complicated. The fundamental forces $d^2\mathbf{F}_{st}$ and $d^2\mathbf{F}_{ts}$ are not central forces but have force components along each of the three natural directions present: $d\mathbf{r}_s$, $d\mathbf{r}_t$, and $\hat{\mathbf{R}}_{st}$. Also, they violate Newton's third law of motion: $d^2\mathbf{F}_{st} \neq -d^2\mathbf{F}_{ts}$. These facts are discussed in detail in Appendix 3 at the end of this chapter.

In electrostatics the source charges and the test charge are all stationary with respect to the observer. It is found experimentally that when the test charge moves but the source charges remain stationary, the electrical force is still the only force present, and the value of **E** is the same as before. When the opposite conditions apply, namely the source charges move but the test charge is stationary, then the electric force is again the only one present; but this time **E** is different from the value in the previous cases. This situation will be considered in Chap. 14. Finally, when both the source charges and the test charge move relative to the observer, then the electric force on the moving test charge is the same as that in the third case; however, it is found experimentally that there is now an additional, velocity-dependent force. This force is the magnetic force. It is seen that this force depends on the observer; for an observer who moves along with the test charge there is no such force.

Examples

1. Two parallel currents Consider the two idealized circuits of Fig. 3-3. The distant batteries, resistors, and lead-in wires have been omitted from the diagram. In the limit

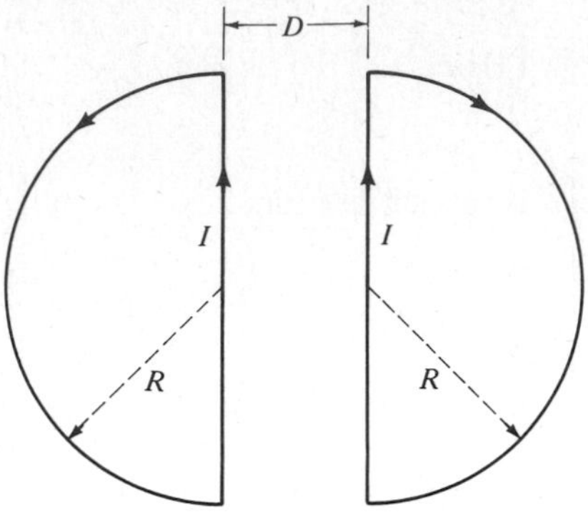

FIGURE 3-3

when $R \to \infty$ this becomes Fig. 3-4. We will find the force between the two straight portions of Fig. 3-4, assuming that all the forces involving the two semicircular sections of Fig. 3-3 become negligible in the limit when $R \to \infty$. This will be the force between the two idealized circuits of Fig. 3-3, in the limit when the radii of the two semicircles are infinite.

The dot product $(d\mathbf{r}_s \cdot d\mathbf{r}_t)$ is here $(dr_s)(dr_t)$. If $y = 0$ is a point on the I_s conductor opposite some convenient $d\mathbf{r}_t$ on the I_t conductor, then setting $r_s = y$ gives

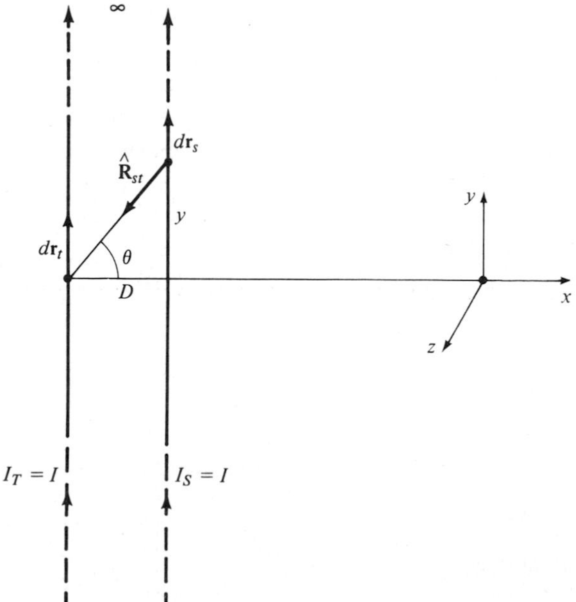

FIGURE 3-4

$$\mathbf{F}_{ST} = -\frac{\mu_0 I^2}{4\pi} \oint_T dr_t \int_{y=-\infty}^{\infty} \frac{dy}{D^2 + y^2} \hat{\mathbf{R}}_{st}$$

By symmetry, $\mathbf{F}_{ST}$ cannot have y or z components. Then

$$(F_{ST})_x = +\frac{\mu_0 I^2}{4\pi} \oint_T dr_t \int_{-\infty}^{\infty} \frac{dy \cos\theta}{D^2 + y^2}$$

Switching, for convenience, to θ as the independent variable, $y = D \tan\theta$ gives

$$dy = \frac{D\, d\theta}{\cos^2\theta}$$

also

$$D^2 + y^2 = \frac{D^2}{\cos^2\theta}$$

So

$$\mathbf{F}_{ST} = \hat{\mathbf{x}} \frac{\mu_0 I^2}{4\pi} \oint_T dr_t \int_{-\pi/2}^{\pi/2} \left(\frac{1}{D}\right) \cos\theta\, d\theta = \hat{\mathbf{x}} \frac{\mu_0 I^2}{2\pi D} \oint_T dr_t$$

Since $\oint_T dr_t \to \int_{-\infty}^{\infty} dy = \infty$, the total force produced by one infinitely long conduction current on another is infinite. If the range of integration of y is restricted to lie between $-\frac{1}{2}$ and $+\frac{1}{2}$, the integral equals unity; then the linear force density, $\mathbf{f}_{ST}$ or $\mathbf{f}$, produced by the entire source conductor on unit length of the test conductor is

$$\mathbf{f} = \hat{\mathbf{x}} \frac{\mu_0 I^2}{2\pi D} \quad \text{N m}^{-1} \quad \text{(attractive)}$$

2. The ampere The results of Example 1 are used, in principle, to define the ampere.

μ_0 is given, by definition, as $4\pi \times 10^{-7}$ m kg s^{-2} A^{-2}. If D and f are measured, I can be calculated from the last equation, above. This measurement may be used to calibrate an ammeter in series with one of the conductors, assuming the same current flows in both lines. The calibrated ammeter, removed and transported elsewhere, serves as a secondary standard.

From the equation above, if $I = 10$ A and $D = 2 \times 10^{-2}$ m then $f = 10^{-3}$ N m^{-1}. This is not a very large force, so the National Bureau of Standards employs a modification of this method in order to increase the sensitivity. Instead of using straight, parallel wires they employ conductors wound as coils, thereby concentrating the force obtainable in a given region. The equation above must, of course, then be modified.

It is interesting to compare the results of the parallel current case here with the case of two parallel charged lines in electrostatics:

$$\mathbf{f} = -\hat{\mathbf{x}} \frac{\lambda^2}{2\pi\epsilon_0 D} \qquad \text{N m}^{-1} \quad \text{(repulsive)}$$

There are three changes: negative to positive, $\lambda \to I$, $\epsilon_0 \to 1/\mu_0$.

3. A filament and a parallel sheath A current I_1 flows in a very long hollow cylinder whose cross section is the area lying between a circle of radius $a + da$ and a circle of radius a. We wish to determine the force that this current exerts on a unit length of a parallel filament of current I_2, at the distance $d > a$ from the I_1 axis. See Fig. 3-5.

The current in the shaded portion of the differential area is

$$\frac{dI_1}{I_1} = \frac{a\,da\,d\phi}{2\pi a\,da} = \frac{d\phi}{2\pi}$$

By the result of Example 1, this differential filament of current exerts the linear force density

$$d\mathbf{f} = -\frac{\mu_0 I_2 dI_1}{2\pi D}$$

on I_2. By symmetry, only an attractive force in the $-x$ direction will exist, so only df_x need be calculated.

$$df_x = -\frac{\mu_0 I_2 I_1 d\phi}{4\pi^2 D} \cos\theta$$

This must be integrated for $0 \leqslant \phi \leqslant 2\pi$. The problem is analogous to that of Fig. 2-12 with

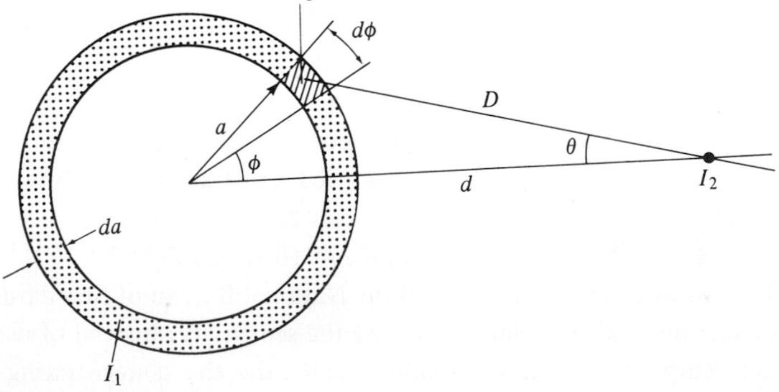

FIGURE 3-5

$$\frac{(1)\Lambda}{\epsilon_0} \to -\mu_0 I_1 I_2$$

So the answer there,

$$\frac{\Lambda}{2\pi\epsilon_0 d}$$

here becomes

$$\mathbf{f} = -\hat{\mathbf{x}}\,\frac{\mu_0 I_1 I_2}{2\pi d}$$

PROBLEMS

1 One ideal circuit is a circular loop of radius R, in the yz plane, centered at $x = 0$; it carries the current I_1. A second circuit is similar and coaxial, but lies in the $x = a$ plane and carries the current I_2. Write the integral expression for the force of the second circuit on the first (both currents flow in the same sense) using the two azimuthal angles as the independent variables.

2 An idealized circuit consists of a circular ring of radius R in the yz plane, with the center of the loop at the origin. A current I_1 flows in the loop. A second circuit consists of the x axis from $-\infty$ to ∞ and carries a current I_2. Assuming that a semicircle of infinite radius, which completes the second circuit, exerts zero force, find the force of the second circuit on the first.

3 Two very long parallel filamentary currents of 100 A each go in the same sense and are separated by 1 cm. Find the force per unit length produced on one by the entire length of the other.

4 Circuit 1 is a circular ring, of radius R, in the xy plane and has its center at the origin. Circuit 2 is a circular ring with the same radius and current. It lies in the xz plane with center at (R,0,0); i.e., the two intertwine. Write the expression for $\mathbf{F}_{21}$ in cartesian coordinates.

5 The cgs form of the basic magnetostatic force equation is

$$\mathbf{F}_{ST} = -\oint_T \oint_S \frac{(I_S\, d\mathbf{r}_s) \cdot (I_T\, d\mathbf{r}_t)}{R^2}\,\hat{\mathbf{R}}_{st}$$

This gives the so-called *emu* system, which differs from the *esu* cgs system described in Chap. 2. The attractive force of an infinitely long conductor on unit length of a second parallel conductor becomes

$$\mathbf{f} = \hat{\mathbf{x}}\,\frac{2I^2}{D} \qquad \text{dyn–cm}^{-1}$$

in the emu system (instead of $\mathbf{f} = \hat{\mathbf{x}}(\mu_0 I^2/2\pi D)$ N m^{-1} in mksa units). The dimensions of I are [A] in mksa and [dyn$^{1/2}$] in *emu*. (1) To make a dimensional adjustment between these two, note that the dimensions of μ_0 are [m kg s^{-2} A^{-2}], so $[I_{emu}/\sqrt{\mu_0}]$ has the same dimensions as $[I_{mksa}]$. (2) The mksa system is rationalized; the *emu* cgs system is unrationalized. The position of the 4π conversion factor relative to the μ_0, combined with the dimensional conversion factor, is incorporated in Reference Table II-B. (3) After the dimensions of the two quantities, I_{emu} and I_{mksa}, have been made the same, further revision is required to compare the centimeters and grams with meters and kilograms. This conversion factor is also listed in Reference Table II-B.

Make the changes required in the three steps above to find the relation between amperes, the unit of current in the mksa system, and abamperes, the unit of current in the *emu* cgs system.

6 If there are q_{mksa} units of charge in the mksa system, how many units of charge q_{esu} are there in the esu system? If there are q_{esu} units in the esu system, how many units of charge q_{emu} are there in the emu system?

7 In Example 3, let I_2 become dI_1 and let d become a. Find $d\mathbf{f}$. (Instead of finding the force produced by the current shell I_1 on a current external to I_1, find the force produced by the shell on a differential part of itself.)

8 In Example 3, let $d < a$. Find **f**.

9 In Fig. 3-6, find the force on the idealized rectangular circuit.

10 An annular cylindrical ring in a current-carrying wire of circular cross section will have magnetic forces on it which are radial inward. Will $\mathbf{J}(r)$ reach a steady-state value which varies with radius? One possibility is that $\rho(r)$, the charge density, varies with r. A radial inward **E** would be established within the metal in the steady state. A second alternative is that $\mathbf{v}(r)$, the drift velocity, varies with r. What actually happens?

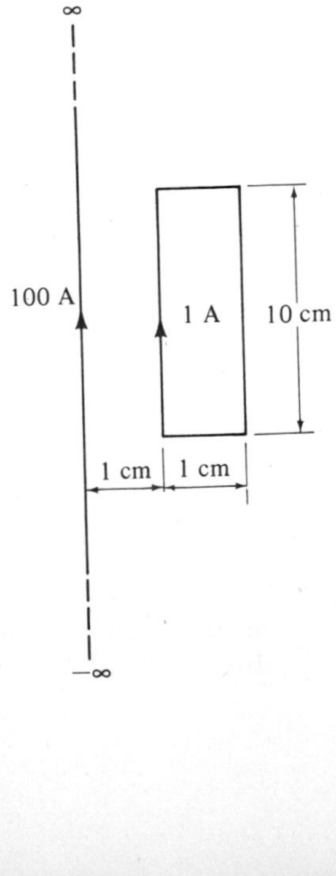

FIGURE 3-6

3-3 THE MAGNETIC FIELD, B

The fundamental force relation between the two filamentary currents of the previous section lends itself to an action-at-a-distance concept. In magnetostatics the velocity of transmission of the force is irrelevant, since it is assumed that a long enough time has elapsed for steady-state conditions to have been attained.

To introduce the concept of a magnetic field, analogous to the manner in which **E** was introduced in electrostatics, it is necessary to convert the equation for the fundamental force relation between currents into another form. For as it stands,

$$\mathbf{F}_{ST} = -\frac{\mu_0 I_S I_T}{4\pi} \oint_T \oint_S \frac{(d\mathbf{r}_s \cdot d\mathbf{r}_t)}{R^2} \hat{\mathbf{R}}_{st}$$

the dot product (which means a cosine factor) prevents the integrand from being written as the product of two factors, one of which depends only on the source circuit and the second of which depends exclusively on the test circuit. The angle of the cosine factor depends on the directions both of $d\mathbf{r}_s$ and of $d\mathbf{r}_t$ so it cannot be grouped with either factor alone. Only if the integrand can be written as such a product can one assign physical significance to the two different parts: (1) the test circuit and (2) the field produced by the source circuit.

Before making the necessary conversion, however, it is worth noting that the fundamental force relation above, because of its symmetry, shows that Newton's third law is obeyed by the two *complete* circuits. Action equals reaction for the complete circuits, unlike the case of two differential current elements. To find $\mathbf{F}_{TS}$, the force of the test circuit on the source circuit, let $s \to t$ and $t \to s$ in the above equation; all factors on the right remain unchanged except that $\hat{\mathbf{R}}_{st} \to \hat{\mathbf{R}}_{ts} = -\hat{\mathbf{R}}_{st}$. So $\mathbf{F}_{TS} = -\mathbf{F}_{ST}$.

To make the conversion to a less symmetric but more useful form, we start with the fact that $du = \nabla u \cdot d\mathbf{r}$. The integral of this expression completely around some closed path and back to the starting point is zero for any u. Let $u = R^{-1}$, where $R = |\mathbf{R}_{st}|$. So

$$\oint_T \nabla\left(\frac{1}{R}\right) \cdot d\mathbf{r}_t = 0$$

Also,

$$\oint_S d\mathbf{r}_s \oint_T \nabla\left(\frac{1}{R}\right) \cdot d\mathbf{r}_t = -\oint_S d\mathbf{r}_s \oint_T \left(\frac{\hat{\mathbf{R}}_{st}}{R^2}\right) \cdot d\mathbf{r}_t = 0$$

Then multiplying this by $-(\mu_0/4\pi)I_S I_T$ and adding the result to the expression for $\mathbf{F}_{ST}$ does not affect the value of F_{ST}. (This is not an obvious procedure; it is justified only by the fact that it leads to the desired result most simply.) So

$$\mathbf{F}_{ST} = \frac{\mu_0 I_S I_T}{4\pi} \oint_S \oint_T \frac{d\mathbf{r}_s(d\mathbf{r}_t \cdot \hat{\mathbf{R}}_{st}) - \hat{\mathbf{R}}_{st}(d\mathbf{r}_s \cdot d\mathbf{r}_t)}{R^2}$$

Now use the vector identity (see Appendix 1-B)

$$\mathbf{A} \times (\mathbf{B} \times \mathbf{C}) = \mathbf{B}(\mathbf{A} \cdot \mathbf{C}) - \mathbf{C}(\mathbf{A} \cdot \mathbf{B})$$

with $\mathbf{A} = d\mathbf{r}_t$, $\mathbf{B} = d\mathbf{r}_s$, and $\mathbf{C} = \hat{\mathbf{R}}_{st}$. This gives

$$\mathbf{F}_{ST} = \frac{\mu_0 I_S I_T}{4\pi} \oint_S \oint_T \frac{d\mathbf{r}_t \times (d\mathbf{r}_s \times \hat{\mathbf{R}}_{st})}{R^2}$$

or

$$\mathbf{F}_{ST} = \oint_T I_t \, d\mathbf{r}_t \times \left[\frac{\mu_0 I_S}{4\pi} \oint_S \frac{d\mathbf{r}_s \times \hat{\mathbf{R}}_{st}}{R^2} \right]$$

In this expression the factor within brackets does not depend on the magnitude or orientation of $I_t \, d\mathbf{r}_t$ (the test element) but only on the position of this element. Omitting the subscripts on $\mathbf{F}_{ST}$,

$$\boxed{\mathbf{F} = \oint_T I_T \, d\mathbf{r}_t \times \mathbf{B}} \qquad \text{(M-2)}$$

where

$$\boxed{\mathbf{B}(\mathbf{r}_t) = \frac{\mu_0}{4\pi} \oint_S \frac{I_S \, d\mathbf{r}_s \times \hat{\mathbf{R}}_{st}}{R^2}} \qquad \text{(M-3)}$$

The vector **B** is usually called the magnetic flux density, but the name is used so often that we will simply call it the magnetic field. There can scarcely be any confusion. Since the dimensions of μ_0 are those of $[\text{m kg s}^{-2}\,\text{A}^{-2}]$, those of **B** must be this multiplied by $[\text{A m}^{-1}]$; i.e., $[\text{kg s}^{-2}\,\text{A}^{-1}]$ or $[\text{V-s m}^{-2}]$. The unit having these dimensions is called the W/m^2 or, recently, the tesla (T). This is the unit that must be employed in mksa equations involving **B**. The size of this unit, however, turns out to be inconvenient for measurements–it is too large a unit–so it is common practice (because of a convenient, power of ten, conversion factor) to use the cgs *emu* magnetic field: 1 gauss. This is equivalent to 10^{-4} T (equivalent, not equal, since the dimensions differ). The size of the cgs measuring unit, the gauss, is much smaller than the mksa unit, so there will be many more such units in the measurement of a given amount of magnetic field.

Equations (M-2) and (M-3), like the other equations in magnetostatics which have been boxed and given numbers, are the mates of their corresponding electrostatic relations.

The equation for **B** is called the law of Biot-Savart. It is a macroscopic law and gives the finite value of **B** produced at some test or field point $\mathbf{r}_t$ by an infinite number of transverse components of the differential current elements at the source points $\mathbf{r}_S$. Note that, like the electrostatic case, the variation is inverse square with distance, but for electrostatics the longitudinal components are integrated. In magnetostatics there is no variation with time, so it does not matter whether one considers the test point and source points simultaneously or not. However, when variations with time are permitted this question assumes importance, and the equation must be modified.

Laplace seems to have been the first to suggest the possibility of going from the macroscopic law of Biot-Savart to its differential form,

$$\boxed{d\mathbf{B} = \frac{\mu_0}{4\pi} \frac{I_S\, d\mathbf{r}_s \times \hat{\mathbf{R}}_{st}}{R^2}} \qquad \text{(M-3, alternate)}$$

The simplicity of this step has great attraction. The resistance to the adoption of this proposal rested on the fact that it is possible to add any term

$$I_S\, d\mathbf{v}_s = I_S(d\mathbf{r}_s \cdot \nabla_s)\mathbf{v}_s$$

to the right-hand side of this equation without affecting the value of **B**, the total magnetic field at a test point. For

$$\oint_S I_S \left(\frac{\partial \mathbf{v}_s}{\partial x_s} dx_s + \frac{\partial \mathbf{v}_s}{\partial y_s} dy_s + \frac{\partial \mathbf{v}_s}{\partial z_s} dz_s \right) = I_S \oint_S d\mathbf{v}_s = 0$$

It is not practical to test this assumption experimentally; in magnetostatics one cannot isolate, effectively enough, the differential current element of interest from all the other current elements present. But it can be shown relativistically (see Chap. 14, Sec. 14-1, Prob. 18) that $d\mathbf{B}$ and $d\mathbf{E}$ are related by the equation $d\mathbf{B} = 1/c^2(\mathbf{v} \times d\mathbf{E})$, and from the expression for $d\mathbf{E}$ one obtains the equation for $d\mathbf{B}$ above, with no added terms. The above expressions for the differential and integral forms of the law of Biot-Savart are only approximations. They are valid when the current I_S is produced by charges which are moving relative to the observer with a velocity that is small compared to c, the speed of light in vacuum. The general formulas are more complicated.

We will anticipate the relativistic derivation and adopt Laplace's suggestion as valid. The differential force produced by a differential source current on a complete test circuit is thus given by

$$d\mathbf{F}_{sT} = \oint_T I_T \, d\mathbf{r}_t \times d\mathbf{B}$$

Similar remarks hold true for $d\mathbf{F}_{St}$, the force produced by a complete source circuit on a differential test current element $I_T \, d\mathbf{r}_t$. The current I_T is the transport per second of charge dq_t past a given point in the circuit–$I_T = dq_t/dt$. Then

$$I_T \, d\mathbf{r}_t = \left(\frac{dq_t}{dt}\right) d\mathbf{r}_t = dq_t \left(\frac{d\mathbf{r}_t}{dt}\right) = \mathbf{v} \, dq_t$$

where $\mathbf{v}$ is the drift velocity of the charge. The relation

$$d\mathbf{F}_{St} = I_T \, d\mathbf{r}_t \times \mathbf{B}$$

obtained from Eq. (M-2) by taking the differential of both sides without adding arbitrary function, then becomes

$$d\mathbf{F}_{St} = dq_t \, \mathbf{v} \times \mathbf{B}$$

The force on a finite charge moving relative to an observer is obtained from this by integrating the charge:

$$\mathbf{F} = q\mathbf{v} \times \mathbf{B}$$

This is called the Lorentz force. In Chap. 14 it will be derived from Coulomb's law; see Sec. 14-1, Examples 1 and 2. Experimentally it has been verified innumerable times.

The connection between the force on the conduction electrons and the force on a wire is not as simple as the one-step process above. The Lorentz force on the moving electrons tends to displace to one side of the wire, and this creates strong electrical attractive forces between the electrons and the positive ions of the metallic lattice. This transfers the magnetic force on the moving electrons to the stationary ions, i.e., the conductor. (For doped semiconductors this phenomenon, called the Hall effect, also occurs, but acting on the fixed impurity charges instead of on the metal ions. For intrinsic semiconductors there is no Hall field and there are no fixed charges–it is necessary to bring diffusion forces and electron gas pressure into the picture.) See "Force on a Wire in a Magnetic Field," A. C. English, *Amer. J. Phys.*, **35**:326 (1967).

Since the above derivation does not depend in any way on whether the charge is q or dq, we will accept the experimental evidence as justifying our use, also, of

$$d\mathbf{F}_{St} = I_T \, d\mathbf{r}_t \times \mathbf{B}$$

What of the second-order differential force $d^2\mathbf{F}_{st}$ exerted by one differential current element on another?

$$d^2\mathbf{F}_{st} = I_T\, d\mathbf{r}_t \times d\mathbf{B} = \frac{\mu_0}{4\pi} \frac{I_T d\mathbf{r}_t \times I_S d\mathbf{r}_s \times \hat{\mathbf{R}}_{st}}{R^2}$$

is essentially deduced in Chap. 14, Sec. 14-1, from Coulomb's law and the Lorentz transformation. This expression is valid for $v \ll c$, like the law of Biot-Savart. By exchanging s and t in this relation it is seen at once that $d^2\mathbf{F}_{st} \neq d^2\mathbf{F}_{ts}$.

In many applications it is useful to consider volume distributions of current rather than current that flows in filamentary circuits. For a current density $\mathbf{J}$ A m^{-2} passing through an area dS perpendicular to the direction $d\mathbf{r}$,

$$I\, d\mathbf{r} = \mathbf{I}\, dr = (\mathbf{J}\, dS)dr = \mathbf{J}\, d\tau$$

So

$$\mathbf{F} = \int_{\mathcal{V}} \mathbf{J} \times \mathbf{B}\, d\tau$$

and

$$\boxed{\mathbf{B}(\mathbf{r}_t) = \frac{\mu_0}{4\pi} \int_{\mathcal{V}} \frac{\mathbf{J} \times \hat{\mathbf{R}}_{st}}{R^2}\, d\tau_s} \qquad \text{(M-4)}$$

This last expression may be compared with the electrostatic expression for $\mathbf{E}$. That field is produced by a continuum charge distribution while the $\mathbf{B}$ field is due to a continuum current distribution. Just as $\mathbf{E}$ does not become infinite, even inside the charge distribution provided the charge density is finite, so $\mathbf{B}$ above does not become infinite, even inside the conductor volume, as long as $\mathbf{J}$ is finite. $\mathbf{E}(\mathbf{r}_t)$ is produced by $\rho(\mathbf{r}_s)\, d\tau$, each contribution $d\mathbf{E}$ being along the line joining $\mathbf{r}_s$ and $\mathbf{r}_t$; $\mathbf{B}(\mathbf{r}_t)$ is produced by $\mathbf{J}(\mathbf{r}_s)\, d\tau$, each contribution $d\mathbf{B}$ being perpendicular to the line joining $\mathbf{r}_s$ and $\mathbf{r}_t$. But $d\mathbf{F}_{St}$, $d\mathbf{F}_{sT}$, and $d^2\mathbf{F}_{st}$ are neither transverse nor longitudinal.

Examples

1. An infinitely long conductor Consider the magnetic field produced at an arbitrary point P by a straight, infinitely long, zero-resistance wire carrying a conduction current I, as shown in Fig. 3-7*a*. This circuit is the limit of the idealized circuit of Fig. 3-7*b* when the radius of the semicircle approaches infinity. The contribution to $\mathbf{B}$ by the semicircular section can be shown to approach zero as r_o grows without bound, and we consider only the contribution from the linear portion.

$I_s\, d\mathbf{r}_s = \hat{\mathbf{z}}I\, dz$ for each differential current element; but $\mathbf{R}_{st} = \mathbf{R}$ varies in magnitude and direction for the different source elements. For different particular differential

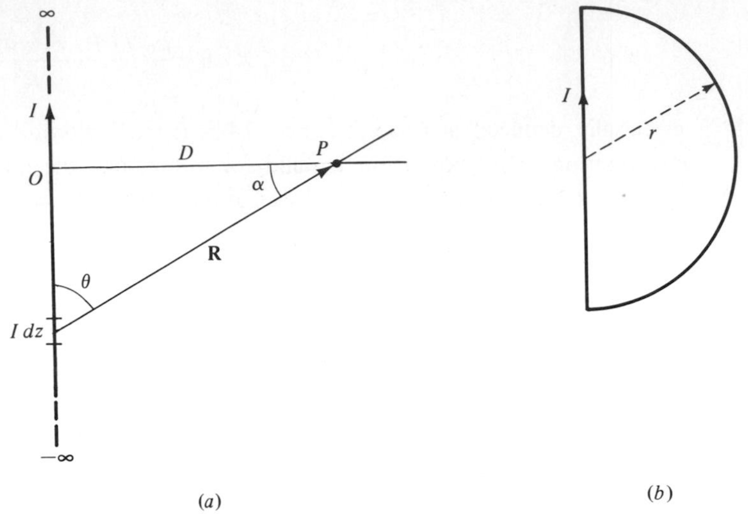

FIGURE 3-7

current elements $I_S\, d\mathbf{r}_s \times \hat{\mathbf{R}}_{st}$ produces different magnitudes at a given P; but whatever the magnitudes, the different differential vectors at P all have the same direction—here, into the plane of the paper. The integral therefore changes from a vector integral, in which directions must be taken into account, to a scalar integral using algebraic addition; the direction of the resultant vector **B** at P is in the same direction as that of each of its constituents. For a point P to the left of the conductor in the plane of the paper, on the other hand, the direction of **B** will be out of the plane of the paper. A similar situation exists in any other plane containing the conductor and, from the cylindrical symmetry, **B** is everywhere in the azimuthal direction: $\mathbf{B} = \hat{\boldsymbol{\phi}} B$.

To obtain B, the magnitude of the magnetic field, one must evaluate

$$\frac{\mu_0}{4\pi} \int_{-\infty}^{\infty} \frac{I\, dz \sin\theta}{R^2}$$

The simplest procedure is to set $z = D \tan\alpha$, i.e.,

$$dz = \frac{D\, d\alpha}{\cos^2\alpha} \qquad \sin\theta = \cos\alpha \qquad R^2 = \frac{D^2}{\cos^2\alpha}$$

Then

$$B = \frac{\mu_0 I}{4\pi} \int_{-\pi/2}^{\pi/2} \frac{\cos\alpha\, d\alpha}{D} = \frac{\mu_0 I}{2\pi D}$$

or
$$\mathbf{B} = \hat{\boldsymbol{\phi}} \frac{\mu_0 I}{2\pi D}$$

This result may be compared with that produced by an infinitely long charged line:

$$\mathbf{E} = \hat{\mathbf{r}} \frac{\lambda}{2\pi\epsilon_0 D} \qquad \text{(cylindrical } \hat{\mathbf{r}}\text{)}$$

2. Convection currents Assume a convection current I obtained, for example, by spraying an external line charge density, λ C m^{-1}, on a very long conductor and then moving the wire past an observer with velocity v m s^{-1}. $I = \lambda v$. We will compare the values of **E** and **B** at an arbitrary point, assuming that **E** has the same value it would have if there were no drift velocity. (This assumption for the very long wire can be shown to be true despite the fact that the **E** produced by a moving point charge differs from that produced by a stationary point charge.)

Here

$$E = \frac{\lambda}{2\pi\epsilon_0 D} \qquad \text{and} \qquad B = \frac{\mu_0 I}{2\pi D}$$

Then
$$\frac{B}{E} = \frac{I}{\left(\dfrac{1}{\epsilon_0\mu_0}\right)\lambda} = \frac{v}{\left(\dfrac{1}{\epsilon_0\mu_0}\right)}$$

Since $\mu_0 = 4\pi \times 10^{-7}$ m kg s^{-2} A^{-2} (by definition) while $\epsilon_0 = 8.87 \times 10^{-12}$ m^{-3} kg^{-1} s^4 A^2 (by experiment), we have

$$\frac{1}{\epsilon_0\mu_0} = 9 \times 10^{16} \text{ m}^2 \text{ s}^{-2}$$

and
$$\frac{B}{E} = \frac{v}{9 \times 10^6} \text{ m}^{-1} \text{ s}$$

B and E have different dimensions here so the magnitude of the numeric is immaterial: the sizes of the two quantities cannot be compared. B and E are different kinds of quantities if they have different dimensions. In Chap. 7, Sec. 7-3, it will be shown that the above experimental value for $(\epsilon_0\mu_0)^{-1}$ is only approximate; on theoretical grounds, $(\epsilon_0\mu_0)^{-1} = c^2$, where c is the speed of light (or any other electromagnetic wave) in vacuum. Since cB and E do have the same dimensions, these *can* be compared:

$$\frac{cB}{E} = \frac{v}{c}$$

The conductor on which the charge λ was sprayed does not actually play a role in this problem. We could just as well have considered a beam of particles moving in space—for instance, in the drift region of a cathode ray tube. cB will be a fraction of E, the size of the fraction being equal to β, the drift velocity normalized to the speed of light.

3. Superconductor currents We will treat a rather lengthy example here. This presents the case of the critical current, one which lies between convection currents (Example 2) and conduction currents (Example 4). Figure 3-8*a* shows a transmission line of two parallel superconducting wires connecting a battery to a resistor. It will be shown in Chap. 16 that when the switch is closed, a wave of electric field starts traveling toward the right with velocity c (very closely, if the medium is air) as shown in Fig. 3-8*b*. We will anticipate a few of the results here.

The leading edge of this wave has a longitudinal component E_ℓ which points to the right at the surface of superconductor *a* and to the left at the surface of superconductor *b*. As the leading edge passes them, electrons on *a* are given an impulse to the left; while electrons on *b* are given an impulse to the right. Soon after the leading edge of the wave has passed, the value of E_ℓ at a given point becomes zero and the steady-state field is then entirely transverse, E_t. On a superconductor the moving electrons encounter zero

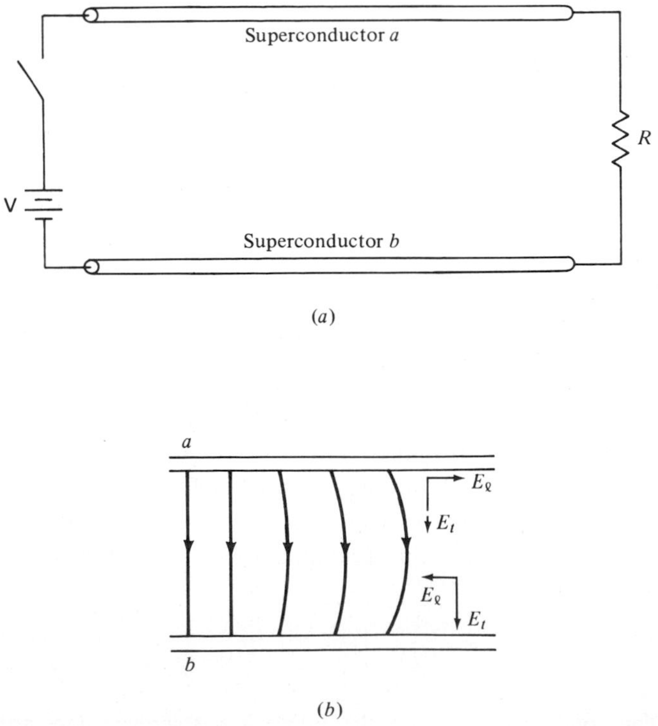

FIGURE 3-8

resistance: there are no collisions. Such conductors are ideal conductors, and the electrons which have been given momentum by the leading-edge impulse of the wave will continue to move at constant speed as long as they are on the superconductor. The situation is similar to that of a vehicle in outer space where there is no air resistance–if the rocket engine is fired for an instant, then the vehicle is given an impulse which alters its momentum and a new value of velocity is obtained and maintained without further use of the engine. Just so on the superconductors–no batteries are required to supply further energy to the electrons and no heat is lost to the wires by collisions between the electrons and the lattice atoms.

Concomitant with the electric field wave which travels to the right, therefore, there is also a wave of current which travels along at the same speed. On superconductor a, at a point to the left of the leading edge (i.e., where the wave has passed) the electrons move to the left with a constant speed v, so there is a steady current to the right (since electrons are negative); to the right of the leading edge (where the wave has not yet reached) there is no current on a. Similarly on superconductor b there will be a steady current, to the left, at any point to the left of the wavefront.

If the load resistor R has a particular value equal to R_0, the characteristic impedance of the transmission line, then no reflected wave occurs when the initial wave reaches the right end. The value of R_0 depends on the diameter and spacing of the wires and on the nature of the medium between them–a typical value is 200 Ω. It can be shown that if the load resistance R has this value, $R = R_0$, then $(c\lambda)^2 - I^2 = 0$. The electrical forces exactly balance the magnetic forces and the current that flows is neither a convection nor a conduction current. In the case of a superconductor there is no charge density, no current, no field in the interior of the superconductor (these actually decay from the surface in a distance of the order of 10^{-6} cm); all effects occur on the surface. If conditions are changed such that $R > R_0$ then the linear charge density on the wires, $\lambda = C_l V$, is unaffected but the current is reduced, $I = V/R$. Consequently the electrical effects will predominate over the magnetic: the current is of the convection type (even though here, unlike in a beam, charges of both signs are present). Similarly, when $R < R_0$ the quantity $(c\lambda)^2 - I^2$ is negative, the magnetic effects predominate, and a conduction current flows.

It should be mentioned that, regardless of the value of R, the original current wave that moves to the right with velocity c when the switch is closed has the value $I_0 = V/R_0$. When $R = R_0$ there is only this one transient wave–the value of the transient current wave equals the steady-state current demanded by the load resistor if there is to be a PD (potential difference) of V volts across its terminals. But when $R \neq R_0$ the value of the initial transient current I_0 is not correct for the steady state, for $V/R_0 \neq V/R$. The changeover from I_0 to I is accomplished by an infinite sequence of left-going and right-going reflected waves which, typically, decay to extremely small values in less than a microsecond.

In the region between the two superconducting wires the value of E is given by

$$E = \frac{\lambda}{2\pi\epsilon_0 x} + \frac{\lambda}{2\pi\epsilon_0 (s-x)}$$

where x is the distance from one wire while s is the separation between the two wires. Similarly,

$$B = \frac{\mu_0 I}{2\pi x} + \frac{\mu_0 I}{2\pi(s-x)}$$

Consequently, $B/E = \epsilon_0\mu_0(I/\lambda)$. Anticipating the future result, as above, that $\epsilon_0\mu_0 = c^{-2}$, this gives $cB/E = (I/\lambda)/c$. Since $I = V/R$ while $\lambda = C_l V$, where C_l is the capacitance between the wires per unit length of the line,

$$\frac{I}{c\lambda} = \frac{1}{cC_l R} = \frac{1}{cC_l R_0}\left(\frac{R_0}{R}\right)$$

The formulas

$$C_l = \frac{\pi\epsilon_0}{\cosh^{-1}(s/d)}$$

(where d is the wire diameter) and

$$R_0 = \frac{1}{\pi}\sqrt{\frac{\mu_0}{\epsilon_0}}\cosh^{-1}(s/d)$$

which are here presented on faith, then give

$$\frac{cB}{E} = \frac{R_0}{R}$$

For

1 $R > R_0$		*1* Convection current
2 $R = R_0$	we have respectively,	*2* Critical current
3 $R < R_0$		*3* Conduction current

and the field which dominates is

1 Electrical
2 Neither
3 Magnetic

[For further details see "A Model for the Current on a Superconducting Transmission Line," A. Shadowitz, *Physica*, **49**:141–152 (1970).]

4. Conduction current If the circuit of Fig. 3-8*a* is modified by making the wires *a* and *b* ordinary conductors instead of superconductors, a number of changes must be made in the preceding analysis. Here the current flows predominantly in the interior of the metal and the volume charge density remains zero for an observer stationary with respect to the ions. Nevertheless, a small fraction of the total current flows on the surface and there is a surface charge density which here, however, decreases in magnitude with distance from the battery. The surface charge is produced by the longitudinal **E** components of the leading edges of the waves after the switch is closed, in a manner analogous to that described above. This surface charge produces the longitudinal field within the wire that is responsible for the flow of the volume current, the predominant part of the total current.

The field within the metal is here necessary to supply the energy to the electrons to compensate for the heat lost in the metal by the electrons in collisions between them and the lattice vibrations, defects, and impurities. The field within the metal that keeps the current flowing despite the resistance cannot be produced directly by the battery. Such an electric field would decrease with distance from the battery while, in actuality if the wire is uniform, the longitudinal field within the wire is a constant independent of distance from the battery. (The surface charge density decreases linearly with distance from the battery, as does the PD between the two conductors.)

It is seen that the common analogy between the flow of current in a wire and the flow of water through a pipe breaks down here. Another indication that this analogy is only vaguely accurate is provided by the effect of adding a resistance. In the case of the flow of water such a resistance may be obtained by restricting the pipe diameter; in such a region the velocity of the water increases. But in the case of the electric circuit the addition of a resistance does not increase the electron drift velocity but, instead, decreases it. The analogy is not too good.

For ordinary conduction currents through ordinary wires there is no fixed relation between the current flowing through the interior and the surface charge on the wires—they may be individually controlled. The current can be varied by changing the value of R, the load resistor, without affecting the surface charge density. Consequently the magnetic and electric fields outside the wires are essentially independent of each other. (Changing V does not affect both, however.)

In ordinary house circuits, e.g., the surface charge density is comparatively small because of large spacing between the wires. The value of cB/E is $I/c\lambda$, as above. Usually this has a magnitude very large compared to unity; e.g., $I = 1$ A, $\lambda = C_l V \approx 10^{-12}\,(10^2)$ or 10^{-10} C m^{-1}, so $I/c\lambda \sim 100$. The drift velocity of the electrons, however, is very small, $v/c \sim 10^{-12}$. So the equation $cB/E = v/c$, which holds true for convection currents is certainly not true for conduction currents (here we would have 10^2 for the left and 10^{-12} for the right).

PROBLEMS

1 (*a*) Find the value of **B** at P when the wire in Example 1 is finite in length as shown in Fig. 3-9. (The rest of the circuit, needed to produce the steady current I, is not shown in the diagram; though present, it is not pertinent to the problem.)

(*b*) Repeat for P'.

2 (*a*) Find the force on a circular filamentary circuit, of radius r, carrying a current I. The circle, in the xy plane, is in a magnetic field, $\hat{\mathbf{z}}B_0$, created by some other circuit.

(*b*) Repeat (*a*) with $\mathbf{B} = \hat{\mathbf{x}}B_0$.

3 Find the value of **B** produced by an idealized circular filamentary current, of radius a, at a point on the z axis of symmetry perpendicular to the plane of the circle. The distance of the point from the center is D.

4 Find the value of **B** produced by the coaxial cable of Fig. 3-10 at P, distant r from the central axis (P is not shown) for $r < a$, $a < r < b$, $b < r < c$, $c < r$. (r in cylindrical coordinates.)

5 Find the value of **B**, in the case of Prob. 3, for an arbitrary point P which is not on the axis of symmetry but whose distance from the center (the origin) r is much larger than a. Use cartesian coordinates. (Note: If r is not much larger than a this problem involves elliptic integrals.)

6 Assume that an electron introduced into a uniform magnetic field $\hat{\mathbf{z}}B_0$ with constant velocity $\hat{\mathbf{x}}v$ moves in a circle of radius r. What is the current? The velocity in the field? The angular velocity? The work done on the electron by the magnetic field? The change in kinetic energy?

7 Obtain the answer to Prob. 5 using spherical coordinates.

8 A steady current I flows in a homogeneous conductor of circular cross section. Is the net magnetic force at a point a spatial constant, or does it vary with distance from the axis? Is the density **J** a constant over the cross section?

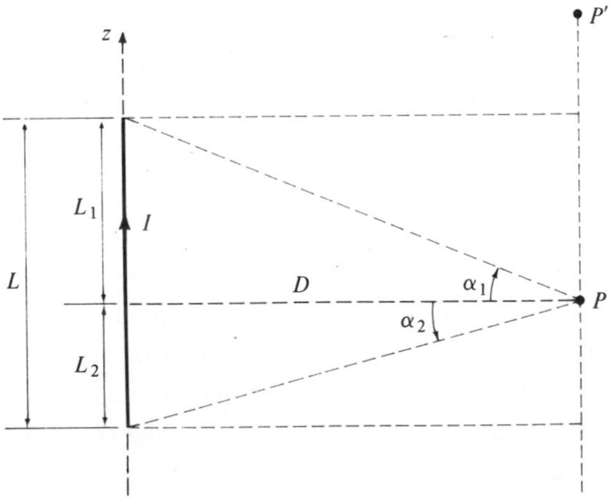

FIGURE 3-9

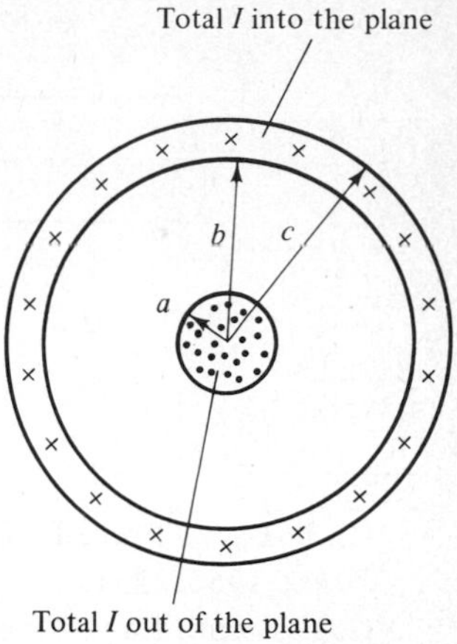

FIGURE 3-10

9 A force-free field is defined as one for which $\mathbf{J} \times \mathbf{B} = 0$. Anticipating a result from the next section, viz., $\nabla \times \mathbf{B} = \mu_0 \mathbf{J}$, derive an equation which $\mathbf{B}$ must satisfy to be force-free. Find a differential equation which must be satisfied by each component of $\mathbf{B}$. Note that this says that $\nabla \times \mathbf{B} = \alpha \mathbf{B}$, i.e., that the curl of a vector can be parallel to the vector.

For further information see:

H. P. Furth, M. A. Levine, and R. W. Waniek, Production and Use of High Transient Magnetic Fields II, *Rev. Sci. Inst.*, 28:949 (1957).

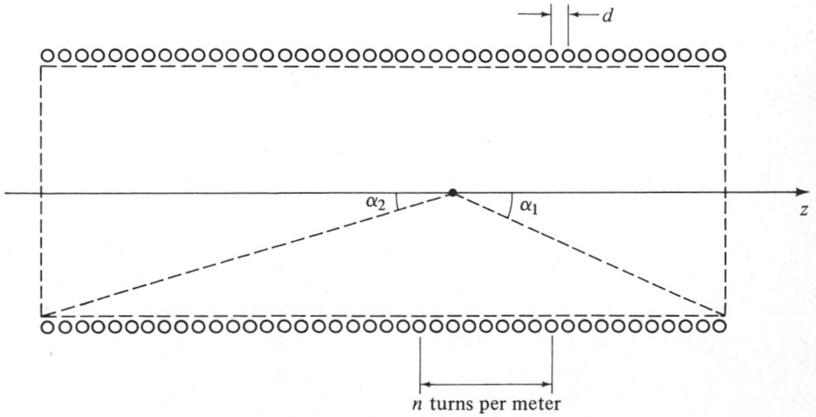

FIGURE 3-11

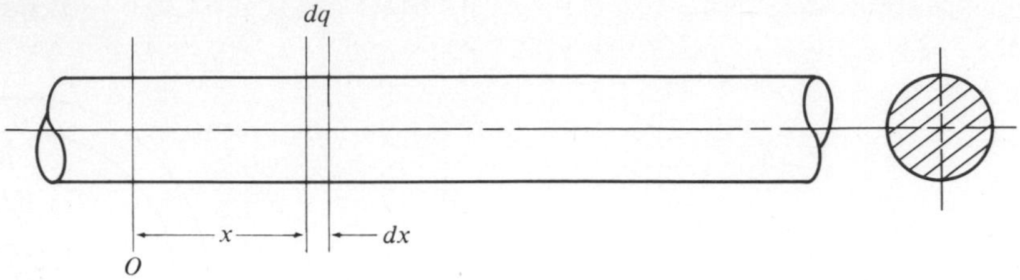

FIGURE 3-12

G. J. Buck, Force-Free Magnetic-Field Solution in Toroidal Coordinates, *J. Appl. Phys.*, **1965**:2231.

G. F. Freire, Force-Free Magnetic-Field Problem, *Amer. J. Phys.*, **1966**:567.

10 A two-wire transmission line consists of two parallel conductors, each of radius a, with a separation between centers of d. Each conductor carries the same current, but in one case the sense is opposite to that in the other. Find the value of **B** along the line of centers in the region outside the conductors.

11 A tightly wound solenoid, with n turns per meter, consists of a helical conductor whose pitch equals the cross-sectional diameter d of the wire; a thin coating on the wire serves for insulation. Use the result of Prob. 3 here by neglecting the pitch of the helix, to find the magnetic field on the axis of the solenoid. See Fig. 3-11. Assume n is very large, so that integration is permissible.

12 Figure 3-12 shows a conductor having a surface charge density given by

$$dq = \sigma_0 \left(1 - \frac{x}{L}\right) dx$$

Find the values of **E** and **J** inside the conductor.

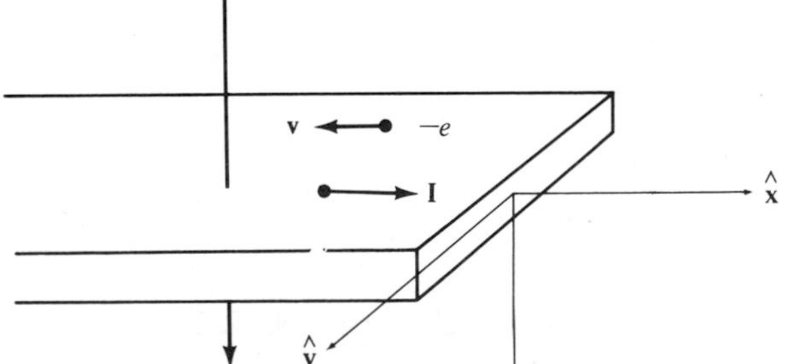

FIGURE 3-13

13 Two infinite parallel planes separated by the distance D have equal but oppositely directed surface current densities $\mathbf{J}_l$ A m^{-1}. Find the force of repulsion per unit area on a plane.

14 Assuming that $\mathbf{B}$ at the surface of a solenoid has the same value as on the axis (an approximation), what would be the magnitude and the direction of the force exerted on the turn at the center of the solenoid winding if the current is 10,000 A? (This problem is not academic; it is a practical consideration in the design of large magnetic field superconducting solenoids.) Compare the result with the tensile strength of brass: 5×10^8 N m^{-2}.

15 Suppose a wire of rectangular cross section, of width w in the $\hat{\mathbf{y}}$ direction, carries a current I in the $\hat{\mathbf{x}}$ direction in a region where there is a magnetic field $+\hat{\mathbf{z}}B$, as in Fig. 3-13. Then the Lorentz force on a moving electron is $\hat{\mathbf{y}}(-e)vB$, and electrons will move in the $-\hat{\mathbf{y}}$ direction until there is an electric field sufficient to halt further electrons from being deflected in this manner. A potential difference is established and can be measured. This is the Hall effect. What is the value of the potential difference? What is the Hall electric field?

3-4 AMPERE'S LAW AND $\nabla \times \mathbf{B}$

The circuital form of the law of Biot-Savart is

$$\mathbf{B}(\mathbf{r}_t) = \frac{\mu_0 I_S}{4\pi} \oint_S \frac{d\mathbf{r}_s \times \hat{\mathbf{R}}_{st}}{R^2}$$

A rather lengthy derivation is required in order to attain from this a simple and useful result: an expression for $\oint \mathbf{B} \cdot d\mathbf{r}$. From this, in magnetostatics, we will find, $\nabla \times \mathbf{B} = \mu_0 \mathbf{J}$. If we know $\mathbf{J}$ everywhere we then know $\nabla \times \mathbf{B}$ everywhere—half the knowledge required by the Helmholtz theorem for the determination of $\mathbf{B}$ itself. So this result is a very useful one.

Figure 3-14 shows an idealized situation to which this law may be applied; the battery and resistors have been omitted for simplicity. It is assumed that a steady, finite current flows around the source circuit S. The magnetic field $\mathbf{B}$ is shown at a test point P. Instead of merely calculating the value of $\mathbf{B}$ at one point P consider the circulation of $\mathbf{B}$ around some closed curve T. Using the value of $\mathbf{B}$ from the law of Biot-Savart,

$$\oint_T \mathbf{B} \cdot d\mathbf{r}_t = \oint_T \left(\frac{\mu_0 I_S}{4\pi} \oint_S \frac{d\mathbf{r}_s \times \hat{\mathbf{R}}_{st}}{R^2} \right) \cdot d\mathbf{r}_t$$

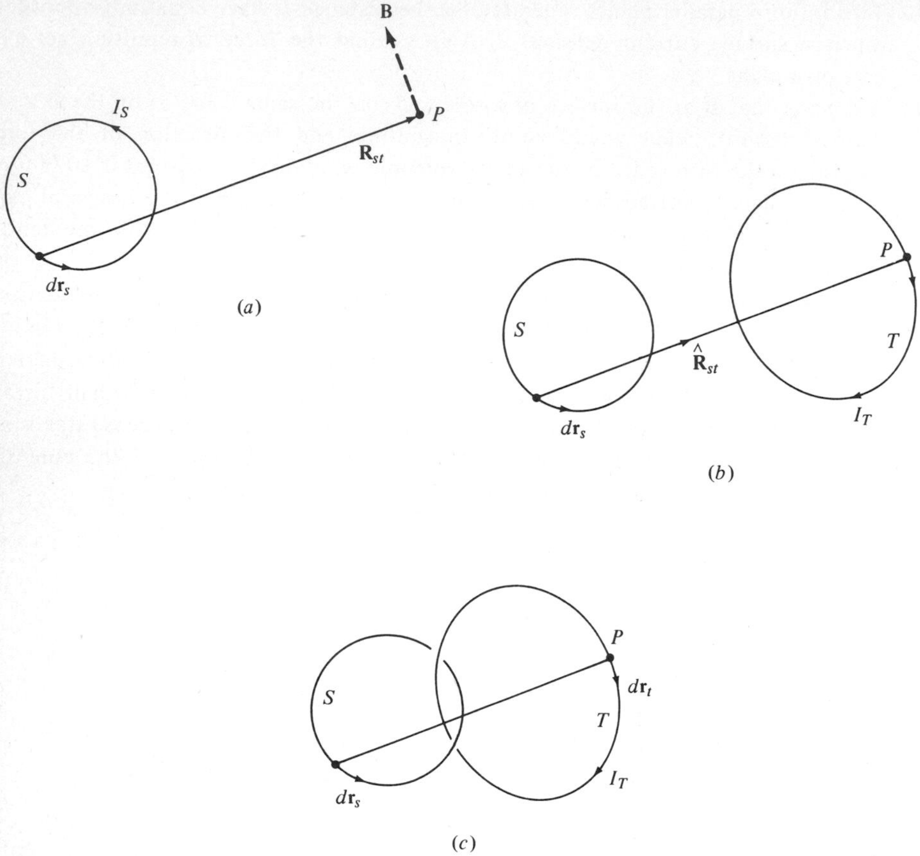

FIGURE 3-14

Figure 3-14 shows this closed path for the test point. Here T has been chosen so that it does not link with S: the two rings do not intertwine.

The double integral may be evaluated by a mathematical trick. For a given differential displacement along the test circuit $d\mathbf{r}_t$, the triple product $d\mathbf{r}_s \times \hat{\mathbf{R}}_{st} \cdot d\mathbf{r}_t$ may be written equally well as $d\mathbf{r}_t \times d\mathbf{r}_s \cdot \hat{\mathbf{R}}_{st}$, both forms representing the volume formed by these three vectors. $d\mathbf{r}_t \times d\mathbf{r}_s$ is an area; it may be visualized more easily if, instead of having the single test point P displaced through $d\mathbf{r}_t$, we consider P held fixed while the entire source circuit is moved through the displacement $-d\mathbf{r}_t$. This is shown in Fig. 3-15. $d\mathbf{S}$, at S, is the negative of $d\mathbf{r}_t \times d\mathbf{r}_s$, at P.

The trick, now, is to replace the two separate line integrals, one over the source circuit and one over the test circuit

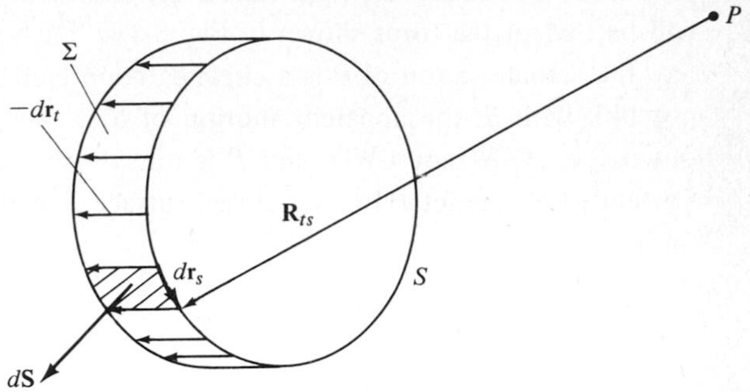

FIGURE 3-15

$$\oint_T \oint_S \frac{d\mathbf{r}_s \times \hat{\mathbf{R}}_{st} \cdot d\mathbf{r}_t}{R^2}$$

by an equivalent double integral over the source circuit alone. Since the independent variables are $\mathbf{r}_s$ and $-\mathbf{r}_s$, this is the equivalent of a surface integration over Σ, the shaded area in Fig. 3-15 that is formed by $-d\mathbf{r}_t \times d\mathbf{r}_s$:

$$\oint_T \mathbf{B} \cdot d\mathbf{r}_t = \frac{-\mu_0 I_S}{4\pi} \iint_\Sigma \frac{-d\mathbf{r}_t \times d\mathbf{r}_s \cdot \hat{\mathbf{R}}_{st}}{R^2}$$

So, with $-\hat{\mathbf{R}}_{st} = \hat{\mathbf{R}}_{ts}$,

$$\oint_T \mathbf{B} \cdot d\mathbf{r}_t = \frac{\mu_0 I_S}{4\pi} \iint_\Sigma \frac{d\mathbf{S} \cdot \hat{\mathbf{R}}_{ts}}{R^2}$$

The integrand here is $d\Omega$, the differential change in the solid angle at P of the entire loop S. This differential change is produced by the apparent displacement $-d\mathbf{r}_t$ of the loop S. Thus

$$\oint_T \mathbf{B} \cdot d\mathbf{r}_t = \frac{\mu_0 I_S}{4\pi} \iint_\Sigma d\Omega$$

In Fig. 3-15 only a differential displacement $-d\mathbf{r}_t$ is shown for the source circuit, corresponding to a differential displacement of the test point $d\mathbf{r}_t$ in Fig. 3-14*b*. When the test point P actually moves around the closed path T of Fig. 3-14*b* and T does not link with S, then the surface Σ becomes a closed surface dividing all space into an inside and an outside. If, e.g., S and T are circles in perpendicular planes the surface Σ will be that of the torus shown in Fig. 3-16*a*. Each point of S describes a closed path. Say the actual motion of P is a circle T, counterclockwise looking down, which does not link with S; the apparent motion of S is then also a circle, again CCW looking down (yes, CCW not CW!); and P is outside the torus. The value of the solid angle subtended at a point P by any closed surface Σ which does not include P within it or on it, is

$$\iint_{\Sigma} d\Omega = 0$$

The proof of this statement is left for one of the problems. The result, for the case when S and T do not link, is a very simple result for such a complicated derivation:

$$\oint_T \mathbf{B} \cdot d\mathbf{r}_t = 0$$

Now consider the case when S and T do link, as shown in Fig. 3-14*c*. Again, take the example above: let S be a circle in the plane of the paper while T, also a circle, is in a plane perpendicular to that of S. Here, also, a torus is formed by the apparent motion of S about P and, again, this apparent motion of S is actually due to the actual motion of P about S; this time, however, P lies within the torus. In this case

$$\iint_{\Sigma} d\Omega = 4\pi$$

This is also left for the student to prove. The result in this case is

$$\oint_T \mathbf{B} \cdot d\mathbf{r}_t = \frac{\mu_0 I_S}{4\pi}(4\pi) = \mu_0 I_S$$

The overall result for both cases is

$$\boxed{\oint_T \mathbf{B} \cdot d\mathbf{r} = \mu_0 I} \qquad \text{(M-5)}$$

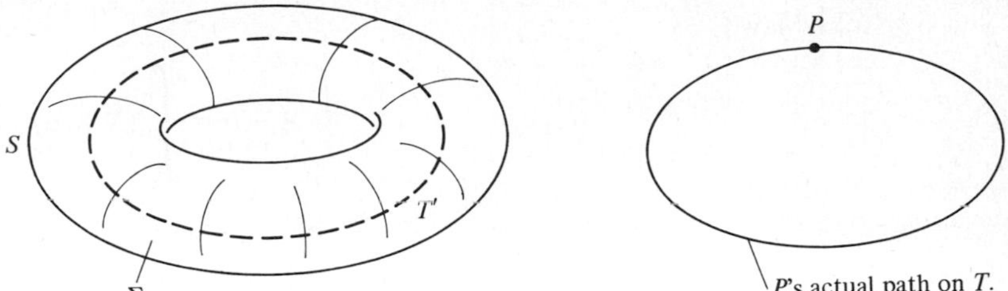
P
S
T'
Σ
P's actual path on T. This is simulated by the path of S along T'. P is outside Σ.

(a)

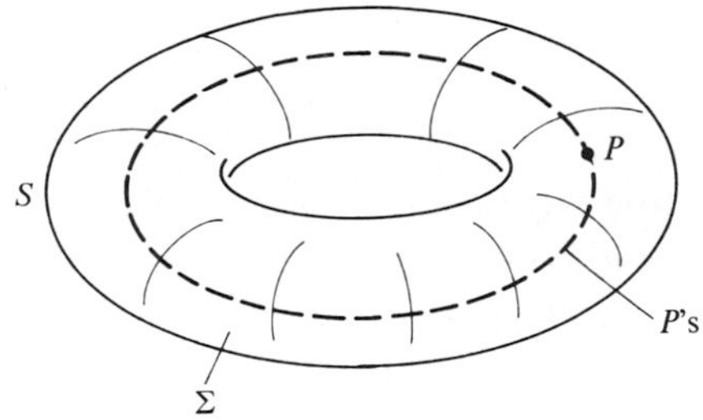
S
P
P's path on T. P is inside Σ.
Σ

(b)

FIGURE 3-16

where I is the current that passes through any surface that is bounded by the curve T. When T does not link with the current loop S, then $I = 0$; when T links once with a current loop carrying a current I_S, then $I = I_S$; if T links twice with a current loop carrying a current I_S, then the value of I in the equation above is either zero or $2I_S$, depending on the sense in which the links are made: opposite or similar; etc.

This result is called Ampere's law. It has a significance for magnetostatics which is similar to that of Gauss' law for electrostatics. Ampere's law relates the circulation of **B** around an arbitrary path bounding some open area to the current through that area. In some cases with a high degree of symmetry it permits a simple determination of **B** but, in any case, it leads to an immediate equation for $\nabla \times \mathbf{B}$.

Let

$$I = \int_\Sigma \mathbf{J} \cdot d\mathbf{S}$$

where Σ now is an open area bounded by T. Then

$$\oint_T \mathbf{B} \cdot d\mathbf{r} = \mu_0 \int_\Sigma \mathbf{J} \cdot d\mathbf{S}$$

But, by Stokes' law,

$$\oint_T \mathbf{B} \cdot d\mathbf{r} = \int_\Sigma (\nabla \times \mathbf{B}) \cdot d\mathbf{S}$$

So

$$\mu_0 \int_\Sigma \mathbf{J} \cdot d\mathbf{S} = \int_\Sigma (\nabla \times \mathbf{B}) \cdot d\mathbf{S}$$

Since this is true for any open area Σ,

$$\boxed{\nabla \times \mathbf{B} = \mu_0 \mathbf{J}} \qquad \text{(M-6)}$$

It turns out that this relation holds true even when v is comparable to c; despite the fact that the law of Biot-Savart, from which it was derived, is only valid for $v \ll c$.

In electrostatics one may say, from $\nabla \cdot \mathbf{E} = (1/\epsilon_0)\rho$, that ρ is the "source" of $\mathbf{E}$; in magnetostatics one may say, from $\nabla \times \mathbf{B} = \mu_0 \mathbf{J}$, that $\mathbf{J}$ is the "vortex source" of $\mathbf{B}$.

Unlike the electrostatic field, the magnetostatic field does have rotation, so it cannot be conservative, even though at any point where $\mathbf{J} = 0$ the value of $\nabla \times \mathbf{B} = 0$. $\mathbf{J}$ cannot be zero everywhere, if a field exists at all; so $\nabla \times \mathbf{B} \neq 0$ everywhere. For $\mathbf{B}$ to be conservative it is not sufficient that $\nabla \times \mathbf{B} = 0$ at an infinity of points—not even all but one point in some region. We require $\nabla \times \mathbf{B} = 0$ everywhere in some singly connected region. If care is taken to restrict oneself to limited singly connected volumes which do not link any currents, it is then possible to treat $\mathbf{B}$ as if $\nabla \times \mathbf{B} = 0$ in that restricted volume. One may then proceed to treat $\mathbf{B}$ in a manner analogous to electrostatics. This procedure is usually followed when dealing with permanent magnets if there are no conventional currents present.

The result above for $\nabla \times \mathbf{B}$ is applicable only to static fields. For the general case of time-varying fields it is necessary to modify this expression. This was also true of the previous result for $\nabla \times \mathbf{E}$.

Examples

1. An infinitely long conductor We will use Ampere's law to find the magnetic field produced by a current I passing through an infinitely long straight conductor, assuming a uniform current density. (In Example 1 of Sec. 3-3 this problem was worked out by direct application of the law of Biot-Savart; in the present case of great symmetry the use of Ampere's law is much simpler.)

Figure 3-17 shows a cross section of the conductor. In applying Ampere's law it is essential to distinguish between two different meanings of the symbol I. (1) In Fig. 3-17 I

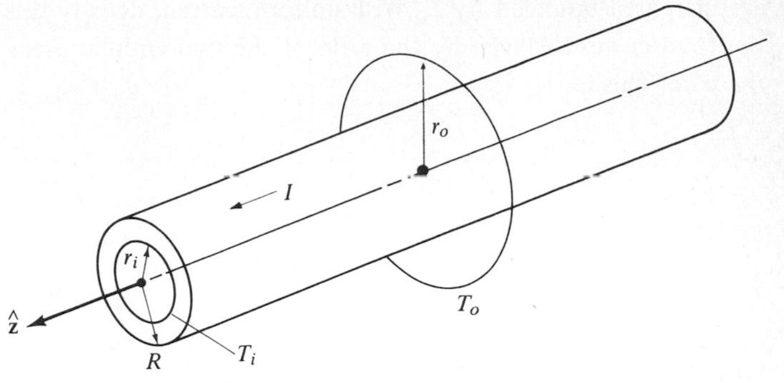

FIGURE 3-17

means the current passing through the conductor. (2) In Ampere's law I means the current passing through the open area bounded by T.

First, consider the test path T_o to find the value of **B** at a point outside the conductor, $r_o > R$. The current I passing through the open area bounded by T_o (that is, the current to be used on the right-hand side of Ampere's law) is equal to the total current I passing through the wire. So

$$\int_{T_0} \mathbf{B}_0 \cdot d\mathbf{r} = \mu_0 I$$

This integral may be evaluated in a manner similar to that employed previously in the case of an infinitely long charged wire. There it was possible to rely on symmetry to rule out any $\hat{\mathbf{z}}$ or $\hat{\boldsymbol{\phi}}$ components of **E**; here the current may flow either in the $+\hat{\mathbf{z}}$ or the $-\hat{\mathbf{z}}$ direction and it is not possible, a priori, to rule out any **B** component on the gounds of symmetry. Recollecting the law of Biot-Savart, however, the factor ($I_S\, d\mathbf{r}_s \times \hat{\mathbf{R}}_{st}$) shows that if **I** flows in the $\hat{\mathbf{z}}$ direction **B** must have only a $+\hat{\boldsymbol{\phi}}$ component. Thus $\mathbf{B}_o = \hat{\boldsymbol{\phi}} B_o$ in the left-hand side of Ampere's law and $d\mathbf{r} = \hat{\boldsymbol{\phi}}\, r\, d\phi$ for the path T_o. Consequently

$$\int_0^{2\pi} B_o r\, d\phi = \mu_0 I$$

$$B_o = \left(\frac{\mu_0 I}{2\pi}\right)\frac{1}{r} \qquad \text{(cylindrical } r\text{)}$$

Within the conductor a slight modification of this procedure is required. Take the test path T_i. The current to be used in Ampere's law is that current which passes through an open area bounded by T_i. With uniform current density this current is that fraction of the total current I given by the ratio of the two circular areas—that within T_i to that of the wire. This is

$$\left(\frac{\pi r_i^2}{\pi R^2}\right) I$$

so

$$\int_0^{2\pi} B_i r \, d\phi = \mu_0 \left(\frac{r_i^2}{R^2} I\right)$$

or

$$B_i = \left(\frac{\mu_0 I}{2\pi R^2}\right) r \qquad \text{(cylindrical } r\text{)}$$

2. A torus Next we find the value of **B** created by a steady current I flowing in a toroidal winding.

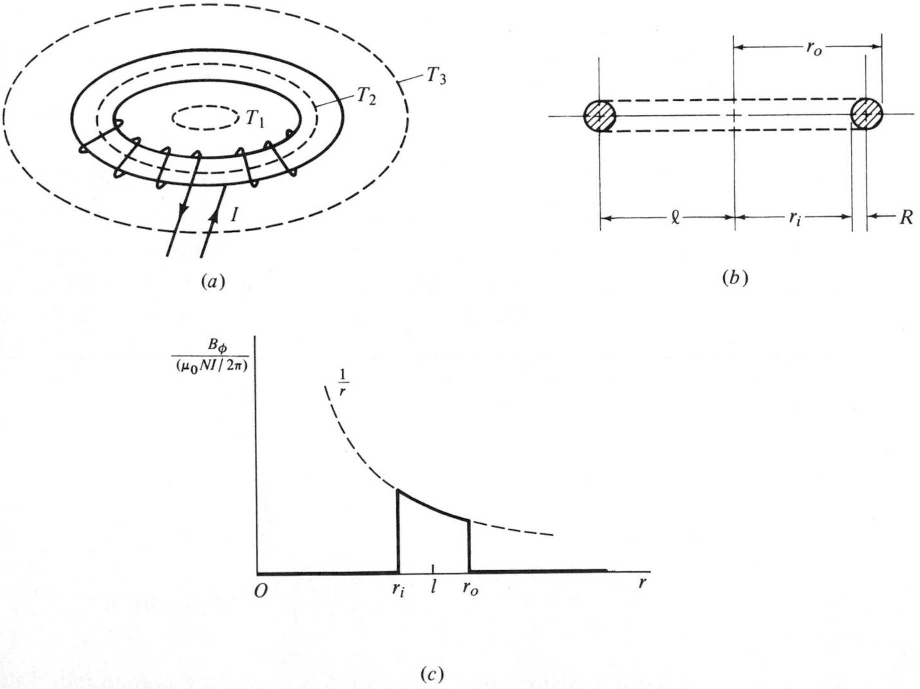

FIGURE 3-18

Figure 3-18a shows the winding, while Fig. 3-18b gives the dimensions. Apply Ampere's law to the circular path T_1 of radius $r < r_i$, which lies in the central plane of the torus. Then

$$\oint_{T_1} \mathbf{B} \cdot d\mathbf{r} = \oint_{T_1} \mathbf{B} \cdot \hat{\boldsymbol{\phi}}\, r\, d\phi = \oint_{T_1} B_\phi r\, d\phi = r \oint_{T_1} B_\phi\, d\phi = 0$$

since the current that passes through the flat circular area bounded by T_1 is zero. From symmetry, this integral is zero not because there are positive parts of B_ϕ which just cancel the negative parts, but because $B_\phi = 0$. This is all one may conclude from this particular path T_1; one cannot conclude that B_z or B_r vanish since such components would not contribute anything with this contour.

Similar results are obtained from a circular path T_3 with $r > r_o$. A flat, circular, area bounded by T_3 will be pierced by (1) N turns of the wire at $r = r_i$, each carrying the current I downward, and (2) N turns of the wire at $r = r_o$, each carrying the current I upward. The net current through the area is zero, so

$$\oint_{T_3} \mathbf{B} \cdot d\mathbf{r} = r \oint_{T_3} B_\phi\, d\phi = 0$$

and B_ϕ is also zero outside the torus.

For $r_i < r < r_o$ this circular path T_2 gives

$$\oint_{T_2} \mathbf{B} \cdot d\mathbf{r} = \mu_0\, (-NI)$$

since now only the N turns at $r = r_i$ contribute to the total current passing through the area bounded by T_2. The minus sign results from the direction taken along T_2: the positive direction of the normal to the open area bounded by T_2 is upward, while the current passing through the area is downward.

$$\int_0^{2\pi} B_\phi r\, d\phi = -\mu_0 NI$$

$$B_\phi = -\frac{\mu_0 NI}{2\pi r}$$

Figure 3-18a shows how the azimuthal component of $\mathbf{B}$ varies with r. When $l \gg R$ the value of B_ϕ is essentially constant for $r_i < r < r_o$ and equals

$$B_\phi \approx -\frac{\mu_0 NI}{2\pi \ell} = -\mu_0 nI$$

where n is the number of turns per unit length of the mean circumference. The minus sign shows only that the azimuthal component of the field is in the $-\hat{\boldsymbol{\phi}}$ direction when the current flows through the winding as indicated. For the opposite sense of current flow the direction of B_ϕ is also opposite, and the minus becomes a plus.

The values of B_z and B_r near to, but outside, the torus (unlike the B_ϕ component) do *not* vanish. They are, however, very small compared to the value of B_ϕ within the torus. They cannot be calculated by Ampere's law but require an integration using the law of Biot-Savart. A rough estimate, made in Prob. 14 below, shows that the field components outside the torus are of the order of $1/N$ times the field inside it.

A toroidal coil is more expensive to manufacture than other types of coils–pancake windings, scramble-wound coils, solenoids–so its use is generally restricted to applications where it is desired to minimize the external fields.

3. A solenoid Find **B** in the region inside and outside an infinitely long solenoid.

If the individual turns of the coil are close-wound and the pitch of the helical winding is small, one may simulate the actual result by assuming a surface current density $\mathbf{J}_l$ flowing either in the $+\hat{\boldsymbol{\phi}}$ or in the $-\hat{\boldsymbol{\phi}}$ direction, somewhat as fhown in Fig. 3-1*b*. In actual practice there are small wiggles in **B** near the individual turns; also, the finite pitch gives a $\hat{\mathbf{z}}$ component to the current, thereby introducing a $\hat{\boldsymbol{\phi}}$ component into **B**.

Problem 11 of the previous section gives $\mathbf{B} = \hat{\mathbf{z}}\mu_0 nI$ on the central axis of an infinitely long solenoid; here it is the value of **B** at a general point that is being considered. Figure 3-19*a* shows a section of the solenoid through the central axis while Fig. 3-19*b* is a cross-sectional view. Application of Ampere's law to a circle in the xy plane shows there is no $\hat{\boldsymbol{\phi}}$ component of **B**, either within or without the solenoid, if the pitch is neglected; i.e., if the solenoid is assumed to consist of rings of current parallel to the xy plane. Any B_z component must be independent of z, because the solenoid is infinitely long in the z direction. Also, there can be no B_r component: this should reverse with a reverse in the direction of the current; which direction should it start with? Thus $\mathbf{B} = \hat{\mathbf{z}}B_z$.

Apply Ampere's law to path 1234, completely outside the solenoid. Sides 12 and 34 contribute nothing to the integral, since $B_r = 0$. Then side 23 must give a contribution equal and opposite to that of side 41. Since these vertical legs may be located anywhere outside, it follows that B_z must also be independent of r; say $\mathbf{B} = \hat{\mathbf{z}}B_o$.

The same reasoning applied to path *abcd* gives a constant field, B_i, inside. To see that B_o and B_i must be different, consider path *ABCD*. If dots indicate current coming out of the paper (the tips of arrows) and crosses represent current into the paper (the tail feathers), then

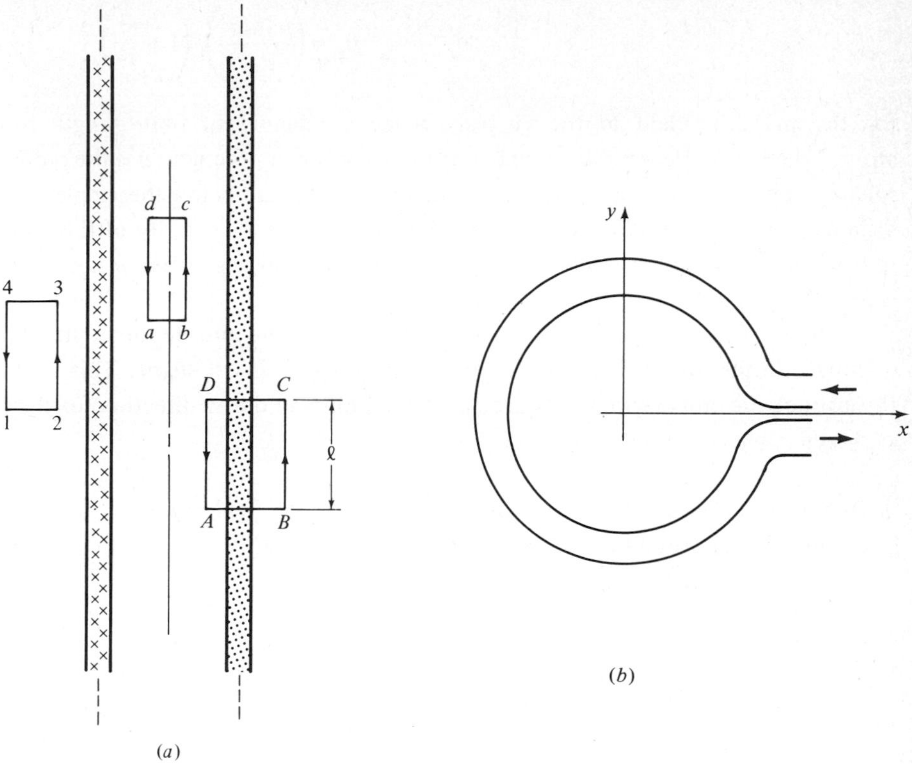

FIGURE 3-19

$$B_o l - B_i l = \mu_0 J_l l$$

or

$$B_o - B_i = \mu_0 J_l$$

Writing $J_l = nI$, where $n = N/l$ is the number of turns per unit length, this may also be given in terms of current and turns.

It now seems reasonable to demand that the magnetic field should approach zero when one goes very far out in the xy plane and, consequently, to assert that the value of the constant B_o must be zero. But this assertion is not really justified: there is no law which forbids B_z from having some finite value at a point arbitrarily far from the infinite solenoid. In Fig. 2-11, e.g., a constant value for **E**, independent of distance, was produced by a charged infinite plane. The justification for here setting $B_o = 0$ must rest on a calculation using the law of Biot-Savart.

D. B. Brick and A. W. Snyder have made such a calculation.* They derive

*D. B. Brick and A. W. Snyder, Field of a Long Solenoid, *Amer. J. Phys.*, **33**:905–909 (1965).

$$B_z = \left(\frac{\mu_0 n I a^2}{2l^2}\right)\left(1 + \frac{r^2}{l^2}\right)^{-3/2}$$

for the magnetic field in the xy plane when a solenoid of finite length is wound around the z axis. Here n is the number of turns per meter of length; a is the radius of the solenoid, and $2l$ is the solenoid's length (half above and half below the xy plane); r is the distance from the z axis in the xy plane. A condition for the validity of this equation is that $r \gg a$, $l \gg a$. When l is allowed to grow without bound, regardless of the value of r, it is seen that $B_z \to 0$.

We are, therefore, justified in taking $B_o = 0$ everywhere outside for an infinitely long solenoid. Inside this solenoid, the results above give $B_i = -\mu_0 nI$. This is constant throughout the cross section of the solenoid and points in the $-z$ direction for the current as shown.

4. $\nabla \times \mathbf{B}$ for an infinitely long conductor The magnetic field produced by an infinitely long conductor, of round cross section, carrying a current $\hat{\mathbf{z}}I$ is

$$\mathbf{B}_o = \hat{\boldsymbol{\phi}}\left(\frac{\mu_0 I}{2\pi}\right)\frac{1}{r} \qquad \text{outside the wire}$$

$$\mathbf{B}_i = \hat{\boldsymbol{\phi}}\left(\frac{\mu_0 I}{2\pi R^2}\right)r \qquad \text{inside the wire}$$

(cylindrical r). We will find $\nabla \times \mathbf{B}$ outside the wire, inside the wire, and on the surface and then check the applicability of Stokes' theorem.

$$\nabla \times \mathbf{B} = \hat{\mathbf{r}}\left[\frac{1}{r}\frac{\partial B_z}{\partial \phi} - \frac{\partial B_\phi}{\partial z}\right] + \hat{\boldsymbol{\phi}}\left[\frac{\partial B_r}{\partial z} - \frac{\partial B_z}{\partial r}\right] + \hat{\mathbf{z}}\left[\frac{1}{r}\frac{\partial}{\partial r}(rB_\phi) - \frac{1}{r}\frac{\partial B_r}{\partial \phi}\right]$$

So

$$(\nabla \times \mathbf{B})_o = 0 \qquad \text{and} \qquad (\nabla \times \mathbf{B})_i = \hat{\mathbf{z}}\frac{\mu_0 I}{\pi R^2}$$

There is a discontinuity, of amount $\mu_0 I/\pi R^2$, in the value of $\partial B_\phi/\partial r$ at the surface of the conductor and the value at the surface depends on the direction from which the surface is approached.

Figure 3-20 shows a cross section of the conductor. A circle C_o is drawn outside the wire. Along C_o,

$$\oint_{C_o} \mathbf{B}_o \cdot d\mathbf{r} = \oint_{C_o} \hat{\boldsymbol{\phi}}\frac{\mu_0 I}{2\pi r} \cdot (\hat{\boldsymbol{\phi}} r\, d\phi) = \frac{\mu_0 I}{2\pi}\int_0^{2\pi} d\phi = \mu_0 I$$

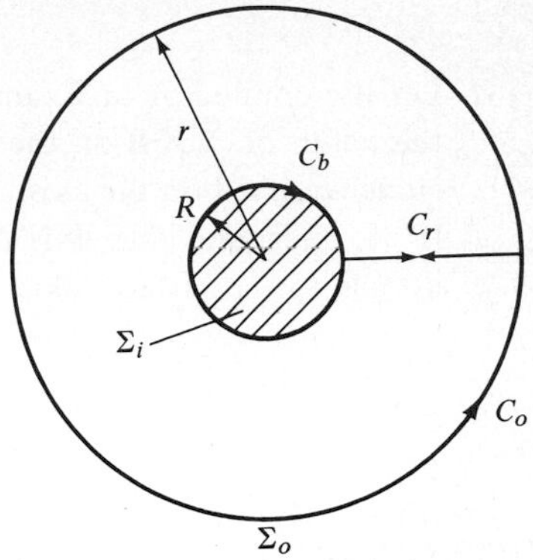

FIGURE 3-20

To check Stokes' theorem for the area bounded by C_o we break up the area Σ enclosed by C_o into Σ_o, the area outside the wire but within C_o; and Σ_i, that inside the wire.

$$\int_{\Sigma} (\nabla \times \mathbf{B}) \cdot d\mathbf{S} = \int_{\Sigma_o} (\nabla \times \mathbf{B})_o \cdot d\mathbf{S} + \int_{\Sigma_i} (\nabla \times \mathbf{B})_i \cdot d\mathbf{S}$$

$$= 0 + \int_0^R \left(\hat{\mathbf{z}}\, \frac{\mu_0 I}{\pi R^2}\right) \cdot (\hat{\mathbf{z}}\, 2\pi r\, dr) = \mu_0 I$$

Thus

$$\oint_{C_o} \mathbf{B}_o \cdot d\mathbf{r} = \int_{\Sigma} (\nabla \times \mathbf{B}) \cdot d\mathbf{S}$$

Stokes' theorem may also be applied to Σ_o alone. C_o is traversed CCW and the boundary of the conductor C_b is traversed CW; a radial segment connecting C_o and C_b is traversed outward and inward. Then

$$\oint_{C_o} \mathbf{B}_o \cdot d\mathbf{r} + \int_{\text{in } C_n} \mathbf{B}_i \cdot d\mathbf{r} + \oint_{C_b} \mathbf{B}_o \cdot d\mathbf{r} + \int_{\text{out } C_n} \mathbf{B}_o \cdot d\mathbf{r} = \mu_0 I + 0 - \mu_0 I + 0 = 0$$

This checks with

$$\int_{\Sigma_o} (\nabla \times \mathbf{B})_o \cdot d\mathbf{S} = 0$$

PROBLEMS

1 Let the conductor of Example 4 become a filamentary wire of zero radius. What is the value of $\nabla \times \mathbf{B}$ on the axis? What is $\mathbf{J}$? Calculate the circulation of $\mathbf{B}$ about a circle centered on the axis.

2 If the magnetic flux Φ is defined as the integral of the magnetic field component normal to the surface taken about some closed surface

$$\Phi = \oint \mathbf{B} \cdot d\mathbf{S}$$

what is the flux, per unit length of the conductor, produced by an infinitely long straight conductor carrying a current I?

3 A current I flows in an idealized filamentary wire (i.e., one with a cross section of zero radius) extending along the z axis from $-\infty$ to $+\infty$. If $\mathbf{B}$ were given by $\hat{\boldsymbol{\phi}} B_o{}^2 r^{-n}$, then for all $n > 0$ the value of B at $r = 0$ would be $+\infty$.

(*a*) Find $\nabla \times \mathbf{B}$.

(*b*) Is this expression valid at $r = 0$?

(*c*) For values of $n > 0$ does $\nabla \times \mathbf{B}$ at $r = 0$ become $+\infty$?

(*d*) Zero?

(*e*) $-\infty$?

4 Prove that, if $d\Omega$ is the solid angle from P to dS on Σ,

$$\int\!\!\int_{\Sigma} d\Omega = 0$$

if Σ is a closed surface that does not include P inside it.

5 In the infinitely long solenoid of Example 3, let the pitch be finite along the $\hat{\mathbf{z}}$ direction. Calculate $B_{\phi(\text{out})}$ by applying Ampere's law to a circle in the xy plane.

6 Prove that, if $d\Omega$ is the solid angle from P to dS on Σ,

$$\int\!\!\int_{\Sigma} d\Omega = 4\pi$$

if Σ is a closed surface that includes P inside it.

7 Use Ampere's law to find the magnetic field produced by a current I flowing along a coaxial circuit having an inner conductor of radius r_0 and an outer conductor of inside radius r_1 and outside radius r_2. Assume a uniform current density and then let $\mathbf{I} = \hat{\mathbf{z}}I$ for the inner conductor, $\mathbf{I} = -\hat{\mathbf{z}}I$ for the outer one.

8 Change Prob. 7 by removing the inner conductor. The return path of the idealized circuit is a circular path at infinity. Using Ampere's law, find $\mathbf{B}$.

9 Given a uniform surface current density $\mathbf{J}_l = \hat{\mathbf{z}}J$ flowing in the yz plane. Direct integration of the law of Biot-Savart shows $B_x = B_z = 0$ everywhere outside the yz

plane. Use Ampere's law to find B_y. Let $\hat{\mathbf{n}}$ be the unit normal to the yz, or $x = 0$, plane.

10 Consider a conducting surface on a nonconducting sphere of radius r. A pigtail at the north pole leads a current I to the sphere; another at the south pole leads the current away. Use Ampere's law to find **B** inside the sphere and outside, in spherical coordinates. This geometry was considered by Pugh and Pugh in an article devoted to the Poynting vector. See *Amer. J. Phys.*, **35**:153–156 (1967). (The Poynting vector will be treated here in Chap. 11.)

11 A circular filamentary ring insulator of 1 m radius lies in the xy plane with its center at the origin. It is given a uniform charge of 10^{-8} C/m. How many revolutions per second must the loop make about the $\hat{\mathbf{z}}$ axis to give a circulating current of 1 μA?

12 The toroidal coil of Example 2 gives a magnetic field, within its interior, of

$$\mathbf{B}_\phi = -\frac{\mu_0 NI}{2\pi r}$$

Suggest a different configuration of the source element which produces this same value of **B**. Can one draw a conclusion from this example about uniqueness of the source geometry to produce a given magnetic field?

13 The **B** produced by the infinitely long conductor of Example 1 is

$$\hat{\boldsymbol{\phi}}\left(\frac{\mu_0 I}{2\pi}\right)\frac{1}{r}$$

outside the wire. Select any closed path not surrounding the wire and find, by explicit integration, the circulation of **B** about the closed path.

14 In Example 2 the azimuthal component of **B** was found for the toroid. Select a path in a plane containing the axis of symmetry and apply Ampere's law to show that the other components of **B**, far from the torus, are approximately $1/N$ times as large as the value of B_ϕ inside the torus; i.e., they are due to only one turn.

15 Find the circulation of **B** along a circle, of radius $r < R$, centered on the axis of the infinitely long round conductor of Example 4. Check the applicability of Stokes' theorem.

16 A necessary condition for the application of Stokes' law is that the function be continuous everywhere within the region considered. Is it also necessary that any first derivatives be continuous?

17 A cone of half-angle 60° about the z axis, extending upward from the origin to an xy plane, such that each element of the cone has a length R, has a uniform volume charge density ρ, and rotates about its axis of symmetry with angular velocity $\hat{\mathbf{z}}\omega$. Find the value of **B** at the apex of the cone.

18 From

$$d\mathbf{B} = \frac{\mu_0}{4\pi}\frac{I_S\, d\mathbf{r}_s \times \hat{\mathbf{R}}_{st}}{R^2}$$

derive $\mathbf{B} = -(\mu_0/4\pi)I_s\nabla\Omega$ for a closed loop, where Ω is the solid angle subtended by the loop at the test point.

Appendix 3

THE FORCE BETWEEN CURRENT ELEMENTS

The momentum of the αth particle of a system may be defined in two different ways. The newtonian momentum is defined by

$$\mathbf{p}_\alpha = m_\alpha \mathbf{v}_\alpha$$

Then $\mathbf{F}_\alpha = d\mathbf{p}_\alpha/dt$ gives the net force acting on this particle. If the system is isolated this force is the resultant of the forces exerted by all the other particles: $\mathbf{F}_\alpha = \Sigma_{\beta\neq\alpha}\mathbf{F}_{\alpha\beta}$. It can then be shown that one must have $\mathbf{F}_{\alpha\beta} = -\mathbf{F}_{\beta\alpha}$ if the angular momentum, $\Sigma_\alpha \mathbf{r}_\alpha \times \mathbf{p}_\alpha$, is to be conserved; also that $\mathbf{F}_{\alpha\beta}$ and $\mathbf{F}_{\beta\alpha}$ must be central, i.e., parallel to $\mathbf{r}_\alpha - \mathbf{r}_\beta$.*

In the lagrangian method one finds a function of the coordinates and velocities called the lagrangian $L(\mathbf{r}_\alpha,\mathbf{v}_\alpha)$. The *canonical* momentum is defined by

$$\mathbf{p}_\alpha = \frac{\partial L}{\partial \mathbf{v}_\alpha}$$

This equation is shorthand for

$$p_{\alpha x} = \frac{\partial L}{\partial v_{\alpha x}} \quad \text{etc.}$$

since division by a vector has no meaning. Taking $\mathbf{F}_\alpha = \partial L/\partial \mathbf{r}_\alpha$ as the *canonical* force it is found here, also, that $\mathbf{F}_\alpha = d\mathbf{p}_\alpha/dt$. In the lagrangian case it can again be shown that $\Sigma_\alpha \mathbf{F}_\alpha = 0$, so $\mathbf{F}_1 = -\mathbf{F}_2$ when only two particles are considered; and, if the potential energy is a function only of the distances between particles, the canonical forces will be central. But, although $\mathbf{F}_1 = -\mathbf{F}_2$ in both methods, the meaning of $\mathbf{F}$ is different in the two cases, even though they have the same symbol and are usually called by the same name: forces.

It is possible for the two definitions to agree:

$$\frac{\partial L}{\partial \mathbf{v}_\alpha} = m_\alpha \mathbf{v}_\alpha$$

Then

$$L = \tfrac{1}{2} \sum_\alpha m_\alpha v_\alpha^{\,2} - V(\mathbf{r}_\alpha)$$

For two particles, e.g.,

$$L = \tfrac{1}{2} m_1 v_1^{\,2} + \tfrac{1}{2} m_2 v_2^{\,2} - V(\mathbf{r}_1, \mathbf{r}_2)$$

*For the proof of these statements see: B. Podolsky, Conservation of Angular Momentum, *Amer. J. Phys.*, **34**:42 (1966).

If L is to remain unchanged either by infinitesimal translations or rotations of the coordinate system, then V must be a function only of $R \equiv |\mathbf{r}_1 - \mathbf{r}_2|$. So

$$F_{1x} = \frac{\partial L}{\partial x_1} = -\frac{\partial V}{\partial x_1} = -\frac{dV}{dR} \cdot \frac{\partial R}{\partial x_1} = -\frac{dV}{dR}\left(\frac{x_1 - x_2}{R}\right)$$

etc., or

$$\mathbf{F}_1 = -\frac{dV}{dR}\left(\frac{\mathbf{r}_1 - \mathbf{r}_2}{R}\right)$$

Also

$$\mathbf{F}_2 = -\frac{dV}{dR}\left(\frac{\mathbf{r}_2 - \mathbf{r}_1}{R}\right)$$

Thus, $\mathbf{F}_1 = -\mathbf{F}_2$; and the two canonical forces, in the lagrangian method, are parallel to $(\mathbf{r}_1 - \mathbf{r}_2)$.

Suppose, instead, that the two definitions do not agree. What then? If there are velocity-dependent terms added to $V(\mathbf{r}_\alpha)$, the canonical forces will still be central but the newtonian forces will not be central. This is the case for an isolated system of two particles with, say,

$$L = \tfrac{1}{2} m_1 v_1^2 + \tfrac{1}{2} m_2 v_2^2 - V(R) + k\mathbf{v}_1 \cdot \mathbf{v}_2$$

The newtonian momenta $m_1\mathbf{v}_1$ and $m_2\mathbf{v}_2$ differ from the canonical momenta, since $\mathbf{p}_1 = m_1\mathbf{v}_1 + k\mathbf{v}_2$ and $p_2 = m_2\mathbf{v}_2 + k\mathbf{v}_1$. The canonical forces, as before

$$\mathbf{F}_1 = -\frac{dV}{dR}\frac{(\mathbf{r}_1 - \mathbf{r}_2)}{R} \quad \text{and} \quad \mathbf{F}_2 = -\frac{dV}{dR}\frac{(\mathbf{r}_2 - \mathbf{r}_1)}{R}$$

will be central, equal in magnitude, and opposite in direction:

$$\frac{d}{dt}(m_1\mathbf{v}_1 + k\mathbf{v}_2) = -\frac{d}{dt}(m_2\mathbf{v}_2 + k\mathbf{v}_1)$$

So,

$$\frac{d}{dt}(m_1\mathbf{v}_1) \neq -\frac{d}{dt}(m_2\mathbf{v}_2)$$

the newtonian forces, the ones ordinarily considered, will not be equal and opposite.

When the two methods do not agree, which is right and which is wrong? The answer to this question is that the two methods are talking about different things—newtonian force and canonical force. It is possible, as in the case above, for one method to give equal and opposite central "forces" while the other does not. Yet both methods are correct. They do not contradict each other because "forces" in one method means something different from "forces" in the other.

For two charged particles moving relative to one another it is found experimentally that the canonical forces are both central and that they obey Newton's third law. Most people intuitively accept the newtonian definition of momentum, $\mathbf{p} = m\mathbf{v}$. It is noteworthy, then, that in the transition to quantum mechanics it is the canonical definition that is employed. Much confusion arises from the fact that the word "forces" is commonly used in both contexts. Actually, however, in quantum mechanics the concept of force is relegated to secondary importance and is only rarely used.

The lagrangian for the case of two charged particles was first derived relativistically in 1920 by C. G. Darwin. A good summary appears in the survey article: Magnetic Interactions between Charged Particles, E. Breitenberger, *Amer. J. Phys.*, **36**:505 (1968). Eliminating the relativistic variation of mass with velocity that appears in Darwin's treatment, the result is:

$$L = \tfrac{1}{2}m_1v_1^2 + \tfrac{1}{2}m_2v_2^2 - \frac{q_1q_2}{4\pi\epsilon_0 R} + \frac{\mu_0 q_1 q_2}{8\pi R}[(\mathbf{v}_1 \cdot \mathbf{v}_2) + (\mathbf{v}_1 \cdot \hat{\mathbf{R}})(\mathbf{v}_2 \cdot \hat{\mathbf{R}})]$$

$$= \tfrac{1}{2}\mathbf{p}_1 \cdot \mathbf{v}_1 + \tfrac{1}{2}\mathbf{p}_2 \cdot \mathbf{v}_2 - \frac{q_1q_2}{4\pi\epsilon_0 R}$$

This gives for the *canonical* momenta, if $\mathbf{R} = \mathbf{r}_1 - \mathbf{r}_2$:

$$\mathbf{p}_1 = m_1\mathbf{v}_1 + \frac{\mu_0 q_1 q_2}{8\pi R}[\mathbf{v}_2 + (\mathbf{v}_2 \cdot \hat{\mathbf{R}})\hat{\mathbf{R}}]$$

$$\mathbf{p}_2 = m_2\mathbf{v}_2 + \frac{\mu_0 q_1 q_2}{8\pi R}[\mathbf{v}_1 + (\mathbf{v}_1 \cdot \hat{\mathbf{R}})\hat{\mathbf{R}}]$$

For the *canonical* forces it gives:

$$\mathbf{F}_1 = \left(\frac{q_1q_2}{4\pi\epsilon_0}\right)\frac{\hat{\mathbf{R}}}{R^2} - \left(\frac{\mu_0 q_1 q_2}{8\pi}\right)[(\mathbf{v}_1 \cdot \mathbf{v}_2) + (\mathbf{v}_1 \cdot \hat{\mathbf{R}})(\mathbf{v}_2 \cdot \hat{\mathbf{R}})]\frac{\hat{\mathbf{R}}}{R^2}$$

and $F_2 = -F_1$. Note that both canonical forces are central. From $\mathbf{F}_2 = (d/dt)\mathbf{p}_2 = \nabla_2 L$ and $F_1 = (d/dt)\mathbf{p}_1 = \nabla_1 L$ it follows, since $\nabla_1 L = -\nabla_2 L$, that $(d/dt)(\mathbf{p}_1 + \mathbf{p}_2) = 0$: *canonical* momentum is conserved. Total canonical angular momentum can also be shown to be conserved.

Although the canonical forces are equal and opposite here, the newtonian forces differ. The center of mass of the system does not move in a straight line; instead, it wiggles around a straight line.

The *newtonian* magnetic force here, obtained in Chap. 3, is

$$\mathbf{F} = \frac{\mu_0}{4\pi}\frac{I_T\,d\mathbf{r}_t \times (I_S\,d\mathbf{r}_s \times \hat{\mathbf{R}})}{R^2} = \frac{\mu_0 q_t q_s}{4\pi R^2}[\mathbf{v}_s(\mathbf{v}_t \cdot \hat{\mathbf{R}}_{st}) - \hat{\mathbf{R}}_{st}(\mathbf{v}_t \cdot \mathbf{v}_s)]$$

It has a central component which is attractive if $\mathbf{v}_t$ and $\mathbf{v}_s$ are in the same sense; it also has a component, at the test point, which is parallel to $\mathbf{v}_s$. It differs considerably from the *canonical* magnetic force,

$$\mathbf{F} = -\frac{\mu_0 q_t q_s}{8\pi R^2}\hat{\mathbf{R}}_{st}[(\mathbf{v}_t \cdot \mathbf{v}_s) + (\mathbf{v}_t \cdot \hat{\mathbf{R}}_{st})(\mathbf{v}_s \cdot \hat{\mathbf{R}}_{st})]$$

Not only is there no component parallel to $\mathbf{v}_s$ here, but the values of the central components in the two expressions differ.

4

THE ELECTROSTATIC CURL IN VACUUM

4-1 $\nabla \times \mathbf{E}$ AND THE POTENTIAL, ϕ

We now return to electrostatics. Knowing that $\nabla \cdot \mathbf{E} = \rho/\epsilon_0$, it is now necessary to find $\nabla \times \mathbf{E}$ in order to determine $\mathbf{E}$. This is required by the Helmholtz theorem, given in Appendix 4 at the end of this chapter. This can be done in short order.

In

$$\mathbf{E}(\mathbf{r}_t) = \frac{1}{4\pi\epsilon_0} \int_{\mathcal{V}} \frac{\rho(\mathbf{r}_s)}{R^2} \hat{\mathbf{R}} \, d\tau_s$$

substitute

$$\nabla \left(\frac{1}{R}\right) = -\frac{1}{R^2} \hat{\mathbf{R}}$$

(See Example 9 of Chap. 1, Sec. 1-1.) Then

$$\mathbf{E}(\mathbf{r}_t) = -\frac{1}{4\pi\epsilon_0} \int_{\mathcal{V}} \rho(\mathbf{r}_s) \nabla \left(\frac{1}{R}\right) d\tau_s$$

In this expression the independent variables under the integral are x_s, y_s, z_s while the variables in the del operator are x_t, y_t, z_t (or x,y,z, for short). The del operator may then be taken out of the integral:

$$\mathbf{E}(\mathbf{r}_t) = \nabla \left[-\frac{1}{4\pi\epsilon_0} \int_{\mathcal{V}} \frac{\rho(\mathbf{r}_s)}{R} \, d\tau_s \right]$$

Take the curl of both sides. From the vector identities listed in Appendix 1-B, the curl of the gradient of any function is zero. So

$$\boxed{\nabla \times \mathbf{E} = 0} \qquad \text{(E-7)}$$

This result, unlike the previous expression for $\nabla \cdot \mathbf{E}$, is only valid in electrostatics. In the general case of time-varying **E**, it will be necessary to modify this equation.

The electric field is an irrotational vector in electrostatic cases and, therefore, also a conservative vector. The remainder of this section is devoted to exploring some of the consequences of this statement. It is worth reiterating that, with $\nabla \cdot \mathbf{E}$ and $\nabla \times \mathbf{E}$ both known in some finite region $\mathcal{V}$ containing all the charges, Helmholtz's theorem determines **E**. The Helmholtz formula consists of two terms, one involving an integral of the divergence and the other an integral of the curl. Here the second term vanishes; the first is identically the same as the expression above:

$$\mathbf{E}(\mathbf{r}_t) = -\nabla \left[\frac{1}{4\pi} \int_{\mathcal{V}} \frac{\rho(\mathbf{r}_s)/\epsilon_0}{R} \, d\tau_s \right]$$

Note that the Helmholtz theorem gives this as the answer only when $\mathcal{V}$ is limited to a finite region.

From the above

$$\boxed{E = -\nabla \phi} \qquad \text{(E-8)}$$

where

$$\boxed{\phi(\mathbf{r}_t) = \frac{1}{4\pi\epsilon_0} \int_{\mathcal{V}} \frac{\rho(\mathbf{r}_s)}{R} \, d\tau_s} \qquad \text{(E-9)}$$

When the charges are to be considered as discrete then, by analogy,

$$\phi(\mathbf{r}_t) = \frac{1}{4\pi\epsilon_0} \sum_{i=1}^{N} \frac{q_i(\mathbf{r}_i)}{R_i}$$

The function $\phi(\mathbf{r})$ is called the potential function. For a single point charge located at the origin this function is

$$\phi(\mathbf{r}) = \frac{1}{4\pi\epsilon_0} \frac{q}{r}$$

The potential function is a scalar function rather than a vector function, so that it may be calculated with, at most, one-third the effort. For a point charge ϕ falls off inversely with distance rather than inversely with square of the distance. The reason why **E** is written as $-\nabla\phi$, with a minus sign, will become clear below.

The dimensions of ϕ in the mksa system, obtained from the defining relation, must be those of inverse permittivity multiplied by those of charge over distance:

$$[\mathrm{m}^{-3}\mathrm{kg}^{-1}\mathrm{s}^4\mathrm{A}^2]^{-1}[\mathrm{sA}]\,[\mathrm{m}]^{-1} = [\mathrm{m}^2\ \mathrm{kg}\ \mathrm{s}^{-3}\mathrm{A}^{-1}]$$

A unit having these dimensions is called the volt, by definition. Comparing these with the dimensions of **E** gives $[\phi] = [E \times \mathrm{m}^{-1}]$. The unit of **E**, the N/C, is, consequently, equivalent to the V/m, which is the designation normally employed.

The potential function was originally introduced as an auxiliary mathematical quantity, useful to obtain **E**. It will turn out, however, that the potential is actually a basic quantity, sometimes more so than the electric field. ϕ will be seen to be related to energy, while **E** is essentially a force; and though force is a simpler, more intuitive, concept than energy it is the latter which is often more significant. We will proceed in the historic fashion and deal with **E** and **B**, the electric and magnetic fields, to obtain and utilize Maxwell's equations. But it is possible to deal only with ϕ and **A**, the electric and magnetic potentials, ignoring **E** and **B** completely if the force is not required explicitly.

It might appear from the equations above that a given charge distribution determines ϕ uniquely, but this is not so. From $\mathbf{E} = -\nabla\phi$ it is seen that $\phi' = \phi + k$, where k is any constant, will give the same **E**. ϕ' is just as valid a potential function as ϕ. For a given charge distribution it is natural to select that value of the additive constant which is most convenient. Physicists usually prefer the choice which makes the potential equal to zero at infinity. It will be seen subsequently that this is not always possible; but it is possible when the charge distribution is limited to a finite domain. When the charge distribution, itself, extends to infinity, e.g., the case of an infinitely long line with a uniform charge distribution, then the zero potential point must be taken somewhere in the finite domain. Electrical engineers usually choose the potential of some convenient body or terminal as zero; the potential at infinity is then different from zero.

That body or object which has been arbitrarily selected as having zero potential is called the ground (in England, the earth). This is true even when the mathematical "ground" has no connection whatever with the actual (earth) ground. In an automobile, e.g., the metallic chassis is taken as the ground, although the auto is insulated from the actual ground by nonconducting rubber tires. If the negative terminal of the battery is electrically connected to the chassis, then the chassis is taken to have zero potential and the potential of the other battery terminal is positive, say +12.6 V, with respect to ground. If instead the positive terminal of the battery is grounded, i.e., connected to the chassis, then the positive terminal is considered to be at zero potential, while the potential of the other terminal would be −12.6 V with respect to ground. *The* potential of a point is arbitrary and there is no unique value; in fact, it cannot be measured. What can be measured is the potential *difference* between two points, and this is a quantity whose value *is* unique and independent of the zero reference point. Sometimes *the* potential is, nevertheless, employed; what is meant is the potential difference between that point and the arbitrarily selected ground.

We will now show why the addition of this arbitrary constant to the potential has no physical significance. Suppose $\mathbf{E} = -\nabla\phi$ is integrated over some path C_1 connecting a and b, as in Fig. 4-1. Then

$$\underset{(C_1)}{\int_a^b} \mathbf{E} \cdot d\mathbf{r} = -\underset{(C_1)}{\int_a^b} \nabla\phi \cdot d\mathbf{r} = -\underset{(C_1)}{\int_a^b} d\phi = \phi_a - \phi_b$$

The quantity ϕ_b is the potential of the point b with respect to ground, which is arbitrarily taken at some convenient point P. Similarly, ϕ_a is taken as the potential of a with respect to this ground point. $\phi_a - \phi_b$ is the potential of a with respect to b, $\phi_b - \phi_a$ is the potential of b with respect to a. The magnitude of this quantity is the potential difference between a and b.

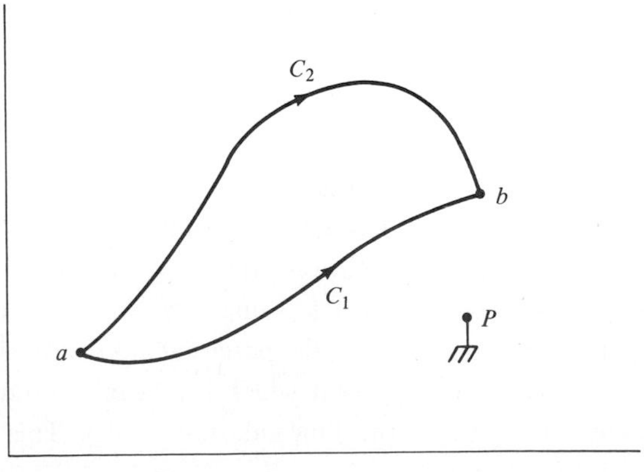

FIGURE 4-1

The right side of the equation above does not depend on the particular path C_1. If the line integral is taken along some other path, such as C_2, then

$$\int_{(C_2)\,a}^{b} \mathbf{E} \cdot d\mathbf{r} = \phi_a - \phi_b$$

So

$$\int_{(C_1)\,a}^{b} \mathbf{E} \cdot d\mathbf{r} - \int_{(C_2)\,a}^{b} \mathbf{E} \cdot d\mathbf{r} = 0$$

$$\int_{(C_1)\,a}^{b} \mathbf{E} \cdot d\mathbf{r} + \int_{(C_2)\,b}^{a} \mathbf{E} \cdot d\mathbf{r} = 0$$

or

$$\oint \mathbf{E} \cdot d\mathbf{r} = 0$$

The circulation of **E** is zero in electrostatics. This result could have been written down immediately from Stokes' law:

$$\oint \mathbf{E} \cdot d\mathbf{r} = \int_{\Sigma} (\nabla \times \mathbf{E}) \cdot d\mathbf{S} = \int_{\Sigma} 0 \cdot d\mathbf{S} = 0$$

In order to attach physical significance to $\phi_a - \phi_b$, consider a line integral over the path C_1 again, but this time take the line integral of $-\mathbf{F}_{\text{elec}} = -q\mathbf{E}$. Let q be transported from a to b over path C_1 at an infinitesimal speed. The journey will then require an infinitely long time for its completion, but this is no problem; it is only an imagined journey, designed simply to obtain a mathematical result. With the speed along the path always infinitesimal, the mechanical force required to move the charge is equal in magnitude but oppoiste in direction to $\mathbf{F}_{\text{elec}} = q\mathbf{E}$, the electrical force exerted on the charge by the field.

If the speed were finite a number of difficulties would be encountered. (1) Coulomb's law gives the value of **F** only for stationary charges. What does **F** become for a test charge which is moving? Actually, the answer is that **F** is given as before; still, it is a complicating factor. (2) If the motion involves acceleration, either along a straight line—to start and stop—or because the path is curved, then inertial forces enter and the electrical and mechanical forces are no longer equal in magnitude. (3) It will be shown subsequently, by relativity theory, that W, the work performed on an object, and ΔU, the resulting energy change, are physical quantities which transform differently between two observers moving relative to one another. Suppose $W_0 = \Delta U_0$ for one observer who is stationary relative to the source charges that are creating the electric field; let another observer move along with the test charge; then if the velocity of the test charge from a to b is finite, the second observer will find $W \neq \Delta U$. Consequently, for finite speeds of the charge—even without acceleration—it cannot be true that energy changes are the equivalent of the work performed. (4) Kinetic energy would have to be taken into account. (5) Magnetic forces would enter the picture as soon as there is a finite velocity.

With infinitesimal speed for the test charge, then, we guarantee that $\mathbf{F}_{\text{mech}}$, the mechanical force required to do the moving, equals the negative of $q\mathbf{E}$, the electrical force on the test charge. So

$$W_{\text{mech}} = \int_{(C_1)\,a}^{b} \mathbf{F}_{\text{mech}} \cdot d\mathbf{r} = \int_{(C_1)\,a}^{b} -q\mathbf{E} \cdot d\mathbf{r}$$

$$= q \int_{(C_1)\,a}^{b} \nabla\phi \cdot d\mathbf{r} = q \int_{(C_1)\,a}^{b} d\phi = q\phi_b - q\phi_a = \Delta U$$

The left side represents the work required to bring the charge at an infinitesimal speed from a to b. The right side gives the change in the potential energy: $\Delta U = U_b - U_a$ or $q\phi_b - q\phi_a$. Just as ϕ, the potential at a point, is not uniquely defined, so U, the potential energy at a point, is not uniquely defined. Only the difference of potential $\phi_b - \phi_a$ and the difference of potential energy $U_b - U_a$ have physical significance. Once an arbitrary point has been picked as the zero point, necessarily of *both* the potential and the potential energy, it is common to refer to *the* potential and *the* potential energy at other points. Then we can set $U_b = q\phi_b$ and $U_a = q\phi_a$.

The reason for the name "potential" for ϕ is now obvious: "potential" means "potential energy per unit charge": $\phi = U/q$. The unit of potential is 1 volt and has the dimensions [m^2 kg s^{-3} A^{-1}]. The unit of potential energy is 1 joule, with dimensions [m^2 kg s^{-2}]. Since the unit of charge is 1 coulomb with dimensions [s A], 1 joule per coulomb is 1 volt.

From $\oint \mathbf{F}_{\text{mech}} \cdot d\mathbf{r} = \oint -q\mathbf{E} \cdot d\mathbf{r} = -q \oint \mathbf{E} \cdot d\mathbf{r} = 0$, the work done in a journey at an infinitesimal rate around a closed path is zero. Energy is conserved, hence the name "conservative" field. Any vector field which does not vary with time and is the gradient of a scalar function at all points of a singly connected region is a conservative field (the earth's gravitational field is another example of a conservative field). Put another way, a conservative field is one, not varying with time, whose curl is zero everywhere. If there is even one point where the curl is not zero the field is no longer conservative, not even if this point is excluded from the region; for then the curl is only zero everywhere in a doubly connected region—the region around the excluded point.

To see why a conservative field must be irrotational at all points in a singly connected region, consider the vector function

$$\mathbf{F} = -\hat{\mathbf{x}}\left[\frac{y}{x^2+y^2}\right] + \hat{\mathbf{y}}\left[\frac{x}{x^2+y^2}\right] = \hat{\boldsymbol{\phi}}\left(\frac{1}{r}\right) \qquad \text{(cylindrical } r\text{)}$$

Here $\nabla \times \mathbf{F} = 0$ everywhere except at $r = 0$, where $\nabla \times \mathbf{F} = \infty$. The work done in a complete journey about the origin, e.g., in a circle of radius r, will not be zero even though $\nabla \times \mathbf{F} = 0$ at all points on the trajectory; consequently, this is not a conservative field. ($\mathbf{F}$, here, actually gives the magnetic field lines produced by an infinitely long filamentary conductor, of zero thickness, carrying a steady current.) If one draws a circle of radius a, where $a < r$, and excludes the area inside a from the region under consideration, the situation is unaltered. The work done in transporting a charge around the circle of radius r will not be zero. The difficulty is not produced by the singularity itself (i.e., by the fact that $F(0) = \infty$), for if a cylinder of radius a is substituted for the filamentary conductor the work done in circulating a charge about the cylinder will still not be zero, even though there is no longer a singularity. The region inside a now has a finite but nonvanishing curl instead of an infinite curl. Excluding this region creates a doubly connected region and the situation is unaltered. The region outside a may be made singly connected by cutting it along a radius from a to ∞.

On the other hand, the vector function

$$\mathbf{F} = \hat{\mathbf{x}}\left[\frac{x}{x^2+y^2}\right] + \hat{\mathbf{y}}\left[\frac{y}{x^2+y^2}\right] = \hat{\mathbf{r}}\left(\frac{1}{r}\right) \qquad \text{(cylindrical } r\text{)}$$

does have $\nabla \times \mathbf{F} = 0$ everywhere, even at $r = 0$, despite the fact that the partial derivatives are infinite at $r = 0$. In spite of the singularity, this is a conservative field, the field of an infinite line charge and no work is required to circumnavigate the z axis.

The concept of a conservative field is an important one. Suppose we have a particle moving with velocity $\mathbf{v}$ at a point in some region of field-free space under the action of a force $\mathbf{F}$. The rate of change of the kinetic energy T with time is

$$\frac{dT}{dt} = \mathbf{F} \cdot \mathbf{v}$$

If there is a field in this region then the potential energy U at the instantaneous point where the particle finds itself can vary for two reasons: either $U = U(t)$ at that fixed point or the potential energy at different points of the particle's trajectory is a function of position. Then, applying the chain rule to $U/x, y, z, t/t$,

$$\frac{dU}{dt} = \frac{\partial U}{\partial t} + (\nabla U) \cdot \mathbf{v}$$

The rate of change with time of the total energy $(T + U)$ is thus

$$\frac{d}{dt}(T + U) = \frac{dT}{dt} + \frac{dU}{dt} = \mathbf{F} \cdot \mathbf{v} + \left[\frac{\partial U}{\partial t} + (\nabla U) \cdot \mathbf{v}\right]$$

$$= (\mathbf{F} + \nabla U) \cdot \mathbf{v} + \frac{\partial U}{\partial t}$$

If both (1) $\partial U/\partial t = 0$ (there is no explicit variation with time of the scalar field U) and (2) $\mathbf{F} = -\nabla U$ (the vector field $\mathbf{F}$ is irrotational), then the quantity $(T + U)$ is conserved. In the electrostatic case $\mathbf{F} = q\mathbf{E}$ and $U = q\phi$ so $\mathbf{F} = -\nabla U$ is equivalent to $\mathbf{E} = -\nabla\phi$; also if $U \neq U(t)$ then $\phi \neq \phi(t)$ and $\mathbf{E} \neq \mathbf{E}(t)$.

The potential ϕ was defined by

$$\phi(\mathbf{r}_t) = \frac{1}{4\pi\epsilon_0}\int_{\mathcal{V}} \frac{\rho(\mathbf{r}_s)}{R}\, d\tau_s$$

with $\mathbf{E} = -\nabla\phi$. It could just as well have been taken

$$\phi'(\mathbf{r}_t) = -\frac{1}{4\pi\epsilon_0}\int_{\mathcal{V}} \frac{\rho(\mathbf{r}_s)}{R}\, d\tau_s$$

Then $\mathbf{E} = +\nabla\phi'$. The reason why the first choice was selected is illustrated in Fig. 4-2. Two large parallel plates are separated by a small distance, with one charged positive and one negative. By integrating $\mathbf{E} = -\nabla\phi$ from a to b along any path one obtains

$$\int_a^b \mathbf{E} \cdot d\mathbf{r} = \phi_a - \phi_b$$

If the path chosen is horizontal from left to right it will coincide with a line of $\mathbf{E}$ and go in the same direction. The left side of the equation is then positive, so ϕ_a will be greater than ϕ_b. Then that plate with the higher potential has a more positive charge; also, a higher potential energy. If the choice had been $\mathbf{E} = +\nabla\phi'$ then ϕ' would be more negative where the charge was more positive; the potential and the potential energy would have

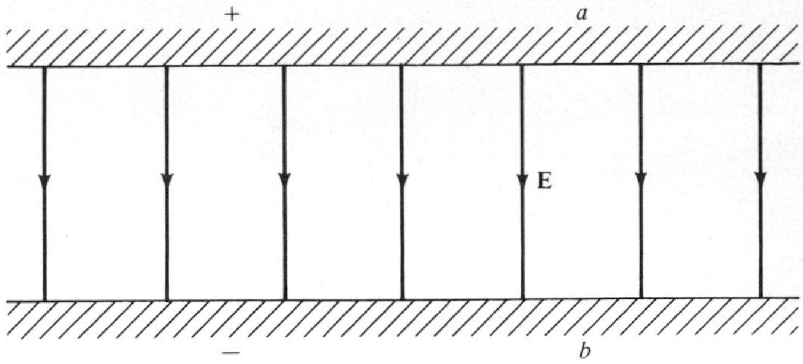

FIGURE 4-2

opposite signs for a positive charge; lines of **E**, proceeding from positive to negative charge, would go from a point of lower potential to a point of higher potential.

Examples

1. The potential function produced by a point charge q at the origin if the point (1,0,0) is chosen to be at ground potential.

Here $\mathbf{E} = (1/4\pi\epsilon_0)\,(q/r^2)\hat{\mathbf{r}}$. In Fig. 4-3 let ϕ_{ab} be the potential of a relative to b. Then

$$\phi_{ab} = \phi_a - \phi_b = \int_a^b \mathbf{E} \cdot d\mathbf{r} = \frac{q}{4\pi\epsilon_0} \int_a^b \frac{\hat{\mathbf{r}}}{r^2} \cdot d\mathbf{r} = \frac{q}{4\pi\epsilon_0} \int_a^b \frac{dr}{r^2}$$

$$= \frac{q}{4\pi\epsilon_0}\left(\frac{1}{r_a} - \frac{1}{r_b}\right)$$

Normally the ground would be taken at infinity; i.e., $\phi_b = 0$ when $r_b = \infty$. So the potential at a would then be

$$\phi_a = \frac{q}{4\pi\epsilon_0}\left(\frac{1}{r_a}\right) \qquad \text{or} \qquad \phi = \frac{q}{4\pi\epsilon_0}\left(\frac{1}{r}\right)$$

The equipotential surfaces, the loci of all points such that ϕ has some constant value, would be spheres about the origin. If for some reason it is more convenient to take another sphere, $r_b = 1$, say, as that equipotential surface for which $\phi_b = 0$,

$$\phi_a - 0 = \frac{q}{4\pi\epsilon_0}\left(\frac{1}{r_a} - \frac{1}{1}\right) \qquad \text{or} \qquad \phi = \frac{q}{4\pi\epsilon_0}\left(\frac{1}{r} - 1\right)$$

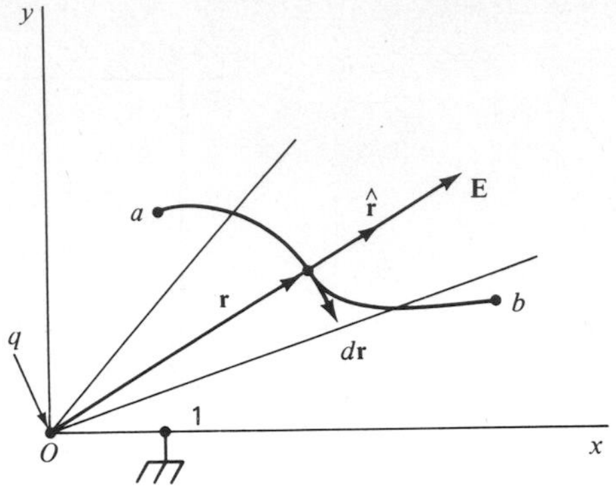

FIGURE 4-3

In effect, the constant $q/4\pi\epsilon_0$ has merely been subtracted from the previous result; the equipotential surfaces are still spheres.

2. The potential function $\phi(\mathbf{r})$ produced by an infinitely long, uniformly charged line.
The value of **E** was obtained for this case in Example 1 of Sec. 3:

$$\mathbf{E} = \frac{\lambda}{2\pi\epsilon_0}\left(\frac{1}{r}\right)\hat{\mathbf{r}}$$

Here the line is along the z axis, as in Fig. 4-4. Cylindrical coordinates are used. Then

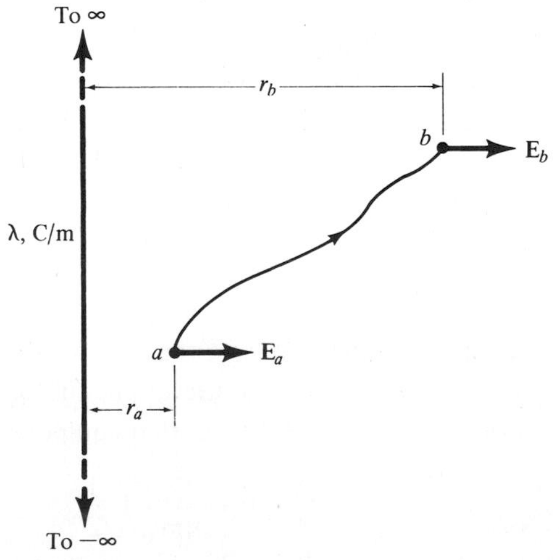

FIGURE 4-4

$$\phi_{ab} = \phi_a - \phi_b = \int_a^b \mathbf{E} \cdot d\mathbf{r} = \frac{\lambda}{2\pi\epsilon_0} \int_a^b \frac{\hat{\mathbf{r}}}{r} \cdot d\mathbf{r}$$

$$= \frac{\lambda}{2\pi\epsilon_0} \int_a^b \frac{dr}{r} = \frac{\lambda}{2\pi\epsilon_0} \ln\left(\frac{r_b}{r_a}\right)$$

If an attempt is made to set $\phi_b = 0$ at $r_b = \infty$ this gives

$$\phi_a - 0 = \frac{\lambda}{2\pi\epsilon_0} \ln\left(\frac{\infty}{r_a}\right) = \infty$$

So $\phi = \infty$ at any point at a finite distance from the z axis. This difficulty is due to the existence of charge at an infinite distance from the origin, that at $z = \pm\infty$.

By symmetry the equipotential surfaces here are cylinders. Suppose the ground reference surface, $\phi_b = 0$, is taken at some finite distance from the z axis, say at $r_b = r_0$. Then

$$\phi_a - 0 = \frac{\lambda}{2\pi\epsilon_0} \ln \frac{r_0}{r_a} \quad \text{or} \quad \phi = -\left(\frac{\lambda}{2\pi\epsilon_0}\right) \ln r + \left[\frac{\lambda}{2\pi\epsilon_0} \ln r_0\right]$$

Figure 4-4 shows the variation of ϕ with r. If r_0 has a finite value there will be some points at a finite distance with a positive potential and others with a negative potential. But if r_0 is taken at $r = \infty$ then the point of intersection of the curve with the horizontal axis moves off to ∞ and all points at a finite distance have a positive, infinite, potential.

At a given distance from the line charge the potential may be made either positive or negative simply by choosing the zero point properly. The fact that the potential at a point is positive simply means that positive work has to be done to transport a positive test charge to that point, at an infinitesimal rate, from the ground point. Selecting a new ground point could make the potential at the test point negative. But the gradient of the equipotential function at a point is unaffected by the choice of the zero point. The negative slope of the tangent to the curve of Fig. 4-5 at $r = 2r_0$, say, gives **E** there. If a new ground surface were to be taken, then the curve of Fig. 4-5 would simply be shifted bodily, up or down, an arbitrary amount, but the slope of the curve at any point would remain unchanged.

3. The potential for the charged line of infinite length by direct integration, using

$$\phi = \frac{1}{4\pi\epsilon_0} \int_{\mathcal{V}} \frac{\rho}{r}\, d\tau$$

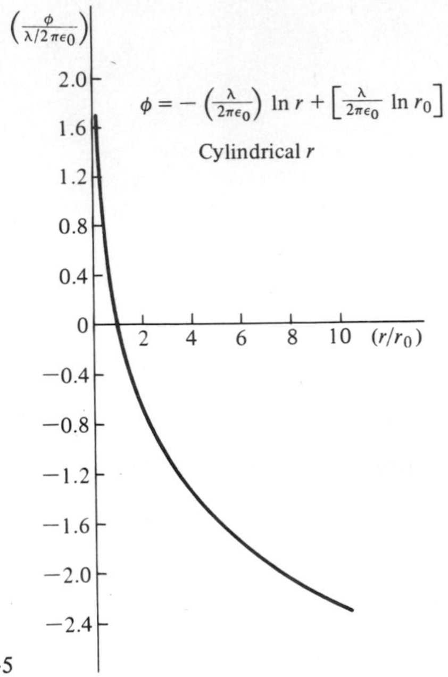

FIGURE 4-5

Here the charge is filamentary, so

$$\phi = \frac{1}{4\pi\epsilon_0} \int_{-\infty}^{\infty} \frac{\lambda}{r}\, dz \qquad \text{(cylindrical } r\text{)}$$

See Fig. 4-6. Since $r = \sqrt{D^2 + z^2}$,

$$\phi = \left(\frac{\lambda}{4\pi\epsilon_0}\right) \sinh^{-1}\left[\left(\frac{z}{D}\right)\right]_{-\infty}^{\infty} = \left(\frac{\lambda}{4\pi\epsilon_0}\right)[\sinh^{-1}(\infty) - \sinh^{-1}(-\infty)]$$

$$= \left(\frac{\lambda}{4\pi\epsilon_0}\right) [\infty-(-\infty)] = \infty$$

This result is meaningless because the equation

$$\phi = \frac{1}{4\pi\epsilon_0}\int_{\mathcal{V}} \frac{\rho}{r}\, d\tau$$

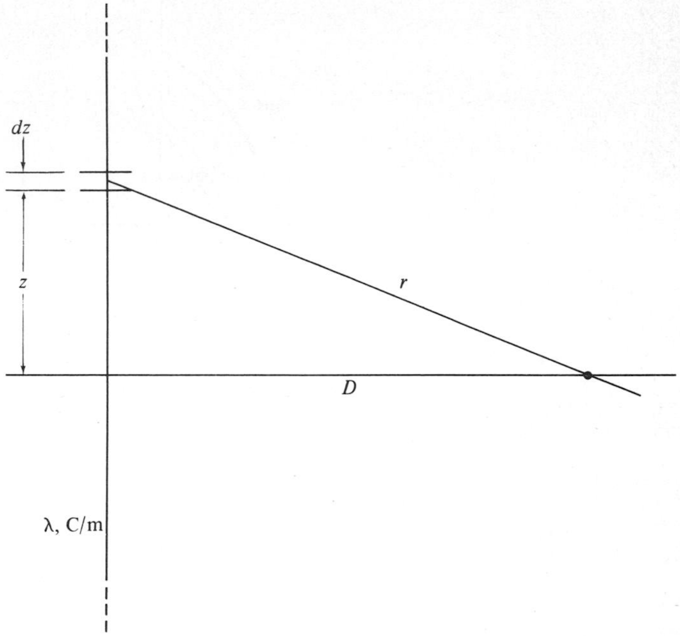

FIGURE 4-6

is only valid when $\mathcal{V}$ is limited to the finite domain. Therefore the potential for the charged line of infinite length cannot be obtained by direct integration of (ρ/r); it must be obtained by integration of $(\mathbf{E} \cdot d\mathbf{r})$. This result is connected with the fact that the Helmholtz theorem is invalid when there is charge at infinity.

4. The potential function produced by a uniformly charged plane circular disk at a point on the axis of symmetry perpendicular to the disk at its center.

Figure 4-7, which illustrates this case, is almost the same as Fig. 2-9, employed for the calculation of **E** in this case. The potential at P may be found from

$$\phi = \frac{1}{4\pi\epsilon_0} \int_{\Sigma} \frac{\sigma}{r}\, dS$$

This integral is a scalar integral, and it is much easier to evaluate than the vector integral that arose in finding **E** directly in connection with Fig. 2-9. An element of charge in the annular ring shown produces the potential

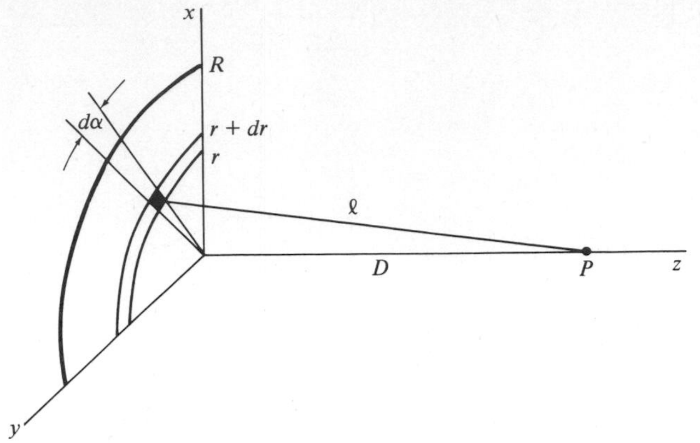

FIGURE 4-7

$$d^2\phi = \frac{1}{4\pi\epsilon_0} \frac{\sigma r \, d\alpha \, dr}{\ell}$$

Since all elements in the ring are at the same distance ℓ from P, the entire annular ring produces the potential, obtained by integrating $d\alpha$ from $\alpha = 0$ to $\alpha = 2\pi$,

$$d\phi = \frac{\sigma r \, dr}{2\epsilon_0 \ell}$$

Integrating next over r:

$$\phi = \frac{\sigma}{2\epsilon_0} \int_0^R \frac{r \, dr}{\sqrt{r^2 + D^2}} = \frac{\sigma}{2\epsilon_0} [\sqrt{D^2 + R^2} - \sqrt{D^2}\,]$$

For P on the $+x$ axis this becomes

$$\phi = \frac{\sigma}{2\epsilon_0} [\sqrt{D^2 + R^2} - D] \qquad D \geqslant 0$$

Here the positive root, $+\sqrt{D^2}$, is taken so that $\phi \to 0$ when $R \to 0$ (i.e., when there is no disk). For P on the $-x$ axis the negative root must be taken if $\sqrt{D^2 + R^2}$ is always taken as positive:

$$\phi = \frac{\sigma}{2\epsilon_0} [\sqrt{D^2 + R^2} + D] \qquad D \leqslant 0$$

A formula applicable to both regions is

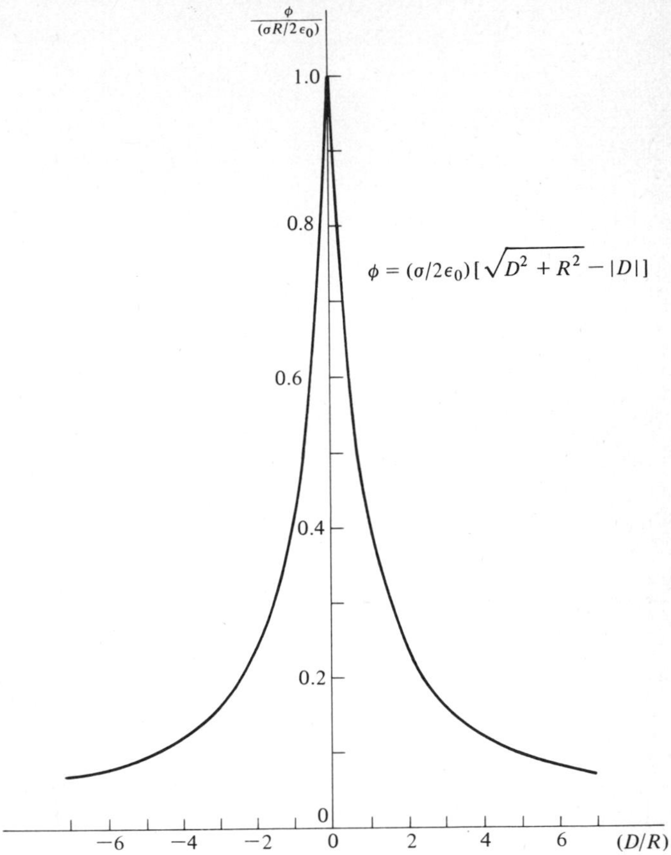

FIGURE 4-8

$$\phi = \frac{\sigma}{2\epsilon_0}[\sqrt{D^2 + R^2} - |D|]$$

Figure 4-8 shows how ϕ varies with distance. The negative of the slope of this curve gives the **E** of Fig. 2-10. While there is a discontinuity in **E** at the origin, there is none in ϕ. Exactly the opposite situation will be found to hold, in a later chapter, for a plane distribution of dipoles.

5. The ϕ produced by a uniformly charged infinite plane.

This result cannot be found immediately from the previous example by letting $R \to \infty$, for in the previous case the point at zero potential was at infinity, and this is not permissible when the charge distribution itself extends to infinity. But a slight change will

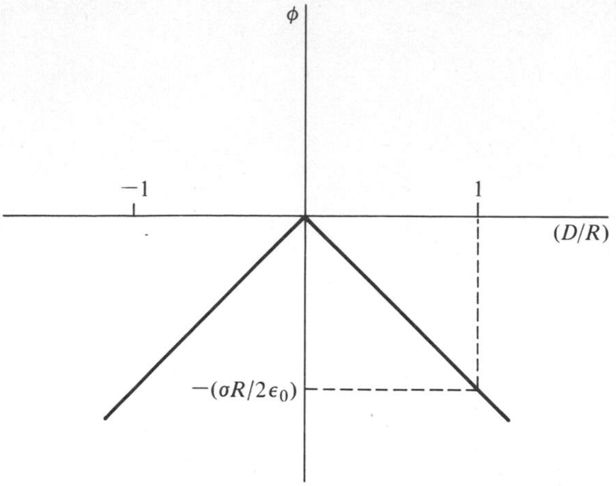

FIGURE 4-9

adapt the previous case to the present one: add a constant to the potential of the previous case to make the zero in potential occur at a finite distance, any constant $\neq 0$. Suppose we select $-\sigma R/(2\epsilon_0)$, so the previous result becomes

$$\phi = \frac{\sigma}{2\epsilon_0}\left[\sqrt{D^2 + R^2} - |D| - R\right]$$

This has no effect on **E** but only shifts the zero of ϕ to the origin. The curve of Fig. 4-8 is shifted downward so that the cusp occurs at the origin and the potential at $D = \infty$ becomes $-\sigma R/2\epsilon_0$.

Letting $R \to \infty$ then goes from the case of the disk to the case of the infinite plane.

$$\phi \to \frac{\sigma}{2\epsilon_0}\left[-|D|\right]$$

This is graphed in Fig. 4-9. The negative of the slope of this curve is the **E** of Fig. 2-11.

It is interesting to compare the **E** and ϕ of this case with those for the infinite uniformly charged line and for the case of the point charge.

	E	ϕ
Point charge	$\frac{1}{r^2}$	$\frac{1}{r}$
Infinite line	$\frac{1}{r}$	$\ln r$
Infinite plane	k	r

PROBLEMS

1 Given the vector field $\mathbf{G} = \hat{\mathbf{x}}x + \hat{\mathbf{y}}y + \hat{\mathbf{z}}z$. Is it possible that this represents a field **E** in electrostatics?

2 Given the vector field $\mathbf{G} = \hat{\mathbf{x}}xyz + \hat{\mathbf{y}}$. Can this represent an electrostatic field?

3 Find the ϕ for Prob. 1 such that $\mathbf{G} = -\nabla\phi$.

4 Given $\mathbf{E} = \hat{\mathbf{x}}(-yz + 2x) - \hat{\mathbf{y}}(zx) - \hat{\mathbf{z}}(xy)$. Is this permissible if **E** is the electrostatic field? If yes, find the ϕ such that $\mathbf{E} = -\nabla\phi$.

5 Find the potential function produced by a spherical distribution of charge having uniform density.

6 Find the potential function produced by a spherical shell of charge, of constant density ρ, extending from r_i to r_o.

7 Find the potential function produced by a spherical surface charge, of constant density σ; use direct integration.

8 A plate of charge of uniform density ρ C/m^3 extends from

$$-L/2 \leqslant x \leqslant L/2, -\infty \leqslant y \leqslant \infty, -\infty \leqslant z \leqslant \infty$$

Find ϕ and **E**.

9 A proton is at the origin. If $\phi = 0$ at $r = \infty$, at what value of r is the potential 1 V?

10 In Prob. 9, how far from the origin would the potential be 1 V if the charge were an electron instead of a proton?

11 In Prob. 9, let another proton be put at the place where the first one produces a potential of 1 V. What is the potential energy of the second proton in the field of the first one?

12 If 1 eV (electron volt) is defined as an energy such that 1 eV is 1.6×10^{-19} J, (*a*) through how many volts must an electron be moved to experience a change in energy of 1 eV? (*b*) Repeat, for a proton. (*c*) How many volts separate the two protons of Prob. 11? (*d*) How many electron volts are needed to take a test charge from $r = 0$ to $r = 1.44 \times 10^{-9}$ m?

13 In Prob. 11, how many joules are expended in moving the second proton from infinity to the point in question at an infinitesimal rate? How many electron volts?

14 An electric charge at one point produces a potential at a second point. Its potential function is also defined at the second point. Is there a distinction between the two concepts? If so, which is a broader one? What use is made of the broader idea?

15 In Prob. 9, suppose $\phi = 0$ at $r = 1$. At what value of r is $\phi = 1$?

16 A charge q is at the origin. What is the potential function produced, expressed in spherical coordinates? In cartesian coordinates?

17 A charge q is at (x_s,y_s,z_s). What is the potential field at (x,y,z)?

18 A charge q_1 is at (x_1,y_1,z_1) and a charge q_2 is at (x_2,y_2,z_2). What is the potential field at (x,y,z)?

19 Given a charge 3.2×10^{-19} C at (2,1), as in Example 1 of Sec. 4-2. (*a*) Find the potential function (the potential field). (*b*) What is the potential at (1,2)? (*c*) Find the field **E**. (*d*) Find the value of **E** at (1,2).

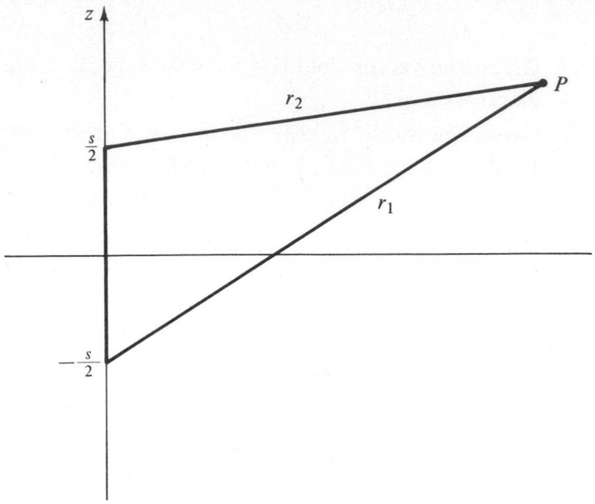

FIGURE 4-10

20 To the charge q_1 of Prob. 19 add a charge $q_2 = -4.8 \times 10^{-19}$ C at (3,2). Find (*a*) the potential field; (*b*) the potential at (1,2); (*c*) the electric field function; (*d*) **E** at (1,2). Compare this method of finding (*d*) with that of Example 4, Sec. 4-2.

21 An infinitely long line charge, with linear density λ C/m, is at $x = -s/2$; another one, with an equal but opposite charge density, is at $x = +s/2$. If $x = 0$ is taken as a ground plane, find the potential field (*a*) in cylindrical coordinates; (*b*) in cartesian coordinates.

22 In Prob. 21, discuss the shape and location of the equipotential surfaces.

23 A line charge of constant density λ extends along the z axis from $z = -s/2$ to $z = s/2$; the total charge is Q. Find the potential function in terms of z and the distances of the test point (field point) from the two ends of the line of charge, r_1 and r_2. See Fig. 4-10.

24 Introduce elliptic coordinates into Prob. 23 by setting

$$u = \tfrac{1}{2}(r_1 + r_2) \qquad v = \tfrac{1}{2}(r_1 - r_2)$$

What are the equipotentials in terms of u and v?

25 Earnshaw's theorem states that an isolated charge cannot be in a state of stable equilibrium in an electrostatic field. Interpret this geometrically for the potential field in the neighborhood of a point P. What about unstable equilibrium?

26 Derive Green's theorem,

$$\int_{\mathcal{V}} (\phi_1 \nabla^2 \phi_2 - \phi_2 \nabla^2 \phi_1)\, d\tau = \int_{\Sigma} (\phi_1 \nabla \phi_2 - \phi_2 \nabla \phi_1) \cdot d\mathbf{S}$$

from Gauss' divergence theorem.

27 If 3 million V/m is the maximum value of **E** in air (average, or rule or thumb) permissible before the onset of ionization, find the maximum potential to which an isolated sphere (*a*) 1 m in radius may be charged. (*b*) 1 mm in radius?

28 A charge $+2q$ is at (0,0) and a charge $-q$ is at $(L,0)$. Find the point on the x axis to the right of L where the potential has a maximum along the x axis. Find $\partial^2 \phi/\partial y^2$ and $\partial \phi/\partial y$ there also. Is the potential a maximum along the y axis?

4-2 CAPACITANCE

Consider a system of two conductors of arbitrary shape, one having a charge $+q$ and the other a charge $-q$, all other bodies being far removed. In electrostatics there is no current flow; so $\mathbf{E}_{\text{tangential}} = 0$ on the surface of the metallic bodies and each of the two conductors has a surface which is an equipotential. The net charge on each body resides entirely on its surface; thus, all the lines of **E** that originate on the positive body terminate on the negative body. By Gauss' law the flux of **E** that leaves the positive body is q/ϵ_0, the same in magnitude as the flux of **E** that terminates on the negative body.

If an additional charge is added to one body then some lines of **E** will be produced that must either terminate on other bodies or else go off to infinity. But if, when an additional charge Δq is given to the positive body, there is an equal and opposite charge $-\Delta q$ given to the negative body, then no additional lines of **E** in other regions will be created; the only effect will be to increase the flux of **E** between the two bodies. It is then reasonable to assume (and this is, indeed, found to be true experimentally) that the added charge Δq will distribute itself on the surface in the same way that the original q was distributed: the σ at any point on the surface will be multiplied by some constant factor, independent of position. This means that the distribution of the lines of **E** will be left unchanged, the intensity being simply multiplied by the same factor f and the difference in potential between the two bodies–the line integral of **E** between them–will also be multiplied by this factor. But if $q \to fq$ and $\Delta\phi \to f\,\Delta\phi$, where f is a constant factor, it follows that $\Delta\phi$ (i.e., $\phi_1 - \phi_2$) must be proportional to q:

$$\boxed{q = C\Delta\phi} \qquad \text{(E-10)}$$

where C is a constant that is applicable to this system.

The constant of proportionality C is called the capacitance of the system and is measured in terms of a unit, the farad, which is simply a short name for 1 C/V. Since 1 C is 1 s A and 1 V = 1 $\text{m}^2\ \text{kg}\ \text{s}^{-3}\ \text{A}^{-1}$, the farad is actually 1 $\text{m}^{-2}\ \text{kg}^{-1}\ \text{s}^4\ \text{A}^2$. Comparing this with the unit of permittivity ϵ_0 it is seen that ϵ_0 is measured in farads per meter.

The farad is too large a unit for most practical applications and it is customary to use either the microfarad (1 μF = 10^{-6} F) or the picofarad (1 pF = 10^{-12} F). In the esu-cgs system (it is only one of three common cgs systems, and there are others) the unit of capacitance has almost the same value as the picofarad, but it has a different dimension, [cm]: the same dimension as that of the cgs unit of length. There is nothing wrong with this—torque and work also have the same dimensions yet are completely distinct concepts—but it is a possible source of confusion. Consequently the cgs unit, the cm, is rarely employed for capacitance.

Electrical devices which are specifically designed to have a given capacitance are known as capacitors, and they play an important role in electronics. Many electric circuits may be analyzed exclusively in terms of capacitance, resistance, and inductance. The underlying reason for this is that all three are connected in one way or another with energy, one of the basic concepts of physics. A capacitor, as will be seen in the next section, stores electrical energy; a resistor dissipates electrical energy as heat when current flows through it; an inductor stores magnetic energy.

The capacitance defined above, that between two conductors in vacuum, depends exclusively on the geometry. When this definition is extended to include other common insulating (i.e., nonconducting) substances in the space between the conductors, it is found that the capacitance depends also on the nature of the medium, in the form of a simple multiplicative factor. Whatever the medium, capacitance is an electrostatic concept. It is a fact, nevertheless, that the concept of capacitance can be linked to that of resistance, which is a dynamic concept connected with the flow of charge. This connection is the basis of the electrolytic tank method for determining capacitance by measuring the resistance between two objects when they are immersed in an electrolyte. And, indeed, it turns out that the differential equations that must be satisfied by **E** and by **I** in the two cases are similar, so that with identical boundary conditions the solutions are also similar. This would indicate only a formal connection between two disparate concepts, dynamic R and static C. But a study of the charging process itself shows that—guided by a series of transient electromagnetic waves propagating between the bodies—the capacitance is only the final result of a dynamic process, and is not the static quantity it appears to be. This is considered in detail in a subsequent chapter.

Examples

1. A charge $+q$ is placed on the surface of a metallic sphere of radius R_1; a charge $-q$ resides on the inner surface of a concentric sphere of larger radius R_2. The capacitance of this spherical capacitor will be found.

Whether the inner sphere is solid, or a shell of finite thickness, or a mathematical surface of zero thickness, the value of **E** at any point with $r < R_1$ is zero under electrostatic conditions. This can be shown by application of Gauss' law to a fictitious sphere of radius sphere of radius $r < R_1$: since the charge inside this sphere is zero, so is the flux of **E** on the surface, and **E** itself must then be zero because of the symmetry.

For the region $r > R_2$ application of Gauss' law again yields the same result, whether the point is within the metal of the outer sphere or completely outside the shell. The electrostatic field therefore exists only between the outer surface of the inner sphere and the inner surface of the outer sphere.

$$\phi_1 - \phi_2 = -\int_{\phi_1}^{\phi_2} d\phi = -\int_{R_1}^{R_2} \nabla\phi \cdot d\mathbf{r} = \int_{R_1}^{R_2} \mathbf{E} \cdot d\mathbf{r}$$

$$= \frac{q}{4\pi\epsilon_0}\int_{R_1}^{R_2} \frac{dr}{r^2} = \frac{q}{4\pi\epsilon_0}\left(\frac{1}{R_1} - \frac{1}{R_2}\right)$$

Then

$$C = \frac{q}{\phi_1 - \phi_2} = \frac{4\pi\epsilon_0}{\left(\frac{1}{R_1} - \frac{1}{R_2}\right)} = 4\pi\epsilon_0\left(\frac{R_1R_2}{R_2 - R_1}\right)$$

2. The capacitance of a system consisting of two plane conductors, each of similar area A, separated by a distance d.

This result may be obtained, approximately, from the previous case. Suppose

$$R_2 = R + \frac{d}{2} \qquad R_1 = R - \frac{d}{2}$$

R is the (arithmetic) mean radius and d is the separation between the plates. Let $R \to \infty$ with d constant. Then $R_2 - R_1 = d$, $R_1R_2 \to d^2$, and $C \to 4\pi\epsilon_0 R^2/d = \epsilon_0 A_{\text{sphere}}/d$. If the areas to be considered are A instead of A_{sphere}, where A is only a part of A_{sphere}, then the capacitance will be reduced proportionately:

$$C = \frac{\epsilon_0 A}{d}$$

The reason this result is only approximately true is that conditions change when the capacitor of area A, originally considered part of a spherical capacitor, is disconnected and removed to exist by itself. As part of the spherical capacitor, all the lines of **E** are radial. This remains true if an infinitesimal gap is cut in the surface of each sphere surrounding the area A, provided the potential of each part of a sphere is kept at the same value. When the capacitor of area A is removed from the two spheres, however, the lines of **E** do not remain unaltered.

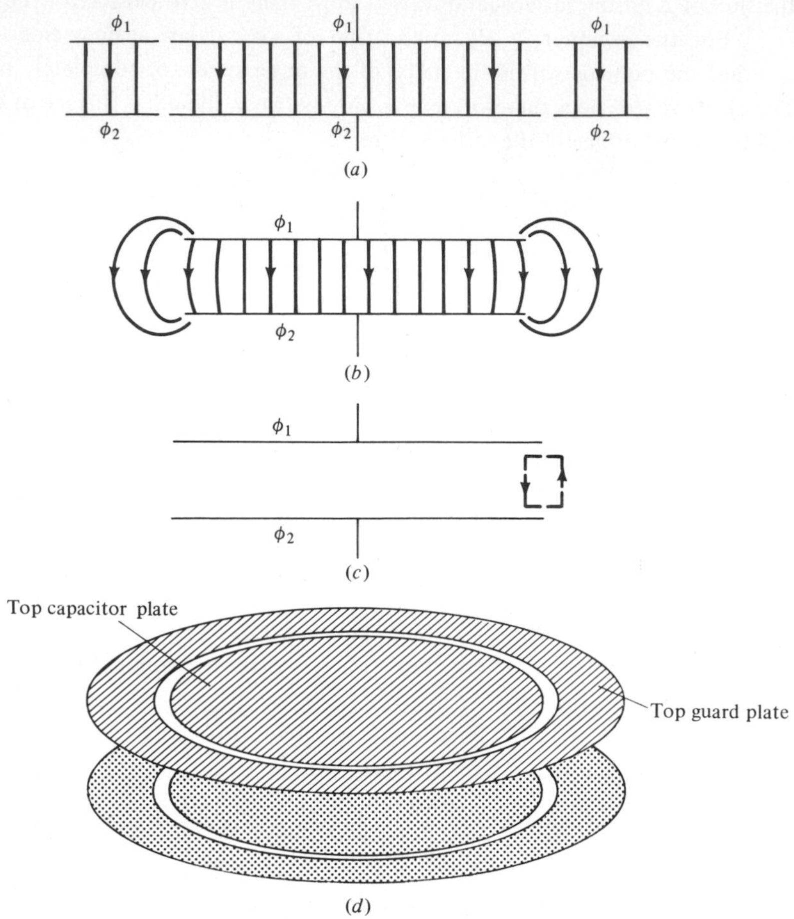

FIGURE 4-11

Figure 4-11*a* shows how the lines of **E** would look if the capacitor, of area A, were part of a spherical capacitor with very large radii. Figure 4-11*b* shows the fringing of the lines of **E** near the edges of the two plates when there is no surrounding metal. Here not all the charge on a plate remains uniformly distributed on the surface facing the other plate; some even resides on the outer surfaces.

The fact that **E** cannot abruptly become zero at the edge of the plates becomes clear if the circulation of **E** is taken along the dotted rectangle of Fig. 4-11*c*. Since $\nabla \times \mathbf{E} = 0$ the integral must vanish. But this condition cannot be satisfied if there is a nonvanishing contribution from the long side of the rectangle inside the capacitor edge and zero contributions from the other sides. Since there is no contribution from $\mathbf{E} \cdot d\mathbf{r}$ along the short sides of the rectangle, where $\mathbf{E}$ is perpendicular to $d\mathbf{r}$, there must be a contribution from the region outside the edge of the capacitor plates.

The formula $\epsilon_0 A/d$ becomes a better and better approximation to the actual results as the separation of the plates gets smaller and smaller compared to the dimensions of the plates. For most applications the simple formula above may be accepted as valid. In accurate experimental work (as in the calibration of standards) guard plates such as those shown in Fig. 4-11*d* are constructed around the actual plates of the capacitor and are kept at the same potentials as those of the plates themselves.

3. The use of capacitance to test Coulomb's law The following abstract, reprinted from *The Physical Review, D* (August 1, 1970), is from the article, "Experimental Test of Coulomb's Law," by D. F. Bartlett, P. E. Goldhagen, and E. A. Phillips.

One of the classic "null experiments" tests the exactness of the electrostatic inverse-square law. The outer shell of a spherical capacitor is raised to a potential V with respect to a distant ground, and the potential difference ΔV induced between the inner and outer shells is measured. If this induced potential difference is not zero, Coulomb's law is violated. For example, if we assume that the force between charges varies as r^{-2+q}, then $\Delta V/V$ is approximately a tenth of q. In our experiment five concentric spheres are used. A potential difference of 40 kV at 2500 Hz is impressed between the outer two spheres. A lock-in detector with a sensitivity of about 0.2 nV measures the potential difference between the inner two spheres. We find $|q| \leqslant 1.3 \times 10^{-13}$. We also find comparable limits on the detected signal when the operating frequency is 250 Hz, and when the detector is synchronized with the charging current rather than with the charge itself.

PROBLEMS

1 What is the capacitance of a parallel-plate capacitor if the area is a square, $1 \times 1\ \text{cm}^2$, and the separation of the plates is 1 mm?

2 What is the capacitance of an isolated metal sphere? (Consider the second body to be a sphere at infinity. The lines of $\mathbf{E}$ which start on the isolated sphere may be considered to terminate on an equal and opposite charge at infinity, to give a capacitance. If, as in the gravitational analog, they are considered to die away, as $r \to \infty$, in such a way that $\nabla \cdot \mathbf{E} = 0$, and there is no charge at infinity, then the capacitance may still be defined by $Q/|\phi_1 - \phi_2|$ even though the Q only exists on one conductor.)

3 Two strips of metal and two strips of insulation, each 2 cm wide by 10 m long, are arranged one on top of the other in the form of a double sandwich–metal A,

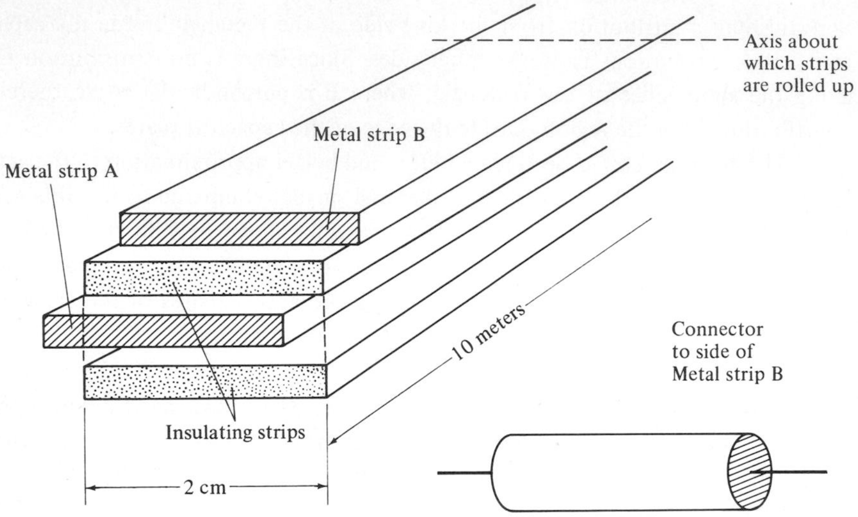

FIGURE 4-12

insulation 1, metal B, insulation 2—and rolled up along an axis parallel to the width. Metals A and B are slightly displaced, relative to each other, in the opposite sense along this axis so that a circular plate may be soldered to each for the electrical connections. See Fig. 4-12. If the insulation thickness is 0.1 mm, what is the capacitance? Assume the effects of the insulators are the same as those of vacuum.

4 A strip of aluminum 2 cm wide by 10 m long has a layer of aluminum oxide formed on both faces, the thickness of the oxide being 10^{-7} m. A layer of aluminum is then evaporated on top of one insulating face and the strip is rolled up to form a capacitor. What is the capacitance?

5 One capacitor, with terminals A and B, has capacitance C_1; another, with terminals D and E, has capacitance C_2. A and D, connected together, serve as one terminal of a combined arrangement; B and E, connected together, serve as the other terminal. Derive the capacitance of this arrangement.

6 In Prob. 5, let A serve as one terminal while B is connected to D, leaving E for the other terminal. What is the capacitance of this arrangement?

7 Find the capacitance of a coaxial cylindrical capacitor of length L and radii R_1 and R_2. Neglect fringing at the ends.

8 Arrangement (a) of three similar capacitors puts a parallel combination of two of them in series with the third one. Arrangement (b) puts a series combination of two of them in parallel with the third one. Which arrangement has the larger capacitance?

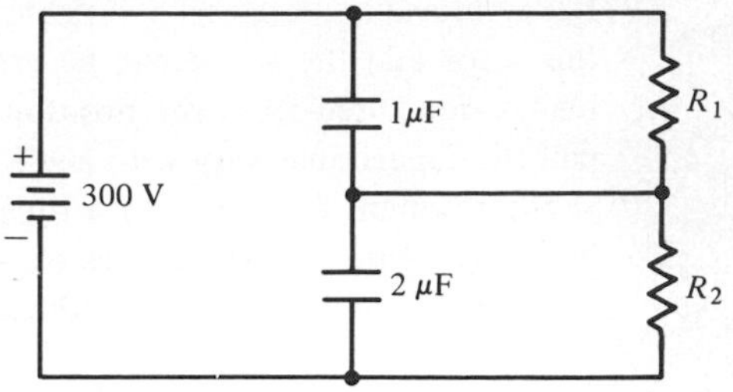

FIGURE 4-13

9 Using the results of Probs. 21 and 22 in the previous section, find the capacitance per unit length between an infinitely long cylinder of diameter d and a parallel plane whose distance from the central axis of the cylinder is $s/2$.

10 Using the results of Prob. 9, find the capacitance per unit length between two infinitely long parallel cylinders, each of diameter d, if the distance between the axes of the cylinders is s.

11 Using the results of Probs. 23 and 24 in Sec. 4-5, find the capacitance of a prolate ellipsoid of revolution, with semimajor axis a, having the distance s between the two foci.

12 A capacitor C_1 is charged to the potential ϕ; another capacitor C_2 is uncharged. C_1 and C_2 are connected together, terminal to terminal. (There is no distinction, here, between series and parallel connections.) Find the resultant charge on each capacitor.

13 A 1-μF and a 2-μF capacitor are connected in series across a potential difference of 300 V. What is the potential difference across each capacitor, if both are considered to be ideal capacitors?

14 Suppose the conditions of Prob. 13 are modified by putting two resistors R_1 and R_2 in parallel with the capacitors, as shown in Fig. 4-13. In actual practice the resistors are not deliberately added but are introduced by the leakage resistances of the two nonideal capacitors. What is the potential difference across each capacitor? Is this answer at all related to that of the previous case? Note that, even if the capacitors are ideal, so that $R_1 = R_2 = \infty$, any attempt to measure the potential differences with vacuum tube voltmeters (having a high but finite input impedance) in effect puts R_1 and R_2 across the capacitors.

15 What is the capacitance of the earth?

16 A parallel-plane capacitor has a spacing d between its plates. A thin metal foil, of thickness t where $t < d$, is inserted between the plates. What is the effect on the capacitance?

17 What is the capacitance of the earth in cgs units? Is there any relation to the radius of the earth?

18 Two parallel plates are 1 mm apart and each is a square of side ℓ m. Find ℓ if the capacitance is 1 F. Give the answer both in meters and miles.

19 If the movable plates on a variable capacitor are semicircular the area common to the rotor and the stator will be proportional to the angle of rotation of the rotor blades, measured from the position where the rotor plates are just disengaged. How will the capacitance vary with angle of rotation?

20 If the resonant frequency of a tuned circuit is $f = 1/2\pi\sqrt{LC}$, where L is a constant in a given case, can the blades of the rotor of a variable capacitor be shaped such that f will vary linearly with θ, the angle of rotation?

4-3 ELECTROSTATIC ENERGY

If the charges on the two conductors of a parallel-plate capacitor are originally q and $-q$, a transfer of charge dq from the second to the first leaves the plates charged $(q + dq)$ and $-(q + dq)$. The potential difference $(\phi_1 - \phi_2)$ between the plates requires the expenditure of work, $dW = dq(\phi_1 - \phi_2)$, to accomplish this transfer. Then

$$dW = \frac{q}{C}\,dq$$

To bring the capacitor from an uncharged state to one with the final charges Q and $-Q$ on the plates requires the work

$$W = \int_0^Q \frac{q}{C}\,dq = \frac{Q^2}{2C}$$

If the transfer is done at an infinitesimal rate this work is all converted into the potential energy

$$U = \frac{Q^2}{2C} = \frac{1}{2}C(\phi_1 - \phi_2)^2 = \frac{1}{2}Q\,|\phi_1 - \phi_2|$$

where ϕ_1 and ϕ_2 are the final potentials of the two conductors. The capacitor can then be said to store the potential energy U. Where does this potential energy reside? In electrostatics various answers have been presented to this question although all agree concerning the result in the case of radiation, where the field in a region in space is effectively uncoupled from the source charges that produced the field. It is instructive to consider the different possibilities that present themselves in electrostatics.

1 One viewpoint is that it is not correct to localize the electrical potential energy in space, any more than one can localize the mechanical potential energy when a ball is lifted above the floor. The potential energy, produced by the relative position of one object with respect to another, is simply due to the configuration and cannot be said to reside in any particular region. Only the total energy can be measured (by means of the work done) and the assignment of certain portions of the total energy to various regions has no physical significance.

2 Quite a different view holds that the energy resides in the charges. This view is reinforced by a calculation of the work needed to assemble a collection of point charges at an infinitesimal rate into some definite configuration. The work W_1 needed to bring the charge Q_1 to $\mathbf{r}_1$ from infinity when no other charges are present is zero. To bring Q_2 to $\mathbf{r}_2$ (when Q_1 is at $\mathbf{r}_1$) requires the work

$$W_2 = \int_\infty^{r_{12}} -\mathbf{F} \cdot d\mathbf{r} = -\int_\infty^{r_{12}} \frac{1}{4\pi\epsilon_0} \frac{Q_1 Q_2}{r^2}\, dr = \frac{1}{4\pi\epsilon_0} \frac{Q_1 Q_2}{r_{12}}$$

where $r_{12} = |\mathbf{r}_1 - \mathbf{r}_2|$. Here $\mathbf{F}$ is the Coulomb force and $-\mathbf{F}$ is the mechanical force just needed to overcome it if the motion is at an infinitesimal rate. To bring Q_3 to $\mathbf{r}_3$ (when Q_1 is at $\mathbf{r}_1$ and Q_2 is at $\mathbf{r}_2$) requires the work

$$W_3 = \frac{Q_3}{4\pi\epsilon_0}\left(\frac{Q_1}{r_{13}} + \frac{Q_2}{r_{23}}\right)$$

Continuing in this fashion, the total work required to assemble the N charges and, consequently, the potential energy of the final configuration is

$$\begin{aligned} U = W &= [0] + \left[\frac{Q_2}{4\pi\epsilon_0}\left(\frac{Q_1}{r_{12}}\right)\right] + \left[\frac{Q_3}{4\pi\epsilon_0}\left(\frac{Q_1}{r_{13}} + \frac{Q_2}{r_{23}}\right)\right] + \cdots \\ &= \frac{Q_1}{4\pi\epsilon_0}\left[\frac{Q_2}{r_{12}} + \frac{Q_3}{r_{13}} + \cdots\right] + \frac{Q_2}{4\pi\epsilon_0}\left[\frac{Q_3}{r_{23}} + \frac{Q_4}{r_{24}} + \cdots\right] + \cdots \\ &= \frac{1}{4\pi\epsilon_0} \sum_{i=1}^{N} \sum_{j>i} \frac{Q_i Q_j}{r_{ij}} \\ &= \frac{1}{8\pi\epsilon_0} \sum_{i=1}^{N} \sum_{j=1}^{N}{}' \frac{Q_i Q_j}{r_{ij}} \end{aligned}$$

where, in the last expression, the prime on the sigma means that the summation is to exclude the term with $j = i$. This arrangement of point charges does not include the self-energies of the individual charges. (The self-energy of the point charge Q_1 is the work needed to bring the constituent individual charges, comprising Q_1, to $\mathbf{r}_1$ from infinity.)

The last two expressions above correspond to the two arrangements shown in the charts below. The $\frac{1}{2}$ factor enters because the value in the ith row, jth column is the same as the in the jth column.

i \ j	1	2	3	4	5	6
1		✓	✓	✓	✓	✓
2			✓	✓	✓	✓
3				✓	✓	✓
4					✓	✓
5						✓
6						

i \ j	1	2	3	4	5	6
1		✓	✓	✓	✓	✓
2	×		✓	✓	✓	✓
3	×	×		✓	✓	✓
4	×	×	×		✓	✓
5	×	×	×	×		✓
6	×	×	×	×	×	

A corresponding expression is obtained for a continuum distribution: The last equation above may be written

$$W = \frac{1}{2} \sum_{i=1}^{N} Q_i \sum_{j=1}^{N}{}' \frac{1}{4\pi\epsilon_0} \frac{Q_j}{r_{ij}}$$

The second sum represents the potential at $\mathbf{r}_1$ produced by all the other charges, not including the effect of Q_i. So

$$W = U = \frac{1}{2} \sum_{i=1}^{N} Q_i \phi_i'$$

where ϕ'_i omits the contribution of Q_i. Going over to the continuum distribution gives

$$W = \frac{1}{2} \int \rho\phi' \, d\tau$$

Now, however, instead of Q_i—a finite charge—at $\mathbf{r}_i$ there is only $\rho \, d\tau$—an infinitesimal charge; consequently, letting $\phi' \to \phi$, where ϕ is the total potential at $\mathbf{r}_i$ (including the effect of $\rho \, d\tau$), does not change the value of the integral:

$$\boxed{U = \frac{1}{2} \int \rho\phi \, d\tau} \qquad \text{(E-11)}$$

At a place where $\rho = 0$ there is, then, no contribution to U; for discrete charges, similarly, at a place where $Q_i = 0$ there is no contribution to U. Thus, if no charge at some point means no contribution to the energy, while a charge there does mean a contribution to the energy, one may say that the energy resides in the charge.

3 On the other hand, from Eq. (E-11) one could equally well say that at a place where $\phi = 0$ there is no contribution to U, while elsewhere the contribution is directly proportional to ϕ. So the potential energy resides in the potential field. Note, however, that without charge there is no energy; just as in the previous case there is no energy without ϕ. The two are intertwined.

4 Still another viewpoint possible in electrostatics is that the potential energy resides in the field. Substitute, for the case of the parallel-plane capacitor,

$$C = \frac{\epsilon_0 A}{d} \quad \text{and} \quad \phi_1 - \phi_2 = E\,d$$

into

$$U = \tfrac{1}{2}\, C(\phi_1 - \phi_2)^2$$

Then

$$U = \tfrac{1}{2}\, \epsilon_0\,(Ad)E^2$$

Letting $u = U/(Ad)$ be the density of the potential energy, this gives

$$u = \tfrac{1}{2}\, \epsilon_0 E^2$$

and

$$U = \int u\, d\tau = \int \frac{1}{2}\, \epsilon_0 E^2\, d\tau$$

A broader significance is obtained for this expression when the general equation for W derived above is found to lead to the same result. Starting with

$$U = \frac{1}{2}\int \rho\phi\, d\tau$$

and substituting for ρ from $\nabla \cdot \mathbf{E} = \rho/\epsilon_0$ gives

$$U = \frac{\epsilon_0}{2}\int \phi\,(\nabla \cdot \mathbf{E})d\tau$$

With the vector identity $\nabla \cdot (\phi\mathbf{E}) = \phi\,(\nabla \cdot \mathbf{E}) + \mathbf{E} \cdot \nabla\phi$ this becomes

$$U = \frac{\epsilon_0}{2}\int \nabla \cdot (\phi\mathbf{E})\, d\tau - \frac{\epsilon_0}{2}\int \mathbf{E} \cdot \nabla\phi\, d\tau$$

By Gauss' divergence theorem the first term equals

$$\frac{\epsilon_0}{2}\oint_{\Sigma} (\phi\mathbf{E}) \cdot d\mathbf{S}$$

where Σ, a surface that encloses all the charge density ρ, may be taken arbitrarily far away. It will be seen later that at great distances $\phi \propto 1/r$ and $E \propto 1/r^2$. (This is certainly plausible, because far from the source charges the distribution behaves essentially like a point charge.) But $dS \propto r^2$, so if Σ is taken to be a sphere the integral approaches the value zero at large distances. Then U is given by the second integral alone:

$$\boxed{U = \frac{1}{2}\int_{\text{all space}} \epsilon_0 E^2 \, d\tau} \tag{E-12}$$

This is the same result obtained above for the special case of the capacitor. From this viewpoint it is seen that there is no contribution to U at points where $E = 0$.

Which viewpoint is preferable? These different views on the localization of the potential energy all lead to the same result in any electrostatic case; so they are all equally valid here, despite the fact that they seem to contradict each other. In electrostatics the field and the sources are effectively tied tightly together and the system must be treated as a whole. In the case of radiation fields, however, a region in space may effectively be considered as separated from its sources, and it will be seen subsequently that the field viewpoint, alone, is the correct one. The simplest view to adopt is the one applicable in all cases: the field viewpoint of Faraday and Maxwell. Accordingly, this is our view: in all cases the electric potential energy density is $\frac{1}{2}\epsilon_0 E^2$ at each point in space.

It should be emphasized that the result $u = \frac{1}{2}\epsilon_0 E^2$ applies only to an isolated region of space, such as a charged parallel-plate capacitor which is not connected to a battery. If the battery is connected the results are different. (Though they can be made, effectively, the same.) A comparison between these two cases is of interest.

ISOLATED CASE Fig. 4-14*a* shows a parallel-plate capacitor in which one plate, originally in equilibrium under the action of two forces, is imagined to undergo a displacement with the battery disconnected, all physical laws being obeyed. The potential electric energy is $U_e = Q^2/2C$, with Q constant. After the virtual displacement (here an increase of the plate separation) the energy is increased by the amount δU_e or $-(Q^2/2C^2)\delta C = \delta u(A\,\delta x)$. It is left to Prob. 14 to show that $\delta u = \frac{1}{2}\epsilon_0 E^2$, in agreement with the previous results. In this case the work δW_{mech} performed in the virtual displacement by F_{mech}—the mechanical restraining force that balances F_{elec}, the electric attractive force—is all converted into δU_e, the increase in the potential electric energy. This is shown diagrammatically in Fig. 4-14*b*; it follows from the conservation of energy. The calculation of F_{mech} and F_{elec} is performed in the next section.

NONISOLATED CASE Suppose the battery is connected to the capacitor before, during, and after the virtual displacement. Here the potential difference rather than the

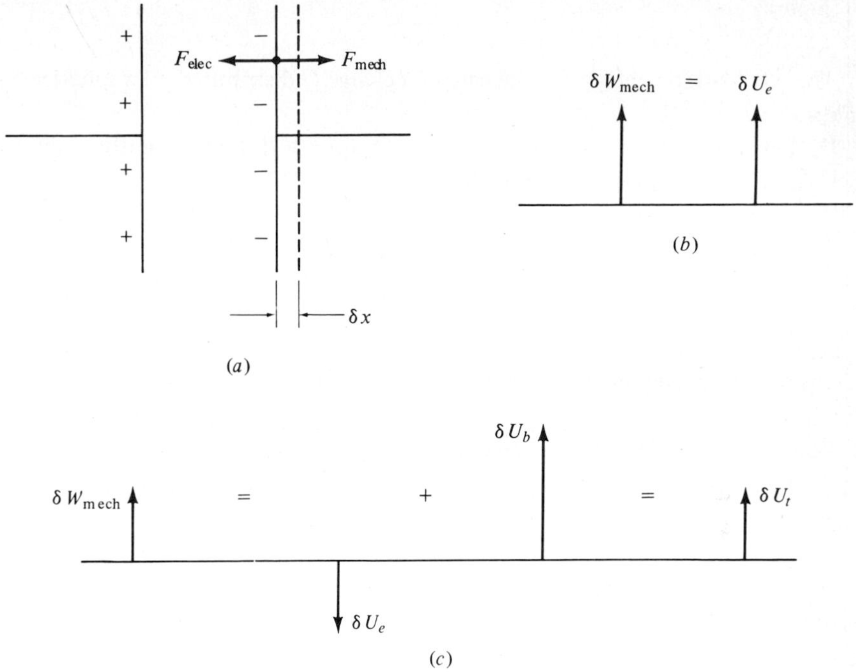

FIGURE 4-14

charge remains the same: $U_e = \frac{1}{2} C(\phi_1 - \phi_2)^2$, with $(\phi_1 - \phi_2)$ constant. Then

$$\delta U_e = \tfrac{1}{2} (\phi_1 - \phi_2)^2 \,\delta C = (Q^2/2C^2)\delta C$$

This has the same magnitude but the opposite sign of the result in the previous case. The potential electric energy of the field decreases after the separation of the plates. To satisfy the conservation of energy it is necessary to take account of the chemical energy of the battery: the battery is charged by the virtual displacement of the plate. The increase in battery energy δU_b may be shown (Prob. 16) to equal $-2\,\delta U_e$. This is the sum of the mechanical work and the magnitude of the change of the field energy: $\delta U_b = -\delta U_e + \delta W = -2\,\delta U_e$, as indicated in Fig. 4-14*c*.

If we set $U_t = U_e + U_b$, then $\delta U_t = \delta U_e + \delta U_b = \delta U_e + (-2\,\delta U_e) = -\delta U_e$ gives the change in the total electric energy: field plus battery. Any changes in the energy density $\frac{1}{2}\epsilon_0 E^2$ apply to this total energy U_t rather than to the field energy U_e alone. The nonisolated system of the capacitor itself is replaced by the isolated system of capacitor and battery, the battery energy being distributed throughout the field. If U_t is substituted for U_e the result here is then similar to that of the previous case.

Examples

1. The electrostatic energy for the case of a charge Q distributed on a spherical surface of radius R.

Here $E = 0$ for $r < R$ while $E = Q/4\pi\epsilon_0 r^2$ for $r \geqslant R$. Consequently, only the volume external to the sphere stores energy. Then

$$U = \frac{\epsilon_0}{2} \int_R^\infty \frac{Q^2}{(4\pi\epsilon_0)^2 r^4} 4\pi r^2 \, dr = \frac{Q^2}{8\pi\epsilon_0 R}$$

This is the self-energy of the spherical surface charge.

2. The electrostatic energy for the case of a point charge.

Here $R = 0$. From the previous example, $U = \infty$. The self-energy of a point charge is infinite.

3. Assume an electron to be a sphere of radius a, with the charge uniformly distributed on the surface. Set the electrostatic energy equal to the rest energy of the electron. What is a?

$$\frac{e^2}{8\pi\epsilon_0 a} = m_0 c^2 \qquad \text{or} \qquad a = \frac{1}{8} \frac{e^2}{\pi\epsilon_0 m_0 c^2}$$

So

$$a = \frac{(9 \times 10^9)(1.6 \times 10^{-19})^2}{2(9.11 \times 10^{-31})(3 \times 10^8)^2} = 1.4 \times 10^{-15} \text{ m}$$

No experimental probing to distances less than this within an electron has yet taken place. It is considered plausible that an electron may be a true fundamental particle—a point particle having no structure. This view is made fuzzy by a quantum fact: the electron is continually surrounded by a cloud of virtual photons. In the case of the proton the results are similar, yet different: the proton is surrounded by a cloud of virtual mesons. But a calculation, similar to the one above, yields unrealistic results for protons. The proton has been probed experimentally, using electrons to explore the structure of the interior.

4. If the energy density is taken to be $u_1 = \frac{1}{2}\epsilon_0 E_1{}^2$ for one charge distribution and

$$u_2 = \tfrac{1}{2}\epsilon_0 E_2{}^2$$

for a second, the energy density when both charge distributions are present would be $u = \frac{1}{2}\epsilon_0(E_1 + E_2)^2$. So $u \neq u_1 + u_2$. Does the superposition principle apply to energy?

$U_1 + U_2 = (\epsilon_0/2) \int (\mathbf{E}_1{}^2 + \mathbf{E}_2{}^2)\, d\tau$ is the total energy stored in the separate, individual fields; here $\mathbf{E}_1$ is produced by $\rho_1(\mathbf{r})$, while $\mathbf{E}_2$ is produced by $\rho_2(\mathbf{r})$. In the combined field the energy stored is $U_{12} = (\epsilon_0/2) \int (\mathbf{E}_1 + \mathbf{E}_2)^2\, d\tau$, since $\mathbf{E}_1 + \mathbf{E}_2$ is the combined field produced by $\rho_1(\mathbf{r})$ and $\rho_2(\mathbf{r})$ acting together. The difference between these two potential energies is

$$\Delta U = U_{12} - (U_1 + U_2) = \epsilon_0 \int \mathbf{E}_1 \cdot \mathbf{E}_2\, d\tau = \epsilon_0 \int \nabla\phi_1 \cdot \nabla\phi_2\, d\tau$$

In the vector identity $\nabla \cdot (a\mathbf{A}) = a\nabla \cdot \mathbf{A} + \nabla a \cdot \mathbf{A}$ set $a = \phi_1$ and $\mathbf{A} = \nabla\phi_2$; then $\nabla \cdot (\phi_1 \nabla\phi_2) = \phi_1 \nabla^2 \phi_2 + \nabla\phi_1 \cdot \nabla\phi_2$. Substituting into ΔU gives

$$\Delta U = \epsilon_0 \int \nabla \cdot (\phi_1 \nabla \phi_2)\, d\tau - \epsilon_0 \int \phi_1 \nabla \cdot \nabla\phi_2\, d\tau$$

$$= \epsilon_0 \oint_\Sigma (\phi_1 \nabla \phi_2) \cdot d\mathbf{S} + \epsilon_0 \int \phi_1 \nabla \cdot \mathbf{E}_2\, d\tau$$

If Σ is taken to be of very large radius, the first integral may be made arbitrarily small. Then

$$\Delta U = \int \phi_1(\mathbf{r})\, \rho_2(\mathbf{r})\, d\tau$$

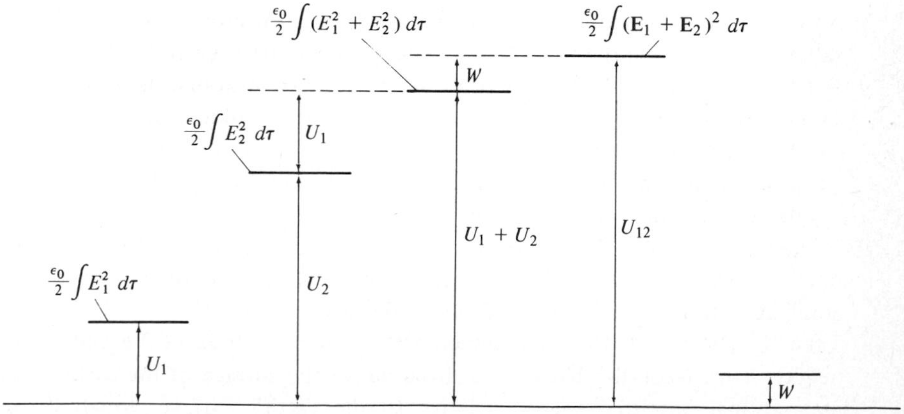

FIGURE 4-15

But this is just the work required to bring the second charge distribution $\rho_2(\mathbf{r})$ into the potential field $\phi_1(\mathbf{r})$ created by the first charge distribution.

That the energy densities of the two fields are not additive is due to the work required to bring the charge density which produces $\mathbf{E}_2$ into the field $\mathbf{E}_1$. This situation is illustrated in Fig. 4-15.

PROBLEMS

1 A battery charges a parallel plate capacitor to potential difference V and is then disconnected from the capacitor. The spacing between the plates is next increased from d to $3d$. By what factor is the potential energy increased?

2 A capacitor C_1 charged to the potential difference ϕ and an uncharged capacitor C_2 are connected together—terminal a_1 to a_2, terminal b_1 to b_2. Compare the initial and final potential energies. Explain this, in view of conservation of energy.

3 How much energy is stored in a 1-μF capacitor charged to a potential difference of 1000 V?

4 If the capacitor of Prob. 3 could be discharged at a constant rate, what would be the behavior of the potential difference as a function of time?

5 There is a maximum value of $\mathbf{E}$ for which an insulator remains nonconducting. Above this value the medium becomes conducting; though possibly only temporarily, as with lightning in air. This maximum value of $\mathbf{E}$ is called the breakdown field. If the electrical breakdown field of air is 3×10^4 V/cm, what is the maximum energy density which an air capacitor may have?

6 In Example 3 assume the charge density of the electron to be uniformly distributed throughout the volume of a sphere instead of being concentrated on the surface only. What is the radius of the electron then?

7 An electron is placed at each corner of a cube of side 1 Å. What is the potential energy?

8 Add a positive charge Ze to the center of the cube of Prob. 7. What is the smallest value of Z for which the total potential energy becomes negative?

9 A 12-V electrochemical battery has a rating of 1000 A-hours. Its size is 8 by 10 by 12 in. How much electrochemical energy is stored per cubic inch, assuming that the battery terminal potential difference (*a*) remains 12 V during the entire time the current is withdrawn, (*b*) drops linearly to 0 V during this period?
(*c*) Which assumption is more nearly correct?

10 If the capacitor of Prob. 3 is a cylinder with a $\frac{1}{2}$-in. diameter and is 2 in. long, how much energy is stored per cubic inch? How does this electrostatic energy density compare with the electrochemical energy density of Prob. 9?

11 Take the density of the lead storage battery of Prob. 9 as 11.3 g cm^{-3}. To what height, h m, must the battery be lifted above the surface of the earth so that its gravitational potential energy, relative to the earth's surface, equals its electrochemical potential energy? Use two different expressions for the gravitational potential energy.

12 What is the electrical energy stored, per unit length, on an infinitely long conductor of radius R carrying λ C/m?

13 Integrate $u\,d\tau$ for a point charge to obtain the self-energy of an isolated point charge. (The energy of a finite point charge here is actually being calculated from the mutual interaction energy of an infinite number of infinitesimal point charges, all at the same point.)

14 Provide the proof, mentioned in the discussion of the isolated capacitor, that if δU is $-(Q^2/2C^2)\delta C = \delta u(A\ \delta x)$ then $\delta u = \frac{1}{2}\epsilon_0 E^2$.

15 The surface of the earth is charged negatively by lightning storms which occur, on the average, somewhere on earth, each 5 min. The potential in the air increases positively, to a height of roughly 50 km, and a typical value of E is 100 V/m. Taking the earth's radius as 6400 km what is the electrostatic potential energy in the earth's field between sea level and 50 km?

16 (*a*) Prove that the increase of the battery energy, in the nonisolated capacitor case discussed in the text, is twice the magnitude of the change of the field energy alone and of opposite sign.

(*b*) An uncharged capacitor is connected to a constant-voltage battery V, thereby transferring a charge Q from battery to capacitor. What is the energy supplied by the battery?

(*c*) What is the energy acquired by the capacitor?

(*d*) What happens to the difference in these two energies? Consider the cases when there is resistance in series with C and when there is no resistance.

4-4 ELECTROSTATIC FORCES

The electric forces exerted by one plate of a parallel-plate capacitor on the second plate may be calculated by Coulomb's law, but a simpler method exists in the concept of virtual displacement. This was also employed in the previous section. Assume a displacement δx of one of the plates, increasing the plate separation from x to $x + \delta x$. The δ indicates an instantaneous displacement, consistent with the constraints, that is virtual–i.e., imagined–rather than real. We are concerned with a comparison of the static conditions before and after the displacement, and are not concerned with the dynamic conditions during the displacement. Suppose the field is confined to the volume τ between the plates; this is not strictly true, but it can be made approximately true to any desired degree of accuracy. Then

$$U = \left(\frac{\epsilon_0}{2}\right)E^2\tau$$

The virtual displacement (to the right) of the right-hand plate in Fig. 4-14*a* leads to a change in this energy. If the capacitor is disconnected from all batteries then the surface charge density σ remains constant, as does $E = \sigma/\epsilon_0$. So $\delta U = (\epsilon_0/2)E^2 A\ \delta x$. In this case the potential energy increases when the plate spacing is made larger.

The right-hand plate of the capacitor is in equilibrium under the action of two forces, an electrical attractive force F_{elec} to the left, and a mechanical restraining force F_{mech} to the right. In a virtual displacement of this plate to the right the mechanical force would perform work $\delta W_{\text{mech}} = \mathbf{F}_{\text{mech}} \cdot \delta \mathbf{x} = F_{\text{mech}}\ \delta x$. Since $\mathbf{F}_{\text{mech}} = -\mathbf{F}_{\text{elec}}$, $F_{\text{mech}} = F_{\text{elec}}$; so $\delta W_{\text{mech}} = F_{\text{elec}}\ \delta x$. Then, by the conservation of energy, the change in the electric field energy δU equals the work performed by the (outside) mechanical force on this isolated system:

$$\frac{\epsilon_0}{2} E^2 A\ \delta x = F_{\text{elec}}\ \delta x$$

The electric force per unit area on the plates is

$$f_{\text{elec}} = \left(\frac{\epsilon_0}{2}\right) E^2$$

If the capacitor is connected to a battery during the virtual displacement the potential difference, rather than the charge, remains constant. From $U = \frac{1}{2} C\ (\phi_1 - \phi_2)^2$ this gives $\delta U = \frac{1}{2}\ (\phi_1 - \phi_2)^2\ \delta C = (Q^2/2C^2)\ \delta C$. In the previous case, with $U = Q^2/2C$ and Q constant, the energy change was $\delta U = -(Q^2/2C^2)\ \delta C$, so here the change in energy is equal in magnitude, but opposite in sign, to the previous case. But now it is not correct to equate δU to δW. As shown in Prob. 16 of the previous section, there is a change in the battery energy here, $\delta U_b = -2\ \delta U$. The total change in the electrical energy is, therefore, $\delta U_t = \delta U + \delta U_b = \delta U - 2\ \delta U = -\delta U$. Setting $\delta W_{\text{mech}} = \delta U_t$ gives the same value for the electrical force per unit area as in the previous case.

The formula for the electrostatic force per unit area of the capacitor plate is the same as the previously obtained expression for electrostatic energy density. The case of the parallel-plate capacitor, however, is only a particularly simple example of the general case. The latter was first worked out by Maxwell. The following derivation of the general case applies to the case of an isolated charge distribution. (The remainder of this section involves matrix algebra; it may be skipped, if so desired, without any loss in continuity.)

The force per unit volume on a charge density ρ is $\rho\mathbf{E}$ or

$$\mathbf{f} = (\epsilon_0\ \nabla \cdot \mathbf{E})\mathbf{E}$$

Note that $\mathbf{f}$ here is not the same as $\mathbf{f}$ in the previous case; here the force is per unit volume. The x component of this equation is

$$f_x = \epsilon_0 \left(E_x \frac{\partial E_x}{\partial x} + E_x \frac{\partial E_y}{\partial y} + E_x \frac{\partial E_z}{\partial z} \right)$$

But

$$E_x \frac{\partial E_x}{\partial x} = \frac{1}{2} \frac{\partial}{\partial x} (E_x{}^2) .$$

Also,

$$E_x \frac{\partial E_y}{\partial y} = \frac{\partial}{\partial y}(E_x E_y) - E_y \frac{\partial E_x}{\partial y}$$

Since $\nabla \times \mathbf{E} = 0$,

$$\frac{\partial E_x}{\partial y} = \frac{\partial E_y}{\partial x}$$

so

$$E_x \frac{\partial E_y}{\partial y} = \frac{\partial}{\partial y}(E_x E_y) - E_y \frac{\partial E_y}{\partial x} = \frac{\partial}{\partial y}(E_x E_y) - \frac{1 \partial}{2 \partial x}(E_y^2)$$

Similarly,

$$E_x \frac{\partial E_z}{\partial z} = \frac{\partial}{\partial z}(E_x E_z) - \frac{1}{2}\frac{\partial}{\partial x}(E_z^2)$$

Therefore

$$f_x = \epsilon_0 \left\{ \frac{\partial}{\partial x}\left[\frac{1}{2}(E_x^2 - E_y^2 - E_z^2)\right] + \frac{\partial}{\partial y}(E_x E_y) + \frac{\partial}{\partial z}(E_x E_z) \right\}$$

Similarly

$$f_y = \epsilon_0 \left\{ \frac{\partial}{\partial x}(E_x E_y) + \frac{\partial}{\partial y}\left[\frac{1}{2}(E_y^2 - E_x^2 - E_z^2)\right] + \frac{\partial}{\partial z}(E_y E_z) \right\}$$

$$f_z = \epsilon_0 \left\{ \frac{\partial}{\partial x}(E_x E_z) + \frac{\partial}{\partial y}(E_x E_z) + \frac{\partial}{\partial z}\left[\tfrac{1}{2}(E_z^2 - E_x^2 - E_y^2)\right] \right\}$$

These three equations can be written

$$f_x = \frac{\partial T_{xx}}{\partial x} + \frac{\partial T_{yx}}{\partial y} + \frac{\partial T_{zx}}{\partial z}$$

$$f_y = \frac{\partial T_{xy}}{\partial x} + \frac{\partial T_{yy}}{\partial y} + \frac{\partial T_{zy}}{\partial z}$$

$$f_z = \frac{\partial T_{xz}}{\partial x} + \frac{\partial T_{yz}}{\partial y} + \frac{\partial T_{zz}}{\partial z}$$

where

$$\{T\} = \left\{ \begin{matrix} T_{xx} & T_{xy} & T_{xz} \\ T_{yx} & T_{yy} & T_{yz} \\ T_{zx} & T_{zy} & T_{zz} \end{matrix} \right\}$$

$$= \epsilon_0 \left\{ \begin{matrix} \tfrac{1}{2}(E_x^2 - E_y^2 - E_z^2) & E_x E_y & E_x E_z \\ E_y E_x & \tfrac{1}{2}(E_y^2 - E_z^2 - E_x^2) & E_y E_z \\ E_z E_x & E_z E_y & \tfrac{1}{2}(E_z^2 - E_x^2 - E_y^2) \end{matrix} \right\}$$

or

$$\{T\} = \epsilon_0 \begin{Bmatrix} E_x^2 & E_xE_y & E_xE_z \\ E_yE_x & E_y^2 & E_yE_z \\ E_zE_x & E_zE_y & E_z^2 \end{Bmatrix} - \frac{\epsilon_0}{2} \begin{Bmatrix} E^2 & 0 & 0 \\ 0 & E^2 & 0 \\ 0 & 0 & E^2 \end{Bmatrix} \tag{E-13}$$

The quantities in braces are matrices. If we define $\{f\} = \{f_x, f_y, f_z\}$ as a row matrix and $\{\nabla\} = \{\partial/\partial x, \partial/\partial y, \partial/\partial z\}$ as an operator row matrix, the three equations may be succinctly written $\{f\} = \{\nabla\}\{T\}$. Here the rule of matrix multiplication is employed. In general $\{A\} = \{B\}\{C\}$ means $A_{ij} = B_{i1}C_{1j} + B_{i2}C_{2j} + B_{i3}C_{3j}$, where i can be 1, 2, or 3 and j can be 1, 2, or 3; but now $\{f\}$ and $\{\nabla\}$, respectively representing a vector and a vector operator, are row matrices. So i can have only the value $i = 1$:

$$f_{1j} = \nabla_{11}T_{1j} + \nabla_{12}T_{2j} + \nabla_{13}T_{3j}$$

The x component of this equation is $f_{11} = \nabla_{11}T_{11} + \nabla_{12}T_{21} + \nabla_{13}T_{31}$; the y and z components are obtained with $j = 2$ and 3.

The matrix $\{T\}$ may be shown to have the transformation properties of a tensor. (It can also be expressed as a dyadic, $\overleftrightarrow{T}_{12} = \hat{x}\hat{y}E_xE_y$.) For this reason, and also because it was first investigated by Maxwell, $\{T\}$ is called the Maxwell stress tensor.

We found the x component of the force density:

$$f_x = \frac{\partial T_{xx}}{\partial x} + \frac{\partial T_{yx}}{\partial y} + \frac{\partial T_{zx}}{\partial z}$$

The total x component of the force on a body is then

$$F_x = \int f_x \, d\tau = \int \left(\frac{\partial T_{xx}}{\partial x} + \frac{\partial T_{yx}}{\partial y} + \frac{\partial T_{zx}}{\partial z} \right) d\tau$$

This can be written

$$F_x = \int \nabla \cdot T_{\alpha x} \, d\tau$$

where $T_{\alpha x}$ is a vector with the components (T_{xx}, T_{yx}, T_{zx}). Apply Gauss' divergence theorem:

$$F_x = \oint_\Sigma T_{\alpha x} \cdot d\mathbf{S} = \oint_\Sigma (n_x T_{xx} + n_y T_{yx} + n_z T_{zx}) \, dS$$

Here n_x, n_y, n_z are the direction cosines of the surface element dS on the surface Σ enclosing the body for which the forces are being calculated. Similar expressions hold for F_y and F_z.

For a body of any shape the knowledge of $\{T\}$ enables one to calculate **F**; e.g., in the capacitor calculation above only $E_x \neq 0$. Then

$$T = \epsilon_0 \begin{Bmatrix} \frac{1}{2} E_x^2 & 0 & 0 \\ 0 & -\frac{1}{2} E_x^2 & 0 \\ 0 & 0 & -\frac{1}{2} E_x^2 \end{Bmatrix}$$

The integral over a closed Σ here becomes one over only a face A in the yz plane; $n_x = 1$ while $n_y = n_z = 0$. Then

$$F_x = \int_A T_{xx}\, dS = \epsilon_0 \frac{1}{2} E_x^2 A$$

while $F_y = F_z = 0$. In general, just as in the case here, a knowledge of $\{T\}$ enables one to calculate **F** on a body of any shape.

Sometimes it is desirable to have the expression for the force per unit volume exerted by a charge distribution on the external world, rather than the expression for the force per unit volume exerted on the charge distribution. This may be true, for example, in the consideration of the conservation of energy on a given volume. That expression is simply the negative of the one considered above.

Examples

1. A capacitor has a difference of potential of 100 V and a separation between plates of 1 mm. Find the electrostatic force density on a plate.

$$E = 10^5 \text{ V/m}$$

so

$$f = \frac{\epsilon_0}{2} E^2 = \frac{8.89 \times 10^{-12}}{2} (10^5)^2 = 0.044 \text{ N/m}^2$$

$$= 0.01 \text{ lb/m}^2 = 6.5 \times 10^{-6} \text{ lb/in}^2$$

By ordinary, macroscopic, mechanical standards this is a rather small tension. This does not mean that electrical forces are weak; it means that the charge which may be put on a conductor is severely limited by another consideration, breakdown potential difference.

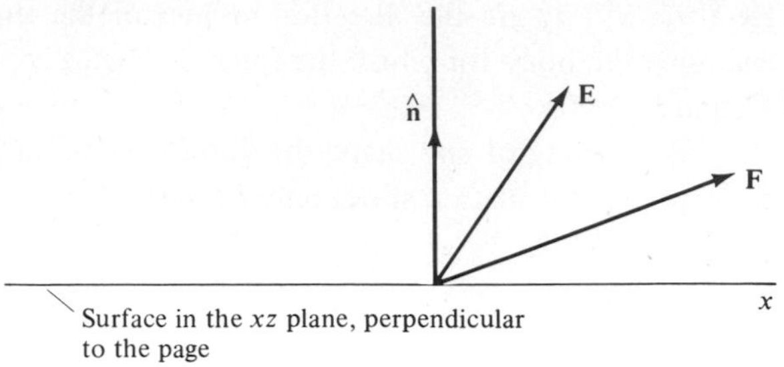

FIGURE 4-16

2. A surface in the xz plane has on it an electric field **E** which makes an angle θ with $\hat{\mathbf{y}}$, as in Fig. 4-16. Using the Maxwell stress tensor, find the force per unit area.

$$\begin{cases} E_x = E \sin\theta \\ E_y = E\cos\theta \\ E_z = 0 \end{cases} \qquad \begin{cases} n_x = 0 \\ n_y = 1 \\ n_z = 0 \end{cases}$$

Then
$$\{T\} = \epsilon_0 \begin{Bmatrix} E^2(\sin^2\theta - \cos^2\theta) & E^2 \sin\theta\cos\theta & 0 \\ E^2\sin\theta\cos\theta & E^2(\cos^2\theta - \sin^2\theta) & 0 \\ 0 & 0 & -\dfrac{E^2}{2} \end{Bmatrix}$$

$$F_x = \int [n_x T_{xx} + n_y T_{yx} + n_z T_{zx}]\,dS = \epsilon_0 \int [0 + E^2 \sin\theta\cos\theta + 0]\,dS$$

$$= \epsilon_0 \frac{E^2}{2} S \sin 2\theta$$

$$F_y = \int [n_x T_{xy} + n_y T_{yy} + n_z T_{zy}]\,dS = \epsilon_0 \int [0 + E^2(\cos^2\theta - \sin^2\theta) + 0]\,dS$$

$$= \epsilon_0 \frac{E^2}{2} S \cos 2\theta$$

$$F_z = \int [n_x T_{xz} + n_y T_{yz} + n_z T_{zz}]\,dS = \epsilon_0 \int [0 + 0 + 0]\,dS = 0$$

The magnitude of the force is $F = \sqrt{F_x{}^2 + F_y{}^2 + F_z{}^2} = \epsilon_0 (E^2/2)S$, so

$$f = \frac{\epsilon_0}{2} E^2 \qquad \text{N/m}^2$$

The force makes an angle α with the normal such that $\tan \alpha = F_x/F_y$, or

$$\tan \alpha = \frac{\sin 2\theta}{\cos 2\theta}$$

i.e., $\alpha = 2\theta$. If **E** is normal to the surface then **F** is a normal pull on the surface; if **E** is at 45° to the surface then **F** is a tangential shear; if **E** is tangential to the surface then **F** is a normal push on the surface.

PROBLEMS

1 In Example 2 is the force a pressure or a tension if **E** is (*a*) $\hat{\mathbf{y}}E$, (*b*) $\hat{\mathbf{x}}E$, (*c*) $-\hat{\mathbf{y}}E$, (*d*) $-\hat{\mathbf{x}}E$?

2 Repeat Prob. 1 when (*a*) $\mathbf{E} = (1/\sqrt{2})\ (\hat{\mathbf{x}} + \hat{\mathbf{y}})$, (*b*) $\mathbf{E} = (1/\sqrt{2})\ (\hat{\mathbf{x}} - \hat{\mathbf{y}})$, (*c*) $\mathbf{E} = (1/\sqrt{2})\ (-\hat{\mathbf{x}} - \hat{\mathbf{y}})$, (*d*) $\mathbf{E} = (1/\sqrt{2})\ (-\hat{\mathbf{x}} + \hat{\mathbf{y}})$.

3 Construct a closed box surrounding one of the plates of a capacitor and apply Maxwell's stress tensor to find the resulting force on the plate, assuming a uniform field with negligible fringing.

4 A charge $+q$ is located at $(a/2,0,0)$, a charge $-q$ at $(-a/2,0,0)$. Construct a closed surface consisting of the infinite yz plane and a hemisphere of infinite radius enclosing the region to the left of the yz plane. Integrate the Maxwell stress tensor over this surface to find the force exerted on $-q$.

5 Change Prob. 4 by making both charges positive. Find the force on q at $(-a/2,0,0)$.

6 Find the force per unit area on one plate of a parallel-plate capacitor in terms of the charge density on the plate.

7 An infinitely large metal plate lies in the yz plane. A charge $+q$ is at $(a,0,0)$. Apply the Maxwell stress tensor to find the force exerted by the charge $+q$ on the metal plate.

8 Change the charge in Prob. 7 to $-q$ at $(a,0,0)$ and repeat.

Appendix 4

THE HELMHOLTZ THEOREM

Given: $\nabla \cdot \mathbf{F} = b(\mathbf{r})$

$\nabla \times \mathbf{F} = \mathbf{c}(\mathbf{r})$

in some bounded region $\mathcal{V}$.

To prove: $\mathbf{F} = (-\nabla\phi) + (\nabla \times \mathbf{A})$

where $$\phi(\mathbf{r}_t) = \frac{1}{4\pi} \int_{\mathcal{V}} \frac{b(\mathbf{r}_s)}{R}\, d\tau_s$$

$$\mathbf{A}(\mathbf{r}_t) = \frac{1}{4\pi} \int_{\mathcal{V}} \frac{c(\mathbf{r}_s)}{R}\, d\tau_s$$

with $$R = |\mathbf{r}_s - \mathbf{r}_t|$$

i.e., to prove: A vector is specified by its divergence and curl. Inversely, any vector ($\mathbf{F}$) may be decomposed into the sum of an irrotational vector ($-\nabla\phi$) and a solenoidal vector ($\nabla \times \mathbf{A}$).

Making use of the properties of the Dirac delta function derived in Appendix 2, we may write

$$\mathbf{F}(\mathbf{r}_t) = \int_{\mathcal{V}} \mathbf{F}(\mathbf{r}_s)\, \delta(R)\, d\tau_s$$

i.e., $$\mathbf{F}(\mathbf{r}_t) = \int_{\mathcal{V}} \mathbf{F}(\mathbf{r}_s) \left[-\frac{1}{4\pi} \nabla^2 \left(\frac{1}{R} \right) \right] d\tau_s$$

Since ∇^2 is a function of $\mathbf{r}_t$ only, and not of $\mathbf{r}_s$,

$$\mathbf{F}(\mathbf{r}_t) = -\nabla^2 \left[\int_{\mathcal{V}} \frac{\mathbf{F}(\mathbf{r}_s)}{4\pi R}\, d\tau_s \right]$$

Then, using the identity $\nabla \times \nabla \times \mathbf{H} = \nabla\nabla \cdot \mathbf{H} - \nabla^2 \mathbf{H}$, we have

$$\mathbf{F}(\mathbf{r}_t) = \nabla \times \nabla \times \left[\int_{\mathcal{V}} \frac{\mathbf{F}(\mathbf{r}_s)}{4\pi R}\, d\tau_s \right] - \nabla\nabla \cdot \left[\int_{\mathcal{V}} \frac{\mathbf{F}(\mathbf{r}_s)}{4\pi R}\, d\tau_s \right] = -\nabla\phi + \nabla \times \mathbf{A}$$

where $$\phi = \nabla \cdot \int_{\mathcal{V}} \frac{\mathbf{F}(\mathbf{r}_s)}{4\pi R}\, d\tau_s \qquad \text{and} \qquad \mathbf{A} = \nabla \times \int_{\mathcal{V}} \frac{\mathbf{F}(\mathbf{r}_s)}{4\pi R}\, d\tau_s$$

In the vector identity $\nabla \cdot (1/R)\mathbf{F} = \mathbf{F} \cdot \nabla(1/R) + (1/R)\nabla \cdot \mathbf{F}$ the last term vanishes here, since $\mathbf{F}$ depends only on $\mathbf{r}_s$ while ∇ involves only $\mathbf{r}_t$. So

$$\phi = \frac{1}{4\pi} \int_{\mathcal{V}} \mathbf{F}(\mathbf{r}_s) \cdot \nabla \left(\frac{1}{R}\right) d\tau_s$$

$$= -\frac{1}{4\pi} \int_{\mathcal{V}} \mathbf{F}(\mathbf{r}_s) \cdot \nabla_s \left(\frac{1}{R}\right) d\tau_s$$

$$= -\frac{1}{4\pi} \int_{\mathcal{V}} \left\{ \nabla_s \cdot \left[\frac{1}{R} \mathbf{F}(\mathbf{r}_s)\right] - \frac{1}{R} \nabla_s \cdot \mathbf{F}(\mathbf{r}_s) \right\} d\tau_s$$

$$= -\oint_{\Sigma} \frac{\mathbf{F}(\mathbf{r}_s) \cdot d\mathbf{S}}{R} + \frac{1}{4\pi} \int_{\mathcal{V}} \frac{\nabla_s \cdot \mathbf{F}(\mathbf{r}_s)}{R} d\tau_s$$

If the sources, i.e., $b(\mathbf{r}_s)$, are all included in some finite region, then $\mathcal{V}$ and Σ may be made so large that the value of $\mathbf{F}$ becomes arbitrarily small on Σ and the first integral may be neglected. In the second integral $b(\mathbf{r}_s)$ may be substituted for $\nabla_s \cdot \mathbf{F}(\mathbf{r}_s)$, so

$$\phi = \frac{1}{4\pi} \int_{\mathcal{V}} \frac{b(\mathbf{r}_s)}{R} d\tau_s$$

Similarly,

$$\mathbf{A} = \frac{1}{4\pi} \int_{\mathcal{V}} \nabla \times \left[\frac{1}{R} \mathbf{F}(\mathbf{r}_s)\right] d\tau_s$$

$$= \frac{1}{4\pi} \int_{\mathcal{V}} \left[\nabla \left(\frac{1}{R}\right) \times \mathbf{F}(\mathbf{r}_s) + \frac{1}{R} \nabla \times \mathbf{F}(\mathbf{r}_s)\right] d\tau_s$$

The second integrand vanishes because $\mathbf{F}$ is not a function of $\mathbf{r}_t$ but only of $\mathbf{r}_s$. $\nabla(1/R) = -\nabla_s(1/R)$, so

$$\mathbf{A} = \frac{1}{4\pi} \int_{\mathcal{V}} \mathbf{F}(\mathbf{r}_s) \times \nabla_s \left(\frac{1}{R}\right) d\tau_s$$

$$= \frac{1}{4\pi} \int_{\mathcal{V}} \left[\frac{1}{R} \nabla_s \times \mathbf{F}(\mathbf{r}_s) - \nabla_s \times \frac{\mathbf{F}(\mathbf{r}_s)}{R} \right] d\tau_s$$

The first integral is
$$\frac{1}{4\pi} \int_{\mathcal{V}} \frac{\mathbf{c}(\mathbf{r}_s)}{R}\, d\tau_s$$

For the second integral we employ a modification of Gauss' divergence theorem:

$$\int_{\mathcal{V}} \nabla \times \mathbf{F}\, d\tau_s = -\oint_{\Sigma} \mathbf{F} \times d\mathbf{S}$$

The derivation follows. Let $\mathbf{K}$ be a constant. Then, by use of a vector identity (see Appendix 1-B), $\nabla \cdot (\mathbf{K} \times \mathbf{F}) = \mathbf{F} \cdot \nabla \times \mathbf{K} - \mathbf{K} \cdot \nabla \times \mathbf{F} = -\mathbf{K} \cdot \nabla \times \mathbf{F}$.

So
$$\int_{\mathcal{V}} \nabla \cdot (\mathbf{K} \times \mathbf{F})\, d\tau = -\mathbf{K} \cdot \int_{\mathcal{V}} \nabla \times \mathbf{F}\, d\tau = \oint_{\Sigma} (\mathbf{K} \times \mathbf{F}) \cdot d\mathbf{S}$$

$$= \oint_{\Sigma} (\mathbf{F} \times d\mathbf{S}) \cdot \mathbf{K} = \mathbf{K} \cdot \oint_{\Sigma} \mathbf{F} \times d\mathbf{S}$$

Since $\mathbf{K}$ is arbitrary, this yields the desired result. Then the second integral becomes $1/4\pi \oint_{\Sigma} (\mathbf{F}/R) \times d\mathbf{S}$. As before, Σ may be made so large, when the vortex sources, i.e., $\mathbf{c}(\mathbf{r}_s)$, are all included in a finite region, that $\mathbf{F}$ becomes negligibly small on Σ and the integral may be made as small as desired. So

$$\mathbf{A} = \frac{1}{4\pi} \int_{\mathcal{V}} \frac{\mathbf{c}(\mathbf{r}_s)}{R}\, d\tau_s$$

This completes the proof.

When there are sources or vortex sources at infinity then the surface integrals do not vanish and the Helmholtz theorem is no longer applicable.

5

THE MAGNETOSTATIC DIVERGENCE IN VACUUM

5-1 $\nabla \cdot \mathbf{B}$ AND THE VECTOR POTENTIAL A

Following our plan of contrasting electrostatic results with those of their magnetostatic mates, we now proceed to a consideration of the corresponding magnetic quantities.

To obtain an expression for **B** it is sufficient, according to the Helmholtz theorem, to find $\nabla \cdot \mathbf{B}$ and $\nabla \times \mathbf{B}$. Having the latter, we now proceed to find $\nabla \cdot \mathbf{B}$. This can be done directly from the law of Biot-Savart.

Starting with the filamentary form of this law,

$$\mathbf{B} = \frac{\mu_0}{4\pi} \oint_S \frac{I_S \, d\mathbf{r}_s \times \hat{\mathbf{R}}_{st}}{R^2}$$

where $R = |\mathbf{R}_{st}|$, take the divergence of both sides. The del operator employed in taking the divergence is a function of the position of the test point $\mathbf{r}_t$ at which the divergence of **B** is being found. But the independent variable under the integral sign is the position of the source point $\mathbf{r}_s$, so the del operator may be moved to the right of the integral sign while I_S, a constant, may be moved to the left:

$$\nabla \cdot \mathbf{B} = \frac{\mu_0 I_S}{4\pi} \oint_S \nabla \cdot \left[d\mathbf{r}_s \times \frac{\hat{\mathbf{R}}_{st}}{R^2} \right]$$

Using the vector identity $\nabla \cdot (\mathbf{A} \times \mathbf{B}) = \mathbf{B} \cdot (\nabla \times \mathbf{A}) - \mathbf{A} \cdot (\nabla \times \mathbf{B})$ with $\mathbf{A} = d\mathbf{r}_s$ and $\mathbf{B} = \hat{\mathbf{R}}_{st}/R^2$,

$$\nabla \cdot \left[d\mathbf{r}_s \times \frac{\hat{\mathbf{R}}_{st}}{R^2} \right] = \frac{\hat{\mathbf{R}}_{st}}{R^2} \cdot (\nabla \times d\mathbf{r}_s) - d\mathbf{r}_s \cdot \left(\nabla \times \frac{\hat{\mathbf{R}}_{st}}{R^2} \right)$$

Since $\mathbf{r}_s$ and $\mathbf{r}_t$ are independent variables, $\nabla \times d\mathbf{r}_s = 0$. For the second term

$$\frac{\hat{\mathbf{R}}_{st}}{R^2} = -\nabla \left(\frac{1}{R} \right)$$

Then

$$\nabla \times \left(\frac{\hat{\mathbf{R}}_{st}}{R^2} \right) = \nabla \times \left[-\nabla \left(\frac{1}{R} \right) \right] = -\nabla \times \nabla \left(\frac{1}{R} \right) = 0$$

since the curl of the gradient of any function is identically zero. Consequently, the entire integrand vanishes, and

$$\boxed{\nabla \cdot \mathbf{B} = 0} \qquad \text{(M-7)}$$

In electrostatics $\mathbf{E}$ is an irrotational vector ($\nabla \times \mathbf{E} = 0$), but the vector $\mathbf{B}$ is found to be solenoidal ($\nabla \cdot \mathbf{B} = 0$). So far $\nabla \cdot \mathbf{B} = 0$ has been shown to hold only for magnetostatics, but this will be found to hold true in general. The physical significance of $\nabla \cdot \mathbf{B} = 0$ is simply this: no sources or sinks of magnetic flux have yet been discovered in nature. There is no theoretical law that forbids the existence of the magnetic equivalent of electrical charge $\nabla \cdot \mathbf{B} = \rho_{\text{mag}}$ where ρ_{mag} is a magnetic charge density. A number of theoretical physicists firmly believe in the existence of such magnetic charges, but despite an intensive search for their existence no experimental evidence has yet been found which shows they do exist. If any sources or sinks of magnetism are ever found it will only be necessary to modify the equation $\nabla \cdot \mathbf{B} = 0$ by adding the term ρ_m to the right and, of course, the law of Biot-Savart will have to be altered, for $\nabla \cdot \mathbf{B} = 0$ was found to be a necessary consequence of this law. Meanwhile the search for these units—magnetic monopoles—goes on.

We list below a few of many recent articles, all from *The American Journal of Physics*, devoted to various aspects of monopoles. References to the basic investigations in the research journals will be found there also.

1. I. R. Lapidus and J. L. Pietenpol, Classical Interaction of an Electric Charge with a Magnetic Monopole, **28**:17 (1969).
2. R. Katz, The Magnetic Pole in the Formulation of Electricity and Magnetism, **30**:41 (1960).
3. H. Harrison, N. A. Krall, O. C. Eldridge, F. Fehsenfeld, W. L. Fite, and W. B. Tentsch, Possibility of Observing the Magnetic Charge of an Electron, **31**:249 (1963).
4. N. Strax, Magnetic Monopoles, Weak Interactions, and Angular Momentum, **33**:102 (1965).
5. R. Mirman, Magnetic Monopoles and Invariance, **34**:70 (1966).
6. H. J. Efinger, An Instructive Model for the Quantization of Magnetic Monopoles, **37**:740 (1969).

Since no free sources or sinks of lines of **B** have yet been found experimentally, a strong temptation exists to think of lines of **B** as closed loops, having neither a beginning nor an end. In some simple cases this interpretation is possible, but this is not true in general.

For example, consider the lines of **B** produced by a current flowing in an infinitely long straight wire along the $\hat{\mathbf{z}}$ axis. The lines of **B** are circles in xy planes centered about the wire. A line of **B** may be drawn through each point in space; but only a few lines are actually drawn, such that the density of lines is proportional to the magnitude of the flux density. A problem exists here, because the three-dimensional density of lines in space does not equal the two-dimensional density which must be employed for a drawing. Neglecting this, as a difficulty in practice but not in principle, there is no hindrance up to this point in imagining closed lines of **B**.

Now let an external field $\hat{\mathbf{z}}B_{\text{ext}}$ be added to the field produced by the current in the wire. Instead of circles, the lines must now become helices; whatever the pitch, even for the weakest B_{ext}, their lengths change from some finite value to infinite value. Instead of being closed loops, the lines of **B** are now open. Though here it is still possible to say that the two ends of a line meet at infinity to form a closed loop. (The convention that the density of lines be proportional to $|\mathbf{B}|$ now becomes impossible to carry out because the smaller B_{ext}, the less the pitch and the greater the density in space of any one line.)

Next, make a slight change in the geometry. Suppose the conductor, instead of being an arbitrarily long straight wire, forms a circle of radius r lying in the xy plane with center on the z axis. Like the infinitely long straight wire, this is an idealized circuit: the battery and circuit resistance are ignored for convenience. The lines of the magnetic field B_1, which are created by the current I_1 in the loop, lie in planes passing through the $\hat{\mathbf{z}}$ axis, as shown in Fig. 5-1, and will all be closed. Now add an additional current $\hat{\mathbf{z}}I_2$ along the main axis of symmetry. This current creates a field $\mathbf{B}_2$ which is azimuthal, and the total magnetic field is the vector sum of $\mathbf{B}_1$ and $\mathbf{B}_2$. The lines of resultant magnetic field are spirals wound about the circle of radius r, and the pitch will depend on the ratio of I_1

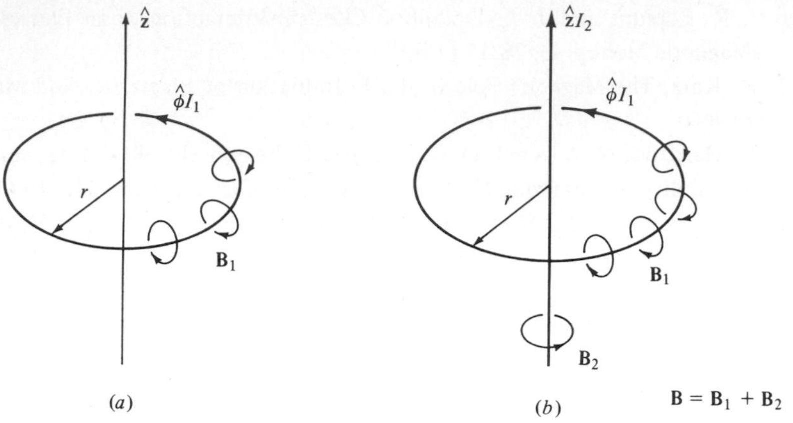

FIGURE 5-1

to I_2. For some value of I_1/I_2 the spiral will close on itself after going around the circumference of the horizontal circle once; for some other value of this ratio the spiral will close after going around twice, etc., but in general, except for certain special values of I_1/I_2, the spiral will not close on itself. The cases where it does close on itself are the exceptional ones of closed loops; in most cases the lines are open.

In summary: closed loops are a sufficient condition, but not a necessary condition, for $\nabla \cdot \mathbf{B} = 0$ to be true. One may have this equation satisfied even with open lines of flux density.

The equation $\nabla \times \mathbf{E} = 0$ in electrostatics had as an immediate consequence the existence of an electric scalar potential ϕ; here the equation $\nabla \cdot \mathbf{B} = 0$ has as an immediate consequence the existence of a magnetic vector potential **A** given by

$$\boxed{\mathbf{B} = \nabla \times \mathbf{A}} \qquad \text{(M-8)}$$

since $\nabla \cdot \nabla \times \mathbf{A} = 0$ for all **A**. That **A** exists, with $\mathbf{B} = \nabla \times \mathbf{A}$, is both a necessary and a sufficient condition for $\nabla \cdot \mathbf{B} = 0$ to be true. We will leave the proof to Prob. 18. The fact that the magnetic potential is a vector rather than a scalar has the consequence that much of the utility of dealing with the potential is lost in magnetostatics. It is often as cumbersome to find first **A** by vector integration and then $\nabla \times \mathbf{A}$ as it is to find **B** directly by vector integration using the law of Biot-Savart. It is introduced, nevertheless, for several reasons: (1) **A** provides tremendous advantages when variations with time are considered; (2) there is an intimate relation between ϕ and **A** which will be brought out when relativity is considered; (3) a strong case can be made for the argument that the potential fields, rather than the electric field and the magnetic field, are the fundamental physical quantities; and (4) sometimes, as in the transformer or the betatron, a knowledge of **A** gives a more direct and physical insight than does a knowledge of **B**. The knowledge of **B** is really only needed when one must know the force. In quantum mechanics, it turns out, ϕ and **A** *must* be used and **E** and **B** cannot be substituted for them.

We now obtain an explicit relation for **A** at a test point in terms of the currents at various source points, starting from the law of Biot-Savart.

$$\mathbf{B}(\mathbf{r}_t) = \frac{\mu_0}{4\pi} \int_{\mathcal{V}} \frac{\mathbf{J}(\mathbf{r}_s) \times \hat{\mathbf{R}}_{st}}{R^2}\, d\tau_s$$

may be written

$$\mathbf{B}(\mathbf{r}_t) = \frac{\mu_0}{4\pi} \int_{\mathcal{V}} \mathbf{J}(\mathbf{r}_s) \times \left[-\nabla \left(\frac{1}{R} \right) \right] d\tau_s = \frac{\mu_0}{4\pi} \int_{\mathcal{V}} \left[\nabla \left(\frac{1}{R} \right) \right] \times \mathbf{J}(\mathbf{r}_s)\, d\tau_s$$

Using the vector space identity $\nabla \times (a\mathbf{A}) = [\nabla a \times \mathbf{A}] + [a \nabla \times \mathbf{A}]$

with

$$a = \frac{1}{R} \quad \text{and} \quad \mathbf{A} = \mathbf{J}(\mathbf{r}_s)$$

gives

$$\nabla \times \left[\frac{\mathbf{J}(\mathbf{r}_s)}{R} \right] = \left[\nabla \left(\frac{1}{R} \right) \right] \times \mathbf{J}(\mathbf{r}_s) + \frac{1}{R} \nabla \times \mathbf{J}(\mathbf{r}_s)$$

The last term vanishes, since **J** is a function only of $\mathbf{r}_s$, while the del operator is a function of $\mathbf{r}_t$. So

$$\mathbf{B}(\mathbf{r}_t) = \frac{\mu_0}{4\pi} \int_{\mathcal{V}} \nabla \times \left[\frac{\mathbf{J}(\mathbf{r}_s)}{R} \right] d\tau_s$$

Since the del operator does not depend on the independent variables of the integrand it may be moved to the left of the integral sign:

$$\mathbf{B}(\mathbf{r}_t) = \frac{\mu_0}{4\pi} \nabla \times \int_{\mathcal{V}} \frac{\mathbf{J}(\mathbf{r}_s)}{R}\, d\tau_s$$

This gives

$$\mathbf{B} = \nabla \times \mathbf{A}$$

where

$$\boxed{\mathbf{A}(\mathbf{r}_t) = \frac{\mu_0}{4\pi} \int_{\mathcal{V}} \frac{\mathbf{J}(\mathbf{r}_s)}{R}\, d\tau_s} \tag{M-9}$$

The equation $\mathbf{B} = \nabla \times \mathbf{A}$ is usually regarded as a formula for finding **B** if **A** is known; but it is also possible to turn this around, as $\nabla \times \mathbf{A} = \mathbf{B}$, and consider this as a prescription for $\nabla \times \mathbf{A}$ in terms of a given **B**. Then the Helmholtz theorem requires only the additional knowledge of $\nabla \cdot \mathbf{A}$ to specify **A**. It is left to Prob. 19 to derive the result that, at least in magnetostatics,

$$\nabla \cdot \mathbf{A} = 0$$

The **A** given by Eq. (M-9), which yields this result, is called the Coulomb gage.

It is instructive to compare this result for **A** with that for ϕ in electrostatics: just let $\phi \to \mathbf{A}$, $\epsilon_0 \to 1/\mu_0$, $\rho \to \mathbf{J}$. As mentioned above, the similarity is not accidental. The three components of **A**, together with the single component of ϕ, will be found to constitute a relativistic four-vector. The characteristics of such quantities will be discussed in Chap. 14. These vectors differ in their properties from ordinary three-dimensional vectors, or three-vectors: it is not simply an extension from three components to four.

One feature of the electric scalar potential is that it is not unique. $\mathbf{E} = -\nabla\phi$ and also

$$\mathbf{E} = -\nabla \left[\frac{1}{4\pi\epsilon_0} \int_{\mathcal{V}} \frac{\rho}{R}\, d\tau_s \right]$$

It follows that

$$\phi(\mathbf{r}_s) = \frac{1}{4\pi\epsilon_0} \int_{\mathcal{V}} \frac{\rho(\mathbf{r}_s)}{R}\, d\tau_s + K$$

where K is any constant. There is an infinity of ϕ's which give the same **E**. The distinction between one ϕ and another, however, is physically unimportant. It is easy to consider the additive constant as trivial: it means only taking a different reference ground point. (Later, when variations with time are permitted, there will still be an infinite number of ϕ's, but then they will differ by more than just a constant.)

Similarly in the case of the vector potential: if $\mathbf{B} = \nabla \times \mathbf{A}$, and if

$$\mathbf{B} = \nabla \times \left[\frac{\mu_0}{4\pi} \int_{\mathcal{V}} \frac{\mathbf{J}(\mathbf{r}_s)}{R}\, d\tau_s \right]$$

then
$$\mathbf{A}(\mathbf{r}_t) = \frac{\mu_0}{4\pi}\int_{\mathcal{V}} \frac{\mathbf{J}(\mathbf{r}_s)}{R}\,d\tau_s + \nabla f(\mathbf{r}_t)$$

where f is any scalar function of $\mathbf{r}_t$. There is an infinity of **A**'s which give the same **B**. The significance of these different **A**'s for a given **B** will be brought out in Example 1.

While it is easy to accept an added constant as unimportant, it is not easy to consider the additive gradient of an arbitrary function as trivial. It is because of this added vector function that **A** has long been considered more a mathematical device than a physical quantity, and quite secondary in importance to **B**. In the subsequent discussion of the Aharanov-Bohm effect (Chap. 14, Sec. 14-2, Example 4), it will be seen that an electron traveling in a region where $\mathbf{B} = 0$ but $\mathbf{A} \neq 0$ is affected experimentally by the value of **A**. Also, the theory of operation of a transformer, Chap. 11, Sec. 11-1, is rather mystifying without a knowledge of **A**. Easy or not, therefore, we must conclude–anticipating these two topics–that **A** has real physical significance.

The argument has been raised that the Aharanov-Bohm effect is a quantum-mechanical effect since it involves Planck's constant h. It is possible to take the position that **A** has physical significance in quantum mechanics but not in classical mechanics, which is obtained by letting $h \to 0$; but this really begs the question. When quantum mechanics and classical mechanics give contradictory answers we must abandon the classical result, for quantum mechanics is a better approximation to the truth, always, than is classical mechanics. Classical mechanics gives pretty pictures that are easier to understand and it is much more convenient, provided it gives the correct answers. In any event, how explain the transformer–a classical device–without a knowlege of **A**? That can only be done mathematically, using **B**, but not physically. Incidentally, note that it is not possible to have a region with $\mathbf{A} = 0$ in which the value of **B** affects the physical result. If $\mathbf{A} = 0$ in some *region* then $\mathbf{B} = \nabla \times \mathbf{A} = 0$ also in that region. (At a single point, however, it *is* possible to have a finite **B** with $\mathbf{A} = 0$.)

The volume $\mathcal{V}$ over which the integral above is to be taken is any volume which includes *all* the current sources. This integral diverges if any current sources are present at infinity, since $d\tau_s/R$ behaves like $(r^2\,dr\,d\Omega)/r$, i.e., like r, for distances very far from the origin. Only if $\mathbf{J} = 0$ at infinity can the integral be finite. In such a case (an infinitely long straight conductor, an infinitely long solenoid, a constant surface current density on an infinite plane), **A** must be determined by other means. A similar situation was seen to exist for ϕ.

When $\mathcal{V}$ is finite it can be shown that the divergence of the integral in the general expression for **A**, above, is zero if there is no variation with time. (The proof of this is left for one of the problems.) Then $\nabla \cdot \mathbf{A} = \nabla^2 f$. In magnetostatics it is always possible to take $f = 0$, in which case $\nabla \cdot \mathbf{A} = 0$ also. With $\nabla \times \mathbf{A} = \mathbf{B}$ and $\nabla \cdot \mathbf{A} = 0$, **A** is then fixed uniquely by Helmholtz's theorem. The choice $f = 0$ is called the Coulomb gage. The reason for this name will be discussed subsequently.

The dimensions of **A**, from Eq. (M-9), are found to be [m kg s^{-2} A^{-1}]. Comparing the unit for **A** having these dimensions with that of **B**—the tesla or the weber per square meter, i.e., [kg s^{-2} A^{-1}]—it is seen that the unit of **A** is 1 T m or 1 Wb m^{-1}. Also, the dimensions of **A** times those of velocity are the same as those of ϕ, which gives 1 Wb s^{-1} = 1 V. So the unit of **A** is also the V-s m^{-1}.

The significance of the fact that the unit of **A** is the weber per meter can be better appreciated by considering the flux through some open surface:

$$\Phi = \int_\Sigma \mathbf{B} \cdot d\mathbf{S} = \int_\Sigma (\nabla \times \mathbf{A}) \cdot d\mathbf{S}$$

Using Stokes' theorem, this give the useful relation

$$\Phi = \oint_C \mathbf{A} \cdot d\mathbf{r}$$

where C is the curve which is the boundary of Σ. Taking the line integral of **A** around C is sometimes a simpler way of finding the flux than is the more obvious determination of the surface integral of **B**; but the chief utilization of this formula will only become apparent later, with time-varying fields, when the induced voltage around a path C is considered.

The vector potential produced by some distribution of source currents is inherently a simpler quantity than the magnetic field produced by those currents. Compare

$$\mathbf{A} = \frac{\mu_0 I_S}{4\pi} \oint \frac{d\mathbf{r}_s}{R} \qquad \text{with} \qquad \mathbf{B} = \frac{\mu_0 I_S}{4\pi} \oint \frac{d\mathbf{r}_s \times \hat{\mathbf{R}}_{st}}{R^2}$$

In classical mechanics one is generally interested in forces; then it is necessary to take the curl of the simpler quantity, and that is no longer so simple. In quantum mechanics one is more interested in the potentials; there the simpler quantity is usually sufficient.

A word of caution must be given concerning the differential $d\mathbf{A}$. Neglecting the arbitrary additive function $\nabla f(r_t)$,

$$d\mathbf{A}(\mathbf{r}_t) = \frac{\mu_0 I_S}{4\pi} \left(\frac{d\mathbf{r}_s}{R}\right)$$

The source element $I_S d\mathbf{r}_s$ makes a contribution $d\mathbf{A}$ which is parallel to $d\mathbf{r}_s$. No problem exists if $d\mathbf{r}_s$ is given in cartesian coordinates, say $\hat{\mathbf{x}}\, dx$, for in the computation of **A** the $\hat{\mathbf{x}}$ can be taken to the left of the integral sign when the x contributions from different source elements are combined. $\hat{\mathbf{x}}$ is not a function of x, the independent variable. But if

$d\mathbf{r}_s$ is given in noncartesian (e.g., cylindrical) coordinates, a $\hat{\boldsymbol{\phi}} d\phi$ contribution, say, from one source element must be combined rather carefully with a $\hat{\boldsymbol{\phi}} d\phi$ contribution from another source element.

In Fig. 5-2 the point P is the test point at which **A** is to be determined. The vector potential is produced by the source current I_S flowing in a filamentary circuit; two differential contributions to the $\hat{\boldsymbol{\phi}}$ component of **A** are shown at P. $(d\mathbf{A}_1)_\phi$ is parallel to the $\hat{\boldsymbol{\phi}}$ component of the current element which produces it, $I_S\, d\mathbf{r}_{s1}$; it does not lie along the $\hat{\boldsymbol{\phi}}$ direction at P. To find the $\hat{\boldsymbol{\phi}}$ component at P it is necessary to take a component of $(d\mathbf{A}_1)_\phi$.

Similarly, $(d\mathbf{A}_2)_\phi$ is parallel to the $\hat{\boldsymbol{\phi}}$ component of the differential source current element which produces it, $\hat{\boldsymbol{\phi}} I_S r_2\, d\phi$; $(d\mathbf{A}_2)_\phi$ lies neither in the same direction as $(d\mathbf{A}_1)_\phi$ nor in the direction of $\hat{\boldsymbol{\phi}}$ at P.

When using noncartesian coordinates to evaluate

$$\mathbf{A}(\mathbf{r}_t) = \frac{\mu_0}{4\pi} \int_{\mathcal{V}} \frac{\mathbf{J}(\mathbf{r}_s)}{R}\, d\tau_s$$

it is necessary to use all the components of **J** to find any one, single, component of **A**. When cartesian coordinates are employed it is only necessary to integrate one component of **J** to find the corresponding component of **A**.

Examples

1. Uniform magnetic field Given: the uniform, constant magnetic field $\mathbf{B} = \hat{\mathbf{z}} B_0$. To find: the vector potential **A**.

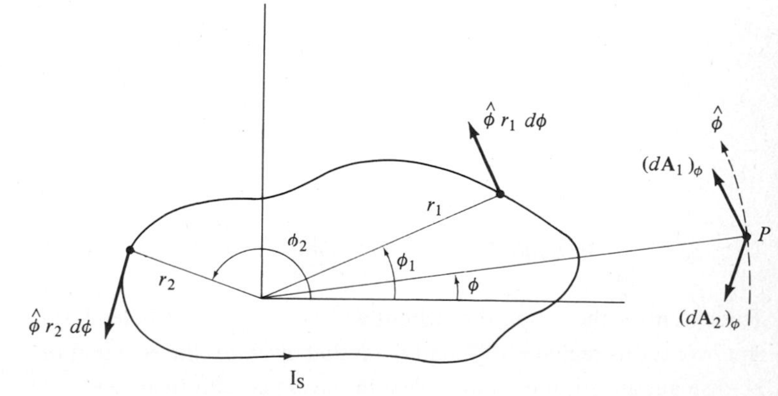

FIGURE 5-2

Any **A** which satisfies $\mathbf{B} = \nabla \times \mathbf{A}$ will be satisfactory. One such **A** is

$$\mathbf{A}_1 = \hat{\mathbf{x}}\left(-\frac{B_0}{2}y\right) + \hat{\mathbf{y}}\left(\frac{B_0}{2}x\right) = \frac{B_0}{2}(\hat{\mathbf{z}} \times \mathbf{r})$$

Several lines of constant $|\mathbf{A}_1|$ are shown in Fig. 5-3*a*. The location of the origin is arbitrary. The lengths of the arrows are proportional to $|\mathbf{A}_1|$.

Another, equally satisfactory **A** is

$$\mathbf{A}_2 = \hat{\mathbf{x}}(-B_0 y)$$

This **A** is shown in Fig. 5-3*b*. The relation between these two cases is

$$\mathbf{A}_1 = \mathbf{A}_2 + \left(\hat{\mathbf{x}}\frac{B_0}{2}y + \hat{\mathbf{y}}\frac{B_0}{2}x\right) = \mathbf{A}_2 + \nabla\left(\frac{1}{2}B_0 xy\right)$$

These are just two of the infinity of **A**'s which satisfy $\hat{\mathbf{z}}B_0 = \nabla \times \mathbf{A}$. The physical significance of **A** is hinted at by these plots. $\mathbf{A}_1$ has circular symmetry, while $\mathbf{A}_2$ has linear symmetry. It is possible to find an **A**, for instance $\mathbf{A}_3 = \hat{\mathbf{x}}\frac{B_0}{2}(x - y) + \hat{\mathbf{y}}\frac{B_0}{2}x$, which has elliptic symmetry, etc. Yet all the **A**'s, which differ from each other only by a gradient of some scalar function, are identical as far as **B** is concerned.

The set of different **A**'s for the same **B** is a reflection of the fact, in magnetostatics, that completely different current sources can produce the given **B** field, though not completely in the same volume. The flux density $\hat{\mathbf{z}}B_0$ can be produced inside an infinitely long solenoid centered about the z axis; the azimuthal symmetry of the current in the solenoid (the boundary of the **B** field) gives the **A** of Fig. 5-3*a*. If, instead, $\hat{\mathbf{z}}B_0$ is caused by two infinite plane current sheets we are, in effect, changing the boundaries. Let the current densities be

$$\mathbf{J}_\ell = \begin{cases} -\hat{\mathbf{x}}\dfrac{B_0}{\mu_0} \text{ in the } \mathbf{xz} \text{ plane at } y = y_0 \\ +\hat{\mathbf{x}}\dfrac{B_0}{\mu_0} \text{ in the } xz \text{ plane at } y = -y_0 \end{cases}$$

Then $\mathbf{A} = \dfrac{\mu_0}{4\pi}\displaystyle\int (\mathbf{J}_\ell/R)\,dS$ has only an x component; this gives $\mathbf{A} = -\hat{\mathbf{x}}B_0 y$, as in Fig. 5-3*b*.

The extent of the region throughout which $\mathbf{B} = \hat{\mathbf{z}}B_0$ is affected by the choice of y_0 but, in any event, this region is different from that given by the solenoid of circular cross section. If, instead, an infinitely long solenoid having an elliptical cross section were used, the **B** field could still be $\hat{\mathbf{z}}B_0$ but the potential field might be, for example,

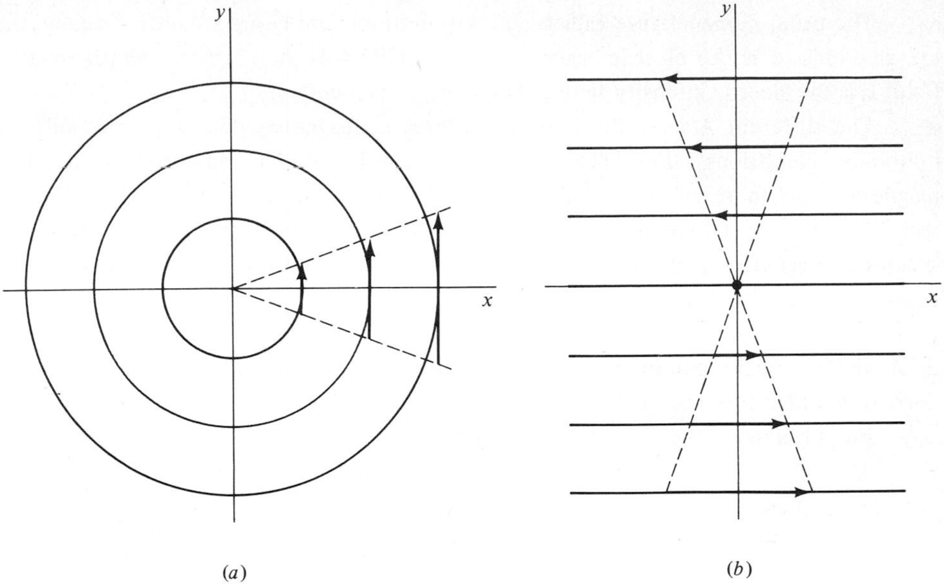

FIGURE 5-3

$$\mathbf{A}_3 = \hat{\mathbf{x}}\left(B_0 \frac{y}{2}\right) + \hat{\mathbf{y}}\left(B_0 \frac{3x}{2}\right) = \mathbf{A}_1 + \nabla(B_0 xy) = \mathbf{A}_2 + \nabla\left(\frac{3}{2} B_0 xy\right)$$

In magnetostatics it is possible to set up a simple rule: There is a unique **A**, for a given **B**, that depends only on the distribution of current density in some finite region. This is given by

$$\mathbf{A}(\mathbf{r}_t) = \frac{\mu_0}{4\pi} \int_{v} \frac{\mathbf{J}(\mathbf{r}_s)}{R}\, d\tau_s$$

without any additive function ∇f. The term ∇f is zero. One distribution of current density $\mathbf{J}_1$ gives $\mathbf{A}_1$; another distribution $\mathbf{J}_2$ gives $\mathbf{A}_2$; etc. All give the same **B** in some subregion. Any **A** can be derived from any other **A** that yields the same **B** merely by adding the proper ∇f. As in electrostatics, it is possible to add an arbitrary constant by selecting the point at which **A** vanishes.

The situation with time-varying fields is different. In radiation fields the sources have essentially become decoupled from the current sources that produced them. The additive function ∇f there can have any value, subject only to some constraints on f imposed by a differential equation.

The usual nomenclature calls ϕ and **A** potentials, and **E** and **B** fields; actually, they are all fields. ϕ is the electric scalar potential field; **A** is the magnetic vector potential field; **E** is the electric intensity field; **B** is the magnetic field.

The different **A**'s for the same **B**, at least in magnetostatics, represent different boundary conditions; thus, different source configurations can produce the same magnetic field in some given region. While the forces on conductors within the field region are unchanged by shifting from one **A** to another, both the region in which the specified magnetic field exists and the source configuration which produces it are connected with the choice of **A**.

2. A finite straight conductor Find the contribution to **A** at P, in Fig. 5-4, of the current $\hat{\mathbf{z}}I$ which flows in the length of the conductor between $z = -L$ and $z = +L$.

For a filamentary circuit the expression

$$\mathbf{A} = \frac{\mu_0}{4\pi}\int \frac{\mathbf{J}}{R}\,d\tau \qquad \text{becomes} \qquad \mathbf{A} = \frac{\mu_0}{4\pi}\int \frac{I\,d\mathbf{r}}{R}$$

So
$$\mathbf{A} = \frac{\mu_0}{4\pi}\int_{-L}^{L} \frac{\hat{\mathbf{z}}I\,dz}{R} = \hat{\mathbf{z}}\,\frac{\mu_0 I}{4\pi}\int_{-L}^{L} \frac{dz}{\sqrt{r^2 + z^2}} = \hat{\mathbf{z}}\,\frac{\mu_0 I}{4\pi}\ln\left[\frac{L + \sqrt{r^2 + L^2}}{-L + \sqrt{r^2 + L^2}}\right]$$

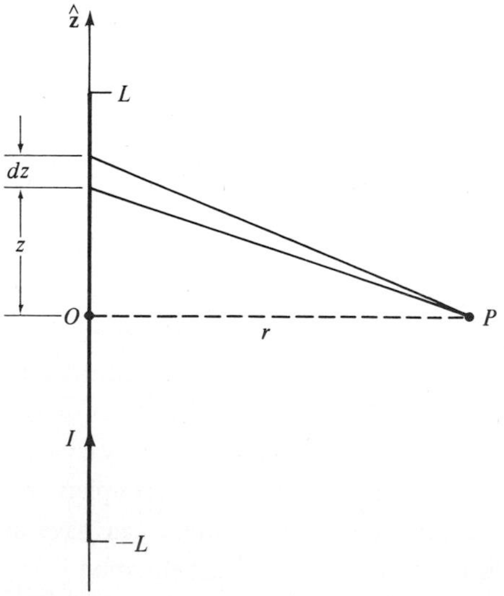

FIGURE 5-4

This is the vector potential produced by the current in a finite straight conductor. If $L \to \infty$ then $A \to \infty$ also, according to

$$\hat{\mathbf{z}}\,\frac{\mu_0 I}{4\pi}\ln\left(\frac{4L^2}{r^2}+1+\cdots\right)\approx\hat{\mathbf{z}}\,\frac{\mu_0 I}{2\pi}\ln\left(\frac{2L}{r}\right)$$

The integral for **A** becomes infinite when current exists at infinity. The value of $\mathbf{B} = \nabla \times \mathbf{A}$ remains finite, however, even though **A** becomes infinite. This was shown in Example 1 of the previous section:

$$\mathbf{B} = \hat{\boldsymbol{\phi}}\,\frac{\mu_0 I}{2\pi r}$$

The integral expressions for **A**, above, are inapplicable when there are current sources at infinity.

3. A for a circular loop Figure 5-5 shows an idealized circuit, a loop of radius $|\mathbf{r}_s| = a$, in the xy plane. The test point P may be first taken in the xz plane; let it have the cartesian coordinates $(r,0,z)$. A current element $I_S\,d\mathbf{r}_s$, distant R from P, produces a contribution $d\mathbf{A}$ at P which has only x and y components since $I_S\,d\mathbf{r}_s$ has no z component. But there is another current element, at an equal but opposite angle from the x axis, which is at the same distance R from P; and it produces a dA_x at P which cancels the first one. Its dA_y there, furthermore, just equals the one above. Since this cancellation of the x component is true for all current elements, there is only an A_y component at P. Generalizing to an arbitrary point, not necessarily in the xz plane: there will only be a component

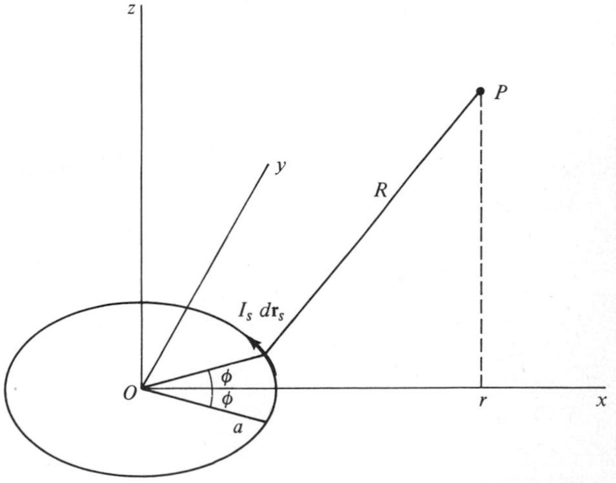

FIGURE 5-5

perpendicular to the plane containing the z axis and the point, i.e., there will only be an A_ϕ component. Once this is established it is possible, without any loss in generality, to pick axes for simplicity, such that P lies in the xz plane.

The distance from the current element $Id\mathbf{r}_s$ to the particular point P in the xz plane is, letting I_S become simply I for convenience,

$$R = [(r - a\cos\phi)^2 + (0 - a\sin\phi)^2 + (z-0)^2]^{1/2}$$
$$= [z^2 + r^2 + a^2 - 2ra\cos\phi]^{1/2}$$

Also $$I\,d\mathbf{r}_s = I\,d(\hat{\mathbf{x}}r_s\cos\phi + \hat{\mathbf{y}}r_s\sin\phi) = Ia(-\hat{\mathbf{x}}\sin\phi + \hat{\mathbf{y}}\cos\phi)d\phi$$

so the component of $I\,d\mathbf{r}_s$ in the y direction is $Ia\cos\phi\,d\phi$. The y direction for the particular point P is perpendicular to the xz plane in which P lies; for an arbitrary point this would be the ϕ component. We may integrate the effects of all the y components of $I\,d\mathbf{r}_s$ to obtain the y component of $\mathbf{A}$ at P; then generalize by calling this the ϕ component. Thus,

$$\mathbf{A} = \hat{\boldsymbol{\phi}}\frac{\mu_0 Ia}{4\pi}(2)\int_0^\pi \frac{\cos\phi\,d\phi}{(z^2 + r^2 + a^2 - 2ra\cos\phi)^{1/2}}$$

This integral cannot be evaluated in terms of elementary functions, but it can be thrown into a form involving elliptic integrals. The latter are so named because they arise in the determination of the length of an ellipse. This derivation follows.

The denominator may be written

$$\left[z^2 + (r+a)^2 - 4ra\left(\frac{1+\cos\phi}{2}\right)\right]^{1/2}$$

Set $\sin^2\alpha = (1 + \cos\phi)/2$; this is equivalent to putting $\phi = \pi - 2\alpha$. Then the denominator becomes

$$[z^2 + (r+a)^2]^{1/2}\left\{1 - \left[\frac{4ra}{z^2 + (r+a)^2}\right]\sin^2\alpha\right\}^{1/2}$$

With $$k \equiv \sqrt{\frac{4ra}{z^2 + (r+a)^2}}$$

the denominator is now

$$\frac{2}{k}\sqrt{ra}\,\sqrt{1 - k^2\sin^2\alpha}$$

The numerator is $2(1 - \sin^2 \alpha)\, d\alpha$; the upper and lower limits of the integral are 0 and $\pi/2$, respectively. Therefore

$$\mathbf{A} = \hat{\boldsymbol{\phi}} \frac{\mu_0 I a}{2\pi} \int_{\pi/2}^{0} \frac{2(1 - 2\sin^2 \alpha) d\alpha}{(2/k)\sqrt{ra}\sqrt{1 - k^2 \sin^2 \alpha}}$$

$$= \hat{\boldsymbol{\phi}} \frac{\mu_0 I k}{2\pi} \sqrt{\frac{a}{r}} \left[2 \int_0^{\pi/2} \frac{\sin^2 \alpha\, d\alpha}{\sqrt{1 - k_2 \sin^2 \alpha}} - \int_0^{\pi/2} \frac{d\alpha}{\sqrt{1 - k^2 \sin^2 \alpha}} \right]$$

The second integral, a function of k, is defined as $K(k)$:

$$K(k) = \int_0^{\pi/2} \frac{d\alpha}{\sqrt{1 - k^2 \sin^2 \alpha}}$$

It is called the elliptic integral (or the complete elliptic function) of the first kind. Its values are tabulated in many places, e.g., Jahnke and Emde, "Table of Functions," or Peirce, "A Short Table of Integrals."[1] For values of the argument ranging from $k = 0$ to $k = 0.9998$ the function $K(k)$ varies monotonically only from $\pi/2$ to 5.44; but as $k \to 1$, beyond this, $K(k) \to \infty$.

The first integral is more complicated. It is given in terms not only of $K(k)$ but also of $E(k)$, the elliptic integral (or the complete elliptic function) of the second kind, where, by definition,

$$E(k) = \int_0^{\pi/2} \sqrt{1 - k^2 \sin^2 \alpha}\, d\alpha$$

This integral function is also tabulated in the reference works above. $E(k)$ is a function which varies only slightly: it goes monotonically from $E(0) = \pi/2$ to $E(1) = 1$. Then

$$\frac{1}{k^2}\left[K(k) - E(k)\right] = \frac{1}{k^2} \int_0^{\pi/2} \left(\frac{1}{\sqrt{1 - k^2 \sin^2 \alpha}} - \sqrt{1 - k^2 \sin^2 \alpha} \right) d\alpha$$

$$= \int_0^{\pi/2} \frac{\sin^2 \alpha\, d\alpha}{\sqrt{1 - k^2 \sin^2 \alpha}}$$

[1] Jahnke and Emde, p. 43, McGraw-Hill, 1960; Peirce, 3d rev. ed., p. 121, Ginn, Boston, 1929.

The final results of the derivation for **A** in terms of the elliptic integrals is

$$\mathbf{A} = \hat{\mathbf{\Phi}} \frac{\mu_0 I}{2\pi} \sqrt{\frac{a}{r}} \left[\left(\frac{2}{k} - k \right) K(k) - \frac{2}{k} E(k) \right]$$

where

$$k = \sqrt{\frac{4ra}{z^2 + (r+a)^2}}$$

For any given value of a, picking a point P fixes r and z; this determines k. $K(k)$ and $E(k)$ may be found from the tables. $\mathbf{A} = \hat{\mathbf{\Phi}} A$ can then be computed. Figure 5-6 shows the variation of A in the planes $z = 0$, $z = a$, and $z = 2a$. The lines of constant A are circles about the z axis.

4. An infinitely long solenoid Find the **A** produced by an infinitely long solenoid, neglecting the pitch of the winding.

The solenoid may be considered to be made up of n loops per unit length along the z axis, each loop producing the **A** of Example 3, above. But now it is possible to integrate the expression for **A** directly, without going to elliptic integrals. First consider a solenoid of finite length, extending from $z = -D$ to $z = +D$.

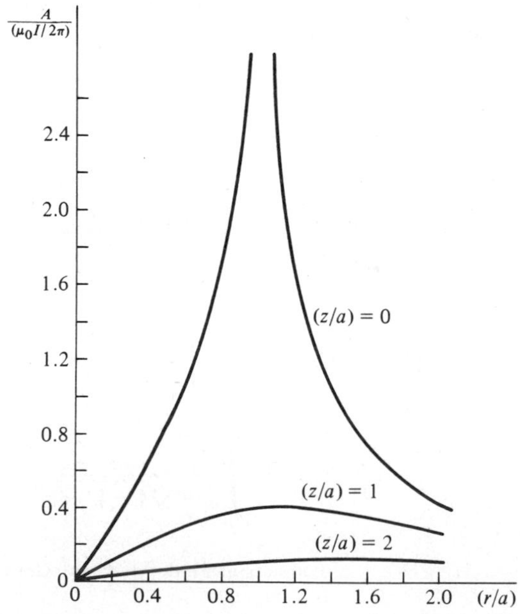

FIGURE 5-6

$$\mathbf{A} = \hat{\boldsymbol{\phi}} \frac{\mu_0 n I a}{2k} \int_{-D}^{D} dz \int_0^{\pi} \frac{\cos \phi \, d\phi}{(z^2 + r^2 + a^2 - 2ra \cos \phi)^{1/2}}$$

Reversing the order of integration gives

$$\mathbf{A} = \hat{\boldsymbol{\phi}} \frac{\mu_0 n I a}{2\pi} \int_0^{\pi} \cos \phi \, d\phi \int_{-D}^{D} \frac{dz}{\sqrt{z^2 + C^2}}$$

where $C^2 = r^2 + a^2 - 2ra \cos \phi$.

As it stands, the second integral will give infinity when $D \to \infty$. This may be overcome, oddly enough, by changing the limits of integration of the first integral. Instead of integrating from 0 to π, suppose we integrate first from 0 to $\pi/2$ and then from $\pi/2$ to π:

$$\mathbf{A} = \hat{\boldsymbol{\phi}} \frac{\mu_0 n I a}{2\pi} \left[\int_0^{\pi/2} \cos \phi \, d\phi \int_{-z}^{z} \frac{dz}{\sqrt{z^2 + C^2}} + \int_{\pi/2}^{\pi} \cos \phi' \, d\phi' \int_{-z}^{z} \frac{dz}{\sqrt{z^2 + C_1^2}} \right]$$

Here also $C_1^2 = r^2 + a^2 - 2ra \cos \phi'$; but in the second term $\cos \phi'$ assumes only the negative of the values $\cos \phi$ takes on in the first term. The result may, then, also be obtained by setting $\phi' = \pi - \phi$. So $\phi' = \pi/2$ and $\phi' = \pi$ (the lower and upper limits, respectively) become $\phi = \pi/2$ and 0, while C_1^2 becomes $r^2 + a^2 + 2ra \cos \phi$. Therefore,

$$\mathbf{A} = \hat{\boldsymbol{\phi}} \frac{\mu_0 n I a}{2k} \int_0^{\pi/2} \cos \phi \, d\phi \left[\int_{-D}^{D} \frac{dz}{\sqrt{z^2 + C^2}} - \int_{-D}^{D} \frac{dz}{\sqrt{z^2 + C_1^2}} \right]$$

Suppose we now consider this in the limit when $D \to \infty$:

$$\mathbf{A} = \hat{\boldsymbol{\phi}} \frac{\mu_0 n I a}{2\pi} \int_0^{\pi/2} \cos \phi \, d\phi \left\{ \lim_{D \to \infty} \left[\ln \frac{D + \sqrt{D^2 + C^2}}{-D + \sqrt{D^2 + C^2}} - \ln \frac{D + \sqrt{D^2 + C_1^2}}{-D + \sqrt{D^2 + C_1^2}} \right] \right\}$$

$$= \hat{\boldsymbol{\phi}} \frac{\mu_0 n I a}{2\pi} \int_0^{\pi/2} \cos \phi \, d\phi \left\{ \lim_{D \to \infty} \ln \left[\frac{D + \sqrt{D^2 + C^2}}{-D + \sqrt{D^2 + C^2}} \, \frac{-D + \sqrt{D^2 + C_1^2}}{D + \sqrt{D^2 + C_1^2}} \right] \right\}$$

Expanding $\sqrt{D^2 + C^2}$ in a power series gives $D(1 + \frac{1}{2} \cdot C^2/D^2 + \cdots)$, with a similar result for $\sqrt{D^2 + C_1{}^2}$. The bracket in the integrand becomes

$$\lim_{D\to\infty} \ln\left[\frac{2D}{(C^2/2D)}\,\frac{(C_1{}^2/2D)}{2D}\right] = \lim_{D\to\infty} \ln\left(\frac{C_1{}^2}{C^2}\right)$$

Therefore when $D \to \infty$ value of the bracket is

$$\ln\left(\frac{C_1{}^2}{C^2}\right) = \ln\left(\frac{r^2 + a^2 + 2ra\cos\phi}{r^2 + a^2 - 2ra\cos\phi}\right)$$

Then

$$\mathbf{A} = \hat{\boldsymbol{\phi}}\,\frac{\mu_0 nIa}{2\pi}\int_0^{\pi/2} \cos\phi \ln\left(\frac{r^2 + a^2 + 2ra\cos\phi}{r^2 + a^2 - 2ra\cos\phi}\right) d\phi$$

Using integration by parts this integral may be evaluated. The details are left for Prob. 6. The calculation yields the result

$$\mathbf{A} = \hat{\boldsymbol{\phi}}\,\frac{\mu_0 nIa}{2\pi}\left[\frac{\pi(r^2 + a^2)}{2ra}\right]\left(1 - \frac{|r^2 - a^2|}{r^2 + a^2}\right)$$

For the region outside the infinitely long solenoid

$$r > a \qquad \text{and} \qquad |r^2 - a^2| = r^2 - a^2$$

so

$$\mathbf{A} = \hat{\boldsymbol{\phi}}\left(\frac{\mu_0 nIa^2}{2}\right)\frac{1}{r}$$

Within the solenoid $|r^2 - a^2| = a^2 - r^2$ and

$$\mathbf{A} = \hat{\boldsymbol{\phi}}\left(\frac{\mu_0 nI}{2}\right) r \qquad (\text{cylindrical } r)$$

Figure 5-7 shows **A** and **B** for the infinite solenoid.

The results of this calculation are significant because there is a region where **B** vanishes but **A** is finite. The Aharanov-Bohm experiment (see Chap. 14, Sec. 14-2, Example 4) utilizes the interference fringes produced by two coherent electron beams (i.e., both beams are produced by one source); the two beams pass around the outside of a solenoid whose length is very large compared to its radius. One beam passes on one side and one on the other, both beams coming from a beam splitter which separates them from each other in the original electron beam. The beams pass only through a region

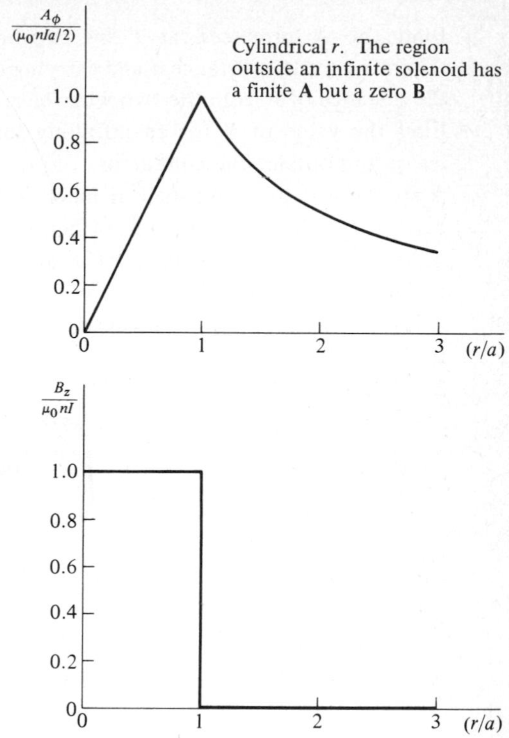

FIGURE 5-7

where $B = 0$. A fringe system is produced in a plane where the beams are brought together again. It is found experimentally that a shift in the fringes occurs, when current passes through the solenoid; though the force on any one electron is unaffected by the absence or presence of **A**. Further, the magnitude of the shift agrees with the prediction of quantum mechanics. This effect has no classical analog and is intrinsically quantum mechanical. The conclusion is nevertheless inescapable: **A** possesses physical significance.

PROBLEMS

1 Using $\Phi = \oint \mathbf{A} \cdot d\mathbf{r}$ find the **A** produced within a straight long conductor of length L and radius R. The conductor carries a current I, which is uniformly distributed throughout the conductor.

2 Repeat Prob. 1, but utilizing the equation

$$\mathbf{A} = \frac{\mu_0}{4\pi} \oint \frac{I\,d\mathbf{r}}{R}$$

3 Find the **A** produced at P by two infinitely long filamentary parallel wires, separated by the distance d and carrying equal currents I in opposite directions. Let the distance of P from the two wires be r_1 and r_2.

4 Find the value of **B** for an infinitely long conductor from the expressions for **A** inside and outside the conductor.

5 A spherical conducting shell of inner radius r_1 and outer radius r_2 conducts a total current I with a uniform current density **J** which flows azimuthally. Find **A** and **B** at the center of the sphere. (The answers obtain only at one point, not over a region. It is not possible to have $\mathbf{A} = 0$, $\mathbf{B} \neq 0$ in a region without violating $\mathbf{B} = \nabla \times \mathbf{A}$.) This problem is unrealistic but has a simple **J**. Use care in choosing dS such that $\mathbf{I} = \mathbf{J}\, dS$.

6 Show that

$$\int_0^{\pi/2} \cos\phi \ln\left(\frac{r^2 + a^2 + 2ra\cos\phi}{r^2 + a^2 - 2ra\cos\phi}\right) d\phi$$

equals

$$\frac{\pi(r^2 + a^2)}{2ra}\left(1 - \frac{|r^2 - a^2|}{r^2 + a^2}\right)$$

This result is utilized in Example 4.

7 Assume the following results for the derivatives of the elliptic integrals:

$$\frac{dK}{dk} = -\frac{K}{k} + \frac{E}{k(1 - k^2)} \qquad \frac{dE}{dk} = -\frac{K}{k} + \frac{E}{k}$$

Find $\mathbf{B} = \nabla \times \mathbf{A}$, in cylindrical coordinates, for the circular loop of Example 3. This problem is fairly long; its only justification is the utility of the answer.

8 Find the value of **B** on the axis of symmetry perpendicular to a circular loop at its center by specializing the results of Prob. 7. (This answer has already been obtained directly from the law of Biot-Savart.) What is the value of **A** there?

9 From the answer to Prob. 7, find **B** in the plane of the loop.

10 From the expression for **A** produced by a circular loop, graph **A** versus (z/a) for $r = 2a$, $r = a$, $r = a/2$. Let z/a go from -4 to $+4$ in unit steps.

11 A Helmholtz coil is constructed as shown in Fig. 5-8. Both coils carry the current I in the same sense. Show that, for the resultant field at the center of the system,

$$\frac{dB_z}{dz} = \frac{d^2 B_z}{dz^2} = \frac{d^3 B_z}{dz^3} = 0$$

Also, find B_z at the center of the system. Such a system is employed either to produce a very uniform field in a small region of space or else to neutralize the earth's magnetic field, plus other stray fields, in a small region.

12 A sphere of radius a, containing a volume charge density ρ = const, rotates with constant angular velocity about a diameter. Neglect any effects of centrifugal force. What is **A** at the center of the sphere?

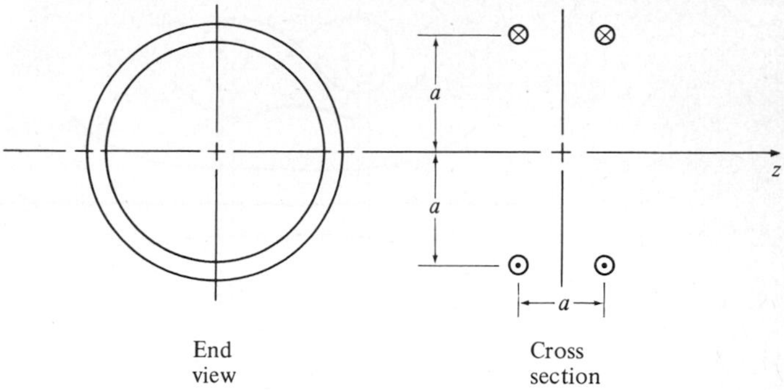

FIGURE 5-8

13 Find the contribution to the **A** at *P* that is due to the current flowing in the straight wire of Fig. 5-9 between *M* and *N*.

14 Figure 5-10 shows some lines of **B** produced by a Helmholtz coil system. Comment on the crossing of the lines of **B** at the point *P*.

15 In Prob. 12, what is the value of **A** at a point z on the axis of rotation when we have (a) $z > a$, (b) $z = a$, (c) $z < a$.

16 The **A** produced by a solenoid of length $2D$ and radius a, from Example 4, is

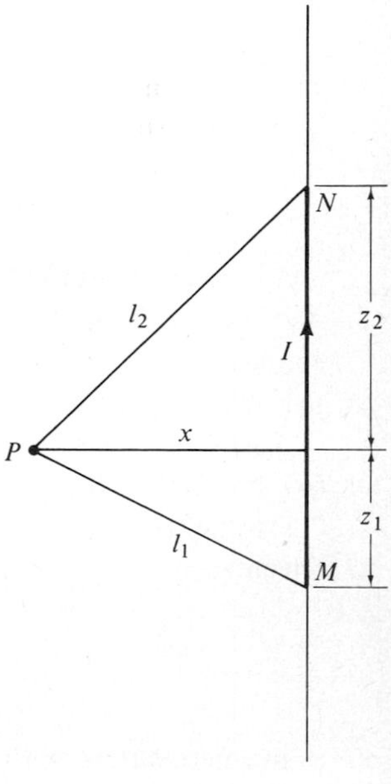

FIGURE 5-9

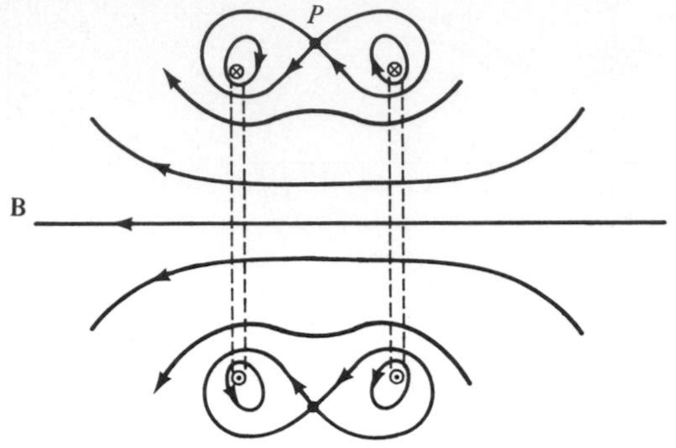

Some of the lines of **B** produced by the two turnes of a Helmholtz coil

FIGURE 5-10
(Modeled after Fig. 7.12b from William T. Scott, "The Physics of Electricity and Magnetism," 2d ed., Wiley, New York, 1966, adapted from E. Durand, "Electrostatique et Magnetostatiquem," Masson et Cie, Paris, 1953.)

$$\hat{\boldsymbol{\phi}}\frac{\mu_0 nIa}{2\pi}\int_0^{\pi/2}\cos\phi\, d\phi\,\ln\left[\frac{-D^2+\sqrt{(D^2+C^2)(D^2+C_1{}^2)}+D(\sqrt{D^2+C_1{}^2}-\sqrt{D^2+C^2})}{-D^2+\sqrt{(D^2+C^2)(D^2+C_1{}^2)}-D(\sqrt{D^2+C_1{}^2}-\sqrt{D^2+C^2})}\right]$$

where $C^2 = L^2 + a^2 - 2La\cos\phi$ and $C_1{}^2 = L^2 + a^2 + 2La\cos\phi$. Reduce this expression to a simpler form for the case $D/a = 3$.

17 Given

$$\mathbf{B}_1 = \hat{\mathbf{z}}B_0\left[-(16+2\sqrt{2})\frac{r}{r_0}+(8+2\sqrt{2})\right] \qquad 0 \leqslant r \leqslant \frac{r_0}{2}$$

$$\mathbf{B}_2 = \hat{\mathbf{z}}B_0\sqrt{\frac{r_0}{r}} \qquad \frac{r_0}{2} \leqslant r \leqslant \frac{3r_0}{2}$$

(cylindrical r). Find A_ϕ in these two regions, eliminating all arbitrary constants. Assume $A_r = 0$, $A_z = 0$, $A_\phi \neq A_\phi(z)$. This case will be considered further in connection with the theory of the betatron.

18 (*a*) What conditions are employed in Prob. 17 that permit the determination of the arbitrary constants?

(*b*) Prove that $\mathbf{B} = \nabla \times \mathbf{A}$ is both a necessary and sufficient condition for $\nabla \cdot \mathbf{B} = 0$ to be true.

19 Given

$$\mathbf{A}(\mathbf{r}_t) = \frac{\mu_0}{4\pi}\int_{\mathcal{V}}\frac{\mathbf{J}(\mathbf{r}_s)}{R}\,d\tau_s$$

in magnetostatics, with $\mathcal{V}$ finite. Find $\nabla \cdot \mathbf{A}$.

20 Instead of employing the vector potential **A** to give the magnetic field **B** it is possible to use the Euler potentials α and β, such that $\mathbf{B} = (\nabla\alpha) \times (\nabla\beta)$. Show that (*a*) every field line of **B** is the intersection of a surface α = const with a surface β = const; (*b*) $\nabla \cdot \mathbf{B} = 0$; (*c*) if $\mathbf{A} \equiv \alpha\nabla\beta$ then $\mathbf{A} \cdot \mathbf{B} = 0$; (*d*) α and β are not unique. For further information see Euler Potentials, D. P. Stern, *Amer. J. Phys.*, **38**:494 (1970).

5-2 INDUCTANCE

In electrostatics a system of two conductors has a magnitude of charge, on either one, which is proportional to the potential difference between the two: $q = C\,\Delta\phi$. The constant of proportionality is a geometric factor—the capacitance. In magnetostatics, similarly, a system of two individual circuits has a magnetic flux passing through one circuit which is produced by, and proportional to, the current in the other circuit. The constant of proportionality is also a geometric factor—the mutual inductance.

Let Φ_{12} be the flux, passing through the circuit 1, that is produced by the current I in circuit 2

$$\Phi_{12} = \int_{\Sigma_1} \mathbf{B} \cdot d\mathbf{S}_1 = \int_{\Sigma_1} (\nabla \times \mathbf{A}) \cdot d\mathbf{S}_1 = \oint_{C_1} \mathbf{A} \cdot d\mathbf{r}_1$$

where C_1 is a contour which coincides with filamentary circuit 1 and Σ_1 is any open area bounded by C_1. This is shown in Fig. 5-11, where the circuits have been idealized. But

$$\mathbf{A} = \frac{\mu_0}{4\pi} \oint_{C_2} \frac{I_2\, d\mathbf{r}_2}{R}$$

where C_2 is the contour of the second filamentary circuit. Then

$$\Phi_{12} = \oint_{C_1} \left[\frac{\mu_0}{4\pi} \oint_{C_2} \frac{I_2\, d\mathbf{r}_2}{R}\right] \cdot d\mathbf{r}_1 = \left[\frac{\mu_0}{4\pi} \oint_{C_1} \oint_{C_2} \frac{d\mathbf{r}_1 \cdot d\mathbf{r}_2}{R}\right] I_2$$

or

$$\Phi_{12} = M_{12} I_2$$

where

$$M_{12} = \frac{\mu_0}{4\pi} \oint_{C_1} \oint_{C_2} \frac{d\mathbf{r}_1 \cdot d\mathbf{r}_2}{R}$$

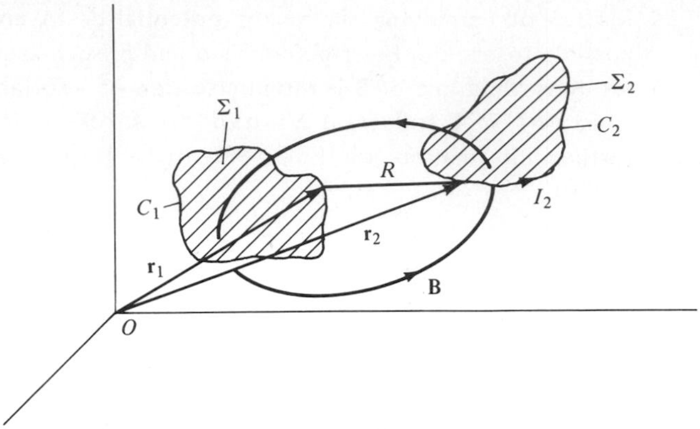

FIGURE 5-11

This is called the Neumann formula. M_{12} is rather difficult to compute; fortunately, it turns out that it can be measured quite simply.

From the symmetry of this expression for M_{12} it is seen that the current $I_1 = I_2$ flowing in circuit 1 would produce identically the same flux passing through circuit 2:

$$\Phi_{21} = M_{21}I_1 = \Phi_{12}$$

Then $M_{12} = M_{21} = M$. In general

$$\Phi_{12} = M_{12}I_2$$
$$\Phi_{21} = M_{21}I_1$$

M is called the mutual inductance between circuits 1 and 2. The unit of mutual inductance, the weber per ampere, is called the henry; its dimensions are those of $[\mathrm{m}^{-2}\ \mathrm{kg}\ \mathrm{s}^{-2}\ \mathrm{A}^{-2}]$.

Suppose we consider one circuit instead of two. The self-produced flux that links the circuit is

$$\Phi_{11} = \oint_C \mathbf{A} \cdot d\mathbf{r} = \oint_C \left[\frac{\mu_0}{4\pi} \oint_C \frac{I\, d\mathbf{r}'}{R} \right] \cdot d\mathbf{r} = \left[\frac{\mu_0}{4\pi} \oint_C \oint_C \frac{d\mathbf{r}' \cdot d\mathbf{r}}{R} \right] I$$

or

$$\Phi_{11} = LI$$

The two equations above may be combined into one expression by setting $L_1 = M_{11}$, $L_2 = M_{22}$, etc. Thus

$$\boxed{\Phi_{ij} = M_{ij} I_j} \qquad \text{(M-10)}$$

To compare (M-10) with its electrostatic mate $q = C\,\Delta\phi$, Eq. (E-10), it is better first to rearrange Eq. (E-10). In $\Phi = MI$ we consider I as the cause and Φ as the effect. Writing Eq. (E-10), correspondingly, as $\Delta\phi = (1/C)q$ would mean that two equal and opposite charges of magnitude q produce the difference of potential $\Delta\phi$. It then becomes evident that M, which is proportional to μ_0, is the analog of $(1/C)$, which is proportional to $1/\epsilon_0$.

The coefficient L measures the ratio of the flux linking a circuit to the current, in that circuit, that produces the flux. L is called the self-inductance of the circuit; like M, it is measured in henrys. The integral in the Neumann expression is taken twice completely around C. It sums the contributions of the differential elements $d\mathbf{r}'$ and $d\mathbf{r}$, both on C, which are separated by the distance R. This integral would diverge even if there were only one pair of elements for which $R = 0$; but there are actually an infinite number of such pairs of elements. For a true filamentary circuit (i.e., one of infinitesimal thickness and infinite current density) L becomes infinite. The reason this happens is that B is proportional to r^{-1} near the conductor. As $r \to 0$, $B \to \infty$ fast enough so that, for an $\ell \times R$ rectangle with one ℓ side parallel to the conductor at a distance a and the other parallel side at distance R,

$$\lim_{a\to 0} \Phi = \lim_{a\to 0} \int \mathbf{B} \cdot d\mathbf{S} \propto \lim_{a\to 0} \int_a^R \frac{1}{r}(\ell\, dr) \to \infty$$

For the more realistic case of conductors of finite thickness this difficulty vanishes. The computation of L, however, is not usually made by a generalization of the double integral above; it is more convenient to work from the definition that L is the self-flux linking a circuit, per unit current producing it. If care is taken in defining what is meant by flux linking the circuit then the self-inductance measures the effect, on each other, of the different infinitesimal filaments in a wire of finite size.

The quantities L and M introduced here will not be further utilized for the moment. They find much use later, however: M in connection with Faraday's law (Chap. 11) and L in connection with electric circuits (Chap. 12).

Examples

1. A cylindrical sheath Using the concept of flux produced per unit current we will calculate the self-inductance per unit length of an infinitely long straight conductor, made of a shell of radius a, which has infinitesimal thickness. The wire is considered part of an idealized coaxial circuit, the return being made by a second concentric, infinitesimally thin, shell of infinite radius.

In cylindrical coordinates the magnetic field is

$$B_\phi = \frac{\mu_0 I}{2\pi r}$$

The total flux crossing a rectangle of length ℓ and width $(R - a)$ between the two shells is, in the limit when $R \to \infty$,

$$\Phi_{11} = \lim_{R\to\infty} \int_a^R \frac{\mu_0 I}{2\pi r} \ell \, dr = \lim_{R\to\infty} \frac{\mu_0 I \ell}{2\pi} \ln\left(\frac{R}{a}\right)$$

Note that this flux is entirely external to the cylindrical sheath that produces it; there is no contribution to Φ_{11} in the region $r < a$ (Ampere's law shows that $\mathbf{B} = 0$ there).

The inductance per unit length L_ℓ is

$$L_\ell = \frac{1}{I}\left(\frac{\Phi_{11}}{\ell}\right) = \frac{1}{I\ell}\left[\lim_{R\to\infty} \frac{\mu_0 I \ell}{2\pi} \ln\left(\frac{R}{a}\right)\right] = \infty$$

This corresponds to the fact that the total flux produced per unit length by a straight wire is infinite.

A simple variation of the example here is the case of a coaxial cable where both conductors are shells of infinitesimal thickness but the outer conductor is at the finite radius b:

$$L_\ell = \frac{\mu_0}{2\pi} \ln\left(\frac{b}{a}\right)$$

2. A coaxial cable What is the self-inductance per unit length of a coaxial cable when the inner conductor is solid, of radius a, and has uniform current density, while the outer conductor is a shell of infinitesimal thickness at radius b?

a Inside the inner conductor: We will calculate the self-inductance in a manner that brings out the fact that it is the result of the mutual inductance between the different differential current elements constituting the current of finite density. L is, therefore, really a special case of M for wires of finite size, quite unlike the case with filamentary circuits.

Consider a differential rectangle of width dr and length ℓ parallel to the cable axis and at a distance r from it, with $r < a$. The current that flows in the cylinder of radius r, i.e., that portion of the total current closer to the axis than the rectangle, is $(\pi r^2/\pi a^2)I$. The magnetic field it produces at the rectangle is $(\mu_0/2\pi)[(r^2/a^2)I]\,1/r$. This current, therefore, generates the flux $[(\mu_0 I/2\pi a^2)r]\,\ell\,dr$ through the rectangle. The total flux linkage throughout the inner conductor–the mutual flux produced at the different places by currents flowing elsewhere (closer to the axis) rather than the self-flux produced by the actual currents flowing at those places–is, then,

$$\int_0^a \left(\frac{\mu_0 I\ell}{2\pi a^2}\right) r\,dr = \frac{\mu_0 I\ell}{4\pi}$$

This is also the total flux.

The flux we have calculated is a mutual flux when we consider that it is due to the interaction between the various component differential elements of the inner conductor. However, we may also consider this as a self-flux Φ_{11} produced by the inner conductor taken as an entity. (From Example 1, there is no contribution to the flux in the inner conductor that is due to the current in the outer conductor.) Then a factor of $\frac{1}{2}$ is necessary (see Example 1 of the next section). So $\frac{1}{2}(\mu_0 I\ell/4\pi) = LI$ and

$$L_{\ell(\text{int})} = \frac{\mu_0}{8\pi}$$

The factor of $\frac{1}{2}$ may also be justified in a slightly different manner. Suppose we integrate the flux through the above rectangle from $r = r_1$ to $r = a$:

$$\int_{r_1}^a \left(\frac{\mu_0 I}{2\pi a^2} r\right) \ell\,dr = \frac{\mu_0 I\ell}{4\pi a^2}(a^2 - r_1{}^2)$$

This flux links a fraction $(2\pi r_1\,dr_1)/\pi a^2$ of the inner conductor (the cylindrical shell between r_1 and $r_1 + dr_1$). The flux linkage to this fraction of the inner conductor is, thus, $[(2\pi r_1\,dr_1)/\pi a^2](\mu_0 I\ell/4\pi a^2)(a^2 - r_1{}^2)$. The flux linkage to the entire conductor is

$$\frac{\mu_0 I\ell}{2\pi a^4}\int_0^a (a^2 r_1 - r_1{}^3)\,dr_1 = \frac{\mu_0 I\ell}{8\pi}$$

So, again,

$$L_{\ell(\text{int})} = \frac{\mu_0}{8\pi}$$

b Outside the inner conductor: All the flux in the annular region between the two conductors links the entire current I of the inner conductor to the outer one. The result here is identical with that of the previous example. This flux per unit current and per unit length is called the per-unit-length external self-conductance. Here

$$L_{\ell(\text{ext})} = \left(\frac{\mu_0}{2\pi}\right) \ln \left(\frac{b}{a}\right)$$

The total per-unit-length self-inductance, including both the internal and external components, is

$$L_\ell = \frac{\mu_0}{8\pi}\left(1 + 4 \ln \frac{b}{a}\right)$$

The two terms are roughly of the same order of magnitude for the usual commercial coaxial cables, where b is only slightly larger than a.

3. **An infinite straight conductor** Find the per-unit-length, internal and external, self-inductances of an infinitely long, straight conductor of radius a.

The per-unit-length internal self-inductance here is the same as that of the coaxial cable of Example 2 with $b \to \infty$:

$$L_{\ell(\text{int})} = \frac{\mu_0}{8\pi}$$

The per-unit-length external self-inductance here is the same as that in Example 1 with $b \to \infty$:

$$L_{\ell(\text{ext})} = \infty$$

The infinite flux per unit length (and the infinite $L_{\ell(\text{ext})}$ which results) is due to the infinite number of in-phase contributions from the other elements of the infinitely long wire.

For wires of radius a which are neither infinitely long nor perfectly straight these results do not apply. The external self-inductance has to be calculated for each particular geometry. The internal self-inductance, however, will be approximately given by the same result

$$L_{(\text{int})} = \frac{\mu_0 \ell}{8\pi}$$

where ℓ is the length of the circuit. This is so because the circuit dimensions are generally much larger than a, so within the wire the flux conditions do not differ much from those for an infinitely long line.

PROBLEMS

1 A long solenoid has n_1 turns per unit length and a cross-sectional area A_1. A short coil of slightly larger radius, placed around the solenoid at its center, has n_2 turns per unit length and a length ℓ_2, as shown in Fig. 5-12. Find the mutual inductance.

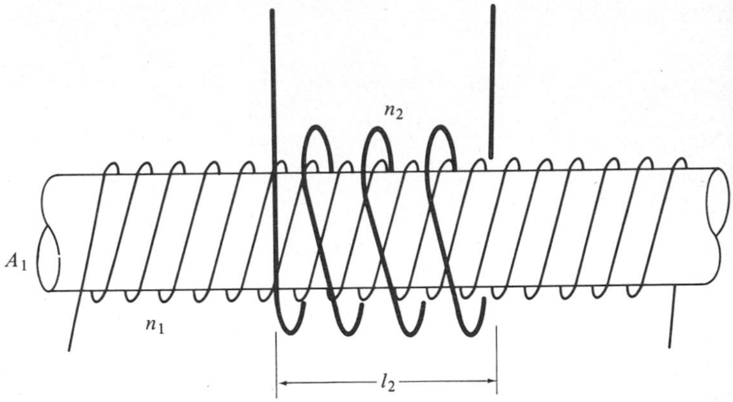

FIGURE 5-12

2 Find the self-inductance of a long solenoid of length ℓ and cross-sectional area A.

3 (*a*) Find the self-inductance of a toroid having N turns if the cross-sectional area is a square of dimensions $d \times d$, and the mean radius is R.
(*b*) Give an approximate answer when $R \gg d$.

4 In Prob. 1, let both windings be long, with lengths ℓ_1 and ℓ_2, and let $n_1 = n_2$. Find the mutual inductance.

5 Derive a relation between the mutual inductance M of two coils and their individual self-inductances, L_1 and L_2, for a specific geometric arrangement where $\Phi_{12} = k_2 \Phi_{22}$ and $\Phi_{21} = k_1 \Phi_{11}$. The k's are called coefficients of coupling.

6 Suppose the outer conductor of the coaxial cable of Example 2 has inner radius b and outer radius c. Find the additional per-unit-length internal self-inductance.

7 An infinitely long filamentary wire lies on the z axis. A second filamentary wire is arranged in a rectangular loop, with dimensions $a \times b$, and lies with the a dimension parallel to the z axis. The distance of the nearest side of the rectangle to the z axis is c, as in Fig. 5-13. Find M.

8 A mutual inductance M exists between two distinct coils when they are not connected together. The two coils, with self-inductance L_1 and L_2, respectively, are then connected together. (*a*) What is the resultant self-inductance when the two coils are connected in series? (*b*) What would the resultant self-inductance be if the two coils were connected in parallel?

9 Suppose the mutual inductance between the two coils of Prob. 8 were zero. What would the resultant self-inductance be? It can be shown, in the parallel case, that each coil produces the same flux; this will be done in Chap. 11.

10 In Prob. 3*a*, let the cross section be circular, of radius r. Find the self-inductance.

11 Give an approximate answer to Prob. 10 when $R \gg r$.

12 Find the external self-inductance of an idealized circuit consisting of a circular loop of mean radius R and cross-sectional radius r. Express the result in terms of the

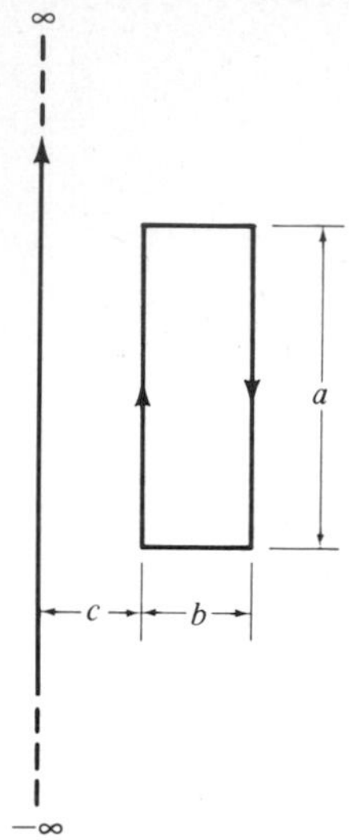

FIGURE 5-13

elliptic integrals of the first and second kind, $K(k)$ and $E(k)$, where k is $(2\sqrt{R(R-r)}/(2R-r)$. The answer bears a marked resemblance to the **A** determined by such a loop, as derived in Example 3 of the previous section.

13 Two idealized circular loops of filamentary conductor are coaxial and are separated by the distance z along the common axis. If the coils have the radii r_1 and r_2, where $r_1 \ll z$ and $r_2 \ll z$, find M.

14 In Prob. 13, suppose r_1 is not very small compared to z but, instead, r_2 is. Find M.

15 Find the total per-unit-length inductance (i.e., internal plus external) of an infinitely long pair of parallel wires of radius r, with a center-to-center separation D, if $D \gg r$.

16 A long solenoid has a tight winding of 30 turns/cm and a radius of 2 cm. A short coil of 100 turns is wound over the center of the solenoid. Find M.

17 In Prob. 13, suppose neither radius is small compared to the axial separation. Find M in terms of the elliptic integrals with argument k, where

$$k = \sqrt{\frac{4r_1 r_2}{z^2 + (r_1 + r_2)^2}}$$

18 In Prob. 17, let $z \ll r_1$, $z \ll r_2$, $r_1 \approx r_2$. Find M.

19 (*a*) Use the Neumann formula to set up the expression for the mutual inductance between two parallel, coaxial, circular, filamentary rings. One has the radius r_1, the other the radius r_2; the separation between their planes is D. Employ the azimuthal angles α_1 and α_2 as the independent variables.

(*b*) With $\alpha_2 = \alpha_1 + \theta$, θ variable, convert the expression above to an integral for θ.

20 Express the second result in Prob. 19 in terms of the elliptic functions of the first and second kind. Use the parameter

$$k = \sqrt{\frac{4r_1 r_2}{D^2 + (r_1 + r_2)^2}}$$

21 (*a*) By setting $\phi = \theta/2 - \pi/2$ in the answer to Prob. 19*b* and using the parameter k of Prob. 20, obtain another integral for M.

(*b*) Expand the denominator in powers of k^2 and find M for small k, i.e., when the separation between the turns is very large.

5-3 MAGNETOSTATIC ENERGY

In dealing with the energy of a system of currents the isolated and nonisolated cases must be considered separately, just as in electrostatics. There the nonisolated case occurs when batteries or power supplies, taken as constant-voltage sources, are connected to the charges. No matter what currents are drawn from these batteries, within their power handling capability, the potential difference between the battery terminals is stipulated as remaining constant. In magnetostatics the nonisolated case corresponds to circuits connected to constant-current supplies. No matter what the potential difference between their supply terminals, within certain limits, the current drain from a supply is defined as remaining constant. But in both instances the isolated case signifies the same thing, that there are no batteries, power supplies, or other sources of electrical energy in the system.

TWO NONISOLATED CIRCUITS We start a derivation which will show that the total energy is subdivided differently in the different cases. Consider the nonisolated case first. Assume a source circuit and a test circuit, as shown in Fig. 5-14*a*, each with a constant-current power supply. For convenience in visualization these may be considered to be far away, connected by twisted pairs to small gaps in idealized circuits, thereby eliminating any magnetic effects of the leads. The supplies will, nevertheless, enter the calculation. The flux $\Phi_S = \Phi_{SS} + \Phi_{ST}$ links the source circuit, where Φ_{SS} is due to the source current and is not shown in Fig. 5-14; Φ_{ST} is produced by the test circuit. Similarly, the flux at the test circuit is $\Phi_T = \Phi_{TT} + \Phi_{TS}$. The magnetic field **B** produced by the source circuit varies from one point to another but is constant in time.

The magnetic force on an element of the test circuit is $d\mathbf{F}_m = I_T\, d\mathbf{r}_t \times \mathbf{B}$. Suppose the test circuit is given a small virtual displacement $\delta\mathbf{l}$, which has the same value at all parts of the test circuit, so that the latter is not distorted. As a result of the displacement there is work done by the magnetic force on a test circuit element:

$$\delta^2 W_m = d\mathbf{F}_m \cdot \delta\mathbf{l} = (I_T\, d\mathbf{r}_t \times \mathbf{B}) \cdot \delta\mathbf{l} = I_T(d\mathbf{l} \times d\mathbf{r}_t) \cdot \mathbf{B}$$

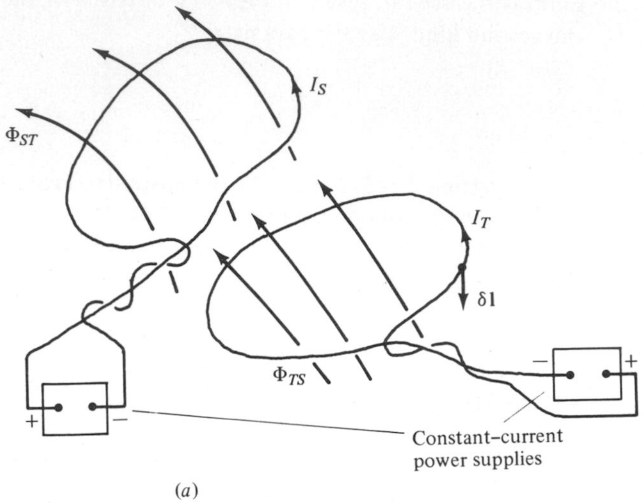

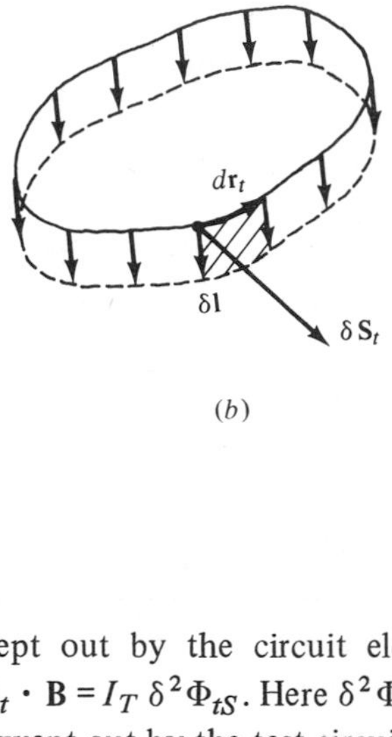

FIGURE 5-14

But $\delta\mathbf{l} \times d\mathbf{r}_t \equiv \delta^2\mathbf{S}_t$ is a differential area swept out by the circuit element in the displacement. See Fig. 5-14*b*. So $\delta^2 W_m = I_t\,\delta^2\mathbf{S}_t \cdot \mathbf{B} = I_T\,\delta^2\Phi_{tS}$. Here $\delta^2\Phi_{tS}$ is the flux, produced by I_S, that links the differential area swept out by the test circuit element $d\mathbf{r}_t$. Taking all the elements of the test circuit into account, the total work done by the magnetic force is

$$\delta W_m = I_T\,\delta\Phi_{TS} = I_T I_S\,\delta M$$

where $\delta\Phi_{TS}$ refers to the change in flux of the whole test circuit. Φ_{TS} can be either positive, as in Fig. 5-14*a*, or negative; $\delta\Phi_{TS}$ can also be positive or negative, independent of the sign of Φ_{TS}. If Φ_{TS} is positive then $\delta\Phi_{TS}$ will be negative when the separation between the circuits increases.

Since δW_m is the work done *by* the magnetic force (or, by the magnetic field), $-\delta W_m$ is the work done *on* the magnetic field. But $-\delta W_m = -I_T I_S \, \delta M$ also equals $+\delta W_{\text{mech}}$, the work done by the mechanical force $\mathbf{F}_{\text{mech}}$ that holds the circuit in its normal position; there $\mathbf{F}_{\text{mech}} = -\mathbf{F}_m$. The mechanical work put into the system by the virtual displacement is $-I_T I_S \, \delta M$; this is positive when δM is negative, as in Fig. 5-14*a*.

It is now necessary to take the constant-current generators into account, for the virtual displacement of the test circuit induces voltage into both circuits. This phenomenon will be treated in Chap. 11 but, for the present, it suffices to accept on faith the experimental evidence of Faraday's law of induction: there is an instantaneous voltage induced in the source circuit whose value is the negative of the rate of change with time of Φ_{ST}:

$$V_S = -\frac{\partial \Phi_{ST}}{\partial t}$$

The instantaneous energy associated with this is $V_S \, dq_S$, where $I_S = dq_S/dt$ is the current in the source circuit; i.e., $dq_S = I_S \, dt$. The total energy induced in the source circuit, over the duration of the displacement, is then

$$\int_{t_i}^{t_f} V_S I_S \, dt = \int_{t_i}^{t_f} -I_S \frac{\partial \Phi_{ST}}{\partial t} \, dt = -I_S \, \delta \Phi_{ST} = -I_S I_T \delta M$$

In the same way, there is an induced voltage in the test circuit: $V_T = -(\partial \Phi_{TS}/\partial t)$. The energy induced in the test circuit is

$$\int_{t_i}^{t_f} V_T I_T \, dt = \int_{t_i}^{t_f} -I_T \frac{\partial \Phi_{TS}}{\partial t} \, dt = -I_T \delta \Phi_{TS} = -I_T I_S \delta M$$

Equal energies are induced into the two circuits by the virtual displacement, the total being $-2I_T I_S \, \delta M$. This is positive if δM is negative, as in Fig. 5-14. In order that the induced power should not change the currents in the two circuits, contrary to our assumption, it is necessary that the constant-current power supplies absorb the energy $\delta U_b = -2I_T I_S \, \delta M$.

Figure 5-15*a* gives an accounting of the energy changes produced by the virtual displacement. To have conservation of energy it is necessary that there be an additional component,

$$\delta U_m = I_T I_S \, \delta M$$

This can only be the change in energy of the magnetic field. The sum of

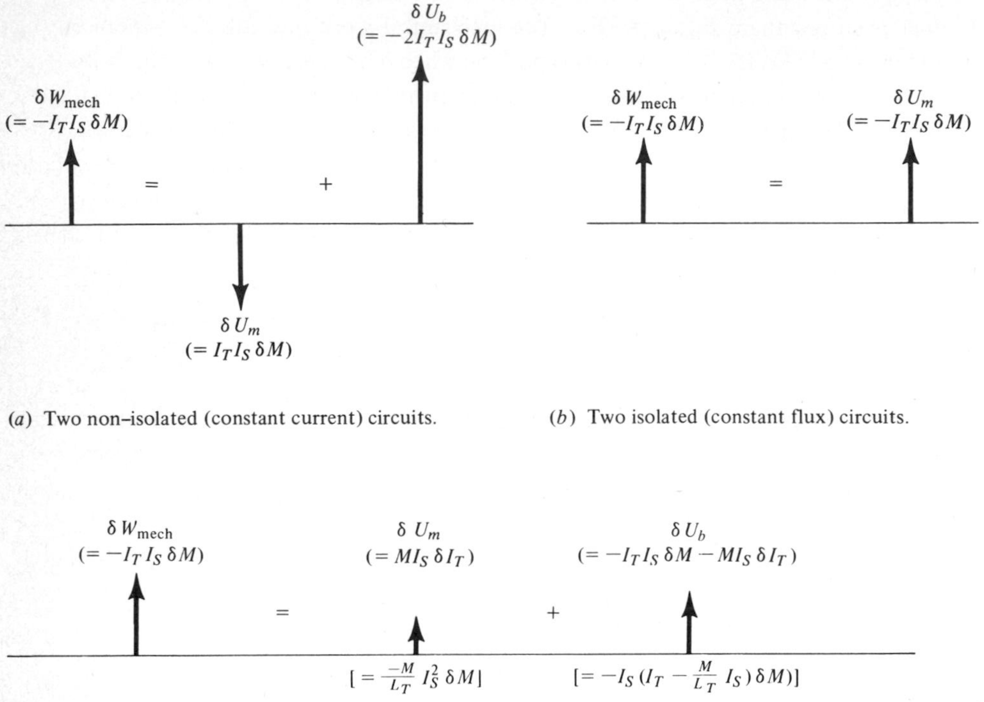

(*a*) Two non-isolated (constant current) circuits.

(*b*) Two isolated (constant flux) circuits.

(*c*) One non-isolated (constant current) circuit and one isolated (constant flux) circuit.

FIGURE 5-15

1 The gain in energy of the two power supplies, *and*
2 The *loss* in magnetic energy of the test circuit *equals*
3 The work done by the mechanical force.

Because (item 1) = 2 × (item 2), it follows that (item 3) = $\frac{1}{2}$ (item 1).

The magnetic energy in this case, or at least the portion of it that is variable, is given by

$$U_m = I_T I_S M = I_T \Phi_{TS} = I_S \Phi_{ST} \qquad (I_T \text{ and } I_S = \text{const})$$

This relation holds only for a steady current flowing in a steady external flux. It is analogous to the potential energy, $U_e = q(\phi_1 - \phi_2)$, of a charge q moved through an external difference of potential $(\phi_1 - \phi_2)$. In the electrical case it is necessary to distinguish such an energy from, e.g., the energy of a capacitor. There the energy of charges in a field which they, themselves, create is given by $U_e = \frac{1}{2}Q|\phi_1 - \phi_2|$. Similarly here we can write $U_m = \frac{1}{2}I_T\Phi_{TS} + \frac{1}{2}I_S\Phi_{ST}$.

The change in the magnetic energy of the currents above is not affected by adding the constant terms $\frac{1}{2}I_T\Phi_{TT}$ and $\frac{1}{2}I_S\Phi_{SS}$ to $U_m = \frac{1}{2}I_T\Phi_{TS} + \frac{1}{2}I_S\Phi_{ST}$. Then

$$U_m = \sum_{k=1}^{2} I_k\Phi_k$$

In general, this will be true also for n circuits. See Example 1 for further justification. The factor of $\frac{1}{2}$ means that we are actually considering the circuits by pairs, the current in one, the flux from another; the number of pairs is half the number of circuits. The electrostatic analog of this expression is $U_e = \frac{1}{2}\Sigma_{k=1}^{n}(q_k\phi_k')$ where ϕ_k' omits the contribution of q_k.

The expression for the total magnetic energy U_m may be written in terms of self-inductance and mutual inductance. Thus

$$\Phi_k = L_kI_k + {\sum_{j=1}^{n}}' M_{kj}I_j$$

where the prime on sigma signifies that $j = k$ is excluded. (Or, without the prime, set $M_{kk} = L_k$; so $\Phi_k = \Sigma_{j=1}^{n} M_{kj}I_j$.) Then

$$U_m = \frac{1}{2}\sum_{k=1}^{n} I_k\left(L_kI_k + {\sum_{j=1}^{n}}' M_{kj}I_j\right)$$

$$= \sum_{k=1}^{n} \frac{1}{2}L_kI_k^2 + \sum_{k=1}^{n}\sum_{j>k}^{n} M_{kj}I_kI_j$$

The first term represents the self-energies of the individual circuits, while the second gives the mutual energies between the various pairs of circuits. $\Big($Using $M_{kk} = L_k$, this is

$$U_m = \frac{1}{2}\sum_{j=1}^{n}\sum_{k=1}^{n} M_{kj}I_kI_j\Big)$$

In the case of the two circuits considered above, the total magnetic energy is

$$U_m = \tfrac{1}{2}(I_T\Phi_T + I_S\Phi_S) = \tfrac{1}{2}I_T(\Phi_{TT} + \Phi_{TS}) + \tfrac{1}{2}I_S(\Phi_{SS} + \Phi_{ST})$$

$$= \frac{I_T}{2}(L_T I_T) + \frac{I_T}{2}(M_{TS} I_S) + \frac{I_S}{2}(L_S I_S) + \frac{I_S}{2}(M_{ST} I_T)$$

$$= \tfrac{1}{2} L_T I_T^2 + \tfrac{1}{2} L_S I_S^2 + M I_T I_S$$

Only the last term varies in the virtual displacement if the currents are held constant and the circuits are not distorted:

$$\delta U_m = I_T I_S \, \delta M$$

TWO ISOLATED CIRCUITS What is the situation when the two circuits are isolated? Without constant-current generators, but with currents that stay constant in the steady state, it follows that no energy is dissipated in the circuits; so the circuits can have no resistance. The most common examples of such currents are those produced by the circulating electrons in any atom. Superconducting residual currents are also of this type. These may have been previously induced in the circuits by changing magnetic fields: the Faraday effect referred to above; or there could have been, in the circuits, batteries and resistors which were subsequently removed, leaving only the superconducting wires. The convection current in a betatron doughnut is also of this type, if collisions with residual gas molecules are neglected. In any event, there are two circulating loops of current in circuits without resistance, each one having some flux produced by itself and some flux produced by the other. If one circuit is given a virtual displacement there will be a voltage induced in each loop, just as before; here, however, the power cannot be absorbed by any power supplies so it must be absorbed by the circulating electrons. The currents must change. We will now show that in this, isolated, case the currents change in such a way that the total flux linking each circuit remains constant.

The magnetic force on the test circuit is the same here as in the constant-current case, for this is a function only of the magnitude and configuration of the currents when the circuits are at rest. It is not a function of the constraints imposed on the currents in the virtual displacement. The work done by the magnetic force in this case, then, is also given by $\delta W_m = I_T I_S \, \delta M$, and the work done by the balancing mechanical force is $\delta W_{\text{mech}} = -I_T I_S \, \delta M$. With no energy to be given to the power supplies, the only other possible change in energy is that in the magnetic field, δU_m. This requires that

$$\delta U_m = -I_T I_S \, \delta M$$

Another expression for δU_m is obtained from $U_m = \frac{1}{2} L_T I_T^2 + \frac{1}{2} L_S I_S^2 + M I_T I_S$, namely $\delta U_m = L_T I_T \, \delta I_T + L_S I_S \, \delta I_S + M I_S \, \delta I_T + M I_T \, \delta I_S + I_T I_S \, \delta M$. Taking half the sum of the two, separate, equations for δU_m:

$$\delta U_m = \tfrac{1}{2}(L_T I_T + M I_S)\delta I_T + \tfrac{1}{2}(L_S I_S + M I_T)\delta I_S$$

$$= \tfrac{1}{2}(\Phi_{TT} + \Phi_{TS})\delta I_T + \tfrac{1}{2}(\Phi_{SS} + \Phi_{ST})\delta I_S$$

$$= \tfrac{1}{2}\,\Phi_T\,\delta I_T + \tfrac{1}{2}\,\Phi_S\,\delta I_S$$

$$= \delta\left[\frac{1}{2}\sum_{k=1}^{2} I_k\Phi_k\right] \qquad (\Phi_k = \text{const})$$

If M changes, Φ_{TS} and Φ_{TT} change in opposite directions, such that $\Phi_T = \Phi_{TS} + \Phi_{TT}$ remains the same. Similarly with Φ_S.

1 In the case of two *isolated* circuits δU_m is obtained from $\frac{1}{2}\Sigma I_k\Phi_k$ by holding Φ_k constant: $\delta U_m = \frac{1}{2}\Sigma\Phi_k\,\delta I_k$. But another method also exists here: δU_m can be obtained by taking $U_m = -\frac{1}{2}\Sigma I_k\Phi_k$ and treating $I_k = \text{const}$; even though these are not actual constraints on the I_k, this procedure yields the equivalent answer $\delta U_m = -\frac{1}{2}\Sigma I_k\,\delta\Phi_k$. Figure 5-15*b* gives an accounting of the energy changes for this case. δU_m represents the entire energy change into which δW_{mech} is converted.

2 In the case of two *nonisolated* circuits δU_m is obtained from $U_m = \frac{1}{2}\Sigma I_k\Phi_k$ by keeping I_k fixed: $\delta U_m = +\frac{1}{2}I_k\,\delta\Phi_k$ and the constraints on the I_k are the actual ones. Here, however, δU_m is not the entire energy change that occurs as a result of δW_{mech}. When $\delta U_b = -\Sigma I_k\,\delta\Phi_k$ is added to δU_m to give the total energy change, we obtain $\delta U_m + \delta U_b = -\frac{1}{2}\Sigma I_k\,\delta\Phi_k$. So $U_m = -\frac{1}{2}\Sigma I_k\Phi_k$ ($I_k = \text{const}$) can here be treated as an effective, or total, magnetic energy just as in the previous case.

3 The case of one isolated and one nonisolated circuit is left to Prob. 19. The result is shown in Fig. 5-15*c*.

The expressions for the magnetostatic energy of a distribution of steady currents, treating the I_k as constant, are as follows.

Convection currents

$$U_{\text{total}} = -\frac{1}{2}\sum I_k\Phi_k \qquad U_m = -\frac{1}{2}\sum I_k\Phi_k \qquad (+\tfrac{1}{2}\text{ if }\Phi_k = \text{const})$$

Conduction currents

$$U_{\text{total}} = -\frac{1}{2}\sum I_k\Phi_k \qquad U_m = +\frac{1}{2}\sum I_k\Phi_k \qquad (U_b\text{ omitted})$$

In any case, for any combination of currents, the total magnetic energy is

$$U_{\text{total}} = -\frac{1}{2}\sum I_k\Phi_k \qquad (I_k\text{ const})$$

This expression can be thrown into a different form by taking

$$\Phi_k = \int_{\Sigma_k} \mathbf{B}\cdot d\mathbf{S} = \int_{\Sigma_k} (\nabla\times\mathbf{A})\cdot d\mathbf{S} = \oint_{C_k} \mathbf{A}\cdot d\mathbf{r}$$

Then
$$U_{\text{total}} = -\frac{1}{2} \sum_k I_k \oint_{C_k} \mathbf{A} \cdot d\mathbf{r} \qquad (I_k = \text{const})$$

Going over to a continuum distribution,

$$\boxed{U_{\text{total}} = -\frac{1}{2} \int_{\text{all space}} \mathbf{J} \cdot \mathbf{A}\, d\tau} \qquad (\mathbf{J} = \text{const}) \qquad \text{(M-11}a\text{)}$$

If all the **J**'s are all conduction current densities we could also write

$$\boxed{U_m = +\frac{1}{2} \int_{\text{all space}} \mathbf{J} \cdot \mathbf{A}\, d\tau} \qquad (\mathbf{J} = \text{const}) \qquad \text{(M-11}b\text{)}$$

with the understanding that the battery energies have been ignored instead of being spread out through all space. As in the electric analog, there is no longer a problem with self-energy; the **A** at any point is that produced by all the **J**'s, including that at the point in question.

Like their electrostatic counterparts, the equations (M-11) turn out to be valid only under static conditions and are not true for time-varying cases. They therefore lose much of their importance thereby. It is possible to modify them slightly in such a way, however, that the results hold true for all cases. Take Eq. (M-11a), e.g.:

$$U_{\text{total}} = -\frac{1}{2} \int_{\mathcal{V}} \left(\frac{1}{\mu_0} \nabla \times \mathbf{B} \right) \cdot \mathbf{A}\, d\tau = -\frac{1}{2\mu_0} \int_{\mathcal{V}} [\nabla \cdot (\mathbf{B} \times \mathbf{A}) + \mathbf{B} \cdot (\nabla \times \mathbf{A})]\, d\tau$$

$$= -\frac{1}{2\mu_0} \int_{\Sigma} (\mathbf{B} \times \mathbf{A}) \cdot d\mathbf{S} - \frac{1}{2\mu_0} \int_{\mathcal{V}} \mathbf{B} \cdot \mathbf{B}\, d\tau$$

If the currents are limited to a finite region of space then Σ, the surface bounding $\mathcal{V}$, can be made large enough so that the first integral becomes arbitrarily small. Then

$$\boxed{U_{\text{total}} = -\frac{1}{2\mu_0} \int_{\text{all space}} B^2\, d\tau} \qquad \text{(M-12}a\text{)}$$

Similarly for U_m, the energy in the magnetostatic field proper, not including any battery energy:

$$U_m = \frac{1}{2\mu_0} \int_{\text{all space}} B^2\, d\tau \qquad \text{(M-12}b\text{)}$$

The following two cases must be distinguished from each other. (1) If the distribution of currents is an isolated one, i.e., if we have convection currents without batteries, we are considering a system without power supplies; U_m then represents the total magnetic energy. This gives a distribution throughout all space of a magnetic energy density $u_m = (1/2\mu_0)B^2$ (Φ_K = const). But this case can also be given by its equivalent, $u_m = (1/2\mu_0)B^2$ (I_k = const) as a mathematical convenience, even though the I_k are not actually constant for isolated currents. Although it occurs widely, this is not the type usually considered. (2) If the distribution of currents is nonisolated, then power supplies must be taken into account. Here $u_m = (1/2\mu_0)B^2$ (I_K = const) represents the energy density of the magnetic field proper, neglecting the energies (and their changes) in the power supplies; this is only a partial energy density. If we wish to consider the total energy changes that occur (as a result of a virtual displacement, say) then we must distribute an additional energy density $u_b = -2u_m$ throughout all space. The total, or effective energy density, becomes

$$u_{\text{total}} = -\frac{1}{2\mu_0} B^2 \qquad (I_k = \text{const})$$

It is not conventional to assign a negative value to the magnetic field energy density; but, if the I_K are to be taken as constant, it is a necessary step. In both the nonisolated and the isolated case this is in the nature of a mathematical ruse, though for different reasons. To our knowledge J. W. Butler was the first, in an article based on relativistic considerations, to point the logical necessity of this step. See A Proposed Electromagnetic Momentum-Energy 4-Vector for Charged Bodies, *Amer. J. Phys.*, **37**:1258 (1969); also Electromagnetic Momentum, Energy, and Mass, by F. Rohrlich, *Amer. J. Phys.*, **38**:1310 (1970).

Examples

1. The factor $\frac{1}{2}$ in the energy formula We will justify the statement in the text that

$$U_m = \frac{1}{2} \sum_{k=1}^{n} I_k \Phi_k$$

instead of $U_m = \Sigma_{k=1}^{n} I_k \Phi_k$ when the Φ are created by the I.

Assume that each I_k, as well as the Φ_k in which it flows, is allowed to grow from its original value to its final value in a common, uniform, manner. At any instant the values of current and flux are αI_k and $\alpha\Phi_k$, where $0 \leqslant \alpha \leqslant 1$. An infinitesimal increase in the currents is given by $I_k \, d\alpha$. To the first order in α the potential energy is

$$U_m = \int_0^1 \sum_{k=1}^{n} (I_k \, d\alpha)(\alpha\Phi_k) = \sum_{k=1}^{n} I_k\Phi_k \int_0^1 \alpha \, d\alpha = \frac{1}{2} \sum_{k=1}^{n} I_k\Phi_k$$

2. Magnetic energy of an infinitely long conductor Next, we find the magnetic energy associated with unit length of an infinitely long straight wire of radius a carrying a current I.

Inside the conductor

$$\mathbf{B} = \hat{\boldsymbol{\phi}} \left(\frac{\mu_0 I}{2\pi a^2}\right) r$$

Then

$$B^2 = \left(\frac{\mu_0{}^2 I^2}{4\pi^2 a^4}\right) r^2$$

and

$$U_{\ell(\text{int})} = \frac{1}{2\pi_0} \int_0^a \left(\frac{\mu_0{}^2 I^2}{4\pi^2 a^4}\right) r^2 \, (2\pi r \, dt) = \frac{\mu_0 I^2}{16\pi}$$

This reveals an interesting feature of the energy distribution: the energy residing within the conductor is independent of the size of the conductor. All infinitely long straight conductors, whether they are thick or thin, have the same internal energy for the same total current.

Outside the conductor

$$\mathbf{B} = \hat{\boldsymbol{\phi}} \left(\frac{\mu_0 I}{2\pi}\right) \frac{1}{r}$$

Then

$$B^2 = \left(\frac{\mu_0{}^2 I^2}{4\pi^2}\right) \frac{1}{r^2}$$

and

$$U_{\ell(\text{out})} = \frac{1}{2\mu_0} \int_a^\infty \left(\frac{\mu_0{}^2 I^2}{4\pi^2}\right) \frac{1}{r^2} \, (2\pi r \, dr) = \infty$$

The magnetic energy per unit length of an infinitely long conductor is infinite. In the previous section it was seen that the flux per unit length and the self-inductance per unit length were also infinite.

3. Self-inductance and energy The equation for the self-energy, $U = \frac{1}{2}LI^2$, provides an alternate method for calculating the self-inductance of a circuit. We will find the self-inductance of a straight, infinitely long conductor by this method.

Using the results of the previous example, the per-unit-length internal self-inductance is

$$L_{\ell(\text{int})} = \frac{2U_{\ell(\text{int})}}{I^2} = \frac{2}{I^2}\left(\frac{\mu_0 I^2}{16\pi}\right) = \frac{\mu_0}{8\pi}$$

The per-unit-length external self-inductance is

$$L_{\ell(\text{ext})} = \frac{2U_{\ell(\text{ext})}}{I^2} = \frac{2}{I^2}(\infty) = \infty$$

These results, obtained previously in Example 3 of the previous section, do not force us to abandon the concept of self-inductance. Given a conductor of finite thickness and a circuit with finite dimension the self-inductance is found to be finite. $L_{\ell(\text{int})}$ is given by $\mu_0/8\pi$ even when the circuit has an arbitrary shape rather than that of an infinitely long conductor; for, if the wire diameter is very small compared to the smallest radius of curvature of the circuit, **B** inside the conductor behaves in essentially the same way as it does for the infinitely long wire. As for the external self-inductance, let contour C_1 coincide with the center of the conductor and contour C_2 with the innermost element of the conductor surface. Then the external self-inductance, the external flux per unit current, can be approximated by calculating the flux contained within the contour C_2 (i.e., the flux outside the wire) on the assumption that the current within the wire (that producing the flux) is concentrated along the contour C_1. Actually, of course, the flux is being produced by a continuum distribution of current throughout the wire; but this simplification gives an excellent approximation to the answer. For an example, the reader is referred to the calculation of $L_{\ell(\text{ext})}$ for a circular current loop in "Introduction to the Principles of Electromagnetism," W. Hauser, p. 347, Addison-Wesley, Reading, Mass., 1971.

PROBLEMS

1 Find the per-unit-length magnetic energy of an infinitely long solenoid having n turns/m, a radius a, and a current I.

2 Compare the electrostatic energy density in a capacitor having $E = 10^5$ V/m with the magnetostatic energy density in a solenoid having $B = 1000$ G. Both field values are reasonable figures and could be normally encountered in the everyday world; but one energy density is 10^5 times as large as the other. Before working the problem, guess which is the larger, then compare with your answer. Explain why one energy density is so much larger than the other. Why isn't it the other way around?

3 Find the per-unit-length magnetic energy of a coaxial cable carrying the current I, if the conductors are shells of infinitesimal thickness with radii a and b, where $b > a$.

4 A cylindrical hole is drilled into an infinitely long straight conductor carrying current I, as shown in Fig. 5-16. Find the per-unit-length magnetic energy stored in the hole.

5 A spherical shell of infinitesimal thickness and radius a has a surface charge density σ C/m^2 and revolves about a diameter with angular velocity ω. Find (*a*) $\mathbf{J}_\ell$ and (*b*) an integral expression for $\mathbf{A}$ at (θ_t, ϕ_t) on the sphere in terms of all the (θ, ϕ) on that surface.

6 Find an equation for the magnetic potential energy of two idealized filamentary circuits in terms of the current elements and the separation between them.

7 Find the magnetic potential energy stored within a toroid having N turns, a central radius a, and a cross-sectional radius b.

8 Find the per-unit-length magnetic potential energy of a parallel wire transmission line. The center-to-center separation of the wires is D, each wire has a radius a, and the wires carry equal but opposite currents I.

9 Compare the magnetostatic and kinetic energy densities in outer space, if, roughly, B is 10^{-6} G and it is assumed that there is one hydrogen atom per cubic centimeter, moving with a speed of 10^5 cm/s.

10 Reconcile the results of Prob. 2 with those of Example 2 of Sec. 5-3, where it is shown that for a convection current $cB = (v/c)E$, with $c = 3 \times 10^8$ m/s. According to the latter, the magnetic energy density should be the smaller, should it not?

11 The mutual inductance between two coils is related to their self-inductances by $M = \pm k\sqrt{L_1 L_2}$ where $0 \leqslant k \leqslant 1$. k is called the coefficient of coupling; the plus is used for currents flowing in a similar sense, the minus is used for currents bucking. If $k = \frac{1}{2}$, $L_1 = 1$ mH, $L_2 = 4$ mH and $I_1 = I_2 = 10$ A, find the total magnetostatic energy for the cases of aiding and bucking.

12 Distinguish between quasistatic and virtual displacements. May each have an arbitrary magnitude? What special forces are required for either one? In how long a time must they be completed?

13 (*a*) Find the energy density per meter stored in the magnetic field of a very long solenoid. The radius of the solenoid is a, the current is I, and there are n turns/m.

(*b*) What is L per meter?

14 What is the effect on a residual superconducting current of an external, randomly fluctuating magnetic field?

15 Rearrange the terms of the expression for the magnetic potential energy

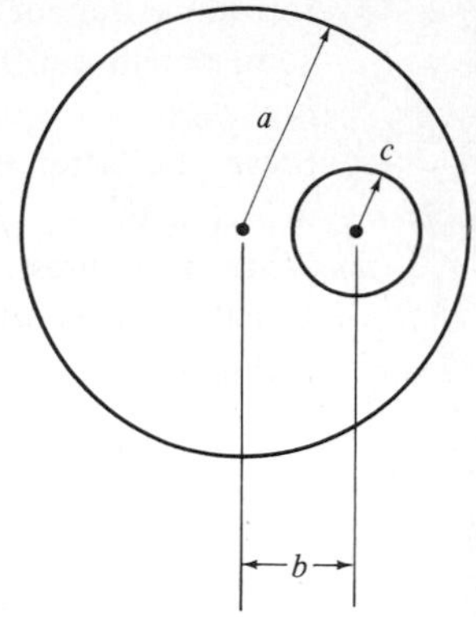

FIGURE 5-16

$$U = \tfrac{1}{2}L_1 I_1{}^2 + MI_1 I_2 + \tfrac{1}{2}L_1 I_1{}^2$$

to show that for $U \geqslant 0$ it is necessary that a relation exist among L_1, L_2, and M^2. What is the relation? (For geometric proofs of this relation see On $L_1 L_2 \geqslant M^2$ by E. A. Power, *Amer. J. Phys.*, **37**:23 (1969).)

16 In Fig. 5-14*a* is the test circuit attracted or repelled by the source circuit? Suppose I_T is changed in direction? Suppose I_S alone is reversed? Suppose I_S and I_T are both reversed? Give a general rule.

17 Derive an expression, by a method similar to that of Example 1, for U_m caused by a collection of I_K's in an external field Φ that is not due to them.

18 The analog of $U_m = \frac{1}{2}\Sigma_{k=1}^{n} I_k \Phi_k$ is $U_e = \frac{1}{2}\Sigma_{k=1}^{n} q_k \phi'_k$. Yet ϕ'_k omits the contributions of q_k while Φ_k includes the contribution of I_k. Explain.

19 Given two circuits. One has a residual superconducting current, the other has a constant-current power supply and ordinary resistive conductors. Find (*a*) δW_{mech}, (*b*) δU_b, and (*c*) δU_m when either circuit is given a virtual displacement producing a change δM in the mutual inductance. In the last case start with the L and M expression for U_m to obtain δU_m; then set $\delta W_{\text{mech}} = \delta U_m + \delta U_b$.

(*d*) From the procedure in (*c*) show that $\Phi_T \equiv \Phi_{TT} + \Phi_{TS}$ is an invariant here also, as well as in the case of two superconducting circuits, if the test circuit is the isolated circuit.

(*e*) Write δU_b and δU_m in terms of δM alone, without δI_T.

20 Figure 5-15*b* shows that $\delta U_m = -I_T I_s \, \delta M$, and this can be either positive or negative. Yet $u_m = (1/2\mu_0)B^2$ is always positive. Are these contradictory?

21 In a change of the mutual inductance between two isolated circuits

$$(d/dM)(L_1 i_1 + Mi_2) = 0 \qquad \text{and} \qquad (d/dM)(L_2 i_2 + Mi_1) = 0$$

(*a*) Obtain from these a differential equation for i_1 in terms of L_1, L_2, and M. (This is also a differential equation for i_2.)

(*b*) Introduce the coefficient of coupling defined by $M = k\sqrt{L_1 L_2}$, where $-1 \leqslant k \leqslant 1$, to obtain a differential equation for y in terms of k, where y is either i_1 or i_2.

(*c*) Solve the latter equation for the current by the power series method, $y = \sum_{m=0}^{\infty} c_m k^m$.

(*d*) Write the expressions for i_1 and i_2 in terms of i_{10}, i_{20}, and f. Here i_{10} is i_1 when $k = 0$; i_{20} is i_2 when $k = 0$ (both are arbitrary initial conditions); and f is the parameter

$$f = \frac{i_{10}}{i_{20}} \sqrt{\frac{L_1}{L_2}}$$

f is the square root of the ratio of the self-energies of the two circuits when they are infinitely far apart. i_1/i_{10} is graphed in Fig. 5-17*a*; i_2/i_{20} is obtained by selecting the curve for $(1/f)$ instead of that for f.

22 Superconducting circuit 1 has a self-inductance of 100 mH, while superconducting circuit 2 has $L_2 = 1$ mH. When widely separated the two circuits each carry a current of 10 A. Find what the two currents become when the circuits are moved together so that (*a*) the mutual inductance is +5 mH and (*b*) $M = -5$ mH. Assume no power loss in any adjacent materials.

23 Given two widely separated circuits: the first superconducting, with self-inductance L_1 and current i_{10}; the second ordinary, with self-inductance L_2 and constant current i_2. When the two circuits are brought closer to each other, with mutual inductance M, the current in the first becomes i_1 while that in the second stays i_2. From the fact that the energy in the magnetic field stays constant (*a*) derive an equation for i_1 in terms of the other quantities mentioned, (*b*) write the expression for i_1 normalized relative to i_{10}, using the parameter

$$f = \frac{i_{10}}{i_2} \sqrt{\frac{L_1}{L_2}}$$

and the coefficient of coupling k. This is graphed in Fig. 5-17*b*.

24 In Probs. 21 and 23 the currents are obtained in normalized form. If $i_{10} = 0$ the normalization is not possible.

(*a*) Obtain expressions for i_1, in these two cases, when $i_{10} = 0$.

(*b*) In Prob. 23 there is another solution: $k/f - \sqrt{1 + (k/f)^2}$. What is its significance?

25 Given one isolated and one nonisolated circuit.

(*a*) Can the results for δU_m and δU_b be made to fall symmetrically between the case of two isolated circuits and that of two nonisolated circuits, $\delta U_m = 0$?

(*b*) Can they be made to behave like two nonisolated circuits?

(*c*) Like two isolated circuits?

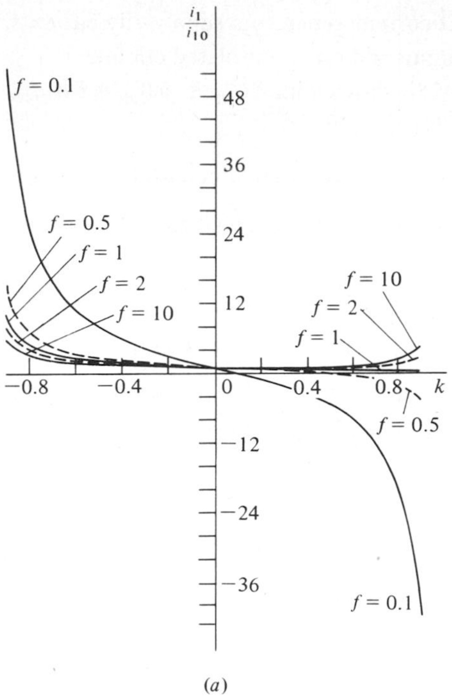

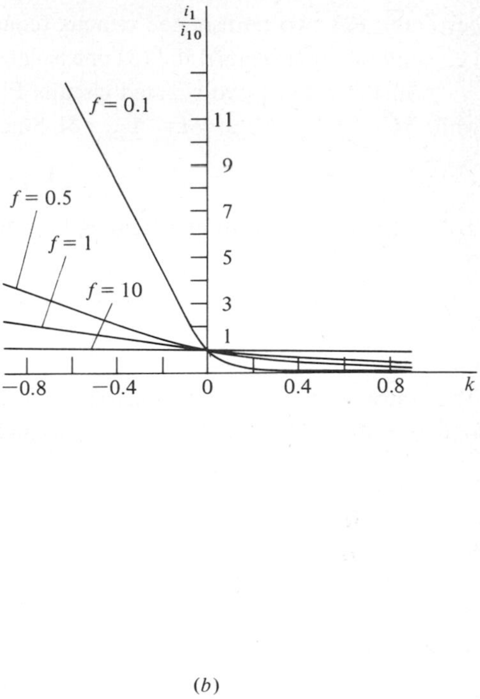

FIGURE 5-17

5-4 MAGNETOSTATIC FORCES

The force which a magnetic field exerts on a circuit carrying a current has a unique value, quite independent of whether the currents or the fields would be kept constant in any displacement of the circuit; the force exists even if there is no displacement. When a filamentary circuit is stationary the magnetic force $I\oint d\mathbf{r}_t \times \mathbf{B}$ is actually exerted on the moving conduction electrons in the wire, the current. This force is transmitted to the wire itself by the attraction which exists between the conduction electrons and the stationary ions of the metal lattice. In effect, the magnetic field then exerts a force on the stationary conductor when it is carrying current.

One can obtain an expression for the magnetic force exerted on a current-carrying circuit in terms of the magnetic energy. As in the electrostatic case there are different possibilities that must be considered separately. Here it is necessary to distinguish among the following: (1) two isolated circuits (no batteries or power supplies, i.e., convection currents); (2) two nonisolated circuits (constant-current generators supply the currents, i.e., conduction currents); and (3) one isolated circuit and one nonisolated circuit.

In the case of two isolated circuits Fig. 5-15*b* shows that $\delta U_m = -\delta W_m = \delta W_{\text{mech}}$ while $\delta U_b = 0$; so $\nabla U_m \cdot \delta \mathbf{l} = -\mathbf{F}_m \cdot \delta \mathbf{l}$. Since $\delta \mathbf{l}$ is arbitrary,

$$\mathbf{F}_m = -\nabla U_m \qquad \text{(two isolated circuits)}$$

This is the normal relation obtaining between a force and the potential energy producing it.

In the case of two nonisolated circuits Fig. 5-15*a* shows that $\delta U_m = +\delta W_m$. Here

$$\mathbf{F}_m = +\nabla U_m \qquad \text{(two nonisolated circuits)}$$

This unusual relation exists because U_m is not the total potential energy in this case, but only part of it. The total or effective, magnetic potential energy here is

$$U_t = U_m + U_b = U_m + (-2U_m) = -U_m$$

so

$$\mathbf{F}_m = -\nabla U_t \qquad \text{(two nonisolated circuits)}$$

A superconducting circuit carrying 10 A in a certain field of flux would experience the same force if 10 A were battery-driven through ordinary wire. In the first case $U_m = 10\Phi$ while in the latter $U_m = -10\Phi$; however, $U_t = 10\Phi$, also, in the nonisolated case.

When one circuit is isolated and one nonisolated, the force—for the same configuration and currents as above—is still the same. Here $\delta U_m = MI_S\,\delta T_T$ in a virtual displacement, and $\mathbf{F}_m$ cannot be written directly as $\pm\nabla U_m$. But here also

$$\mathbf{F}_m = -\nabla U_t \qquad \text{(one isolated circuit, one nonisolated circuit)}$$

$\delta U_t = \delta U_m + \delta U_b = -I_T I_S\,\delta M$ in this mixed case has the same value as

$$\delta U_t = \delta U_m + \delta U_b = -I_T I_S\,\delta M + 0$$

in the doubly isolated case.

It is possible, then, to say that $\mathbf{F} = -\nabla U_t$ in all cases. To express $\mathbf{F}_m$ in terms of a gradient of U_m alone is to risk confusion.

It is often desirable to express the magnetostatic force on an object in terms of the field which exists on the surface of the object. This is similar to the case with electrostatic forces; and the derivation of the result, as well as the result, itself, are quite similar to the electrostatic case. Starting with

$$\mathbf{F} = \int_v \mathbf{J} \times \mathbf{B}\, d\tau$$

we identify the integrand as the force density, $\mathbf{f} = \mathbf{J} \times \mathbf{B}$. Utilizing $\nabla \times \mathbf{B} = \mu_0 \mathbf{J}$ this becomes

$$\mathbf{f} = \frac{1}{\mu_0} (\nabla \times \mathbf{B}) \times \mathbf{B}$$

Then, e.g., the x component of $\mathbf{f}$ is

$$f_x = \frac{1}{\mu_0} \left\{ B_z \left[\frac{\partial B_x}{\partial z} - \frac{\partial B_z}{\partial x} \right] - B_y \left[\frac{\partial B_y}{\partial x} - \frac{\partial B_x}{\partial y} \right] \right\}$$

$$= \frac{1}{\mu_0} \left\{ \left[\frac{\partial}{\partial z} (B_z B_x) - B_x \frac{\partial B_z}{\partial z} - \frac{1}{2} \frac{\partial}{\partial x} B_z^2 \right] + \left[\frac{\partial}{\partial y} (B_x B_y) - B_x \frac{\partial B_y}{\partial y} - \frac{1}{2} \frac{\partial}{\partial x} B_y^2 \right] \right\}$$

We may add

$$-B_x \frac{\partial B_x}{\partial x} + \frac{1}{2} \frac{\partial}{\partial x} B_x^2 = 0$$

without altering the value of f_x, so

$$f_x = \frac{1}{\mu_0} \left\{ \frac{\partial}{\partial x} \left[\frac{1}{2} (B_x^2 - B_y^2 - B_z^2) \right] + \frac{\partial}{\partial y} (B_x B_y) + \frac{\partial}{\partial z} (B_z B_x) - B_x (\nabla \cdot \mathbf{B}) \right\}$$

Since $\nabla \cdot \mathbf{B} = 0$, the last term vanishes. The rest of the expression is similar to the previous result in electrostatics, and the magnetostatic stress tensor may be written down at once by analogy:

$$\{T\} = \frac{1}{\mu_0} \begin{Bmatrix} \frac{1}{2} (B_x^2 - B_y^2 - B_z^2) & B_x B_y & B_x B_z \\ B_y B_x & \frac{1}{2} (-B_x^2 + B_y^2 - B_z^2) & B_y B_z \\ B_z B_x & B_z B_y & \frac{1}{2} (-B_x^2 - B_y^2 + B_z^2) \end{Bmatrix} \qquad \text{(M-13)}$$

The remarks which apply to the stress tensor in electrostatics all have similar validity here.

Examples

1. An air gap between two nonmagnetic plates is 1 mm thick. (The meaning of the term nonmagnetic will become clear in Chap. 7. For the moment it is sufficient to state that **B** within and on the surface of the plate behaves essentially the same way as it does in a vacuum.) If the magnetic field is normal to the surface of the gap and has the value **B** = 1000 G, find the magnetostatic tension on the plates.

$$f = \frac{1}{2\mu_0}B^2 = \frac{(0.1)^2}{2(4\pi \times 10^{-7})} = 3980 \text{ N/m}^2$$

Since 1 pound-force = 4.45 N, f = 894 lb/m^2 = 0.58 lb/in^2. This is not an insignificant tension by ordinary, macroscopic, mechanical standards. It is small compared to the pressure at the surface of the earth which is caused by the weight of the atmosphere above the surface: 14.7 lb/in^2. However, it is large compared to the electrostatic tension in a capacitor: 6.5×10^{-6} lb/in^2 in Example 1 of Sec. 4-4 of the previous chapter.

2. A surface in the xz plane has a field **B** which makes an angle θ with $\hat{\mathbf{y}}$, similar to the case of Fig. 4-16 in electrostatics. Using the Maxwell stress tensor, find the force per unit area.

This is exactly similar to the case of Example 2 of Sec. 4-8 in the previous chapter, and the results are similar to the ones there. The force makes the angle 2θ with the normal to the surface. If **B** is normal to the surface then **F** is a normal pull on it, regardless of whether **B** points toward or away from the surface. If **B** is at an angle of 45° to the surface then **F** is a tangential shear, in the same direction as the tangential component of **B**. If **B** is tangential to the surface then **F** is a normal push on the surface.

PROBLEMS

1 At the Canadian National Magnet Laboratory a magnetic field of 250,000 G is created (for 10-min intervals) on the axis of a solenoid having 6,000 turns. It is a hybrid solenoid, consisting of two solenoids, one within the other. The inner one is merely cryogenically coiled to lower its resistance, but the outer one is superconducting (which also requires liquid helium cooling) and has zero resistance. Figure 5-18 shows the design. A cylindrical air core is surrounded by a multilayer inner solenoid, tight wound with aluminum ribbon, 0.045 by 0.34 in; a coating of AlO acts as insulation. The current of 5,000 A is made possible by the cryogenic cooling with liquid helium, thereby reducing the resultant power consumption and heat dissipation. The outer solenoid carries 1000 A through a niobium-tin superconducting alloy ribbon (higher currents and fields would destroy the superconducting state).

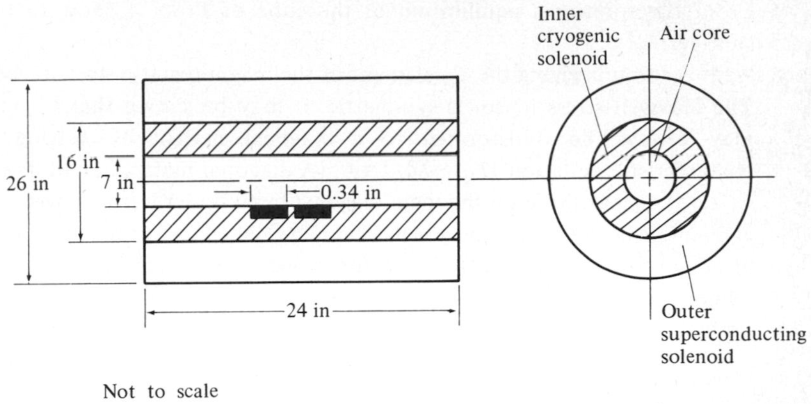

FIGURE 5-18

Because of the large mechanical stresses it is necessary to encase each layer of ribbon in stainless steel, despite the fact that this greatly increases the problem of removal of heat. Calculate the radial force density (*a*) in newtons per square centimeter and (*b*) in pounds per square inch, on a ribbon in the innermost layer of the cryogenic solenoid, midway between the ends, assuming **B** has a constant value in the air core.

(*c*) How does this compare with the tensile strength of aluminum? Why isn't the conductor itself made of stainless steel rather than aluminum?

(*d*) Can the formula $f = B^2/2\mu_0$, found in some texts, be used here? This would be a much faster way of making the calculation. (See Chap. 4, Sec. 4-4 for the electrical analog.)

2 Write an expression for the Maxwell stress tensor that includes both the electrostatic and magnetostatic forces. Supply a meaning to the subscripts of T_{xx} and T_{xy}.

3 (*a*) In Prob. 1, is the axial force on a ribbon of aluminum a push or a pull with respect to a neighboring turn?

(*b*) Is the radial force outward or inward?

(*c*) Does either direction reverse if the current is reversed?

(*d*) Which is larger, the radial or axial force? Reconcile this with the Maxwell stress tensor.

4 Apply the outward force $T_{xx}\,dy\,dz$ to the left face of an infinitesimal cube and the outward force $(T_{xx} + \partial T_{xx}/\partial x)\,dy\,dz$ to the right face. Add the shear forces $T_{xy}\,dy\,dz$ and $T_{xz}\,dy\,dz$ to the left face, with corresponding forces $-(T_{xy} + \partial T_{xy}/\partial x)\,dy\,dz$ and $-(T_{xz} + \partial T_{xz}/\partial x)\,dy\,dz$ to the right face; then add analogous forces for the other sides. Find the net force density f_x. What is the significance of the first and second subscripts of T_{ji}?

5 From the rotational equilibrium of the cube of Prob. 4 show that $\{T\}$ must be symmetric.

6 Write a formula giving the T_{ij} element of the magnetostatic stress tensor.

7 The Maxwell stress tensor is symmetric. It may be shown that a symmetric matrix may always be transformed to a diagonal matrix by solving the so-called characteristic equation $|T_{ij} - \lambda\delta_{ij}| = 0$. (A diagonal matrix has all its elements equal to zero, except those on the main diagonal from upper left to lower right.) Find the elements that lie on the diagonal. These are called the principal values of $\{T\}$. The principal axes are so oriented that the single root of λ is the value T_{ii} parallel to **B**, while the double root of λ gives the values T_{ii} on the axes orthogonal to this direction.

Appendix 5

ALTERNATE DERIVATION OF THE $\nabla \times \mathbf{B}$ EXPRESSION

A rather lengthy derivation of the result $\nabla \times \mathbf{B} = \mu_0 \mathbf{J}$ for magnetostatics was given in Chap. 3, Sec. 3-4. Now that the vector potential has been introduced it is possible to supply a much simpler proof, and we proceed to do so here.

We have $\nabla \times \mathbf{A} = \mathbf{B}$ and $\nabla \cdot \mathbf{A} = 0$. (The latter is not necessarily true in cases where conditions change with time.) Then $\nabla \times \mathbf{B}$ is given by

$$\nabla \times \mathbf{B} = \nabla \times \nabla \times \mathbf{A} = \nabla\nabla \cdot \mathbf{A} - \nabla^2 \mathbf{A} = -\nabla^2 \mathbf{A}$$

But

$$\mathbf{A} = \frac{\mu_0}{4\pi} \int_{\mathcal{V}} \frac{\mathbf{J}}{R}\, d\tau_s$$

so

$$\nabla^2 \mathbf{A} = \frac{\mu_0}{4\pi} \int_{\mathcal{V}} \nabla^2 \left(\frac{\mathbf{J}}{R}\right) d\tau_s$$

However, **J** is a function of $\mathbf{r}_s$, not of **r**, and ∇ is dependent on **r**, not on $\mathbf{r}_s$. Then

$$\nabla^2 \mathbf{A} = \frac{\mu_0}{4\pi} \int_{\mathcal{V}} \mathbf{J} \nabla^2 \left(\frac{1}{R}\right) d\tau_s$$

We now utilize the result, shown in Appendix 1, that

$$\nabla^2\left(\frac{1}{R}\right) = -4\pi\,\delta(\mathbf{r}_s - \mathbf{r}_t)$$

Thus $$\nabla^2 \mathbf{A}(\mathbf{r}_t) = \frac{\mu_0}{4\pi}\int_{\mathcal{V}} \mathbf{J}(\mathbf{r}_s)[-4\pi\,\delta(\mathbf{r}_s - \mathbf{r}_t)]\,d\tau_s = -\mu_0 \mathbf{J}(\mathbf{r}_t)$$

and $$\nabla \times \mathbf{B}(\mathbf{r}_t) = \mu_0 \mathbf{J}(\mathbf{r}_t)$$

or, simply, $$\nabla \times \mathbf{B} = \mu_0 \mathbf{J}$$

6

ELECTROSTATICS IN MATTER

6-1 THE ELECTRIC DIPOLE

We now return once more to electrostatics, this time when matter is present. Suppose we have a collection of N source charges q_s at various points $\mathbf{r}_s$ in some region $\mathcal{V}$. The potential ϕ produced by this assembly at some point P is

$$\phi(\mathbf{r}) = \frac{1}{4\pi\epsilon_0} \sum_{s=1}^{N} \frac{q_s}{R_s}$$

where each R_s is the distance between $\mathbf{r}$ and the corresponding $\mathbf{r}_s$, as shown in Fig. 6-1. The origin O has been taken at some arbitrary point in or near $\mathcal{V}$. If P as well as O has also been taken near $\mathcal{V}$, then this expression for ϕ is as simple as any that can be found; but if P is far from $\mathcal{V}$ it is possible to convert this expression into another, more meaningful, form which employs the distances r_s and r instead of R_s. The result then becomes dependent only on the nature of the charge distribution within $\mathcal{V}$, the factor r remaining independent of the particular charge being considered. The meaning of the words *near* and *far* will become clear in the development that follows.

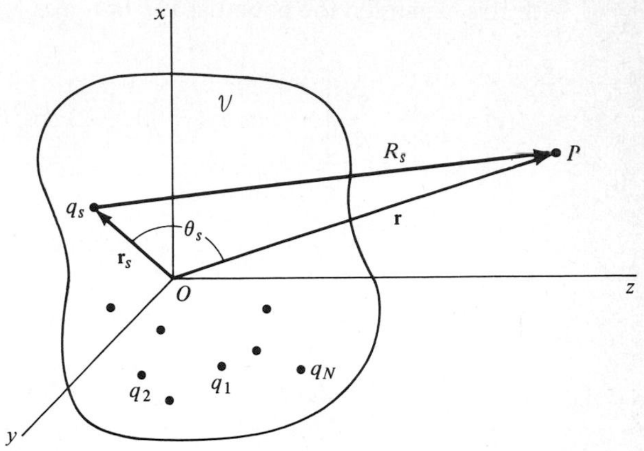

FIGURE 6-1

By the law of cosines, considering a charge q_s at r_s,

$$R_s^2 = r_s^2 + r^2 - 2r_s r \cos\theta_s$$

$$= r^2\left[1 - (2\cos\theta_s)\left(\frac{r_s}{r}\right) + \left(\frac{r_s}{r}\right)^2\right]$$

(Note: θ_s is measured from $\mathbf{r}$ rather than from an axis.) Therefore

$$\frac{1}{R_s} = \frac{1}{r}\left[1 - (2\cos\theta_s)\left(\frac{r_s}{r}\right) + \left(\frac{r_s}{r}\right)^2\right]^{-1/2}$$

Using the series expansion

$$(1+x)^n = 1 + nx + \frac{n(n-1)}{2!}x^2 + \cdots$$

with $$n = -\frac{1}{2} \quad \text{and} \quad x = -2\cos\theta_s\left(\frac{r_s}{r}\right) + \left(\frac{r_s}{r}\right)^2$$

gives $$\frac{1}{R_s} = \frac{1}{r}\left[1 + \cos\theta_s\left(\frac{r_s}{r}\right) + \frac{1}{2}(3\cos^2\theta_s - 1)\left(\frac{r_s}{r}\right)^2 + \cdots\right]$$

The restriction that P be far from $\mathcal{V}$ guarantees that the higher-order terms in (r_s/r) are negligible.

With this expansion the potential at P becomes

$$\phi(\mathbf{r}) = \frac{1}{4\pi\epsilon_0 r} \sum_{s=1}^{N} \left[1 + \cos\theta_s \left(\frac{r_s}{r}\right) + \frac{1}{2}(3\cos^2\theta_s - 1)\left(\frac{r_s}{r}\right)^2 + \cdots\right] q_s$$

$$= \frac{1}{4\pi\epsilon_0 r} \sum_{s=1}^{N} q_s + \frac{1}{4\pi\epsilon_0 r^2} \sum_{s=1}^{N} q_s r_s \cos\theta_s$$

$$+ \frac{1}{4\pi\epsilon_0 r^3} \sum_{s=1}^{N} q_s r_s^2 \frac{3\cos^2\theta_s - 1}{2} + \mathcal{O}\left(\frac{1}{r^4}\right)$$

$$= \sum_{n=0}^{\infty} \frac{A_n(\theta_s, r_s)}{r^{n+1}}$$

For a continuous charge distribution, of density ρ throughout some volume $\mathcal{V}$, the corresponding equation is

$$\phi(\mathbf{r}) = \frac{1}{4\pi\epsilon_0 r} \int_{\mathcal{V}} \rho(\mathbf{r}_s)\, d\tau_s + \frac{1}{4\pi\epsilon_0 r^2} \int_{\mathcal{V}} \rho(\mathbf{r}_s) r_s \cos\theta_s\, d\tau_s$$

$$+ \frac{1}{4\pi\epsilon_0 r^3} \int_{\mathcal{V}} \rho(\mathbf{r}_s) \left(\frac{3\cos^2\theta_s - 1}{2}\right) d\tau_s + \mathcal{O}\left(\frac{1}{r^4}\right)$$

The symbol $\mathcal{O}(1/r^4)$ means that the remaining terms are of order $1/r^n$ with $n = 4$ or higher.

These expressions are called the multipole expansion of the potential. The first term, of order $(1/r)(1/r^n) = 1/r$ when $n = 0$, is called the monopole term. The prefix mono comes from $2^n = 1$ with $n = 0$. The second term $\mathcal{O}[(1/r)(1/r^n)]$ with $n = 1$, is called the dipole term ($2^n = 2$ with $n = 1$); the third, having $\phi \propto (1/r)(1/r^n)$ with $n = 2$, is the $2^n = 4$ or quadrupole term; next the octupole, $2^3 = 8$; etc.

The coefficient in the numerator of the monopole term is Q, the total charge within $\mathcal{V}$; its value does not depend on the choice of origin. This term is the leading one in the expansion; if $Q \neq 0$ then all the other terms may be made negligibly small by going to a point P sufficiently far removed from O. At such a point the distribution within produces a potential essentially the same as that which would be produced by a point charge of the same total charge, located at O.

The numerator of the second term is the $\hat{\mathbf{r}}$ (i.e., longitudinal) component of a vector quantity, $\Sigma_{s=1}^{N}(q_s\mathbf{r}_s)$ or $\int_{\mathcal{V}} \rho(\mathbf{r}_s)\mathbf{r}_s\,d\tau_s$. For any particular case there is a unit vector $\hat{\mathbf{t}}$ transverse to $\hat{\mathbf{r}}$ such that, e.g.,

$$\sum_{s=1}^{N} q_s\mathbf{r}_s = \hat{\mathbf{t}}\left(\sum_{s=1}^{N} q_s r_s \sin\theta_s\right) + \hat{\mathbf{r}}\left(\sum_{s=1}^{N} q_s r_s \cos\theta_s\right)$$

It is left to Prob. 2 to show that when $Q = 0$ the quantity $\Sigma_{s=1}^{N} q_s\mathbf{r}_s$ is independent of the choice of the origin. (The component $\Sigma\, q_s r_s \cos\theta_s$ is only independent of the placement of the origin along the $\hat{\mathbf{r}}$ direction.) This vector quantity, $\Sigma_{s=1}^{N} q_s\mathbf{r}_s$ or $\int_{\mathcal{V}} \rho\mathbf{r}_s\,d\tau_s$, is called *the* dipole moment of the distribution in all cases. In the case when $Q = 0$, that is when the first term of the multipole expansion vanishes and the dipole term becomes the leading term, the dipole moment of the distribution acquires a special, invariant, importance.

The simplest nontrivial case with $Q = 0$ occurs with a distribution of only two charges. Here one charge must be the negative of the other. Such a distribution of two charges, one the negative of the other, is called an electric dipole. Note the different, distinct, ways in which the term dipole is used here. (1) The dipole *term* is the r^{-2} term in the multipole expansion of an arbitrary charge distribution. (2) The dipole *moment* is the magnitude of a vector, of which the numerator of the dipole term is the longitudinal component. (3) *The* electric dipole is a specific distribution of two charges, equal but opposite, placed at two different points; it is implicit that the separation between the charges is much smaller than the distance to the field point.

The potential $\phi \propto r^{-2}$ produced at distant points by a dipole falls off much more rapidly with increasing distance than that produced by a single point charge $\phi \propto r^{-1}$. That of the dipole is the residue remaining when two equal but opposite charges almost, but not quite, cancel each other. The name dipole stems from the fact that two opposite charges are required for an invariant dipole moment. The dipole distribution is an extremely important one because it provides the key to the explanation of the problems raised by the behavior of matter in the presence of an electric field. The remainder of this chapter is essentially devoted to the further exploration of this subject.

If the dipole moment is zero then the quadrupole term becomes the leading term in the expansion of the potential. This term falls off with distance much more rapidly than the term preceding it, the dipole term. At a point far from the charges it is the remainder that is left when two equal but opposite dipoles almost, but not exactly, cancel each other. The name "quadrupole" comes from $2^n = 4$ with $n = 2$. Two *pairs* of equal and opposite charges are required to have both the total charge and *the* dipole moment vanish. This requires four charges, therefore.

Figure 6-2*a* shows a simple dipole charge distribution. Here

$$\sum_{s=1}^{2} q_s r_s \cos\theta_s = q\left(\frac{\ell}{2}\right)\cos\theta + (-q)\left(\frac{\ell}{2}\right)\cos(\pi-\theta)$$

$$= q\ell\cos\theta = p\cos\theta = \mathbf{p}\cdot\hat{\mathbf{r}}$$

$\mathbf{p}$, the dipole moment, is a vector of magnitude $p = q\ell$ which points from the negative charge to the positive charge. This convention is opposite to the one which determines $\mathbf{E}$, the electric field. If the sense of the vector $\mathbf{p}$ were taken the same as that of $\mathbf{E}$, from positive to negative charge, then the coefficient $\Sigma\, q_s r_s \cos\theta_s$ would become $-\mathbf{p}\cdot\hat{\mathbf{r}}$. There is nothing wrong with this; but the other convention is the one that has been universally adopted.

The field of the charge q at P is almost, but not quite, canceled by the field of the charge $-q$ when $r \gg \ell$. The greater r is, compared to ℓ, the more nearly do the effects of the two charges cancel each other. Closer to the dipole, where r is not large compared to ℓ, the potentials of the two charges do not cancel each other, even approximately. The dipole term, alone, is then not sufficient to give the potential ϕ.

In Fig. 6-2*b* several examples are given of electric quadrupoles, obtained by taking the dipole of Fig. 6-2*a* and adding to it another, reversed, dipole slightly displaced from the former. Three configurations are shown; there is actually an infinite variety of arrangements possible, all of them equally valid as quadrupoles. In all cases the origin has been taken, for simplicity, at the center of charge. Since the total charge is zero, the dipole moment would have the same value for any other origin.

The monopole term cannot vanish when a charge distribution consists of only one charge. The dipole term does not vanish either, except for the special case when the origin and charge coincide. Though the sum of all terms beyond the first equals zero, the individual multipole terms are considered to exist. The multipole terms, however, have values which depend on the origin and one cannot speak of *the* dipole moment of this distribution. At large enough distances from the charge, however, all the terms beyond the monopole term become insignificant, even if one overlooks the fact that their collective sum is zero, and only the monopole term remains.

For a distribution of two charges where the monopole term does not vanish similar remarks apply, except that now the sum of the terms beyond the first one does not vanish. There is, for example, a dipole term; again, the value of the dipole moment depends on the location of the charges relative to the origin. At far distances one can say that only a monopole term exists.

The case of two charges which are equal in value but opposite in sign (the monopole term vanishes) gives the smallest number of charges for which the dipole moment of the distribution has invariant significance. Similarly, though a quadrupole term can arise in distributions with three or even two charges, the minimum number of

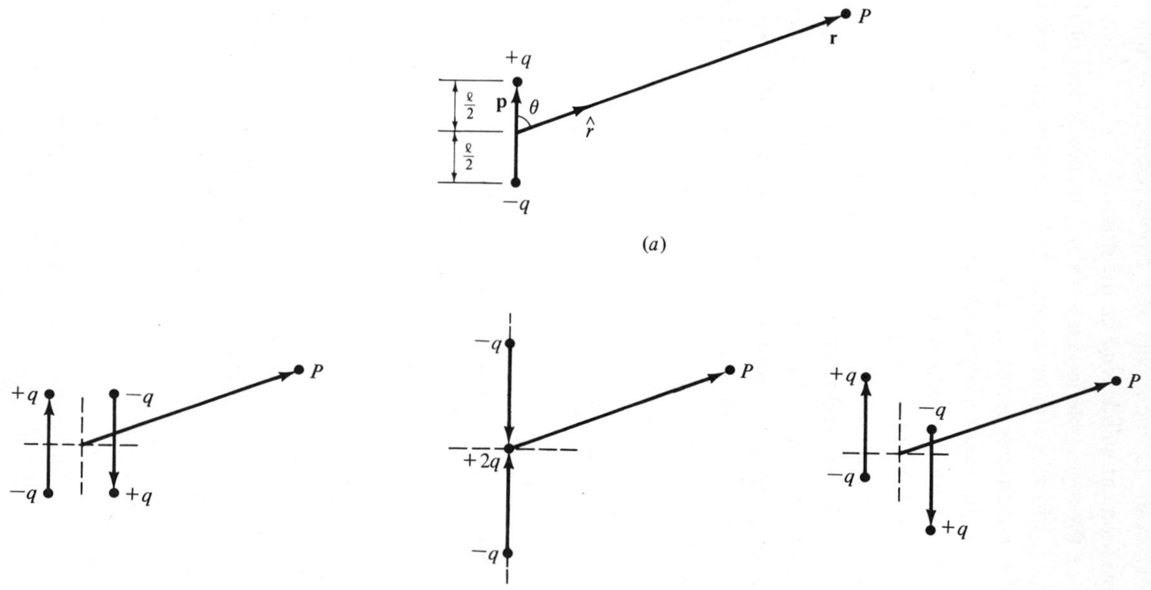

FIGURE 6-2

charges which can give a distribution with a unique value of the quadrupole moment is four. This requires that both the monopole and dipole moments vanish; otherwise the value of the quadrupole moment will depend on the origin.

The coefficient of the monopole term is a scalar. The coefficient of the dipole term is a function of a vector, which is a more complicated quantity than a scalar. Similarly, the coefficient of the quadrupole term is a function of a quantity that is more complicated than a vector. In tensor analysis it is shown that this is a tensor of rank two. A vector is a tensor of rank one, a scalar is a tensor of rank zero. The word *tensor* has been used once before in this text in a purely formal way, in connection with the Maxwell stress tensor; its definition will be deferred until Chap. 14.

The multipole expansion of the potential, which has been carried out to three terms above, applies to an arbitrary charge distribution. In particular, it is applicable to the distribution of charge which resides on any material body. When the body is a dielectric, i.e., one which is a nonconductor rather than a conductor or a semiconductor, then the first term in the multipole expansion generally vanishes, since the total charge on the body is zero unless special pains are taken to produce and maintain a net nonvanishing charge on it. The second term is also zero for most dielectrics in their normal state, i.e., when the bodies are not placed in an electric field. Most dielectrics do not possess a permanent dipole moment of their own. When a dielectric is placed in an electric field, however, the atoms of the dielectric are distorted and develop an electric dipole moment. The dielectric, itself, is said to acquire an induced dipole moment.

A similar statement is also true, in general, for the third term of the multipole expansion. The quadrupole term, however, has an effect which drops off with distance much more quickly than does the dipole term; and this applies with even more force to higher-ordered terms. Consequently, for the study of dielectric materials in electric fields the electric dipole acquires a predominant importance, while the quadrupole and higher moments play a trivial role.

Examples

1. The point dipole What is the distinction between the potential of a point electric dipole and that of the dipole of Fig. 6-2*a*.

The monopole term of the multipole expansion of the potential is equal to zero for the dipole of Fig. 6-2*a*:

$$\sum_{s=1}^{2} q_s = (-q) + (q) = 0$$

When $r \gg \ell$ the quadrupole term, which varies as r^{-3}, is much smaller than the dipole term, which varies as r^{-2}. With the condition that the result is only applicable far from the dipole, the potential for this configuration is

$$\boxed{\phi_{\text{dipole}} = \frac{1}{4\pi\epsilon_0}\frac{\mathbf{p}\cdot\hat{\mathbf{r}}}{r^2}} \tag{E-14}$$

This is called *the* dipole potential. Note that the dipole potential varies as r^{-2}, unlike the potential of a point charge, which varies as r^{-1}.

In actuality, for a dipole—one which consists of two equal but opposite finite charges, separated by a finite distance—this potential is only applicable at points far from the dipole. For points close to the dipole this result is quite wrong. It is then necessary to include the higher order terms of the multipole expansion; so it is simpler to employ

$$\phi = \frac{q}{4\pi\epsilon_0}\left(\frac{1}{r_a} - \frac{1}{r_b}\right)$$

where r_a and r_b are the distance from P to the upper and lower charge, respectively.

It is convenient to invent a fictitious entity called the point dipole. This is a charge distribution for which the dipole potential

$$\left(\frac{1}{4\pi\epsilon_0}\right)\frac{p\cos\theta}{r^2}$$

is valid for all values of r, no matter how close they are to the point dipole. The point dipole can be thought of as a limiting case of the actual dipole of Fig. 6-2*a*, one for which:

1 $q \to \infty$ and $-q \to -\infty$
2 $\ell \to 0$
3 but q grows and ℓ shrinks in such a way that $p = q\ell$ remains constant during the process

Since there is no infinite charge in nature this concept is an artificial one, introduced merely to eliminate the requirement limiting the validity only to distant points. It is common to say dipole when one really means either point dipole or *far* field of a real dipole.

By setting the dipole potential of a given dipole equal to a constant,

$$\left(\frac{p}{4\pi\epsilon_0}\right)\frac{\cos\theta}{r^2} = \phi_0$$

we obtain the expression for the family of equipotentials of the dipole:

$$r^2 = \left(\frac{p}{4\pi\epsilon_0\phi_0}\right)\cos\theta = C_1 \cos\theta$$

Figure 6-3 shows several equipotentials drawn for a point dipole on the $+z$ axis. The constant C_1 which is applicable to any one equipotential is a parameter connected with the dipole moment and the potential by the equation

$$C_1 = \frac{p}{4\pi\epsilon_0\phi_0}$$

The curves shown are actually the intersections of surfaces of revolution about the z axis with the plane of the page. The top half of the drawing is drawn with solid lines, the bottom half with dotted lines, to indicate that the two halves of the drawing are not identical. The upper curves correspond to $C_1 > 0$, i.e., to a positive potential; the lower curves correspond to a negative potential. (The curves may also be thought of as being obtained by taking horizontal xy slices of a three-dimensional figure in which ϕ is plotted along the z axis normal to the plane of the paper.) In the upper half of Fig. 6-3 ϕ gives the contours of a hill; in the lower half ϕ indicates a valley. Otherwise the curves are symmetric: one is the negative of the other in magnitude. Note that $\phi > 0$ for points closer to the $+q$ charge than to the $-q$ charge.

These curves apply either to (1) a point dipole or (2) a dipole with $r \gg \ell$.

2. The field of a point dipole We now find the **E** produced by the point electric dipole of Example 1.

In spherical coordinates, using

$$\Phi = \left(\frac{p}{4\pi\epsilon_0}\right)\frac{\cos\theta}{r^2}$$

for potential and ϕ for azimuthal angle,

$$\mathbf{E} = -\nabla\Phi = \hat{\mathbf{r}}\left(-\frac{\partial\Phi}{\partial r}\right) + \hat{\boldsymbol{\theta}}\left(-\frac{1}{r}\frac{\partial\Phi}{\partial\theta}\right) + \hat{\boldsymbol{\phi}}\left(-\frac{1}{r\sin\theta}\frac{\partial\Phi}{\partial\theta}\right)$$

$$= \hat{\mathbf{r}}\left(\frac{p}{4\pi\epsilon_0}\right)\frac{2\cos\theta}{r^3} + \hat{\boldsymbol{\theta}}\left(\frac{p}{4\pi\epsilon_0}\right)\frac{\sin\theta}{r^3}$$

Note how much more complicated this is than the dipole potential. One may draw curves for the lines of force by comparing the E_r and E_θ components with the r and θ components of $d\mathbf{r}$: dr and $r\,d\theta$. Thus

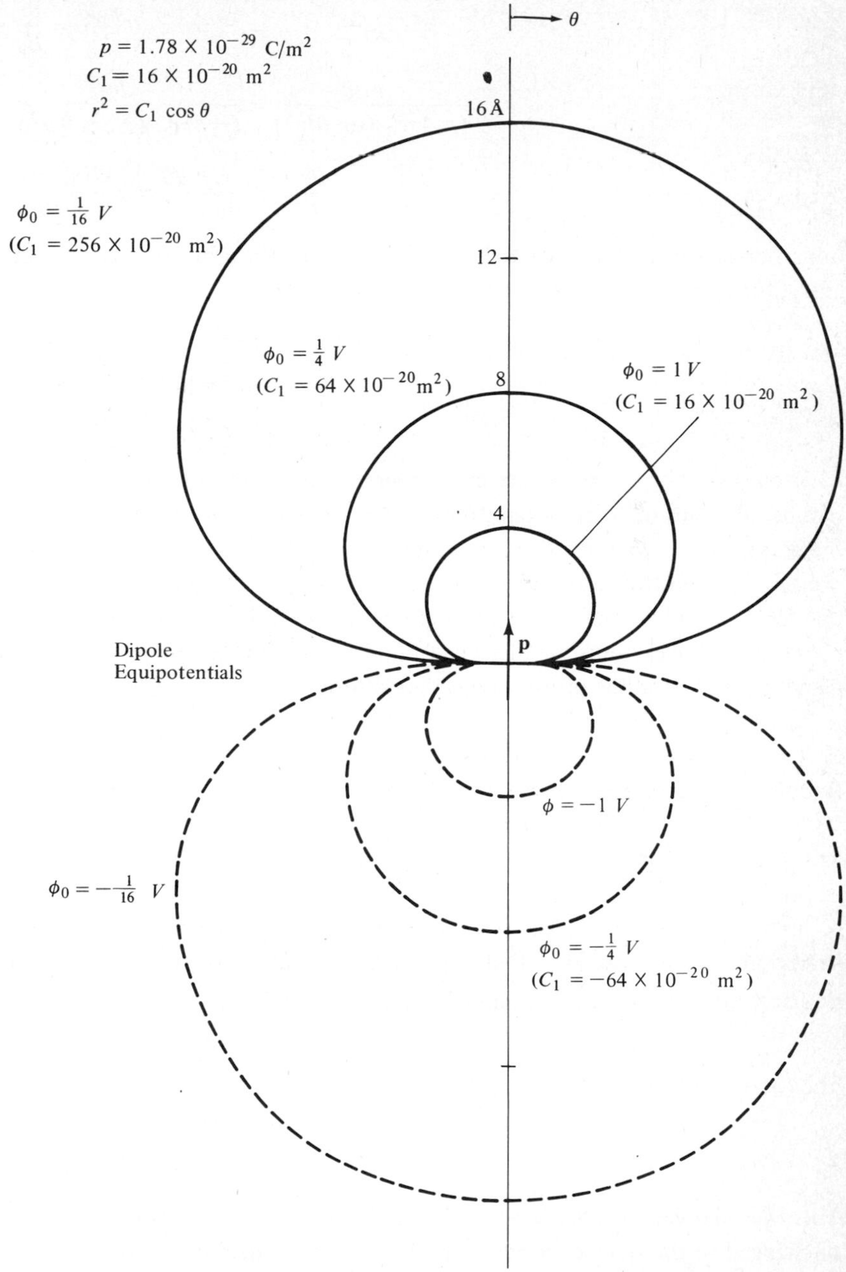

$p = 1.78 \times 10^{-29}$ C/m^2
$C_1 = 16 \times 10^{-20}$ m^2
$r^2 = C_1 \cos\theta$
θ
16 Å
12
8
4
$\phi_0 = \frac{1}{16}$ V
($C_1 = 256 \times 10^{-20}$ m^2)
$\phi_0 = \frac{1}{4}$ V
($C_1 = 64 \times 10^{-20}$m^2)
$\phi_0 = 1$ V
($C_1 = 16 \times 10^{-20}$ m^2)
p
Dipole
Equipotentials
$\phi = -1$ V
$\phi_0 = -\frac{1}{16}$ V
$\phi_0 = -\frac{1}{4}$ V
($C_1 = -64 \times 10^{-20}$ m^2)

FIGURE 6-3

$$\frac{r\,d\theta}{E_\theta} = \frac{dr}{E_r}$$

$$\frac{r\,d\theta}{(p/4\pi\epsilon_0)(\sin\theta/r^3)} = \frac{dr}{(p/4\pi\epsilon_0)(2\cos\theta/r^3)}$$

$$\frac{dr}{r} = \frac{2\cos\theta\,d\theta}{\sin\theta} = 2\,\frac{d(\sin\theta)}{\sin\theta}$$

Integrating, one may simplify the result by choosing the arbitrary constant in the form of $\ln C_2$, for then

$$\ln r = 2\ln\sin\theta + \ln C_2$$

$$\ln r = \ln(C_2\sin^2\theta)$$

$$r = C_2\sin^2\theta$$

A given value of C_2 gives one curve representing a line of force. Figure 6-4 shows a number of lines of force. Again, these are really figures of revolution, symmetric about the z axis. The arrows on the curves indicate the direction of the force which would be exerted on a positive test charge located at that point. At the (r,θ) shown, for example, the force would be mainly to the right but a little downward. The significance of the constant C_2 is clear: it is the distance from the origin to the line of force in the equatorial plane. $C_2 = OR$ for line of force which goes through P.

3. The force on a dipole A completely separate problem from that in Example 2 is this: find the force exerted on an electric dipole by an external electric field **E**.

Figure 6-5 illustrates this case. Suppose we consider the ith component ($i = 1,2,3$) of the force $\mathbf{F}^+$ on the positive charge $+q$ of a dipole. Let the ith electric field component at the center of the dipole be E_i; then at the point $\frac{1}{2}\,\mathbf{l}$ where the charge $+q$ is located, this component, by a Taylor expansion, is $E_i + (\nabla E_i)\cdot(\mathbf{l}/2)$. The higher-order terms have been neglected here by assuming that ℓ is small. So, $F_i^+ = (+q)\,[E_i + (\nabla E_i)\cdot(+\mathbf{l}/2)]$, while similarly, the ith force component on the negative charge is

$$F_i^- = (-q)[E_i + (\nabla E_i)\cdot(-\mathbf{l}/2)]$$

The total ith component force on the dipole, itself, will then be

$$F_i = F_i^+ + F_i^- = q\,[(\nabla E_i)\cdot\mathbf{l}] = (\nabla E_i)\cdot\mathbf{p} = (\mathbf{p}\cdot\nabla)E_i$$

So

$$\mathbf{F} = (\mathbf{p}\cdot\nabla)\mathbf{E}$$

This was derived for the case of a dipole, with small but finite ℓ; but the result is unchanged in the limit of a point dipole with the same dipole moment **p**. Consequently, this is called the dipole force; though, for a real dipole it is only true if the field does not vary too fast with distance. Otherwise one would require higher-order terms in the expansion. The meaning of the expression for F_i can be seen more clearly by explicit expansion, say, of the x component.

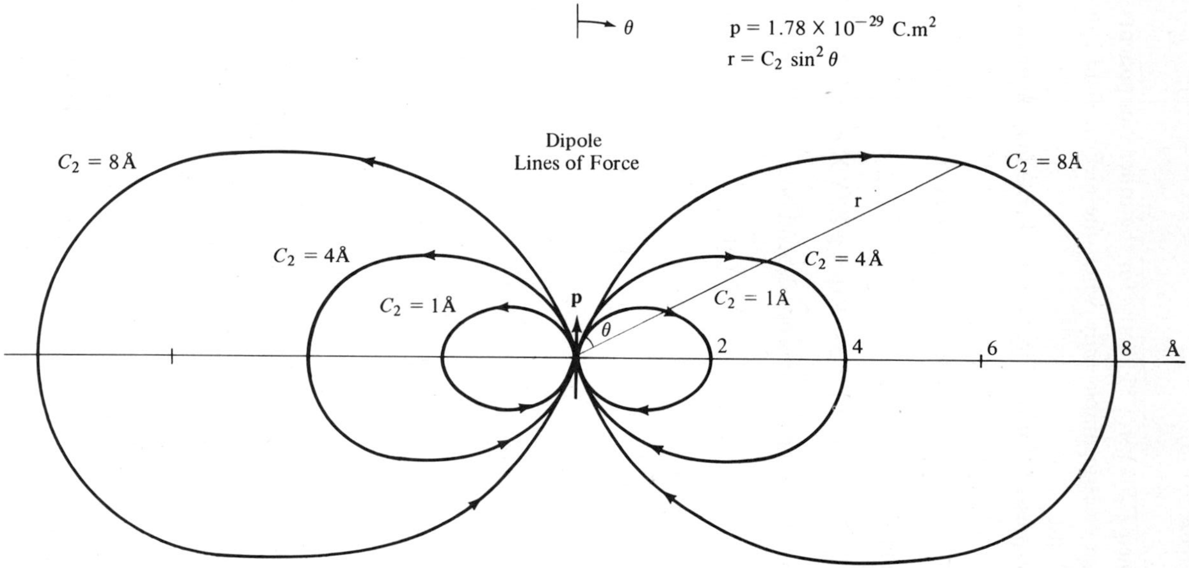
θ
p = 1.78 X 10^−29 C.m^2
r = C2 sin^2 θ
Dipole
Lines of Force
C2 = 8Å
C2 = 4Å
C2 = 1Å
p
θ
r
2
4
6
8
Å
FIGURE 6-4

$$F_x = \left(p_x \frac{\partial}{\partial x} + p_y \frac{\partial}{\partial y} + p_z \frac{\partial}{\partial z}\right) E_x = p_x \frac{\partial E_z}{\partial x} + p_y \frac{\partial E_x}{\partial y} + p_z \frac{\partial E_x}{\partial z}$$

If E_x is a constant then F_x is zero. Similarly for the other components. A translational force is exerted on an electric dipole only when the electric field is not constant, but varies from point to point.

An alternative expression for the force on the dipole may be obtained from a vector identity. In

$$\nabla(\mathbf{A} \cdot \mathbf{B}) = (\mathbf{A} \cdot \nabla)\mathbf{B} + (\mathbf{B} \cdot \nabla)\mathbf{A} + \mathbf{A} \times (\nabla \times \mathbf{B}) + \mathbf{B} \times (\nabla \times \mathbf{A})$$

let $\mathbf{A} = \mathbf{p}$ and $\mathbf{B} = \mathbf{E}$. Then

$$\mathbf{F} = \nabla(\mathbf{p} \cdot \mathbf{E}) - (\mathbf{E} \cdot \nabla)\mathbf{p} - \mathbf{p} \times (\nabla \times \mathbf{E}) - \mathbf{E} \times (\nabla \times \mathbf{p})$$

The dipole moment $\mathbf{p}$ is given here; it does not depend on $x, y,$ or z. Consequently

$$(\mathbf{E} \cdot \nabla)\mathbf{p} = 0 \qquad \text{and} \qquad \nabla \times \mathbf{p} = 0$$

so the electrical force on the dipole is

$$\mathbf{F} = \nabla(\mathbf{p} \cdot \mathbf{E}) - \mathbf{p} \times (\nabla \times \mathbf{E})$$

But in electrostatics

$$\nabla \times \mathbf{E} = 0$$

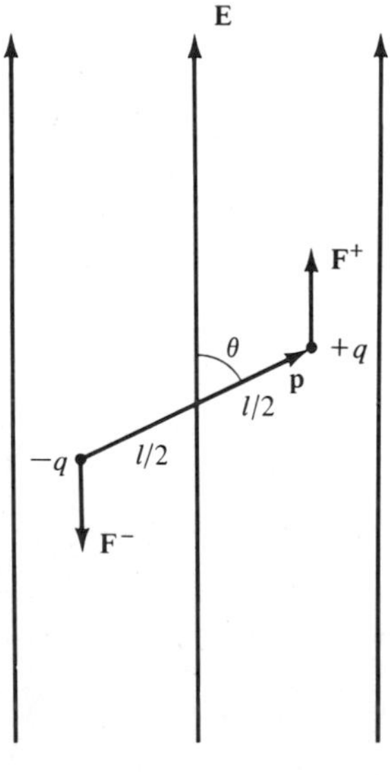

FIGURE 6-5

so $$\mathbf{F} = \nabla(\mathbf{p} \cdot \mathbf{E}) = -\nabla(-\mathbf{p} \cdot \mathbf{E})$$

This result is only valid in electrostatics. When $\mathbf{E}$ varies with time it will be found subsequently that $\nabla \times \mathbf{E} \neq 0$; then the expression for $\mathbf{F}$ also contains the term $-\mathbf{p} \times (\nabla \times \mathbf{E})$.

Electrical force $\mathbf{F}_e$ and potential electrical energy U_e are related by $\mathbf{F}_e = -\nabla U_e$. For a single charge q, e.g., dividing this equation by q gives $\mathbf{E} = -\nabla\phi$. For a dipole, then, comparing $\mathbf{F} = -\nabla U_e$ with $\mathbf{F} = -\nabla(-\mathbf{p} \cdot \mathbf{E})$ gives

$$\boxed{U_e = -\mathbf{p} \cdot \mathbf{E}} \qquad \text{(E-15)}$$

We have arbitrarily taken the additive constant in U_e equal to zero in order to have U_e vanish when p and $\mathbf{E}$ are perpendicular to each other. The potential energy is a minimum when $\mathbf{p}$ and $\mathbf{E}$ are parallel, i.e., when the $+q$ of a dipole is turned as far as it can go along the positive direction of $\mathbf{E}$ and the $-q$ is toward the tail of $\mathbf{E}$. When $\mathbf{p}$ and $\mathbf{E}$ are antiparallel the potential energy is a maximum. When $\mathbf{p}$ and $\mathbf{E}$ are perpendicular to each other the potential energy is halfway between its maximum and minimum values. The choice of this attitude as the zero is convenient but arbitrary.

PROBLEMS

1 Find $\mathbf{E}$ in cartesian coordinates for an ideal electric dipole which lies at the origin with $\mathbf{p} = \hat{\mathbf{z}}p$.

2 (*a*) Show that when the monopole term of the multipole expansion of the potential vanishes, i.e., when $Q = 0$, then the value of the dipole moment of the charge distribution does not depend on the origin.
(*b*) Show that then the numerator of the dipole term depends on the transverse position of the origin, but not on longitudinal position.

3 From $\mathbf{E} = -\nabla\phi$ find a vector expression for the $\mathbf{E}$ produced by an electric dipole in terms of $\mathbf{p}$ and $\mathbf{r}$.

4 If two curves are so related that the slopes $m_1 \equiv (dy/dx)_1$ and $m_2 \equiv (dy/dx)_2$ satisfy $m_2 = -1/m_1$, then the curves are orthogonal to each other. Show that the families of equipotentials and lines of force of the ideal dipole are orthogonal to each other.

5 In Example 3 the (translational) force on a real dipole was found by making a Taylor expansion of $\mathbf{F}$ on the two charges in terms of $\mathbf{E}$ at the center of the dipole. Using the expression for the torque caused by these two forces,

$$\mathbf{T} = (\mathbf{r}^+ \times \mathbf{F}^+) + (\mathbf{r}^- \times \mathbf{F}^-)$$

find the torque on the dipole in terms of $\mathbf{p}$ (and $\mathbf{E}$ at the center of the dipole). *Hint*: Take $\mathbf{T} = \hat{\mathbf{z}}T_z = \hat{\mathbf{z}}(T_z^{\ +} + T_z^{\ -})$.

6 The potential energy of a dipole in an external electrostatic field is $U = (+q)\phi_+ + (-q)\phi_-$, where ϕ_+ and ϕ_- are the potentials at the points a and b occupied by $+q$ and $-q$, respectively. Let $\boldsymbol{\ell}$ be the vector from a to b. Find ϕ_+ and ϕ_- in terms of ϕ, the potential at the center of dipole, by means of a Taylor expansion, to obtain the expression $U = -\mathbf{p} \cdot \mathbf{E}$, given in Example 3.

7 Find the torque exerted by $\mathbf{E}$ on an electric dipole if the torque is taken about some point distant $\mathbf{r}$ from the center of the dipole. Use the Taylor expansions for $\mathbf{F}^+$ and $\mathbf{F}^-$ in Example 3.

8 Write $U = -\mathbf{p} \cdot \mathbf{E}$ as $U = -pE \cos\theta$. By analogy with the expression for the force exerted on the dipole by an external field $\mathbf{F} = -\nabla U$, find the expression relating the torque (exerted on the dipole, at its center, by the external field) and the potential energy U in terms of θ and $\hat{\mathbf{n}}$. Here θ is as shown in Fig. 6-5 and $\hat{\mathbf{n}}$ is a unit vector perpendicular (down) into the plane of the page.

9 Calculate the quadrupole moment of a real electric dipole with respect to an origin O off the center of symmetry but on the dipole axis. The designation *real* means two equal, but opposite, finite charges displaced from each other by a finite distance—in distinction from a point dipole. Let r_+ and r_- be the distances from O, off the center of symmetry, to q_+ and q_-; while θ_+ and θ_- are the angles between the two source vectors and $\mathbf{r}$.

10 A spherical surface of infinitesimal thickness has a surface charge distribution $\sigma = \sigma_0 \cos\theta$. For such a distribution the dipole term of the multipole expansion of the potential is

$$\frac{1}{4\pi\epsilon_0 r^2} \int_\Sigma \sigma(\mathbf{r}_s)\mathbf{r}_s \cos\theta_s \, dS$$

and the integral, which depends only on the nature of the charge distribution, is the dipole moment of the surface charge. Find the dipole moment of this distribution.

11 Calculate the dipole moments of the three quadrupole charge distributions shown in Fig. 6-2*b*.

12 An equilateral triangle has a charge q at each apex. Find the dipole moment and the quadrupole moment of the distribution.

13 In Example 3, the expression for the force of an electrostatic field on an electric dipole has $\nabla \times \mathbf{E} = 0$. In quantum mechanics the angle θ between $\mathbf{p}$ and $\mathbf{E}$ must remain constant. Obtain a simplified equation for $\mathbf{F}$ in that case.

14 Express the force on a dipole, $F = (\mathbf{p} \cdot \nabla)\mathbf{E}$, in cylindrical coordinates.

15 Two dipoles, $\mathbf{p}_1$ and $\mathbf{p}_2$, have arbitrary direction and separation. See Fig. 6-6. Find the expression for the energy of either one in the field of the other. This is called the dipole-dipole interaction energy.

16 Suppose the first charge distribution in Fig. 6-2*b* is modified by changing each of the two $+q$ charges to $+2q$ charges. Let the distance from each of the four charges to the center be a. Calculate the distance along the x axis at which the dipole and quadrupole terms of the potential multipole expansion are equal.

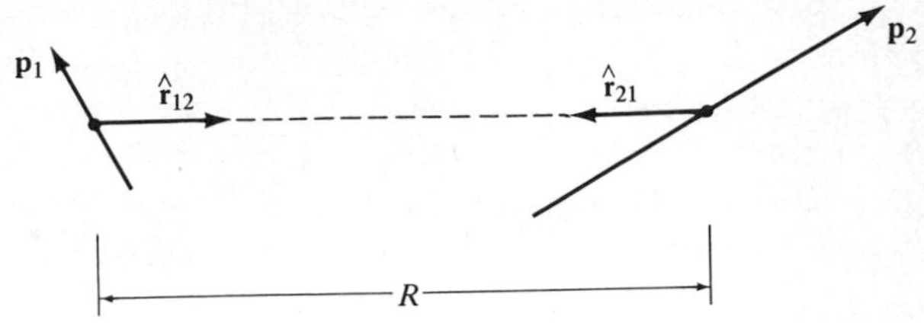

FIGURE 6-6

17 From the dipole-dipole interaction energy of Prob. 15 obtain an explicit expression for the so-called tensor force $\mathbf{F}_{12} = -\nabla_2 U$, the force exerted by one dipole on another, in terms of $\mathbf{p}_1$, $\mathbf{p}_2$, and $\hat{\mathbf{R}}$.

18 From the dipole-dipole interaction energy of Prob. 15 obtain an explicit expression for the torque exerted by $\mathbf{p}_1$ on $\mathbf{p}_2$. What is the torque on $\mathbf{p}_2$ for the configuration of Fig. 6-7*a*? Is the torque of $\mathbf{p}_2$ on $\mathbf{p}_1$ equal and opposite to this?

19 Two dipoles are arranged as in Fig. 6-7*b* and *c*. What is the magnitude and direction of the force between them? (A typical value for the dipole moment of an atom would be 2×10^{-18} esu-cm or 6×10^{-30} C m.)

20 The quadrupole term of the potential multipole expansion of the charge distribution of Fig. 6-1 is

$$\frac{1}{4\pi\epsilon_0 r^3} \sum_{s=1}^{N} q_s r_s^2 \left(\frac{3\cos^2\theta_s - 1}{2} \right)$$

(*Note*: Consideration of this term and, therefore, this problem, involves tensor language. This should be skipped if the reader is unfamiliar with tensors.)

(*a*) Show that this may also be written as

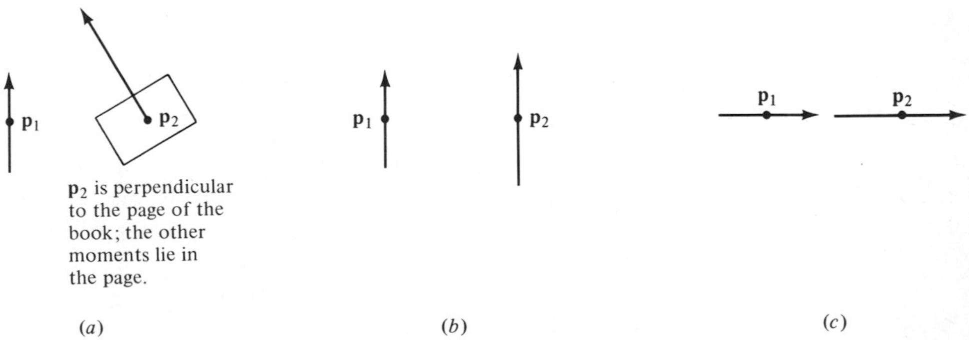

FIGURE 6-7

$$\frac{1}{4\pi\epsilon_0 r^3}\sum_{\substack{i=1\\j=1}}^{3} Q_{ij}\left(\frac{r_i r_j}{2r^2}\right)$$

where
$$Q_{ij} = \sum_{s=1}^{N} q_s[3(\mathbf{r}_s)_i(\mathbf{r}_s)_j - (\mathbf{r}_s)^2\delta_{ij}]$$

Q_{ij} is the quadrupole moment of the charge distribution. Written in matrix form it looks like this:

$$\{Q_{ij}\} = \sum_{s=1}^{N} q_s \begin{Bmatrix} 3x_s^2 - r_s^2 & 3x_s y_s & 3x_s z_s \\ 3y_s x_s & 3y_s^2 - r_s^2 & 3y_s z_s \\ 3z_s x_s & 3z_s y_s & 3z_s^2 - r_s^2 \end{Bmatrix}$$

Q_{ij} is a function only of the source charge distribution (amount of charge and its position). The other factor in the quadrupole term of the potential is a function only of the test point position. A similar situation exists in the dipole term, where $\mathbf{p}$ gives the effect of the distribution and $\cdot\,\hat{\mathbf{r}}$ modifies this effect in relation to the test point.

For matrix manipulations it is convenient to rewrite the test point factor in a form which is similar to that of Q_{ij}.

(*b*) Show that, when used in the quadrupole term, this factor may be written

$$a_{ij} = \frac{1}{6r^2}\,[3r_i r_j - r^2\,\delta_{ij}]$$

i.e.,
$$\{a_{ij}\} = \frac{1}{6r^2}\begin{Bmatrix} 3x^2 - r^2 & 3xy & 3xz \\ 3yx & 3y^2 - r^2 & 3yz \\ 3zx & 3zy & 3z^2 - r^2 \end{Bmatrix}$$

The quadrupole term may then be written, since this factor is symmetric, as

$$\frac{1}{4\pi\epsilon_0 r^3}\sum_{\substack{i=1\\j=1}}^{3} Q_{ij}a_{ji}$$

$\{a_{ij}\}$ is a function only of the test point at which the effect of the quadrupole term is being evaluated. $\{a_{ij}\}$ gives a direction along which to measure the effect of $\{Q_{ij}\}$, just as in the dipole term $\hat{\mathbf{r}}$ gives a direction along which to measure the effect of $\mathbf{p}$.

In deriving the result here, the following will be found useful:

$$\cos^2\theta = \frac{(\mathbf{r}_s\cdot\mathbf{r})^2}{r_s^2 r^2} = \frac{(x_s x + y_s y + z_s z)^2}{r_s^2 r^2}$$

21 How many independent components does $\{Q_{ij}\}$ have when referred to arbitrary axes? Suppose $\{Q_{ij}\}$ is diagonalized, referring it to the principal axes: how many independent components does $\{Q_{ij}\}$ have then? Can this number be reduced to one?

22 Plot the +10-V equipotential for a dipole with $\mathbf{p} = \hat{\mathbf{z}}10^{-30}$. Add one line of field intensity to the drawing, to show the orthogonality.

6-2 THE POLARIZATION P

When an atom which normally has no dipole moment is placed in an electric field, the nucleus of the atom tends to move along the direction of **E** while the electrons in the atom tend to go in the opposite direction. As soon as such movements (caused by these disruptive forces) occur, restoring electric forces between the opposite charges are set up which tend to bring conditions back to their original status. These restoring forces bring the atom into equilibrium, with the center of positive charge displaced from the center of negative charge by a very small fraction of an angstrom unit, $\sim 10^{-5}$ Å; so the atom acquires a dipole moment, induced by the external electric field. If **p** is this induced electric moment per atom, and if there are N such atoms per unit volume of some fixed piece of nonconductive matter, then the polarization vector is defined by the relation

$$\mathbf{P} = N\mathbf{p}$$

Thus, the polarization **P** is the dipole moment density in the material.

If there are n kinds of atoms present in the dielectric then similar results, with individual N_i and $\mathbf{p}_i$, hold for each of them and $\mathbf{P} = \Sigma_{i=1}^{N} N_i\mathbf{p}_i$.

In addition to induced polarization of the atoms, many substances have a permanent polarization caused, e.g., by the fact that their *molecules* are asymmetric. (*Atoms*, on the other hand, always have a symmetric distribution in their ground state.) When two neutral atoms combine to form a molecule the electrons of one atom quite often shift over to the second atom by approximately 0.1 Å, thereby giving the molecule a permanent dipole moment. This molecular permanent dipole moment is usually considerably larger than the induced atomic dipole moment. Dielectrics with molecules having permanent dipole moments are called polar substances. In the normal state, when there is no external electric field **E**, most polar substances do not have any polarization since the temperature-dependent molecular agitation results in a random distribution. There are some substances, however, called ferroelectrics by analogy with the permanent magnetism of iron, in which the alignment force is strong enough to make itself felt despite the randomizing effect of the heat.

Polar molecules tend to align themselves in the external electric field. But nonpolar molecules also develop an induced dipole moment in an external field; there are different mechanisms by which a body may become polarized. No matter how it is produced, polarization is in any case measured in units of C m^{-2}; then $P\,d\tau$ has the units C m, just like those of a single dipole. It is customary to refer to such a nonconducting substance as a dielectric. In practically all dielectrics the polarization induced by the field is not a permanent property but one which exists only when **E** is present. When the field is removed the dielectric reverts to its normal unpolarized state. There are substances, however, in which the polarization may be caused to remain permanently, even after the field is removed. When molten carnauba wax is inserted into an electric field and then solidified by cooling, e.g., a permanent **P** remains after **E** is reduced to zero; for thin slabs and large **E** in fact, even the melting is not required. Such a substance is called an electret. It is the electrical analog of the permanent magnet; but, unlike the latter, it has not achieved much technical importance.

Suppose we have a dielectric, of volume $\mathcal{V}$, for which (1) the electric charge density is zero throughout the volume ($\rho = 0$) and (2) there is no external charge density on its surface ($\sigma = 0$). When this dielectric is placed in an external electric field **E**, the individual atoms acquire induced dipole moments—the dielectric acquires a polarization. These moments produce a potential at any point and thereby modify the potential and electric field which would exist there if the dielectric were not present. This potential will now be calculated.

From the potential produced by one point electric dipole,

$$\phi = \frac{1}{4\pi\epsilon_0}\frac{\mathbf{p}\cdot\hat{\mathbf{r}}}{r^2}$$

it is an obvious generalization to go to the potential produced at a point *outside* the dielectric by a polarized dielectric consisting of many dipoles. This is:

$$\phi(\mathbf{r}_t) = \frac{1}{4\pi\epsilon_0}\int_{\mathcal{V}} \frac{\mathbf{P}\cdot\hat{\mathbf{R}}}{R^2}\,d\tau_s$$

The reason it is necessary to specify $\mathbf{r}_t$ as outside $\mathcal{V}$ is that this equation for the dipole potential is only valid either: (1) for point dipoles or (2) for real dipoles when $\mathbf{r}_t$ is far from the dipoles. But within the dielectric there are 10^{19}, or more, dipoles per cubic centimeter and it becomes necessary to distinguish points where the equation is valid from those where it is not—a difficult task. (Strictly speaking, $\mathbf{r}_t$ must be not only outside $\mathcal{V}$ but also not too close to the surface. But the dimensions of the dipoles in the dielectric are only of the order of 1 Å = 10^{-8} cm; so it would only be necessary to restrict $\mathbf{r}_t$ such that it did not come within 10 Å of the surface. This distance is so small that we neglect it.)

Taking $\mathbf{r}_t$ as outside $\mathcal{V}$, the last result may be transformed to

$$\phi(\mathbf{r}_t) = -\frac{1}{4\pi\epsilon_0}\int_{\mathcal{V}} \mathbf{P}\cdot\nabla\left(\frac{1}{R}\right)d\tau_s = +\frac{1}{4\pi\epsilon_0}\int_{\mathcal{V}} \mathbf{P}\cdot\nabla_s\left(\frac{1}{R}\right)d\tau_s$$

where ∇_s is the del operator taken with respect to the source point variables (x_s, y_s, z_s) of the particular $d\tau_s$ under consideration. The vector identity

$$\nabla_s\cdot\left(\frac{\mathbf{P}}{R}\right) = \frac{1}{R}\nabla_s\cdot\mathbf{P} + \mathbf{P}\cdot\nabla_s\left(\frac{1}{R}\right)$$

then, gives

$$\phi(\mathbf{r}_t) = \frac{1}{4\pi\epsilon_0}\left[\int_{\mathcal{V}} \nabla_s\cdot\left(\frac{\mathbf{P}}{R}\right)d\tau_s + \int_{\mathcal{V}} \frac{-\nabla_s\cdot\mathbf{P}}{R}\,d\tau_s\right]$$

By Gauss' theorem the first integral may be rewritten as a surface integral over Σ, the surface bounding $\mathcal{V}$:

$$\oint_{\Sigma} \frac{\mathbf{P}\cdot d\mathbf{S}}{R} = \oint_{\Sigma} \frac{P_n\,dS}{R}$$

Finally, removing the subscript in the second integral for simplicity,

$$\phi(\mathbf{r}_t) = \frac{1}{4\pi\epsilon_0}\oint_{\Sigma} \frac{\sigma_t}{R}\,dS + \frac{1}{4\pi\epsilon_0}\oint_{\mathcal{V}} \frac{\rho_p}{R}\,d\tau$$

where

$$\boxed{\begin{aligned}\sigma_p &= P_n\\ \rho_p &= -\nabla\cdot\mathbf{P}\end{aligned}} \qquad \text{(E-15)}$$

σ_p is called the polarization surface charge density; ρ_p is the polarization volume charge density. Both σ_p and ρ_p have direct physical significance, as shown in Fig. 6-8. A cube of matter, exposed to the field **E**, develops a polarization **P**. The normal component of **P** exists only on two faces: at the right and at the left of the cube in Fig. 6-8*a*. On the former, P_n is outward, or positive, so σ_p is positive; at the left P_n is inward, or negative, so σ_p is negative. If the dielectric is homogeneous throughout its volume $\nabla\cdot\mathbf{P} = 0$, then the cube may be replaced, as far as its effect on the region outside it is concerned, by the two sheets of charge shown in Fig. 6-8*b*. The potential and field produced by these

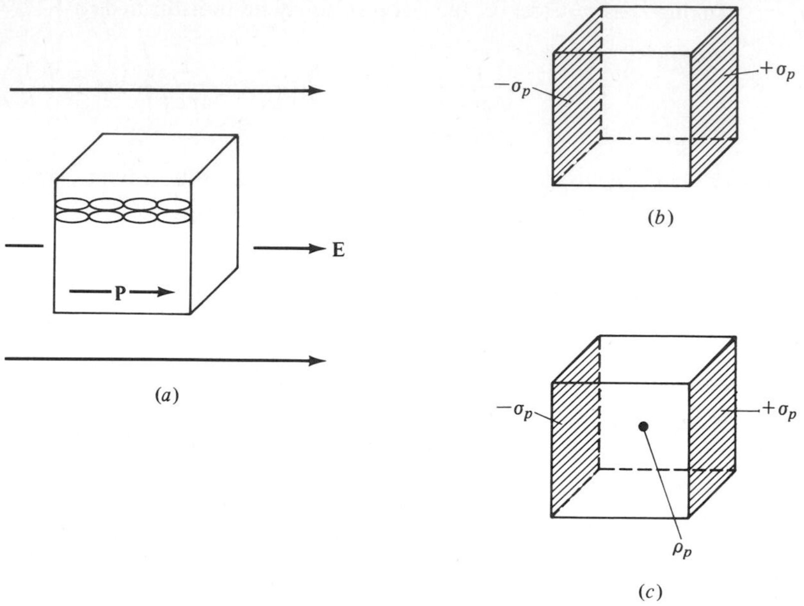

FIGURE 6-8

effective surface charge densities add (vectorially, for the field) to the external values produced by the charges in the external sources.

Let the dielectric be homogeneous. If one considers a small volume within the dielectric, the effect of the external field is to remove some positive charges from this small volume but to introduce an equal number of negative charges, leaving the charge of the small volume unchanged. If the dielectric is inhomogeneous, however, $\nabla \cdot \mathbf{P} \neq 0$; and there will be some regions in the interior which acquire more negative charges than they lose (or vice versa) and thereby become charged. Then it becomes necessary to add ρ_p to the region between the plates, as in Fig. 6-8*c*. This total effective charge density, surface plus volume, replaces the actual cube of dielectric, in determining the field outside the cube. ρ_p and σ_p are real charges, produced by the polarization of the dielectric, and the potential and field produced by these polarization charge densities are also real. Nevertheless, it is convenient to distinguish them from ρ_f and σ_f, the usual charge densities; i.e., those produced by charges which are free to flow. The means for making this distinction will be discussed in the next section.

What is the potential at a point *within* the dielectric? The real polarization charge densities, ρ_p and σ_p of Fig. 6-8*c*, determine potentials at all points inside the cube as well as at all points outside it. But inside the dielectric this is not the only component of the potential: there must also be added the intense contribution of the atomic nuclei and

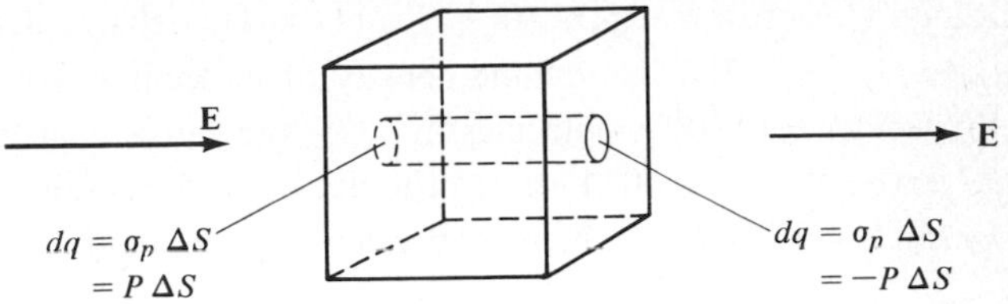

FIGURE 6-9

electrons, with electric fields of the order of hundreds of millions of volts per meter. The latter, microscopic, contribution varies both with space and with time, however, in such a way that for any macroscopic sample its average value is zero, leaving only the contribution of σ_p and ρ_p. Consequently, the equation

$$\phi(\mathbf{r}_t) = \frac{1}{4\pi\epsilon_0} \oint_{\Sigma} \frac{\sigma_p}{R} \, dS + \frac{1}{4\pi\epsilon_0} \int_{\mathcal{V}} \frac{\rho_p}{R} \, d\tau$$

gives the potential ϕ at any point $\mathbf{r}_t$, whether this point is within or without the dielectric; in the former case the potential is the macroscopic time-averaged and space-averaged potential.

The fact that σ_p and ρ_p give the potential inside $\mathcal{V}$ may also be seen from Fig. 6-9, where a needlelike cavity has been drilled in the dielectric. ΔS, the cross-sectional area of this cylinder, has a maximum dimension much less than the length; the latter is parallel to **E**, the externally produced electric field. A point at the center of the needle, being far enough from the dielectric, would have a potential $\phi(\mathbf{r}_t)$ given by the equation above. The presence of this hole produces a negligible alteration in the average electric field and potential which would otherwise exist for the charges introduced at the ends of the cavity are far away and very small in magnitude. So the $\phi(\mathbf{r}_t)$ of the equation above also gives the space-averaged and time-averaged potential within the dielectric itself.

Examples

1. Surface charge density Suppose the effect of an electric field on a dielectric is to induce an electric dipole moment of 10^{-30} C m^{-1} per molecule. What is the surface polarization charge density on a surface making an angle of 45° with **E** if there are 10^{21} molecules/cm^3?

$$\sigma_p = P_n = P \cos 45° = Np \cos 45°$$

$$= \left(10^{21} \frac{\text{mol}}{\text{cm}^3} \times \frac{10^6 \, \text{cm}^3}{\text{m}^3}\right) (10^{-30} \text{ cm}) \, (0.707)$$

$$= 0.0007 \text{ C m}^{-2}$$

Since 1 electron = 1.6×10^{-19} C, this surface charge density is equivalent to 4.4×10^{15} electrons/m^2. But the volume density of molecules, 10^{27} m^{-3}, gives a surface density of molecules of 10^{18} molecules/m^2; so the surface charge density is produced by an effective shift of 0.0044 electron/molecule to the surface. For each 1000 molecules at the surface, 4.4 develop a charge of 1 electron each, in effect.

2. A circular polarized disk We will find the potential and the electric field at a point on the axis of symmetry perpendicular to a circular disk which has a uniform polarization perpendicular to the surface of the disk. The disk electret has infinitesimal thickness.

Figure 6-10*a* shows the geometry. In Fig. 6-10*b* the previous results of ϕ and **E** for a charged disk are reproduced, from Fig. 4-8 and Fig. 2-10, for comparison. The equations that apply to the *charged* disk of Fig. 6-10*b* are:

$$\phi = \frac{\sigma}{2\epsilon_0}\left[\sqrt{z^2 + R^2} - |z|\right]$$

$$\mathbf{E} = \pm\frac{\sigma}{2\epsilon_0}\left[1 - \frac{|z|}{\sqrt{z^2 + R^2}}\right]\hat{\mathbf{z}}$$

Here, for the *polarized* disk, the situation is that which would obtain if two oppositely charged disks were to be placed coaxially, spaced apart by a distance ℓ, giving a dipole plate with $P = \sigma\ell$. Letting $\sigma \to \infty$ and $\ell \to 0$ such that P stays fixed then gives a dipole plane. The potential for a surface distribution of dipoles is given by modifying $P\,d\tau$ (where $P = p/\upsilon$ and $[P] = [\mathrm{C\,m^{-2}}]$) for a volume distribution to $P_{\text{surf}}\,dS$ for a surface distribution, with $P_{\text{surf}} = p/S$ and $[P_{\text{surf}}] = [\mathrm{C\,m^{-1}}]$. Thus,

$$\phi(\mathbf{r}_t) = \frac{1}{4\pi\epsilon_0}\int_\Sigma \frac{\mathbf{P}_{\text{surf}} \cdot \hat{\mathbf{R}}}{R^2}\,dS_s$$

But $\hat{\mathbf{R}}\,dS_s/R^2 = d\mathbf{\Omega}_s$, a differential vector solid angle subtended at the test point by the source area but pointing, with $\hat{\mathbf{R}}$, from source to test point. Taking $d\mathbf{\Omega} = -d\mathbf{\Omega}_s$ as this solid angle but pointing from the test point,

$$\phi(\mathbf{r}_t) = \frac{1}{4\pi\epsilon_0}\int_\Sigma \mathbf{P}_{\text{surf}} \cdot d\mathbf{\Omega}_s = -\frac{1}{4\pi\epsilon_0}\mathbf{P}_{\text{surf}} \cdot \int_\Sigma d\mathbf{\Omega} = -\frac{\mathbf{P}_{\text{surf}} \cdot \mathbf{\Omega}}{4\pi\epsilon_0}$$

The solid angle subtended by the disk at the point $z = z$ on the axis is $2\pi(1 - \cos\alpha)$, where α is the half-angle of the cone. This result is readily obtained by integrating the area of a zone on the surface of a sphere of radius r: $d\Omega = dS/r^2 = 2\pi r^2 \sin\theta\, d\theta/r^2$ and $\Omega = \int_0^\alpha 2\pi \sin\theta\, d\theta$. So $\phi = \mp(P_{\text{surf}}/2\epsilon_0)(1 - \cos\alpha) = \mp(\mathbf{P}_{\text{surf}}/\epsilon_0)\sin^2(\alpha/2)$. The minus is employed when **P** and **Ω** are parallel. Here **Ω** points to $-\hat{\mathbf{z}}$ for points on the $+\hat{\mathbf{z}}$ axis and $\mathbf{P}_{\text{surf}}$ points to $+\hat{\mathbf{z}}$; so $\mathbf{P}_{\text{surf}}$ and **Ω** are antiparallel and the plus is used for $\mathbf{z} > 0$. Then

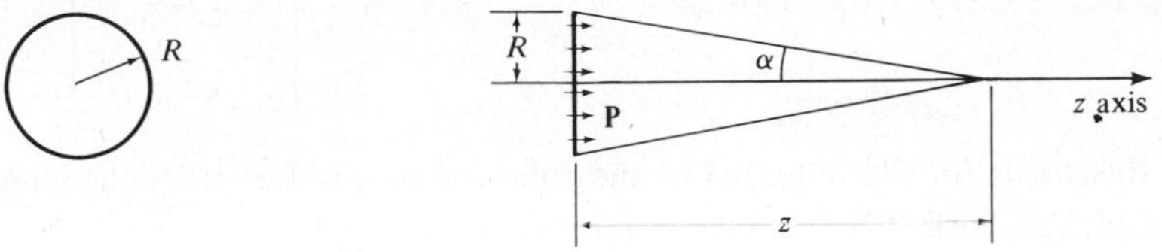

(a) Geometry.

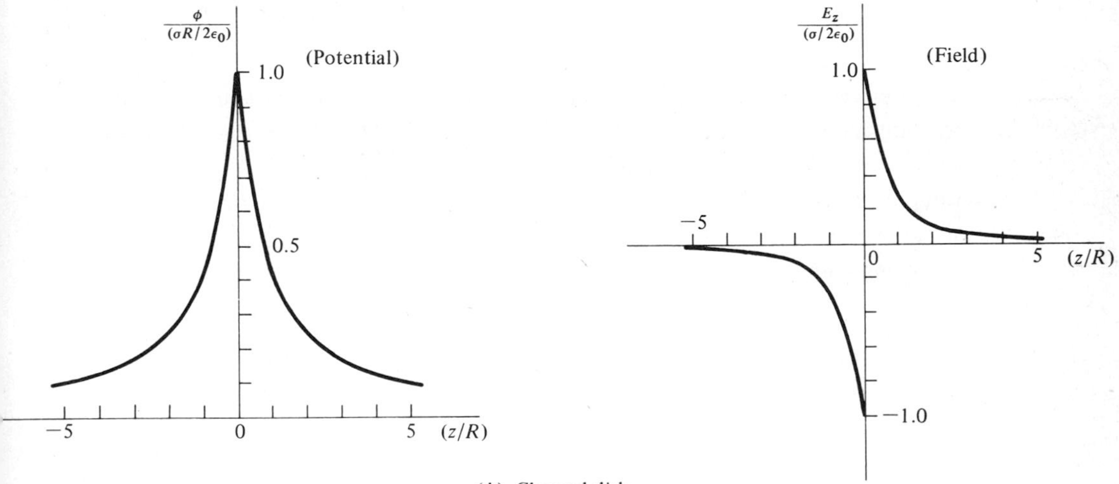

(b) Charged disk

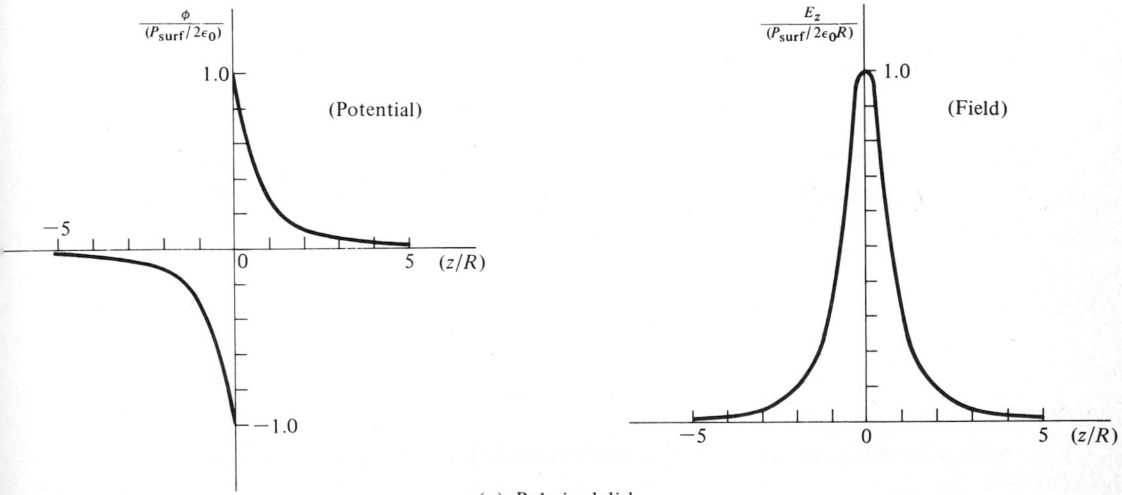

(c) Polarized disk

FIGURE 6-10

$$\phi = \pm \frac{\mathbf{P}_{\text{surf}}}{2\epsilon_0}\left[1 - \frac{|z|}{\sqrt{z^2 + R^2}}\right]$$

Note that this result for the potential of the polarized disk is essentially the same as that for **E** of the charged disk. This is shown in Fig. 6-10*c*.

The **E** for the polarized disk is obtained from ϕ by taking $\mathbf{E} = -\nabla\phi$:

$$\mathbf{E} = \frac{\mathbf{P}_{\text{surf}}}{2\epsilon_0}\left[\frac{R^2}{(z^2 + R^2)^{3/2}}\right]\hat{\mathbf{z}}$$

3. A cylindrical electret Let the disk of Exercise 2 become a cylinder of length L. We will find **E** at any point on the axis of the cylinder. This is the field on the axis of a cylindrical electret with uniform axial polarization.

The value of **E** may be obtained from a uniform surface charge distribution $\sigma_p = -P$ on a disk of radius R at $z = 0$ and a distribution with surface density $\sigma_p = +P$ at $z = L$, the cylinder itself being eliminated. The value of **E** for a point on the axis inside the cylinder is

$$\mathbf{E} = \left\{\frac{-P}{2\epsilon_0}\left[1 - \frac{z}{\sqrt{z^2 + R^2}}\right] - \frac{(+P)}{2\epsilon_0}\left[1 - \frac{L - z}{\sqrt{(L - z)^2 + R^2}}\right]\right\}\hat{\mathbf{z}}$$

The minus outside the second bracket comes from the fact that this point is to the left of the right end of the cylinder; the first minus is due to the negative charge density on the end at $z = 0$.

For a point on the axis outside the cylinder with $z > L$,

$$\mathbf{E} = \left\{\frac{-P}{2\epsilon_0}\left[1 - \frac{z}{\sqrt{z^2 + R^2}}\right] + \frac{+P}{2\epsilon_0}\left[1 - \frac{z - L}{\sqrt{(L - z)^2 + R^2}}\right]\right\}\hat{\mathbf{z}}$$

For a point on the axis outside the cylinder with $z < 0$,

$$\mathbf{E} = \left\{-\frac{-P}{2\epsilon_0}\left[1 - \frac{|z|}{\sqrt{z^2 + R^2}}\right] - \frac{+P}{2\epsilon_0}\left[1 - \frac{|z| + L}{\sqrt{(|z| + L)^2 + R^2}}\right]\right\}\hat{\mathbf{z}}$$

Note that, within the cylinder, **E** points left while **P** points right on the axis.

4. A spherical electret A spherical electret is uniformly polarized: $\mathbf{P} = \hat{\mathbf{z}}P$. Find **E** at the center of the sphere.

There is no polarization volume charge density since $\rho_p = -\nabla \cdot \mathbf{P} = 0$. The polarization surface charge density is $\sigma_p = P_n = \mathbf{P} \cdot \hat{\mathbf{r}} = P \cos\theta$. The $\mathbf{E}$ at the center may be found from σ_p. Consider an annular ring on the surface, between θ and $\theta + d\theta$. All the charge in this ring is at the distance r from the center; by symmetry, this charge will produce only a $-\hat{\mathbf{z}}$ component. So

$$\mathbf{E}_{\text{in}} = \frac{1}{4\pi\epsilon_0} \oint (-\hat{\mathbf{z}}) \frac{\sigma_p \, dS}{r^2} \cos\theta = -\frac{\hat{\mathbf{z}}}{4\pi\epsilon_0} \int_0^{\pi} (P\cos\theta)(2\pi \sin\theta \, d\theta) \cos\theta$$

$$= -\hat{\mathbf{z}} \frac{P}{2\epsilon_0} \int_{\pi}^{0} \cos^2\theta \, d(\cos\theta) = -\hat{\mathbf{z}} \frac{P}{3\epsilon_0} = -\frac{1}{3\epsilon_0}\mathbf{P}$$

This value, calculated this way, pertains to the center of the sphere only, not necessarily to other points in the electret. Nevertheless, the electric field throughout the entire polarized sphere turns out to be a constant. The proof of this is left to Prob. 16. The potential inside the electret that gives this electric field is $\phi_{\text{in}} = (1/3\epsilon_0)(\mathbf{P} \cdot \mathbf{r}) = (Pz/3\epsilon_0)$.

PROBLEMS

1 What is the polarization volume charge density in a uniform dielectric surrounding a charged metal sphere?

2 Given a region in which $\mathbf{P} = \hat{\mathbf{z}}P(1 + bz)$. Find the polarization charge density ρ_p. Repeat this for $P = \hat{\mathbf{z}}P(1 + bx)$. Suppose there is a slab of this material of thickness d along the $\hat{\mathbf{z}}$ axis. Find the surface charge density. What is the total bound charge of the slab?

3 Given the spherical electret of Example 4. Let the radius of the sphere be R. Find $\mathbf{E}$ and ϕ outside the sphere.

4 Find the polarization surface charge densities, for the two polarizations of Prob. 2, on a cube of side a with center at the origin and edges parallel to the coordinate axes. Show, also, that the net total charge is zero.

5 A spherical hole is cut in an electret having uniform polarization. Find (a) ϕ_h and (b) $\mathbf{E}_h$ at the center of the hole in terms of $\mathbf{P}$ and $\mathbf{E}$, the values within the electret before the hole was cut.
(c) Are $\mathbf{P}'$ and $\mathbf{E}'$, the values after the hole is cut, the same as $\mathbf{P}$ and $\mathbf{E}$?

6 Is the electric field within a dielectric in an external $\mathbf{E}$ greater or smaller than it would be at that point if there were no dielectric? Explain.

7 Figure 6-11 shows an atom that has been polarized by an external field. If a typical value of $\mathbf{p}$ is 10^{-30} C m, what is x in angstrom units?

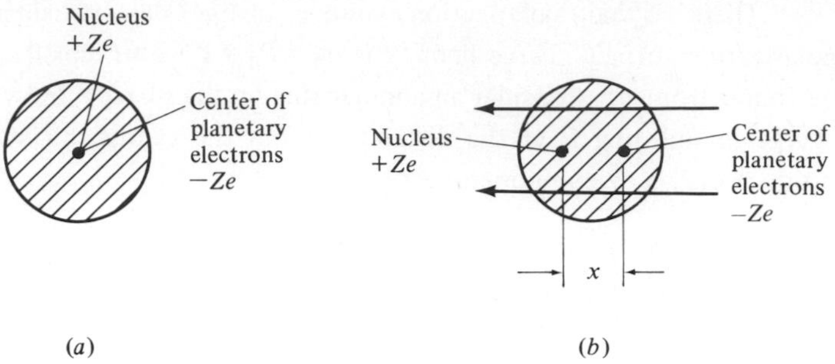

FIGURE 6-11

8 A cube of dielectric, having finite **E** dimensions, is uniformly polarized: $\mathbf{P} = \hat{\mathbf{z}}P$. Find **E** at the center of the cube.

9 Find the electric field and the potential produced by the $z = 0$ plane, of infinite extent in the x and y directions, when this plane has a uniform polarization $\mathbf{P} = \hat{\mathbf{z}}P$ perpendicular to the plane. Can ϕ here be derived from the case of the polarized disk by letting $R \to \infty$? Why?

10 Suppose a homogeneous, isotropic dielectric of arbitrary shape is inserted into a uniform external field. The bound charges produced on the surface will in general produce a nonuniform field inside the dielectric. Show that for ellipsoidal bodies the field inside will be uniform. (This is a complicated problem.)

11 A spherical dielectric of radius a has $\mathbf{P} = \hat{\mathbf{z}}P$. What is the total polarization charge for the upper hemisphere? For the whole sphere?

12 A comb is charged by passing it through one's hair. A small piece of electrically neutral tissue paper will be attracted to the comb. Why?

13 What are ϕ and **E** produced by the $z = 0$ plane, polarized uniformly in the z direction?

14 A vector is determined by its divergence and curl. In $\mathbf{P} = N\mathbf{p}$ let N and **p** depend on position. What is $\nabla \cdot \mathbf{P}$ and $\nabla \times \mathbf{P}$?

15 The number density of dipoles in a gas varies with height according to $N = N_0 \exp(-w_z/kT)$, where w is the weight of the molecule. Find $\nabla \cdot \mathbf{P}$ and $\nabla \times \mathbf{P}$ if $\mathbf{p} = \alpha\mathbf{E}$ and (a) $\mathbf{E} = \hat{\mathbf{x}}E_0$ and (b) $\mathbf{E} = \hat{\mathbf{z}}E_0$.

16 Show that the electric field at all points within a uniformly polarized spherical electret is equal to the value at the center of the sphere. *Hint*: Consider the individual fields of two spheres of uniform charge density, one positive and one negative, separated by **l**. Then, if N is the number of double spheres per unit volume and q is the charge, $\mathbf{P} = Nq\,\mathbf{l}$. The surface charge density given by these two displaced spheres is just that of the spherical electret.

17 Compare the values of ϕ just inside and just outside a spherical electret at $\theta = 0$. Repeat for **E**.

6-3 THE D FIELD

The equation $\nabla \cdot \mathbf{E} = (1/\epsilon_0)\rho$ was derived for source charges of density ρ in a vacuum. But in the preceding section a bound polarization charge, produced in a dielectric medium, was assumed to create a field in exactly the same way as one created by free charge. Accordingly, in the equation above for $\nabla \cdot \mathbf{E}$ we will now consider that $\mathbf{E}$ is produced by any ρ, whether the charge is free or bound. Suppose we divide ρ explicitly into free charge density and polarization charge density: $\rho = \rho_f + \rho_p$. Then

$$\boxed{\nabla \cdot \mathbf{E} = \frac{1}{\epsilon_0}(\rho_f + \rho_p)} \qquad \text{(E-17)}$$

$$\nabla \cdot \mathbf{E} = \frac{1}{\epsilon_0}\rho_f - \frac{1}{\epsilon_0}\nabla \cdot \mathbf{P}$$

Setting

$$\nabla \cdot (\epsilon_0 \mathbf{E} + \mathbf{P}) = \rho_f$$

$$\boxed{\mathbf{D} = \epsilon_0 \mathbf{E} + \mathbf{P}} \qquad \text{(E-18)}$$

gives

$$\boxed{\nabla \cdot \mathbf{D} = \rho_f} \qquad \text{(E-19)}$$

The vector **D** has generally been called the displacement, after an idea of Maxwell's concerning displacement of the ether, and the name has stuck for want of a better one. It is sometimes referred to as electric flux density, but this is incorrect. What is actually meant is free electric flux density, and this is cumbersome. We will sometimes use the term displacement if we have to; but mostly, following a recent healthy trend to deemphasize both the word and the vector, we will simply call it the **D** field or the **D** vector.

Equations (E-17) and (E-19) display an essential distinction between **E** and **D**. Any kind of charge density, whether ρ_f or ρ_p, acts as a source for **E**; but only free charge density ρ_f is a source of **D**. However, **D** is not uniquely determined by ρ_f, though **E** is determined uniquely by $\rho_f + \rho_p$. For $\nabla \times \mathbf{E} = 0$ in electrostatics; $\nabla \cdot \mathbf{E}$ and $\nabla \times \mathbf{E}$ being given, **E** is determined by the Helmholtz theorem. But $\nabla \times \mathbf{D} = \nabla \times \mathbf{P}$ and this can have a wide variety of values; so, for a given ρ_f, **D** is not determined until $\nabla \times \mathbf{P}$ is given.

In Example 3 below, we discuss the case of an electret for which $\rho_f = 0$ and **P** is uniform. Then $\nabla \times \mathbf{P} = 0$ outside the electret (because $\mathbf{P} = 0$ there) and $\nabla \times \mathbf{P} = 0$ inside the electret (because $\mathbf{P} = \hat{z}P_0$ there). If we could say $\nabla \times \mathbf{P} = 0$ everywhere then $\nabla \times \mathbf{D} = 0$ everywhere also, and since $\nabla \cdot \mathbf{D} = 0$ if $\rho_f = 0$, the Helmholtz theorem would give $\mathbf{D} = 0$ everywhere. But $\nabla \times \mathbf{P} \neq 0$ everywhere, for on the sides of the electret $\nabla \times \mathbf{P}$

does not vanish. The result is that $\mathbf{D} \neq 0$ for the case of the electret, even though $\rho_f = 0$. This contribution from $\nabla \times \mathbf{P}$ on the sides of the electret cannot be overlooked; without it one would have $\mathbf{D} = 0$ and $\mathbf{E} = -1/\epsilon_0 \mathbf{P}$, so that $\mathbf{E}$ would vanish also outside the electret. Of course, instead of calculating $\mathbf{D}$ from $\nabla \cdot \mathbf{D} = 0$, $\nabla \times \mathbf{D} = \nabla \times \mathbf{P}$ ($\neq 0$ on the sides), we could find $\mathbf{E}$ from $\nabla \cdot \mathbf{E} = -\nabla \cdot \mathbf{P}$ ($\neq 0$ on the ends) and $\nabla \times \mathbf{E} = 0$. The latter method, using two disks of charge, is the usual one.

The vector $\mathbf{D}$ is not as important as the simplicity of $\nabla \cdot \mathbf{D} = \rho_f$ might suggest. The reason for this is that ρ_f is a difficult quantity either to measure or to calculate.

The dimensions of $\mathbf{D}$ are $[\mathrm{m}^{-2}\ \mathrm{C}]$, just like those of σ and $\mathbf{P}$; but $\mathbf{E}$ and $\mathbf{D}$ differ dimensionally. Consequently $\mathbf{E}$ and $\mathbf{D}$ cannot be compared directly, and it is first necessary to multiply the first (or divide the second) by ϵ_0.

Equations (E-17) and (E-19) are alternative forms of one of Maxwell's equations. The right sides give the cause, the particular source charges that are relevant; the left sides give the effect, the field responses. These equations will be found to be true even after time variations enter the picture.

$\mathbf{D} = \epsilon_0 \mathbf{E} + \mathbf{P}$ connects $\mathbf{D}$ and $\mathbf{E}$ by means of an additive term. It is also possible to connect $\mathbf{D}$ and $\mathbf{E}$ by a multiplicative factor which is not necessarily a constant. Only the criterion of convenience decides the question of which one to use. Thus, assume that the polarization is proportional to the electric field. There are some substances for which this is not true—for some crystalline substances $\mathbf{P}$ and $\mathbf{E}$ are not even in the same direction; and for all substances $\mathbf{P}$ will vary as a higher-than-linear power of $\mathbf{E}$ if the latter is made sufficiently large; but, for most substances at ordinary field strengths this assumption is valid. Then

$$\boxed{\mathbf{P} = \chi(\epsilon_0 \mathbf{E})} \qquad \text{(E-20)}$$

where χ is a numerical factor called the electric susceptibility. The constant ϵ_0 appears, if χ is to be dimensionless, in order to make the equation correct dimensionally. So

$$\mathbf{D} = \epsilon_0 \mathbf{E} + \chi \epsilon_0 \mathbf{E} = (1 + \chi)\epsilon_0 \mathbf{E} = \kappa \epsilon_0 \mathbf{E}$$

Setting $\epsilon = \kappa \epsilon_0$,

$$\boxed{\mathbf{D} = \epsilon \mathbf{E}} \qquad \text{(E-21)}$$

This equation, necessary as an adjunct to Maxwell's equations if they are written in terms of both $\mathbf{E}$ and $\mathbf{D}$, is called a constitutive equation since it attempts to characterize the constitution of the material through ϵ.

The multiplying factor ϵ that connects $\mathbf{D}$ and $\mathbf{E}$ is called the permittivity of the substance. It has the same dimensions as ϵ_0; the quantity κ is a scalar called the dielectric constant and for all known substances $\kappa > 1$ in electrostatic conditions. The method of definition employed here permits κ to have the same values in the cgs and mksa systems. Most substances have values of κ which fall between 1 and 10. Water is quite unusual, with $\kappa = 80$ at room temperature (for steam, κ is only slightly larger than unity). There are other substances with very high dielectric constants, e.g., barium titanate with $\kappa > 1000$; these really constitute a different class of substances, since **P** and **E** are not simply proportional for them. In all cases the value of κ varies with frequency in a complicated fashion. κ for water, for example, drops from the static value of 80, roughly, to a value of 1.8 at frequencies ($\sim 10^{15}$ Hz) characteristic of light waves.

At low frequencies, say from zero to 10^9 Hz (the latter is roughly the beginning of the microwave region of the spectrum), the predominant contribution to κ comes from the fact that the water molecule is polar; i.e., it has a permanent dipole moment. This is due to the fact that the oxygen atom and the two hydrogen atoms do not align themselves in a straight line. Instead, if the O atom is thought of as up, or north, one H atom would be southwest and another would be southeast. The angle between the two lines connecting the H atoms to the O atom is approximately 105°; the permanent dipole moment, 5.5×10^{-30} m C^{-1}, points symmetrically due south. Without any **E** the moments of individual molecules point in random directions, neutralizing themselves; but the presence of a field tends to align **p** parallel to **E**.

For frequencies that are increasingly higher than the microwave frequencies, water molecules find it increasingly difficult to align themselves with the field, and the polar contribution soon becomes negligible. Somewhat above 10^{12} Hz the κ for water becomes about 5. Other contributions drop out at even higher frequencies. For any substance, at high enough frequencies, the value of κ is very close to unity. Our interest at present is in the static value, i.e., the value at zero frequency.

The replacement of the vacuum as the medium by a nonpolarized dielectric results only in the substitution of ϵ for ϵ_0 in any equation with which we will be concerned: homogeneous and isotropic media, not-too-large electric fields. Thus, a calculation of the potential energy of a charge distribution, embedded in a dielectric with permittivity ϵ, yields $dU = \frac{1}{2}\epsilon E^2\, d\tau = \frac{1}{2}\mathbf{D} \cdot \mathbf{E}\, d\tau$ instead of $\frac{1}{2}\epsilon_0 E^2\, d\tau$. Similarly, the capacitance of a parallel-plane capacitor with dielectric insulation becomes $\epsilon A/d$ instead of $\epsilon_0 A/d$, the value for vacuum insulation. Several examples of this general statement are left for the problems.

Is **D** more fundamental a vector than **E**? Considerable controversy existed among some authors in the past concerning this question. Similarly, in magnetism, is **B** more important than the still-to-be-introduced **H**? There no longer seems to be much difference of opinion in electricity: **E** is generally accepted as primary, but the difference of opinion still exists concerning **B** versus **H**. The distinction between **E** and **D** is clear—the divergence of one is produced by any source charges, of the other only by free source charge. A similar, though not identical, distinction exists in magnetism, where it concerns the curl and vortex sources. For us the fundamental quantities are **E** and **B**, when we wish to use fields; but sometimes they are the potentials, ϕ and **A** (ϕ and **A** are actually fields just as much as any one of **E**, **D**, **B**, or **H**). The potential ϕ can exist in some region even though both **E** and **D** vanish there, and its presence or absence can produce a measurable physical effect. Similar remarks apply to **A** versus **B** or to **A** versus **H**. But the question of which quantities, fields or potentials, are the fundamental ones is actually a matter of personal preference. Incidentally, the distinction between free and polarization charges cannot always be made uniquely.

Examples

1. Polarization versus free charge density A parallel-plate capacitor has an insulating plate with dielectric constant κ. We will find an expression for the bound surface charge density σ_p on a face of the insulator in terms of the free charge density on the metal plates.

For the face of the dielectric adjacent to the positive metal plate

$$\sigma_p = \mathbf{P} \cdot \hat{\mathbf{n}} = -P$$

If the plate separation is small compared to the lateral dimensions then **D**, **E**, and **P** are collinear and all point in the same direction:

$$\mathbf{D} = \epsilon_0 \mathbf{E} + \mathbf{P} \qquad \text{becomes} \qquad D = \epsilon_0 E + P$$

Then

$$\sigma_p = \epsilon_0 E - D = D\left(\frac{1}{\kappa} - 1\right) = -D\left(\frac{\kappa - 1}{\kappa}\right)$$

To find D in terms of σ, apply Gauss' law to the cylinder of infinitesimal height in Fig. 6-12, partly in and partly outside the metal. A free surface charge density σ_f exists on the surface. Then

$$\nabla \cdot \mathbf{D} = \rho_f \qquad \text{and} \qquad \int \nabla \cdot \mathbf{D}\, d\tau = \int \rho_f\, d\tau$$

$$\oint \mathbf{D} \cdot d\mathbf{S} = \int \sigma_f\, dS \qquad \int_{\text{top}} \mathbf{D} \cdot dS = \int \sigma_f\, dS$$

since $\mathbf{D} = 0$ on the bottom, inside the metal, and the sides $\to 0$. So

$$D = \sigma_f$$

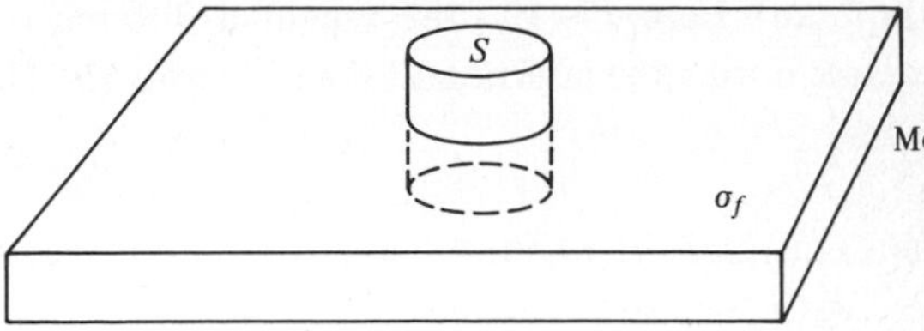

FIGURE 6-12

This is a very useful relation, not only for this problem but in its own right: on the surface of a metal the free surface charge density equals D. For other than metals, in the more general case, this becomes $\sigma_f = D_n$, where D_n is the normal component of $\mathbf{D}$.

For the capacitor, then, $\sigma_p = -\sigma_f\,[(\kappa - 1)/\kappa]$. This equation is useful in visualizing the change in the lines of flux produced by the introduction of a dielectric. If σ_p is added to σ_f to give $\sigma = \sigma_f/\kappa$ we find that $E = \sigma_f/\epsilon$ instead of $E = \sigma/\epsilon_0$, the relation with vacuum. This illustrates a rule: if we use ϵ instead of ϵ_0, we must only employ σ_f (and disregard σ_p). If we use σ_p then we must continue to use ϵ_0. Thus, in the case of the spherical electret of Example 4 in the previous section, ϵ_0 was used rather than ϵ. The effect of the dielectric was already accounted for in finding σ_p.

2. **The effect of a dielectric on E and D** A parallel-plate capacitor, with vacuum as the insulation, has a charge of magnitude Q_0 on each plate; a capacitance C_0; a potential difference $V_0 = \phi_1 - \phi_2$ between its plates; an electric field $\mathbf{E}_0$; and a $\mathbf{D}$ field vector $\mathbf{D}_0$. An insulating slab with dielectric constant $\kappa = 2$ is inserted into the region between the plates, filling the region completely. Find the new values–Q, C, V, E, D–for the following two cases. (1) The capacitor is not connected to a battery or any other element during the insertion of the slab. (2) The capacitor is connected to a battery while the dielectric is placed in position.

CASE 1 (*a*) Here $Q = Q_0$: the charge remains the same, before and after insertion of the dielectric, since current cannot flow anywhere. (*b*) The capacitance

$$C_0 = \epsilon_0 A/d \rightarrow C = \epsilon A/d = \kappa\epsilon_0 A/d = 2C_0$$

(*c*) The equation $Q = CV$ becomes $Q_0 = (2C_0)V$ so $V = Q_0/2C_0 = V_0/2$. (*d*) Since $E = V/d$, and d remains the same, then $E = (V_0/2)/d = E_0/2$. (*e*) $D_0 = \epsilon_0 E_0$ originally; $D = \epsilon E = (\kappa\epsilon_0)E = (2\epsilon_0)(E_0/2) = \epsilon_0 E_0 = D_0$. In this case the free charge remains the same, so D remains the same. The polarization charge density (see the previous example) is $-\frac{1}{2}$ the free charge density so the total charge density is reduced by a factor of $\frac{1}{2}$. Therefore the electric field is reduced to $\frac{1}{2}$ its original value.

CASE 2 (*a*) Here $V = V_0$: the potential difference, determined by the battery, remains the same during the insertion of the dielectric. (*b*) The capacitance

$$C_0 = \epsilon_0 A/d \rightarrow C = \epsilon A/d = 2C_0$$

again, regardless of whether the battery is connected or not. (*c*) $Q = CV = (2C_0)(V_0) = 2Q_0$. With the battery connected to the plates of the capacitor, current can flow while the dielectric is inserted; it does flow, and the free charge on the plates of the capacitor is doubled. (*d*) From $E = V/d = V_0/d = E_0$ it is seen that the electric field remains the same. Though the free charge on the capacitor plates has doubled, a charge of opposite polarity (with magnitude equal to the original free charge) has developed on the surfaces of the dielectric, leaving the net charge unaltered. (*e*) $D = \epsilon E = (2\epsilon_0)E_0 = 2D_0$. This is to be expected, since the free charge has doubled.

The following table summarizes the results:

Original values	Final values: 1. Isolated case	Final values: 2. Nonisolated case
Q_0	Q_0	$2Q_0$
C_0	$2C_0$	$2C_0$
V_0	$\frac{1}{2}V_0$	V_0
E_0	$\frac{1}{2}E_0$	E_0
D_0	D_0	$2D_0$

When the dielectric is changed it is not possible to say whether D increases or E decreases unless one specifies the conditions during the change.

3. A bar electret Suppose we have an electret, which has bound charge on its surfaces but no free charge. Since **D** is produced only by free charge it might seem that there is no **D** anywhere, though there is an **E**. But $\mathbf{D} = \epsilon\mathbf{E}$ gives a finite **D** for a finite **E** unless $\epsilon = 0$. Does this imply $\epsilon = 0$ for an electret?

The equation $\mathbf{D} = \epsilon_0\mathbf{E} + \mathbf{P}$ applies to an electret as well as to an ordinary dielectric. The equation $\mathbf{D} = \epsilon\mathbf{E}$, on the other hand, does not apply to an electret: it was derived on the assumption that $\mathbf{P} = \chi(\epsilon_0\mathbf{E})$. For an electret the relation between **P** and **E** is not a linear one.

Suppose the electret, with no free charge, has the shape of a long, thin, cylinder with uniform polarization pointing along the axis. At all points within the rod, as at all points outside it, the derivatives of **E**, **P**, and **D** are continuous. And there, since $\nabla\cdot\mathbf{P} = 0$, $\nabla\times\mathbf{P} = 0$:

$$\begin{cases} \nabla\cdot\mathbf{E} = \dfrac{1}{\epsilon_0}(\rho_f + \rho_p) = 0 \\ \nabla\times\mathbf{E} = 0 \end{cases} \qquad \begin{cases} \nabla\cdot\mathbf{D} = \epsilon_0(\nabla\cdot\mathbf{E}) + \nabla\cdot\mathbf{P} = 0 \\ \nabla\times\mathbf{D} = \epsilon_0(\nabla\times\mathbf{E}) + \nabla\times\mathbf{P} = 0 \end{cases}$$

If these were the only points then, by the Helmholtz theorem, **E** and **D** would have to be constant everywhere. This result is inconsistent with $\mathbf{D} = \epsilon_0 \mathbf{E} + \mathbf{P}$ since **P** vanishes outside the rod but not inside it. The reason this result for **E** and **D** is incorrect is that there are points where the derivatives of **E**, **P**, and **D** are not continuous: various points on the surface of the electret. It is necessary to employ care in the definition of divergence and curl at such points. In Appendix 6, at the end of this chapter, we show how to modify the usual definitions of divergence and curl in order to take care of these discontinuities. Figure 6-13*a* and *b* is utilized there in order to provide the extended definitions of the divergence and curl. Here, however, we will only treat this problem in the conventional manner.

Figure 6-13*c* shows the lines of **P**, assumed uniform throughout the electret. From $\sigma_p = \mathbf{P}$ we see that there is a positive surface polarization charge density on the right end of the bar electret and a negative σ_p of equal magnitude at the left end of the bar. There is no σ_p on the long surface, nor is there any volume polarization charge density: $\rho_p = -\nabla \cdot \mathbf{P} = 0$. Since **E** is produced by any charge, free or bound, the bound charges at

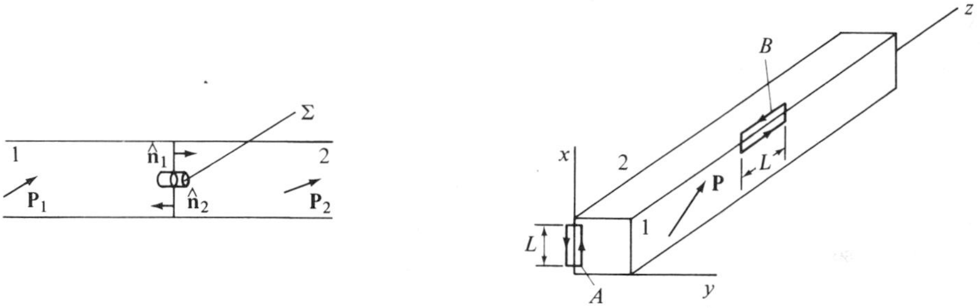

(*a*) Divergence. (*b*) Curl.

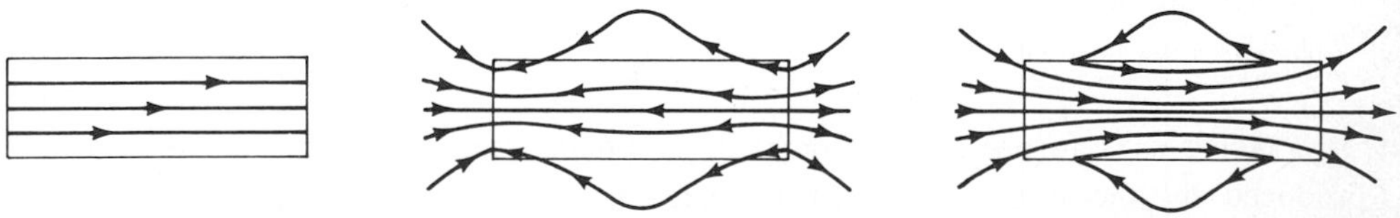

(*c*) **P** + (*d*) $\epsilon_0 \mathbf{E}$ = (*e*) **D**.

FIGURE 6-13

the two ends act to produce an **E**. The actual calculation of **E** is complicated (see Example 3 of the previous section) but Fig. 6-13*d* shows the general appearance of either **E** or $\epsilon_0\mathbf{E}$. Then **D** is simply obtained by adding $\epsilon_0\mathbf{E}$ and **P**, giving the result shown in Fig. 6-13*e*. Note that **D** is not zero, even though there is no free charge; it is $\nabla \cdot \mathbf{D}$ that vanishes, not **D**.

The directions of **E** and **D** within the bar are opposite to each other.

If the bar is very short, so that it is really a puck or disk, then **E** within it will be very nearly uniform. **P** and $\epsilon_0\mathbf{E}$ would be practically equal in magnitude but opposite in direction, so that $\mathbf{D} \approx 0$. If the bar is long then **E** has radial components even if **P** is strictly axial, so there must be a **D** inside the bar. The lines of **D** pass through either end in one continuous sense; but the lines of **E** diverge from one face and converge on the other. **P** and **E** have discontinuous derivatives on the ends; **D** has discontinuous derivatives on the sides.

4. A spherical dielectric in an external uniform E Let us take a region in space where we can produce a uniform electric field, $\mathbf{E}_{ext} = \hat{\mathbf{z}}E_{ext}$ say, such that the charges which cause $\mathbf{E}_{ext}$ are not appreciably affected by the presence or absence of a dielectric introduced into this region. Then it is plausible to assume that, whatever its shape, the dielectric will be uniformly polarized. Suppose, in particular, that we consider a spherical dielectric; this is not the simplest geometry to take, but it is an important one and it is also not too difficult to test the hypothesis in this case. It turns out that the hypothesis of uniform polarization is true for a sphere; indeed, it is true for all ellipsoids–of which spheres are only one special category. But it is *not true* for any other shape. Plausible or not, an assumption must be verified.

The dielectric acquires an induced polarization because of $\mathbf{E}_{ext}$ so it, itself, will produce an electric intensity $\mathbf{E}_{sph}$. This will be comparatively large in and near the sphere, and comparatively small far from the sphere. At any point *the* electric field, the *total* electric field, will be given by $\mathbf{E}_{tot} = \mathbf{E}_{ext} + \mathbf{E}_{sph}$. For an ordinary dielectric the polarization will then be $\mathbf{P} = \chi(\epsilon_0\mathbf{E}_{tot})$. Note that when the electric susceptibility χ is employed then the proportionality is between $\mathbf{E}_{tot}$ (the *total* field) and **P** (which contributes to $\mathbf{E}_{tot}$), not between $\mathbf{E}_{ext}$ (the cause) and **P** (the effect). In the next section we will introduce α, the polarizability, which is the factor employed when one wants to emphasize the cause-effect relationship. Here, using $\chi = \kappa - 1$, we can write

$$\mathbf{P} = (\kappa - 1)\epsilon_0\mathbf{E}_{tot}$$

Going back to our assumption that the polarization is uniform, the $\mathbf{E}_{sph}$ that is produced by the induced **P** must be indistinguishable from the **E** produced by a permanently polarized spherical electret. For, the derivation of $\mathbf{E}_{sph} = -(1/3\epsilon_0)\,\mathbf{P}$ in the previous section did not in any way depend on the manner in which the dipoles were produced. So here $\mathbf{E}_{sph} = -(1/3\epsilon_0)\mathbf{P}$ inside the sphere and

$$\mathbf{E}_{tot} = \mathbf{E}_{ext} - (1/3\epsilon_0)\,[(\kappa - 1)\epsilon_0 \mathbf{E}_{tot}]$$

Then $\mathbf{E}_{tot} = [3/(\kappa + 2)]\,\mathbf{E}_{ext}$ inside the dielectric. $\mathbf{E}_{ext}$ will now be the field outside, and far from, the sphere.

The consistency of the above results with the original assumption of uniform **P** is shown by now solving for **P**:

$$\mathbf{P} = (\kappa - 1)\epsilon_0 \mathbf{E}_{tot} = (\kappa - 1)\epsilon_0 \left[\left(\frac{3}{\kappa + 2}\right)\mathbf{E}_{ext}\right] = \left[3\left(\frac{\kappa - 1}{\kappa + 2}\right)\epsilon_0\right]\mathbf{E}_{ext}$$

Since $\mathbf{E}_{ext}$ is uniform, **P** is also uniform. Note that **P** and $\mathbf{E}_{tot}$ inside the sphere are both in the $+\hat{\mathbf{z}}$ direction, unlike the case of the spherical electret. Figure 6-14*a* shows the lines of $\mathbf{E}_{tot}$ inside the sphere. They are similar to $\mathbf{E}_{ext}$ but, because $3/(\kappa + 2) < 1$, the field is smaller and the lines are drawn farther apart.

Outside the sphere $\mathbf{E}_{tot} = \mathbf{E}_{ext} + \mathbf{E}_{sph}$ also, but here we have

$$\mathbf{E}_{sph} = \mathbf{E}_{out} = \left(\frac{PR^3}{3\epsilon_0}\right)\left[\hat{\mathbf{r}}\,\frac{2\cos\theta}{r^3} + \hat{\boldsymbol{\theta}}\,\frac{\sin\theta}{r^3}\right]$$

(See Prob. 3 of the previous section.)

Thus

$$\mathbf{E}_{out} = \frac{PR^3}{3\epsilon_0}\left[\hat{\mathbf{r}}\,\frac{2\cos\theta}{r^3} + \hat{\boldsymbol{\theta}}\,\frac{\sin\theta}{r^3}\right] + \left[\hat{\mathbf{r}}(\mathbf{E}_{ext}\cos\theta) - \hat{\boldsymbol{\theta}}(\mathbf{E}_{ext}\sin\theta)\right]$$

With

$$P = 3\left(\frac{\kappa - 1}{\kappa + 2}\right)\epsilon_0 E_{ext}$$

this gives

$$\mathbf{E}_{out} = \hat{\mathbf{r}}\,E_{ext}\cos\theta\left[\left(\frac{\kappa - 1}{\kappa + 2}\right)\frac{2R^3}{r^3} + 1\right]$$

$$+\,\hat{\boldsymbol{\theta}}\,E_{ext}\sin\theta\left[\left(\frac{\kappa - 1}{\kappa + 2}\right)\frac{R^3}{r^3} - 1\right]$$

Inside the dielectric sphere

$$\mathbf{E}_{in} = \left(\frac{3}{\kappa + 2}\right)\mathbf{E}_{ext}$$

Figure 6-14*a* shows the resulting lines for $\mathbf{E}_{tot}$ outside the sphere, as well as those inside. When $r \gg R$ the electric field $\mathbf{E}_{tot}$ approaches $\mathbf{E}_{ext}$. The electric field outside the sphere here is the resultant of the dipole field of the polarized sphere, $\mathbf{E}_{sph}$ above, and the linear, uniform, external field $\mathbf{E}_{ext}$.

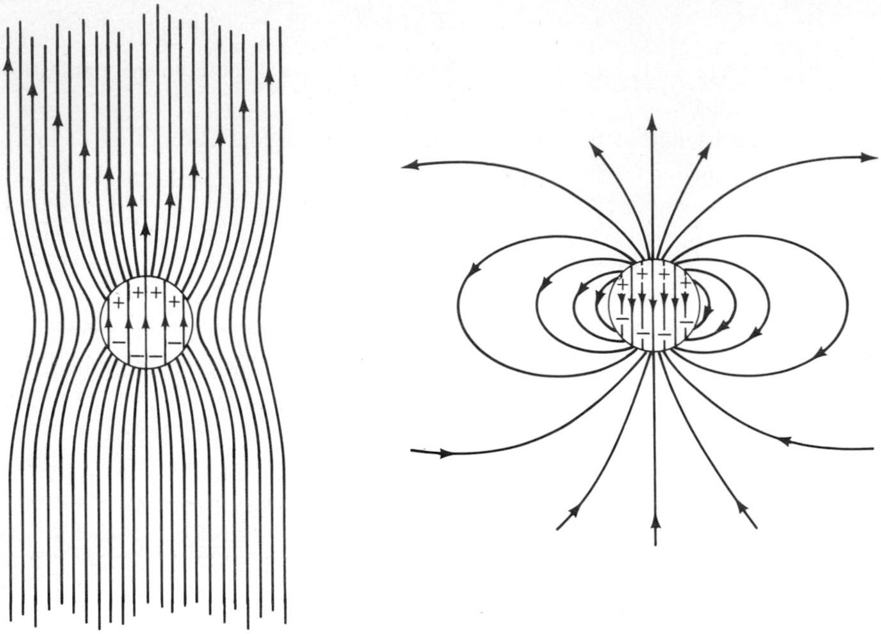

(*a*) Dielectric. This shows **E**. Within the sphere, **P** is uniform and up.

(*b*) Electret. This shows **E**. Within the sphere, **P** is uniform and up.

FIGURE 6-14

Figure 6-14*b* shows the electric field of the spherical electret, for contrast with that of the spherical dielectric of Fig. 6-14*a*; this is given by

$$\mathbf{E}_{\text{out}} = \frac{PR^3}{3\epsilon_0}\left[\hat{\mathbf{r}}\,\frac{2\cos\theta}{r^3} + \hat{\boldsymbol{\theta}}\,\frac{\sin\theta}{r^3}\right] \qquad \text{and} \qquad \mathbf{E}_{\text{in}} = -\frac{1}{3\epsilon_0}\,\mathbf{P}$$

PROBLEMS

1 A sphere of radius a has a uniform dielectric constant κ. A charge Q is at its center. Find the total bound charge on the surface of the sphere.

2 Find the polarization and volume charge density in Prob. 1.

3 A substance with dielectric constant κ has a spherical cavity of radius r within it, with a charge Q at the center of the hole. Find the total bound charge on the cavity's surface.

4 In Example 2 above, suppose the dielectric inserted into the capacitor has $\kappa = 1.5$. Find the values of Q, C, V, E, and D for the two cases considered.

5 A plane capacitor with square plate area A has an insulator with dielectric constant κ filling the right half of the volume between the plates. The left half is vacuum. If the plate separation is d what is the capacitance?

6 Show that the capacitance of a parallel-plate capacitor with a dielectric having permittivity ϵ is given by $\epsilon A/d$.

7 The capacitor of Prob. 5 has, instead, the dielectric filling the bottom half (thickness $d/2$) of the volume between the plates. What is the capacitance?

8 The capacitor of Prob. 5 has a metal plate, of thickness $d/2$ and area A, inserted into the region between the plates. Let a be the separation between the inner surface of one capacitor plate and the nearer surface of the inserted metal plate. What is the capacitance? Does it depend on a? Compare with Prob. 7 to find the effective κ that applies for a metal if we wish to take over the equations that apply to dielectrics.

9 In comparison with the spherical electret and the spherical dielectric in a uniform $\mathbf{E}_{\text{ext}}$, consider the case of a metal sphere, of radius R placed in a uniform external electric field, $\mathbf{E}_{\text{ext}} = \hat{\mathbf{z}}E_{\text{ext}}$. Take the potential to be zero at the surface of the sphere.

(*a*) Apply the results of Prob. 8, for the effective value of κ for a metal, to the results of Example 4 for a dielectric to obtain the electric field everywhere for the case of the metal sphere.

(*b*) What is the potential everywhere? Figure 6-15 shows the total electric field in the case of the metal sphere; compare with Fig. 6-14*a*, which is drawn for the case of $\kappa = 3$.

10 Show that the energy density of a parallel-plate capacitor having an insulator with dielectric constant κ is $\frac{1}{2}\epsilon E^2 = \frac{1}{2}\kappa\epsilon_0 E^2$.

11 If the bottom plate of the capacitor of Fig. 6-14*c* is at zero potential (it is ground) and the top plate is at potential ϕ, what is the potential of the metal plate in the interior?

12 In Example 2 above, suppose the insulating slab is inserted halfway into the space between the capacitor plates with the capacitor connected to a battery. Then the capacitor is isolated from the battery and the insulating slab is inserted the remaining distance into the capacitor. What are the final values of Q, C, V, E, and D?

13 (*a*) Considering a dielectric sphere in an $\mathbf{E}_{\text{ext}}$, assign a value of κ when the dielectric becomes an ideal conductor. Why is this so?

(*b*) What are the values of E, P, and D within the conductor? Why?

(*c*) Repeat for the dielectric sphere analog of the metal sphere.

14 In Example 2 a comparison was made between the electric field inside the dielectric and that which existed before the dielectric was inserted. $\mathbf{E}$ could equal $\mathbf{E}_0$ or $\mathbf{E}$ could be less than $\mathbf{E}_0$, depending on the circumstances of insertion. Suppose a dielectric is surrounded by air. If the electric field just outside a surface is $\mathbf{E}_{\text{out}}$, what

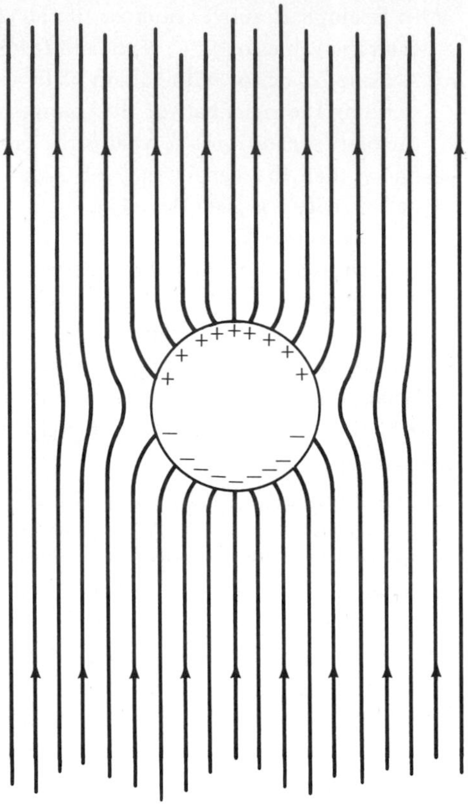

FIGURE 6-15

Metal. This shows **E**. Within the sphere, **E** is zero.

is the electric field $\mathbf{E}_{\text{in}}$ just inside the surface, in terms of $\mathbf{E}_{\text{out}}$ and κ? Can $\mathbf{E}_{\text{in}}$ ever equal $\mathbf{E}_{\text{out}}$?

15 A dielectric slab is inserted into the region between the plates of a parallel-plate capacitor, filling the region completely. The capacitor is connected to a constant-voltage battery during the insertion so the original electric field within the material $\mathbf{E}$ is equal to the original uniform electric field in vacuum, $\mathbf{E}_{\text{ext}} = \hat{\mathbf{z}}E_{\text{ext}}$. (See Example 2.) A uniform polarization, $\mathbf{P} = \hat{\mathbf{z}}P$, within the material is also obtained. A hole is then cut in the dielectric. Find $\mathbf{E}_h$ at the center of the hole in terms of $\mathbf{P}$ and $\mathbf{E}$ if the hole is (*a*) a needle, a long cylinder with its axis parallel to $\mathbf{P}$ and an infinitesimal cross section; (*b*) a sphere; (*c*) a disk, a squat cylinder with its axis parallel to $\mathbf{P}$ and having infinitesimal height.

Now take an electret slab having the same shape as the dielectric slab above. Let it have the same uniform polarization $\mathbf{P}$ as above but discard the capacitor; so there is no $\mathbf{E}_{ext}$ in this case. Find $\mathbf{E}_h$ at the center of a hole cut in the electret when the hole is (*d*) a needle, (*e*) a sphere, (*f*) a disk; in the same directions as above.

16 From $\mathbf{D}_{in} = \epsilon_0 \mathbf{E}_{in} + \mathbf{P}$ and $\mathbf{D}_{in} = \mathbf{D}_{out}$, at a surface normal to a uniform external field $\mathbf{E}_{out}$, we obtain $\epsilon_0 \mathbf{E}_{out} = \epsilon_0 \mathbf{E}_{in} + \mathbf{P}$ and $\mathbf{E}_{in} = \mathbf{E}_{out} - \mathbf{P}/\epsilon_0$. The second term on the right is known as the depolarizing field. Find its value for a long cylinder, of infinitesimal cross section, parallel to $\mathbf{E}_{in}$. Find its value for a cylinder of finite cross section, but infinitesimal length, parallel to $\mathbf{E}_{in}$.

17 A spherical cavity is made in the dielectric of Example 4, not necessarily at the center of the sphere. Find $\mathbf{E}_h$ at the center of the hole (*a*) in terms of $\mathbf{E}$, the total electric field in the dielectric *before the cavity was made*; (*b*) in terms of $\mathbf{E}_{ext}$.

18 Let the electret of Example 3 have $\mathbf{P} = \hat{\mathbf{z}}P_0$. The electret extends between $x = 0$ and $x = w$; $y = 0$ and $y = h$; and $z = 0$ and $z = \ell$. Write $\nabla \times \mathbf{P}$ on the sides in terms of the Dirac delta function.

19 A thin, large dielectric slab with $\kappa = 1.5$ is placed in an external field $\mathbf{E} = \hat{\mathbf{z}}10^5$ V/m. $\hat{\mathbf{n}}$, a normal to a large face, makes the angle of 30° with $\hat{\mathbf{z}}$. Find σ_p and $\mathbf{E}_{slab}$, the field induced within the slab.

20 Is there a contradiction between the statement (in Prob. 15) that **P** is uniform in a slab and the statement (in Example 4) that **P** is only uniform in an ellipsoid of revolution?

6-4 INDUCED AND PERMANENT P

Polar substances have molecules with an inherent electric dipole moment; the atoms, themselves, are unpolarized in their ground states. If an external electric field is introduced the atoms acquire an induced electric dipole moment, as do the molecules; the two are roughly of the same order of magnitude. We consider first only the induced dipoles.

The induced electric dipole moment per atom **p** will be proportional to $\mathbf{E}_h$ the local electric field that acts on that atom, where this field does not include the contribution of the atom itself. $\mathbf{E}_h$ is due to the other induced dipoles of the dielectric. Taking the volume of an atom as spherical, the local field is that produced at the center of a spherical cavity in a dielectric: $\mathbf{E}_h = \mathbf{E} + \mathbf{P}/3\epsilon_0$. Here **E** is the *total* space and time average of the rapidly fluctuating microscopic electric field and includes the effect of the atom itself. $+\mathbf{P}/3\epsilon_0$ is the effect of the induced surface charges on the wall of the cavity (see Sec. 6-2, Prob. 5, as well as Sec. 3, Prob. 15).

Another way of looking at this is to rearrange the equation above:

$$\mathbf{E} = \mathbf{E}_h + \left(-\frac{\mathbf{P}}{3\epsilon_0}\right)$$

$\mathbf{E}$ is the total macroscopic (i.e., average) field at a point; $\mathbf{E}_h$ is the field that would exist there if a spherical piece of dielectric were removed, the point being at the center of the sphere; $-\mathbf{P}/3\epsilon_0$ is the field at the center of the removed spherical piece. See Sec. 6-2, Example 4.

Then $\mathbf{p} = \alpha\epsilon_0 \mathbf{E}_h$ can be taken as the relation between the cause $\mathbf{E}_h$ and the effect $\mathbf{p}$. The proportionality factor α is called the polarizability. Since $\mathbf{P} = N\mathbf{p}$, where N is the number of atoms per unit volume,

$$\mathbf{P} = N\alpha\epsilon_0 \mathbf{E}_h = N\alpha\epsilon_0 \left(\mathbf{E} + \frac{\mathbf{P}}{3\epsilon_0}\right)$$

$$\left(1 - \frac{N\alpha}{3}\right)\mathbf{P} = N\alpha\epsilon_0 \mathbf{E}$$

$$\mathbf{P} = \frac{N\alpha}{1 - \frac{1}{3}N\alpha}\,\epsilon_0 \mathbf{E} = \chi\epsilon_0 \mathbf{E}$$

The susceptibility is

$$\chi = \frac{N\alpha}{1 - \frac{1}{3}N\alpha}$$

Then

$$N\alpha = \frac{\chi}{1 + \frac{1}{3}\chi}$$

For gases, where N is much smaller than for liquids or solids and $\chi \ll 1$, we have $\chi = \kappa - 1 \approx N\alpha$. For liquids,

$$\kappa - 1 = \frac{N\alpha}{1 - \frac{1}{3}N\alpha}$$

$$\frac{1}{3}N\alpha = \frac{\kappa - 1}{\kappa + 2}$$

This is called the Clausius-Mosotti equation. For $\kappa \approx 1$ this again becomes $N\alpha = \kappa - 1$. There is good agreement between these equations and experimental results. For solids the results are more complicated. In addition to the type of polarizability discussed above, the so-called electronic polarizability, there is also an ionic polarizability in crystals having a heteropolar bond (the molecules are held together by Coulomb forces between positive and negative ions as, e.g., in NaCl). There is also an orientational type of polarizability. The effects of all three types become visible in the variation of the dielectric constant with frequency.

Suppose we now neglect the induced dipole moment, which always exists, and consider only the permanent dipole moments of certain types of molecules; e.g., H_2O.

When such a substance, in the fluid phase, is placed in an external electric field there is a torque produced on the individual dipole molecules which tends to align them with the field. The collisions between molecules, a temperature-dependent effect, opposes the alignment process and, with normal fields and temperatures, the alignment is slight. The potential energy of the dipole in the field is $U = -\mathbf{p} \cdot \mathbf{E}$ and the torque on the randomly oriented molecules is in such a direction as to tend to decrease this energy.

Statistical mechanics provides the answer to the next question: What are the relative numbers of molecules with a given orientation? In thermal equilibrium the relative number having the potential energy U is proportional to $\exp(-U/kT)$, where $k = 1.380 \times 10^{-23}$ J/K and T is the temperature in kelvins. As a function of the angle θ between $\mathbf{p}$ and $\mathbf{E}$ the number of dipoles, per unit solid angle, having the orientation θ is $n = n_0 \exp(pE \cos\theta/kT) \approx n_0(1 + pE \cos\theta/kT)$. The expansion of the exponential is justified by the fact that kT is roughly 4×10^{-21} J at room temperature and $p \sim 10^{-30}$ C m, so even if $E = 4 \times 10^7$ V/m the exponent equals only 10^{-2}. Integrating the expression for n over all angles and setting the result equal to N, the number density of molecules (Loschmidt number), gives $n_0 = N/4\pi$.

The net polarization will have only a z component if $\mathbf{E}$, for example, is in this direction. Then

$$P = \int_0^\pi n(p\cos\theta)\, 2\pi \sin\theta\, d\theta$$

$$P = \frac{Np^2E}{3kT}$$

This is called Curie's law. The inverse behavior with temperature agrees very well with experimental results and serves as a means of distinguishing between the effects of permanent and induced dipoles.

Examples

1. **Oscillation** A stationary molecule with a permanent electric dipole moment $\mathbf{p}$ is released in a uniform electrostatic field. We will describe the motion of the molecule.

The initial kinetic energy is zero; the initial potential energy is $U = -\mathbf{p} \cdot \mathbf{E}$ and depends on the angle θ of the initial attitude; the initial total energy is $W = T + U = -\mathbf{p} \cdot \mathbf{E}$. The force on the dipole, $\mathbf{F} = -\nabla(-\mathbf{p} \cdot \mathbf{E})$, is initially zero and remains zero, so the center of mass of the molecules moves with constant velocity. The torque on the dipole, $\mathbf{T} = \mathbf{p} \times \mathbf{E}$, is not zero unless the dipole moment is initially aligned parallel or antiparallel to $\mathbf{E}$.

A molecule will tend to turn in such a direction as to reduce U and align $\mathbf{p}$ with $\mathbf{E}$. If there are no collisions with other molecules, however, the law of conservation of energy does not permit this alignment. At the equilibrium attitude the potential energy has been converted to rotational kinetic energy and the dipole will overshoot the equilibrium angle. The molecule will oscillate like a pendulum.

From $T = I\alpha$, where I is the moment of inertia about the center of mass and α is the angular acceleration, we obtain $-pE \sin\theta = I\ddot{\theta}$. For small enough θ this gives a sinusoidal oscillation with frequency

$$f = \frac{1}{2\pi}\sqrt{\frac{pE}{I}}$$

Collisions with other molecules play a double role here. (1) They are necessary to enable a molecule to lose energy, thereby permitting it to align itself more nearly along $\mathbf{E}$, in a position of lower energy. (2) They have a randomizing effect on the angle between a given molecule and $\mathbf{E}$, for the angle with $\mathbf{E}$ after a collision does not depend on the angle with $\mathbf{E}$ before the collision. Instead of all the dipoles either oscillating about $\mathbf{E}$ or aligning themselves along $\mathbf{E}$, the effect of the collisions is to establish an equilibrium Boltzmann distribution such that the relative number of molecules making a given angle with $\mathbf{E}$ is determined by exp $(-U/kt)$, where U is the total energy and k is Boltzmann's constant.

The motion of permanent electric dipoles here is quite different from that of permanent magnetic dipoles in a magnetic field, discussed in the next chapter. There, because of the presence of spin angular momentum associated with the magnetic dipole moment, gyroscopic forces are brought into play and then precession, rather than oscillation, occurs in the absence of collisions.

PROBLEMS

1 Consider the Z electrons of an atom to constitute a sphere of radius r with uniformly distributed charge. The nucleus, of charge $+Ze$, is pulled over to a position, distant d from the center ($d < r$), by an external field. Find the force on the nucleus that is due to the electron cloud (a) within the sphere of radius d about the center of the cloud, and (b) outside the sphere of radius d.

2 What is the structure of the CO_2 molecule? Does it have a permanent electric dipole moment?

3 By equating the total restoring force of Prob. 1 to the force on the nucleus that is due to the external field, find the induced electric dipole moment and the polarizability.

4 What is the dielectric constant of a gas with atoms similar to those in Prob. 3, containing N atoms/m^3?

5 Standard conditions are defined as 0°C and 1 atm. A kg-mol of gas (defined as an amount whose mass in kilograms equals, numerically, the molecular weight on a scale with the atomic weight of C^{12} taken as 12.000000) selects a sample with the same number of molecules (N_A = Avogadro's number) for any gas at standard conditions. If 1 kg-mol of a gas has a volume of 22.4 m^3 at standard conditions, find N, the number of molecules per cubic centimeter at 0°C. This is known as Loschmidt's number. What is the number of molecules/m^3 at 27°C, approximately room temperature? (In both bases let $p = 1$ atm.)

6 What relations, if any, exist among (*a*) N_A, Avogadro's number: 6.025×10^{26} $(\text{kg-mol})^{-1}$, (*b*) N, Loschmidt's number: 2.7×10^{25} m^{-3}, (*c*) R, the gas constant: 8310 J K^{-1} (kg mol)$^{-1}$, and (*d*) k, Boltzmann's constant: 1.38×10^{-23} J K^{-1}?

7 The quantity $\frac{1}{3}N\alpha$ in the Clausius-Mosotti equation depends on the room temperature and pressure. At standard conditions this would depend, instead, on $N_A \cdot \alpha_M = N_A \alpha/3$ is defined as the molar polarizability. Use the results of Prob. 6 to find the molar polarizability for a nonpolar substance from the induced polarizability.

8 For a polar substance, is the field that induces polarization the same as that which turns the permanent dipole? Is the first field exactly $\mathbf{E} + \mathbf{P}/3\epsilon_0$?

9 Find the molar polarizability for a molar substance in terms of N_A, assuming a positive answer to Prob. 8. The result, known as the Debye equation, does not correspond too closely to experimental fact because the effect of the polarization of a molecule on its neighbors has been neglected.

10 In the derivation of Curie's law an exponential factor was expanded in a Taylor series. Complete the derivation without the expansion, to obtain the Langevin equation

$$P = Np\left[\coth\left(\frac{pE}{kT}\right) - \frac{1}{(pE/kT)}\right]$$

What is the value of the quantity in the brackets for $(pE/kT) \ll 1$? This condition is always true in practice.

11 Use the dipole-dipole interaction energy formula to calculate the field at an atom (at the origin) in a cubic crystal. Each atom has a permanent polarization $p = \hat{\mathbf{z}}p$. Make the calculation for the 6 nearest neighbors, 12 second-nearest neighbors, and 8 third-nearest neighbors and add the results.

12 Barium titanate is a ferroelectric crystal whose unit cell has a Ti^{+4} ion at its center; six O^{-2} ions—one on each face; and eight Ba^{+2} ions—one at each corner. Show that the cell is electrically neutral. (Above 118°C $BaTiO_3$ is extraordinary because, while it has no permanent electric dipole moment, κ has a value in the thousands. Below 118°C it is extraordinary because it becomes an electret.)

13 Apply $\mathbf{F} = (\mathbf{p} \cdot \nabla)\mathbf{E}$ to the case with $\mathbf{p} = (\chi\epsilon_0 E)\mathcal{V}$, where $\mathcal{V}$ is the volume of a small dielectric. Will $\mathbf{F}$ be toward a stronger or weaker part of the field? Does $\mathbf{F}$ depend on the sign of $\mathbf{E}$?

14 Write a matrix equation $\{P\} = \{\chi\}\{\epsilon_0 E\}$ for the case of a crystalline substance when χ cannot be represented simply as a single scalar constant.

6-5 BOUNDARY CONDITIONS

There are a number of conditions which must be satisfied by ϕ, **E**, and **D** at the interface surface between two different dielectrics, say, characterized by κ_1 and κ_2.

The potential ϕ at a point on the surface of discontinuity between two dielectrics represents the work per unit charge needed to bring a charge from the point of zero potential (usually infinity) to the point on the surface. In electrostatics the potential difference between two points has a unique value, independent of the path taken in calculating it. The potential at any point on the boundary surface would, therefore, usually be taken to be single-valued; the value on the boundary would be independent of the direction from which it is approached. But this is contrary to the result of Fig. 6-10*c*, which shows that the potential on a polarized disk is double-valued. Two values of ϕ are obtained there, of equal magnitude but opposite sign.

One reason for this is that the polarized disk (a plane area with a circular boundary) is actually the limiting case of a polarized plate or puck. Figure 6-16 shows the potential and field distribution for this case. When D, the thickness of the puck (the distance between the two plane surfaces of the plate), approaches zero, the results of Fig. 6-16 approach those of Fig. 6-10*c*; E_{int} approaches $-\infty$, but for an ever-smaller linear extent, one approaching zero thickness. Figure 6-16 applies for a real dipole distribution; the potential has a unique value. Figure 6-10*c* applies for ideal dipoles only—a mathematical figment of the imagination—and the potential is not unique.

There is another way of seeing that the potential can be double-valued on a boundary surface of infinitesimal thickness: consider such a closed boundary surface. Then, if the two values of potential differ by a fixed amount everywhere on the closed boundary surface, the line integral of **E** around any closed path will still be zero, even though ϕ is not unique on the closed boundary surface.

A condition on **E** at the interface between dielectrics is obtained from the fact that $\nabla \times \mathbf{E} = 0$ in electrostatics, whether one considers vacuum or dielectrics. Figure 6-17 shows the vector **E** in medium 1 decomposed into a component $\mathbf{E}_{1n}$ normal to the boundary surface between the two dielectrics and a second component $\mathbf{E}_{1t}$, which is tangent to the border. The vectors $\mathbf{E}_1$ and $\mathbf{E}_2$ determine a plane, and in that plane we consider the rectangle Σ bounded by *abcd* (length L, width w).

$$\oint_{\Sigma} (\nabla \times \mathbf{E}) \cdot d\mathbf{S} = 0$$

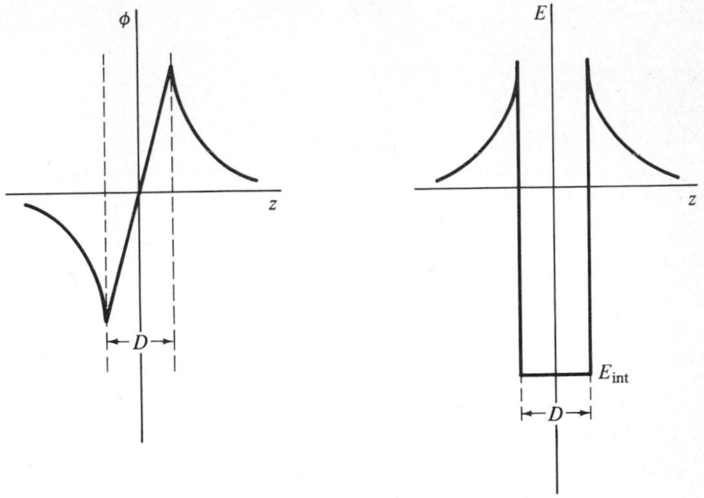

FIGURE 6-16

$$\oint_{abcd} \mathbf{E} \cdot d\mathbf{r} = 0$$

$$\int_a^b \mathbf{E} \cdot d\mathbf{r} + \int_b^c \mathbf{E} \cdot d\mathbf{r} + \int_c^d \mathbf{E} \cdot d\mathbf{r} + \int_d^a \mathbf{E} \cdot d\mathbf{r} = 0$$

The second and fourth integrals may be made to vanish by taking w to be infinitesimal. Then

$$\int_a^b E_{1t}\, dr + \int_c^d E_{2t}\, dr = 0$$

(The limits in the latter integral are the reverse of those in the former when w is infinitesimal; since these have been reversed on the integral sign, the independent variable has not been reversed.)

$$\int_a^b E_{1t}\, dr - \int_d^c E_{2t}\, dr = 0$$

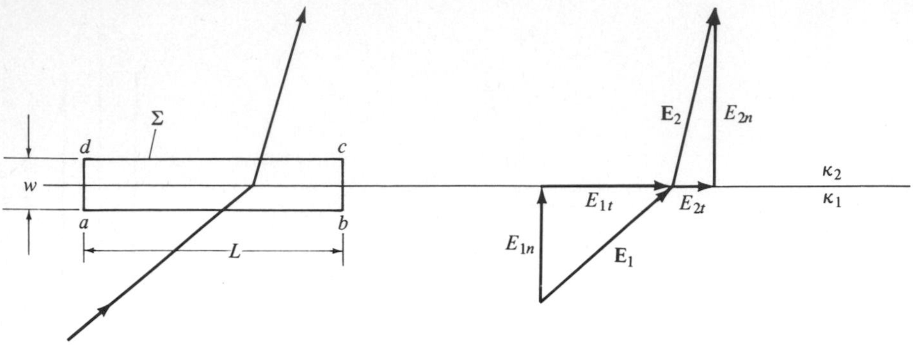

FIGURE 6-17

Taking L as finite but arbitrarily small makes E_{1t} and E_{2t} constant for the interval of integration and gives

$$E_{1t}L - E_{2t}L = 0$$

so

$$E_{1t} = E_{2t}$$

Consequently: the tangential component of **E** is continuous across the dielectric boundary.

To obtain a condition on **D** at the interface consider the sandwich of two dielectrics shown in Fig. 6-18. A cylinder (cross-sectional area A, length L) lies partially in the two dielectrics. Suppose the interface between the two insulators has a free charge density σ_F. Then Gauss' law gives

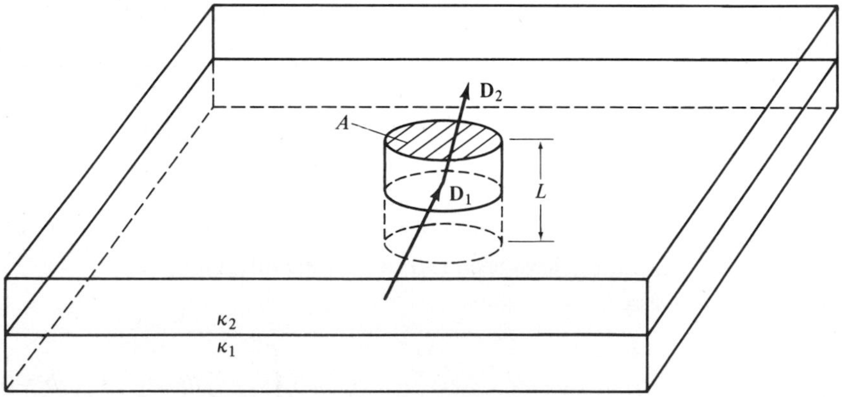

FIGURE 6-18

$$\oint \mathbf{D} \cdot d\mathbf{S} = Q$$

or

$$\int_{\text{top}} \mathbf{D} \cdot d\mathbf{S} + \int_{\text{bottom}} \mathbf{D} \cdot d\mathbf{S} + \int_{\text{sides}} \mathbf{D} \cdot d\mathbf{S} = \int_{\substack{\text{cross}\\\text{section}}} \sigma_f \, dS$$

By making L infinitesimal the integral over the curved side can be made to vanish, so

$$\int_{\text{top}} D_{2n} \, dS - \int_{\text{bottom}} D_{1n} \, dS = \int_{\substack{\text{cross}\\\text{section}}} \sigma_f \, dS$$

Taking A as finite but small makes D_{2n} and D_{1n} constant over A. Then

$$D_{2n} - D_{1n} = \sigma_f$$

If $\sigma_f = 0$, as is usual,

$$D_{1n} = D_{2n}$$

The normal component of **D** is continuous across a dielectric boundary if there is no free surface charge on it.

Examples

1. Two dielectrics–no surface charge Medium 1 has $\kappa_1 = 1.5$ and extends to the left of the yz plane. Medium 2 has $\kappa_2 = 2.5$ and extends to the right of the yz plane.

$$\mathbf{E}_1 = (\hat{\mathbf{x}}2 - \hat{\mathbf{y}}3 + \hat{\mathbf{z}}) \quad \text{V m}^{-1}$$

We will find $\mathbf{D}_1$, $\mathbf{E}_2$, and $\mathbf{D}_2$.

$$\mathbf{D} = \epsilon\mathbf{E} = \kappa\epsilon_0\mathbf{E} \quad \text{Wb m}^{-2}$$

So

$$\mathbf{D}_1 = \kappa_1\epsilon_0\mathbf{E}_1$$

$$\mathbf{D}_1 = 1.5\epsilon_0(\hat{\mathbf{x}}2 - \hat{\mathbf{y}}3 + \hat{\mathbf{z}}) = \hat{\mathbf{x}}3\epsilon_0 - \hat{\mathbf{y}}4.5\epsilon_0 + \hat{\mathbf{z}}1.5\epsilon_0$$

The normal direction at this interface is the x direction, so

$$\mathbf{E}_{1n} = \hat{\mathbf{x}}2$$

$$\mathbf{E}_{1t} = -\hat{\mathbf{y}}3 + \hat{\mathbf{z}}$$

$$\mathbf{D}_{1n} = \hat{\mathbf{x}}3\epsilon_0$$

$$\mathbf{D}_{1t} = -\hat{\mathbf{y}}4.5\epsilon_0 + \hat{\mathbf{z}}1.5\epsilon_0$$

From $E_{1t} = E_{2t}$ and $D_{1n} = D_{2n}$

$$E_{2t} = -\hat{y}3 + \hat{z} \qquad D_{2n} = \hat{x}3\epsilon_0$$

$\mathbf{E}_{2n}$ is found from $\mathbf{D}_{2n}$: $\mathbf{D}_{2n} = \kappa_2\epsilon_0\mathbf{E}_{2n}$. So

$$\mathbf{E}_{2n} = \frac{1}{2.5\epsilon_0}(\hat{x}3\epsilon_0) = \hat{x}1.2$$

Similarly $\mathbf{D}_{2t}$ is found from $\mathbf{E}_{2t}$: $\mathbf{D}_{2t} = \kappa_2\epsilon_0\mathbf{E}_{2t}$,

$$\mathbf{D}_{2t} = 2.5\epsilon_0(-\hat{y}3 + \hat{z}) = -\hat{y}7.5\epsilon_0 + \hat{z}2.5\epsilon_0$$

Summarizing,

$$\mathbf{E}_1 = \hat{x}2 - \hat{y}3 + \hat{z} \qquad \mathbf{E}_2 = \hat{x}1.2 - \hat{y}3 + \hat{z}$$

$$\mathbf{D}_1 = \epsilon_0(\hat{x}3 - \hat{y}4.5 + \hat{z}1.5) \qquad \mathbf{D}_2 = \epsilon_0(\hat{x}3 - \hat{y}7.5 + \hat{z}2.5)$$

The laws for the behavior of the components of **E** are different from those for **D** but the behavior of **E** and **D** are similar in this respect—$\mathbf{E}_1$ is parallel to $\mathbf{D}_1$ and $\mathbf{E}_2$ is parallel to $\mathbf{D}_2$.

PROBLEMS

1 $\mathbf{E}_1$ and $\mathbf{D}_1$ make the angle θ_1, with the normal to the boundary between two dielectrics. Find the angle θ_2 made by $\mathbf{E}_2$ and $\mathbf{D}_2$ with this normal.

2 $\mathbf{E}_1$ and $\mathbf{D}_1$ make the angle θ_1 with the normal to a dielectric interface on which a free surface charge density σ_f exists. Prove $\mathbf{E}_2$ and $\mathbf{D}_2$ in medium 2 are parallel to each other.

3 Medium 1 is air (i.e., approximately vacuum) and occupies the region $z < 2$. Medium 2 has $\kappa = 2$ and lies in $z > 2$. If $E_1 = \hat{x} + \hat{y}2 + \hat{z}3$ and there is no free charge on the interface, find $\mathbf{D}_2$.

4 In the derivation of the equation $E_{1t} = E_{2t}$ the line integral of **E** around a rectangle was utilized. Suppose we had taken the line integral of **D** instead. What equation would result?

5 The plane interface between two dielectrics has the equation $x + y + z = 1$. What is the normal component of $\mathbf{E}_1 = \hat{x} + \hat{y}2 + \hat{z}3$? What is the tangential component of $\mathbf{E}_1$?

6 Given $\mathbf{E}_1 = \hat{x} + \hat{y}3 + \hat{z}$, is it possible to have an interface such that the component of $\mathbf{E}_1$ normal to the interface has its x component equal to 2 (i.e., larger than the x component of E_1)? Can it be equal to 3? Can it be equal to 4? Explain.

7 In Prob. 5 consider the two dielectrics to be in the first quadrant. One dielectric lies between the origin and the interface and has $\kappa_1 = 1$. The other has $\kappa_2 = 2$. Find $\mathbf{E}_{2n}$, $\mathbf{E}_{2t}$, and $\mathbf{E}_2$.

8 $\mathbf{E} = \hat{\mathbf{r}}E_r + \boldsymbol{\theta}E_\theta + \hat{\boldsymbol{\phi}}E_\phi$ in spherical coordinates; an arbitrary surface is given, $F(r,\theta,\phi) = 0$. Find the components of **E** which are normal and tangential to the given surface.

9 Are lines of **E** bent toward or away from the normal to the interface when going into a dielectric in which κ is larger?

10 In Fig. 6-13 show that $\theta_1 = \theta_2$ for (*a*) **E** on the curved cylindrical surface, and (*b*) **D** on the flat end surfaces.

11 Medium 1 has $\kappa = 1$ and occupies the region $z > 0$. Medium 2 has $\kappa = 2$ and occupies the region $z < 0$. $\mathbf{E}_1 = \hat{\mathbf{x}} - \hat{\mathbf{y}}2 + \hat{\mathbf{z}}3$. Find $\mathbf{D}_1$, $\mathbf{E}_2$, and $\mathbf{D}_2$.

12 In Prob. 1 let dielectric 1 have $\kappa_1 = 1$ and let dielectric 2 become a metal. Find θ_1 and θ_2.

13 Given a metal sphere with total charge Q and radius a, surrounded by an insulator with $\kappa = \kappa_0 r/a$, a function only of distance from the center. Find **E**, **P**, and **D** inside the dielectric. Do the **E** and **D** curves fall off similarly with distance?

14 Compare the equation of Prob. 1 with Snell's law for the refraction of light rays.

(*a*) Write the two expressions for paraxial conditions, i.e., θ_1 and $\theta_2 \ll 1$.

(*b*) Design a lens that will focus parallel lines of **E** to a point, similar to the way a convex lens does this for light.

(*c*) What difference exists between the two cases?

15 Write an expression for $E_{1n} - E_{2n}$ at the interface between two dielectrics.

16 Are the boundary conditions listed above for E_t and D_n valid in the case of an electret? Explain.

17 In Prob. 1, is it possible to have a phenomenon, similar to total internal reflection in light, where $\theta_2 > 90°$?

Appendix 6

SURFACE DIVERGENCE AND SURFACE CURL

Here we will reconsider the bar electret treated in Sec. 6-3, Example 3, of the text. As pointed out there, the derivatives of **E**, **P**, and **D** are discontinuous at various places on the surface of the electret. This affects the definitions of the divergence and the curl.

We will proceed as follows in connection with divergence. Instead of employing the ordinary definition involving the volume τ bounded by Σ,

$$\nabla \cdot \mathbf{P} = \lim_{\tau \to 0} \left[\frac{1}{\tau} \oint_\Sigma \mathbf{P} \cdot d\mathbf{S} \right]$$

we will define a surface divergence given by

$$\nabla_{\text{surf}} \cdot \mathbf{P} = \lim_{\Sigma\to 0}\left[\frac{1}{\Sigma}\oint_{\Sigma} \mathbf{P}\cdot d\mathbf{S}\right]$$

Figure 6-13*a* shows the situation in general, at a boundary between two media. Thus

$$\frac{1}{\Sigma}\oint_{\Sigma} \mathbf{P}\cdot d\mathbf{S} = \frac{1}{\Sigma}\left[\int_{\Sigma} \mathbf{P}_1 \cdot (\hat{\mathbf{n}}_2\, dS) + \int_{\Sigma} \mathbf{P}_2 \cdot \hat{\mathbf{n}}_1\, dS\right]$$
$$= -(\mathbf{P}_1 - \mathbf{P}_2)\cdot \hat{\mathbf{n}}_1$$

When $\mathbf{P}_2 = 0$ we may take $\hat{\mathbf{n}}_1 = \hat{\mathbf{n}}$. Then

$$\nabla_{\text{surf}} \cdot \mathbf{P} = -\mathbf{P}_1 \cdot \hat{\mathbf{n}} = -P_{1n}$$

In the present case $\sigma_p = \mathbf{P}\cdot\hat{\mathbf{n}}$. So $\nabla_{\text{surf}} \cdot \mathbf{P} = -\sigma_p$. Note that the surface divergence of $\mathbf{P}$, $\nabla_{\text{surf}} \cdot \mathbf{P}$, has different dimensions [C m^{-2}] than the divergence of $\mathbf{P}$, $\nabla\cdot\mathbf{P}$, whose dimensions are [C m^{-3}].

In a similar fashion we will define a surface curl of $\mathbf{P}$, designated $\nabla_{\text{surf}} \times \mathbf{P}$. Figure 6-13*b* shows a rod of rectangular cross section; the polarization, in general, will have x, y, and z components. Let the rod be medium 1, the outside medium 2, and consider the top surface so that $\hat{\mathbf{n}}_1 = \hat{\mathbf{x}}$ and $\hat{\mathbf{n}}_2 = -x$. Draw rectangle A, in the xy plane, having two infinitesimal sides and two sides of length L, where one is in the rod and the other outside it. We define

$$(\nabla_{\text{surf}} \times \mathbf{P})_z = \lim_{L\to 0}\left[\frac{1}{L}\oint_{C_{xy}} \mathbf{P}\cdot d\mathbf{r}\right]$$

The circulation is taken in the direction shown in accordance with the right-hand screw convention. Then $(\nabla_{\text{surf}} \times \mathbf{P})_z = P_{2y} - P_{1y}$.

Next consider rectangle B in an xz plane and take

$$(\nabla_{\text{surf}} \times \mathbf{P})_y = \lim_{L\to 0}\left[\frac{1}{L}\oint_{C_{zx}} \mathbf{P}\cdot d\mathbf{r}\right]$$

Then here

$$(\nabla_{\text{surf}} \times \mathbf{P})_y = -P_{2z} + P_{1z}$$

These two components specify $\nabla_{surf} \times \mathbf{P}$ on the top surface:

$$\begin{aligned}\nabla_{surf} \times \mathbf{P} &= \hat{z}(P_{2y} - P_{1y}) + \hat{y}(-P_{2z} + P_{1z}) \\ &= (-\hat{z}P_{1y} + \hat{y}P_{1z}) + (\hat{z}P_{2y} - \hat{y}P_{2z}) \\ &= (\mathbf{P}_1 \times \hat{\mathbf{n}}_1) + (\mathbf{P}_2 \times \hat{\mathbf{n}}_2) \\ &= (\mathbf{P}_1 - \mathbf{P}_2) \times \hat{\mathbf{n}}_1\end{aligned}$$

When $\mathbf{P}_2 = 0$ we may take $\hat{\mathbf{n}}_1 = \hat{\mathbf{n}}$. Then

$$\nabla_{surf} \times \mathbf{P} = \mathbf{P} \times \hat{\mathbf{n}}$$

Using these definitions we may now write for the electret

$$\nabla \cdot \mathbf{E} = \frac{1}{\epsilon_0}(\rho_f + \rho_p) = 0$$

$$\nabla \times \mathbf{E} = 0$$

and

$$\nabla_{surf} \cdot \mathbf{E} = \frac{1}{\epsilon_0}(\sigma_f + \sigma_p) \qquad \text{(by Gauss' law applied to the uncharged pillbox of Fig. 6-13}a\text{)}$$

$$= \frac{1}{\epsilon_0}\sigma_p$$

$$\nabla_{surf} \times \mathbf{E} = 0 \qquad \text{(since the circulation of } \mathbf{E} \text{ is zero, everywhere, in electrostatics)}$$

So

$$\nabla_{surf} \cdot \mathbf{D} = \epsilon_0 \nabla_{surf} \cdot \mathbf{E} + \nabla_{surf} \cdot \mathbf{P} = \epsilon_0 \left(\frac{1}{\epsilon_0}\sigma_p\right) + (-\sigma_p) = 0$$

$$\nabla_{surf} \times \mathbf{D} = \epsilon_0 \nabla_{surf} \times \mathbf{E} + \nabla_{surf} \times \mathbf{P} = 0 + (\mathbf{P} \times \hat{\mathbf{n}}) = \mathbf{P} \times \hat{\mathbf{n}}$$

Taking discontinuous derivatives into account, therefore, we have arrived at the result that, for an electret, **E** has a divergence (considering both the usual divergence and the surface divergence) which is not zero everywhere. Similarly, the generalized curl of **D** is not zero everywhere. Just as $(1/\epsilon_0)\mathbf{P} \cdot \hat{\mathbf{n}}$ produces the surface divergence of **E**, $\mathbf{P} \times \hat{\mathbf{n}}$ produces the surface curl of **D**.

7

MAGNETOSTATICS IN MATTER

7-1 THE MAGNETIC DIPOLE

We now have to treat the effect of magnetic dipoles in matter, analogous to the way we dealt with the effects of electric dipoles there. The magnetic vector potential produced at a point P, having coordinates $\mathbf{r}$, is

$$\mathbf{A}(\mathbf{r}) = \frac{\mu_0}{4\pi} \int_{\mathcal{V}} \frac{\mathbf{J}(\mathbf{r}_s)}{R} \, d\tau_s$$

here $\mathbf{J}(\mathbf{r}_s)$ is the current density in a source element $d\tau_s$ at $\mathbf{r}_s$, as in Fig. 7-1. All the source currents are contained in some finite region $\mathcal{V}$; the origin is, indiscriminately, either in $\mathcal{V}$ or out of $\mathcal{V}$. As in the corresponding expression for the electric potential ϕ, if P is close to $\mathcal{V}$ then this formula is as simple as any other. But if P is far from $\mathcal{V}$ an expansion for $1/R$ can be made in terms of $\mathbf{r}$ and $\mathbf{r}_s$:

$$\frac{1}{R} = \frac{1}{r} + \frac{\hat{\mathbf{r}} \cdot \mathbf{r}_s}{r^2} + \frac{3(\hat{\mathbf{r}} \cdot \mathbf{r}_s)^2 - r_s^2}{2r^3} + \cdots$$

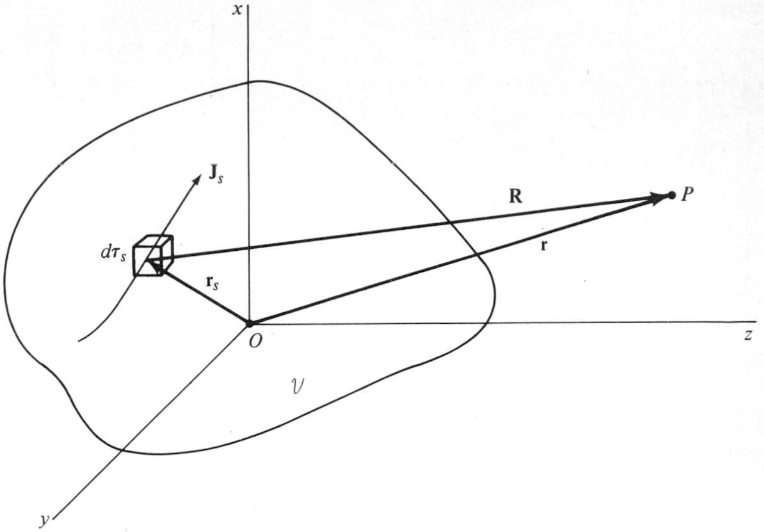

FIGURE 7-1

and
$$\mathbf{A}(\mathbf{r}) = \frac{\mu_0}{4\pi r}\left[\int_{\mathcal{V}} \mathbf{J}(\mathbf{r}_s)d\tau_s + \frac{1}{r}\int_{\mathcal{V}} \mathbf{J}(\mathbf{r}_s)(\hat{\mathbf{r}}\cdot\mathbf{r}_s)d\tau_s + \cdots\right]$$

The first term in this expansion, proportional to $(1/r)1/r^0$, is the monopole $(2^0 = 1)$ term as with the corresponding electrostatic expansion; but here it is always equal to zero.

To see this, consider the fact that $\oint_\Sigma \mathbf{J}\cdot d\mathbf{S} = 0$ for any closed surface in magnetostatics, as the total charge within any Σ remains constant. Then $\int_{\mathcal{V}} \nabla\cdot\mathbf{J} = 0$ for an arbitrary volume $\mathcal{V}$, so $\nabla\cdot\mathbf{J} = 0$. Within Σ we can then write $\mathbf{J} = \nabla\times\mathbf{N}$.

Take Σ large enough so that all the currents are contained within it. Then on Σ we can write $\mathbf{N} = \nabla\psi$ so that $\mathbf{J} = \nabla\times\mathbf{N} = \nabla\times\nabla\psi = 0$. Thus

$$\int_{\mathcal{V}} \mathbf{J}\,d\tau = \int_{\mathcal{V}} \nabla\times\mathbf{N}\,d\tau$$

Now we make use of a vector identity which is obtained by an extension of Stokes' theorem (see Chap. 1, Sec. 1-4, Prob. 12):

$$\int_{\mathcal{V}} \nabla \times \mathbf{N}\, d\tau = \oint_{\Sigma} \hat{\mathbf{n}} \times \mathbf{N}\, d\mathbf{S}$$

Then

$$\int_{\mathcal{V}} \mathbf{J}\, d\tau = \oint_{\Sigma} \hat{\mathbf{n}} \times \mathbf{N}\, dS$$

Next, cut $\mathcal{V}$ into two by a closed curve C in the surface, dividing Σ into Σ_1 and Σ_2. This gives

$$\int_{\mathcal{V}} \mathbf{J}\, d\tau = \int_{\Sigma_1} \hat{\mathbf{n}} \times \nabla\psi\, dS + \int_{\Sigma_2} \hat{\mathbf{n}} \times \nabla\psi\, dS$$

We now make use of another extension of Stokes' theorem (see Chap. 1, Sec. 1-7, Prob. 11):

$$\int_{\mathcal{V}} \mathbf{J}\, d\tau = \oint_{C} \psi\, d\mathbf{r} + \oint_{C} \psi\, d\mathbf{r} = 0 \qquad \text{Q.E.D.}$$

The second term in the multipole expansion, proportional to $(1/r)1/r^1$, is the dipole (2^1) term

$$\frac{\mu_0}{4\pi r^2} \int_{\mathcal{V}} \mathbf{J}(\mathbf{r}_s)(\hat{\mathbf{r}} \cdot \mathbf{r}_s)\, d\tau_s$$

This can be thrown into a different form by a somewhat lengthy manipulation. This derivation is relegated to Appendix 7*a*, at the end of this chapter.

The net result of the entire complicated calculation there is that the second term of the multipole expansion for **A** is

$$\frac{\mu_0}{4\pi r^2} \left\{ \int_{\mathcal{V}} \left[\frac{1}{2} \mathbf{r}_s \times \mathbf{J}(\mathbf{r}_s) \right] d\tau_s \right\} \times \hat{\mathbf{r}}$$

The quantity in braces is defined as the magnetic dipole moment of the current distribution, **m**. Neglecting higher-ordered terms in the multipole expansion,

$$\boxed{\mathbf{A}_{\text{dipole}} = \left(\frac{\mu_0}{4\pi}\right)\frac{\mathbf{m} \times \hat{\mathbf{r}}}{r^2}} \tag{M-14}$$

This result shows that **A** depends only on the transverse component of **m**. The corresponding formula for ϕ_{dipole} shows it to depend only on the longitudinal component of **p**.

The integrand in the expression for **m** is defined as the magnetization: $\mathbf{M} = \frac{1}{2}\mathbf{r} \times \mathbf{J}$. Just as $\mathbf{p} = \int \mathbf{P}\, d\tau$ in electrostatics, so $\mathbf{m} = \int \mathbf{M}\, d\tau$ in magnetostatics. The magnetization, **M**, is thus the density of magnetic dipole moments in the material.

A more physical aspect of the definition of the magnetic dipole moment becomes apparent if we consider a filamentary circuit instead of a volume distribution of currents. Then

$$\mathbf{m} = \int_{\mathcal{V}} \frac{1}{2}\mathbf{r} \times \mathbf{J}\, d\tau \rightarrow I\left[\frac{1}{2}\oint_C \mathbf{r} \times d\mathbf{r}\right]$$

Consider an idealized current-carrying loop of arbitrary shape; for simplicity take the loop to be in the xy plane, as in Fig. 7-2. Then

$$\frac{1}{2}\oint_C \mathbf{r} \times d\mathbf{r} = \frac{\hat{\mathbf{z}}}{2}\oint_C (x\, dy - y\, dx) = \hat{\mathbf{z}}\oint_C \mathbf{G} \cdot d\mathbf{r}$$

with

$$G_x = -\tfrac{1}{2}y \qquad G_y = \tfrac{1}{2}x \qquad G_z = 0$$

Now use Stokes' theorem:

$$\frac{1}{2}\oint_C \mathbf{r} \times d\mathbf{r} = \hat{\mathbf{z}}\int_\Sigma (\nabla \times \mathbf{G}) \cdot d\mathbf{S} = \hat{\mathbf{z}}\int (\hat{\mathbf{z}}) \cdot (\hat{\mathbf{z}}\, dS) = \hat{\mathbf{z}}S$$

where S is the area of the loop. Therefore

$$\mathbf{m} = \hat{\mathbf{z}}IS$$

If the loop lies in some other plane, the direction of the magnetic dipole moment will be perpendicular to that other plane. The direction of the current determines the sense of the vector representing the moment according to the usual rule; the direction of advance is that of a right-hand screw turned in the sense of the current. If the loop does not lie in

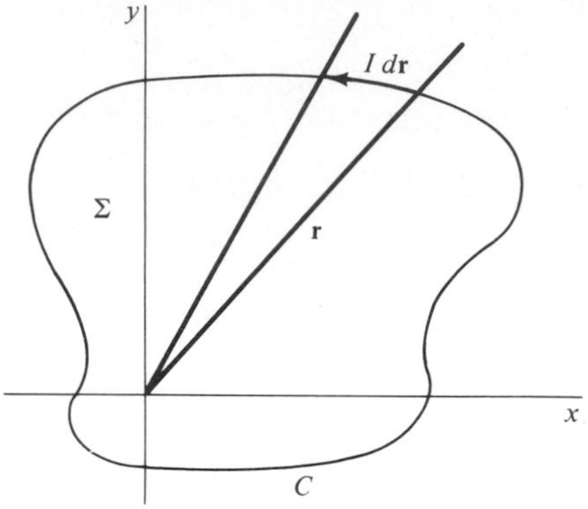

FIGURE 7-2

a plane $\mathbf{m} = \hat{\mathbf{n}}IS$ would still result from the integration; but $\hat{\mathbf{n}}$ would not necessarily have an obvious direction.

The term magnetic dipole, like its electrical counterpart, is used in several different ways. (1) The magnetic dipole term in the multipole expansion of $\mathbf{A}$ for an arbitrary distribution of many currents is the term which varies as r^{-2}. (2) That term may be written $(\mu_0/4\pi)(\mathbf{m} \times \hat{\mathbf{r}}/r^2)$; $\mathbf{m}$ is then called the magnetic dipole moment of this particular current distribution. Only the transverse component of $\mathbf{m}$ is effective in producing a contribution to $\mathbf{A}$; in contrast, only the longitudinal component of $\mathbf{p}$ is effective in contributing to ϕ. (3) A magnetic dipole is a specific current distribution. There are two versions. (a) Any real, filamentary, current-carrying loop (neglecting batteries or lead-in wires) which has dimensions that are small compared to the distance from the loop to a field point is a real magnetic dipole in the far-field approximation. (b) A point magnetic dipole is a useful fiction. It is obtained from (a) by letting the dimensions of the loop approach zero while letting the current approach infinity in such a way that the product IS is constant. The purpose of these limiting processes is to make the higher-order terms in the expansion for $\mathbf{A}$ negligible for all values of $\mathbf{r}$, near as well as far.

The third term in the multipole expansion of $\mathbf{A}$ is a $(1/r)1/r^2$, or quadrupole, term $(2^2 = 4)$. Let the current distribution be taken, for simplicity, to consist of one filamentary circuit. Then, if the circuit dimensions are small compared to the distances from the circuit elements to the test point, the quadrupole and higher-order terms become negligible compared to the dipole term. The value of the latter then becomes independent of the shape of the circuit. On the other hand, close to the coil $\mathbf{A}$ has a significant quadrupole contribution—even if the coil is a circular loop; and the value of the dipole term will be very dependent on the circuit shape. Our interest is in the gross behavior of magnetic materials rather than in the fine details close to the atoms and their circulating currents. So the quadrupole term becomes of negligible importance.

Examples

1. Comparison of the real electric dipole and the real magnetic dipole We will compare the appearance and behavior of the lines of **E** for an electric dipole with those of **B** for a magnetic dipole. Figure 7-3*a* shows the configuration of the electric dipole. Figure 7-3*b, c, d* shows the lines of **E** as they appear at successively increasing distances from the dipole. In the far-field approximation, $r \gg \ell$, the electric field is given by

$$\mathbf{E} = \hat{\mathbf{r}}\left(\frac{p}{4\pi\epsilon_0}\right)\frac{2\cos\theta}{r^3} + \hat{\boldsymbol{\theta}}\left(\frac{p}{4\pi\epsilon_0}\right)\frac{\sin\theta}{r^3}$$

in spherical coordinates. This corresponds to Fig. 7-3*d*.

In the near-field approximation, $r \ll \ell$,

$$\mathbf{E} = \hat{\mathbf{r}}\left(\frac{-2q}{\pi\epsilon_0\ell^2}\right)\cos\theta + \hat{\boldsymbol{\theta}}\left(\frac{2q}{\pi\epsilon_0\ell^2}\right)\sin\theta$$

This expression corresponds to the central region of Fig. 7-3*b* and is not particularly useful. Although it is an electric field that is produced by a so-called dipole (one positive charge together with one negative charge of equal magnitude) it is not what is referred to as *the* dipole field. For a point electric dipole the **E** is given, at *all* points, by the equation that gives Fig. 7-3*d* for a real dipole.

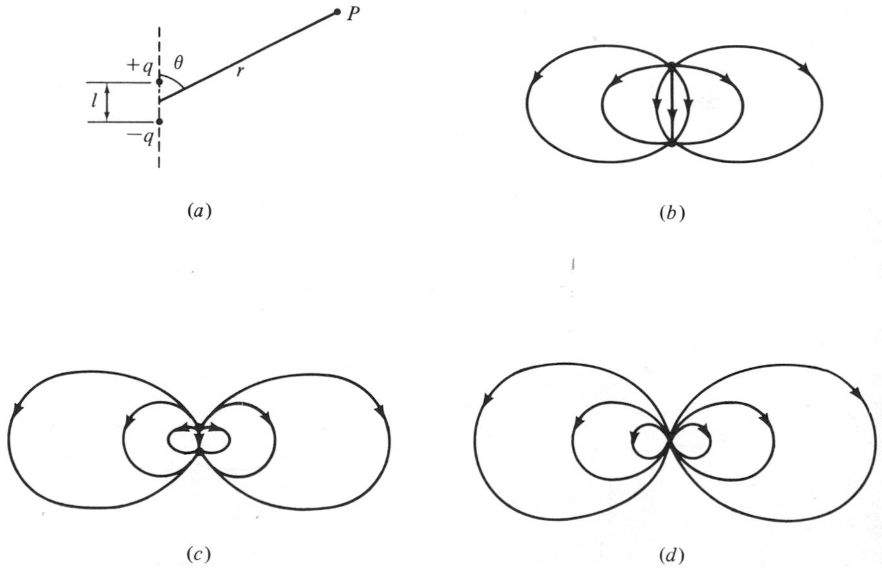

FIGURE 7-3

Figure 7-4*a* shows one configuration of a magnetic dipole: a circular loop with P far from O. This is the same as that of Fig. 5-5, which was used to obtain the **A** of a circular loop. This dipole is real in the sense that it has finite current and dimensions; even so, it is idealized in the sense that the wire resistance is zero and the battery, current limiting resistor, and lead-in wires have been neglected. Any other loop, e.g., a rectangular coil, could serve equally well—in its far-field approximation—for a magnetic dipole. The **A** obtained previously (Chap. 5, Sec. 5-1, Example 3) for the circular loop is

$$\mathbf{A} = \hat{\boldsymbol{\phi}}\,\frac{\mu_0 I}{2\pi}\sqrt{\frac{a}{L}}\left[\left(\frac{2}{k} - k\right)K(k) - \frac{2}{k}E(k)\right]$$

Here

$$k = \sqrt{\frac{4La}{z^2 + (L + a)^2}}$$

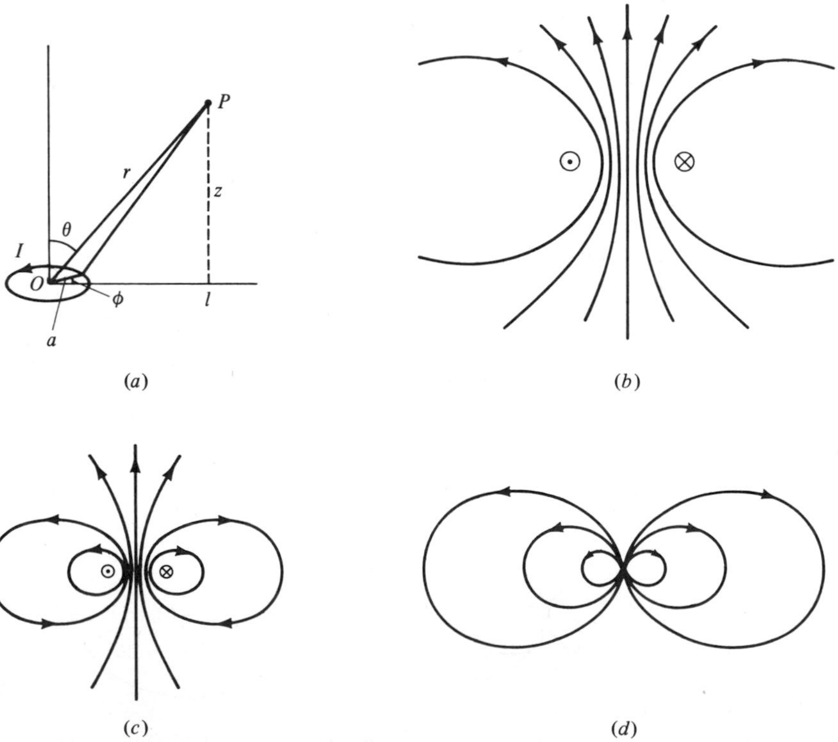

FIGURE 7-4

$$K(k) = \int_0^{\pi/2} \frac{d\alpha}{\sqrt{1 - k^2 \sin^2 \alpha}}$$

is the elliptic integral of the first kind, while

$$E(k) = \int_0^{\pi/2} \sqrt{1 - k^2 \sin^2 \alpha} \cdot d\alpha$$

is the elliptic integral of the second kind. $K(k)$ and $E(k)$ are functions which are tabulated. Knowing k, one can find values for K and E, just as one can find $\sin \theta$ and $\cos \theta$ as soon as one knows θ.

We will now obtain $\mathbf{B} = \nabla \times \mathbf{A}$ in spherical coordinates, for comparison with the $\mathbf{E}$ of the electric dipole. The derivation is a bit lengthy and can be skipped without loss of continuity.

First, replace L by $r \sin \theta$ in the expressions for $\mathbf{A}$ and k:

$$A_\phi = \left(\frac{\mu_0 I \sqrt{a}}{2\pi}\right) \frac{1}{\sqrt{r \sin \theta}} \left[\left(\frac{2}{k} - k\right) K - \frac{2}{k} E\right]$$

where

$$k = \sqrt{\frac{4ar \sin \theta}{a^2 + r^2 + 2ar \sin \theta}}$$

The general expression for $\mathbf{B}$ may be found from these. But this is too complicated a formula for our purpose and it is more meaningful to obtain the near-field and far-field approximations.

The formula for k gives

$$\sqrt{r \sin \theta} = \frac{k\sqrt{a^2 + r^2 + 2ar \sin \theta}}{2\sqrt{a}}$$

Then

$$A_\phi = \left(\frac{\mu_0 I \sqrt{a}}{2\pi}\right)\left(\frac{2\sqrt{a}}{k\sqrt{a^2 + r^2 + 2ar \sin \theta}}\right)\left[\frac{(2 - k^2)K - 2E}{k}\right]$$

$$= \frac{\mu_0 I a}{\pi}\left(\frac{1}{\sqrt{a^2 + r^2 + 2ar \sin \theta}}\right)\left[\frac{(2 - k^2)K - 2E}{k^2}\right]$$

This formula for k can be simplified in special cases.

For $r \ll a$: $\quad k^2 = [4 \sin \theta] \left(\frac{r}{a}\right) - [8 \sin^2 \theta] \left(\frac{r}{a}\right)^2 + \cdots$

$$\text{for } r \gg a: \quad k^2 = [4 \sin\theta]\left(\frac{a}{r}\right) - [8 \sin^2\theta]\left(\frac{a}{r}\right)^2 + \cdots$$

in either case $k^2 \ll 1$. So both the near-field and the far-field approximations for **A** and **B** correspond to letting $k \to 0$ in the expression for A_ϕ. Expand $\sqrt{1 - k^2 \sin^2\alpha}$ in a Taylor series in k:

$$\sqrt{1 - k^2 \sin^2\alpha} \approx 1 - \tfrac{1}{2} k^2 \sin^2\alpha - \tfrac{1}{8} k^4 \sin^4\alpha$$

Then

$$K(k) \approx \int_0^{\pi/2} d\alpha \left[1 + \frac{1}{2} k^2 \sin^2\alpha + \frac{1}{8} k^4 \sin^4\alpha\right] = \frac{\pi}{2} + \frac{k^2}{2}\left(\frac{\pi}{2}\right) + \frac{3k^4}{8}\left(\frac{3\pi}{16}\right)$$

Similarly

$$E(k) \approx \int_0^{\pi/2} d\alpha \left[1 - \frac{1}{2} k^2 \sin^2\alpha - \frac{1}{8} k^4 \sin^4\alpha\right] = \frac{\pi}{2} - \frac{k^2}{2}\left(\frac{\pi}{4}\right) - \frac{k^4}{8}\left(\frac{3\pi}{16}\right)$$

Then

$$\left[\frac{(2 - k^2)K - 2E}{k^2}\right] \approx \frac{\pi k^2}{16}$$

when k is small, and

$$A_\phi = \frac{\mu_0 I a}{\pi}\left(\frac{1}{\sqrt{a^2 + r^2 + 2ar\sin\theta}}\right)\left[\left(\frac{\pi}{16}\right)\frac{4ar\sin\theta}{a^2 + r^2 + 2ar\sin\theta}\right]$$

$$= \left(\frac{\mu_0 I a^2}{4}\right)\frac{r\sin\theta}{(a^2 + r^2 + 2ar\sin\theta)^{3/2}}$$

[Curves for several lines of constant A_ϕ given previously (Fig. 5-6) were calculated from the general formula for **A**.] The expression for $\mathbf{A}_\phi$ given here is valid not only near the center of the loop ($r/a \ll 1$) and anywhere far from the loop ($r/a \gg 1$) but also, since $k \to 0$ when $\theta \ll 1$ for any r, for any point (near, far, or in between) that lies close to the central axis of symmetry. $\mathbf{B} = \nabla \times \mathbf{A}$ now gives the desired result:

$$B_r = \frac{1}{r\sin\theta}\frac{\partial}{\partial\theta}(\sin\theta\, A_\phi) = \frac{\mu_0 I a^2}{4}\,\frac{(2a^2 + 2r^2 + ar\sin\theta)\cos\theta}{(a^2 + r^2 + 2ar\sin\theta)^{5/2}}$$

$$B_\theta = -\frac{1}{r}\frac{\partial}{\partial r}(rA_\phi) = -\left(\frac{\mu_0 I a^2}{4}\right)\frac{(2a^2 - r^2 + ar\sin\theta)\sin\theta}{(a^2 + r^2 + 2ar\sin\theta)^{5/2}}$$

In the far-field approximation $r \gg a$ and

$$\mathbf{B} = \hat{\mathbf{r}} \left(\frac{\mu_0 I \pi a^2}{4\pi} \right) \frac{2 \cos \theta}{r^3} + \hat{\boldsymbol{\theta}} \left(\frac{\mu_0 I \pi a^2}{4\pi} \right) \frac{\sin \theta}{r^3} \qquad \text{(far field)}$$

Comparing this with the far-field approximation for the **E** produced by an electric dipole, it is seen that the behavior with r and θ is identical in the two cases. Instead of $p/4\pi\epsilon_0$ one has here $(\mu_0/4\pi)(I\pi a^2) = (\mu_0/4\pi)m$, where $m = I(\pi a^2)$ is the magnetic dipole moment of the loop. The curves for **B** in this case, shown in Fig. 7-4*d*, are identically the same as those for **E** in Fig. 7-3*d*.

In the near-field approximation, valid near the center of the loop,

$$\mathbf{B} = \hat{\mathbf{r}} \left(\frac{\mu_0 I}{2a} \right) \cos \theta - \hat{\boldsymbol{\theta}} \left(\frac{\mu_0 I}{2a} \right) \sin \theta \qquad \text{(near field)}$$

Not only does this differ from the far-field approximation for **B** but it also differs (in the sign of both components) from the corresponding expression for **E**. The curves for this case are shown in Fig. 7-4*b*. The reason for the difference in sign is seen to rest on the fact that the lines of **B** continue always in the same sense while the lines of **E** reverse in passing near one of the charges.

The case intermediate between $r \ll a$ and $r \gg a$, shown in Fig. 7-4*c*, cannot be obtained from the formulas for B_r and B_θ above but must be worked out from the general expression.

2. Force on a magnetic dipole What is the force exerted on a magnetic dipole by a magnetic field? This result is used in the interpretation of the Stern-Gerlach experiment, where the force of a magnetic field on a beam of neutral atoms enables one to measure the magnetic moment of the beam atoms. This experiment is significant historically because it first established the spatial quantization of angular momentum.

The magnetic force on a circuit carrying a current I_T is given, for a nonisolated circuit, by

$$\mathbf{F}_m = +\nabla U_m = -\nabla U_{\text{total}} = -\nabla(I_T \Phi_{TS})_{I_T=\text{const}}$$

For an isolated circuit, it is given by

$$\mathbf{F}_m = -\nabla U_m = -\nabla(I_T \Phi_{TS})_{\Phi_T=\text{const}}$$

where $\Phi_T = \Phi_{TS} + \Phi_{TT}$ (see Sec. 5-4, Chap. 5). An orbital electron circulating within an atom constitutes an isolated circuit: no energy is required to maintain the current. When such an atom is introduced into a magnetic field the current changes; this will be discussed in connection with diamagnetism, later in this chapter.

The magnetic force for the isolated case may also be obtained from $\mathbf{F}_m = -\nabla U_m$ taking $U_m = -I_T \Phi_{TS}$ and treating I_T as constant, even though I_T is not constant. Also, the force obtained on an isolated dipole will be the same as that on an ordinary nonisolated dipole, since the force on a circuit in any position is independent of whether it is isolated.

If **S** is the area of the dipole, considering the isolated case, then

$$\mathbf{F}_m = -\nabla U_m = -\nabla\left[-I(\mathbf{B}_{av}\cdot\mathbf{S})\right]_{I=\text{const}} = -\nabla[-(I\mathbf{S})\cdot\mathbf{B}_{av}] = -\nabla(-\mathbf{m}\cdot\mathbf{B}_{av})$$

The magnetic dipole is not necessarily a circular loop but can have any shape provided only that it is small, far away, or both, so that the quadrupole term is negligible. The value of $\mathbf{B}_{av}$ is then essentially the value of **B** at the center of the dipole and

$$\mathbf{F}_m = -\nabla(-\mathbf{m}\cdot\mathbf{B})$$

The potential energy of the magnetic dipole in the external magnetic field **B**, for the isolated case, is

$$\boxed{U_m = -(\mathbf{m}\cdot\mathbf{B})} \qquad \text{(M-15)}$$

There is an arbitrary constant which can be added to the right side but this may be taken to be zero for convenience. It is worth noting the similarity with $U_e = -(\mathbf{p}\cdot\mathbf{E})$.

By use of a vector identity we may obtain a different expression for $\mathbf{F}_m$:

$$\mathbf{F}_m = \nabla(\mathbf{m}\cdot\mathbf{B}) = (\mathbf{m}\cdot\nabla)\mathbf{B} + (\mathbf{B}\cdot\nabla)\mathbf{m} + [\mathbf{m}\times(\nabla\times\mathbf{B})] + [\mathbf{B}\times(\nabla\times\mathbf{m})]$$

The second and fourth terms on the right vanish, since **m** does not depend on position. The third term vanishes also, both for a real magnetic dipole and for a point magnetic dipole, but the reason for this is different in these two cases; we leave the proof to Prob. 2. Then

$$\mathbf{F} = (\mathbf{m}\cdot\nabla)\mathbf{B}$$

PROBLEMS

1 A small rectangular loop lies at the origin, as shown in Fig. 7-5. Find $\mathbf{A}(\mathbf{r})$ if $r \gg a$, $r \gg b$. It is interesting to compare the curves of constant A with those of constant ϕ for the electric dipole shown in Fig. 6-3.

2 The derivation of the expression for the force on a dipole in an external **B** contained the term $\mathbf{m}\times(\nabla\times\mathbf{B})$. Prove that this vanishes for a real dipole. Why does it equal zero for a point dipole?

3 Find the value of $\mathbf{B}(\mathbf{r})$ for the rectangular magnetic dipole of Prob. 1.

4 Compare the answer to Prob. 3 with the **B** obtained for a circular dipole loop and the **E** obtained from an electric dipole.

5 From $\mathbf{T}_m = -\hat{\boldsymbol{\theta}}(\partial U_m/\partial\theta)$ find the torque exerted on a magnetic dipole in an external magnetic field **B**. Here θ goes from **B** toward **m** while $\hat{\boldsymbol{\theta}}$ is in the direction, perpendicular to the plane of **B** and **m**, determined by a right-hand screw.

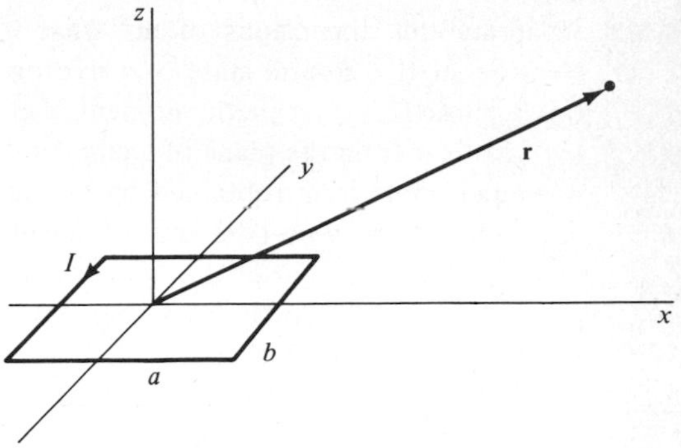

FIGURE 7-5

6 What is the force on a magnetic dipole created by $\mathbf{B} = \hat{\mathbf{z}}B$, where B is a constant? If a magnetic field $\mathbf{B}$ is not uniform will the force on a magnetic dipole placed in $\mathbf{B}$ be toward a region of greater B or smaller B?

7 Two isolated magnetic dipoles, $\mathbf{m}_1$ and $\mathbf{m}_2$, are separated by the distance R.

(*a*) What is the magnetic potential energy of either one in the field of the other one?

(*b*) What is U_m if the dipoles are nonisolated?

(*c*) How is the force found on each one, if one is isolated and the other not?

(*d*) What is U_{tot} in the last two cases?

8 Is the torque on a magnetic dipole in such a sense that it tends to align $\mathbf{m}$ with $\mathbf{B}$, $\mathbf{m}$ with $-\mathbf{B}$, or $\mathbf{m}$ perpendicular to $\mathbf{B}$?

9 Given two magnetic dipoles, $\mathbf{m}_1 = \hat{\mathbf{z}}m_1$ and $\mathbf{m}_2 = \hat{\mathbf{x}}m_2$, with the first at the origin and the second at $y = r$. What is the force exerted by the first on the second?

10 Give an expression for the torque on a magnetic dipole in a field $\mathbf{B}$ in terms of U_{tot}. How does this compare with the corresponding formula in electrostatics for an electric dipole in an electric field $\mathbf{E}$?

11 In Prob. 9, what is the torque exerted by $\mathbf{m}_1$ on $\mathbf{m}_2$?

12 Write an explicit expression for the quadrupole terms in the multipole expansion of $\mathbf{A}$.

13 A magnetic dipole is situated on the axis of a solenoid and just outside it. If $\mathbf{m}$ is parallel to $\mathbf{B}$ would the dipole be attracted to the solenoid or repelled from it? Suppose $\mathbf{m}$ is antiparallel to $\mathbf{B}$?

14 Explain the answers to Prob. 13 in terms of flux density and potential energy.

15 Change Prob. 9 by making $\mathbf{m}_2 = \hat{\mathbf{z}}m_2$. What is the torque exerted by $\mathbf{m}_1$ on $\mathbf{m}_2$? Is this a condition of stable equilibrium?

16 The behavior of electric dipoles and magnetic dipoles are remarkably similar in comparing the equations for $\mathbf{E}$ and $\mathbf{B}$. Yet, the potential ϕ is proportional to $\mathbf{p} \cdot \hat{\mathbf{r}}$ while $\mathbf{A}$ depends on $\mathbf{m} \times \hat{\mathbf{r}}$. How, then, do the forces and torques come to be equal?

17 In quantum mechanics the angle between $\mathbf{m}$ and $\mathbf{B}$ can only assume certain fixed values. Find an expression for the force on a magnetic dipole in an inhomogeneous magnetic field in that case. This is used, e.g., in the Stern-Gerlach experiment.

18 What are the dimensions of **m**? What is the magnitude of the orbital m for an electron in the ground state of a hydrogen atom? Compare this with the value of the intrinsic (spin) magnetic moment of the electron.

19 (*a*) How far from the plane of a circular dipole, on the central axis, does **B** have the quadrupole term (obtained by taking the curl of the **A** quadrupole term) equal to the dipole term (similarly obtained from the dipole terms of **A**)?

(*b*) Equal to 1 percent of the dipole term?

(*c*) How far away on the axis does **A** have its quadrupole term equal to 0.01 of its dipole term? (Another way to obtain the results is to find an expression for **A** near, but not on, the axis and take its curl; then expand in a power series. See J. D. Jackson, "Classical Electrodynamics," Wiley, New York, 1962.)

20 The force on a magnetic loop current may be obtained from $\mathbf{F} = I\oint d\mathbf{r}_s \times \mathbf{B}$ if **B** is obtained by a Taylor expansion from $\mathbf{B}_0$, the value at the center. Derive $\mathbf{F} = (\mathbf{m} \cdot \nabla)\mathbf{B}$.

7-2 THE MAGNETIZATION M

Suppose that an atom or molecule of some substance has a magnetic dipole moment **m**, one which is either characteristic of the species or one which has been induced by the introduction of an external **B.** If there are N such atoms or molecules per cubic meter we define the magnetization by

$$\mathbf{M} = N\mathbf{m}$$

The dimensions of **M** are $[\text{m}^{-3}\ (\text{m}^2\ \text{A})]$, i.e., $[\text{m}^{-1}\ \text{A}]$. This is usually given in the reverse order, $[\text{A m}^{-1}]$. The magnetic vector potential that is due to one of these magnetic dipoles is

$$\mathbf{A} = \frac{\mu_0}{4\pi} \frac{\mathbf{m} \times \hat{\mathbf{r}}}{r^2}$$

the **A** that is due to all of them is

$$\mathbf{A}(\mathbf{r}_t) = \frac{\mu_0}{4\pi} \int_{\mathcal{V}} \frac{\mathbf{M} \times \hat{\mathbf{R}}}{R^2}\, d\tau_s$$

Here $\mathbf{r}_t$ is a test point and $\mathcal{V}$ is the volume of the magnetic material. As in the case of dielectrics, $\mathbf{r}_t$ is first taken outside $\mathcal{V}$ since the formula for **A** is the far-field approximation and it is necessary that $\mathbf{r}_t$ be far from any dipoles. Then

$$\mathbf{A}(\mathbf{r}_t) = -\frac{\mu_0}{4\pi}\int_{\mathcal{V}} \mathbf{M} \times \nabla\left(\frac{1}{R}\right) d\tau_s = \frac{\mu_0}{4\pi}\int_{\mathcal{V}} \mathbf{M} \times \nabla_s\left(\frac{1}{R}\right) d\tau_s$$

where ∇_s operates with respect to a source point $\mathbf{r}_s$. Using the vector identity

$$\nabla_s \times \left(\frac{1}{R}\mathbf{M}\right) = \frac{1}{R}\nabla_s \times \mathbf{M} - \mathbf{M} \times \nabla_s\left(\frac{1}{R}\right)$$

gives
$$\mathbf{A}(\mathbf{r}_t) = \frac{\mu_0}{4\pi}\int_{\mathcal{V}} \frac{\nabla_s \times \mathbf{M}}{R} d\tau_s + \frac{\mu_0}{4\pi}\int_{\mathcal{V}} -\nabla_s \times \left(\frac{1}{R}\mathbf{M}\right) d\tau_s$$

There is a theorem (see Chap. 1, Sec. 1-4, Prob. 12) that for a vector **F**,

$$\int_{\mathcal{V}} \nabla \times \mathbf{F}\, d\tau_s = -\oint_{\Sigma} \mathbf{F} \times d\mathbf{S}$$

where Σ is the surface bounding $\mathcal{V}$. This is analogous to Gauss' theorem. Substituting this for the last term,

$$\mathbf{A}(\mathbf{r}_t) = \frac{\mu_0}{4\pi}\int_{\mathcal{V}} \frac{\nabla_s \times \mathbf{M}}{R} d\tau_s + \frac{\mu_0}{4\pi}\oint_{\Sigma} \frac{\mathbf{M} \times \hat{\mathbf{n}}}{R} d\mathbf{S}_s$$

Then, dropping the subscript for simplicity,

$$\mathbf{A}(\mathbf{r}_t) = \frac{\mu_0}{4\pi}\int_{\mathcal{V}} \frac{\mathbf{J}_m}{R} d\tau_s + \frac{\mu_0}{4\pi}\oint_{\Sigma} \frac{\mathbf{j}_m}{R} d\mathbf{S}_s$$

where
$$\boxed{\begin{aligned}\mathbf{j}_m &= \mathbf{M} \times \hat{\mathbf{n}}\\ \mathbf{J}_m &= \nabla \times \mathbf{M}\end{aligned}} \qquad \text{(M-16)}$$

$\mathbf{j}_m$ is the magnetization surface current density; $\mathbf{J}_m$ is the magnetization volume current density. (Our standard nomenclature would employ $\mathbf{J}_{\ell m}$ for magnetization surface current density, but for the sake of economy in subscripts we are using $\mathbf{j}_m$ here.)

The significance of the two magnetization current densities is similar to that of the two polarization charge densities in electrostatics: the actual effect of the magnetic

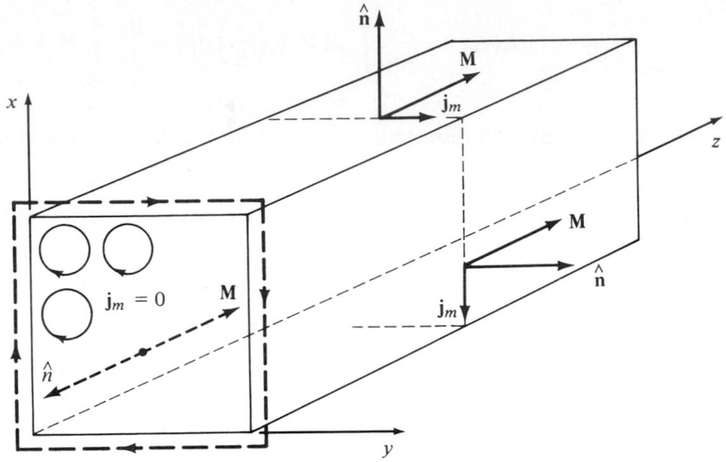

FIGURE 7-6

material can be simulated by removing the material and distributing $\mathbf{J}_m$ throughout the volume $\mathcal{V}$, with $\mathbf{j}_m$ on its surface. Figure 7-6 shows these so-called amperian surface currents for a rectangular rod with a uniform longitudinal magnetization $\mathbf{M} = \hat{\mathbf{z}}M$. Only the surface current appears, transverse to $\mathbf{M}$, when M is a constant. There is no surface contribution on the front and rear faces, just the opposite of the case with $\mathbf{P}$. If $\mathbf{M}$ varies with r, however, a volume contribution is added (and the magnitude of the surface current densities are also affected).

$\mathbf{J}_m$ and $\mathbf{j}_m$ are real, not fictitious, current densities. Though they do not involve the physical transport of charge, unlike conduction or convection current densities, they are the cause of an $\mathbf{A}$ and a $\mathbf{B}$. A view of the end face of the rod of Fig. 7-6 shows how they arise from the incomplete cancellation of the microscopic dipole currents. If the material is homogeneous the dipoles cancel each other at all points in the interior; but at the surface there is an uncanceled residue, like the surface tension on a liquid. This is $\mathbf{j}_m$. If the material is inhomogeneous there are interior points, in addition, where the microscopic currents do not cancel each other, leaving a resultant $\mathbf{J}_m$. The microscopic dipole currents, themselves, are caused either by the orbital circulating motions of the planetary electrons of an atom or by the inherent magnetic moments of these electrons.

Although the expression for $\mathbf{A}$ in terms of $\mathbf{J}_m$ and $\mathbf{j}_m$ was derived only for points outside $\mathcal{V}$, it is also valid inside $\mathcal{V}$. This follows at once from our view that $\mathbf{J}_m$ and $\mathbf{j}_m$ are real, and that they completely replace the magnetic material; provided only that we take $\mathbf{A}$, at an interior point, to be time-averaged and space-averaged sufficiently to leave a zero contribution for the intense fluctuating microscopic $\mathbf{A}$.

Examples

1. **A longitudinally magnetized rod** Let $\mathbf{M} = \hat{\mathbf{z}}M_0 y$ in Fig. 7-7a. M_0 is a constant, with dimensions [m^{-2} A]. This shows a rod, of length L into the paper, made of many parallel slabs, the magnetization increasing linearly in the slabs toward the right. We will find $\mathbf{J}_m$ and $\mathbf{j}_m$.

$$\mathbf{j}_m = \mathbf{M} \times \hat{\mathbf{n}} = (\hat{\mathbf{z}}M_0 y) \times \hat{\mathbf{n}}$$

$$\begin{array}{llll} \text{Top:} & [(\hat{\mathbf{z}}M_0 y) \times \hat{\mathbf{x}}]_{x=a} & = (\hat{\mathbf{y}}M_0 y)_{x=a} & = \hat{\mathbf{y}}M_0 y \\ \text{Right:} & [(\hat{\mathbf{z}}M_0 y) \times \hat{\mathbf{y}}]_{y=b} & = (-\hat{\mathbf{x}}M_0 y)_{y=b} & = -\hat{\mathbf{x}}M_0 b \\ \text{Bottom:} & [(\hat{\mathbf{z}}M_0 y) \times (-\hat{\mathbf{x}})]_{x=0} & = (-\hat{\mathbf{y}}M_0 y)_{x=0} & = -\hat{\mathbf{y}}M_0 y \\ \text{Left:} & [(\hat{\mathbf{z}}M_0 y) \times (-\hat{\mathbf{y}})]_{y=0} & = (\hat{\mathbf{x}}M_0 y)_{y=0} & = 0 \end{array}$$

Figure 7-7b shows the surface current around the outside of the rod. $\mathbf{j}_m$ has an unusual property. Although steady with time, its divergence does not vanish on the top and bottom surfaces.

$$\mathbf{J}_m = \nabla \times \mathbf{M} = \hat{\mathbf{x}}M_0$$

(The dimensions of $\mathbf{J}_m$ are $[M_0]$ = [A m^{-2}], while $\mathbf{j}_m$ has dimensions [A m^{-1}].) $\mathbf{J}_m$ has a divergence of zero within the rod. When $\mathbf{J}_m$ is added to $\mathbf{j}_m$ the picture of Fig. 7-7c is obtained. The two contributions on the right side, at $y = b$, do not cancel each other: they have different dimensions so they are different kinds of quantities (but see Prob. 6).

2. **Orbital and spin magnetic moment** What is the magnetic dipole moment of an electron (a) circulating about the nucleus of an atom, and (b) spinning about its axis?

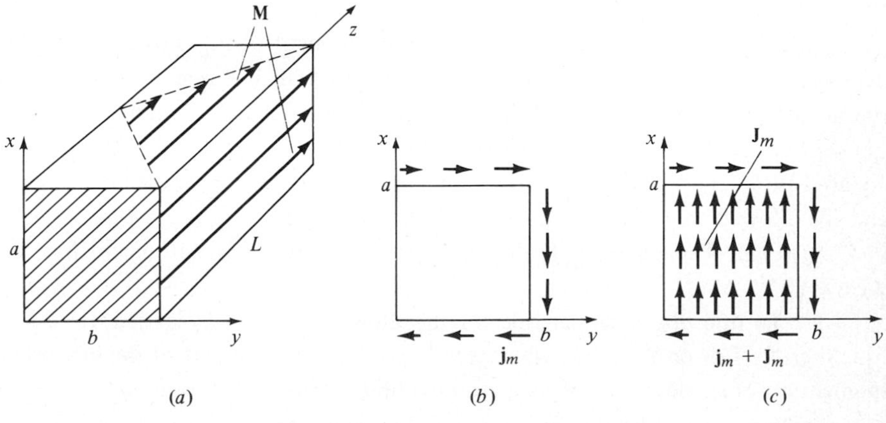

FIGURE 7-7

a. $\mathbf{m}_L = I\mathbf{S}$, where the subscript L is the conventional designation for orbital. The current is the charge passing a given point in the orbit per second: the charge magnitude times the frequency, $I = (-e)f = -e(v/2\pi r)$. This assumes a circular orbit of radius r with electron velocity v. The area $\mathbf{S}$ is $\hat{\mathbf{n}}\pi r^2$, where $\hat{\mathbf{n}}$ is the unit vector in the direction of advance of a right-hand screw turning in the sense of the electron velocity. Then $\mathbf{m}_L = -\hat{\mathbf{n}}evr/2$. The angular momentum is $\mathbf{L} = \hat{\mathbf{n}}m_e vr$, where m_e, the mass, is not to be confused with m, the moment. Substituting,

$$\mathbf{m}_L = \left(-\frac{e}{2m_e}\right)\mathbf{L}$$

b. If, instead of using the orbital angular momentum, one considered the spin angular momentum $\mathbf{S}$, then it would be necessary to apply the quantum-mechanical result (here given on faith, without further justification) that the magnetic moment is twice as large for a given spin angular momentum as it would be for that value of orbital angular momentum:

$$\mathbf{m}_S = 2\left(-\frac{e}{2m_e}\right)\mathbf{S}$$

The factor multiplying the angular momentum to give $\mathbf{m}$ is called the gyromagnetic ratio.

For a system in which both orbital and spin angular moments are considered, the general expression becomes

$$\mathbf{m} = g\left(-\frac{e}{2m_e}\right)\mathbf{J}$$

where g is the Lande g-factor and $\mathbf{J} = \mathbf{L} + \mathbf{S}$ is the total angular momentum. ($\mathbf{J}$ is not to be confused with current density.) Note that the orbital magnetic moment $\mathbf{m}_L$ is antiparallel to $\mathbf{L}$, and the spin magnetic moment is antiparallel to $\mathbf{S}$, but $\mathbf{m} = \mathbf{m}_L + \mathbf{m}_S$ is not antiparallel to $\mathbf{J}$.

a′. Suppose we wish to find $\mathbf{m}_{Lz}$, the component of $\mathbf{m}_L$ along a given direction, here the z axis. Then another quantum rule tells us that, if ℓ is an integer characterizing the angular momentum, the only permissible values of the z component of the angular momentum L_z are given by $L_z = \ell\hbar, (\ell - 1)\hbar, (\ell - 2)\hbar, \ldots, -(\ell - 1)\hbar, -\ell\hbar$ where $\hbar \equiv h/2\pi$ equals 1.06×10^{-34} J s. Thus, for a circulating electron in its lowest energy level, $\ell = 1$ and the allowed values for L_z are $\hbar$, 0, $-\hbar$. So $m_{Lz} = 1(-e/2m_e)L_z = (-e/2m_e)\ell\hbar$ or 0 or $(-e/2m_e)(-\ell\hbar)$. The quantity $(-e\hbar/2m_e)$ is called a Bohr magneton and has the value 9.27×10^{-24} A m^2.

b′. To find m_{Sz} a similar rule for the allowed values of S_z is used. Here S_z is given by either $S_z = s\hbar$ or $S_z = -s\hbar$, where $s = \frac{1}{2}$. So, for a component of the intrinsic angular momentum of an electron along a given direction, say the z direction, $m_{Sz} = 2(-e/2m_e)\frac{1}{2}\hbar$ or $m_{Sz} = 2(-e/2m_e)(-\frac{1}{2}\hbar)$, i.e., $m_{Sz} = \pm$ one Bohr magneton.

The rules given above are all ad hoc here; in quantum mechanics they are derived as necessary consequences of the differential equations employed there.

The Bohr magneton, $\sim 10^{-23}$ A m^2, is a round, order-of-magnitude, figure for atomic magnetic moments. The explanation of why this particular combination of constants is used is our only justification for bringing in the quantum-mechanical rules into this classical treatment.

3. Surface magnetization current density Suppose the effect of a magnetic field is to induce a magnetic dipole moment of 10^{-23} A m^2 per atom. Let there be 10^{27} atoms/m^3. Find the surface current density at a surface making an angle of 45° with **M**.

$$\mathbf{j}_m = \mathbf{M} \times \hat{\mathbf{n}} = Nm \sin 45^\circ$$

The tangential component of **M** is

$$(10^{27})(10^{-23})(0.707) = 7070 \text{ A m}^{-1}$$

and this is the surface current density.

10^{27} atoms/m^3 is equivalent to taking 10^9 atoms/m. Then the contribution of each atom to $\mathbf{j}_m$ is

$$7.07 \times 10^{-6} \text{ A/atom}$$

This seems like a tremendously large value for the current contribution from a single atom. But while 0.7 μA is a current of macroscopic size, it is several orders of magnitude smaller than the current produced in an atom by one electron circulating with a frequency of 10^{15} Hz. The currents in ordinary matter are very, very large; we do not notice them because they cancel each other out—there is no preferential direction.

PROBLEMS

1 Calculate the current of an electron circulating in a circular orbit in a plane, with a radius equal to that of an electron in the lowest energy orbit of a hydrogen atom. Find the Bohr radius; the velocity; and the frequency. Use

$$m_e vr = \hbar \quad \text{and} \quad m_e v^2/r = (1/4\pi\epsilon_0)\, e^2/r^2$$

2 Assume 10 electrons of each atom of Example 3 to be responsible for the induced moment. Suppose they all have a Bohr radius, in order to get a crude estimate. What fraction of the circulating current is represented by the induced equivalent surface current?

3 Derive an expression for the magnetic potential energy of an isolated dipole in a magnetic field in terms of the divergence of a quantity involving **m** and **A**. ("Isolated" means, e.g., a convection current without a battery.)

4 Given a region with $\mathbf{M} = \hat{\mathbf{z}}M(1 + bz)$. Find the magnetization current density $\mathbf{J}_m$. Repeat, with $\mathbf{M} = \hat{\mathbf{z}}M(1 + bx)$. Take a slab of thickness d along the $\hat{\mathbf{z}}$ axis. Find $\mathbf{j}_m$.

5 The interaction energy of two electric dipoles is proportional to $(1/4\pi\epsilon_0)(p^2/R^3)$, with p of the order of 10^{-30} C m for a typical polar molecule. The interaction energy of two magnetic dipoles $\propto (\mu_0/4\pi)(m^2/R^3)$, with m of the order of 10^{-23} A m^2 for a typical molecule. Compare the electric dipole-dipole energy with the magnetic dipole-dipole energy by considering two molecules, separated by R, which have typical values of both types of dipoles.

6 Show that the net transport of $\mathbf{J}_m$ plus $\mathbf{j}_m$ in Fig. 7-6 is zero in the x, y, and z directions. It would, in fact, be possible to have physical currents behave like the magnetization current $\mathbf{J}_m + \mathbf{j}_m$ here. How do you adapt to the fact that $\mathbf{J}_m$ and $\mathbf{j}_m$ have different dimensions?

7 A spherical piece of magnetic material is uniformly magnetized: $\mathbf{M} = \hat{\mathbf{z}}M$. Find $\mathbf{J}_m$ and $\mathbf{j}_m$, using spherical coordinates.

8 A right circular cylinder, symmetric with respect to the $\hat{\mathbf{z}}$ axis, is uniformly magnetized with $\mathbf{M} = \hat{\mathbf{z}}M$. Find $\mathbf{J}_m$ and $\mathbf{j}_m$ using cylindrical coordinates.

9 In Prob. 8, let $\mathbf{M} = \hat{\mathbf{x}}M$. Find $\mathbf{J}_m$ and $\mathbf{j}_m$ using cylindrical coordinates.

10 In Prob. 8, let $\mathbf{M} = \hat{\mathbf{r}}M$ (cylindrical $\hat{\mathbf{r}}$). Find $\mathbf{J}_m$ and $\mathbf{j}_m$.

11 The ellipse $x^2/a^2 + y^2/b^2 = 1$ is rotated about the line of the $\pm\hat{\mathbf{x}}$ axes to give a three-dimensional figure of revolution. If $\mathbf{M} = \hat{\mathbf{x}}M$, find $\mathbf{J}_m$ and $\mathbf{j}_m$.

12 Is it possible to have $\mathbf{J}_m$ and $\mathbf{j}_m$ in a body such that there are some directions with a net transport of magnetizing current?

13 Suppose the dipole density in a gas varies with height according to $N = N_0 \exp(-wz/kT)$ where w, k, and T are constant. In $\mathbf{M} = N\mathbf{m}$ let $\mathbf{m} = \alpha_m \mathbf{B}$ where α_m is a magnetizability. Find $\nabla \cdot \mathbf{M}$ and $\nabla \times \mathbf{M}$ when (*a*) $\mathbf{B} = \hat{\mathbf{x}}B_0$ and (*b*) $\mathbf{B} = \hat{\mathbf{z}}B_0$. (*c*) Compare with the electrostatic case.

14 The expression for $\mathbf{A}$ in terms of $\mathbf{J}_m$ and $\mathbf{j}_m$, applicable both outside and inside V, was based on the far-field approximation. Will the result be valid very close to the amperian surface currents?

15 In Prob. 13, let the gas be contained in a vertical cylinder, between $z = 0$ and $z = h$, of radius a. Find $\mathbf{j}_m$.

7-3 THE H FIELD

The equation $\nabla \times \mathbf{B} = \mu_0 \mathbf{J}$ is applicable for a $\mathbf{J}$ of any kind, no matter how it is produced. It is similar to $\nabla \cdot \mathbf{E} = (1/\epsilon_0)\rho$ in this respect; in fact, this is the reason $\mathbf{E}$ and $\mathbf{B}$ are the vectors of primary interest. $\mathbf{J}$ can be divided explicitly into a conventional current density component $\mathbf{J}_c$ (i.e., conduction plus convection) and a magnetization current density component $\mathbf{J}_m$:

$$\boxed{\nabla \times \mathbf{B} = \mu_0(\mathbf{J}_c + \mathbf{J}_m)} \qquad \text{(M-17)}$$

Then

$$\nabla \times \frac{1}{\mu_0} \mathbf{B} = \mathbf{J}_c + \nabla \times \mathbf{M}$$

$$\nabla \times \left(\frac{1}{\mu_0} \mathbf{B} - \mathbf{M} \right) = \mathbf{J}_c$$

Setting

$$\boxed{\mathbf{H} = \left(\frac{1}{\mu_0} \right) \mathbf{B} - \mathbf{M}} \tag{M-18}$$

gives

$$\boxed{\nabla \times \mathbf{H} = \mathbf{J}_c} \tag{M-19}$$

The vector **H** has its vortex sources only in the conventional current density—the conduction and convection densities; **B** has its vortex sources not only in these but in the magnetization current density as well.

Even if $\mathbf{J}_c = 0$ everywhere, this equation does not imply that $\mathbf{H} = 0$. For $\nabla \cdot \mathbf{H} = \nabla \cdot [(1/\mu_0)\mathbf{B} - \mathbf{M}] = -\nabla \cdot \mathbf{M}$, so inhomogeneities in **M** are sources or sinks of **H**. (The boundary surface between two regions of homogeneous but different **M** is also an inhomogeneity. Called a pole, such a surface also acts as a source or sink of lines of **H**.) Note that though conventional current acts as a vortex source for **H**, it does not act as an ordinary source for **H**.

The dimensions of **H** are $[\mathrm{m}^{-1}\ \mathrm{A}]$. A useful unit that is often used in connection with solenoids is the A-turns/m, where the number of solenoid turns is a dimensionless numeric.

Just as **E** and **D** may be connected either by addition ($\mathbf{D} = \epsilon_0 \mathbf{E} + \mathbf{P}$) or by multiplication ($\mathbf{D} = \epsilon \mathbf{E}$), so may **B** and **H**. Thus, since **M** and **H** have the same dimensions, set

$$\boxed{\mathbf{M} = \chi_m' \left(\frac{1}{\mu_0} \mathbf{B} \right)} \tag{M-20}$$

This is directly analogous to the procedure with **P**. It is also possible to set $\mathbf{M} = \chi_m \mathbf{H}$; indeed, that is the more common assumption, though both procedures are found in the literature. Since χ_m, the magnetic susceptibility, is extremely small for nonferromagnetic substances, the distinction here is academic; it is simply a matter of taste. For ferromagnetic substances it is more convenient to take $\mathbf{M} = \chi_m \mathbf{H}$ since, from Eq. (M-19),

$\oint \mathbf{H} \cdot d\mathbf{r}$ is independent of the material and is directly proportional to the conduction current I, which is easily measured. $\mathbf{B}$, on the other hand, depends on the material and is also more difficult to measure.

Substituting Eq. (M-19) in (M-17),

$$\mathbf{H} = \frac{1}{\mu_0}(1 - \chi'_m)\mathbf{B}$$

This gives

$$\mathbf{H} = \left(\frac{1}{\mu}\right)\mathbf{B} \qquad \text{(M-21)}$$

where

$$\frac{1}{\mu} = (1 - \chi'_m)\,\frac{1}{\mu_0}$$

or

$$\mu = \left(\frac{1}{1 - \chi'_m}\right)\mu_0 \approx (1 + \chi'_m)\mu_0$$

Equation (M-21) is usually written $\mathbf{B} = \mu\mathbf{H}$. This, of course, is correct but it tends to be misleading if it is compared with $\mathbf{D} = \epsilon\mathbf{E}$, making $\epsilon \leftrightarrow \mu$ instead of $\epsilon \leftrightarrow 1/\mu$. One difference between μ and its electrical analog is that χ'_m can be either positive or negative. This will be discussed further in the next section.

μ is the permeability of the magnetic substance and has the dimensions of μ_0. The quantity $1/(1 - \chi'_m) \approx 1 + \chi'_m$ is a numeric; if we set it equal to κ_m and call it the relative permeability constant it would be similar to κ, the dielectric constant; then $\mathbf{B} = \kappa_m \mu_0 \mathbf{H}$. This would be appropriate for most substances; but for iron, e.g., κ_m is not a constant but is a function of $\mathbf{B}$; in fact, a multivalued function. Leaving aside the ferromagnetic substances, the replacement of the vacuum by a medium results, for us, in the replacement of μ_0 by μ in the equations that apply to the vacuum. The reverse procedure, letting $\mu \to \mu_0$, will also be correct; but sometimes this will give the trivial, if true, result $0 = 0$. The equation $\mathbf{H} = (1/\mu_0)\mathbf{B} - \mathbf{M}$ is always true; the equation $\mathbf{H} = (1/\mu)\mathbf{B}$ with constant μ is only true when $\mathbf{M}$ varies linearly with $\mathbf{B}$ or with $\mathbf{H}$.

With the amperian current approach to magnetic materials, $\nabla \times \mathbf{B}$ is produced by any $\mathbf{J}$, whether $\mathbf{J}_c$ or $\mathbf{J}_m$, while $\nabla \times \mathbf{H}$ is produced only by $\mathbf{J}_c$. The effect of the material is left out of $\mathbf{H}$—this is what is taken care of by $\mathbf{M}$. A different approach to magnetic materials was formerly popular: that of magnetic poles. Here the formalism is exactly equivalent to that of electrostatics, with the free pole replacing the free charge and a scalar magnetic potential replacing the vector magnetic potential. This method is still used where only $\mathbf{J}_m$, but not $\mathbf{J}_c$, is present in some region; but the risks of going astray with this method are too great when not only materials but also conduction or convection currents are present. Consequently, this approach has lost its appeal. Especially since it is founded on a fiction—there are no free magnetic poles, at least not in our everyday world.

In the magnetic pole approach the vector **H** is exactly equivalent to the vector **E** in electrostatics (originally **E** was called the electric field intensity and **H** was called the magnetic field intensity) and all the equations in magnetostatics could be obtained by letting **E** → **H**, $\epsilon \to \mu$, etc. Therefore, **H** was at one time considered, like **E**, its namesake and prototype, as the more important vector. This position was reinforced by numerous similarities between **E** and **H**, and between **D** and **B**. Thus, in Example 1 immediately following, this similarity will be found in a comparison between an electret and a bar magnet; again, in Sec. 7-5 of this chapter it will be found that the boundary conditions between magnetic materials are such that **B** and **D** behave similarly, as do **E** and **H**. In all such cases, however, there is an important restriction–that there be no ρ_f or $\mathbf{J}_c$; so the comparison holds only in certain special cases. "Magnetic intensity" is misleading, since **H** gives the force only on free poles. We will try to ignore the need for naming **H**, thereby treating it like **D**. Recently P. Penfield, Jr. and H. A. Haus, in their book, "Electrodynamics of Moving Media" (M.I.T. Press, Cambridge, Mass., 1967), have claimed that Prof. L. J. Chu of M.I.T. found an effect that is correctly explained by the **H** approach but not by that of **B**; but this does not seem to be accepted fact. In a fairly fundamental way, the natural grouping is (**E**,**B**) and (**D**,**H**); very occasionally it is (**E**,**H**) and (**D**,**B**).

Still another view is possible: to treat (ϕ,**A**) as the basic quantities. (1) The latter two, together, have four components and this equals the number of degrees of freedom. Either (**E**,**B**) or (**D**,**H**) have six components. They are not all independent–there are two connecting equations which must be satisfied. (2) The potential fields are simpler quantities relativistically. (3) The potentials deal with energy, which is more important than force, the concept connected with **E**, **D**, **B**, and **H**. (4) The potentials explain the Aharanov-Bohm effect, which is inexplicable on the basis of the magnetic field and electric field.

Examples

1. A bar magnet In Fig. 6-13 a comparison was made between the lines of **E** and **D** for a cylindrical bar electret with uniform longitudinal **P**. We will make a similar comparison between the lines of **B** and **H** for a cylindrical bar of magnetic material with a uniform longitudinal **M**. We will consider all conduction and convection currents to vanish.

In the case of the uncharged electret we had $\mathbf{D} = \epsilon_0 \mathbf{E} + \mathbf{P}$, with **P** pointing uniformly from left to right. At all points within and without the rod, but not including points on the surface (where the derivatives are discontinuous), we had

$$\nabla \cdot \mathbf{E} = 0 \qquad \nabla \times \mathbf{E} = 0$$

$$\nabla \cdot \mathbf{D} = 0 \qquad \nabla \times \mathbf{D} = 0$$

$$\nabla \cdot \mathbf{P} = 0 \qquad \nabla \times \mathbf{P} = 0$$

On the surface of the electret we had, on the other hand (see Appendix 6)

$$\nabla_{\text{surf}} \cdot \mathbf{E} = \frac{1}{\epsilon_0}\sigma_p \qquad \nabla_{\text{surf}} \times \mathbf{E} = 0$$

$$\nabla_{\text{surf}} \cdot \mathbf{D} = 0 \qquad \nabla_{\text{surf}} \times \mathbf{D} = \mathbf{P} \times \hat{\mathbf{n}}$$

$$\nabla_{\text{surf}} \cdot \mathbf{P} = -\mathbf{P} \cdot \hat{\mathbf{n}} \qquad \nabla_{\text{surf}} \times \mathbf{P} = \mathbf{P} \times \hat{\mathbf{n}}$$

Suppose now, in the case of the permanent magnet, that we make the transition: $\mathbf{E} \to \mathbf{B}$, $\mathbf{D} \to \mathbf{H}$, $\mathbf{P} \to -\mathbf{M}$, $\epsilon_0 \to 1/\mu_0$. This gives $\mathbf{H} = (1/\mu_0)\mathbf{B} - \mathbf{M}$. We will here, similarly, take **M** pointing from left to right, or $-\mathbf{M}$ pointing uniformly from right to left. At all points within and without the bar, but not including points on the surface, we have $\nabla \cdot \mathbf{B} = 0$, $\nabla \times \mathbf{B} = 0$; $\nabla \cdot \mathbf{H} = 0$, $\nabla \times \mathbf{H} = 0$; $\nabla \cdot \mathbf{M} = 0$, $\nabla \times \mathbf{M} = 0$. On the surface we have

$$\nabla_{\text{surf}} \cdot \mathbf{M} = -\mathbf{M} \cdot \hat{\mathbf{n}}$$

$$\nabla_{\text{surf}} \times \mathbf{M} = \mathbf{M} \times \hat{\mathbf{n}}$$

as well as $\nabla_{\text{surf}} \cdot \mathbf{B} = 0$ and $\nabla_{\text{surf}} \times \mathbf{H} = 0$. From $\mathbf{H} = (1/\mu_0)\mathbf{B} - \mathbf{M}$ we then obtain $\nabla_{\text{surf}} \cdot \mathbf{H} = +\mathbf{M} \cdot \hat{\mathbf{n}}$. Similarly, $\nabla_{\text{surf}} \times \mathbf{B} = \mu_0 \mathbf{j}_m = \mu_0 \mathbf{M} \times \hat{\mathbf{n}}$.

If we choose the fundamental vector here to be **B** we see that the vortex sources of **B** (actually, surface vortex sources) are the $\mu_0 \mathbf{j}_m$, the amperian currents, on the sides of the bar. This is quite different from the electret where, with **E** the fundamental vector, the surface sources of **E** were the $(1/\epsilon_0)\sigma_p$ on the ends of the bar. From the knowlege of $\nabla \cdot \mathbf{B}$ and $\nabla \times \mathbf{B}$ *everywhere* we can now find **B**, by the Helmholtz theorem, and then calculate $\mathbf{H} = (1/\mu_0)\mathbf{B} - \mathbf{M}$.

By focusing our attention on **H**, rather than **B**, we obtain a closer comparison between the electret and the permanent magnet. Here we write $\mathbf{B} = \mu_0 \mathbf{H} + \mu_0 \mathbf{M}$, with $\mathbf{D} \leftrightarrow \mathbf{B}$, $\mathbf{E} \leftrightarrow \mathbf{H}$, $\mathbf{P} \leftrightarrow \mu_0 \mathbf{M}$, $\epsilon_0 \leftrightarrow \mu_0$. The fundamental vector now is **H** and its surface sources are the poles $+\mathbf{M} \cdot \hat{\mathbf{n}}$ on the ends of the magnet. From the knowledge of $\nabla \cdot \mathbf{H}$ and $\nabla \times \mathbf{H}$ *everywhere* we can now find **H** by the Helmholtz theorem, and then calculate $\mathbf{B} = \mu_0 \mathbf{H} + \mu_0 \mathbf{M}$.

A third alternative exists: we could assign the primary roles to **M** (or **P**). But **M** (like **P**) has both surface sources and surface vortex sources. Though this does not necessarily make **M** more complicated than **B** or **H** it is usual to take either one of the other two vectors as the one of primary interest; they exist even in the absence of matter. In any case, all three approaches would give the same results. Figure 7-8*a* gives the amperian current point of view; Fig. 7-8*b* gives the magnetic pole point of view. The latter is seen to look essentially the same as Fig. 6-13 for the electret; but both points of view are actually identical.

2. Bar magnet and solenoid Next, we will compare the case of Example 1 with a close-wound solenoid having the same dimensions as the bar magnet.

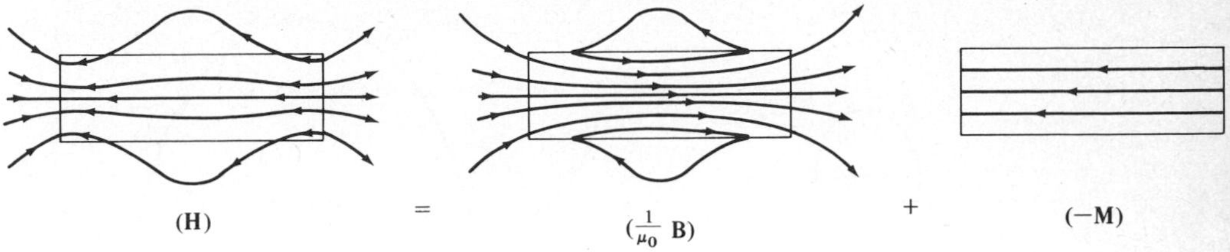

(*a*) Amperian current viewpoint.

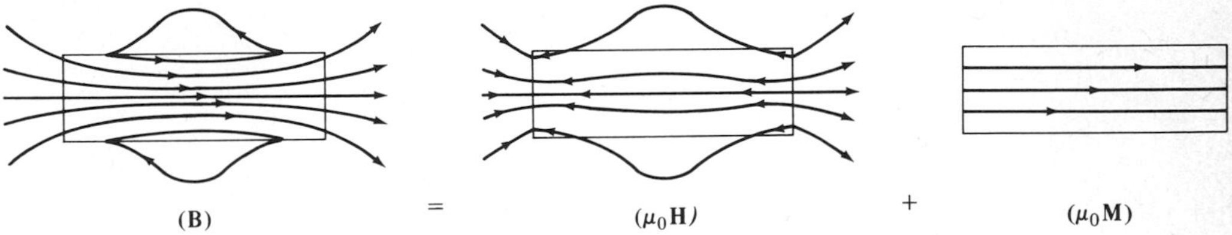

(*b*) Magnetic pole viewpoint.

FIGURE 7-8

$\mathbf{M} = 0$ here, so $\mathbf{B} = \mu_0 \mathbf{H}$. Figure 7-9 shows the lines of **B** (or **H**) for a short solenoid and a long one. When there is no matter present, one of the two vectors is superfluous. Note that the direction of **H** inside the solenoid is reversed compared to the direction of **H** inside the bar magnet. The value of **B** inside a solenoid whose length is very much larger than its diameter is $\mu_0 nI$ so **H** is nI where n is the number of turns of the winding per unit axial length. This is the reason the mksa unit for **H** is the A-turn/m, though the dimensions are $[\text{m}^{-1}\ \text{A}]$.

Since $\mathbf{j}_m = \mathbf{M} \times \hat{\mathbf{n}}$ for the magnet, $|\mathbf{j}_m| = |\mathbf{M}|$. The **B** produced by the magnet will be equal to that of the solenoid if $|\mathbf{M}| = nI$. Each turn gives the surface current per unit length per n.

3. Dimensions From $\mathbf{H} = (1/\mu_0)\mathbf{B} - \mathbf{M}$, the mksa dimensions of **H** must be equal to those of **M**: $[\text{m/V}] = [(\text{m}^2\ \text{A})\,\text{m}^{-3}] = [\text{m}^{-1}\ \text{A}]$. This agrees with the formula nI of Example 2. Designers of solenoids and other coils use the term ampere-turns per unit length for this quantity. If turns is considered a numeric, like revolutions in radial measure, the unit and dimensions of **H** can be conveniently taken as the [A-turns/m].

A complication arises when one wishes to employ both **B** and **H**. The mksa unit for **B**, the tesla (T), is inconveniently large; so the cgs unit of **B**, the gauss (G) $\equiv$ (1 G $\leftrightarrow$ 10^{-4}

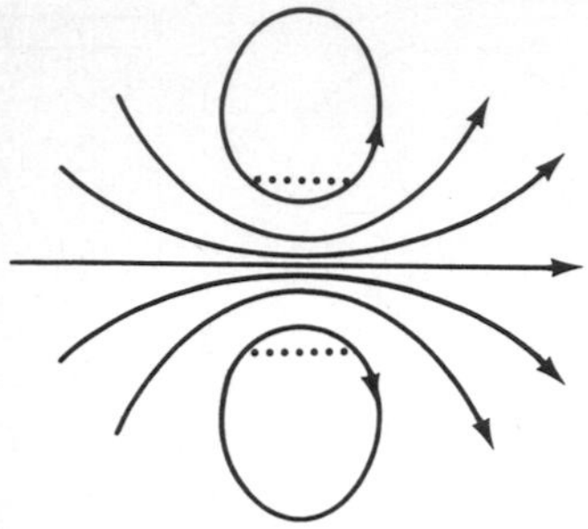

(a) $l = r$.

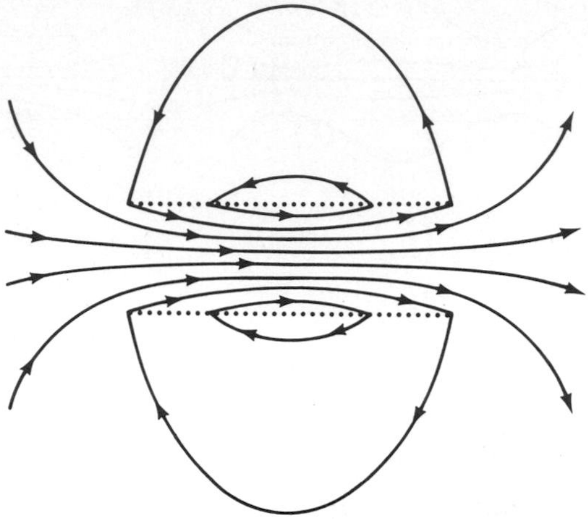

(b) $l = 6r$.

FIGURE 7-9

T; see below) is commonly used. Then it is also convenient to use the cgs unit for **H**. The mksa equation $\mathbf{B} = \mu_0 \mathbf{H}$ in vacuum becomes the cgs equation $\mathbf{B} = \mathbf{H}$, with **B** measured in gauss and **H** measured in oersteds (Oe). For, starting with

$$\mathbf{B}(\mathrm{T}) = \mu_0 \mathbf{H}\left(\frac{\text{A-turns}}{\text{m}}\right)$$

Appendix 7 (at the end of this chapter) gives

$$\left(\sqrt{\frac{\mu_0}{4\pi}}\,\mathbf{B}_{\text{cgs}}\right) = \mu_0\left(\frac{1}{\sqrt{4\pi\mu_0}}\,\mathbf{H}_{\text{cgs}}\right)$$

or

$$\mathbf{B}_{\text{cgs}} = \mathbf{H}_{\text{cgs}}$$

In vacuum the number of gauss equals the number of oersteds.

To prove 1 G $\leftrightarrow 10^{-4}$ T, we start with $B = \mu_0 I/2\pi r$, letting $I = 1$ A, $r = 1$ m. Then $\mathbf{B} = 2 \times 10^{-7}$ T. Now transform the equation to cgs (emu) form, using Appendix 7: $\mathrm{B} = 2I/r$ with $I = 0.1$ abA, $r = 100$ cm. Then $B = 2 \times 10^{-3}$ G. So 1 G $\leftrightarrow 10^{-4}$ T.

$\mathbf{B} = \mu_0 \mathbf{H}$ shows that if

$$\mathbf{H}_{\text{mksa}} = 1\left(\frac{\text{A-turn}}{\text{m}}\right)$$

then

$$\mathbf{B}_{\text{mksa}} = 4\pi \times 10^{-7}\ \mathrm{T}$$

So

$$B_{cgs} = 12.56 \times 10^{-7}\ \text{T}\ \frac{10^4\ \text{G}}{1\ \text{T}}$$

$$= 0.01256\ \text{G} \rightarrow \mathbf{H}_{cgs} = 0.01256\ \text{Oe}$$

Conversely

$$1\ \text{Oe} \leftrightarrow 79.6\left(\frac{\text{A-turn}}{\text{m}}\right)$$

$\oint \mathbf{H} \cdot d\mathbf{r}$ is a quantity that is useful occasionally. Its mksa units would be A-turns, its cgs units Oe cm. This cgs unit is called the gilbert (Gb). Since 1 A-turn/m $\leftrightarrow 4\pi \times 10^{-3}$ Oe, 1 A-turn $\leftrightarrow 4\pi \times 10^{-3}$ Oe $\times\ 10^2$ cm, or 1 A-turn $\leftrightarrow 4\pi/10$ Gb. This quantity $\oint \mathbf{H} \cdot d\mathbf{r}$ is often called mmf (for magnetomotive force) by analogy with the misnomer emf for $\oint \mathbf{E} \cdot d\mathbf{r}$. When we need to use this quantity we will call it magnetomotance.

4. ϵ versus μ When the theory of electricity and magnetism was first being developed it was believed that the vectors **E** and **H** were analogous to each other. **H**, a quantity to which we have deliberately assigned no name, was originally called the magnetic field intensity. By placing ϵ_0 in the denominator of the Coulomb equation and μ_0 in the numerator of the basic magnetostatic force equation the following parallel results were then achieved:

$\nabla \cdot \mathbf{E} = \frac{1}{\epsilon_0}(\rho_f + \rho_p)$	$\nabla \times \mathbf{B} = \mu_0(\mathbf{J}_c + \mathbf{J}_m)$
$= \frac{1}{\epsilon_0}\rho_f - \frac{1}{\epsilon_0}\nabla \cdot \mathbf{P}$	$= \mu_0 \mathbf{J}_c + \mu_0 \nabla \times \mathbf{M}$
$\nabla \cdot (\epsilon_0 \mathbf{E} + \mathbf{P}) = \rho_f$	$\nabla \times \left(\frac{1}{\mu_0}\mathbf{B} - \mathbf{M}\right) = \mathbf{J}_c$
$\mathbf{D} = \epsilon_0 \mathbf{E} + \mathbf{P}$	$\mathbf{H} = \frac{1}{\mu_0}\mathbf{B} - \mathbf{M}$
$\mathbf{P} = \chi(\epsilon_0 \mathbf{E})$	$\mathbf{M} = \chi_m(\mathbf{H})$
$\mathbf{D} = (1 + \chi)\epsilon_0 \mathbf{E} = \kappa\epsilon_0 \mathbf{E}$	$\mathbf{B} = (1 + \chi_m)\mu_0 \mathbf{H} = \kappa_m \mu_0 \mathbf{H}$
$\mathbf{D} = \epsilon\mathbf{E}$ (*Note:* $\epsilon = \kappa\epsilon_0$ for substances such that **P** is proportional to **E**)	$\mathbf{B} = \mu\mathbf{H}$ (*Note:* $\mu = \kappa_m \mu_0$ for substances such that **M** is proportional to **H**)

The last two equations are known as the constitutive equations. They accomplished the objective of making **B** and **D** analogous; likewise **E** and **H**. In fact, **D** is sometimes referred to, erroneously, as the electric flux density.

Today it is generally accepted that **E** and **B** are the analogous quantities, rather than **E** and **H**. For **E** is produced by any kind of charge, free or bound, just as **B** is caused by any kind of current, conventional or bound. **D** and **H**, on the other hand, are only produced by free charge and conventional current, respectively. From this point of view it is unfortunate that ϵ_0 and μ_0 were placed in opposite positions in the two defining equations.

Suppose, for example, that ϵ_0 were replaced by $\epsilon_0' = 1/\epsilon_0$ in Coulomb's equation:

$$\mathbf{F}_{st} = \frac{\epsilon_0'}{4\pi} \frac{q_s q_t}{R_{st}^2} \hat{\mathbf{R}}_{st}$$

The constant of proportionality is then in the same position, the numerator, as the factor μ_0 in the basic magnetostatic force equation:

$$\mathbf{F}_{ST} = -\frac{\mu_0}{4\pi} \oint_T \oint_S \frac{(I_S\, d\mathbf{r}_s) \cdot (I_T\, d\mathbf{r}_t)}{R_{st}^2} \hat{\mathbf{R}}_{st}$$

The following equations then result.

$$\nabla \cdot \mathbf{E} = \epsilon_0'(\rho_f + \rho_p) \qquad \nabla \times \mathbf{B} = \mu_0(\mathbf{J}_c + \mathbf{J}_m)$$

$$= \epsilon_0'\rho_f - \epsilon_0' \nabla \cdot \mathbf{P} \qquad = \mu_0 \mathbf{J}_c + \mu_0 \nabla \times \mathbf{M}$$

$$\nabla \cdot \left(\frac{1}{\epsilon_0'} \mathbf{E} + \mathbf{P}\right) = \rho_f \qquad \nabla \times \left(\frac{1}{\mu_0} \mathbf{B} - \mathbf{M}\right) = \mathbf{J}_c$$

$$\mathbf{D} = \frac{1}{\epsilon_0'} \mathbf{E} + \mathbf{P} \qquad \mathbf{H} = \frac{1}{\mu_0} \mathbf{B} - \mathbf{M}$$

$$\mathbf{P} = \chi \frac{1}{\epsilon_0'} \mathbf{E} \qquad \mathbf{M} = \chi_m' \left(\frac{1}{\mu_0} \mathbf{B}\right)$$

$$\mathbf{D} = (1 + \chi) \frac{1}{\epsilon_0'} \mathbf{E} = \frac{\kappa}{\epsilon_0'} \mathbf{E} \qquad \mathbf{H} = (1 - \chi_m') \frac{1}{\mu_0} \mathbf{B} = \frac{\kappa_m}{\mu_0} \mathbf{B}$$

$$\mathbf{E} = \epsilon' \mathbf{D} \left(\epsilon' = \frac{\epsilon_0'}{\kappa}\right) \qquad \mathbf{B} = \mu' \mathbf{H} \left(\mu' = \frac{\mu_0}{\kappa_m}\right)$$

It is apparent that here $\mathbf{D} \leftrightarrow \mathbf{H}$, $\epsilon_0' \leftrightarrow \mu_0$, $\mathbf{E} \leftrightarrow \mathbf{B}$, $\kappa \leftrightarrow \kappa_m$. But both κ and κ_m appear in the denominator, unlike the situation existing at present.

Another alternative exists: instead of letting $\epsilon_0' = 1/\epsilon_0$ we may choose to set $\mu_0' = 1/\mu_0$. Then, again, we obtain similar behavior in the defining equations:

$$\mathbf{F}_{st} = \frac{1}{4\pi\epsilon_0} \frac{q_s q_t}{R_{st}^2} \hat{\mathbf{R}}_{st}$$

$$\mathbf{F}_{ST} = -\frac{1}{4\pi\mu_0'} \oint_T \oint_S \frac{(I_S d\mathbf{r}_s) \cdot (I_T d\mathbf{r}_t)}{R_{st}^2} \hat{\mathbf{R}}_{st}$$

The other equations have a slightly different parallelism now.

$$\nabla\cdot\mathbf{E} = \frac{1}{\epsilon_0}(\rho_f + \rho_p) \qquad \nabla\times\mathbf{B} = \frac{1}{\mu_0'}(\mathbf{J}_c + \mathbf{J}_m)$$

$$= \frac{1}{\epsilon_0}\rho_f - \frac{1}{\epsilon_0}\nabla\cdot\mathbf{P} \qquad = \frac{1}{\mu_0'}\mathbf{J}_c + \frac{1}{\mu_0'}\nabla\times\mathbf{M}$$

$$\nabla\cdot(\epsilon_0\mathbf{E} + \mathbf{P}) = \rho_f \qquad \nabla\times(\mu_0'\mathbf{B} - \mathbf{M}) = \mathbf{J}_c$$

$$\mathbf{D} = \epsilon_0\mathbf{E} + \mathbf{P} \qquad \mathbf{H} = \mu_0'\mathbf{B} - \mathbf{M}$$

$$\mathbf{P} = \chi(\epsilon_0\mathbf{E}) \qquad \mathbf{M} = \chi_m'(\mu_0'\mathbf{B})$$

$$\mathbf{D} = (1+\chi)\epsilon_0\mathbf{E} = \kappa\epsilon_0\mathbf{E} \qquad \mathbf{H} = (1-\chi_m')\mu_0'\mathbf{B} = \kappa_m\mu_0'\mathbf{B}$$

$$\mathbf{D} = \epsilon\mathbf{E}(\epsilon = \kappa\epsilon_0 \text{ when } \mathbf{P} = \chi\epsilon_0\mathbf{E}) \qquad \mathbf{H} = \mu'\mathbf{B}(\mu' = \kappa_m\mu_0' \text{ when } \mathbf{M} = \chi_m\mu_0'\mathbf{B}$$

Here also, we have $\mathbf{D} \leftrightarrow \mathbf{H}$ and $\mathbf{E} \leftrightarrow \mathbf{B}$. But now κ and κ_m are multiplying factors, just as in the present system.

By the arbitrary choice of where we wish to put the proportionality constants in the defining equations we can emphasize one correspondence or another. It is the physics of the situation that should determine the choice.

PROBLEMS

1 A material has uniform magnetization. Three different cavities exist in the material: (*a*) a long, thin needle parallel to **M**, (*b*) a wide, squat cylinder parallel to **M**, (*c*) a spherical cavity. If **B** is the value of the magnetic field in the medium, find **B** at the centers of the cavities.

2 In Prob. 1, let **H** be the value in the material. Find $\mathbf{H}_{\text{cav}}$, the value of **H** at the centers of the cavities, in terms of **H** and **M**. Compare with **D** and **E** for a dielectric. These results misled many to make a correspondence between **E** and **H**; the results, however, are only valid for $\rho = 0$ and $\mathbf{J} = 0$.

3 Using the definition of the self-inductance of a circuit as the flux produced per unit current, calculate the self-inductance of the following circuits when a substance of permeability μ is the medium.

(*a*) An infinitely long, straight conductor in an infinite medium.
(*b*) A coaxial cable with the medium between the conductors.
(*c*) A coaxial cable with the medium outside the cable.
(*d*) A long solenoid of small radius a with the medium inside it.
(*e*) The same, but with the medium outside it.
(*f*) State the general rule.

4 In Prob. 1*a* account for the $(-\mu_0\mathbf{M})$ by finding $\mathbf{j}_m$. Repeat for 1*b*. The result for 1*c* lies intermediate between these two.

5 A long rod of length ℓ and small circular cross section is magnetized axially: $\mathbf{M} = \hat{\mathbf{z}}M$. Find **B** on the axis (*a*) at the end of the rod, (*b*) at the center of the rod.

6 Example 2, Sec. 5-3, of Chap. 5 gave a table summarizing the changes in the values of Q, C, $V = \phi_1 - \phi_2$, E, D when a dielectric with $\kappa = 2$ was inserted into a capacitor. Two possibilities were considered: (1) The capacitor was isolated from any battery; (2) it was connected to a battery. Prepare a similar table giving the new values of I, L, Φ, B, H (as multiples of the old ones) when a magnetic material having a relative permeability constant κ_m is inserted inside a long solenoid of small radius. Neglect end effects. Consider the two cases. (1) Isolated: the coil is a superconductor and requires no battery. When the material is inserted the flux remains constant but the current changes. (2) Nonisolated: the coil is connected to a battery which keeps the current constant after the transients attending the insertion have died away, but the flux changes.

7 Given a cylinder with $\mathbf{M} = \hat{\boldsymbol{\phi}}M_0(r/a)$, where M_0 is a constant and a is the radius. Find **B** and **H** in the cylinder.

8 The equation $\nabla \times \mathbf{H} = \mathbf{J}_c$ is equivalent to $\oint_C \mathbf{H} \cdot d\mathbf{r} = I_c$ where C is a circle of radius a and I_c is the conventional current passing through the circle. μ does not appear in this equation, and if the medium is symmetric about the center then, for a given I_c, **H** has a value **C** which is not only constant but is independent of the material. Suppose the medium is not symmetric about the origin: let the upper half circle have κ_{m1} and the lower half $\kappa_{m1} \neq \kappa_{m2}$. Is the value of **H** on C dependent on the material? Is $\oint_C \mathbf{H} \cdot d\mathbf{r}$ then dependent on the material?

9 Find **H** at a point P on the axis of a long solenoid with length ℓ and radius a. The distance from P to O, the center of the solenoid, is r, where $r \gg \ell$.

10 The vector potential of a small loop, a magnetic dipole, is

$$\mathbf{A} = \frac{\mu_0}{4\pi}\left[\frac{\mathbf{m} \times \hat{\mathbf{r}}}{r^2}\right]$$

then $\mathbf{B} = \nabla \times \mathbf{A}$. Show that one can obtain **B** also from a magnetic scalar potential ϕ_m, where $\mathbf{B} = \mu_0\mathbf{H}$, $\mathbf{H} = -\nabla\phi_m$, and

$$\phi_m = \frac{1}{4\pi}\left[\frac{\mathbf{m} \cdot \hat{\mathbf{r}}}{r^2}\right]$$

in analogy with electrostatics. (This scalar potential is multiple-valued in general. But a real circuit or loop is a dipole only far from the loop; and there ϕ_m is single-valued.)

11 Find **B** inside and outside a sphere of radius R having $\mathbf{M} = \hat{\mathbf{z}}M_0$.

12 Find **H** in Prob. 11.

13 Compare the answer to Prob. 9 with the **H** produced at P by a magnetic dipole at O to find the value of the dipole moment that would give the same **H** at very large distances. Beyond what value of r would the difference between the two formulas become less than 10 percent of the dipole formula?

14 A superconductor is an ideal conductor (its resistance to the flow of current is zero) which necessarily has the additional property that the magnetic field within it is zero. Find the susceptibility and the permeability μ for a superconductor when the magnetization is defined as (*a*) $\mathbf{M} = \chi'_m[(1/\mu_0)\mathbf{B}]$, as we did above and (*b*) $\mathbf{M} = \chi_m \mathbf{H}$, as it is usually done.

15 An idealized circular turn of wire has a radius of 4 cm to the axis of symmetry, the *z* axis. Find **H** at a point on this axis 3 cm from the plane of the ring if the current is 2 A.

16 Any vector may be written as the sum of a solenoidal vector and an irrotational vector; but sometimes one or the other is zero and the vector becomes a bit simpler to manipulate though not necessarily to understand. Given electrostatic and magnetostatic conditions, which of the vectors **E**, **D**, **B**, **H**, **P**, **M** simplifies to (*a*) a solenoidal vector, (*b*) an irrotational vector, (*c*) neither? This will give one an idea of which vectors to avoid, if possible.

7-4 INDUCED AND PERMANENT M

All atoms acquire an induced magnetic moment when they are placed in a magnetic field. This behavior is similar to that in electrostatics; the effect here, however, is very small. A second difference between the electric and magnetic cases is that the induced magnetic moment tends to line up antiparallel to the external field. This phenomenon is called diamagnetism, and a substance which has only this type of moment is called diamagnetic. Such a body has a negative magnetic susceptibility χ_m, and tends to be repelled from regions of higher **B** toward regions of lower **B**.

Some atoms or molecules have a permanent magnetic moment. When such an atom is placed in an external magnetic field it tends to align its magnetic moment parallel to the field. In the corresponding electric case the moment would actually execute a pendulum motion, back and forth about the field, were it not for collisions with other atoms. The collisions enable an atom to go into a lower energy state, where its moment is more nearly aligned with the direction of **E**. In the magnetic case the moments, instead of oscillating, would actually precess about the external field, except to the extent that collisions with other atoms permit them to lose energy. The explanation for the different behavior here lies in the angular momentum which is always a concomitant of the magnetic moment. (See Example 2, Sec. 7-2.) The phenomenon involving the permanent moments is called paramagnetism. χ_m here is positive. Its value falls in a broad range, from slightly larger than the induced χ_m of diamagnetism to two orders of magnitude larger than this. A substance which has an inherent (positive) χ_m that predominates over the induced (negative) χ_m is called paramagnetic. Such a body tends to be attracted to a region having a large field.

There are many other types of magnetic materials but by far the most important is ferromagnetism, the type of magnetism which first attracted the attention of man thousands of years ago. It is the only type whose effect is large enough to make it noticeable in our everyday lives. Ferromagnetism is found in cases where atoms with a permanent magnetic moment tend to align their moments parallel to each other, not just singly but in groups. In other words, it is a cooperative phenomenon. The aligning force here is not the magnetic dipole-dipole force between neighboring atoms but a very much stronger quantum-mechanical *electrical* force between them. We will discuss only the first two types, starting with the induced moment, since all the other types require a knowledge of quantum mechanics.

DIAMAGNETISM Consider an electron executing a circular trajectory in the xy plane about an atomic nucleus in a region originally having no external field. This is the Bohr semiclassical view. It is employed here, although it does not correspond to reality, because (1) it is simple and (2) a quantum-mechanical treatment yields the same answers. Next, let us establish a uniform magnetic field directed, say, along the $\hat{\mathbf{z}}$ axis. Anticipating a result from Chap. 11 to which we have referred perviously, we will assume that a changing magnetic field is accompanied by an induced electric field. Faraday's law gives the integral $\oint \mathbf{E} \cdot d\mathbf{r}$ of the electric field around the circular orbit here: it equals the negative rate of change of the magnetic flux through the orbit, $-(d\Phi/dt)$. Thus

$$2\pi r E = -\frac{d}{dt}(B\pi r^2)$$

In Example 1 a calculation is made to justify the assertion that r, the radius of the orbit, will remain constant. Accepting this initially on faith,

$$\mathbf{E} = -\hat{\boldsymbol{\phi}}\,\frac{r}{2}\frac{dB}{dt}$$

If the orbiting electron constitutes a $+\hat{\boldsymbol{\phi}}$ current and B is increasing in the $+\hat{\mathbf{z}}$ direction, as in Fig. 7-10, then $\mathbf{E}$ will be in the $-\hat{\boldsymbol{\phi}}$ direction, tending to decrease the current by slowing down the electron. The slowing-down force on the electron during the time that the field is being established is

$$\mathbf{F} = (-e)\mathbf{E} = \hat{\boldsymbol{\phi}}\,\frac{er}{2}\frac{dB}{dt}$$

So there is a torque about the circuit, $\mathbf{T} = \mathbf{r} \times \mathbf{F}$, about the origin:

$$\mathbf{T} = \hat{\mathbf{z}}\,\frac{er^2}{2}\frac{dB}{dt}$$

(The current in the circuit may be taken as continuous, equal in value to the average of the fluctuating current produced by the circulation of the point electron.) Corresponding to this torque there must be a rate of change of angular momentum:

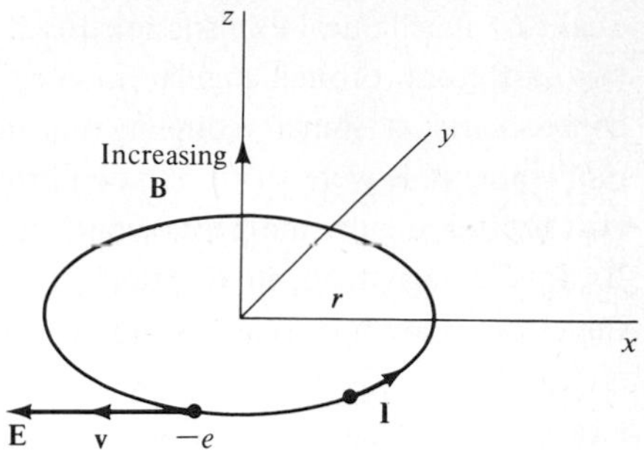

FIGURE 7-10

$$\frac{d\mathbf{L}}{dt} = \mathbf{T} = \hat{\mathbf{z}}\,\frac{er^2}{2}\,\frac{dB}{dt}$$

Since r is constant, this may be integrated, giving

$$\Delta\mathbf{L} = \hat{\mathbf{z}}\tfrac{1}{2}eBr^2$$

Time does not enter this expression so $\Delta\mathbf{L}$ is independent of the rate of change of **B** or of the duration of the transient period. It is only the final value of **B** that counts.

Introducing the gyromagnetic ratio of Example 2 in Sec. 7-2, $(-e/2m_e)$, the changing magnetic field causes a change in the magnetic moment of the circulating electron. This change of **m** is given by

$$\Delta\mathbf{m} = \left(\frac{e}{2m_e}\right)\Delta\mathbf{L} = -\,\hat{\mathbf{z}}\,\frac{e^2Br^2}{4m_e}$$

(Here, again, **m** is moment, m_e is mass, to avoid confusion.) The magnetic moment has thus been reduced, and this corresponds to a negative magnetic susceptibility χ_m. Typical values of χ_m for solids are: Cu, -0.94×10^{-5}; Ag, -2.6×10^{-5}; Sb, -7.0×10^{-5}. The effect is very small.

We have only considered the case where the plane of the orbit is perpendicular to the direction of **B**. If all directions are allowed then the formula for $\Delta\mathbf{m}$ is modified slightly: a different numerical factor appears and $\langle r^2 \rangle$, the average value of r^2, must be used. This applies to atoms in which the average magnetic moment for all the circulating electrons is zero.

Suppose we consider an atom in which a diamagnetic moment has been induced for an electron orbit in a plane not perpendicular to **B**. Classical physics would now have **B** act on this moment in a manner similar to the way it acts on a permanent magnetic dipole: **B** would tend to tip the moment into alignment with itself, but the **L** would cause precession to occur (see below). However, this is not what actually happens in nature. **B** creates the induced moment but does not try to tip it. This is inexplicable classically. The quantum-mechanical explanation for this behavior depends on the necessary coupling of two different orbital angular momenta into a single pair having oppositely directed components; a similar coupling occurs between two oppositely directed spin angular momenta. If **B** were to tip the two magnetic moments into alignment with itself then the two orbital angular momenta would be parallel to each other, in the direction opposite to **B**. So diamagnetism, in the final analysis, is basically a quantum-mechanical effect and the classical arguments above only tell half the story.

PARAMAGNETISM We now neglect all diamagnetic effects and turn to those atoms which possess a permanent magnetic moment **m**. As shown in Fig. 7-11*a*, **m** is in the opposite direction from the orbital angular momentum, alone, of the atom: $\mathbf{m} = g(-e/2m_e)\mathbf{L}$. A similar picture would hold if the spin, alone, were being considered. For an electron, the Landé *g*-factor is 2 for intrinsic angular momentum (spin) and 1 for orbital angular momentum. (For a proton the minus is replaced by a plus, so that **m** and **L** are parallel instead of antiparallel; and $g = 2 \times 2.79$ for spin angular momentum but $g = 1$ for orbital angular momentum. For a neutron the minus holds again, just as it does for the electron, even though the neutron has no charge, so that **m** is antiparallel to **L** for the neutron also. Here $g = -2 \times 1.93$ for spin, while $g = 1$ for orbital angular momentum.)

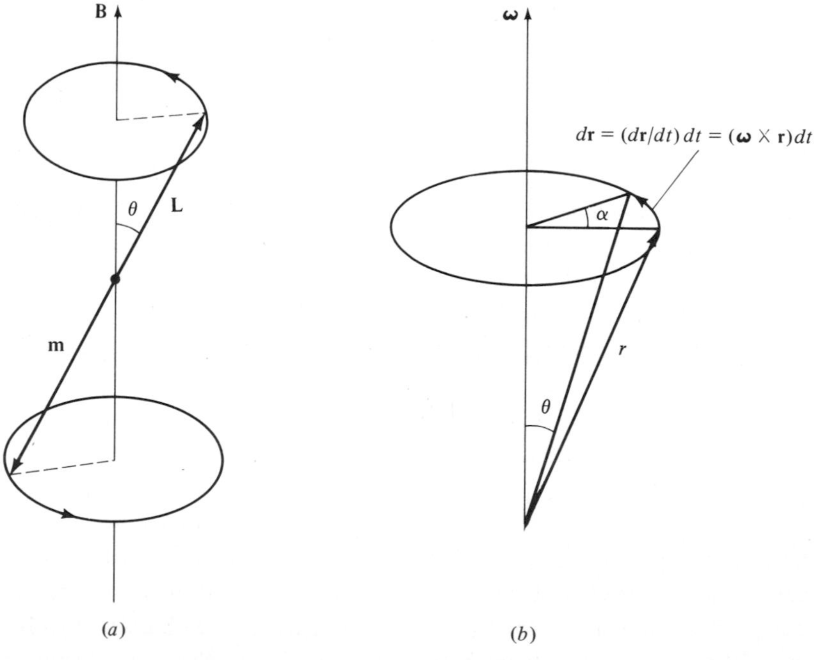

FIGURE 7-11

The torque applied to the moment by the field is $\mathbf{T} = \mathbf{m} \times \mathbf{B}$ (see Sec. 7-1, Prob. 5). But

$$\mathbf{T} = \frac{d\mathbf{L}}{dt} \quad \text{and} \quad \mathbf{m} = g\left(-\frac{e}{2m_e}\right)\mathbf{L}$$

for electrons (the effects that are due to the nuclei in atoms are very much smaller than this because of the greatly increased mass factor in the denominator). Then

$$\frac{d\mathbf{L}}{dt} = \left(\frac{-ge}{2m_e}\right)\mathbf{L} \times \mathbf{B} = \left(\frac{ge}{2m_e}\right)\mathbf{B} \times \mathbf{L}$$

The rate of change of $\mathbf{L}$ is perpendicular to $\mathbf{L}$, itself. To determine the significance of this, consider Fig. 7-11*b*, which shows a vector $\mathbf{r}$ precessing with angular velocity $\omega = d\alpha/dt$. Here $dr = \alpha(r \sin\theta)$ and $dr/dt = \omega r \sin\theta$. Checking the sense, we see that $d\mathbf{r}/dt = \boldsymbol{\omega} \times \mathbf{r}$. Comparing $\mathbf{r}$ here with $\mathbf{L}$ in the equation above,

$$\boldsymbol{\omega} = \frac{g}{2}\left(\frac{e}{m_e}\right)\mathbf{B}$$

The angular precession frequency becomes, in the case when $g = 1$, the Larmor frequency ω_L. This is connected with Larmor's theorem, which states that the final motion of any one of a system of electrons (without spin) after the introduction of a weak magnetic field is the original motion plus the angular precession frequency ω_L about the field axis.

If this atom were an isolated one, the magnetic moment $\mathbf{m}$ would never change the angular attitude it made with $\mathbf{B}$; $\mathbf{m}$ would continuously precess around $\mathbf{B}$ with the angular frequency $\boldsymbol{\omega}$. The magnetic field does no work in causing this precession and the energy of the atom would remain constant: $U = -\mathbf{m} \cdot \mathbf{B}$. The isolated atom cannot lose energy to the field. When the atom is not isolated, however, and collisions do occur with neighboring atoms (in gases; in liquids and in solids the interaction is more complicated, but the results are similar) then U can change. One atom gains energy at the expense of the other, and it becomes possible for θ to decrease. Incidentally, in the case of permanent electric dipole moments both atoms and molecules play a role, but in the magnetic case molecules rarely have a magnetic moment. This is so because of the coupling of the spins of the outer electrons into pairs with oppositely directed components. There are exceptions to this rule, however, such as free radicals, which have an odd number of valence electrons.

The relative number of atoms having a given orientation in thermal equilibrium is found by statistical mechanics, in exactly the same way as with permanent electric dipoles, to be proportional to $\exp(-U/kT)$. There is an apparent difference between the electric and magnetic cases, however, since the magnetic field cannot absorb any energy from the atoms while the electric field can (Prob. 16). It would seem, then, that while individual atoms might change their energies when colliding in the presence of a magnetic field the overall distribution would necessarily have the same relative numbers at various energies with or without the magnetic field. But this is not so if the collision rate is so high that numerous collisions can occur *during the time that the field is being applied.* The energy of the assembly can, and does, change because of the induced voltage while the field is being applied.

Examples

1. Constancy of the radius A magnetic field is created perpendicular to the circular orbit of a circulating electron in an atom. Does the radius stay constant?

Let the original speed be v, so the original centripetal force is $m_e v^2/r$. For a change in speed from v to $v + \Delta v$, there is an increase in the centripetal force by the amount

$$\frac{2m_e v\,\Delta v}{r} + \frac{m_e(\Delta v)^2}{r}$$

The original centripetal force is equal to the Coulomb attraction between the electron and the nucleus; the addition of the magnetic field creates the additional inward radial force. $|-e|(v + \Delta v)B = +eBv + eB\,\Delta v$. We now make use of the result obtained above, in this section, that $E = -(r/2)(dB/dt)$ gives the induced electric field which is a concomitant of a changing magnetic field; this assumes a constant radius. Substituting this in $F = (-e)E = m_e(dv/dt)$ gives $dv = +(e/m_e)(r/2)\,dB$ or $\Delta v = +(e/m_e)(r/2)B$. The first term of the additional centripetal force, $2m_e v\,\Delta v/r$, is seen to be equal to the first term of the additional magnetic force, eBv. The second term of the latter, $eB\,\Delta v$, however, is $e(2\,\Delta v m_e/er)\,\Delta v = [2m_e(\Delta v)^2]/r$. This is just twice the second term of the additional centrifugal force.

To the extent that $\Delta v \ll v$, therefore, the additional forces–centrifugal and magnetic–just match. For an extremely powerful magnetic field, however, the value of r would change.

2. Paramagnetic magnetization Assume that the energy of an atom is changed by the presence of a magnetic field only by the amounts of $\pm m_0 B$ from the original value. Quantum mechanically, $m_0 = g(-e/2m_e)(\hbar/2)$ is the m_{S_z} component of m_S in the field direction. $\hbar/2$ is the allowed magnitude of S_z for spin angular momentum. We will find the magnetization $\mathbf{M}$ of a collection of such atoms possessing the permanent magnetic moment $\mathbf{m}_S$.

The number of atoms per unit volume with spin angular momentum having a z component along the direction of $\mathbf{B}$ (or a magnetic moment component opposite to that of $\mathbf{B}$) will be called N_{par}. Note that the designation "parallel" is with respect to the spin and $\mathbf{B}$, not the moment and **B**. The magnetic energy is $U_{par} = -\mathbf{m}_S \cdot \mathbf{B} = -m_S B \cos\theta$. Setting $\theta = \pi - \theta_0, U_{par} = m_S B \cos\theta_0 = m_{S_z} B = m_0 B$. Statistically, then, $N = N_0 e^{-(U/kT)}$ gives the distribution of dipoles according to their potential energy, and $N_{par} = N_0 e^{-m_0 B/kT}$.

The number of atoms per unit volume with spin component opposite to **B** (**m** component parallel to **B**) will be called N_{anti}. Then $U_{anti} = -m_0 B$ and $N_{anti} = N_0 e^{m_0 B/kT}$. Here $m_0 = m_S \cos\theta_0$ with $\theta_0 = (\mathbf{m}_S, \mathbf{B})$. θ_0 and $\theta = \pi - \theta_0$ are the only two possibilities, quantum mechanically, for the attitude of $\mathbf{m}_S$ relative to **B**. If N is the number density of all the atoms then

$$N = N_0(e^{m_0 B/kT} + e^{-m_0 B/kT})$$

and

$$N_0 = \frac{N}{e^{m_0 B/kT} + e^{-m_0 B/kT}}$$

The N_{par} atoms per unit volume contribute a magnetic moment along $\hat{\mathbf{z}}$ of $N_{par}(-m_0)$. The N_{anti} atoms per unit volume contribute a magnetic moment along $\hat{\mathbf{z}}$ of $N_{anti}(+m_0)$. Together they contribute

$$-N_{par}(m_0) + N_{anti}(m_0) = m_0 N_0 (e^{m_0 B/kT} - e^{-m_0 B/kT})$$

and the average magnetic moment per atom along $+\hat{\mathbf{z}}$ is this expression divided by N:

$$\langle m_z \rangle = m_0 \frac{(e^{m_0 B/kT} - e^{-m_0 B/kT})}{(e^{m_0 B/kT} + e^{-m_0 B/kT})}$$

Then

$$M = N\langle m_z \rangle = N m_0 \tanh\left(\frac{m_0 B}{kT}\right)$$

This is the quantum-mechanical result of the calculation for the paramagnetic magnetization. It has the same general behavior as the Langevin equation, the classical result, but differs from it. At low temperatures $M \approx N m_0$; i.e., the dipoles are nearly all aligned. At room temperature and normally attainable fields, the argument of the tanh is very small, so

$$M = \frac{N m_0^2 B}{kT}$$

Except that there is no factor 3 in the denominator here, this is analogous to the result derived classically with electric moments. It is left to Prob. 2 to insert this factor.

PROBLEMS

1 In Example 1 above, the radius of the orbit was shown to stay constant upon application of a magnetic field if $\Delta v/v \ll 1$. Another criterion could be used: compare the applied external field to that created by the circulating electron.

(*a*) What is the magnitude of $\mathbf{B}_{\text{self}}$ at the center of the orbit of Prob. 1 in Sec. 7-2 produced by the circulating electron?

(*b*) What is $B_{\text{ext}}/B_{\text{self}}$ for an external magnetic field of 10,000 G?

(*c*) Would the radius stay constant, to within 1 percent, on application of this $\mathbf{B}_{\text{ext}}$?

2 The formula for paramagnetic magnetization at normal temperatures differs by a factor of 3 from the formula for polarization caused by permanent electric dipole moments. The quantum development for M utilizes an m_0 factor, while the classical derivation for P used the factor p. Show quantum mechanically that $m_0^2 = \frac{1}{3}m^2$, where m is the magnitude of $\mathbf{m}$. The electric and magnetic cases are, therefore, exactly similar.

3 (*a*) The formula for the paramagnetic magnetization gives what value for M when the temperature is very low?

(*b*) For a magnetic field of 10^4 G and room temperature, taking $m_0 = 10^{-23}$ m^2 A, what is $m_0 B/kT$?

(*c*) Suppose T is changed to 6 K. What is M?

4 In the derivation of the formula for diamagnetism an electron was considered orbiting in the $+\hat{\boldsymbol{\phi}}$ direction with $\mathbf{B}$ increasing toward $+\hat{\mathbf{z}}$. Consider the alternative possibilities (*a*) orbiting $+\hat{\boldsymbol{\phi}}$ but field increasing toward $-\hat{\mathbf{z}}$; (*b*) orbiting $-\hat{\boldsymbol{\phi}}$ with field increasing toward $+\hat{\mathbf{z}}$; (*c*) orbiting $-\hat{\boldsymbol{\phi}}$ with field increasing toward $-\hat{\mathbf{z}}$. Make a general statement covering all four possibilities.

5 The formula for the induced moment of an atom, considering only orbits perpendicular to the direction of the applied field, has the factor $r^2/4$ where $r^2 = x^2 + y^2$. What does this factor become when we consider $\langle r^2 \rangle$, the average value of $(x^2 + y^2 + z^2)$ for all possible orbit planes?

6 Neglect the fact that classically the equilibrium distribution of directions of permanent dipoles cannot be altered by a magnetic field, and proceed formally with the derivation based on $e^{-U/kT}$, where $U = -\mathbf{m} \cdot \mathbf{B}$. What is the resultant classical formula for M? The result, called the Langevin equation, is similar to the one obtained for electric dipoles.

7 Derive an expression for the diamagnetic susceptibility χ_m. Assume N atoms/m^3, each with Z electrons, so that $Z\langle r^2 \rangle = \Sigma_i \langle r_i^2 \rangle$. Take $\chi_m \ll 1$.

8 The answer to Prob. 3*a* depicts the condition of saturation. Draw a curve for M versus $m_0 B/kT$ and explain the physical meaning of the saturation value.

9 What is the precession frequency, in hertz, for an electron in a magnetic field of 1000 G for (*a*) orbital angular momentum?

(*b*) Spin angular momentum?

(*c*) What part of the electromagnetic frequency spectrum do these lie in?

10 The classical equilibrium distribution of permanent magnetic dipoles enables one to find the probability that the dipole makes an angle lying between θ and $\theta + d\theta$ with $\mathbf{B}$. Plot a curve of probability versus angle $(0 \leqslant \theta \leqslant \pi)$ for (*a*) $mB/kT = 0.02$, (*b*) $mB/kT = 20$.

11 When $mB/kT = 0.02$, how many times more likely is it that a dipole will make (*a*) an angle $\theta = 0°$ with $\mathbf{B}$ as it is that the angle is $\theta = 90°$?

(*b*) $\theta = 0°$ versus $\theta = 180°$?

(*c*) Is the variation linear with θ?

(*d*) Is the variation large or small?

12 Repeat Prob. 11*a* through *d* when $mB/kT = 20$.

(*e*) Where will the vast preponderance of the moments lie? Does this have a connection with saturation?

13 How large a value of **B** would be needed, in diamagnetism, to create a difference of 0.1 percent of the magnetic force between the two requirements on the circulating electron: magnetic force and centrifugal force? Use the conditions of Prob. 1, Sec. 7-2.

14 Write a formula for the paramagnetic susceptibility, χ_m, at room temperature.

15 (*a*) Using the results of Prob. 6, to what does the classical (Langevin) equation for **M** reduce, when $y = M/Nm$ and $x = mB/kT$, for $x \ll 1$?

(*b*) Using the results of Prob. 2, repeat (*a*) for the quantum-mechanical equation.

(*c*) Find y for $x = 2$ from the classical equation.

(*d*) Repeat for the quantum case.

(*e*) What are the asymptotic values for y in the two cases when $x \to \infty$?

16 From $\mathbf{F} = -e\mathbf{v} \times \mathbf{B}$, prove the work done by the magnetic field on an atom in a displacement ds is zero.

7-5 BOUNDARY CONDITIONS

At the interface between two different magnetic materials the value of the magnetic vector potential **A** must be single-valued. This restriction is similar to that imposed on ϕ, the electric scalar potential, and it follows from the same consideration—the connection with energy. For the magnetic energy of a circuit C carrying the current I in a region where the vector potential is **A** is given by $U = I\Phi = I\oint_C \mathbf{A} \cdot d\mathbf{r}$; to give a U that is unique, A should be taken as single-valued everywhere. In particular, **A** should be taken as single-valued on the interface between two different magnetic materials, so that U should have a unique value.

This restriction, however, is not absolutely necessary. Consider a closed surface. If **A** is double-valued on this surface in such a way that at any point the difference between the two values is a constant vector, both in magnitude and direction, then $\oint_C \mathbf{A} \cdot d\mathbf{r}$ about any path C will still be unique. A plate, magnetized parallel to its surface as in Fig. 7-12*a*, would behave this way in the limit of infinitesimal thickness, Fig. 7-12*b*. Instead of real magnetic poles, producing oppositely directed magnetization surface current densities on the two sides, the plate would become a disk, and would have point magnetic dipoles with two oppositely directed $\mathbf{j}_m$'s, of the same magnitude, on a single surface. This would give an infinite value for the curl of **A**, and for **B** also, within the infinitesimal thickness of the disk. This is an artificial situation, but it is not impossible in principle. In actual practice it is sufficient to consider **A** as single-valued.

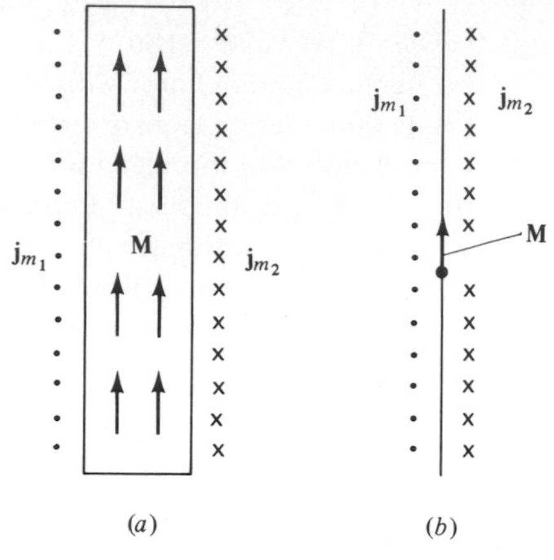

FIGURE 7-12

The boundary condition on **B** at an interface between two magnetic media is derived from the equation $\nabla \cdot \mathbf{B} = 0$. Applying this to a pillbox of infinitesimal height, partly in both media, leads at once to the relation

$$B_{1n} = B_{2n}$$

This is directly analogous to the boundary condition on **D** and tends to make the association (**B**,**D**); especially since it will be seen shortly that the boundary condition on **H** is similar to that on **E**, so that we tend to make the association (**H**,**E**). It is, therefore, necessary to stress a distinction to loosen this misleading analogy. $B_{1n} = B_{2n}$ always, since $\nabla \cdot \mathbf{B} = 0$ always; but $D_{1n} = D_{2n}$ only where there is no free charge on the interface, for $\nabla \cdot \mathbf{D} = \rho_f$. Where there is free charge on this surface then $D_{1n} \neq D_{2n}$.

The boundary condition on **H** at an interface derives from the equation $\nabla \times \mathbf{H} = \mathbf{J}_c$ when $\mathbf{J}_c$, the conventional current density, vanishes on the surface (using our nomenclature this should actually be $\mathbf{j}_c = 0$). Applying $\nabla \times \mathbf{H} = \mathbf{J}_c$ to a rectangular path with two infinitesimal sides and two parallel, finite sides—one in each medium—gives

$$H_{2t} - H_{1t} = j_c$$

where j_c is the component of $\mathbf{j}_c$ that is perpendicular to $\mathbf{H}_{1t}$ and $\mathbf{H}_{2t}$. When there is no conduction or convection surface current on the interface then

$$H_{1t} = H_{2t}$$

a similar boundary condition to that which applies for **E**. But when there is a finite $\mathbf{j}_c$ then the boundary condition for **H** is not at all the same as that for **E**.

The conditions on **D** and **B**, whether similar or not, turn out to be valid even when variations with time are permitted. Those on **E** will be valid, with time variations, only when there is no $\mathbf{j}_c$ on the surface; those on **H** will be valid then only when there is no ρ_f on the surface.

Examples

1. Tangential B **a.** What are the boundary conditions on the tangential component of **B**?

For the tangential component of **H** we have $\mathbf{H}_{2t} - \mathbf{H}_{1t} = \mathbf{j}_c$. The positive direction of these quantities is defined as in Fig. 7-13. Reversing the positive direction of $\mathbf{H}_{1t}$ and $\mathbf{H}_{2t}$ reverses the positive direction of conventional surface current density into the plane of the page. If there are components of $\mathbf{j}_c$ parallel to the page they do not affect the tangential components of **H** in the plane of the page; they would affect H_t components perpendicular to the plane of the page.

The equation $\mu_0 M_{2t} = \mu_0 \mathbf{M}_2 \times \hat{\mathbf{n}}_2$ can be added to μ_0 times the equation above. We can also add the equation $0 = \mu_0 M_{1t} - \mu_0 \mathbf{M}_1 \times \hat{\mathbf{n}}_1$. Then

$$\mu_0 H_{2t} + \mu_0 M_{2t} = \mu_0 H_{1t} + \mu_0 M_{1t} + \mu_0 j_c + \mu_0 \mathbf{M}_2 \times \hat{\mathbf{n}}_2 - \mu_0 \mathbf{M}_1 \times \hat{\mathbf{n}}_1$$

or

$$B_{2t} = B_{1t} + \mu_0 j_c + \mu_0 (j_{2m} - j_{1m})$$

The tangential component of **B** is discontinuous across the boundary surface by an amount proportional to the total surface current density (conventional and magnetization).

b. The boundary conditions on the normal and tangential components of **A**.

The boundary conditions on the components of **E**, **D**, **B**, and **H** are obtained from a knowledge of the divergence and curl of each. We have $\nabla \times \mathbf{A} = \mathbf{B}$ if we wish to find similar conditions on the components of **A**, but so far we do not know $\nabla \cdot \mathbf{A}$. We must first find $\nabla \cdot \mathbf{A}$; this result has self-importance, aside from the particular problem here.

For **A** in a medium we let $\mu_0 \to \mu$ in the equation for **A** in vacuum:

$$\mathbf{A} = \frac{\mu}{4\pi} \int_{\mathcal{V}} \frac{\mathbf{J}}{R}\, d\tau_s$$

Then

$$\nabla \cdot \mathbf{A}(\mathbf{r}_t) = \frac{\mu}{4\pi} \int_{\mathcal{V}} \nabla \cdot \left[\frac{\mathbf{J}(\mathbf{r}_s)}{R}\right] d\tau_s$$

Using

$$\nabla \cdot (f\mathbf{F}) = f\, \nabla \cdot \mathbf{F} + \mathbf{F} \cdot (\nabla f)$$

this becomes

$$\nabla \cdot \mathbf{A}(\mathbf{r}_t) = \frac{\mu}{4\pi} \int_{\mathcal{V}} \frac{1}{R}\, \nabla \cdot \mathbf{J}(\mathbf{r}_s)\, d\tau_s + \frac{\mu}{4\pi} \int_{\mathcal{V}} \mathbf{J} \cdot \nabla \left(\frac{1}{R}\right) d\tau_s$$

The first integrand vanishes since $\mathbf{J} \neq \mathbf{J}(x_t, y_t, z_t)$; in the second we substitute $\nabla(1/R) = -\nabla_s(1/R)$, where ∇_s is obtained by varying the source point rather than the test point. So

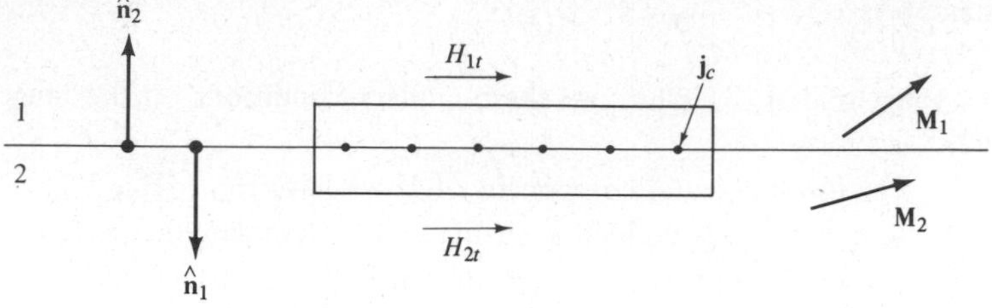

FIGURE 7-13

$$\nabla \cdot \mathbf{A}(\mathbf{r}_t) = -\frac{\mu}{4\pi}\int_{\mathcal{V}} \mathbf{J} \cdot \nabla_s\left(\frac{1}{R}\right) d\tau_s$$

Now we go round the mulberry bush, using the vector identity above in the reverse sense:

$$\nabla \cdot \mathbf{A}(\mathbf{r}_t) = \frac{\mu}{4\pi}\int_{\mathcal{V}} \frac{1}{R}\nabla_s \cdot \mathbf{J}(\mathbf{r}_s) d\tau_s - \frac{\mu}{4\pi}\int_{\mathcal{V}} \nabla_s \cdot \left[\frac{\mathbf{J}(\mathbf{r}_s)}{R}\right] d\tau_s$$

Since $\nabla_s \times \mathbf{H} = \mathbf{J}(\mathbf{r}_s)$ at any source point and $\nabla_s \cdot (\nabla_s \times \mathbf{H}) = 0$ identically, $\nabla_s \cdot \mathbf{J}(\mathbf{r}_s) = 0$ also. Applying Gauss' divergence theorem to the second integral,

$$\nabla \cdot \mathbf{A}(\mathbf{r}_t) = -\frac{\mu}{4\pi}\iint_{\Sigma} \left[\frac{\mathbf{J}(\mathbf{r}_s)}{R}\right] \cdot d\mathbf{S}$$

In the steady state, if Σ includes all the **J**'s, there can be no normal component of **J** on Σ: **J** is either zero or tangential there. In either case, the integral vanishes:

$$\nabla \cdot \mathbf{A} = 0$$

For the vector potential in magnetostatics we now have

$$\begin{cases} \nabla \cdot \mathbf{A} = 0 \\ \nabla \times \mathbf{A} = \mathbf{B} \end{cases}$$

The first equation, while valid here, will not be true in general when variations with time are permitted. The second equation is usually written the other way around; it is written this way here to emphasize the Helmholtz theorem aspect; **A** is now uniquely determined.

Turning to the boundary conditions on **A**, the first equation is true only in the media on either side of an interface but also on the boundary surface itself, since

$(\mathbf{M} \times \hat{\mathbf{n}}) \cdot \hat{\mathbf{n}} = 0$. Considering a differential pill box partly in each of the two media gives $A_{1n} = A_{2n}$: the normal component of **A** is continuous.

Integrating the second equation over a differential rectangle partly in each of the two media and using Stokes' theorem for the left side gives $(A_{2t} - A_{1t})L$, where L is the length of the rectangle; the right side becomes zero because **B** stays finite at the interface, even if there is a $\mathbf{j}_c$ there. So $A_{1t} = A_{2t}$: the tangential component of **A** is also continuous.

Consequently, **A** has no discontinuity in either component. Note the exception previously mentioned: a constant vector discontinuity, everywhere on a closed surface, is permissible.

PROBLEMS

1 Find the boundary conditions on the normal component of **H**. Take all components as positive when parallel to $\hat{\mathbf{n}}_1$.

2 Can the normal component of **H** change its sense in going from one medium to another? The normal component of **B**? The tangential component of **H**? The tangential component of **B**?

3 The value of the magnetic field in vacuum is 1000 G in the $\hat{\mathbf{x}}$ direction. The flux enters a substance, having $\chi_m = 10^{-5}$, at a plane face making an angle of 45° with the horizontal (southwest-northeast). Find the normal and tangential components of $\mathbf{B}_1$, $\mathbf{H}_1$, $\mathbf{B}_2$, and $\mathbf{H}_2$ in mksa units.

4 Find the answers to Prob. 3 with **B** in gauss and **H** in oersteds.

5 **B** and **H** in the second medium of Prob. 3 are bent from the horizontal. Is this upward or downward? Find the angle of deviation in degrees; in minutes of arc; in seconds of arc.

6 Suppose χ_m' in Prob. 3 was -10^{-5} and the second medium was diamagnetic instead of paramagnetic. Find the normal and tangential components of **B** and **H** in the second medium then. Also find the angle of deviation and the direction.

7 The extremely small angles of deflection which are characteristic of diamagnetism and paramagnetism make it easy to calculate the following. In Prob. 3 let $\chi_m' = +10^{-3}$, an increase by a factor of 100. Find the angle of deviation.

8 Repeat Prob. 7 for the diamagnetic case $\chi_m' = -10^{-3}$.

9 Using $\mathbf{M} = \chi_m'[(1/\mu_0)\mathbf{B}]$, let $\chi_m' = -9$. To what value of χ_m does this correspond, using the usual definition $\mathbf{M} = \chi_m \mathbf{H}$? Find the normal and tangential components of **B** and **H** in Prob. 3 with this χ_m'. What is μ in the diamagnetic medium?

10 In the case of a superconductor the increase in $|\chi_m'|$ from Prob. 6 to Prob. 9 is continued to its limit: set $\mu = 0$. Find the normal and tangential components of **B** and **H** for Prob. 3 in this case.

11 Going in the other direction, take $\mathbf{M} = \chi_m \mathbf{H}$ with $\chi_m = +99$. What would be the value of χ_m' corresponding to this? Find the normal and tangential components for **B** and **H** in this case, which would correspond to iron.

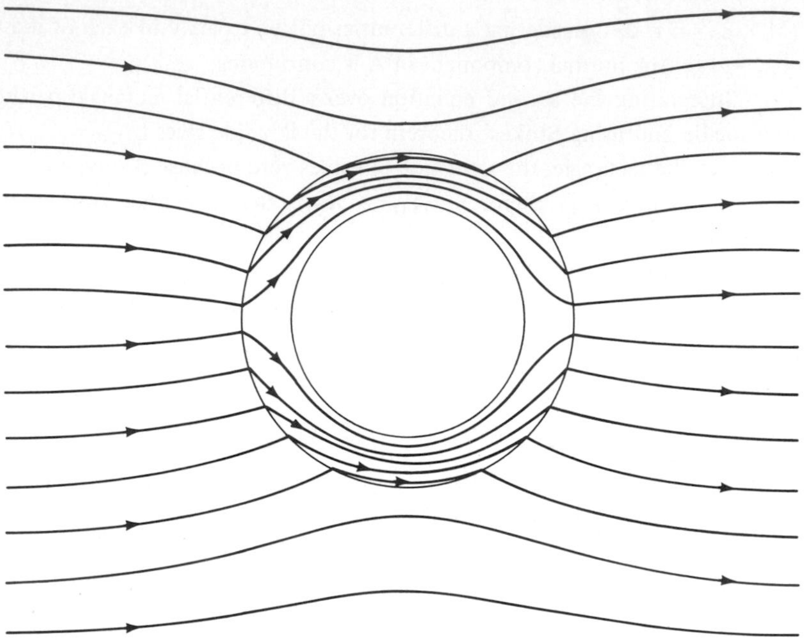

FIGURE 7-14
(*From J. D. Jackson, "Classical Electrodynamics," Wiley, New York, 1962.*)

12 Compare the results of Probs. 10 and 11, generalizing the latter to the case $\chi_m = +\infty$, to make a statement about the direction of the lines of **B** and **H** for the extreme values of susceptibility.

13 If χ_m is positive and very large but finite, is it possible for this material to act as a magnetostatic shield? Perfectly? Compare with electrostatics where a metal can act as a perfect shield even though its conductivity is finite though large. Figure 7-14 shows the shielding action where χ_m is very large but finite.

14 Make a diagram of the cross section of an interface between two magnetic materials having a perpendicular surface current. Draw lines of **B** for this current. Add them to the continuous lines of **B** produced by external means to illustrate why the total tangential **B** is discontinuous.

15 A very small, wedge-shaped gap of mean length L is made in a torus, of mean radius R, having a circular cross section with radius $a \ll L$. Take **B** to be uniform in the torus, with negligible fringing near the edges of the gap, and let there be N turns of wire around the torus with current I. Find **B**, $\mathbf{H}_m$ in the material, and **H** in the air gap. What is the direction of $\mathbf{H}_m$ relative to **H**?

16 Show that there must be fringing in the lines of **B** or **H** near the edges of the gap in Prob. 15 for any finite gap but not for an infinitesimal gap. Though actually present, the fringing may be neglected if $L \ll a$.

17 In Prob. 15 let $\mu = 100\,\mu_0$, $L = 1$ mm, and $R = 10$ cm. Using Ampere's law, find the portion of the ampere-turns that are used in getting B across the air gap and the portion of NI used in getting B through the material.

18 Let $\mu = 1000\,\mu_0$ in Prob. 17. What is the percentage of NI used in getting flux across the gap then?

Appendix 7

A. DERIVATION OF THE EXPRESSION FOR THE DIPOLE TERM IN THE MULTIPOLE EXPANSION FOR A

Using the vector identity $\mathbf{A} \times (\mathbf{B} \times \mathbf{C}) = \mathbf{B}(\mathbf{A} \cdot \mathbf{C}) - \mathbf{C}(\mathbf{A} \cdot \mathbf{B})$ with $\mathbf{B} \to \mathbf{J}$, $\mathbf{A} \to \mathbf{r}$, $\mathbf{C} \to \mathbf{r}_s$, the dipole term becomes, using the symbol (II) for simplicity,

$$(\mathrm{II}) \equiv \frac{\mu_0}{4\pi r^2} \int_{\mathcal{V}} \hat{\mathbf{r}} \times (\mathbf{J} \times \mathbf{r}_s)\, d\tau_s + \frac{\mu_0}{4\pi r^2} \int_{\mathcal{V}} \mathbf{r}_s(\hat{\mathbf{r}} \cdot \mathbf{J})\, d\tau_s$$

$$\equiv (\mathrm{II}_A) + (\mathrm{II}_B)$$

where (II_A) is the first term and (II_B) is the second term.

Consider the x component of the last term: $(\mathrm{II}_B)_x$

$$(\mathrm{II}_B)_x = \frac{\mu_0}{4\pi r^2} \int_{\mathcal{V}} x_s \left(\frac{xJ_x + yJ_y + zJ_z}{r} \right) d\tau_s$$

$$= \frac{\mu_0}{4\pi r^3} \left[x \int_{\mathcal{V}} x_s J_x\, d\tau_s + y \int_{\mathcal{V}} x_s J_y\, d\tau_s + z \int_{\mathcal{V}} x_s J_z\, d\tau_s \right]$$

In order to simplify this it is, unfortunately, first necessary to complicate it, using a vector identity. Thus

$$\nabla_s \cdot (x_s J) = x_s(\nabla_s \cdot \mathbf{J}) + \mathbf{J} \cdot (\nabla_s x_s) = 0 + (\mathbf{J} \cdot \hat{\mathbf{x}}) = J_x$$

Similarly,

$$J_y = \nabla_s \cdot (y_s \mathbf{J}) \quad \text{and} \quad \mathbf{J}_z = \nabla_s \cdot (z_s \mathbf{J})$$

So

$$(\text{II}_B)_x = \frac{\mu_0}{4\pi r^3}\left[x \int_{\mathcal{V}} x_s \nabla_s \cdot (x_s \mathbf{J})\, d\tau_s + y \int_{\mathcal{V}} x_s \nabla_s \cdot (y_s \mathbf{J})\, d\tau_s + z \int_{\mathcal{V}} x_s \nabla_s \cdot (z_s \mathbf{J})\, d\tau_s\right]$$

$$\equiv (\text{II}_B)_{x_1} + (\text{II}_B)_{x_2} + (\text{II}_B)_{x_3}$$

Take the first term, $(\text{II}_B)_{x_1}$.

$$(\text{II}_B)_{x_1} = \frac{\mu_0}{4\pi r^3}\, x \int_{\mathcal{V}} x_s \left[\frac{\partial}{\partial x_s}(x_s J_x) + \frac{\partial}{\partial y_s}(x_s J_y) + \frac{\partial}{\partial z_s}(x_s J_z)\right] d\tau_s$$

Designating the first of these three components by $(\text{II}_B)_{x_{1a}}$ and writing out the triple integral:

$$(\text{II}_B)_{x_{1a}} = \frac{\mu_0}{4\pi r^3}\, x \int_{\mathcal{V}} dz_s \int_{\mathcal{V}} dy_s \int_{\mathcal{V}} x_s \frac{\partial}{\partial x_s}(x_s J_x)\, dx_s$$

This may be integrated by parts:

$$(\text{II}_B)_{x_{1a}} = \frac{\mu_0}{4\pi r^3}\, x \int_{\mathcal{V}} dz_s \int_{\mathcal{V}} dy_s \left\{ x_s(x_s J_x) \Big|_{x_{\min}}^{x_{\max}} - \int_{\mathcal{V}} x_s J_x \frac{\partial}{\partial x_s}(x_s)\, dx_s \right\}$$

$J_x = 0$ at the boundaries of $\mathcal{V}$, so the integrated term vanishes and

$$(\text{II}_B)_{x_{1a}} = -\frac{\mu_0 x}{4\pi r^3} \int_{\mathcal{V}} x_s J_x\, d\tau_s$$

Next we proceed in a similar fashion with $(\text{II}_B)_{x_{1b}}$, the second of the three components. The integration by parts involves a term which vanishes at the boundaries, plus the integral of $(x_s J_y)(\partial/\partial y_s)(x_s)$; so this component vanishes. The third of the three components, $(\text{II}_B)_{x_{1c}}$, is also equal to zero. So $(\text{II}_B)_{x_1}$ is the same as $(\text{II}_B)_{x_{1a}}$:

$$(\text{II}_B)_{x_1} = -\frac{\mu_0 x}{4\pi r^3} \int_{\mathcal{V}} x_s J_x\, d\tau_s$$

The second and third terms of the expression for $(\mathrm{II}_B)_x$ are found in analogous fashion:

$$(\mathrm{II}_B)_{x_2} = -\frac{\mu_0 y}{4\pi r^3} \int_{\mathcal{V}} y_s J_x \, d\tau_s$$

$$(\mathrm{II}_B)_{x_3} = -\frac{\mu_0 z}{4\pi r^3} \int_{\mathcal{V}} z_s J_x \, d\tau_s$$

The equation for $(\mathrm{II}_B)_x$ is then

$$(\mathrm{II}_B)_x = -\frac{\mu_0}{4\pi r^3}\left[x \int_{\mathcal{V}} x_s J_x \, d\tau_s + y \int_{\mathcal{V}} y_s J_x \, d\tau_s + z \int_{\mathcal{V}} z_s J_x \, d\tau_s\right]$$

So the result for (II_B), itself, is

$$(\mathrm{II}_B) = -\frac{\mu_0}{4\pi r^3} \int_{\mathcal{V}} (\mathbf{r} \cdot \mathbf{r}_s)\mathbf{J} \, d\tau_s$$

$$= -\frac{\mu_0}{4\pi r^2} \int_{\mathcal{V}} (\hat{\mathbf{r}} \cdot \mathbf{r}_s)\mathbf{J} \, d\tau_s$$

But this is the negative of the complete second term II of the multipole expansion:

$$(\mathrm{II}) = (\mathrm{II}_A) + (\mathrm{II}_B) = (\mathrm{II}_A) - (\mathrm{II})$$

Then

$$(\mathrm{II}) = \tfrac{1}{2}(\mathrm{II}_A)$$

Therefore the second term of the multipole expansion for **A**, the dipole term we have designated by (II), is

$$(\mathrm{II}) = \frac{\mu_0}{8\pi r^2} \int_{\mathcal{V}} \hat{\mathbf{r}} \times (\mathbf{J} \times \mathbf{r}_s) d\tau_s$$

$$= \frac{\mu_0}{4\pi}\left\{\frac{1}{r^2}\left[\int_{\mathcal{V}} \left(\tfrac{1}{2}\mathbf{r}_s \times \mathbf{J}\right) d\tau_s\right] \times \hat{\mathbf{r}}\right\}$$

B. COMPARISON OF THE EQUATIONS OF ELECTROSTATICS AND MAGNETOSTATICS

Electrostatics	Magnetostatics
(E-1) $\mathbf{F}_{st} = \dfrac{1}{4\pi\epsilon_0} \dfrac{q_s q_t}{R^2} \hat{\mathbf{R}}_{st}$	(M-1) $\mathbf{F}_{ST} = -\dfrac{\mu_0}{4\pi} \int_T \int_S \dfrac{(I_S\, d\mathbf{r}_s) \cdot (I_T d\mathbf{r}_t)}{R^2} \hat{\mathbf{R}}_{st}$
(E-2) $\mathbf{F} = q\mathbf{E}$	(M-2) $\mathbf{F} = \oint_T I_T\, d\mathbf{r}_t \times \mathbf{B}$
(E-3) $\mathbf{E}(\mathbf{r}_t) = \dfrac{1}{4\pi\epsilon_0} \dfrac{q_s}{R^2} \hat{\mathbf{R}}_{st}$	(M-3) $\mathbf{B}(\mathbf{r}_t) = \dfrac{\mu_0}{4\pi} \oint_S \dfrac{I_S d\mathbf{r}_s \times \hat{\mathbf{R}}_{st}}{R^2}$
(E-4) $\mathbf{E}(\mathbf{r}_t) = \dfrac{1}{4\pi\epsilon_0} \int_{\mathcal{V}} \dfrac{\rho(\mathbf{r}_s)}{R^2} \hat{\mathbf{R}}_{st}\, d\tau_s$	(M-4) $\mathbf{B}(\mathbf{r}_t) = \dfrac{\mu_0}{4\pi} \int_{\mathcal{V}} \dfrac{\mathbf{J}(\mathbf{r}_s) \times \hat{\mathbf{R}}_{st}}{R^2} d\tau_s$
(E-5) $\oint_\Sigma \mathbf{E} \cdot d\mathbf{S} = \dfrac{1}{\epsilon_0} Q$	(M-5) $\oint_T \mathbf{B} \cdot d\mathbf{r} = \mu_0 I$
(E-6) $\nabla \cdot \mathbf{E} = \dfrac{1}{\epsilon_0} \rho$	(M-6) $\nabla \times \mathbf{B} = \mu_0 \mathbf{J}$
(E-7) $\nabla \times \mathbf{E} = 0$	(M-7) $\nabla \cdot \mathbf{B} = 0$
(E-8) $\mathbf{E} = -\nabla \phi$	(M-8) $\mathbf{B} = \nabla \times \mathbf{A}$
(E-9) $\phi(\mathbf{r}_t) = \dfrac{1}{4\pi\epsilon_0} \int_{\mathcal{V}} \dfrac{\rho(\mathbf{r}_s)}{R} d\tau_s$	(M-9) $\mathbf{A}(\mathbf{r}_t) = \dfrac{\mu_0}{4\pi} \int_{\mathcal{V}} \dfrac{\mathbf{J}(\mathbf{r}_s)}{R} d\tau_s$
(E-10) $q = C\, \Delta\phi = CV$	(M-10) $\Phi = MI$
(E-11) $U_e = \dfrac{1}{2} \int_{\mathcal{V}} \rho\phi\, d\tau$	(M-11) $U_m = \dfrac{1}{2} \int_{\mathcal{V}} \mathbf{J} \cdot \mathbf{A}\, d\tau$
(E-12) $U_e = \dfrac{\epsilon_0}{2} \int E^2\, d\tau$	(M-12) $U_m = \dfrac{1}{2\mu_0} \int_{\mathcal{V}} B^2\, d\tau$

(E-13) $\frac{1}{\epsilon_0}\{T\} =$

$$\begin{Bmatrix} \frac{E_x^2 - E_y^2 - E_z^2}{2} & E_xE_y & E_xE_z \\ E_yE_x & \frac{-E_x^2 + E_y^2 - E_z^2}{2} & E_yE_z \\ E_zE_x & E_zE_y & \frac{-E_x^2 - E_y^2 + E_z^2}{2} \end{Bmatrix}$$

(E-14) $\phi = \frac{1}{4\pi\epsilon_0}\frac{\mathbf{p}\cdot\hat{\mathbf{r}}}{r^2}$

(E-15) $U_e = -\mathbf{p}\cdot\mathbf{E}$

(E-16) $\sigma_p = P_n \qquad \rho_p = -\nabla\cdot\mathbf{P}$

(E-17) $\nabla\cdot\mathbf{E} = \frac{1}{\epsilon_0}(\rho_f + \rho_p)$

(E-18) $\mathbf{D} = \epsilon_0\mathbf{E} + \mathbf{P}$

(E-19) $\nabla\cdot\mathbf{D} = \rho_f$

(E-20) $\mathbf{P} = \chi(\epsilon_0\mathbf{E})$

(E-21) $\mathbf{D} = \epsilon\mathbf{E}$

(M-13) $\mu_0\{T\} =$

$$\begin{Bmatrix} \frac{B_x^2 - B_y^2 - B_z^2}{2} & B_xB_y & B_xB_z \\ B_yB_x & \frac{-B_x^2 + B_y^2 - B_z^2}{2} & B_yB_z \\ B_zB_x & B_zB_y & \frac{-B_x^2 - B_y^2 + B_z^2}{2} \end{Bmatrix}$$

(M-14) $\mathbf{A} = \frac{\mu_0}{4\pi}\frac{\mathbf{m}\times\hat{\mathbf{r}}}{r^2}$

(M-15) $U_m = -\mathbf{m}\cdot\mathbf{B}$

(M-16) $\mathbf{j}_m = \mathbf{M}\times\hat{\mathbf{n}} \qquad \mathbf{J}_m = \nabla\times\mathbf{M}$

(M-17) $\nabla\times\mathbf{B} = \mu_0(\mathbf{J}_c + \mathbf{J}_m)$

(M-18) $\mathbf{H} = \frac{1}{\mu_0}\mathbf{B} - \mathbf{M}$

(M-19) $\nabla\times\mathbf{H} = \mathbf{J}_c$

(M-20) $\mathbf{M} = \chi_m'\left(\frac{1}{\mu_0}\mathbf{B}\right) = \chi_m\mathbf{H}$

(M-21) $\mathbf{H} = \frac{1}{\mu}\mathbf{B}$

8

SPECIAL SOLUTIONS IN ELECTROSTATICS

8-1 THE METHOD OF IMAGES

In Chap. 2 we drew a number of conclusions concerning the behavior of metals in electrostatics. These results were obtained by the application of Gauss' law. However, the connection between metals and electric fields, or between metals and electric currents, is so important from a practical standpoint that it is easy to justify several chapters devoted to this subject. It would be nice, in fact, if we could follow the pattern we have established above: a chapter, say, on special solutions for the electric potential ϕ in a region having metal boundaries; followed by a chapter on special solutions for the vector potential **A** in such a region; but in that second chapter the pattern would be found to cause difficulties. This is true despite the fact that both ϕ and **A** obey Laplace's equation, $\nabla^2 \phi = 0$ and $\nabla^2 \mathbf{A} = 0$, respectively, in a region where $\rho = 0$ and $\mathbf{J} = 0$.

The reason the equation is applicable to electrostatics is that the metal boundaries to a region are readily held at constant potentials by connecting them to batteries. One then has to solve a classical boundary-value problem: the solution, in some region, to a differential equation having given values on the boundaries of the region. But in magnetostatics it is not easy—in fact, it is very difficult—to give the boundaries a fixed value of **A**. The currents do not flow on the surfaces only, except in the case of superconductors. (The charges do reside only on the surfaces in electrostatics.) But even if we only consider superconductors there are difficulties: $\nabla^2 \mathbf{A} = 0$ is a vector equation. It can only be broken up into three independent one-dimensional equations in a cartesian coordinate system; otherwise each component equation contains some of the other variables. Unless the boundaries naturally fit a cartesian system, then, the solutions are bound to be extremely complicated.

For these reasons, therefore, it is best to halt the comparison between the two disciplines at this point. Only this one chapter is devoted to metals. The three sections are concerned with several different types of electrostatic solutions to Laplace's equation, $\nabla^2 \phi = 0$, subject to certain boundary conditions.

The electric potential ϕ is known as soon as the charge density is specified:

$$\phi = \frac{1}{4\pi\epsilon_0} \int_{\mathcal{V}} \frac{\rho}{R}\, d\tau$$

Given $\rho(\mathbf{r}_s)$ we can find $\phi(\mathbf{r}_t)$ by a triple integration and, in principle, the problem is solved. Unfortunately, ρ is scarcely ever known. The measurement of ρ would be extremely difficult, and, even if ρ could be determined, its mere specification would be very tedious. But if the boundaries of the region are conductors it is usually not too difficult to determine their potential differences with respect to some reference ground. So it becomes desirable to have ways of finding $\phi(\mathbf{r}_t)$ if ϕ on the boundaries is given. A number of different types of attack can be made on this problem. One method is very general but requires the aid of a computer; other methods are rather specialized but are easy to apply. The first method we will consider is the method of images.

Suppose we consider the case of two equal but opposite charges shown in Fig. 8-1*a*. The potential at P is

$$\phi = \frac{q}{4\pi\epsilon_0}\left(\frac{1}{r_+} - \frac{1}{r_-}\right)$$

The diagram shows some of the equipotentials, which are actually surfaces of revolution obtained by rotating the figure about the $-x,x$ axis. The equipotential which passes through O, midway between the charges, is a plane and has the potential zero by symmetry. No work is required to bring a test charge from infinity, where the potential is zero, to O or any other point on the 0-V equipotential; if the path is along the y axis, for example, the force on the test particle is always in the $-x$ direction at each point in the path.

Suppose a metal plate of negligible thickness is placed in the $x = 0$ plane and connected to ground–its potential difference with respect to the reference is made to be zero volts. Then nothing will be altered. What happens in the region $x > 0$ if we now

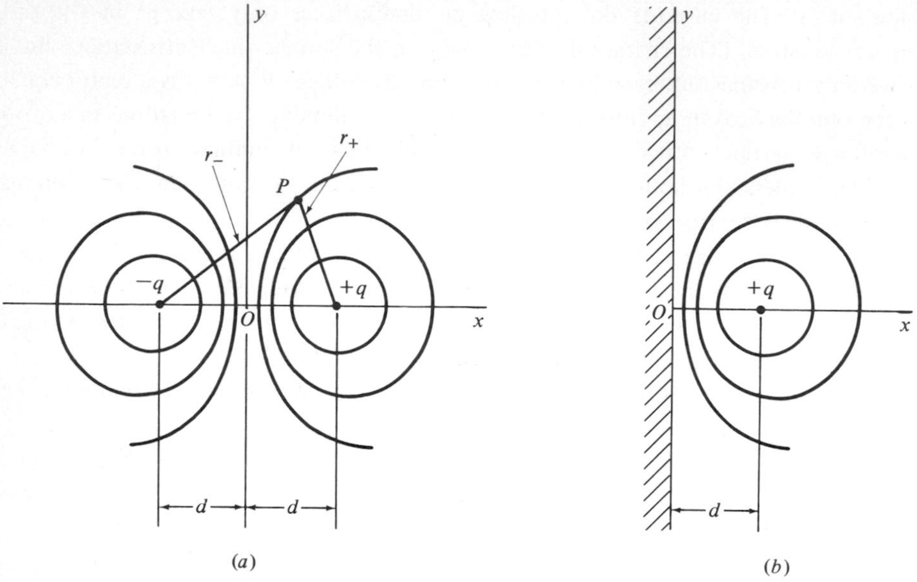

FIGURE 8-1

remove the charge $-q$ altogether but leave the metal plate, grounded, at $x = 0$? Again, nothing. The potentials at all the points in any region (here, $x > 0$) are determined by the charges within that region [$+q$ at $(d,0,0)$] and the potentials at all points on the boundary (the plane plus the hemisphere $r = \infty$). There is a theorem, which we will not prove, to the effect that the potentials are then unique. The point here is that it does not matter whether $\phi = 0$ along the $x = 0$ plane is obtained by utilizing a charge $-q$ at $(-d,0,0)$ or by using a grounded metal plane.

Now look at the problem backward: given the geometry of Fig. 8-1b, what are the equipotentials in the region $x > 0$? The situation may be made even more general: the region $x < 0$ can be all metal, or part metal, or have any behavior whatever, just so long as $x = 0$ is kept at potential 0 V. A charge $+q$ is placed near this metal; what is the potential distribution? The answer, good only in the region $x > 0$, is: the distribution is the same as that obtained with two charges alone, and no metal.

This is, possibly, the simplest example of the method of images. Knowing the potential produced by a given charge distribution, we seek some geometry of boundary metals that will give the same potentials within the region. This is analogous to finding a medicine which we suspect is a cure; all we need to know is the specific disease which it cures. In both cases a similar procedure would be followed in practice: we would carefully file away the knowledge of the potentials (cure) until one day we find this is just what is needed for the given boundary conditions (disease). So utilization of this method involves mostly art, luck, and memory.

Examples

1. Point charge and metal plate Find the charge density induced on an infinite metal plate by a charge q distant d from the plate.

Our file on "images" reveals that two charges, $-q$ at $x = -d$ and $+q$ at $x = d$, give the value $\phi = 0$ for $x = 0$. So does the metal sheet and the charge q at d. We therefore substitute the former for the latter.

In the case of the two charges the electric field at the $x = 0$ plane is in the $-x$ direction.

$$(E_x)_{x=0} = -\left(\frac{\partial \phi}{\partial x}\right)_{x=0} = \frac{-q}{4\pi\epsilon_0}\frac{\partial}{\partial x}\Big\{[(x-d)^2 + y^2 + z^2]^{-1/2} - [(x+d)^2 + y^2 + z^2]^{-1/2}\Big\}_{x=0}$$

$$= \frac{q}{4\pi\epsilon_0}\left\{\frac{x-d}{[(x-d)^2 + y^2 + z^2]^{3/2}} - \frac{x+d}{[(x+d)^2 + y^2 + z^2]^{3/2}}\right\}_{x=0}$$

$$= \frac{-qd}{2\pi\epsilon_0 r^3}$$

where $r = (d^2 + y^2 + z^2)^{1/2}$ is the distance from a point in the plane $x = 0$ to either charge. Then D_x, in the case of the two charges, is $-qd/2\pi r^3$ on the $x = 0$ plane. So D_x has this value also in the case of one charge and the metal plate. But $D_x = D_n$ in the latter case: it is normal to the plane $x = 0$. On the surface of a metal D_n equals σ, the surface charge density; so

$$\sigma = -\left(\frac{qd}{2\pi}\right)\frac{1}{r^3}$$

The negative charge on the metal plate is said to have been induced by the charge q. The plate is charged by induction. Electrons are attracted to the metal plate from the ground source by the Coulomb force of the $+q$ charge at $(d,0,0)$. If, with the charge $+q$ at $(d,0,0)$, the electrical connection between the plate and ground is broken then the originally neutral plate will be found to be charged. The total induced charge, by integration, is $-q$.

2. Point charge and metal sphere A point charge $+q$ is located a distance d from the surface of a grounded metal sphere of radius a. Find the potential at an arbitrary point outside the sphere. See Fig. 8-2a.

We again refer to the file on solutions for ϕ, to find that configuration of charges which gives a spherical equipotential surface with $\phi = 0$. Two equal but opposite charges do not have this property; the only spherical equipotential there has infinite radius. It

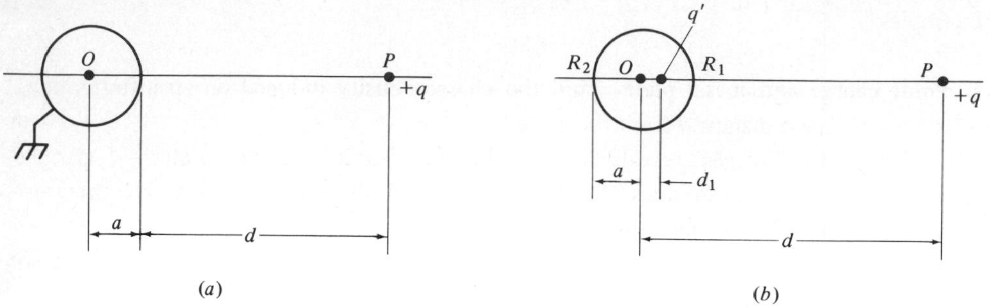

FIGURE 8-2

turns out, however, that in this case of particularly high symmetry we can figure the answer out even without referring to the file. As shown in Fig. 8-2*b*, we replace the grounded sphere with a charge q' at a point, distant d_1 from O, along the line OP; and we demand that the sphere of radius a have $\phi = 0$. There are two unknowns, q' and d_1; two easy conditions are present to enable us to find them: $\phi(R_1) = 0$ and $\phi(R_2) = 0$. Thus,

$$\text{at } R_1: \quad \frac{1}{4\pi\epsilon_0}\left(\frac{q}{d-a} + \frac{q'}{a-d_1}\right) = 0$$

$$\text{and at } R_2: \quad \frac{1}{4\pi\epsilon_0}\left(\frac{q}{a+d} + \frac{q'}{a+d_1}\right) = 0$$

These two linear simultaneous equations have the solutions

$$q' = -\left(\frac{a}{d}\right)q$$

$$d_1 = \frac{a^2}{d}$$

The image charge has the sign opposite to that of the source charge q. It is located at a point, $d_1 = a^2/d_2$, which is called the inversion point of P in the sphere.

We are almost done, although all that has been shown so far is that the potential at the points R_1 and R_2 is zero when the grounded sphere is replaced by the image charge $-(a/d)q$ at $d_1 = a^2/d$. But q and q' give equipotentials which are all circles, and the circle through R_1 and R_2 is such that all points on it have $\phi = 0$. We leave for Prob. 2 an explicit demonstration that this is true.

PROBLEMS

1 Let a point charge be brought near a conducting sphere as in Fig. 8-2*a* but let there be no connection from the sphere to ground, and assume the sphere is uncharged. Find the configuration of image changes which replaces the sphere.

2 (*a*) Show explicitly, in Example 2, that the potential at an arbitrary point on the surface of the sphere $r = a$ is zero.

(*b*) Show that the $\phi = 0$ equipotential surface produced by two unequal charges is a sphere.

3 Change Prob. 1 so that the sphere, while ungrounded, is charged with a total charge Q. What is the image charge configuration?

4 What is the total force exerted by $+q$ on the metal plane at $x = 0$ in Example 1? What is the force exerted by the metal plane on q? What is the force exerted by the image charge q' on q?

5 In Example 1, what is the induced charge contained in an annular ring of radius s and width ds centered at O on the plate? At what normalized distance, $R \equiv s/d$, is this a maximum?

6 A charge $+q$ is distant d from the surface of an ungrounded metal sphere of radius a. The sphere is uncharged, but q exerts a force on the sphere nevertheless. Explain this. Also, calculate the force.

7 A charge is located near two grounded metal planes, intersecting at right angles, as shown in Fig. 8-3. Find the configuration of image charges which, together with $+q$ but without the metal plates, gives the potential distribution in the first quadrant of Fig. 8-3.

8 In Example 2 find the surface charge density induced on the sphere. What is the total charge induced on the sphere?

9 Replace the point charge q of Fig. 8-1*b* by a metal sphere of radius $a < d$ having an unknown charge. The potential distribution can be found by an infinite sequence of images. To start, we replace the sphere by a point charge q, to retrieve the situation of Fig. 8-1*b* again. This makes the sphere an equipotential, but not the plane.

(*a*) Find q', the image of this charge in the plane. This makes the plane an equipotential but makes the sphere nonequipotential.

(*b*) By inversion, find q'', the image of q' within the sphere. This makes the sphere equipotential but undoes the plane.

(*c*) Find q''', the image of q'' in the plane. This fixes the plane but spoils the sphere. Find q', q'', q'''.

10 In Prob. 9 find (*a*) $q + q''$, the first two terms of the infinite series giving the total charge within the sphere, i.e., on the actual sphere; (*b*) the potential of the sphere, i.e., the potential difference between it and a point at infinity; (*c*) the capacitance between the sphere and the plane, to the accuracy of the first two terms of an infinite series.

11 The combination of one infinitely long filament at $(d/2,0)$ having λ C m^{-1} and another, parallel to it, at $(-d/2,0)$ with $-\lambda$ C m^{-1} produces the potential

$$\phi = (\lambda/2\pi\epsilon_0) \ln (r_-/r_+)$$

at a point, where r_- and r_+ are the distances to the negative and positive filaments from the point in question. The z axis through the origin has $\phi = 0$. Find the equations of the equipotential surfaces. These are cylinders of circular cross section with radius r, centered at $(a,0)$. Here

$$r = \frac{(d)r_+r_-}{|r_+^2 - r_-^2|}$$

while

$$a = \frac{d}{2}\left(\frac{r_-^2 + r_+^2}{r_-^2 - r_+^2}\right)$$

12 Two metal plates intersect at an angle of 120° and $+q$ is located d units from the line of intersection on the bisector of the wedge. Find the images.

13 Two infinitely long, parallel cylinders of circular cross section each have a radius r. The center-to-center spacing is D. From the results of Prob. 11 calculate d, the spacing between two equivalent filaments (one inside each cylinder) such that if they were given equal and opposite linear charge densities the surfaces of the cylinders would be equipotentials; i.e., find d in terms of D and r. Also, find r_-/r_+ in terms of D and r.

14 Repeat Prob. 12 for 72° and 60°. Generalize to $\theta = 360°/n$, where n is an integer. What if n is not an integer? What if the source charge is not on the bisector of the wedge?

15 In Prob. 13, find the per-unit-length capacitance of the parallel pair, from $C_\ell = \lambda/\Delta\phi$, in terms of D and r.

16 Interpret Fig. 8-2 as a cross section of an infinitely long circular cylinder and line charge, with $q \to \lambda$ C m^{-1}. Find λ' and d_1, the image charge density and inversion line in the cylinder.

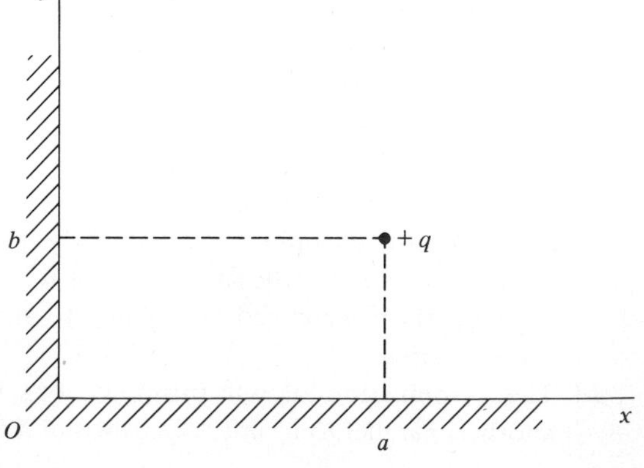

FIGURE 8-3

17 In Prob. 4, calculate the work required to carry the charge from its initial point to infinity.

18 Is the force on the charge in Fig. 8-3 directed toward O? What is its magnitude and direction?

8-2 METHOD OF COMPLEX VARIABLES

The potential ϕ satisfies $E = -\nabla\phi$; and $\mathbf{E}$ satisfies $\nabla \cdot \mathbf{E} = (1/\epsilon_0)\rho$. So the following differential equation is obtained:

$$\nabla^2\phi = -(1/\epsilon_0)\rho$$

This is called Poisson's equation. Its solution in any given case depends on the boundary conditions. The various surfaces enclosing the region of interest $\mathcal{V}$ have certain specified potentials. Often Laplace's equation, the simpler homogeneous equation obtained by setting $\rho = 0$, is the equation of interest. The latter equation arises in many branches of physics and has been the object of much investigation. Solutions to this boundary-value problem in terms of elementary functions have only been obtained in a few cases with a high degree of symmetry; generally it is necessary to resort to computers, either digital or analog, to obtain answers. Then, of course, the solution is an approximation.

In this section we discuss a method for the solution of Laplace's equation which is applicable when the boundaries of $\mathcal{V}$ are conductors which vary in potential only in two dimensions. The third dimension, generally taken along the z axis, extends from $-\infty$ to $+\infty$ and does not enter the problem. The method is based on a property possessed by complex functions of a complex variable.

Consider the complex variable $z = x + iy$; x and iy are the real and imaginary coordinates of a point in the complex xy plane, while y, itself, is real. z is the complex coordinate of the point; it has no connection with the usual z, the third dimension that has no bearing here. At each point of this complex z plane it is possible to define a field, $w(x,y)$. If w is complex then it, too, can be expressed in terms of a real part u, and an imaginary part v: $w(x,y) = u(x,y) + iv(x,y)$. In Chap. 1, Sec. 1-8, we showed that if dw/dz is to have a unique meaning (in which case w is said to be analytic) it is both necessary and sufficient that the Cauchy-Riemann equations be true:

$$\frac{\partial u}{\partial x} = \frac{\partial v}{\partial y} \qquad \frac{\partial u}{\partial y} = -\frac{\partial v}{\partial x}$$

By taking $\partial/\partial x$ of the first and $\partial/\partial y$ of the second it follows that $\nabla^2 u = 0$. Similarly, $\nabla^2 v = 0$. Since the real and imaginary components of w individually satisfy Laplace's equation, w also satisfies Laplace's equation. Further, we saw in Chap. 1 that the two families of curves, u = const and v= const, were orthogonal to each other.

Suppose, now, that we are given the problem of determining a potential function $\phi(x,y)$ in some region $\mathcal{V}$ such that ϕ assumes certain values on the conducting boundaries of $\mathcal{V}$. If we can find a complex function w whose real part u assumes the given values on the boundaries then u gives ϕ in $\mathcal{V}$; v then represents the lines of **E**, the electric field. If we can find a w whose imaginary part v assumes the given values on the boundaries then v gives ϕ; in this case u gives the lines of **E**. So, like the method of images, this method requires a catalog of solutions–in this case two-dimensional–to unknown problems.

Examples

1. Vicinity of crossed metal plates Figure 8-4*a* shows a line charge, λ C m^{-1}, at (a,b) in the first quadrant. The walls of the first quadrant are made of metal and have the potential zero. Taking the region $\mathcal{V}$ to be the first quadrant, find ϕ at any point in $\mathcal{V}$ by the method of complex variables.

Going to the catalog of complex functions we find one that proves to be useful: $w = z^2$. For $w = (x + iy)^2$ or $w = (x^2 - y^2) + i(2xy)$, so $u = x^2 - y^2$, $v = 2xy$. Figure 8-4, similar to Fig. 1-26, shows that the imaginary part of $w = z^2$ assumes the value zero along the boundary xz and yz planes. Therefore $kv = 2kxy$ can represent the function ϕ, where k is an undetermined constant.

The curves of Fig. 8-4*c* are the solutions of $\nabla^2\phi = 0$ subject to the boundary conditions $\phi = 0$ for $x = 0$ or $y = 0$; they are not the solutions we want–those of $\nabla^2\phi = -(1/\epsilon_0)\rho$, where $\rho = \infty$ at (a,b) and $\rho = 0$ elsewhere, with $\int_{\mathcal{V}} \rho\, d\tau = q$. But if the line charge is taken far away, i.e., if $x \ll a$ and $y \ll b$, the curves of Fig. 8-4*c* will represent approximations to the potential. For any given position of the line charge these curves give ϕ in a region sufficiently close to the edge of intersection of the two planes; the solution shown is not valid throughout all $\mathcal{V}$.

To find the value of the constant k we go back to the method of images. Then

$$\phi = \frac{\lambda}{4\pi\epsilon_0}\left[\ln\left(\frac{r_0^2}{r_1^2}\right) - \ln\left(\frac{r_0^2}{r_2^2}\right) + \ln\left(\frac{r_0^2}{r_3^2}\right) - \ln\left(\frac{r_0^2}{r_4^2}\right)\right]$$

where

$$r_1^2 = (a-x)^2 + (b-y)^2 \qquad r_2^2 = (a+x)^2 + (b-y)^2$$
$$r_3^2 = (a+x)^2 + (b+y)^2 \qquad r_4^2 = (a-x)^2 + (b+y)^2 \qquad r_0^2 = a^2 + b^2$$

We leave it to Prob. 2 to expand this expression. For example,

$$\ln\left(\frac{r_0^2}{r_1^2}\right) = \ln\frac{a^2+b^2}{(a^2+b^2)\left\{1 - \dfrac{1}{a^2+b^2}[2ax + 2by - x^2 - y^2]\right\}}$$

$$\approx \ln \left\{ 1 + \frac{1}{a^2 + b^2} [2ax + 2by - x^2 - y^2] \right\}$$

When $f \ll 1$, $\ln(1 + f) \approx f + \frac{1}{2} f^2$. The result, here, is

$$\phi = \left(\frac{2\lambda}{\pi\epsilon_0}\right) \frac{abxy}{(a^2 + b^2)^2}$$

Then

$$k = \left(\frac{\lambda}{\pi\epsilon_0}\right) \frac{ab}{(a^2 + b^2)^2}$$

and

$$w = kz^2 = \left(\frac{\lambda}{\pi\epsilon_0}\right) \frac{ab}{(a^2 + b^2)^2} z^2$$

The curves of Fig. 8-4*b*,

$$k(x^2 - y^2) = \left(\frac{\lambda}{\pi\epsilon_0}\right) \frac{ab}{(a^2 + b^2)^2} (x^2 - y^2)$$

represent the lines of electric field in a region close enough to the walls and far enough from the line charge.

2. Two adjacent metal cylinders Two metal cylinders are separated by an infinitesimal gap so that, while each is an equipotential surface, the two potentials are not necessarily the same. We wish to find formulas for the potential and field in the neighborhood of the cylinders.

We must look for a complex function, $w = f(z) = u + iv$, such that $u = K_1$ is a circle (in the $z = 0$ plane) and $u = K_2$ is another circle tangent to the first; it could also be v that behaves this way. A function that possesses this property is

$$w = \frac{1}{z}$$

Then

$$\frac{1}{z} = \frac{1}{x + iy} = \left(\frac{x}{x^2 + y^2}\right) + i\left(\frac{-y}{x^2 + y^2}\right)$$

so, using the cylindrical r,

$$u = \frac{x}{x^2 + y^2} = \frac{\cos\phi}{r} \qquad v = \frac{-y}{x^2 + y^2} = \frac{-\sin\phi}{r}$$

Consider $u = \text{const} = k_1$. Then $x/(x^2 + y^2) = k_1$ gives

$$\left(x - \frac{1}{2k_1}\right)^2 + y^2 = \left(\frac{1}{2k_1}\right)^2$$

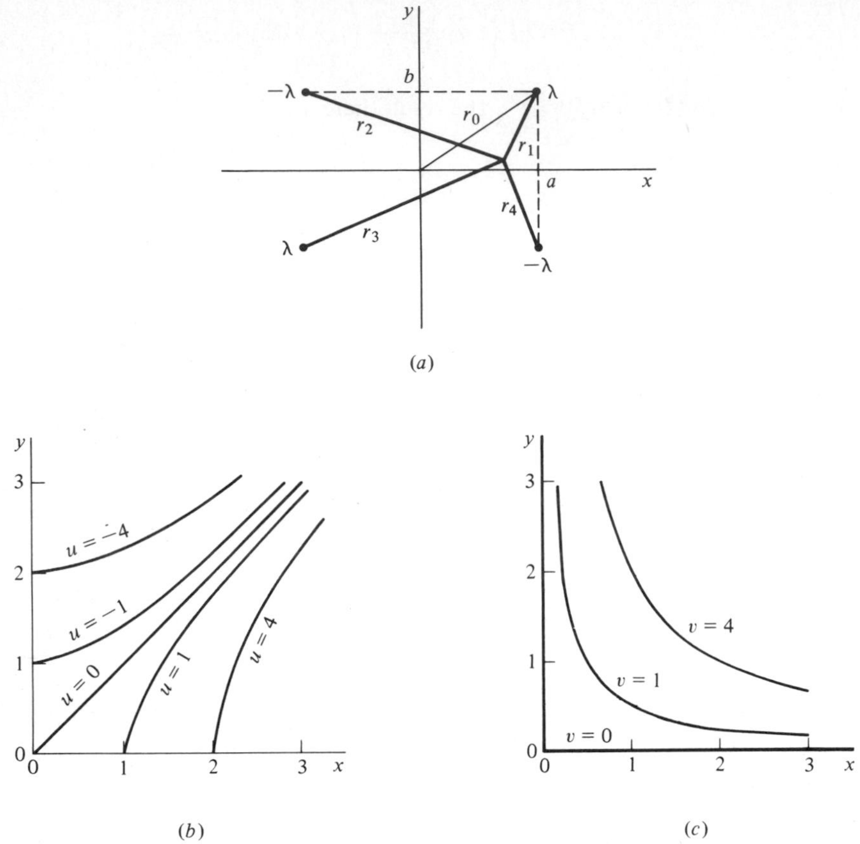

FIGURE 8-4

This represents a family of circles, all tangent to the y axis, with centers on the x axis. If $k_1 > 0$ the circle is to the right, if $k_1 < 0$ it is to the left as shown in Fig. 8-5*a*.

Similarly, Fig. 8-5*b* shows several curves for constant v: $-(\sin\phi)/r = k_2$. These are circles, all tangent to the x axis, with centers on the y axis.

Suppose we take u as the equipotentials. There are three possibilities, with the following representative cases:

1 The cylinder shown for $u = -1$, together with that for $u = +3$

2 The cylinder shown for $u = -2$, together with the cylinder (of infinite radius, i.e., the $x = 0$ plane) for $u = 0$.

3 The cylinder shown for $u = 3$ with that for $u = 1$. The curves $v = \text{const}$, in the region between the cylinders in each case, give the lines of the electric field. If the

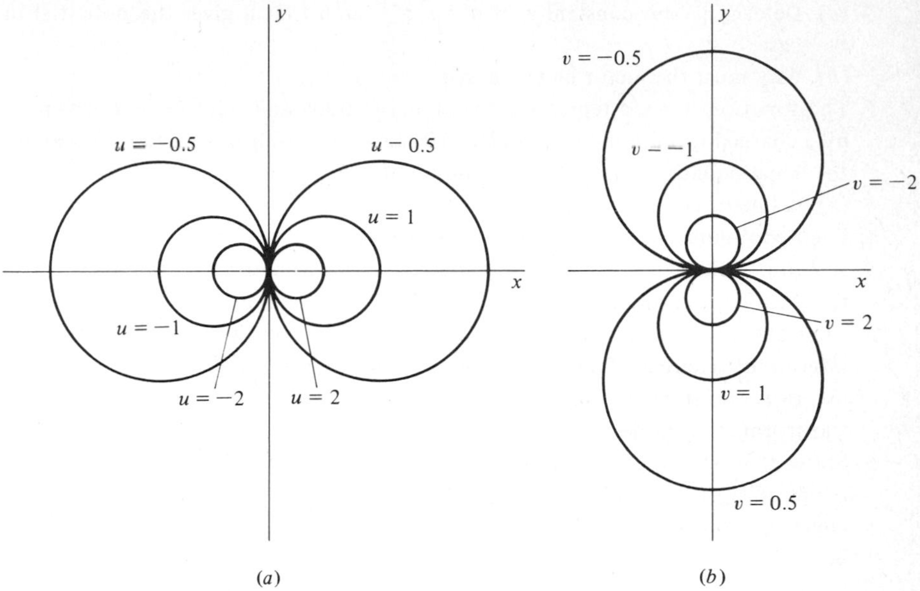

FIGURE 8-5

potentials of the two cylinders are the same there must be a parallel line charge, far away, to produce a nontrivial solution; this method is then approximate.

PROBLEMS

1 In Example 1 find the induced linear charge density, on the surface of either plane, at a distance r from the origin.

2 Complete the expansion of ϕ in Example 1 to derive the result

$$\phi = \frac{2\lambda}{\pi\epsilon_0}\left[\frac{ab}{(a^2+b^2)^2}\right]xy$$

3 The $x = 0$ plane is at ground potential. A cylinder of radius 10 cm, with its axis at $x = 10$ cm, has a potential of +100 V. Find the electric field along the top of the cylinder.

4 The curves of Fig. 8-5*a* can also represent the equipotential surfaces of a linear array of point dipoles, spread with uniform density along an axis perpendicular to the paper through the origin. The dipole moments would all be directed along the x axis, giving a dipole moment per unit length $\mathbf{p}_\ell = \hat{\mathbf{x}}p_\ell$.

(*a*) Determine the constant a in $w = a^2 z^{-1}$ such that u gives the potential in that case.

(*b*) Why must the factor be taken as a^2, i.e., positive?

5 The function $w = \sqrt{z}$ represents the equipotentials and electric field lines produced by a charged infinite metal half-plane ($0 \leqslant x \leqslant \infty$, $y = 0$) having zero potential.

(*a*) What equation represents an equipotential?

(*b*) A line of electric field?

6 What configuration is represented by $w = \ln z$?

7 In Prob. 5 there is a surface linear charge density λ induced on the plane. How does this vary with x, the distance from the edge?

8 Consider the transformation $w = a^2/z$ in polar coordinates. To what does the interior of a circle of radius a, center at the origin, transform? To what does the exterior transform? Show that a circle of radius a with center at $z = b \neq 0$ transforms to a circle.

9 Show that $w = z^{2/3}$ represents the equipotentials and field lines near the outside of an edge of a metallic quadrant. (The metal would be on the inside of the right angle and the field would be on the outside; instead of, as in Fig. 8-4, the metal being on the outside.)

10 Two metal planes intersect at an angle α, the edge being perpendicular to the paper through the origin. A parallel line charge, λ C m^{-1}, lies in the wedge of angle α far from the origin. Find the function w which gives the equipotentials for its real part u. This is superior to the method of images in that it works for any angle; but it is inferior in that the line charge must be far away.

11 Show that the transformation $w = i(a - z)/(a + z)$ maps the inside of the circle $x^2 + y^2 = a^2$ onto an infinite upper half-plane.

12 In Prob. 11 take the following points in the z plane and find the points into which they map in the w plane: (0,0) and $re^{i\phi}$ with $r = 1$ and $\phi = 0, \pi/2, \pi, 3\pi/2$, respectively. What is the pattern?

13 How does the induced surface linear charge density vary with R_1, the distance from the edge in Prob. 9?

14 In Prob. 6 find z as a function of w and express x and y as functions of u and v. With an orthogonal set of (u,v) axes, what does a line of constant u map into in the z plane? What becomes of a line of constant v? What is the domain of variation of u? Of v?

15 Given

$$\cos\theta = \tfrac{1}{2}(\epsilon^{i\theta} + \epsilon^{-i\theta}) \qquad \sin\theta = \frac{1}{2i}(\epsilon^{i\theta} - \epsilon^{-i\theta})$$

$$\cosh\theta = \tfrac{1}{2}(\epsilon^{\theta} + \epsilon^{-\theta}) \qquad \sinh\theta = \tfrac{1}{2}(\epsilon^{\theta} - \epsilon^{-\theta})$$

$$\sin(\alpha + \beta) = \sin\alpha\cos\beta + \cos\beta\sin\alpha$$

If $w = \sin z$, find the expression and nature of (*a*) the curves, in the w plane, of constant x; (*b*) the curves of constant y.

8-3 METHOD OF FINITE DIFFERENCES

Laplace's equation for the potential plays a central role in the solution of electrostatic problems. The two methods discussed in the previous sections represent special attacks of limited applicability. There are a number of other such methods. The presence of one or more of them in all textbooks devoted to electrostatics is testimony to the importance of the concept of potential, though many authors contend these methods are only a step on the path to finding **E**. Instead of pursuing these methods further, we will next consider an approximation method employing the modern techniques of the digital computer. This method is universally applicable to any configuration. The method of finite differences, also called the iteration method, has been known for a long time but it was used with reluctance in the past because of the great computational labor involved to obtain any appreciable accuracy. With the present widespread knowledge of digital programming this method promises to become more popular.

The basis of this treatment consists of the observation, made previously in the last section of Chap. 1, that the laplacian of the potential at a point is approximately proportional to the difference between (1) the average potential at six equidistant neighboring points and (2) the potential at the point in question:

$$(\nabla^2 \phi)_0 \approx \frac{6}{\ell^2} (\bar{\phi} - \phi_0)$$

where ℓ is the distance from O to each of the six points. The accuracy of the approximation increases as ℓ decreases, for the method of finite differences substitutes ratios of finite quantities for derivatives. In a region of space without charge $\nabla^2 \phi = 0$ and we have, simply, that ϕ_0, the potential at O, equals the average potential of other points equally spaced from O: $\bar{\phi} = \phi_0$. Example 1 will show how this is worked out in a systematic fashion.

Examples

1. ϕ at the points of intersection of the horizontal and vertical lines of Fig. 8-6 It is assumed that (1) very small insulating gaps separate the top plate from the two side plates and (2) that no variation occurs in the direction perpendicular to this cross section. Take $\phi_c = \frac{1}{2}(0 + 100.0) = 50.0$ for the potential at a gap. By symmetry about the vertical line through A, the values obtained on the right-hand side will be identical with those on the left side. Using the diagonals through A, we first calculate the potential at A from the average of the potentials at the four corners of the boundary: $\phi_A = \frac{1}{4}(50 + 50 + 0 + 0) = 25.0$.

FIRST RUN Using the diagonals, we now calculate the potential at B from the average of the potentials at C, A_4, A, and A_1:

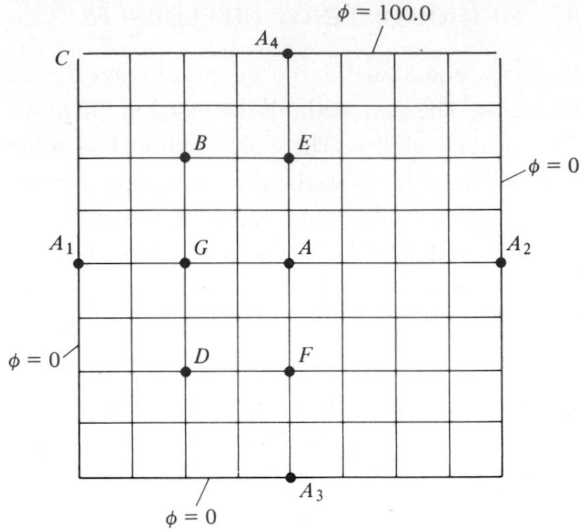

FIGURE 8-6

$$\phi_B = \tfrac{1}{4}(50.0 + 100.0 + 25.0 + 0) \qquad = 43.8$$

Again on diagonals, we next find the potential at D, in a similar manner, from A_1, A, A_3, and the bottom left corner:

$$\phi_D = \tfrac{1}{4}(0 + 25.0 + 0 + 0) \qquad = \ 6.2$$

Continuing, horizontally and vertically, we find ϕ_E as the average of the potentials at B, A_4, the point on the right symmetric to B, and A:

$$\phi_E = \tfrac{1}{4}(43.8 + 100.0 + 43.8 + 25.0 \quad = 53.2$$

Similarly for the potentials at F and G:

$$\phi_F = \tfrac{1}{4}(6.2 + 25.0 + 6.2 + 0) \qquad = \ 9.4$$

$$\phi_G = \tfrac{1}{4}(0 + 43.8 + 25.0 + 6.2) \qquad = 18.8$$

SECOND RUN We now repeat the procedure, using points to the left and right or above and below the actual point we are interested in. Thus, for ϕ_B we would find the average of the potentials at the points spaced two boxes apart to the left, above, to the right, and below point B, always using the last previous values found for the potentials at the various points.

$$\phi_B = \tfrac{1}{4}(0 + 100.0 + 53.2 + 18.8) \qquad = 43.0$$

This is a change from the previous value

Similarly, for ϕ_E we use the potentials at B, A_4, the point two boxes to the right of E, and A.

$\phi_E = \frac{1}{4}(43.0 + 100.0 + 43.0 + 25.0)$	$= 52.8$	change
$\phi_G = \frac{1}{4}(0 + 43.0 + 25.0 + 6.2)$	$= 18.6$	change
$\phi_A = \frac{1}{4}(18.6 + 52.8 + 18.6 + 9.4)$	$= 24.8$	change
$\phi_D = \frac{1}{4}(0 + 18.6 + 9.4 + 0)$	$= 7.0$	change
$\phi_F = \frac{1}{4}(7.0 + 24.8 + 7.0 + 0)$	$= 9.7$	change
		6 changes

THIRD RUN

$\phi_B = \frac{1}{4}(0 + 100.0 + 52.8 + 18.6)$	$= 42.8$	change
$\phi_E = \frac{1}{4}(42.8 + 100.0 + 42.8 + 24.8)$	$= 52.6$	change
$\phi_G = \frac{1}{4}(0 + 42.8 + 24.8 + 7.0)$	$= 18.6$	. . . . .
$\phi_A = \frac{1}{4}(18.6 + 52.6 + 18.6 + 9.7)$	$= 24.9$	change
$\phi_D = \frac{1}{4}(0 + 18.6 + 9.7 + 0)$	$= 7.1$	change
$\phi_F = \frac{1}{4}(7.1 + 24.9 + 7.1 + 0)$	$= 9.8$	change
		5 changes

FOURTH RUN

$\phi_B = \frac{1}{4}(0 + 100.0 + 52.6 + 18.6)$	$= 42.8$	. . . . .
$\phi_E = \frac{1}{4}(42.8 + 100.0 + 42.8 + 24.9)$	$= 52.6$	. . . . .
$\phi_G = \frac{1}{4}(0 + 42.8 + 24.9 + 7.1)$	$= 18.7$	change
$\phi_A = \frac{1}{4}(18.7 + 52.6 + 18.7 + 9.8)$	$= 25.0$	change
$\phi_D = \frac{1}{4}(0 + 18.7 + 9.8 + 0)$	$= 7.1$	. . . . .
$\phi_F = \frac{1}{4}(7.1 + 25.0 + 7.1 + 0)$	$= 9.8$	. . . . .
		2 changes

FIFTH RUN

$\phi_B = \frac{1}{4}(0 + 100.0 + 52.6 + 18.7)$	$= 42.8$	. . . . .
$\phi_E = \frac{1}{4}(42.8 + 100.0 + 42.8 + 25.0)$	$= 52.6$	. . . . .
$\phi_G = \frac{1}{4}(0 + 42.8 + 25.0 + 7.1)$	$= 18.7$	. . . . .
$\phi_A = \frac{1}{4}(18.7 + 52.6 + 18.7 + 9.8)$	$= 25.0$	. . . . .
$\phi_D = \frac{1}{4}(0 + 18.7 + 9.8 + 0)$	$= 7.1$	. . . . .
$\phi_F = \frac{1}{4}(7.1 + 25.0 + 7.1 + 0)$	$= 9.8$	. . . . .
		no changes

Even though the fifth run has produced no changes from the previous one the results are only approximately true, when compared to values calculated from a mathematical formula. This is typical of the method of finite differences, which replaces a continuum by a discrete set of points. To obtain more accuracy it is necessary to use more points; using the same number of points but increasing the number of significant figures in the calculations gives increased precision but not necessarily increased accuracy.

PROBLEMS

1 Put twice as many rows and twice as many columns into the square of Fig. 8-6. Repeat the iterative process of Example 1. What are the new values of the potential at B, E, G, A, D, F?

2 Which of the following two procedures is preferable? Why? (1) Carry out the computation to more decimal points? (2) Add more lines to the grid?

3 What are the values of ϕ midway between B and E? Midway between G and A? Midway between D and F?

4 Suppose the potential, $\phi = 100.0$, of the top plate in Fig. 8-6 is changed to 17.3 V. What is the potential at D?

5 Find the potentials at the points of intersection of the rows and columns in Fig. 8-7 using the method of finite differences for the grid lines shown. The distance between all adjacent grid lines is the same. About a dozen runs are required; if done by hand this is a tedious procedure.

6 Change the slot of Fig. 8-7 so that instead of being four grid spaces deep it is eight grid spaces deep. Find ϕ at E, F, G, H, I, J, K—all in column 6.

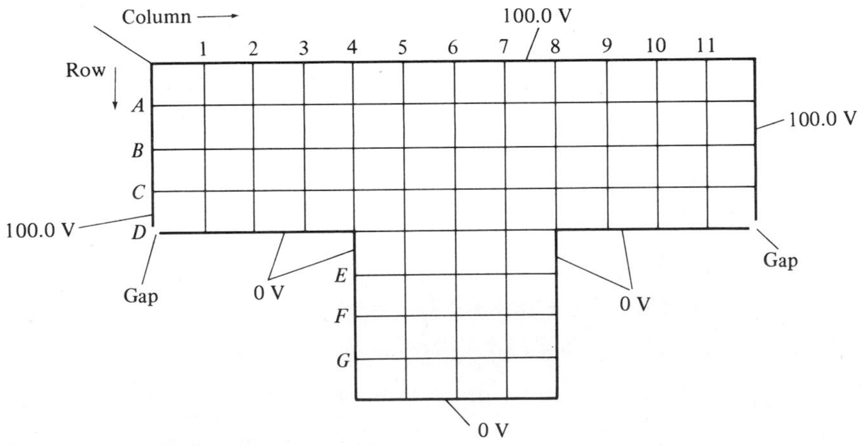

FIGURE 8-7

7 Change the slot of Fig. 8-7 so that instead of being four grid spaces deep, it is two grid spaces deep. Find ϕ at the points of intersection.

8 Change the two side walls of Fig. 8-7 from 100.0 V to 0 V. Repeat Prob. 5.

9 The method of finite differences is particularly suitable for digital computers. There are also methods using analog computers, in effect. These methods are based on measuring an analog of the potential, rather than the potential itself. The analog is a quantity which also obeys Poisson's or Laplace's equation. For example, ϕ is analogous to the height in a gravitational field.

Give the equation for a gravitational analog such that Δh, the difference in height between any two points, is proportional to $\Delta\phi$ between those points for the case of the field of an infinite line charge of density λ. Obtain the model by equating changes in the energy of an electrostatic field to those in a gravitational field. A rubber membrane may be stretched to give the required heights at the boundaries. These correspond to the boundary potentials in an electric field. A ball rolling on the membrane, in a gravitational field, corresponds to a charged particle moving through an analogous electric field.

10 Make a sketch of the model of Prob. 9 showing several equipotentials. Assume some convenient r_0 to have zero potential and zero height. Is this analog method applicable only to two-dimensional variations in ϕ or can it also be used for three-dimensional cases?

11 Assume a particle of mass m is placed on the rigid surface of an analog such as that of Prob. 9.

(*a*) What is the effective force on the particle that could cause any motion? Let θ be the angle the surface makes with the horizontal.

(*b*) What would be the gravitational analog of **E** expressed in terms of θ?

(*c*) What condition must be satisfied to make the effective gravitational force in the analog equal the electrical force in the actual case?

12 Prove that the vertical displacement of an elastic sheet satisfies the two-dimensional Laplace equation if the slope is small. Here the surface is not made to fit an equation; instead, the surface adapts itself to Laplace's equation for a given setting of boundary conditions. Such models have long been used to determine the trajectories of electrons in vacuum tubes. See "Vacuum Tubes," by K. R. Spangenberg (McGraw-Hill, New York, 1948), or "Introduction to Electric Fields," by W. E. Rogers (McGraw-Hill, New York, 1954). The models are gravitational analog computers, in effect, for the electrical forces and trajectories.

13 Another analog is provided by a comparison between $\mathbf{D} = \epsilon\mathbf{E}$ and the microscopic form of Ohm's law, $\mathbf{J} = g\mathbf{E}$, where g is the conductivity. Assuming (*a*) Ohm's law, $R = \Delta\phi/I$, for the resistance between two metal objects of arbitrary shapes immersed in an electrolyte and (*b*) the usual equation for C, the capacitance between these objects in terms of the magnitude of charge on either and $\Delta\phi$, find an equation for RC. This enables the capacitance to be determined by a resistance measurement in an electrolytic trough.

14 A very much simpler variation of the electrolytic trough method is available in two dimensions to plot equipotential surfaces. Paper is available which conducts current.

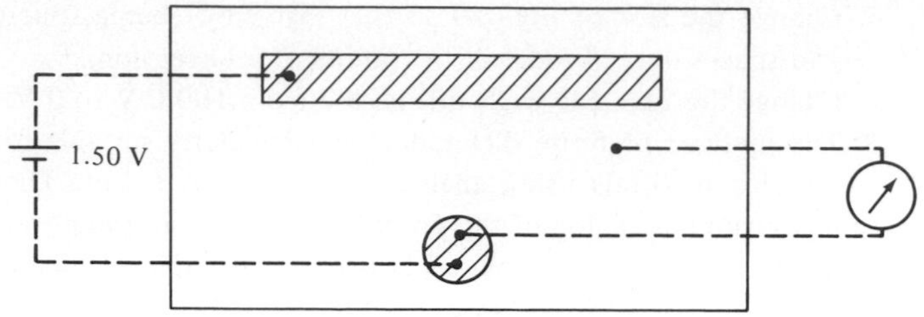

FIGURE 8-8

The resistance between any two points is small compared to the input resistance of a voltmeter whose leads are connected to these points, but is high compared to the resistance between these points if silver paint connects them. If the terminals of a dry cell are connected to the two silver areas (shown shaded) in Fig. 8-8, then a voltmeter probe can be used to determine the equipotentials in the neighborhood of the conductors. Determine the 0.25 V, 0.50 V, 0.75 V, 1.00 V, and 1.25 V equipotentials for this configuration. The paper and silver paint may be obtained from many suppliers of laboratory equipment.

15 Two thin metal plates, 10 cm wide and 10 m long, are placed with their flat faces parallel, facing each other and 10 cm apart in air.

(*a*) What is the capacitance?

(*b*) Suppose the plates are made 30 cm wide and placed 30 cm apart, what is the capacitance?

(*c*) How would the resistance between these plates in an electrolytic bath compare in the two situations above?

16 Suppose a square slice of the above two configurations was made, in effect, and one measured the resistance between two silver-coated edges of a square of conducting paper. How would the resistance for the 10 by 10 cm square compare with that of the 30 by 30 cm square?

17 Use the method of finite differences to obtain the potentials at various points in the two-dimensional cross section of a very long, square metal tube. The bottom $(0 \leqslant x \leqslant a)$ and two sides $(0 \leqslant y \leqslant a)$ have $\phi = 0$; the top is made of many narrow insulated metal strips with $\phi = 100 \sin \pi x/a$. Find ϕ at $x = a/6$, $2a/6$, $3a/6$, $4a/6$, $5a/6$; and similar values of y.

9
METALLIC CONDUCTION

The previous chapter considered metals as the boundaries enclosing a region in which we desired the solution of a problem in electrostatics. Even more important, from a practical point of view, is the case of a metal considered as the region of interest in a problem in magnetostatics. What actually happens, from a microscopic point of view, when a steady current flows in a metal? We will only attempt to supply a few broad answers to this question; for a more detailed account it is necessary to use quantum mechanics.

The current in an ordinary metal is due to the motion of the so-called conduction electrons through the interstitial region of the lattice atoms and ions. The mean lattice spacing is, very roughly, of the order of 1 Å; there is approximately one conduction electron per atom; it is found there are $n \approx 10^{29}$ conduction electrons per cubic meter. When there is no electric field applied to the metal, the average velocity of these charges $\langle \mathbf{v} \rangle$ is zero. The current density at any point, $\mathbf{J} = N(-e)\mathbf{v}$, is then also zero on the average; though at any instant its value will not generally vanish but will fluctuate in a random fashion in direction and magnitude. The mean speed with which the electrons move about in their thermal agitation is quite high, $v_T \approx 2 \times 10^6$ m s^{-1} (almost 5 million miles per hour!). But they do not move very far in any given direction, for they collide with the phonons of vibrational energy of the crystal atoms and ions. The phonons are similar in many ways to the quantized photons of electromagnetic energy. The mean free time between collisions, $\tau \approx 2 \times 10^{-14}$ s, and the mean free path, $\lambda \approx 4 \times 10^{-8}$ m, are both quite small. The collisions have a randomizing effect on the velocity: there is no correlation between the initial direction before a collision and the final direction afterward.

Suppose an electric field **E** is now applied to the interior of the metal. This is no longer electrostatics; here it is quite permissible that $\mathbf{E} \neq 0$, so that a current flows. Let an electron's velocity immediately after a collision be $\mathbf{v}_0$; its initial momentum is then $m\mathbf{v}_0$. The electric force provides an increment in this momentum of amount $(-e)\mathbf{E}t$ at the end of a time t; so the momentum at time t is $(m\mathbf{v}_0 - e\mathbf{E}t)$. The average momentum at the end of a time interval equal to the mean time between collisions is $\langle \mathbf{p} \rangle_\tau = -e\mathbf{E}\tau$, and the average velocity at the end of this time is $\langle \mathbf{v} \rangle_\tau = -(e/m)\mathbf{E}\tau$. We leave it to Prob. 2 to show that the average time since the last collision $\langle t \rangle$ equals τ, the average time between collisions (for random collisions). With this so, $\langle \mathbf{v} \rangle = -(e/m)\mathbf{E}\tau$ gives the average velocity of the electron, the average for all times t. Note that there is no factor $\frac{1}{2}$ in this expression. If this were motion under constant acceleration, the velocity would increase linearly with time and $\langle \mathbf{v} \rangle_\tau$ would equal $2\langle \mathbf{v} \rangle$; instead, here we have a constant average velocity, $\langle \mathbf{v} \rangle_\tau = \langle \mathbf{v} \rangle$.

The current density is now $\mathbf{J} = -ne\langle \mathbf{v} \rangle = n(e^2/m)\tau\mathbf{E} = g\mathbf{E}$, where $g = n(e^2/m)\tau$ is called the conductivity of the material. (The common symbol is σ but this involves the possibility of confusion with surface charge density. Like some other authors, we prefer g; G is the accepted symbol for the conductance of a bulk piece of matter.) Let us apply $\mathbf{J} = g\mathbf{E}$ to the uniform metallic rod of Fig. 9-1. The current density is $\mathbf{J} = (1/A)\mathbf{I}$; the electric field is $E = (1/\ell)V$. Then

$$I = GV$$

where $G = gA/\ell$ is the conductance

$$V = RI$$

where $R = \ell/gA$ is the resistance.

This relation, with g a constant independent of E, is called Ohm's law. The relation $\mathbf{J} = g\mathbf{E}$ is the microscopic equivalent of Ohm's law.

R and G are bulk quantities and depend on the shape of the material; g and $1/g$ (equal to the resistivity ρ) are functions only of the nature of the material, not of its shape. The dimensions of resistance are $[\mathrm{m^2\ kg\ s^{-3}\ A^{-2}}]$, and the unit having these dimensions is called the ohm. The dimensions of conductance are the inverse of those of resistance: $[\mathrm{m^{-2}\ kg^{-1}\ s^3\ A^2}]$ but the name of the unit having these dimensions is, figuratively, the inverse of the ohm: the mho. Resistivity is then given in ohm-meters and conductivity is in mho per meter.

Not all substances obey Ohm's law: in general, g is a function of **E** rather than a constant. This occurs in the following way. The mean drift speed per unit applied electric

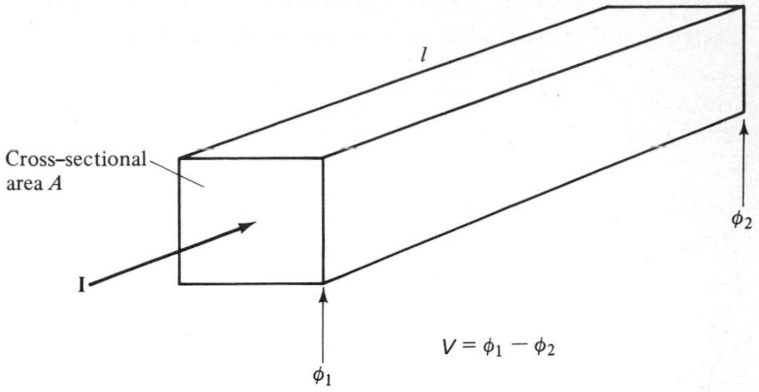

FIGURE 9-1

field is called mobility. In Prob. 1 the student is asked to derive a formula for this mobility, g/ne, which is applicable for the usual values of **E**. For copper, e.g., this formula yields a mobility of 4×10^{-3} m s^{-1}/V-m^{-1}, so the drift speed is perhaps 10^{-5} times the thermal speed. The addition of the drift speed to the thermal speed can scarcely affect τ; so g is constant. But suppose **E** is made so large that the drift speed becomes comparable to the thermal speed; then τ changes and g becomes $g(\mathbf{E})$. This would require about 10^8 V/m, a value much greater than those that occur in practice.

Resistors made of a substance called thyrite are nonohmic even with ordinary values of E. They are employed as protective devices for the field windings of dynamos, among other uses. A diode vacuum tube is a nonohmic device in the sense that the curve of the convection current versus the cathode-to-anode potential difference is not a straight line. The resistance of such a device is given either by the slope of (1) the line connecting a given point to the origin (dc resistance), or (2) the tangent to the curve at that point (ac resistance); neither slope is constant here. Many other examples of nonohmic resistors can be found, but the most common types of resistors and conductors are ohmic. The reasons for the nonohmic behavior of many substances—gases, semiconductor junctions, electrolytes—are many and varied. The mechanism for transport of the charge is the important factor. When this is nothing like that in metallic conduction there is no reason to expect linearity between current and potential difference.

The resistivity of materials shows a tremendous range of variation. Even for metals there is a considerable spread: 1.5×10^{-8} for silver, 10^{-6} for Nichrome (an alloy used for the heating element in electric irons). Semiconductors have much higher resistivities: 0.45 for germanium, 640 for silicon. Insulators have resistivities which are vastly greater than these: 10^{10} for glass, 7.5×10^{17} for quartz. The fact that the resistivities over this tremendous range of 10^{26}:1 fall predominantly into three groups of substances is the basis for the classification of substances into conductors, semiconductors, and insulators.

Examples

1. Power dissipation in resistors When the current-voltage relation in an electric circuit parameter is a straight line for all values of current or voltage in the range used, the element is said to be a linear one. Resistors constitute one of the three linear types of parameters available to circuit designers; inductors and capacitors are the other two types. (The really interesting circuit parameters are the nonlinear elements: the vacuum tubes, transistors, diodes, etc.) For resistors there are two significant operating specifications: the resistance, in ohms, and the power dissipation, in watts. That the power dissipation is a vital feature becomes evident if one considers the fact that a 10-Ω resistor rated at $\frac{1}{4}$ W is, perhaps, 1 cm in length. A 50,000-W searchlight may also be considered a resistor; if it, too, is rated at 10 Ω it will fill a space whose shortest dimension is of the order of 10 ft.

The power rating, or power handling capacity, of a resistor is determined by the manufacturer; $\frac{1}{4}$ W, $\frac{1}{2}$ W, 1 W, 2 W, 5 W, 10 W are the common sizes. The actual power dissipation, determined by the user in a particular operating condition, should be less than the power rating.

To determine the power dissipation in a resistor, suppose the current through it is I while the difference of potential between its ends is $V = \phi_1 - \phi_2$. The work done on a charge q that drifts through the resistor is qV. The rate at which this work is performed is the rate at which work done by the field is being converted to heat dissipated by the resistor.

$$P = \frac{d}{dt}(qV) = V\frac{dq}{dt} = VI \qquad \text{watts}$$

Using Ohm's law, this can also be written

$$P = I^2 R = \frac{V^2}{R}$$

These are three different forms of Joule's law. The microscopic equivalent of this is, if τ represents volume,

$$\frac{dP}{dt} = \mathbf{E} \cdot \mathbf{J}$$

Aside from superconductors, all practical wires used for the conduction of power from one point to another are made of metals having a finite resistance. There is, therefore, a dissipation of heat energy in them. This is called the Joule heat. The commercial code which limits the current-carrying capacity of a wire of given composition and diameter is based on the temperature rise which is admissible.

The Joule heat in a wire is, in general, wasted energy; though, this is not always true—in a toaster it is the desired result. Usually a wire is employed for the transfer of energy which is several orders of magnitude larger than the Joule energy. It is commonly thought that the metallic wires transfer this energy within their interiors, much as a hollow pipe transfers water. But this analogy is a poor one. In Chap. 11, when discussing the Poynting vector, it will be seen that the longitudinal transfer of energy along the wires occurs completely outside the wires; none is transferred longitudinally within the wires. The Joule heat energy, which is dissipated within the wires, is supplied to the interior, transversely, from the outside. Both the transferred power and the Joule heat are supplied by the fields outside the conductors.

In a superconducting wire, again, all the longitudinal transfer of energy occurs external to the wire. Here there is no Joule heat and no transverse transfer of energy from outside the wire to inside it. In a superconducting wire, in fact, both the electric and magnetic fields within the wire are zero.

2. **Relaxation time** In electrostatics we have seen that the equilibrium condition within a metal is $\rho = 0$ and $\mathbf{E} = 0$. Suppose we deliberately disturb this condition by somehow placing a charge within the interior of the metal. How long does it take for equilibrium conditions to return?

Let us apply the law of conservation of charge to some volume within the metal. Then $\partial Q/\partial t + I = 0$, where Q is the charge within the volume and I is the current flowing outward across the surface of that volume. So, if $\mathcal{V}$ and Σ are the volume and surface,

$$\int_{\mathcal{V}} \frac{\partial \rho}{\partial t}\, d\tau + \oint_{\Sigma} \mathbf{J} \cdot d\mathbf{S} = 0$$

or

$$\int_{\mathcal{V}} \frac{\partial \rho}{\partial t}\, d\tau + \int_{\mathcal{V}} \nabla \cdot \mathbf{J}\, d\tau = 0$$

Since $\mathcal{V}$ is arbitrary,

$$\boxed{\frac{\partial \rho}{\partial t} + \nabla \cdot \mathbf{J} = 0}$$

This is called the equation of continuity. It is the microscopic equivalent of the very fundamental law of conservation of charge.

For the second term we substitute for $\mathbf{J}$ from $\mathbf{J} = g\mathbf{E}$:

$$\frac{\partial \rho}{\partial t} + g\, \nabla \cdot \mathbf{E} = 0$$

But $\nabla \cdot \mathbf{E} = (1/\epsilon)\rho$; so

$$\frac{\partial \rho}{\partial t} + \frac{g}{\epsilon}\rho = 0$$

Then
$$\int_{\rho_0}^{\rho} \frac{d\rho}{\rho} = -\frac{g}{\epsilon}\int_0^t dt \qquad \text{and} \qquad \rho = \rho_0\, e^{-(g/\epsilon)t}$$

The quantity ϵ/g has the dimensions of time. It is called the relaxation time of the substance. For copper $\epsilon/g = 1.5 \times 10^{-19}$ s; for distilled water the value is 10^{-6} s; for quartz it is 10^6 s. The relaxation time for all metals is extremely short. To answer the question: it takes an infinite time for equilibrium to be fully restored; but it only takes the time τ for any disturbing charge density to fall to $1/e$ of its original value.

3. Analogy between J and D Many problems in steady-state current flow **J** may be solved by transposing them into problems involving **D** in electrostatics. (There is also an analogy to fluid dynamics, which is even suggested by the names.) This rests on the following mathematical similarities:

$$\begin{cases} \nabla \cdot \mathbf{J} = 0 \quad \left(\text{since } \dfrac{\partial \rho}{\partial t} = 0\right) \\ \nabla \times \mathbf{J} = 0 \quad (\text{since } \nabla \times \mathbf{J} = \nabla \times \mathbf{E}) \\ \mathbf{J} = g\mathbf{E} \ (\text{Ohm's law}) \end{cases} \qquad \begin{cases} \nabla \cdot \mathbf{D} = 0 \\ \nabla \times \mathbf{D} = 0 \\ \mathbf{D} = \epsilon \mathbf{E} \end{cases}$$

As a result of these equations both **J** and **D** are given by $-\nabla\psi$ where ψ, a potential function, satisfies Laplace's equation $\nabla^2\psi = 0$ in some finite region, subject to given boundary conditions. (Incidentally, the **J** here can refer to fluid dynamics as well as to electricity.)

The analogy between the two cases is, however, not exact because g can vary down to a value of zero while ϵ can only attain a minimum value of ϵ_0. In Fig. 9-2*a* the space between the plates is filled with a substance of conductivity g. The current flow between the two plates is limited to the region between them, and the current is uniform in this region. In Fig. 9-2*b* the space between the plates is a dielectric. Some lines of **D** fringe out into the outer region because $\epsilon \neq 0$ there. The analogy between **J** and **D** here is good to the extent that the fringing may be ignored.

Suppose the conditions of Fig. 9-2 are altered so that, in each case, the two metal plates are embedded in uniform media; i.e., in Fig. 9-2*a* $g = g$ both between and outside the plates, while in Fig. 9-2*b* $\epsilon = \epsilon$ everywhere outside the plates. Then

$$R = \frac{V}{I} = \frac{\int_1^2 \mathbf{E} \cdot d\mathbf{r}}{\int_\Sigma (g\mathbf{E}) \cdot d\mathbf{S}}$$

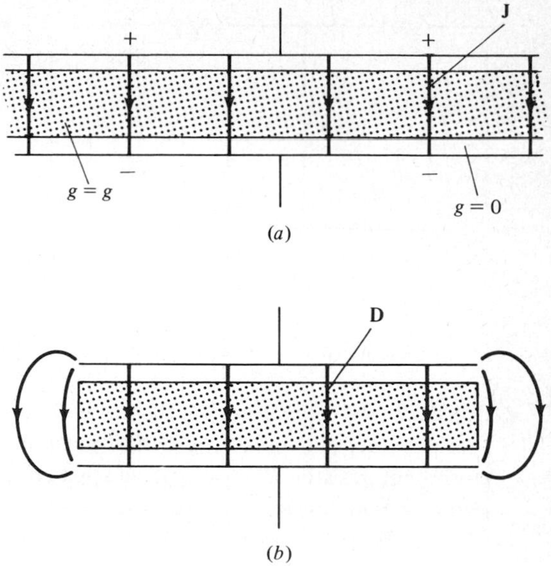

FIGURE 9-2

while

$$C = \frac{Q}{V} = \frac{\int_{\Sigma}(\epsilon\mathbf{E}) \cdot d\mathbf{S}}{\int_1^2 \mathbf{E} \cdot d\mathbf{r}}$$

Thus $RC = \epsilon/g$. A measurement of R will, therefore, suffice to determine C—to the extent that the fringing of the lines of **D** does not require the Σ of the surface integral for C to differ from the Σ of the surface integral for R.

PROBLEMS

1 Derive a formula for the mobility of the conduction electrons in metallic conduction from the definition of mobility and the relation between current density and drift velocity.

2 For collisions spaced at regular and equal intervals, (average time between collisions) equals (average time since the previous collision) plus (average time until the next collision). In symbols: $\tau = \langle t_{\leftarrow} \rangle + \langle t_{\rightarrow} \rangle$. Show that for collisions occurring at random intervals $\tau = \langle t_{\leftarrow} \rangle = \langle t_{\rightarrow} \rangle$.

3 Determine the equivalent resistance when two resistors, R_1 and R_2, are put (*a*) in series and (*b*) in parallel. *Series* means the same current flows through both; *parallel* means the same potential difference is applied to both.

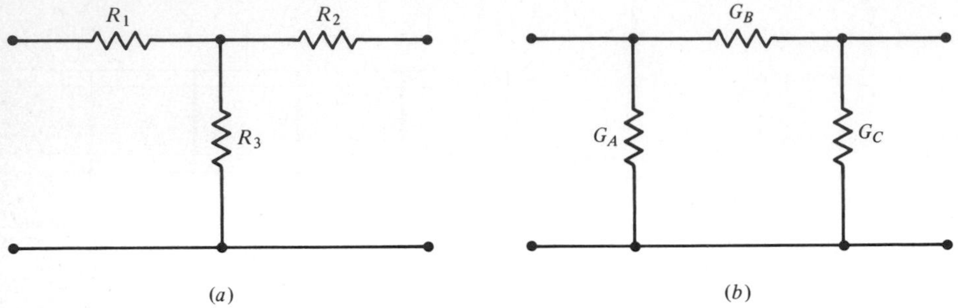

FIGURE 9-3

4 Consider two cases. In (*a*) there are two resistors such that $R_1 = 2R_2$; and the two resistors are put in series. What portion of the total heat dissipated by the equivalent resistor R is actually dissipated in the smaller resistor? In (*b*) there are two different resistors such that $R'_1 = 2R'_2$; and the two resistors are put in parallel. Let the equivalent resistor R in this case be the same as in (*a*). What portion of the total heat dissipated by R, in the two cases, is actually dissipated in the smaller resistor?

5 Figure 9-3*a* shows a three-terminal network: two input terminals at the left, two output terminals at the right, with one terminal common to both. This configuration is called a T (tee) or a Y (wye). Let

R_{O1} = the input resistance at the left end with the terminals at the right end open-circuited

R_{O2} = the input resistance at the right end with the terminals at the left end open-circuited

R_{S1} = the input resistance at the left end with the terminals at the right end short-circuited

R_{S2} = the input resistance at the right end with the terminals at the left end short-circuited

Find R_1, R_2, R_3 in terms of these four measured quantities.

6 Fig. 9-3*b* shows a π (pi) or Δ (delta) configuration for a three-terminal network. By measuring the four quantities listed in Prob. 5 it is possible to represent any unknown three-terminal network either in terms of the equivalent π network or the equivalent T network. Here it is easier to deal with the conductances G, rather than resistances R. Show that the result for the π is that obtained by letting

$$\begin{cases} R_1 \to G_A \\ R_2 \to G_C \\ R_3 \to G_B \end{cases} \qquad \begin{cases} R_{O1} \to G_{S1} \\ R_{O2} \to G_{S2} \end{cases} \qquad \begin{cases} R_{S1} \to G_{O1} \\ R_{S2} \to G_{O2} \end{cases}$$

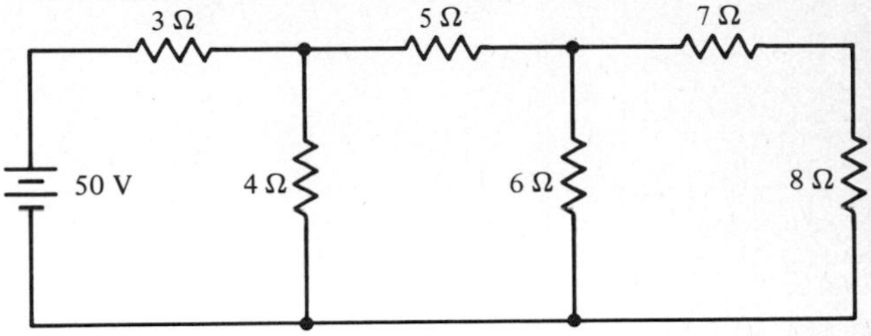

FIGURE 9-4

7 Three resistors are in parallel, having 2 Ω, 3 Ω, and 6 Ω, respectively (Ω = ohms). A 2-Ω resistor is in series with the parallel combination. What is the equivalent resistance?

8 Suppose, in Prob. 5 or 6, that a term with a minus sign exceeds a term with plus sign to which it is added. What would be the significance of this?

9 Find the current in the 8-Ω load resistor of Fig. 9-4.

10 Figure 9-5 shows a battery, whose internal resistance is r ohms, supplying current to a load resistor of R ohms.

(*a*) Assuming r is fixed, what value of R will give the maximum power in R?

(*b*) Assuming R is fixed, what value of r will give the maximum power in R?

11 If R_A is the resistance of the resistor in Fig. 9-3 whose conductance is G_A, etc., find R_A, R_B, R_C in terms of R_1, R_2, R_3.

12 In Prob. 11, find the resistors of the star configuration in terms of those of the delta.

13 A spherical resistor consists of an inner metal sphere of radius a and an outer metal sphere of radius b with a medium having conductivity g between them. Find the resistance.

14 Suppose a conical sector, of half-angle θ, is taken from the spherical resistor of Prob. 13. What is its resistance?

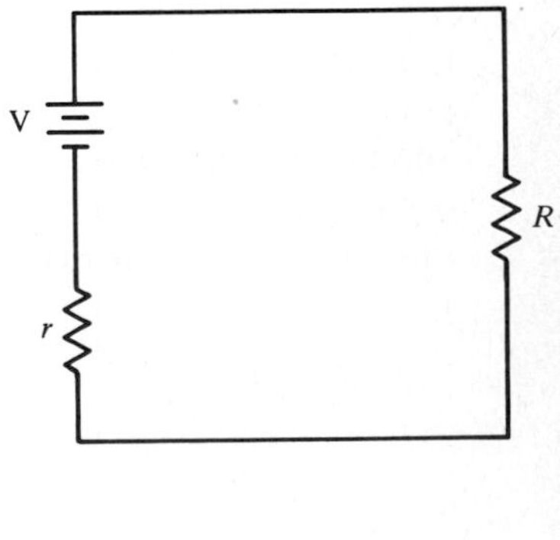

FIGURE 9-5

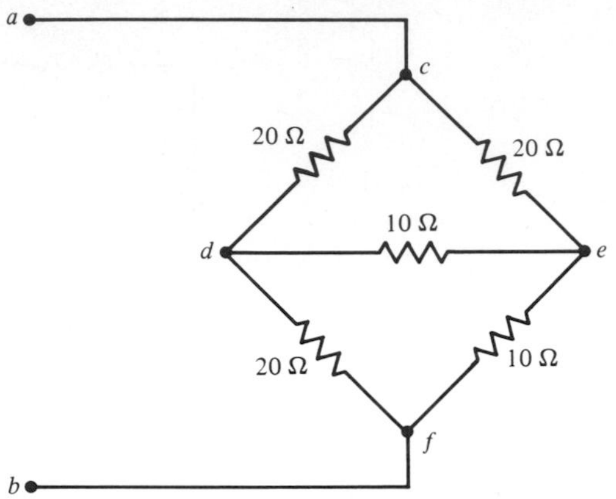

FIGURE 9-6

15 Figure 9-6 illustrates a widely used configuration, the bridge. Find the wye equivalent to the delta section *cde*. Then compute the equivalent resistance presented by the entire bridge circuit between the terminals *a* and *b*.

16 Take the resistivity of silver to be 1.47×10^{-8} Ω-m.

(*a*) Suppose 1 cm^3 of silver is formed into a wire of square cross section 1 by 1 mm. What will be its resistance between the two ends of the wire?

(*b*) What will the resistance be between two opposite faces, separated by 1 mm?

17 The density of copper is 8.92 g cm^{-3}; its atomic weight is 63.5. Using Avogadro's number, (*a*) find n, the number of atoms/m^3.

(*b*) If the current density is $J = 10^3$ A m^{-2}, find the drift velocity, assuming there is one conduction electron per atom.

18 If the resistance of metals is caused primarily by the collisions of the conduction electrons with the vibrating lattice atoms, what effect on the resistance would you expect from an increase of the temperature? Write a law, giving resistance versus temperature, that would be reasonable for not too large a temperature change.

19 Find the resistance of the insulation of a coaxial cable of length ℓ, inner radius r, and outer radius R to the radial flow of current. Assume the insulation has finite conductivity g.

20 If the drift velocity of electrons, as found in Prob. 17, is so small, then how is it possible to send a telephone call from New York to San Francisco—a distance of 3000 miles—with delays in the time of transmission which are only a small part of a second? This delay time is so small that the ear cannot detect it in the answer, by the person at the other end, to one's question.

10
FERROMAGNETISM

We have considered metals in two ways: as boundaries to a region in which we wish to find the electrostatic field, and as the region itself, in which there exists a magnetostatic current. We now treat a particular group of metals–iron, nickel, and cobalt–from a completely different viewpoint. The magnetic effects of these elements are very different from that of other substances and the subject, therefore, warrants a special study, which is that of ferromagnetism. It is odd that this should be the last topic we consider under magnetostatics, for ferromagnetism is the type of magnetism responsible for the properties of the lodestones found in Magnesia and this started the study of magnetism. Yet it is to this day still not completely understood, and that which is understood requires the sophisticated approach of quantum mechanics for explanation. The ironic feature is that ferromagnetism is many orders of magnitude stronger than either diamagnetism or paramagnetism, and is the only type easily noticed.

In electrostatics the process of electrification by rubbing, which started the experimental observations of electricity, is also rather complicated and not too well understood. But ferromagnetism is much more important, both theoretically and practically, than electrification. It is also much more difficult to explain. The effects are certainly quantum mechanical rather than classical; but quantum mechanics sometimes gives not only wrong magnitudes but wrong signs. We will limit ourselves to a description of some of the phenomena.

In treating paramagnetism and diamagnetism we defined the magnetic susceptibility as $\mathbf{M} = \chi'_m [1/\mu_0)\mathbf{B}]$, by analogy with the definition $\mathbf{P} = \chi(\epsilon_0 \mathbf{E})$ in electrostatics. Because it is easier to determine **H** experimentally than it is **B**, it is usual to define the magnetic susceptibility as

$$\mathbf{M} = \chi_m \mathbf{H}$$

in dealing with ferromagnetism. For paramagnetism and diamagnetism χ_m and χ'_m are very small (e.g., $+2.3 \times 10^{-5}$ for Al and -1.7×10^{-5} for Bi) and the value of **M** does not appreciably alter the value of total magnetic field. This is not the case with ferromagnetism. Here the contribution made by **M** can not only outweigh that of the external **H** or **B**, but a large **M** can exist even when there is no external field. This is the case with magnetite, and with other permanent magnets. Using $\mathbf{M} = \chi_m \mathbf{H}$, it is found that the quantity χ_m is not a constant in ferromagnetism; there is no linear relation here between magnetization and any current producing an external field or an external magnetic field. As long as the variability of χ_m is recognized, it is then possible to use not only $\mathbf{M} = \chi_m \mathbf{H}$ but also

$$\mathbf{B} = \mu_0 \mathbf{H} + \mu_0 \mathbf{M} = \mu_0 \mathbf{H} + \mu_0 \chi_m \mathbf{H} = (1 + \chi_m)\mu_0 \mathbf{H} = \mu \mathbf{H}$$

The permeability μ and the relative permeability μ/μ_0 are then also not constants. Figure 10-1*a* shows the relation between B and H for a sample of iron. This is called the magnetization curve of the material despite the fact that it plots B versus H rather than M versus H. It is important to note that the points on this curve are obtained by *monotonically* increasing the values of H, starting from zero; the reason for this restriction will appear shortly. B may be labeled in teslas or, more usually, in gauss; H would be given in amperes per meter (alternatively, ampere-turns per meter) or in oersteds. Figure 10-1*b* shows the slope of the B–H curve (the so-called magnetization curve) versus H. A relative permeability of 5000 is not unusual; the χ_m of 5000 is larger than a paramagnetic or diamagnetic χ_m by the factor of 10^8!

Suppose that, instead of traversing all the successive points of the "magnetization" curve, one stopped increasing the H at H_1 and started decreasing it, as indicated in Figure 10-2*a*. Then when H was reduced to zero the value B_r would be found to be finite. The value of B_r (B with $H = 0$) is called the retentivity: a permanent magnet has been attained. If the current in the solenoid surrounding the sample of iron is now increased in the opposite direction the value $B = 0$ will be reached. Continuing to a symmetric value of $-H_1$ and then stopping and returning gives the complete hysteresis curve of Fig. 10-2*a*. As long as one goes between the extremes $-H_1$ and H_1 this closed loop is repeated over and over again. Figure 10-2*b* shows the corresponding initial tracking of the magnetization curve, followed by subsequent hysteresis loops, for larger values of the maximum H field, one for H_2 and one for H_3.

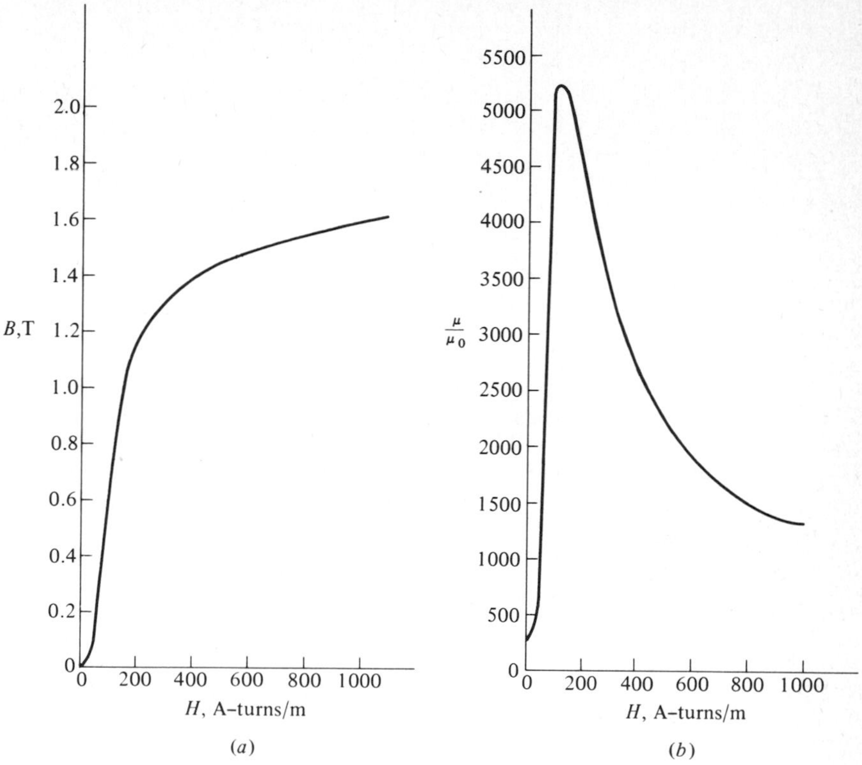

FIGURE 10-1

It is worth emphasizing that the behavior outlined in Fig. 10-2 is quite different from any encountered previously in this book. For example, for a given hysteresis curve the value of H is not sufficient to fix the value of B uniquely: the curve is double-valued. For a given hysteresis curve the magnetic field is determined by H only when one specifies whether H is increasing or decreasing. But, further, if the cycle is changed by altering the amplitude of the variation of H, i.e., if one goes from one hysteresis curve to another, then the two values of B that correspond to a given H are changed also. The values of B are determined not only by H but by H_{max}; i.e., the past history of the specimen enters the picture. Mathematically this problem is very difficult and it is normal to perform the calculations graphically.

The actual testing of ferromagnetic specimens is performed most conveniently by using a toroidal sample. This geometry, with N turns of wire carrying a current I, gives an azimuthal $\mathbf{H} = \hat{\boldsymbol{\phi}}(NI/2\pi R)$, where R is the radius of the central axis of the torus. Then $\mathbf{B} = \hat{\boldsymbol{\phi}}\mu_0\,[(NI/2\pi R) \pm M]$. If R is large compared to r, the radius of the circular cross section, then B will be almost constant over the cross-sectional area of the torus. Further, there will be no surfaces that spoil the symmetry by acting as poles. If an ordinary

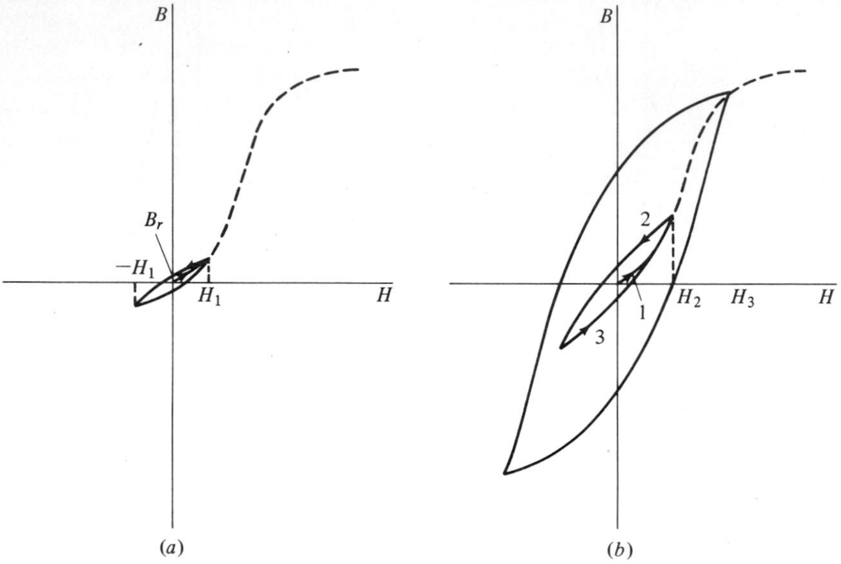

FIGURE 10-2

solenoid were used, with a straight bar of given length inside it, complications produced by the ends would have to be taken into account.

Examples

1. Toroid with an air gap Suppose that our toroidal sample has an air gap cut into it, so that the lines of flux go through both iron and air. Ampere's law gives

$$\oint \mathbf{H} \cdot d\mathbf{r} = NI$$

where N is the total number of turns of wire around the torus. $\mathbf{H} \cdot d\mathbf{r} = H\,dr$ here; using

$$H = \frac{B}{\mu} = \frac{1}{\mu}\left(\frac{\Phi}{A}\right) \qquad \oint \frac{\Phi}{\mu A}\,dr = NI$$

Let the air gap be not too big, so that there is negligible fringing of the lines of flux and let the permeability of the torus be much greater than that of the surrounding air, so that there is negligible leakage flux. Then Φ will be constant within the circular cross section in traversing the line integral and it can be taken to the left of the integral sign. So

$$\Phi \mathcal{R} = NI$$

where $\mathcal{R} = \oint dr/\mu A$ is called the reluctance of the magnetic circuit.

This equation is the magnetic equivalent of Ohm's law. NI, the ampere-turns or magnetomotance, corresponds to the voltage (usually this is called the magnetomotive force or mmf); Φ, the flux, corresponds to the current; $\mathcal{R}$, the reluctance, corresponds to the resistance but here $\mathcal{R}$ is not a constant. In the present case the reluctance becomes the sum of two terms,

$$\mathcal{R} = \frac{\ell_i}{\mu_i A} + \frac{\ell_a}{\mu_a A}$$

where ℓ_i is the length in the iron along the mean radius and μ_i is the permeability of the iron; ℓ_a is the width of the air gap and μ_a is the permeability of air.

The analogy with Ohm's law suffers from a defect. To find

$$\Phi = \frac{NI}{\mathcal{R}}$$

we need to know $\mathcal{R}$; to find $\mathcal{R}$ we need to know μ_i; to find μ_i we need to know B as well as H, i.e., we need to know Φ. This dilemma is solved in Example 2.

Figure 10-3 shows a torus with a gap. Here the gap is shown with radial sides but if the gap is small, the result is essentially the same as if the sides were parallel. Suppose the current in the winding is reduced to zero. Ampere's law,

$$\oint \mathbf{H} \cdot d\mathbf{r} = NI$$

becomes

$$H_i(2\pi R - \ell_a) + H_a \ell_a = 0$$

where H_i and H_a pertain to iron and air, respectively. Then

$$H_i = -\left(\frac{\ell_a}{2\pi R - \ell_a}\right) H_a$$

shows that the $\mathbf{H}$ fields in the iron and in the air are oppositely directed. Suppose $\mathbf{H}$ is CCW in the air gap; then $\mathbf{H}$ must be CW in the iron. In Fig. 10-3 the surface 1-1 will have lines of $\mathbf{H}$ leaving it in the iron and leaving it in the air; such a surface is called a north pole, by definition. Surface 2-2 will have lines of $\mathbf{H}$ coming to it, both in the air and in the iron; this is called a south pole. The lines of $\mathbf{B}$ must run the same way as $\mathbf{H}$ in the air gap, where $\mathbf{M} = 0$; so $\mathbf{B}$ is CCW there. But $\nabla \cdot \mathbf{B} = 0$, so the lines of $\mathbf{B}$ do not change direction at the pole faces but continue CCW. Inside the iron, then, $\mathbf{B}$ and $\mathbf{H}$ go in opposite directions.

The magnetization of the iron is given by $\mathbf{B} = \mu_0 \mathbf{H} + \mu_0 \mathbf{M}$. Thus

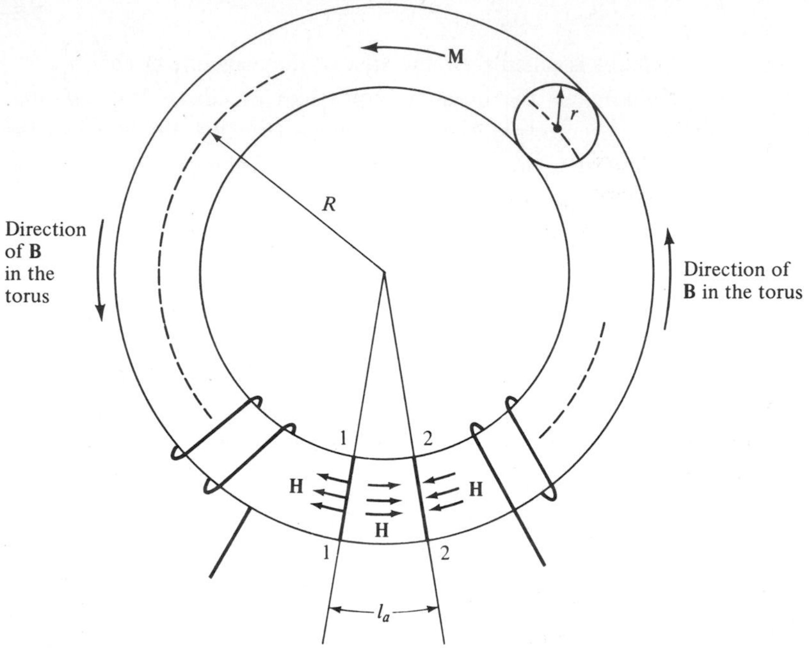

FIGURE 10-3

$$\hat{\boldsymbol{\phi}}B = \mu_0(-\hat{\boldsymbol{\phi}}H) + \mu_0\mathbf{M}$$

$$\mathbf{M} = \hat{\boldsymbol{\phi}}\left(\frac{B}{\mu_0} + H\right)$$

The directions of **H**, **B**, and **M** are shown in the figure.

By assuming $R \gg r$ we have essentially taken **B** to be uniform. Then **M** is uniform also and $\mathbf{J}_m = \nabla \times \mathbf{M} = 0$: there is zero magnetization volume current density. But $\mathbf{j}_m = \mathbf{M} \times \hat{\mathbf{n}}$; at a point on the outside rim of the torus $\hat{\mathbf{n}} = \hat{\mathbf{r}}$ and $\mathbf{j}_m = -\hat{\mathbf{z}}M$ there. There is a nonvanishing magnetization surface current density around the doughnut producing the azimuthal **M**. This is the amperian current view. The pole, alternative, viewpoint takes magnetic poles produced on 1-1 and 2-2 with surface density $\sigma_{\text{mag}} = \mathbf{M} \cdot \hat{\mathbf{n}}$ there. When **M** is uniform $\rho_{\text{mag}} = -\nabla \cdot \mathbf{M} = 0$; in general, however, a volume magnetic charge density must also be taken into account.

2. Magnetic circuits The iron torus with air gap is typical of the simple type of magnetic circuit having two reluctances in series. Since the B–H "magnetization" curve for iron is nonlinear it is necessary to have actual values of B versus H for the specimen if, e.g., it is desired to find the total ampere-turns necessary to produce a given flux. This is one type of problem. Another is the inverse problem: given the ampere-turns, find the flux; this

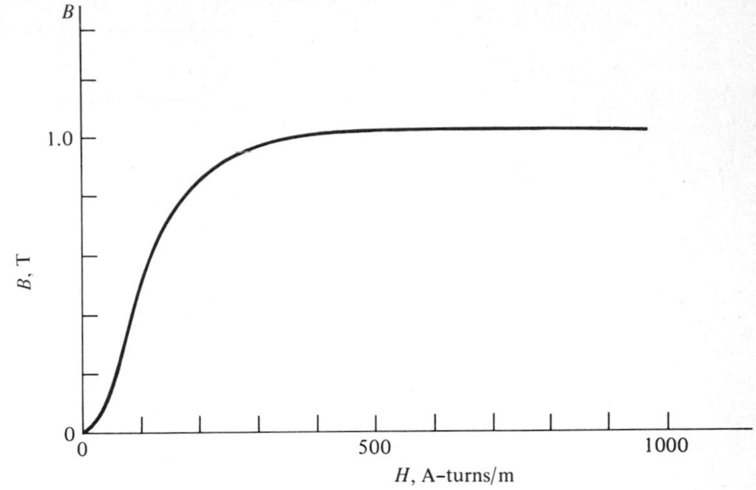

FIGURE 10-4

requires a different procedure. In either case, we will assume the B-H curve of Fig. 10-4. (It is more common practice in such work to use inches for length and maxwells for flux, where 10^8 maxwells are the equivalent of 1 weber.) We will take the torus to have the dimensions shown in Fig. 10-5; and we assume B uniform, with negligible fringing in the air gap.

a. Given: $\Phi = 2.5 \times 10^{-4}$ Wb. Find: the ampere-turns required in the winding.

The area of the core is $\pi(10^{-2})^2 = 3.14 \times 10^{-4}$ m^2. $B = \Phi/S = 0.8$ T $= 8000$ G. From Fig. 10-4 the value of H in the core is 130 A-turns/m. The mean length of the core is $[2\pi(10) - 0.05]\,10^{-2} \approx 0.628$ m. The ampere-turns required for the iron core is then $(NI)_i = 130 \times 0.628 = 81.6$ A-turns.

The B in the air gap is also 0.8 T. It is left to Prob. 1 to show that for air (actually, vacuum), if B is given in teslas and H is given in ampere-turns per meter, then

$$H = 8 \times 10^5\, B$$

Then H here is 6.4×10^5 A-turns/m. The ampere-turns required for the air gap is $(NI)_2 = (6.4 \times 10^5)(5 \times 10^{-4}) = 320$ A-turns.

The total number of ampere-turns that must be supplied by the current is $NI = (NI)_i + (NI)_a = 81.6 + 320$ or 402 A-turns. The air gap needs 80 percent of this even though its width is less than 0.1 of 1 percent of the length of iron.

b. Given: $NI = 400$ A-turns in the winding. Find: the flux produced in the core.

This case is more complicated than the previous one, for we do not know, offhand, how to apportion the given ampere-turns between the iron and the air gap. One approach is the use of a trial-and-error method. A flux is assumed and the ampere-turns required are

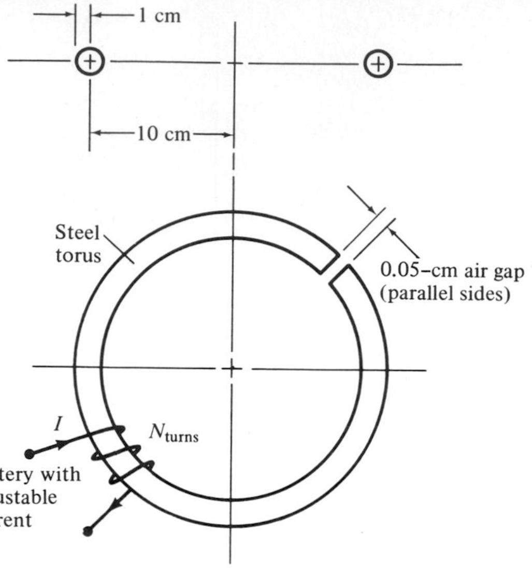

FIGURE 10-5

found by the method above. Comparing this with the given value of ampere-turns, a new value of flux is assumed, etc. We will employ a second approach which utilizes the intersection between the curve of Fig. 10-4 and a second curve—a straight line.

To obtain the equation of the straight line we first write $(NI)_{\text{total}} = (NI)_i + (NI)_a$, where the subscripts i and a mean iron and air. $(NI)_a = H_a \ell_a$ as in Example 1. But for air we know that $H_a = 8 \times 10^5 B_a$ if B_a is in teslas; so $(NI)_a = 8 \times 10^5 \ell_a B_a$. Similarly $(NI)_i = H_i(2\pi R - \ell_a) \approx 2\pi R H_i$. Since we are taking $B_i = B_a$, $(NI)_{\text{total}} \approx (2\pi R)H_i + (8 \times 10^5 \ell_a)B_i$. In the present case $R = 0.1$, $\ell_a = 5 \times 10^{-4}$, and $(NI)_{\text{total}} = 400$ so $400 = 0.63H_i + 400B_i$. In Fig. 10-6 this curve is shown superimposed on the "magnetization" curve of Fig. 10-4. The point of intersection has $H_i = 130$ A-turns/m, $B_i = 0.79$ T. These values compare quite well with those in part *a* of this example. More accurate results may be obtained by using a hysteresis curve, in the proper quadrant, instead of the "magnetization" curve.

PROBLEMS

1 If H is in ampere-turns per meter and B is in teslas, find H in terms of B in air.

2 (*a*) Suppose H is given in ampere-turns per inch while B is given in maxwells per square inch, what is the relation between B and H in air?

(*b*) Suppose H is in oersteds and B is in gauss, what is the relation?

3 In Example 2*a* above, how many ampere-turns would be required, instead of 402, if the flux requirement were (*a*) doubled to 5×10^{-4} Wb or (*b*) changed to 3.75×10^{-4} Wb? These results illustrate the effects of saturation.

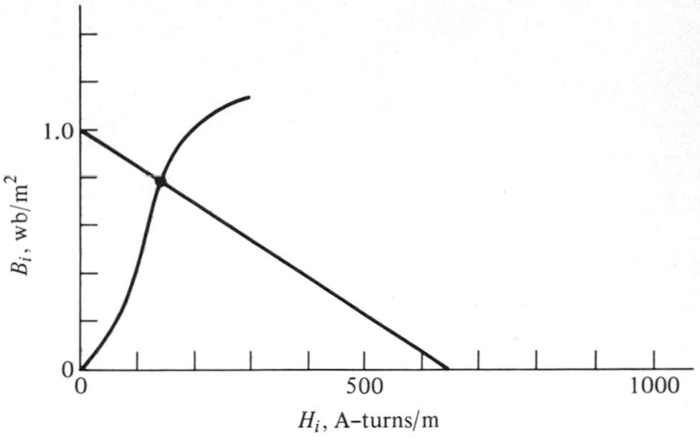

FIGURE 10-6

4 What modifications need be made in a series magnetic circuit of three reluctances instead of two? Are any further conditions imposed if the cross-sectional areas in different parts are unequal?

5 Suppose that in Example 2*b* the given number of ampere-turns were doubled to 800. To what value would the flux (originally 2.5×10^{-4} Wb) change?

6 What are the relative merits for the use of oersteds vs. ampere-turns per meter as units for H?

7 In the magnetized but nonenergized toroid with air gap of Fig. 10-3 the lines of B are azimuthal CCW; those of H are CW in the iron. Suppose that, leaving the **B** direction unaltered, the gap is replaced with iron to make a homogeneous torus. What will be the direction of the lines of **H** in the torus now—the same as, or opposite to, the previous case? Explain.

8 The lines of flux in air from a north pole leave the surface. Consequently two north poles repel each other. Considering the fact that the northern end of a compass needle seeks to set itself closer than the southern end to the earth's north magnetic pole, is the northern end of the compass needle a north pole or a south pole?

9 The work done in taking a sample of iron from one point on its hysteresis curve to another is $NI\,d\Phi$; also, Ampere's law gives

$$\oint_C \mathbf{H} \cdot d\mathbf{r} = NI$$

if C, say, is a closed path in the iron. From these two equations find the work required, per unit volume of iron, to go (*a*) from one point on the hysteresis curve to another; and (*b*) completely around the hysteresis curve. Give a graphical explanation of the results (curves, not adjectives).

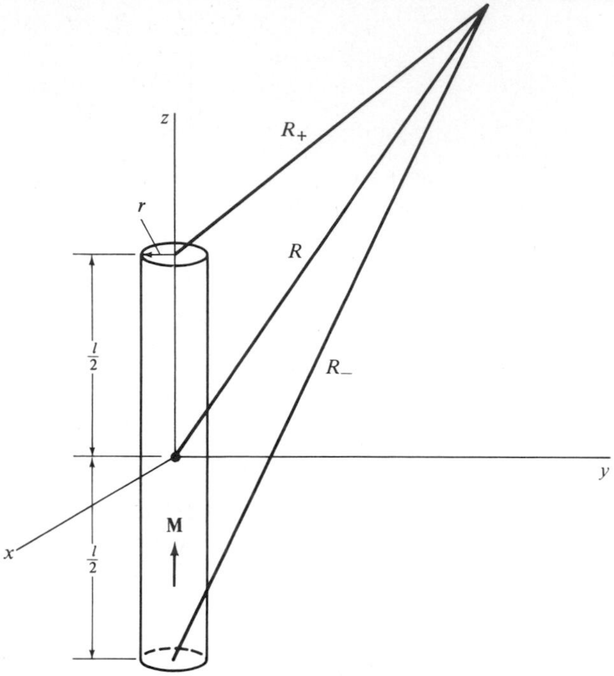

FIGURE 10-7

10 Reconcile ampere-turns = flux times reluctance with Example 1, where a torus (with air gap, but without excitation current in the winding) had flux but did not have ampere-turns.

11 In Example 1 the direction of **B** and **H** are opposite to each other in the iron. Does this imply a negative permeability?

12 (*a*) If there are no free magnetic poles (i.e., no free sources of lines of flux) and div **B** = 0 is a measure of the source strength (e.g., in electrostatics $\nabla \cdot \mathbf{E} = (1/\epsilon_0)\rho_{\text{total}}$ and $\nabla \cdot \mathbf{D} = \rho_{\text{free}}$), how is it possible for **B** to have a nonvanishing value?

(*b*) In Example 1 there is zero conventional current. If the (vortex) sources of **H** are only conventional currents, not magnetization currents, how can **H** have nonvanishing values there?

13 A toroidal winding in air has a self-inductance L_0. When a current I_0 flows, it produces a flux Φ_0, a flux density B_0, and an H field H_0.

(*a*) Suppose the current is produced by a battery and an initially unmagnetized core of soft iron of relative permeability κ_m is inserted into the torus. In terms of the original values and κ_m, find the new values of Φ, L, I, B, and H.

(*b*) Repeat with a superconducting torus and no battery. (*a*) and (*b*) are the nonisolated and isolated cases, respectively.

14 Describe how you would, experimentally, measure H and B in the core of the torus of Fig. 10-3.

15 Would the results of Prob. 13 hold true if a small air gap were made in the core? Using the symbols of Example 1, find expressions for H_a and H_i if H_0 is the value of the H field within the torus when there is no iron.

16 If the conventional current in and near a piece of iron is zero, then $\nabla \times \mathbf{H} = \mathbf{J}_C = 0$ and $\mathbf{H}$ may be obtained from a scalar magnetic potential: $\mathbf{H} = -\nabla\phi_m$. If the magnetization of the iron is $\mathbf{M}$, find the integral expression for ϕ_m at a field point in terms of $\mathbf{M}$ and R, the distance between source points and field point.

17 A cylindrical rod of length ℓ and radius r centered on the z axis has a uniform magnetization parallel to its axis, $\mathbf{M} = \hat{\mathbf{z}}M$ as in Fig. 10-7. Find (*a*) ρ_m, the volume magnetic charge density; (*b*) σ_m, the surface magnetic charge density; (*c*) the magnetic "charge" on the poles; (*d*) ϕ_m, the magnetic scalar potential, at a point whose distances from top and bottom are R_+ and R_-, where these are very large compared to r.

18 Under what conditions could the scalar magnetic potential be employed in a region near filamentary currents? Consider both singly connected and doubly connected regions. Then suppose the region contains a volume distribution of currents.

19 A ring of iron 2 cm wide has a rectangular cross section, the inner diameter being 14 cm and the outer diameter 16 cm. A 1000-turn coil produces a flux of 10^{-3} Wb in the ring when the coil current is 3 A. Find (*a*) B, (*b*) H, (*c*) NI, (*d*) $\mathcal{R}$, (*e*) μ, (*f*) κ_m.

20 In measurements on magnetic materials one always prefers to use a torus having uniform properties rather than a long, thin rod, for example. Why?

11

MAXWELL'S EQUATIONS

11-1 FARADAY'S LAW

When static conditions hold, i.e., when time does not enter the picture, the results of the previous chapters show that electricity and magnetism are two separate, somewhat parallel, disciplines:

$$\begin{cases} \nabla \cdot \mathbf{E} = \dfrac{1}{\epsilon_0}\rho \\ \nabla \times \mathbf{E} = 0 \end{cases} \qquad \begin{cases} \nabla \cdot \mathbf{B} = 0 \\ \nabla \times \mathbf{B} = \mu_0 \mathbf{J} \end{cases}$$

Expressed in terms of the potential fields, these equations are equivalent to

$$\phi = \frac{1}{4\pi\epsilon_0}\int_{\mathcal{V}} \frac{\rho}{R}\,d\tau \qquad \mathbf{A} = \frac{\mu_0}{4\pi}\int_{\mathcal{V}} \frac{\mathbf{J}}{R}\,d\tau$$

ρ (both free and polarization type) is the cause; ϕ and $\mathbf{E}$ are the results. Similarly $\mathbf{J}$ (both conventional and magnetization type) is the cause; $\mathbf{A}$ and $\mathbf{B}$ are the results. Nothing has been said of the action-at-a-distance mechanism by which the fields are propagated from one point to another. The forces are assumed to be transmitted either with infinite speed or with finite speed such that sufficient time is allowed for an equilibrium situation to develop.

Now let us take up the time-varying situation. It is found experimentally that the equations for $\nabla \cdot \mathbf{E}$ and $\nabla \cdot \mathbf{B}$ remain the same; but the other four equations require modification. In the relations for $\nabla \times \mathbf{E}$ and $\nabla \times \mathbf{B}$ the change consists in the addition of extra terms; in the case of the potential fields it is necessary to interpret the integrands differently. It is convenient to distinguish the quasistatic case, when conditions are changing only slowly with time, from the case of radiation, when the changes occur quickly. Except when otherwise mentioned we will assume that quasistatic conditions exist—radiation will be considered separately. We treat, first, the modifications that are necessary in the static equation $\nabla \times \mathbf{E} = 0$.

The force on a charge moving in a region of magnetic flux is $\mathbf{F} = q\mathbf{v} \times \mathbf{B}$. Consider the conductor *ab* of Fig. 11-1*a*; it moves with velocity $\mathbf{v}$ through a region of uniform magnetic field $\mathbf{B}$, perpendicular to $\mathbf{v}$. The Lorentz forces on conduction electrons are here downward, so they will flow toward *a* and build up a negative charge there, leaving *b* positive. An electrostatic field in the rod $\mathbf{E}_{elec}$ is then produced from *b* to *a*. For a steady rod velocity $\mathbf{v}$, an equilibrium situation would be established very quickly, with $\mathbf{E}_{elec} = -\mathbf{v} \times \mathbf{B}$. The magnetic force that is due to the motion of the rod may be thought of as equivalent to a motionally induced electric field, $\mathbf{E}_{ind} = \mathbf{v} \times \mathbf{B}$, which goes from *a* to *b*. $\mathbf{E}_{ind}$ is the cause here. $\mathbf{E}_{elec}$ is an effect that is produced by $\mathbf{E}_{ind}$ because there is no closed circuit around which the electrons can flow; instead, the electrons pile up at one end of the rod, causing a deficiency at the other end.

In Fig. 11-1*b* the moving metallic rod *ab* makes contact with the two stationary conducting rails, connected together at one end as shown. Again the electrons in the rod *ab* will tend to go toward *a* as a result of $\mathbf{E}_{ind}$, but now they can continue to circulate at a steady rate around *adcba* instead of piling up at *a*. A CW steady flow of electrons results, equivalent to a CCW current. The Lorentz force between *a* and *b* acts like a battery between those two points in the sense that it produces a current flow in the closed circuit. The positive terminal of the battery would be at *b*, the negative at *a*.

How does a battery cause a constant current to flow in a circuit? An electrostatic $\mathbf{E}$ will cause an electron to move in the direction opposite to that of E; and, in fact, an electrostatic $\mathbf{E}$ can produce a large current by causing many electrons to move. But such an $\mathbf{E}$ cannot cause a *steady* current to flow, not even an exceedingly small one. If there were zero resistance in the circuit the electrons would accelerate, so the current density, $\mathbf{J} = -ne\mathbf{v}$, would increase with time instead of remaining steady, contrary to the assumption; if there were nonzero resistance the electrons would impart some kinetic energy to the lattice atoms in collisions, and this energy would appear as the irreversible joule heat loss in the material. Where would this heat energy come from? An electrostatic field is a conservative field. Any electron circumnavigating a circuit in such a field would

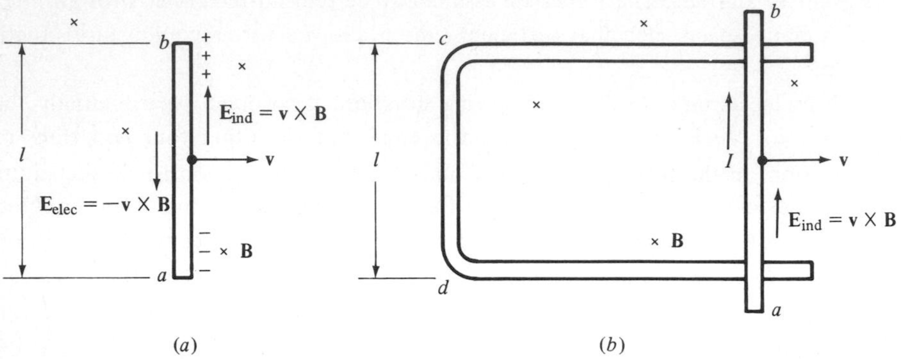

FIGURE 11-1

come back to its starting point with the same potential energy it had when it started, so the electrostatic field cannot be the source of the joule heat.

The conclusion is clear—a steady flow of current can only be produced by a nonelectrostatic **E**. Such an electric field can be produced in many ways: by the conversion of mechanical work, or of chemical energy, or by the action of light, etc. It can even be produced by electrical means. But in all cases the electric field must be one that cannot be derived from a potential function. $\mathbf{E}_{ind} = \mathbf{v} \times \mathbf{B}$, the motionally induced electric field of Fig. 11-1, is one such nonelectrostatic **E**. Technologically it is the most important one, for it is the prototype of the **E** produced in dynamos, dc generators, and ac alternators. The chemically produced **E** within a battery is another such nonelectrostatic example.

It is worth pointing out an important feature of Fig. 11-1. (This case illustrates the conversion of mechanical work into a nonelectrostatic **E**, but the feature illustrated is true regardless of the type of conversion.) The nonconservative $\mathbf{E}_{ind}$ of Fig. 11-1*b*, which is said to exist when the steady current flows, is opposite in direction to the open-circuit conservative $\mathbf{E}_{elec}$ of Fig. 11-1*a*, which exists when no current is permitted to flow. The flow of current around the circuit is in that direction dictated by $\mathbf{E}_{ind}$—in the moving rod *ab* it is upward, from *a* to *b*—despite the fact the conservative $\mathbf{E}_{elec}$ of Fig. 11-1*a* is downward. An analogy with water being circulated by a pump through a closed system is appropriate here. Everywhere outside the pump region *ab* of Fig. 11-2 the water flows downhill, to a region of lower potential energy, but inside *ab* the pump causes the water to flow uphill. It takes external energy, supplied to the pump, to accomplish this. If a valve were shut at *b*, say, so the water could not circulate, then gravity (a conservative field) would tend to move a water particle at *b* downward toward *a*, in the direction opposite to that in which it actually moves when the pump is working. An analogous

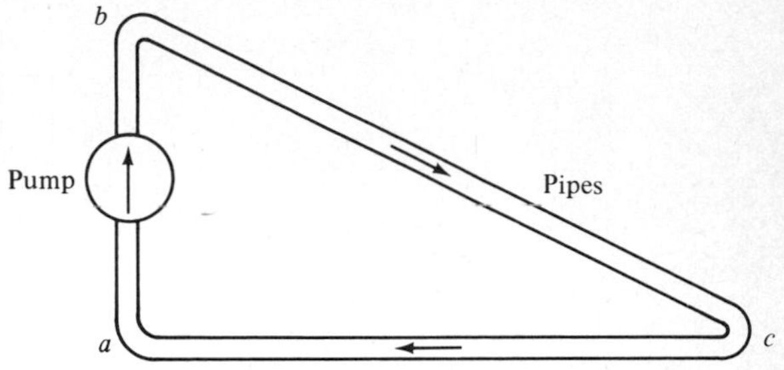

FIGURE 11-2

situation occurs in the flow of current from a battery, where the nonelectrostatic **E** is produced in a rather complex, quantum mechanical manner by chemical means.

The nonconservative electric field $\mathbf{E}_{\text{ind}}$ induced in one part of a circuit by some means is not usually the primary object of interest; that role is played by

$$V_{ba} = \int_a^b \mathbf{E}_{\text{ind}} \cdot d\mathbf{r}$$

a quantity called the induced voltage. Most people call this quantity the electromotive force, or the emf, but we will use the term voltage exclusively. The dimensions of V_{ba} are the same as those of ϕ or of $\Delta\phi$—volts. The integral above gives the voltage of b with respect to a. Because this nonconservative voltage is restricted to one region of the complete circuit, that of the source of energy, no change takes place in its value if the line integral is changed to one around the complete circuit:

$$V = \oint \mathbf{E}_{\text{ind}} \cdot d\mathbf{r}$$

Substituting $\mathbf{E}_{\text{ind}} = \mathbf{F}/q$, where q is a charge and $\mathbf{F}$ is the nonconservative force on that charge, then gives the induced voltage in the circuit,

$$V = \oint \frac{1}{q} \mathbf{F} \cdot d\mathbf{r}$$

In the case of Fig. 11-1, $(1/q)\mathbf{F} = \mathbf{v} \times \mathbf{B}$. But, in general, there may also be an induced-voltage component created electrically, so we may write

$$V = \oint [\mathbf{E} + (\mathbf{v} \times \mathbf{B})] \cdot d\mathbf{r}$$

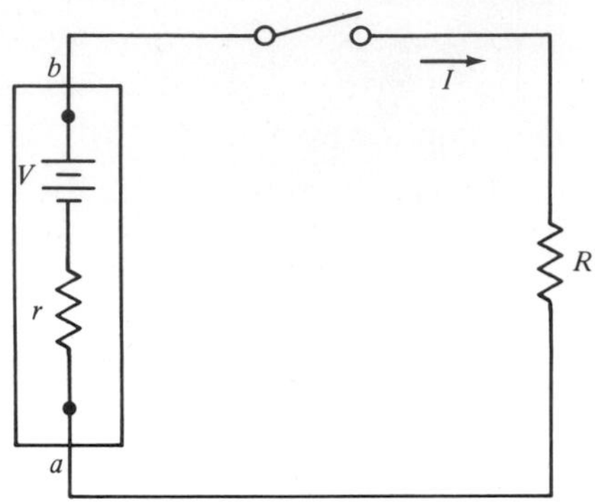

FIGURE 11-3

to take care of all cases. Here $\mathbf{E}$ is either the nonconservative induced field $\mathbf{E}_{\text{ind}}$ or that plus the conservative electric field, since the line integral of the latter alone is zero.

The quantity we have called the voltage is, as we have mentioned, often called the emf (pronounced ee-em-ef), but that name has fallen into disrepute because the quantity is work per unit charge, not force. Note that potential difference is also work per unit charge.

Take a circuit consisting of a battery (open-circuit voltage V, internal resistance r), a load resistor R, and a switch as in Fig. 11-3. When the switch is open the positive charge on the upper terminal of the battery and the negative charge on the lower terminal produce $\mathbf{E}_{\text{elec}}$, a conservative electric field downward, from b to a, inside the battery as well as outside it. When the switch is closed, nevertheless, the current flows upward from a to b inside the battery. It is not produced by the conservative electric field, $\mathbf{E}_{\text{elec}}$ or $\oint \mathbf{E}_{\text{elec}} \cdot d\mathbf{r}$, but by the nonconservative voltage

$$V = \oint (\mathbf{E}_{\text{elec}} + \mathbf{E}_{\text{ind}}) \cdot d\mathbf{r} = \int_a^b \mathbf{E}_{\text{ind}} \cdot d\mathbf{r} = -\int_a^b \mathbf{E}_{\text{elec}} \cdot d\mathbf{r}$$

Ohm's law then gives $I = V/R$ for the current that comes from the battery terminals. $\mathbf{E}_{\text{ind}}$ is the cause, produced by chemical action; $\mathbf{E}_{\text{elec}}$ is the effect, produced by $\mathbf{E}_{\text{ind}}$ when the circuit is open.

Despite the fact that voltage and potential difference have the same dimensions and units, they are distinctly different concepts. Potential difference applies to potential, a quantity which gives a conservative $\mathbf{E}$: $\mathbf{E}_{\text{elec}} = -\nabla\phi$, $\nabla \times \mathbf{E}_{\text{elec}} = -\nabla \times \nabla\phi = 0$. A charge taken around any closed path in such a field has the same potential energy after the trip as it had before, and no energy is either required or released as a result of the journey. On the other hand, voltage is not a result of a conservative field. V, like $\mathbf{E}_{\text{ind}}$ in Fig. 11-1b, can cause a steady current to flow in a closed circuit. The energy required from the heat loss in the conductors is supplied, in the present instance, by the mechanical work necessary to move the rod ab. (The current is given by V divided by the circuit resistance.) Nevertheless, while V and $\Delta\phi = \phi_1 - \phi_2$ are different quantities, they are both measured in volts and they are both measured by a voltmeter.

To apply our definition of voltage to the circuit of Fig. 11-1b we note that $\oint \mathbf{E}_{\text{elec}} \cdot d\mathbf{r} = 0$. (For Fig. 11-1$a$ take any closed path; then $\oint \mathbf{E}_{\text{elec}} \cdot d\mathbf{r} = 0$ for it also.) So

$$V = \oint (\mathbf{v} \times \mathbf{B}) \cdot d\mathbf{r} = \int_a^b + \int_b^c + \int_c^d + \int_d^a$$

where the integrand in the last four cases is the same as that on the left. But in the last three cases $\mathbf{v} = 0$, so the integrals vanish. ($\mathbf{v}$ is the velocity of the conductors relative to the stationary observer, for whom the magnetic field is $\mathbf{B}$. It has no connection with the speed with which the charge is considered to move around the circuit. The time for that displacement may be taken as zero: the displacement is then virtual rather than real.) So, if the $\hat{\mathbf{y}}$-direction is upward in the plane of the page,

$$V = \int_a^b (\hat{\mathbf{y}}vB) \cdot (\hat{\mathbf{y}}\, dy) = vB\ell$$

This result applies when $\mathbf{B}$ is uniform in space and constant in time.

When $\mathbf{B}$ is nonuniform we can write, if $\mathbf{v} = d\mathbf{s}/dt$,

$$V = \oint_C (\mathbf{v} \times \mathbf{B}) \cdot d\mathbf{r} = \oint_C (d\mathbf{r} \times \mathbf{v}) \cdot \mathbf{B} = -\frac{d}{dt} \oint_C (\mathbf{s} \times d\mathbf{r}) \cdot \mathbf{B}$$

But $\mathbf{s} \times d\mathbf{r} = d\mathbf{S}$, an element of area swept out by C in its displacement; and $\int \mathbf{B} \cdot d\mathbf{S} = \Phi$, the flux through the area bounded by C; so

$$V = \oint_C \mathbf{E}_{\text{ind}} \cdot d\mathbf{r} = -\frac{d\Phi}{dt}$$

Let Φ be positive if it comes out of the page. This is, then, the positive direction of $\mathbf{S}$, and the circumnavigation of C is to be made CCW.

The induced motional voltage $V = -d\Phi/dt$, derived above for a $\mathbf{B}$ which is constant in time, is a special case of Faraday's law. This law gives the induced voltage, in the general case, when there is any rate of change of magnetic flux, regardless of the cause. Instead of only a part of the circuit moving relative to the rest, the entire circuit may move; or the flux itself, produced by some other circuit, may change with time. The voltage in the latter case is called a transformer induced voltage. Discovered in 1840 by Michael Faraday, it is important theoretically; it gave an intimate connection between the hitherto separate fields of electricity and magnetism. However, it is also fundamental technologically: it serves to this day as the basis for the conversion of mechanical to electrical energy for the vast power needs of our society.

The differential equivalent of Faraday's law is easily obtained. Let $\mathbf{E}_{\text{ind}}$ be the total induced electric field, both motional and transformer induced. Then the general case of Faraday's law gives

$$V = \oint_C \mathbf{E}_{\text{ind}} \cdot d\mathbf{r} = -\frac{d}{dt}\Phi$$

$$\int_\Sigma (\nabla \times \mathbf{E}_{\text{ind}}) \cdot d\mathbf{S} = -\frac{d}{dt}\int_\Sigma \mathbf{B} \cdot d\mathbf{S}$$

For a fixed C, which is the case most often encountered,

$$\nabla \times \mathbf{E}_{\text{ind}} = -\frac{d\mathbf{B}}{dt} = -\frac{\partial \mathbf{B}}{\partial t}$$

Suppose we let $\mathbf{E}_{\text{elec}}$ be the conservative electrostatic field obtained from a potential function. So $\nabla \times \mathbf{E}_{\text{elec}} = 0$. Then, if $\mathbf{E} = \mathbf{E}_{\text{elec}} + \mathbf{E}_{\text{ind}}$ is the total electric field,

$$\nabla \times \mathbf{E} = (\nabla \times \mathbf{E}_{\text{elec}}) + (\nabla \times \mathbf{E}_{\text{ind}}) = \nabla \times \mathbf{E}_{\text{ind}}$$

Thus

$$\boxed{\nabla \times \mathbf{E} = -\frac{\partial \mathbf{B}}{\partial t}}$$

This is the law that replaces $\nabla \times \mathbf{E} = 0$. It is important in its own right: when it is integrated over an area bounding a circuit it yields the voltage induced in that circuit. But it is even more important in a broader sense because, together with $\nabla \cdot \mathbf{E} = (1/\epsilon_0)\rho$, it satisfies in the general case the requirements of the Helmholtz theorem for the divergence and the curl. Note, however, that $\nabla \times \mathbf{E}$ is given not in terms of the source charges and currents but, instead, in terms of $\mathbf{B}$.

It is a mistake to interpret this equation as meaning that $-\partial \mathbf{B}/\partial t$ is cause and $\nabla \times \mathbf{E}$ is effect. $\mathbf{E}$ and $\nabla \times \mathbf{E}$, as well as $\mathbf{B}$ and $-\partial \mathbf{B}/\partial t$, are all effects; the causes are ρ and $\mathbf{J}$. From this point of view the equation is best written

$$\nabla \times \mathbf{E} + \frac{\partial \mathbf{B}}{\partial t} = 0$$

The reason it is generally written as an equation for $\nabla \times \mathbf{E}$ is that, as one of the four Maxwell equations (equations for $\nabla \cdot \mathbf{E}$, $\nabla \times \mathbf{E}$, $\nabla \cdot \mathbf{B}$, $\nabla \times \mathbf{B}$), it apparently follows the dictates of the Helmholtz theorem though, in actuality, the vortex sources are not given.

A microscopic form of Faraday's law may also be obtained in terms of the potential fields. Substituting $\mathbf{B} = \nabla \times \mathbf{A}$ into $\nabla \times \mathbf{E} = -\partial \mathbf{B}/\partial t$ gives $\nabla \times (\mathbf{E} + \partial \mathbf{A}/\partial t) = 0$. Therefore $\mathbf{E} + \partial \mathbf{A}/\partial t = -\nabla \phi$, where ϕ is a potential function. So

$$\mathbf{E} = -\nabla \phi - \frac{\partial \mathbf{A}}{\partial t}$$

where $-\nabla \phi$ gives the conservative contribution to the total electric field and $-\partial \mathbf{A}/\partial t$ gives the so-called transformer induced electric field, $\mathbf{E}_{\text{ind}}$—the Faraday term. Here $\mathbf{E}_{\text{ind}}$ refers to a fixed point in space, so there is no motional induced term. The two equations

$$\boxed{\begin{aligned} \mathbf{E} &= -\nabla \phi - \frac{\partial \mathbf{A}}{\partial t} \\ \mathbf{B} &= \nabla \times \mathbf{A} \end{aligned}}$$

give $\mathbf{E}$ and $\mathbf{B}$ in the general case if ϕ and $\mathbf{A}$ are known. Note that ϕ and $\mathbf{A}$ together have only four independent components while $\mathbf{E}$ and $\mathbf{B}$ together have six components. It will be shown later that there are two relations among the components of $\mathbf{E}$ and $\mathbf{B}$.

In the case of transformer induced voltages

$$V = \oint_C \mathbf{E}_{\text{ind}} \cdot d\mathbf{r} = \oint_C \left(-\frac{\partial}{\partial t}\mathbf{A}\right) \cdot d\mathbf{r} = -\frac{\partial}{\partial t}\oint_C \mathbf{A} \cdot d\mathbf{r}$$

This expression is often more significant physically than the equivalent

$$V = -\frac{\partial}{\partial t}\Phi$$

For example, Fig. 11-4 shows a very long solenoid carrying an alternating current (of rms value I_p) in the primary winding, the one containing the source generator; here the instantaneous current is $i_p = I_{p0} \sin \omega t$, where I_{p0} is the amplitude, and $I_p = (\sqrt{2}/2)I_{p0}$. If the coil is tightly wound and the length is very long compared to the radius, then Φ is confined to the inside of the solenoid and $\mathbf{B} = 0$ outside it. When a secondary circuit is wound around and outside the primary circuit as shown, it is difficult to see physically how $V = -\partial\Phi/\partial t$ can be induced in the secondary winding to supply power to the load; the $\mathbf{B}$ produced in the region of the secondary wires by the current in the primary winding is zero. However, $\mathbf{A}$ outside the primary solenoid is not zero; and $V = -(\partial/\partial t) \oint_C \mathbf{A} \cdot d\mathbf{r}$ is readily understandable if C is identified with the secondary circuit. Here $\mathbf{A}$ has direct physical significance while $\mathbf{B}$ is, more or less, a mathematical aid.

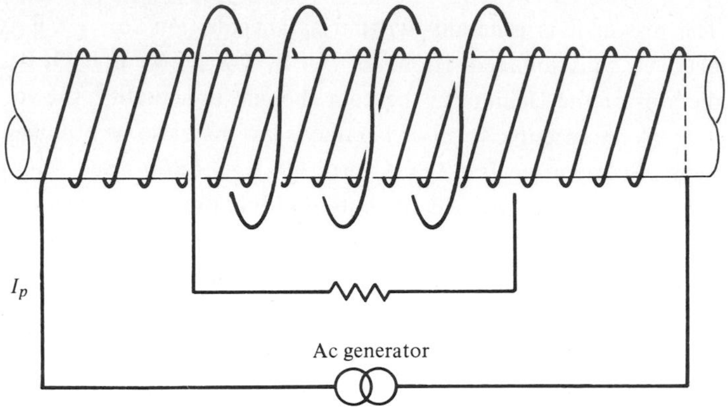

FIGURE 11-4

Faraday's law is usually expressed as $V = -\partial\Phi/\partial t$, applicable to a stationary circuit. When the circuit is moving the flux law becomes $V = -d\Phi/dt$. This is merely a summary of Faraday's experimental results that the cause of the rate of change of Φ with time is irrelevant: whether it is due to circuit motion or to changing current in another circuit, the same V results for the same rate of change of Φ. We may obtain the individual transformer and motional components of this voltage as follows. Let $\int_{\Sigma(t)} \mathbf{B} \cdot d\mathbf{S}$ be the flux at time t, when the circuit is in position C; and let $\int_{\Sigma'(t+\Delta t)} \mathbf{B} \cdot d\mathbf{S}$ be the flux at a slightly later time $t + \Delta t$, when the circuit is in position $C + \Delta C$. The area S swept out by the circuit during the time interval Δt is $\oint_C (\mathbf{v}\,\Delta t) \times d\mathbf{r}$, where $\mathbf{v}$ is the circuit velocity. See Fig. 11-5.

Then

$$V = -\frac{d\Phi}{dt} = -\lim_{\Delta t \to 0} \left(\frac{\Delta\Phi}{\Delta t}\right) = -\lim_{\Delta t \to 0} \left[\frac{1}{\Delta t}\left(\int_{\Sigma'(t+\Delta t)} \mathbf{B} \cdot d\mathbf{S} - \int_{\Sigma(t)} \mathbf{B} \cdot d\mathbf{S}\right)\right]$$

The first integral on the right, the flux through Σ' at time $t + \Delta t$, equals the original flux through Σ at time t *plus* the gain in the flux through Σ during Δt *plus* the gain in the flux in an instantaneous virtual displacement from $\Sigma(t)$ to $\Sigma'(t)$ *plus* higher-ordered terms. So, neglecting the latter,

$$V = -\lim_{\Delta t \to 0} \frac{1}{\Delta t}\left\{\left[\int_{\Sigma(t)} \mathbf{B} \cdot d\mathbf{S} + \int_{\Sigma(t)} \left(\frac{\partial}{\partial t}\mathbf{B}\,\Delta t\right) \cdot d\mathbf{S}\right.\right.$$

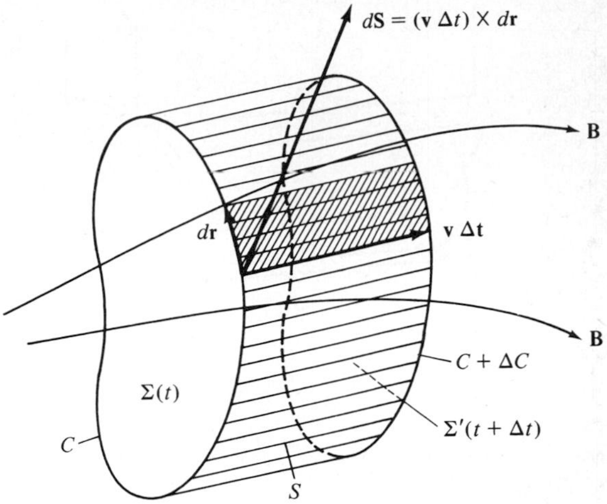

FIGURE 11-5

$$+ \oint_{C(t)} \mathbf{B} \cdot (\mathbf{v}\,\Delta t \times d\mathbf{r})\Bigg] - \int_{\Sigma(t)} \mathbf{B} \cdot d\mathbf{S}\Bigg\}$$

The first and fourth integrals cancel; in the second integral, the derivative can go to the left of the integral; in the third, the integrand may be written $d\mathbf{r} \cdot (\mathbf{B} \times \mathbf{v}\,\Delta t)$. Then

$$V = -\frac{\partial}{\partial t}\int_{\Sigma} \mathbf{B} \cdot d\mathbf{S} + \oint_{C} (\mathbf{v} \times \mathbf{B}) \cdot d\mathbf{r} = -\frac{\partial}{\partial t}\int_{\Sigma} (\nabla \times \mathbf{A}) \cdot d\mathbf{S} + \oint_{C} (\mathbf{v} \times \mathbf{B}) \cdot d\mathbf{r}$$

$$= \oint_{C} \left[-\frac{\partial \mathbf{A}}{\partial t} + (\mathbf{v} \times \mathbf{B})\right] \cdot d\mathbf{r} = \oint_{C} (\mathbf{E}_t + \mathbf{E}_m) \cdot d\mathbf{r}$$

where $\mathbf{E}_t = -\partial\mathbf{A}/\partial t$ is the transformer induced field and $\mathbf{E}_m = \mathbf{v} \times \mathbf{B}$ is the motional induced field. The integrals in these expressions for V are all to be evaluated at time t. We may write

$$V = V_{\text{transformer induced}} + V_{\text{motional induced}} = V_t + V_m$$

where

$$V_t = -\int_{\Sigma} \frac{\partial \mathbf{B}}{\partial t} \cdot d\mathbf{S} = -\oint_{C} \frac{\partial \mathbf{A}}{\partial t} \cdot d\mathbf{r} \qquad \text{and} \qquad V_m = \oint_{C} (\mathbf{v} \times \mathbf{B}) \cdot d\mathbf{r}$$

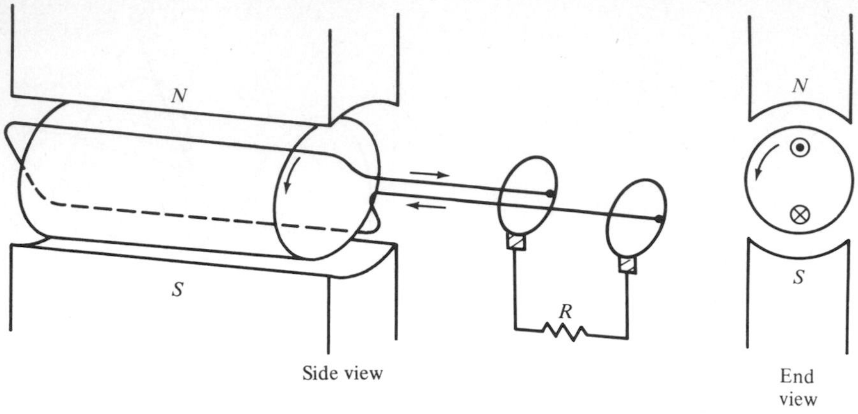

FIGURE 11-6

The induced electric field $\mathbf{E}_{\text{ind}}$ becomes the sum of the transformer induced field $\mathbf{E}_t = -\partial\mathbf{A}/\partial t$ and the motional induced field $\mathbf{E}_m = \mathbf{v} \times \mathbf{B}$, where $\mathbf{v}$ refers to the physical motion of the circuit. In the case of Fig. 11-1*b*, the transformer induced electric field $\mathbf{E}_t = 0$, and only the motional induced voltage is present; in the case of Fig. 11-4 the second term is zero; but in general both effects may occur.

Examples

1. The alternator The usual source of electric power both for home and for industry, by an overwhelming margin, is a dynamo called the alternator. Shown schematically in Fig. 11-6, it is a rotor made of laminated iron sheets with slots carrying insulated copper coils. Only one coil is shown, but the entire circumference actually contains similar coils. These traverse a region of magnetic flux created by two iron magnetic poles, labeled N for north and S for south. When the rotor is turned mechanically, either by a steam or hydroelectric turbine, a motional induced voltage is produced in the coil which causes a current to flow in the direction shown at one particular moment. A half-revolution later, however, a given coil side finds itself under a pole of opposite polarity and the current is reversed in direction. The ends of the coil are permanently fastened to slip rings which rotate with the rotor; stationary carbon brushes make electrical contact with the rings and carry an alternating current to the load resistance R, which may be an incandescent lamp, a motor, a factory, etc.

The conversion of energy here, from mechanical to electrical, is accomplished by means of the motional induced voltage $V = \oint (\mathbf{v} \times \mathbf{B}) \cdot d\mathbf{r}$ in each coil. The induced fields $\mathbf{E}_{\text{ind}}$ in the two sides of a coil are in opposite directions for a stationary observer; but they are in series, aiding, for the coil as a whole. The alternating voltage that is obtained is

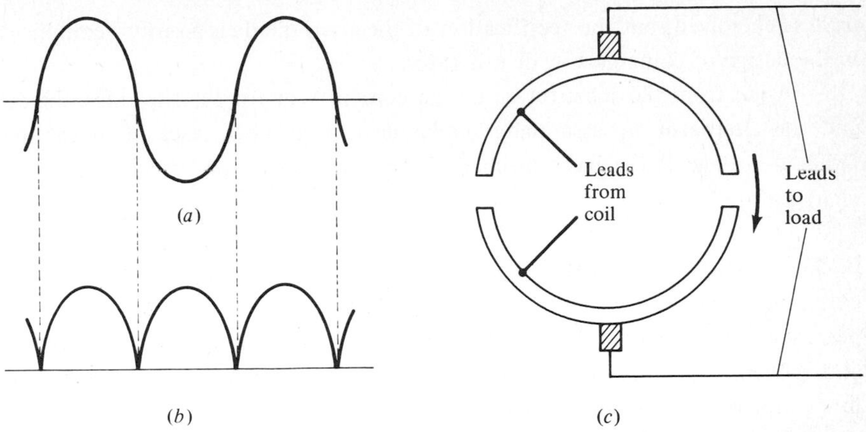

FIGURE 11-7

desirable because of the efficiency and simplicity with which the amplitude may be raised or lowered, using transformers. The sliding contacts between the stationary carbon brushes and the rotating slip rings of an alternator are a source of problems, but these are not too serious.

2. The dc generator For some applications, such as electroplating, it is necessary to have a unidirectional current flowing through the load resistor. A device that changes the sinusoidal (ac) voltage of Fig. 11-7*a* to the fluctuating, but unidirectional (dc) voltage of Fig. 11-7*b* is called a rectifier. There are many types of electronic rectifiers; but the common type that is actually employed in power stations is a mechanical rectifier, called a commutator, that is attached to the rotor of the alternator and revolves with it. As shown in Fig. 11-7*c*, the commutator consists of two semicircular cylinders with small insulating gaps between them; two brushes make contact with the separate halves. This device replaces the two slip rings of the alternator. A given brush makes contact with a coil end when that end is positive, say, with respect to the other end; when the rotor turns sufficiently so that this coil end is about to become negative rather than positive with respect to the other end, then the brush automatically shifts over to the other commutator segment. By this means a given brush, and a given end of the load resistor, always has one sign of voltage with respect to the other end.

In some applications the fluctuations of the load voltage and current are undesirable, even if these are unidirectional. Devices called filters are then employed to remove the unwanted frequency components and leave only a steady dc component.

The time intervals when the brushes straddle two segments of a commutator are periods of difficulty for the design engineer. Serious sparking problems, short-circuit conditions, high transient voltages, and many other complications arise. The problems have been solved, but the rectification of the ac to the dc is a serious complicating factor in the design of commutating dc generators.

Aside from the substitution of the commutator for the slip rings, the dc generator and the alternator are essentially similar devices. In both cases an alternating motional induced voltage is the basic mechanism for the conversion of mechanical to electrical energy.

3. The homopolar generator Faraday invented a dc generator which requires neither a commutator nor a filter yet gives a steady dc voltage, with no ripple components. It has been employed successfully in a few instances but, by and large, its inherent simplicity and efficiency have been vitiated by the necessity of high rotary speeds to produce even low voltages (with extremely high currents, however).

Figure 11-8 shows a circular metal disk revolving about a metal shaft through its center. One stationary brush makes contact with the rim of the revolving disk, another with the shaft. Two stationary permanent magnets (also circular disks) sandwich the metal disk; one magnet has a north pole (a source of **H**) on the face toward the disk; the other magnet has a south pole (a sink for lines of **H**) on the face toward the disk.

If one considers the path *abcda*, with *da* turning at angular velocity ω, the voltage V induced in it is made up, by the general Faraday law, of a transformer induced voltage $\oint -(\partial \mathbf{A}/\partial t) \cdot d\mathbf{r}$ and a motional induced voltage $\oint (\mathbf{v} \times \mathbf{B}) \cdot d\mathbf{r}$. But the former is zero here, for **B** and **A** are constant with time at all points of space if the magnets and the disk are uniform. The motional voltage is zero for *ab*, *bc*, and *cd*. But for *da* the circuit path is moving instantaneously upward with the speed $v = \omega r$; then

$$V = \int_d^a (\mathbf{v} \times \mathbf{B}) \cdot d\mathbf{r} = \int_0^R (-\hat{\mathbf{r}}\omega r B) \cdot (\hat{\mathbf{r}}\, dr) = -\frac{\omega B R^2}{2}$$

The minus sign shows that the voltage is directed opposite to the sense from d to a, i.e., it is directed from a to d; the resultant current therefore flows CW in Fig. 11-8*b*. Note the absence of slots, coils, and commutators.

Two interesting questions arise: (1) what happens if the path for calculating V is taken with *da* stationary in space and (2) what happens if the metal disk is held stationary while the two magnets are both revolved with the same angular speed in the opposite sense? The answer to (1) is that $V = 0$ for such a path, one which is simply not of physical interest. It is the path in the turning metal disk that counts, not the path in the space in which the disk is imbedded. In (2), when the disk is held stationary, whether or not the uniform magnets revolve, the motional induced voltage in *abcd* is zero. If one thinks of the lines of **B** as being rigidly attached to the magnets it might seem that there would be a

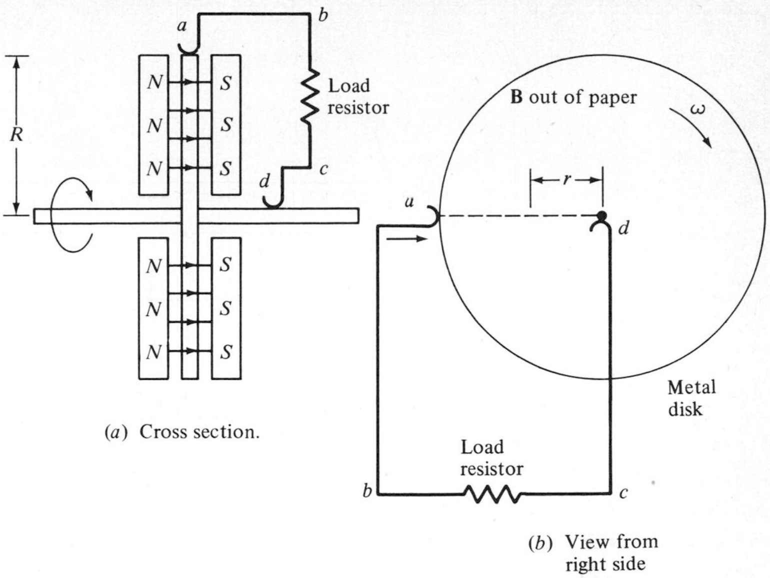

(*a*) Cross section.

(*b*) View from right side of (*a*).

FIGURE 11-8

transformer induced voltage; but this is incorrect. The lines of **B** are not defined with respect to the magnets but are defined with respect to a stationary observer. If **B** is absolutely uniform then the turning of the magnets has absolutely no effect on the **B** at any point: **B** is independent of time. So there is no transformer induced voltage in this case.

4. The linear homopolar generator The conclusions of the previous example often prove to be disturbing to students. The following variation on the theme may help to clarify the situation. Figure 11-9 shows a very long conducting ribbon, of width ℓ, which may move with constant velocity v parallel to its length. There is a uniform constant magnetic field B perpendicular both to **v** and to the width. A voltmeter makes contact with the moving conductor via two slide wires. The voltmeter may also move to the right with velocity **v**. Finally, the north pole magnet above the rod, as well as the south pole magnet below the rod, are absolutely uniform; they, too, may move in unison with velocity **v** to the right. Given all possible combinations of motion to the right—of rod, of magnets, and of voltmeter—the problem is to determine the voltmeter reading.

The table below shows the results. The velocity of the magnets is irrelevant if **B** is uniform.

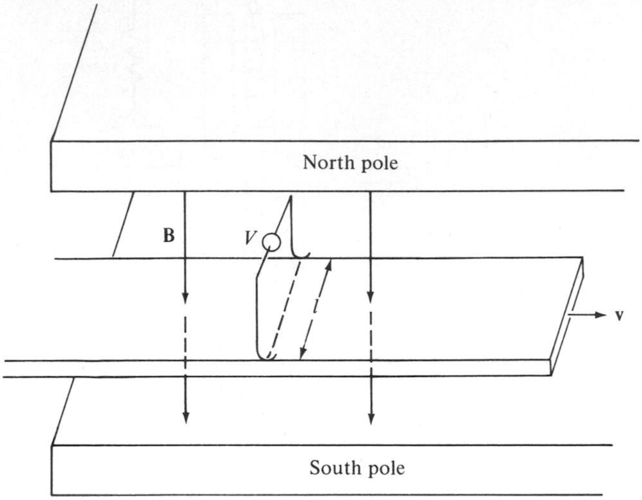

FIGURE 11-9

Velocity of rod	Velocity of magnets	Velocity of voltmeter	Reading of voltmeter
0	0	0	0
0	0	v	$vB\ell$
0	v	0	0
0	v	v	$vB\ell$
v	0	0	$-vB\ell$
v	0	v	0
v	v	0	$-vB\ell$
v	v	v	0

5. Lenz's law The minus sign in Faraday's law, $V = -d\Phi/dt$, gives the direction of the induced voltage. In Fig. 11-10*a* let the circle represent a circuit in which a voltage is being induced by a changing $\mathbf{B}_{\text{ext}}$. Suppose $\mathbf{B}_{\text{ext}}$ is out of the page. Then the area of the circle is represented by a vector which is out of the page. If $\mathbf{B}_{\text{ext}}$ increases, a voltage will be induced in the circuit, and this will cause a current to flow. A CW current flow would itself produce a $\mathbf{B}_{\text{ind}}$, within the circle, which is directed into the page; such a current would tend to oppose the increase of $\mathbf{B}_{\text{ext}}$ out of the page. Lenz's law says that the induced voltage would be CW, in order that the current should minimize the change in $\mathbf{B}_{\text{ext}}$. That is the meaning of the minus sign in Faraday's law. In Fig. 11-10*b* $\mathbf{B}_{\text{ext}}$ is again outward, but now it is decreasing. Then the sign of V is such as to produce an I_{ind}, here CCW, which produces a $\mathbf{B}_{\text{ind}}$ in the outward sense, to minimize the decrease.

6. Voltage vs. potential difference Figure 11-11 shows a circuit diagram which brings out the distinction between potential difference and voltage. The potential difference between two points is independent of the path employed in going between the two

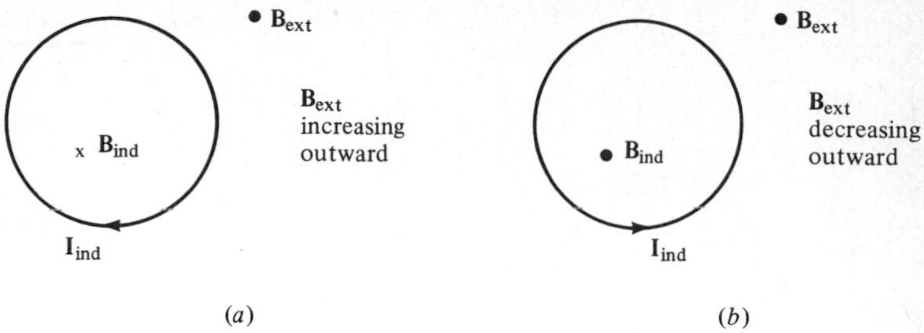

FIGURE 11-10

points. In measuring the potential difference between two points, therefore, the position of the voltmeter is immaterial. Voltage, however, is not a conservative function of position and it is quite possible for the induced voltage between two points to depend on the path between them, as we will now see.

In Fig. 11-11*a* a current produces a sinusoidal magnetic field $\mathbf{B} = \hat{\mathbf{z}}B_0 \sin \omega t$ in the interior of the circular cross section of a very long solenoid, where we take $\hat{\mathbf{z}}$ as the direction out of the page toward the reader. The alternating flux $\Phi = B_0 \pi a^2 \sin \omega t$ produces the transformer-induced voltage $V = -\omega B_0 \pi a^2 \cos \omega t$ around the circuit $A - 100\ \Omega - B - 200\ \Omega - A$. As a result, the current $I = V/R = -(\omega B_0 \pi a^2/300) \cos \omega t$ flows around this circuit and, at any instant, the voltage drop from A to B across the 100-Ω resistor is $I(100) = -(\frac{1}{3}\omega B_0 \pi a^2) \cos \omega t$; while the voltage drop, from B to A, across the 200-Ω resistor is $I(200) = -(\frac{2}{3}\omega B_0 \pi a^2) \cos \omega t$. Now, then, what does the ac voltmeter read in the configuration of Fig. 11-11*a*? Assume the voltmeter has a very high impedance and therefore draws a negligible current.

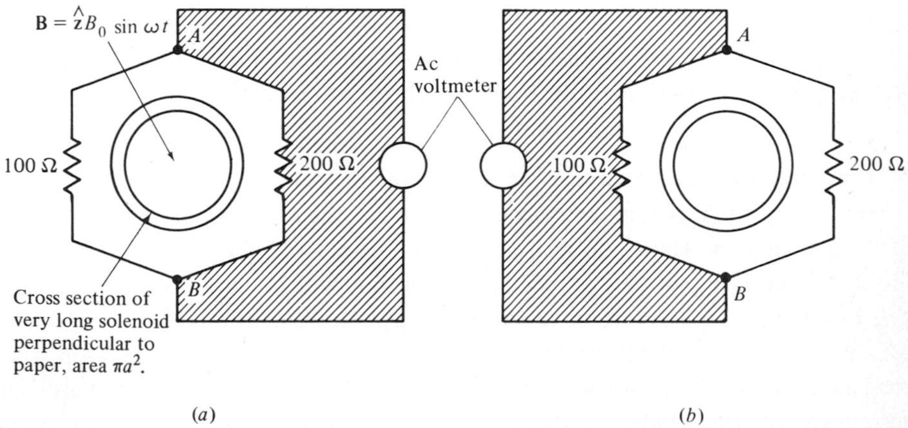

FIGURE 11-11

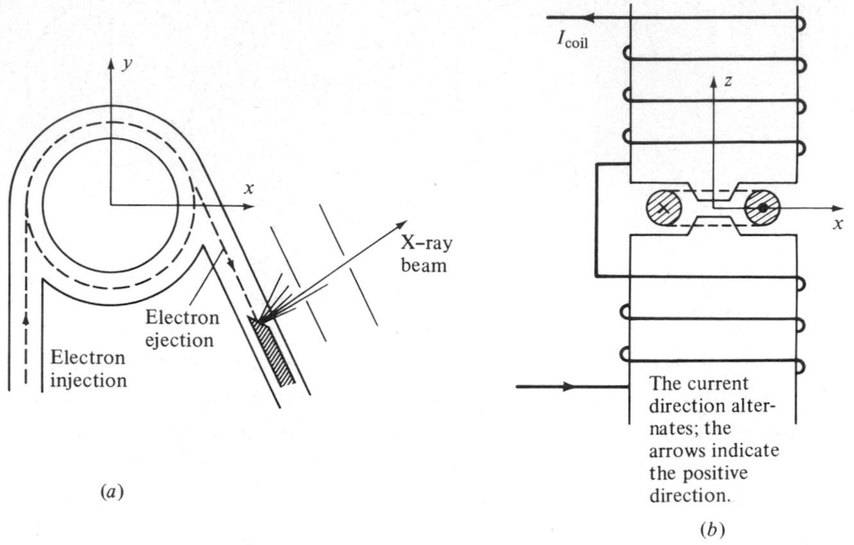

FIGURE 11-12

In the shaded area of Fig. 11-11*a* there is no flux; neither is there a rate of change of flux in this area. So there is zero induced voltage along the boundary of this area. But the voltage drop from A to B across the 200-Ω resistor is $(\frac{2}{3}\omega B_0 \pi a^2)\cos \omega t$; so the voltage drop from B to A across the voltmeter must be $-(\frac{2}{3}\omega B_0 \pi a^2)\cos \omega t$. The voltmeter needle only indicates the root-mean-square (rms) value of the amplitude of this sinusoidal voltage, $(\sqrt{2}/2)(\frac{2}{3}\omega B_0 \pi a^2)$, without regard to polarity.

In Fig. 11-11*b* the ac voltmeter's position has been shifted as shown. No voltage is induced in the shaded area here. Therefore the voltage from B to A across the voltmeter is $(\frac{1}{3}\omega B_0 \pi a^2)\cos \omega t$. The voltmeter needle now indicates the rms voltage, which is $(\sqrt{2}/2)(\frac{1}{3}\omega B_0 \pi a^2)$. This is half the reading of the previous case.

7. The betatron The betatron is an accelerator in which a narrow beam of electrons circles around the $\hat{\mathbf{z}}$ axis, in the xy plane, within an evacuated toroidal tube. Figure 11-12 shows a plan view. (Injection and ejection of the electron beam takes place following the introduction of special transient pulse currents which will not be considered here.) The bending of the beam is due to a vertical magnetic field. The electrons are accelerated by the azimuthal $\mathbf{E}$ field associated with $d\mathbf{B}/dt$.) As shown in Fig. 11-12*b*, the $\mathbf{B}$ is produced by an electromagnet. The first line of Fig. 11-13 shows the sinusoidal coil current in the field winding of the electromagnet, usually at 400 Hz. The second line of Fig. 11-13 shows that, at most, only 50 percent of a cycle can be utilized; during the

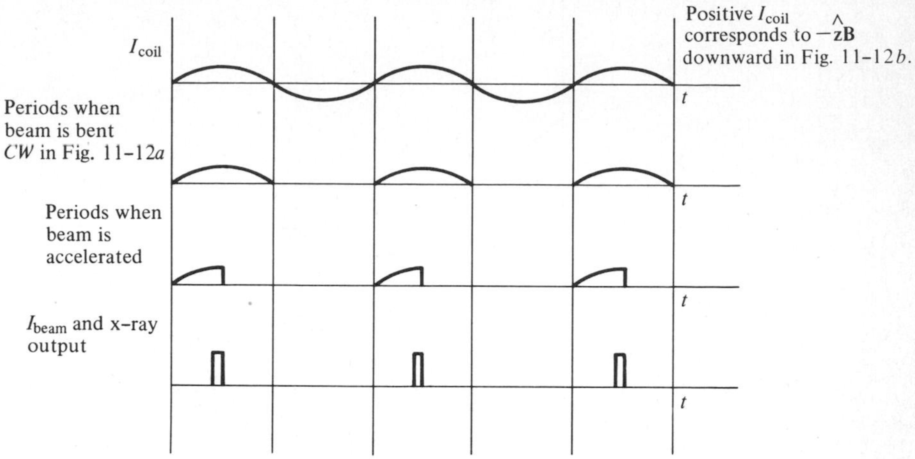

FIGURE 11-13

other half-cycle the magnetic field is along the $+\hat{z}$ direction and the injected electrons would be bent the other way, hitting the glass wall of the tube.

Even this time cannot be completely used, however. While the magnetic flux is changing there is a transformer induced voltage along the path of the mean radius of the torus, the path followed by the electron beam. But when the flux is increasing the voltage will be in one sense, and when the flux is decreasing the voltage will be in the opposite sense. The electron speed is so great that roughly a hundred thousand revolutions take place during the comparatively long time of one half-period of the coil current, but if the whole half-period were used the electrons would first be accelerated in each revolution and then be decelerated back to their original injection velocity. Consequently, only a half of each half-period is available, as shown in the third line of Fig. 11-13. The last line shows that only a small portion of this quarter cycle is actually used; the ejected electrons are then at maximum energy and almost monoenergetic. After ejection the accelerated electron beam is stopped by a metal block, thereby producing an intense beam of pulsed x-rays.

In order to make the induction accelerator a practical device there are two details that must be considered rather carefully. The first concerns focusing. Since the beam makes of the order of a hundred thousand revolutions before ejection it is essential that forces should be provided that tend to move the circulating electrons back to the mean orbit; for collisions with remanent air molecules, Coulomb repulsion of the beam electrons, and imperfect collimation of the injected beam all tend to produce a broadening beam. It is found that such focusing forces exist if **B** in the beam region varies as r^{-n} with $0 < n < 1$; and this focusing occurs both radially, in the xy plane, and axially, along the z direction. In the present example we will assume that

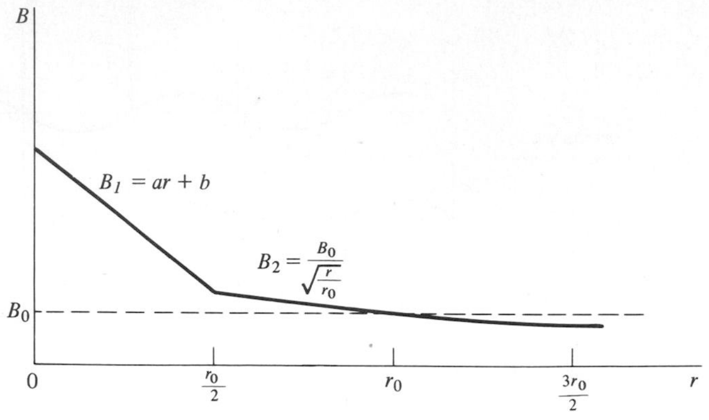

FIGURE 11-14

$$\mathbf{B}_2 = \hat{\mathbf{z}} \frac{B_0}{\sqrt{r/r_0}} \sin \omega t$$

The beam would, therefore, be a focused one. As shown in Fig. 11-14, this magnetic field will be taken to extend from a radius $r_0/2$ outward.

In what follows, the time factor sin ωt will be incorporated in the B_0 factor; all the equations involving **B** or **A** contain this implicit factor, and it can merely be explicitly inserted at the end. The subscript on **B** shows that the behavior of **B** is not the same everywhere: there is an inner region in which **B** varies differently from that in the beam region. This is the second detail. It was only the recognition of this factor by D. W. Kerst[1] in 1941 that enabled this device to become practical.

The all-important second detail deals with the fact that one wishes a stable circular orbit of fixed radius when the induced voltage causes the electrons to accelerate. One does not wish to obtain a spiral orbit, which would greatly increase the size and cost of the magnets.

For an electron to move in a fixed circle of radius r_0, given a radially symmetric magnetic field $\mathbf{B}(r) = \hat{\mathbf{z}}B(r)$, the centripetal force must equal the Lorentz force:

$$-\hat{\mathbf{r}} \frac{mv^2}{r_0} = (-e)(\hat{\boldsymbol{\phi}}v) \times (\hat{\mathbf{z}}B_0)$$

$$\frac{mv^2}{r_0} = evB_0$$

Then $p = mv = er_0B_0$. When Φ, the flux enclosed by the orbit, changes with time, an induced azimuthal force will be produced, $F_\phi = dp/dt = er_0\dot{B}_0 = (-e)E_\phi$, corresponding to

[1] D. W. Kerst and R. Serber, *Phys. Rev.*, **60**:53 (1941).

a transformer induced electric field $E_\phi = -r_0\dot{B}_0$. But the induced voltage, by the Faraday law, is $V = -\dot{\Phi} = E_\phi(2\pi r_0)$; so $-\dot{\Phi}/2\pi r_0 = -r_0\dot{B}_0$. Integrating this gives

$$\Phi = 2\pi r_0^2 B_0 + k$$

Taking $\Phi = 0$ when $B_0 = 0$ makes k vanish, so

$$\Phi = 2\pi r_0^2 B_0$$

This is called the Kerst condition, or the betatron condition.

Suppose we had $B = B_0$ everywhere, from $r = 0$ to beyond $r = r_0$; let us call the enclosed flux in that case Φ_0; i.e., $\Phi_0 = \pi r_0^2 B_0$. Then the Kerst condition

$$\Phi = 2\Phi_0$$

must be satisfied in a region of changing flux in order that the orbit radius stay constant. That is why Fig. 11-12*b* shows the poles closer together in the inner region, so that B is sufficiently greater for $r < r_0$ than its value at $r = r_0$. If we restrict our attention to **B**, then it is not clear why the value of **B** at points which the electrons never reach should affect the trajectory. The remainder of this example will show how a knowledge of **A** does make this evident; **A** gives the physical insight here, not **B**.

We will arbitrarily adopt a linear variation of **B** within a circle of radius $r_0/2$ (region 1): $B_1 = ar + b$. Then

$$\Phi_1 = \int_0^{r_0/2} (ar + b)\, 2\pi r\, dr = \pi r_0^2 \left(\frac{ar_0}{12} + \frac{b}{4}\right)$$

In the annular ring $r_0/2 \leqslant r \leqslant r_0$ the flux is

$$\Phi_2 = \int_{r_0/2}^{r_0} \frac{B_0}{\sqrt{r/r_0}}\, 2\pi r\, dr = \frac{4\pi B_0 r_0^2}{3}\left(1 - \frac{1}{2\sqrt{2}}\right)$$

Then the Kerst condition gives

$$\frac{ar_0}{12} + \frac{b}{4} = \left(\frac{2}{3} + \frac{2}{3\sqrt{2}}\right) B_0$$

Another equation connecting a and b is obtained from the fact that B is continuous at $r = r_0/2$:

$$\frac{ar_0}{2} + b = \sqrt{2}\, B_0$$

Solving the two simultaneous equations gives

$$\begin{cases} a = -2(8+\sqrt{2})\dfrac{B_0}{r_0} \\ b = (8+2\sqrt{2})B_0 \end{cases}$$

so $B_1 = -(16+2\sqrt{2})B_0(r/r_0) + (8+2\sqrt{2})B_0$.

Now assume that the frequency of the current in the coils around the magnets is so low that, essentially, **A** is connected with I by the same formula as in magnetostatics, with no additive arbitrary gradient function. This is the quasistatic approximation. For high frequencies **A** and **I** need not have the same symmetry. Since I is azimuthal (neglecting the winding pitch) then **A** must be azimuthal also. So $\mathbf{B} = \nabla \times \mathbf{A}$ gives

$$B_z = \frac{1}{r}\frac{d}{dr}(rA)$$

In region 1 this gives $(1/r)[d/dr]rA_1 = ar + b$. Integrating:

$$A_1 = \frac{a}{3}r^2 + \frac{b}{2}r + \frac{c_1}{r}$$

Here a and b have the values obtained above while c_1 is a constant of integration. c_1 may be evaluated from $\Phi = \oint_C \mathbf{A} \cdot d\mathbf{r}$ if C is taken to be a circle of infinitesimal radius about the z axis in the $z = 0$ plane. For

$$\int_0^{2\pi} \left(\frac{a}{3}r^2 + \frac{b}{2}r + \frac{c_1}{r}\right) r\, d\phi = 2\pi\left(\frac{a}{3}r^3 + \frac{b}{2}r^2 + c_1\right)$$

can approach zero as $r \to 0$ only if $c_1 = 0$. So

$$\mathbf{A}_1 = \hat{\boldsymbol{\phi}} B_0 r_0 \left[-\frac{2}{3}(8+\sqrt{2})\left(\frac{r}{r_0}\right)^2 + (4+\sqrt{2})\left(\frac{r}{r_0}\right)\right]$$

In region 2 the equation $\mathbf{B} = \nabla \times \mathbf{A}$ gives

$$\frac{1}{r}\frac{d}{dr}(rA_2) = \frac{\sqrt{r_0}B_0}{r^{1/2}}$$

Then

$$A_2 = \frac{2}{3}\sqrt{r_0}B_0 r^{1/2} + \frac{c_2}{r}$$

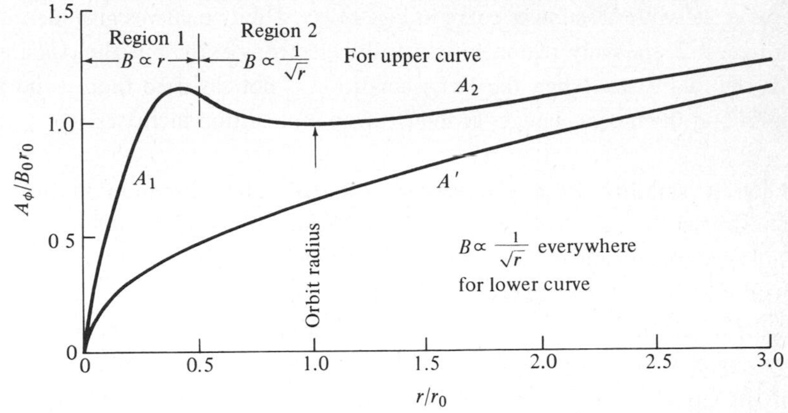

FIGURE 11-15

where c_2 is another constant of integration. Since the flux across the boundary between the two regions is a continuous function of r, c_2 may be evaluated by equating A_1 to A_2 at $r = r_0/2$. This gives

$$\mathbf{A}_2 = \hat{\boldsymbol{\phi}} \frac{B_0 r_0}{3} \left[2\sqrt{\frac{r}{r_0}} + \frac{1}{(r/r_0)} \right]$$

The top curve in Fig. 11-15 shows the resultant **A** for this assumed variation of **B** in the two regions, satisfying the Kerst condition. The stable radius $r = r_0$ corresponds to a minimum point in this curve, just as a stable equilibrium point in mechanics is given by a minimum point in the potential function. Note that there is an implicit $\sin \omega t$ factor here and the induced intensity is given by $-\dot{\mathbf{A}}$, i.e., by $(-\omega \cos \omega t)\, \mathbf{A}$; so the azimuthal force on the electrons is also a minimum at $r = r_0$.

What would be the picture if the Kerst condition were not satisfied? Suppose, for example, that there was no region 1 and that $\mathbf{B} = \hat{\mathbf{z}} B_0 / \sqrt{r/r_0}$ for all values of r. Then we would obtain

$$\mathbf{A}' = \hat{\boldsymbol{\phi}} \;\; \tfrac{2}{3} \sqrt{r_0} B_0 r^{1/2} + \frac{c_3}{r}$$

Here we would use $\Phi = \oint_C \mathbf{A} \cdot d\mathbf{r}$ about an infinitesimal circle to find that, for $\Phi \to 0$ as $r \to 0$, it is necessary to have $c_3 = 0$. So

$$A' = \tfrac{2}{3} B_0 r_0 \sqrt{\frac{r}{r_0}}$$

This is shown by the lower curve in Fig. 11-15. The two curves give identically the same **B** in region 2, the only region in which the electrons exist; but the **A**'s are quite different for the two cases. When the Kerst condition is not satisfied there is no stable minimum point and the radius changes as the electron momentum increases.

8. Orbit stability from the vector potential The minimum in the A-versus-r curve determines the value of r corresponding to the radius of the stable orbit in the betatron. But this conclusion is based on reasoning different from that employed in the similar interpretation of the minimum in either a ϕ-versus-r curve or a U-versus-r curve as a point of stability. In these latter cases $E = -d\phi/dr$ or $F = -dU/dr$, and the positive slope of the curves for $r > r_0$ gives a force directed toward smaller r: toward r_0. The negative slope of the curves for $r < r_0$ gives a force directed toward larger r: again toward r_0. So r_0 is a point of stable equilibrium.

In the case of an A-versus-r curve the significance of a minimum may be obtained from the principle of least action. It would lead us too far astray to employ this principle here in detail, but it is feasible to outline the method briefly.

As an alternative to solving various differential equations, one can find the actual trajectory followed by a particle by considering the action, a quantity defined by the integral

$$\int_{t_i}^{t_f} \mathcal{L}\,dt$$

This integral is taken from an initial point at some time t_i to a final point at another time t_f. The integrand, called the lagrangian, is given by

$$\mathcal{L} = -m_0c^2\sqrt{1 - \frac{v^2}{c^2}} - q(\phi - \mathbf{v} \cdot \mathbf{A})$$

Here m_0 and q are the particle's rest mass and charge, while **v** is the velocity with which it moves through an electromagnetic field (ϕ, A).

The action is evaluated over any path connecting the two end points at their fixed times. This value is then compared with that obtained over all other differentially different paths between the same two points and times. In general, the action depends on the choice of the path. But it is found that when the action over each of all the neighboring paths has the same value as that over any particular one, i.e., when the value of the action is stationary while the path is varied differentially, then that particular path is the one actually followed by the particle. This stationary value of the action corresponds either to a maximum, a minimum, or an inflection point in the value of the action when compared over neighboring paths. Usually it is a minimum, so the method is called the principle of least action.

In our case ϕ is zero while A is azimuthal. We wish the orbit to lie along a circular arc, so the initial and final points must be at the distance r from the center of the arc. Also, A is azimuthal here so

$$\mathcal{L} = -m_0 c^2 \sqrt{1 - \frac{v^2}{c^2}} + qvA$$

The curve for A which satisfies the Kerst condition has a minimum at r_0. It gives a stationary value for the action if the path is the circular arc $r = r_0$ with v constant. This is so because the slope of the A-versus-r curve is zero at $r = r_0$ and a differential displacement of r to $r_0 + dr$ does not, to the first order, change the value of A. If, however, the other curve for A is used, the one not satisfying the Kerst condition, its slope is not zero at $r = r_0$. In a variation of the path from a circular arc $r = r_0$ to some other path connecting the two end points it is easy to see that the value of the action changes as A increases if v is constant. It is more difficult to tell what happens to the action if v also changes as A changes, in order to keep the elapsed time over different paths the same; but it turns out that the value of the action then varies with the path selected. The value of the action along the circular arc is not stationary when compared with the value along other paths, so the actual trajectory followed by an electron in such a field would not be a circular arc $r = r_0$.

The lagrangian contains the potentials alone, not the fields. We see that a knowledge of the fields alone is not sufficient to tell us about stable orbits in the betatron. We are driven to the conclusion that the potentials here are more important than the fields. Cases arise, however, where the reverse is true. A very simple expression is obtained for the energy flux, the Poynting vector, when the fields are used; but when expressed in terms of potentials this becomes a most formidable, difficult-to-understand formula.

PROBLEMS

1 A rectangular loop of length ℓ and width w lies in a region of magnetic field $\mathbf{B}$. The angle between a normal to the plane of the loop and the direction of $\mathbf{B}$ is $(\hat{\mathbf{n}},\mathbf{B}) = \theta$, and $\mathbf{B}$ is a sinusoidal function of the time: $\mathbf{B} = \hat{\mathbf{z}}B \sin \omega t$. Find the transformer induced voltage in the loop.

2 In Prob. 1 let the sinusoidal frequency and amplitude be 60 Hz and 500 G, respectively, and let w and ℓ be 5 cm and 10 cm, respectively, and $\theta = 0°$. Find the amplitude of the induced voltage.

3 In Prob. 1 let B remain constant instead of varying with time but suppose the plane of the coil revolves, so that $\theta = \omega t$. Find the motional induced voltage in the loop.

4 In Prob. 3, how fast would the loop have to revolve to give an amplitude of the induced voltage equal to that of Prob. 2?

5 Let the loop of Probs. 1 and 3 be such that $\mathbf{B} = \hat{\mathbf{z}}B \sin \omega_1 t$ while $\theta = \omega_2 t$. Find the total, transformer plus motional, induced voltage in the loop.

6 Is it possible to choose such angular frequencies, ω_1 and ω_2, in Prob. 5 that, by selecting a particular phase angle between them, the total induced voltage in the coil will be zero?

7 A metal rod lies parallel to the y direction and is 1 m long. It moves in the $+\hat{\mathbf{x}}$ direction with a speed of 30 mi/hr. A magnetic field points in the $+\hat{\mathbf{z}}$ direction, with $B = 1000$ G. What is the motional induced voltage, in volts? In what sense does it point?

8 Suppose a voltmeter is connected to the ends of the rod of Prob. 7, to measure the voltage.

(*a*) Let the voltmeter be of the electrostatic type. This has an infinite input impedance, i.e., it draws no current. Which end of the rod will the voltmeter indicate as having the higher voltage?

(*b*) Let the voltmeter be of the d'Arsonval type. This has a high, but finite, input impedance, i.e., it permits a small current to flow in the closed circuit of voltmeter and rod. Which end of the rod will the voltmeter indicate as having the higher voltage?

(*c*) If the voltmeter resistance is r, what will be the value of the current in (*b*)?

9 What is the voltage obtained in a homopolar generator if the disk has a radius of 50 cm and turns at 3600 rev/min in a field with $B = 1000$ G? (This value is considerably smaller than the voltages obtained in standard alternators or dc generators, where the voltages induced in individual loops are added in series.)

10 Modify the conditions in Example 4 illustrating the linear homopolar generator so that

(*a*) the voltmeter contacts are made with the sides of the conducting bar rather than with the upper surface and

(*b*) the voltmeter is above the upper surface of the N magnet, rather than below the lower surface of the N magnet as in Fig. 11-9. Make a table similar to that in Example 4.

11 A particle moving in response to a central force obeys the equation $\mathbf{F}_i = m\mathbf{a}_i$, where the subscript i refers to an inertial reference frame. In a noninertial frame rotating about the z axis with constant angular velocity $\boldsymbol{\omega} = \hat{\mathbf{z}}\omega$ it is necessary to add two so-called pseudoforces to $\mathbf{F}_i$. One is the Coriolis force $-2m(\boldsymbol{\omega} \times \mathbf{v}_n)$, where $\mathbf{v}_n$ is the velocity of the particle relative to the noninertial rotating frame; if $\mathbf{v}_i$ is the velocity with respect to the inertial frame then $\mathbf{v}_n = \mathbf{v}_i - (\boldsymbol{\omega} \times \mathbf{r})$, so the Coriolis force is

$$-2m(\boldsymbol{\omega} \times \mathbf{v}_i) + 2m\boldsymbol{\omega} \times (\boldsymbol{\omega} \times \mathbf{r})$$

The second pseudoforce is the centrifugal force $-m\boldsymbol{\omega} \times (\boldsymbol{\omega} \times \mathbf{r})$. Thus, in the rotating frame $\mathbf{F}_i - 2m(\boldsymbol{\omega} \times \mathbf{v}_i) + m\boldsymbol{\omega} \times (\boldsymbol{\omega} \times \mathbf{r}) = m\mathbf{a}_n$, where $\mathbf{a}_n$ is the acceleration of the particle referred to the noninertial frame.

I. Let $\mathbf{F}_{i0} = (-e)\mathbf{E}$ be the original central Coulomb force of attraction of a nucleus for a circulating electron (where $\mathbf{E} = (Ze/4\pi\epsilon_0 r^2)\hat{\mathbf{r}}$); and suppose that a magnetic field $\mathbf{B} = \hat{\mathbf{z}}B$ is added to the region about the atom. The total force in the inertial frame is thus $\mathbf{F}_i = -e\mathbf{E} - e(v_i \times \mathbf{B})$. Then the equation for the motion relative to the rotating frame is

$$[-e\mathbf{E} - e(\mathbf{v}_i \times \mathbf{B})] - 2m(\boldsymbol{\omega} \times \mathbf{v}_i) + m\,\boldsymbol{\omega} \times (\boldsymbol{\omega} \times \mathbf{r}) = m\mathbf{a}_n$$

Assume that $m\,\boldsymbol{\omega} \times (\boldsymbol{\omega} \times \mathbf{r}) \ll 2m(\boldsymbol{\omega} \times \mathbf{v}_i)$.

(*a*) For what value of $\boldsymbol{\omega}$ does this reduce to $-e\mathbf{E} = m\mathbf{a}_n$? This equation utilizes the same force as that in the inertial frame. This value of $\boldsymbol{\omega}$ is called the Larmor angular frequency, $\boldsymbol{\omega}_L$. In this rotating frame the magnetic field has been removed from the problem.

(*b*) Is the direction of the vector representing the frame rotation parallel or antiparallel to $\mathbf{B}$ for an electron? For a proton?

(*c*) How is the magnitude and sense of $\boldsymbol{\omega}_L$ related to the magnitude and sense of $\boldsymbol{\omega}_c$, the cyclotron angular frequency with which an electron rotates in a magnetic field $\mathbf{B}$?

(*d*) In the lowest, circular, Bohr orbit of the hydrogen atom the central force is $e^2/4\pi\epsilon_0 r_0^{\,2}$; $mvr = \hbar$; and $r_0 = 4\pi\epsilon_0\hbar^2/me^2$. If we take a $\mathbf{B}$ to be no longer small when $m\omega^2 r \approx 2m\omega v_i$, for what value of $\mathbf{B}$ is the Larmor frequency no longer correct?

Larmor's theorem is a generalization of this result to the case of many electrons attracted to one nucleus. Whatever their motions (including interactions with each other), the effect of an added $\mathbf{B}$ can be eliminated in a frame rotating at $\boldsymbol{\omega}_L$, provided the $\mathbf{B}$ is not too large and provided the spins of the electrons are ignored (the Landé g factor is different for spin and orbital moments). The Larmor frequency of the rotating frame is the same as the precession frequency with which a magnetic moment revolves about a magnetic field, so the latter is often called the Larmor precession frequency.

II. The Larmor frequency does not apply to an electron circulating in a magnetic field (which in a sense may be said to be providing a central force) when the latter is increased from B to δB. This is so because here it is not true that $m\,\boldsymbol{\omega} \times (\boldsymbol{\omega} \times \mathbf{r})$ is small compared to $2m(\boldsymbol{\omega} \times \mathbf{v}_i)$.

Suppose an electron's circular orbit is due to its motion in a magnetic field $\mathbf{B} = z\mathbf{B}$ instead of, as above, being due to a central electrical force. We seek, again, the $\boldsymbol{\omega}$ of a rotating frame such that, when an extra magnetic field $\hat{\mathbf{z}}\,\delta B$ is added to the original $\hat{\mathbf{z}}B$, the total force in this frame is the same as the original one. Then

$$\mathbf{v}_i \mathrm{x}(\delta \mathbf{B}) = \hat{\mathbf{r}}\mathbf{v}_\phi\,\delta B = \hat{\mathbf{r}}(\omega_c r)\delta B = \hat{\mathbf{r}}(\omega_c r)\left(\frac{m}{e}\,\delta\omega_c\right)$$

$$\boldsymbol{\omega} \times \mathbf{v}_\phi = -\hat{\mathbf{r}}v_\phi\omega = -\hat{\mathbf{r}}r\omega_c\omega$$

$$\boldsymbol{\omega} \times (\boldsymbol{\omega} \times \mathbf{r}) = -\hat{\mathbf{r}}\omega^2 r$$

So

$$\mathbf{F}_i = \mathbf{F}_{io} - e(\mathbf{v}_i \times \delta\mathbf{B}) = -e\,\mathbf{v}_i \times \mathbf{B} - e\,\mathbf{v}_i \times \delta\mathbf{B}$$

Then $-e(\mathbf{v}_i \times \mathbf{B}) + [-e(\mathbf{v}_i \times \delta\mathbf{B}) - 2m(\boldsymbol{\omega} \times \mathbf{v}_i) + m\,\boldsymbol{\omega} \times (\boldsymbol{\omega} \times \mathbf{r})] = m\mathbf{a}_n$

Find (*a*) the equation giving ω in terms of ω_c and $\delta\omega_c$, and (*b*) the value of ω such that the quantity in brackets equals zero, (*c*) the ratio of centrifugal to Coriolis force here.

12 Suppose the shaded plane of Fig. 11-11 in Example 6 (the plane containing the voltmeter and the end points *A* and *B*) is swung around through an angle of almost 180°. This would be between the two extreme positions illustrated there, but excluding a small range of angles near the center, where the solenoid would interfere with the voltmeter.

(*a*) How would the voltmeter readings vary between the two extremes?

(*b*) Suppose the solenoid had a finite length, less than that of the voltmeter leads; how would the readings vary as the plane was turned through 180°?

13 Suppose the rod and the rails of Fig. 11-1 are ordinary conductors with a total circuit resistance of *R* ohms at a given time.

(*a*) What is the circuit current at that time?

(*b*) What is the power dissipated in the resistor?

(*c*) What is Fv, the rate at which work is being done on the rod?

14 Repeat Prob. 13 when the rod and rails are superconductors.

15 (*a*) Find the energy gained per revolution by an electron in the betatron, in joules.

(*b*) Find the energy gained, in electron volts, during the first revolution. Assume $B = B_0 \sin 2\pi f t$ at the orbit radius and the Kerst condition is satisfied; and also assume that the time per revolution $\ll 1/f$.

16 Two circuits, each consisting of a single circular turn, are parallel to each other and coaxial. If the currents in the two flow in the same sense and one of the circuits is moved axially away from the other, what happens to the current in the second circuit when (*a*) the circuits carry conduction currents supplied by batteries, (*b*) the circuits are both superconducting, (*c*) the circuits carry convection currents, such as those in two betatron doughnuts?

17 Assume that a betatron doughnut has a mean radius of 0.5 m, operates at 400 Hz, and has an injection energy so high (e.g., 10^4 eV) that the electrons travel at essentially a constant speed of 2.9×10^8 m/s. (This is a relativistic effect, the increase in energy resulting from an increase of mass rather than of speed.) If a quarter of a cycle is utilized between injection and ejection, find the number of revolutions.

18 Figure 11-16 shows a circuit, which surrounds a very long solenoid with some given, constant, magnetic flux inside it, shown shaded in cross section. There is negligible flux outside the solenoid. Two single-pole switches, S_1 and S_2, are ganged so that S_1 is originally open and S_2 originally closed. When S_1 is closed and S_2 is opened the voltmeter circuit area is changed from *ABCD*, which has no flux, to *AEFD*,

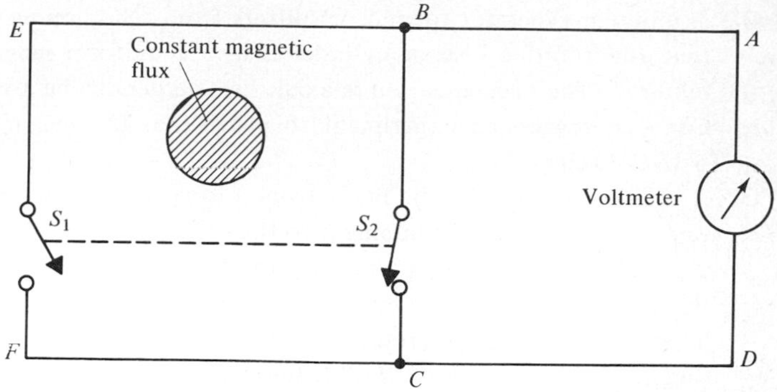

FIGURE 11-16

which does have flux. There is no motional induced voltage and no transformer induced voltage, but there is a finite $d\Phi/dt$. Will the voltmeter read when the switch bar is pressed?

19 In a diathermy machine, living human tissue is exposed to an alternating magnetic field created by ac current in a solenoid winding. If the tissue conductivity is g, the power density given to the tissue is $P = jE = gE^2$. It is often desirable that this be uniform within the solenoid so that, e.g., an arm would be uniformly heated. Taking **B** to be uniform inside the solenoid, find P as a function of r, the distance from the axis. Let the tissue currents be assumed so small that **B** is unaffected by them. Is the heating uniform?

20 Figure 11-17 shows two metal plates so constructed that with very slight angular motion the point of contact between them changes by the large distance from P to P'. The flux through the shaded area changes drastically. Will the voltmeter read?

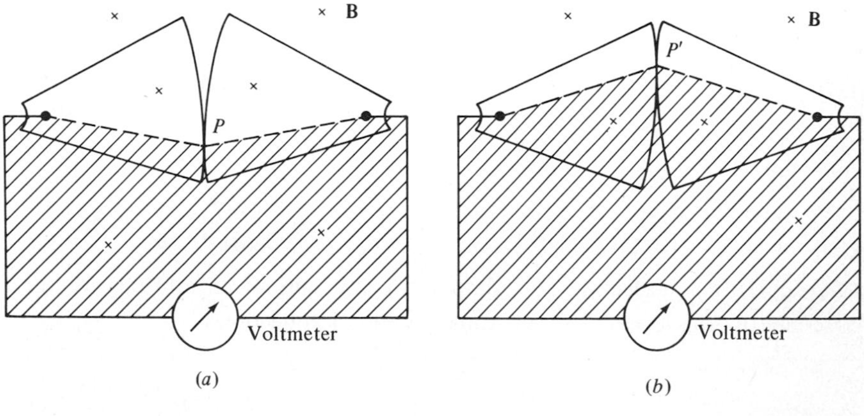

FIGURE 11-17

21 A unipolar generator (or motor) differs from a homopolar generator (or motor) in that the rotating disk or cylinder is also a uniform magnet; no other magnet is required. The magnetization is axial, i.e., perpendicular to the plane circular faces. Can you suggest an experiment to prove that the magnetic field does not rotate with the magnet?

22 A Möbius surface may be made from a long narrow rectangle of metal by twisting one end 180° and then joining it to the other end. Such a surface has only one side: a pencil, starting at any point, may reach any other point without being lifted from the metal or crossing the edge. Does Faraday's law apply to such a piece of metal? That is, is the voltage around its edge given by the negative rate of change of the magnetic flux through its area? If there is a voltage, what is its value?

11-2 THE SO-CALLED DISPLACEMENT CURRENT

Maxwell was the first to see that Ampere's law applied only to static conditions and would have to be modified in order to hold for time-varying conditions also. This followed at once from the equation of continuity,

$$\frac{\partial \rho}{\partial t} + \nabla \cdot \mathbf{J} = 0$$

The equation $\nabla \times \mathbf{B} = \mu \mathbf{J}$, the microscopic form of Ampere's law, gives $\nabla \cdot \mathbf{J} = (1/\mu)\ \nabla \cdot \nabla \times \mathbf{B} = 0$. Then the equation of continuity yields $\partial\rho/\partial t = 0$. This is certainly true for static conditions and it is certainly not true for nonstatic conditions.

Maxwell was also the first to see how to remedy the situation. Suppose we add an unknown term $\mathbf{X}$ to the right-hand side of the microscopic Ampere equation:

$$\nabla \times \mathbf{B} = \mu \mathbf{J} + \mathbf{X}$$

Then

$$\nabla \cdot \mathbf{J} = \frac{1}{\mu}\, \nabla \cdot [(\nabla \times \mathbf{B}) - \mathbf{X}] = -\frac{1}{\mu}\, \nabla \cdot \mathbf{X}$$

Also,

$$\frac{\partial \rho}{\partial t} = \frac{\partial}{\partial t}\, \nabla \cdot \mathbf{D} = \nabla \cdot \frac{\partial \mathbf{D}}{\partial t}$$

so the equation of continuity gives

$$\nabla \cdot \left(\frac{\partial \mathbf{D}}{\partial t}\right) - \frac{1}{\mu}\, \nabla \cdot \mathbf{X} = 0$$

or $\mathbf{X} = \mu(\partial \mathbf{D}/\partial t)$, neglecting (for simplicity) any curl of an arbitrary vector function. Consequently, the revised form of Ampere's equation becomes

$$\boxed{\nabla \times \mathbf{B} = \mu \left(\mathbf{J} + \frac{\partial \mathbf{D}}{\partial t}\right)}$$

The addition of the term $\partial \mathbf{D}/\partial t$ to the $\mathbf{J}$ on the right-hand side of Ampere's equation has the most far-reaching consequences. In fact, it is probably Maxwell's most important contribution to the study of electricity and magnetism. The physical reality it is describing is simply this: a magnetic field can be associated not only with a conventional current but also with a changing electric field. Thus, as we will see below, in the region between the plates of a capacitor which is either charging or discharging there will also exist a magnetic field; this is absent under electrostatic conditions.

Under static conditions $\dot{\mathbf{D}} = 0$ and the equation becomes the usual Ampere's law. Maxwell, unfortunately, gave the added term a name: the displacement current density, by analogy with the conventional (i.e., conduction and convection) current density $\mathbf{J}$. This name has caused considerable confusion because the displacement current density was subsequently interpreted to be a vortex source (i.e., a cause of $\nabla \times \mathbf{B}$) just as $\mathbf{J}$ was a vortex source, or cause. The situation is somewhat similar to that with Faraday's law.

We will show in Example 3, below, that the value of $\mathbf{B}$ determined at any field point by the contribution of $\dot{\mathbf{D}}$ to the law of Biot-Savart, integrated over all space, is zero. Consequently, the addition of $\dot{\mathbf{D}}$ to the $\mathbf{J}$ in the integrand of the law of Biot-Savart is immaterial and $\dot{\mathbf{D}}$ cannot be regarded as a cause of $\mathbf{B}$. Only ρ and $\mathbf{J}$ are the causes; all the fields—$\mathbf{E}$, ϕ, $\mathbf{B}$, $\mathbf{A}$—and their space and time derivatives are effects. From this point of view the equation above should be written

$$\nabla \times \mathbf{B} - \mu \frac{\partial \mathbf{D}}{\partial t} = \mu \mathbf{J}$$

while Faraday's law becomes

$$\nabla \times \mathbf{E} + \frac{\partial \mathbf{B}}{\partial t} = 0$$

The latter equation simply states that there are two effects which cancel each other. Similarly, when $\mathbf{J} = 0$ the former equation states that there are two effects which cancel each other; when $\mathbf{J} \neq 0$ the two effects do not cancel each other. But, in any event, they are all effects, not causes.

In this text we have avoided the terms "displacement," "displacement current," and "displacement current density." We will stay with the two vectors, $\mathbf{E}$ and $\mathbf{B}$, which are produced by all the charges, free or polarization, and by all the currents, conventional or magnetization. The two auxiliary vectors, $\mathbf{D}$ and $\mathbf{H}$, offer only illusory simplicity, for the problems are merely transferred: to determine $\mathbf{P}$ and $\mathbf{M}$. Even then, there are still problems with $\mathbf{D}$ and $\mathbf{H}$. True $\nabla \cdot \mathbf{D}$ will be given only by the free charge density; but even if $\rho_{free} = 0$, a nonvanishing solenoidal $\mathbf{D}$ can still exist, as with an electret. So $\mathbf{D}$ is not so simple. As for $\mathbf{H}$, which is neither irrotational nor solenoidal, it has not only vortex sources (conventional current) but also ordinary sources (magnetization inhomogeneities), so it is certainly not simple.

We are, finally, in a position to write the divergence and curl expressions for the vectors **E** and **B**, the famous Maxwell equations. For simplicity, if it is assumed that ϵ is not a function of time, then

$$\nabla \cdot \mathbf{E} = \frac{1}{\epsilon}\rho \qquad \nabla \cdot \mathbf{B} = 0$$

$$\nabla \times \mathbf{E} = -\frac{\partial \mathbf{B}}{\partial t} \qquad \nabla \times \mathbf{B} = \epsilon\mu\frac{\partial \mathbf{E}}{\partial t} + \mu\mathbf{J}$$

It is seen that they really do not satisfy the requirements of the Helmholtz theorem, for the right-hand sides of the two curl equations are unknown, instead of given, functions of space and time. This is too bad insofar as the simplicity of finding the answers is lost, but it is very good indeed insofar as finding the succinct differential equations that summarize *all* classical electrodynamics. The only factors that change, from one problem to another, are either the boundary conditions or the initial conditions.

Examples

1. The (A,ϕ) analog of Maxwell's equations The four Maxwell equations may be converted to differential equations giving the two potential functions or fields in terms of ρ and **J**. This is done by means of the equations

$$\mathbf{E} = -\nabla\phi - \frac{\partial \mathbf{A}}{\partial t} \qquad \text{and} \qquad \mathbf{B} = \nabla \times \mathbf{A}$$

It is interesting to note that only two of the four Maxwell equations are employed in this conversion. These are the ones for $\nabla \cdot \mathbf{E}$ and $\nabla \times \mathbf{B}$, the two equations involving ρ and **J**. The other two Maxwell equations are consequences of the defining equations, above, for **E** and **B** as functions of ϕ and A. We will restrict ourselves to a vacuum.

Putting $\mathbf{E} = -\nabla\phi - \partial\mathbf{A}/\partial t$ into $\nabla \cdot \mathbf{E} = (1/\epsilon_0)\rho$ gives

$$\nabla^2\phi + \frac{\partial}{\partial t}(\nabla \cdot \mathbf{A}) = -\frac{1}{\epsilon_0}\rho$$

Similarly, $\nabla \times \mathbf{B} = \epsilon_0\mu_0(\partial\mathbf{E}/\partial t) + \mu_0\mathbf{J}$ gives $\nabla \times \nabla \times \mathbf{A} = \epsilon_0\mu_0(\partial/\partial t)(-\nabla\phi - \partial\mathbf{A}/\partial t) + \mu_0\mathbf{J}$ or

$$\nabla\nabla \cdot \mathbf{A} - \nabla^2\mathbf{A} = -\epsilon_0\mu_0\nabla\dot{\phi} - \epsilon_0\mu_0\ddot{\mathbf{A}} + \mu_0\mathbf{J}$$

where a dot represents a derivative with respect to time. So

$$(\nabla^2\mathbf{A} - \epsilon_0\mu_0\ddot{\mathbf{A}}) - \nabla(\nabla \cdot \mathbf{A} + \epsilon_0\mu_0\dot{\phi}) = -\mu_0\mathbf{J}$$

The two partial differential equations in the boxes above are the equivalents, using the potentials ϕ and **A**, of the Maxwell equations using **E** and **B**. Like Maxwell's equations, these equations are mixed: each equation contains both ϕ and **A**.

The value of $\nabla \cdot \mathbf{A}$ in magnetostatics is zero (Chap. 3, Sec. 5, Prob. 19). Inserting this into the two equations above gives

$$\begin{cases} \nabla^2 \phi = -\dfrac{1}{\epsilon_0}\rho \\ \nabla^2 \mathbf{A} - \epsilon_0 \mu_0 (\ddot{\mathbf{A}} + \nabla\dot{\phi}) = -\mu_0 \mathbf{J} \end{cases}$$

One of the equations still seems to be mixed. If the currents are constant in time, however, it follows that $\dot{\phi} = 0$ and $\ddot{\mathbf{A}} = 0$; $-\epsilon_0(\ddot{\mathbf{A}} + \nabla\dot{\phi})$ is, in fact, the so-called displacement current density. So in statics

$$\nabla^2 \phi = -\frac{1}{\epsilon_0}\rho \qquad \nabla^2 \mathbf{A} = -\mu_0 \mathbf{J}$$

We have already obtained these independent equations.

Lorentz found a more general condition for making the equations separable, namely

$$\boxed{\nabla \cdot \mathbf{A} + \epsilon_0 \mu_0 \dot{\phi} = 0}$$

The two equations then become

$$\boxed{\begin{aligned} \nabla^2 \phi - \epsilon_0 \mu_0 \ddot{\phi} &= -\frac{1}{\epsilon_0}\rho \\ \nabla^2 \mathbf{A} - \epsilon_0 \mu_0 \ddot{\mathbf{A}} &= -\mu_0 \mathbf{J} \end{aligned}}$$

It will be shown subsequently that

$$\epsilon_0 \mu_0 = \frac{1}{c^2}$$

where c is the speed of any electromagnetic wave in vacuum, relative to any observer. Then the two equations may be written in terms of the d'Alembertian operator, $\Box^2 \equiv \nabla^2 - (1/c^2)(\partial^2/\partial t^2)$, as follows:

$$\boxed{\begin{aligned} \Box^2 \phi &= -\frac{1}{\epsilon_0}\rho \\ \Box^2 \mathbf{A} &= -\mu_0 \mathbf{J} \end{aligned}}$$

All that happens formally with the Lorentz condition is that the laplacian operator is replaced by the d'Alembertian.

Up to this point the Lorentz condition is quite arbitrary. It is a convenience, because it enables us to deal with one equation for ϕ and another for **A**. In Chap. 14, however, it will be seen that the Lorentz condition is much more than merely a simplifying choice. $(\mathbf{A},\phi)$ constitute a four-vector in space-time; there is an operator in four-dimensional space-time which corresponds to the divergence in the ordinary three-dimensional space of the real world—let us call it the four-divergence; then the Lorentz condition says that the four-divergence of the four-vector $(\mathbf{A},\phi)$ is zero. This is a very basic statement. When the $(\mathbf{A},\phi)$ four-vector is "solenoidal" in space-time, i.e., when its four-divergence vanishes, then the differential equations for **A** and ϕ become independent of each other; but this does not imply that **A** and ϕ are independent of each other, in general. That is only true under electrostatic and magnetostatic conditions.

2. The charging capacitor Maxwell's equation $\nabla \times \mathbf{B} = \mu_0(\mathbf{J} + \dot{\mathbf{D}})$, when integrated over an open surface Σ which is bounded by a curve C, gives

$$\oint_C \mathbf{B} \cdot d\mathbf{r} = \mu_0 \int_\Sigma (\mathbf{J} + \dot{\mathbf{D}}) \cdot d\mathbf{S}$$

We will write this as

$$\oint_C \mathbf{B} \cdot d\mathbf{r} - \epsilon_0\mu_0 \frac{\partial}{\partial t} \int_\Sigma \mathbf{E} \cdot d\mathbf{S} = \mu_0 \int_\Sigma \mathbf{J} \cdot d\mathbf{S}$$

to emphasize the cause-effect relationship. Suppose we apply this equation to various surfaces Σ in turn, in the vicinity of a vacuum capacitor with circular plane plates which is being charged by a time-dependent current.

Figure 11-18a shows the situation when the surface Σ_1 is taken far from the capacitor. The lines of the electric field lie predominantly in the region between the capacitor plates, but there are also a few fringing lines. By symmetry,

$$\oint_{C_1} \mathbf{B} \cdot d\mathbf{r} = 2\pi r B$$

If we let $\Phi_{\text{elec}} = \int_{\Sigma_1} \mathbf{E} \cdot d\mathbf{S}$ be the electric flux through Σ_1, the second term in the equation is $-\epsilon_0\mu_0 \dot{\Phi}_{\text{elec}}$; $\dot{\Phi}_{\text{elec}}$ is negative with the given Σ_1. The last term in the equation is $+\mu_0 I$ since, with the positive direction of C_1 as shown, the $+d\mathbf{S}$ direction is to the right. So

$$2\pi r B - \epsilon_0\mu_0 \dot{\Phi}_{\text{elec}} = \mu_0 I$$

Very far from the capacitor $\dot{\Phi}_{\text{elec}} \approx 0$ and **B** is given essentially by

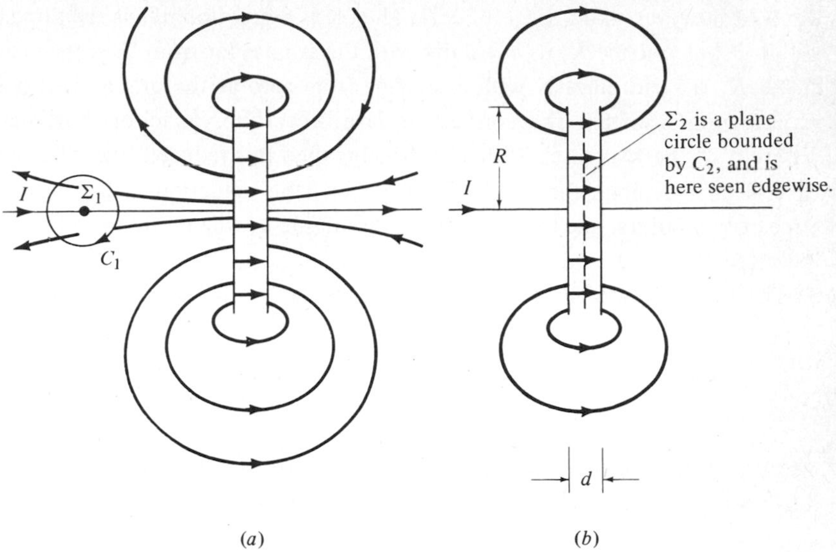

FIGURE 11-18

$$\mathbf{B} = \hat{\boldsymbol{\phi}} \frac{\mu_0 I}{2\pi r}$$

This is also the value obtained by Ampere's law or the law of Biot-Savart. For the latter two it is true, exactly, for an infinitely long conductor; the approximation here is caused by the effect of a small, faraway gap in the circuit.

If C_1 is moved closer and closer to the capacitor the value of B at a given value of r decreases more and more below the Biot-Savart value for the infinitely long conductor. This is reasonable, since the effect of the gap becomes more and more important. As one approaches the capacitor at constant r two effects, the decrease in $2\pi rB$ and the increase in $-\epsilon_0\mu_0\dot{\Phi}_{\text{elec}}$, add together in such a way that the sum is the constant cause (at a given time) $\mu_0 I$.

In Fig. 11-18b the surface Σ_2 is between the capacitor plates. The line integral again gives $2\pi rB$; $\dot{\Phi}_{\text{elec}} > 0$ here, so the second term is negative; the last term is zero. Then

$$2\pi rB - \epsilon_0\mu_0\dot{\Phi}_{\text{elec}} = 0$$

$$\mathbf{B} = \hat{\boldsymbol{\phi}} \frac{\mu_0}{2\pi r}(\epsilon_0\dot{\Phi}_{\text{elec}})$$

Within the capacitor plates two effects occur as one goes out radially: a change in $2\pi rB$ and a change in $-\epsilon_0\mu_0\dot{\Phi}_{\text{elec}}$. These two effects, when added, cancel each other exactly. We leave it to Prob. 20 to show that, when the fringing effect is small, $\epsilon_0\dot{\Phi}_{\text{elec}} = I(r^2/R^2)$

if $r \leqslant R$; however, $\epsilon_0 \dot{\Phi}_{\text{elec}} = I$ if $r \geqslant R$. Then **B** as a function of r is the same (in the region within d but outside R) as at a point with the same r, far from the capacitor. Within both d and R, **B** varies linearly with r, starting from zero at the origin. This is similar to the variation of **B** within the conductors leading to the capacitor, but the range of r is different in the two cases. If there is fringing then **B** is reduced from the value above, just as it is farther along the conductor; however, the reduction in **B** is not necessarily the same at two points, one inside and the other inside d, when both have the same r.

3. The J in the law of Biot-Savart In magnetostatics the law of Biot-Savart is

$$\mathbf{B}(\mathbf{r}_t) = \frac{\mu_0}{4\pi} \int_{\mathcal{V}} \frac{\mathbf{J}(\mathbf{r}_s) \times \hat{\mathbf{R}}_{st}}{R^2} \, d\tau_s$$

where $\mathcal{V}$ is all space. A question arises: if we were to permit variations with time, would this law have the same form or rather should it become

$$\mathbf{B}(\mathbf{r}_t) = \frac{\mu_0}{4\pi} \int_{\mathcal{V}} \frac{(\mathbf{J} + \dot{\mathbf{D}}) \times \hat{\mathbf{R}}_{st}}{R^2} \, d\tau_s$$

We shall now show that

$$\int_{\mathcal{V}} \frac{\dot{\mathbf{D}} \times \hat{\mathbf{R}}_{st}}{R^2} \, d\tau_s = 0$$

if the integration is to be taken over all space.

Let

$$\mathbf{B}'(\mathbf{r}_t) = \frac{\mu_0}{4\pi} \int_{\mathcal{V}} \frac{\dot{\mathbf{D}} \times \hat{\mathbf{R}}_{st}}{R^2} \, d\tau_s = \frac{\mu_0}{4\pi} \int_{\mathcal{V}} \dot{\mathbf{D}} \times \nabla_s \left(\frac{1}{R} \right) d\tau_s$$

Using

$$\nabla_s \times \left(\frac{1}{R} \dot{\mathbf{D}} \right) = \left[\frac{1}{R} (\nabla_s \times \dot{\mathbf{D}}) \right] + \left[\left(\nabla_s \frac{1}{R} \right) \times \dot{\mathbf{D}} \right]$$

and considering empty space, for convenience,

$$\nabla_s \times \dot{\mathbf{D}} = \epsilon_0 \nabla_s \times \dot{\mathbf{E}} = \epsilon_0 \frac{\partial}{\partial t} (\nabla_s \times \mathbf{E})$$

Then

$$\mathbf{B}'(\mathbf{r}_t) = \frac{\epsilon_0 \mu_0}{4\pi} \frac{\partial}{\partial t} \int_{\mathcal{V}} \frac{1}{R} (\nabla_s \times \mathbf{E}) d\tau_s - \frac{\mu_0}{4\pi} \int_{\mathcal{V}} \nabla_s \times \left(\frac{1}{R} \dot{\mathbf{D}} \right) d\tau_s$$

But $\mathbf{E} = -\nabla_s \phi - (\partial \mathbf{A}/\partial t)$; defining quasistatic conditions as those for which $\dot{\mathbf{A}} \ll \nabla_s \phi$ (radiation effects negligible) and assuming such conditions to hold, this gives

$$\nabla_s \times \mathbf{E} \approx -\nabla_s \times \nabla_s \phi = 0$$

Then

$$\mathbf{B}'(\mathbf{r}_t) = -\frac{\mu_0}{4\pi} \int_{\mathcal{V}} \nabla_s \times \left(\frac{\dot{\mathbf{D}}}{R} \right) d\tau_s$$

Now we make use of a theorem, analogous to Gauss' theorem, which we have employed in Chap. 1, viz., for any vector function $\mathbf{F}$,

$$\int_{\mathcal{V}} \nabla \times \mathbf{F}\, d\tau = -\oint_{\Sigma} \mathbf{F} \times d\mathbf{S}$$

So

$$\mathbf{B}'(\mathbf{r}_t) = \frac{\mu_0}{4\pi} \oint_{\Sigma} \frac{\dot{\mathbf{D}}}{R} \times d\mathbf{S}_s = \frac{\epsilon_0 \mu_0}{4\pi} \frac{\partial}{\partial t} \oint_{\Sigma} \frac{\mathbf{E}(\mathbf{r}_s) \times d\mathbf{S}_s}{R}$$

(Note that $\dot{\mathbf{D}}$ is considered at $\mathbf{r}_s$, just like $\mathbf{J}$, since we are attempting to think of these as causes. In Maxwell's laws, on the other hand, all the quantities– $\nabla \times \mathbf{B}$, $\mathbf{J}$, and $\dot{\mathbf{D}}$–are considered at one point; in the generalized Ampere's law (see the previous example) both $\mathbf{J}$ and $\dot{\mathbf{D}}$ are considered on an open surface while $\mathbf{B}$ is taken on its boundary.) When $\mathcal{V}$ includes all space, Σ is at infinity. There $\mathbf{E} \propto R^{-2}$ while $dS_s \propto R^2$, so the integrand approaches zero as R^{-1}. Therefore

$$\mathbf{B}'(\mathbf{r}_t) = 0$$

The $\mathbf{J}$ that is to be employed in the law of Biot-Savart, even when there is variation with time (at least in the quasistatic approximation), should not include $\dot{\mathbf{D}}$ if the integration is over all the sources. We may, however, find it either inconvenient or impossible to apply this law over all the currents. We may then use the alternative approach of applying the generalized Ampere's law to some finite open surface. Here we disregard the $\mathbf{J}$'s not passing through this surface, but we take them into account with the $\dot{\mathbf{D}}$ term through the surface. This allows us to compute $\oint_C \mathbf{B} \cdot d\mathbf{r}$ by discarding all $\mathbf{J}$'s not passing through a Σ bounded by C. Except in a few cases of high symmetry, however, this procedure does not give **B** at a point, but only gives a line integral of **B** over a curve.

The derivation above does not hold for nonquasistatic conditions, since $\nabla_s \times \mathbf{E}$ is then not equal to zero. But this does not matter, since the law of Biot-Savart, itself, turns out to be applicable only under quasistatic conditions. When there is radiation it is necessary to take the time of travel of the energy into account.

PROBLEMS

1 Let a medium have conductivity g. If an alternating **E** is applied, with angular frequency ω, find $|\mathbf{J}/\dot{\mathbf{D}}|$.

2 In a vacuum the **J** that should be used in the law of Biot-Savart is $\mathbf{J}_c$—conventional current density associated with the flow of free charge, conduction plus convection. Show that when matter is present it is also necessary to add a term $\dot{\mathbf{P}}$ to $\mathbf{J}_c$, since a varying polarization corresponds to a movement of bound charge which produces a **B**; and it is also necessary to add a term $\nabla \times \mathbf{M}$ for the amperian currents which simulate the magnetization of the matter.

3 The plates of a parallel-plate capacitor with vacuum dielectric have a separation which varies sinusoidally about a mean value: $d = 0.001 + 0.0001 \sin 2\pi(400)t$ meters. Let the area of each plate be $A = 0.01$ m^2. If the capacitor, connected to a constant voltage power supply, has a potential difference of 1000 V between the plates, what is the maximum value of the current that flows from the battery? Is the current sinusoidal?

4 Suppose one attempts to interpret the equation

$$\frac{\partial \mathbf{D}}{\partial t} = -\mathbf{J} + \frac{1}{\mu_0} \nabla \times \mathbf{B}$$

as cause and effect: the term $(1/\mu_0)\, \nabla \times \mathbf{B}$ could be called, e.g., the replacement current density, and could be looked on as independent of, and similar to, $-\mathbf{J}$ in producing $\dot{\mathbf{D}}$. What additional argument can be given against this view if **J** and $\nabla \times \mathbf{B}$ are truly independent quantities?

5 Let the medium of Prob. 1 be the dielectric insulation in a parallel-plate capacitor having a capacitance C, with a high but finite resistance R, in parallel with C, between the plates. Find $\mathbf{J}/\dot{\mathbf{D}}$ in terms of C and R.

6 Suppose the plates of the mechanically vibrated capacitor of Prob. 3 are first connected to the battery, when the plate separation is $d = 10^{-3}$ m and not vibrating, thereby charging the capacitor. Then the battery is disconnected and the plates are vibrated as previously. What is **J**? What is $\dot{\mathbf{D}}$?

7 In Fig. 11-19 let a current element $I\, ds$ be represented by a charge q and another, $-q$, separated by ds, and let the charge magnitudes vary—$\dot{q}$ and $-\dot{q}$, respectively.

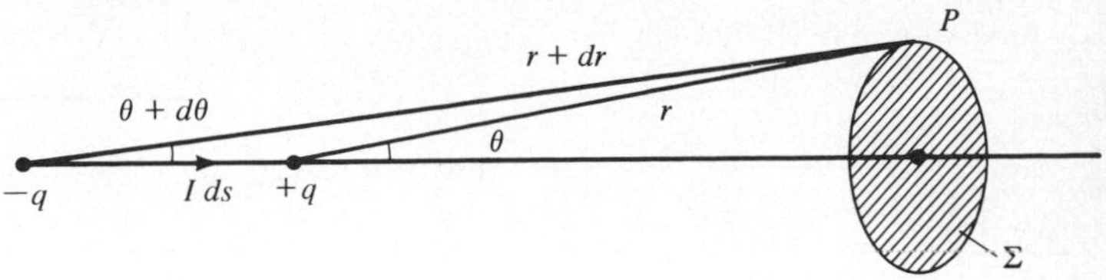

FIGURE 11-19

(a) Find $\int_\Sigma \dot{\mathbf{D}} \cdot d\mathbf{S}$ caused by each charge.
(b) Find the sum of these two.
(c) Use the generalized Ampere's law to find **B** at P.
(d) Compare this with the value obtained by using $I\,ds$ in the law of Biot-Savart.

8 (a) Write Maxwell's equations in terms of **E**, **B**, **D**, **H**, using the auxiliary constitutive equations $\mathbf{D} = \epsilon\mathbf{E}$ and $\mathbf{H} = (1/\mu)B$.
(b) Write Maxwell's equations in terms of **E**, **B**, **P**, **M**.

9 Figure 11-20 [like Fig. 11-19, taken from A. P. French and J. R. Ressman, "Displacement Currents and Magnetic Fields," *Amer. J. Phys.*, **31**:201 (1963)] shows a capacitor being discharged via an axial wire connected between the plates. The generalized Ampere's law gives the B_ϕ at P from $2\pi RB = \mu_0(I + \int_\Sigma \dot{\mathbf{D}} \cdot d\mathbf{S})$, where Σ is the circle of radius R in the equatorial plane. The two terms on the right are almost equal but opposite. Use the law of Biot-Savart, with conventional current density only, to find **B** at P.

10 $\mathbf{B} = \nabla \times \mathbf{A}$ where $\mathbf{A}(\mathbf{r}_t) = \mu_0/4\pi \int_\mathcal{V} (\mathbf{J}(\mathbf{r}_s)/R)d\tau_s$. Take $\mathbf{J} = \mathbf{J}_C + \mathbf{J}_D$, where $\mathbf{J}_C$ is the conventional current density and $\mathbf{J}_D = \dot{\mathbf{D}}$, and assume quasistatic conditions. Show that the $\mathbf{B}(\mathbf{r}_t)$ produced by $\mathbf{J}_D$ is zero. This is an alternative derivation to that in Example 3.

11 Let a parallel-plate capacitor with circular plates have a sinusoidally varying, free surface charge density that is uniform over the surface. Assume quasistatic conditions. Modify the static **E** and **B** by time-dependent factors. Show that the assumption that σ is uniform leads to a contradiction: there must be a nonuniform σ.

12 Show that $\mathbf{B} = \hat{\mathbf{z}}B_0 \sin \omega t$ with $\rho = 0$ and $\mathbf{J} = 0$ is a set of conditions that does not satisfy Maxwell's equations. (See Prob. 15 of the previous section.)

13 Any real parallel-plate capacitor has a dielectric, of permittivity ϵ, which conducts current slightly: $\mathbf{J} = g\mathbf{E}$. Assume the conductivity g is independent of position. If

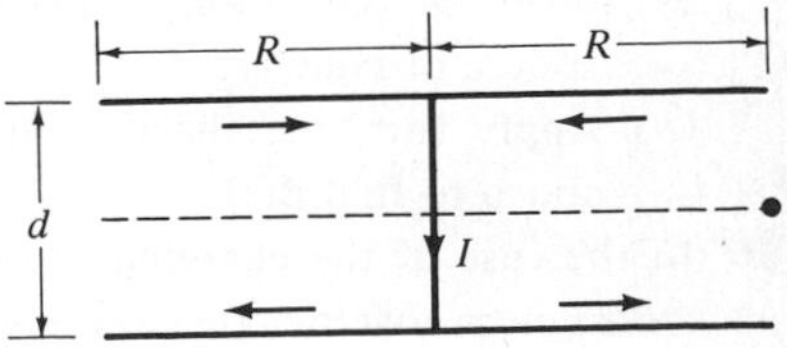

FIGURE 11-20

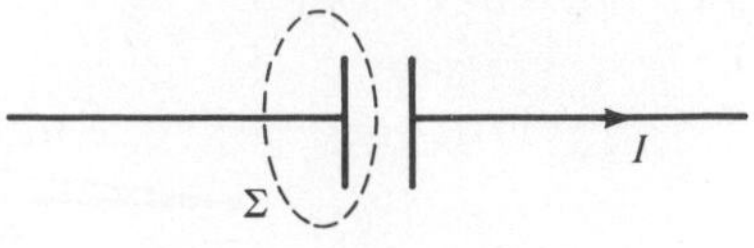

FIGURE 11-21

the capacitor is first charged to the potential difference $\phi_1 - \phi_2$ and is then isolated from the battery, find (*a*) the free charge on a plate as a function of time, (*b*) $\dot{\mathbf{D}}$ as a function of time, assuming all the charges reside on the adjacent (inner) surfaces, and (*c*) **B** between the plates as a function of time. (Compare with Prob. 9 to see that the assumption of charge on the inner surfaces only is not true.)

14 Put $\mathbf{J} = g\mathbf{E}$ (if g is a constant this is the microscopic form of Ohm's law) into the equation of continuity; then, using one of Maxwell's laws, obtain a differential equation for ρ as a function of time. Solve this for ρ. Provide an explanation for the relaxation time, a quantity that appears in this answer.

15 In Example 1 we saw that Maxwell's equations are equivalent to $\Box^2\phi = 0$, $\Box^2\mathbf{A} = 0$, and $\nabla \cdot \mathbf{A} + \epsilon_0\mu_0\dot{\phi} = 0$ in free space—a vacuum region—if $\rho = 0$ and $\mathbf{J} = 0$. Suppose that, instead, the region considered is a dielectric where $\mu = \mu_0$, as before, but $\mathbf{D} = \epsilon_0 E + \mathbf{P}$ with $\mathbf{P} = \mathbf{P}(\mathbf{r},t)$. The Hertz vector π_e, defined by $\phi = -\nabla \cdot \boldsymbol{\pi}_e$ and $\mathbf{A} = \epsilon_0\mu_0\dot{\boldsymbol{\pi}}_e$, satisfies the differential equation $\Box^2\boldsymbol{\pi}_e = -(1/\epsilon_0)\mathbf{P}$. Find **E** and **B** as functions of $\boldsymbol{\pi}_e$.

16 Apply the equation for conservation of charge to the closed surface Σ of Fig. 11-21 which encloses one plate of a charging capacitor. Σ is pierced by one conductor. Use one of Maxwell's to obtain an expression for a surface integral over Σ and interpret the result physically.

17 (*a*) Suppose we have a uniform, sinusoidally varying **B**, of amplitude 10^{-1} T, passing through a circular area of 20 cm radius. What would be the minimum frequency such that the amplitude of the sinusoidally varying **E** at the circumference of the circle would be 10^2 V/m?

(*b*) Suppose we have a uniform, sinusoidally varying **E**, of amplitude 10^2 V/m, passing through a circular area of 20 cm radius. What would be the minimum frequency such that the amplitude of the sinusoidally varying **B** at the circumference of the circle would be 10^{-1} T?

18 Analogous to $\boldsymbol{\pi}_e$ in Prob. 15, suppose that we have a magnetic material with $\mathbf{M}(\mathbf{r},t)$ instead of a dielectric. Assume $\rho = 0$, $\mathbf{J} = 0$, $\epsilon = \epsilon_0$. If the Hertz function $\boldsymbol{\pi}_m$ satisfies $\Box^2\boldsymbol{\pi}_m = -\mu_0\mathbf{M}$, find ϕ, **A**, **E**, and **B** as functions of $\boldsymbol{\pi}_m$.

19 Let $\mathbf{J} = \mathbf{r}J_0$ in spherical coordinates.

(*a*) From the continuity equation find the rate of change of the charge within a sphere of radius r.

(*b*) Apply the generalized Ampere's law to a circle on the surface of the sphere above to find $\mathbf{B}(r)$.

20 In the case of the charging capacitor of Example 2, assume the flat circular plates are so close together that there is negligible fringing of the flux. Show that

$$\epsilon_0 \dot{\Phi}_{\text{elec}} \begin{cases} I\left(\dfrac{r^2}{R^2}\right) & r \leqslant R \\ I & r \geqslant R \end{cases}$$

11-3 THE POYNTING VECTOR

With Maxwell's equations as a starting point we now derive expressions for several quantities of interest. The first of these are for energy flux and momentum flux. The word "flux" here means the rate of flow of energy or momentum across a surface; this meaning, which is derived from the Latin word *fluxus*, differs from that meant when the word "flux" is used in connection with magnetism, e.g., magnetic flux. There "flux" means the surface integral of the magnetic field vector distributed over a surface, $\Phi = \int \mathbf{B} \cdot d\mathbf{S}$ and no rate is implied. Both uses are by now very well established and it is probably too late to change either one.

We start with energy flux, limiting ourselves to a vacuum for simplicity. First, multiply $\nabla \times \mathbf{E} = -\dot{\mathbf{B}}$ by $\mathbf{B} \cdot$ to give $\mathbf{B} \cdot (\nabla \times \mathbf{E}) = -\mathbf{B} \cdot \dot{\mathbf{B}}$; then we will multiply $\nabla \times \mathbf{B} = \mu_0 \dot{\mathbf{D}} + \mu_0 \mathbf{J}$ by $\mathbf{E} \cdot$ to give $\mathbf{E} \cdot (\nabla \times \mathbf{B}) = \mu_0(\mathbf{E} \cdot \dot{\mathbf{D}}) + \mu_0(\mathbf{J} \cdot \mathbf{E})$. Subtract the second from the first and use the vector identity

$$\nabla \cdot (\mathbf{E} \times \mathbf{B}) = \mathbf{B} \cdot (\nabla \times \mathbf{E}) - \mathbf{E} \cdot (\nabla \times \mathbf{B})$$

to give

$$\nabla \cdot (\mathbf{E} \times \mathbf{B}) = -\frac{\partial}{\partial t}(\tfrac{1}{2}\mathbf{B} \cdot \mathbf{B}) - \frac{\partial}{\partial t}(\mu_0 \epsilon_0 \mathbf{E} \cdot \mathbf{E}) - \mu_0 \mathbf{J} \cdot \mathbf{E}$$

or

$$\nabla \cdot \left[\frac{1}{\mu_0}(\mathbf{E} \times \mathbf{B})\right] + \frac{\partial}{\partial t}\left[\frac{\epsilon_0}{2}E^2 + \frac{1}{2\mu_0}B^2\right] + \mathbf{J} \cdot \mathbf{E} = 0$$

Integrate this over some finite volume $\mathcal{V}$ within a boundary area Σ to give

$$\int_\Sigma \frac{1}{\mu_0}(\mathbf{E} \times \mathbf{B}) \cdot d\mathbf{S} + \frac{\partial}{\partial t}\int_{\mathcal{V}} \left(\frac{\epsilon_0}{2}E^2 + \frac{1}{2\mu_0}B^2\right) d\tau + \int_{\mathcal{V}} \mathbf{J} \cdot \mathbf{E}\, d\tau = 0$$

This is the equation we desire. We have to interpret it. Suppose we start with the second term. Under static conditions the integrand would represent the sum of the energies in the electric and magnetic fields. If we make the *assumption* that its significance is unchanged under time-varying conditions then the second term represents the rate of change of energy stored in the electromagnetic field within $\mathcal{V}$. That this is an assumption, to be justified by comparison with experience, is brought out by the fact that there are other expressions for energy density under static conditions, viz., $\frac{1}{2}\rho\phi$ and $\frac{1}{2}\mathbf{J} \cdot \mathbf{A}$, which lose this interpretation under dynamic conditions. The assumption made here has always, so far, proven justified.

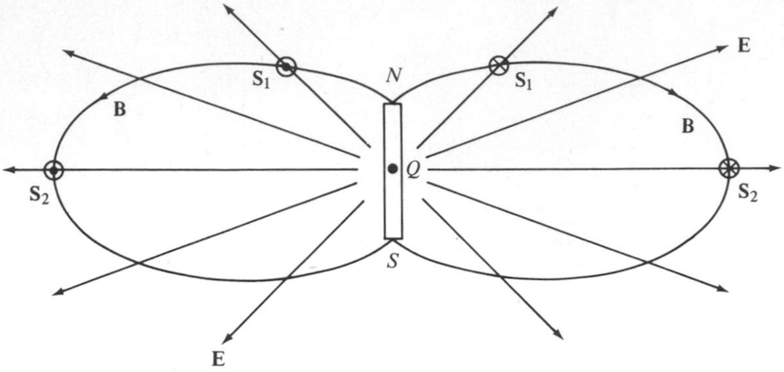

FIGURE 11-22

The third term has a simple meaning; it is the rate at which the field does work on the charges within $\mathcal{V}$. For $d\mathbf{F}$, the force on a charge $\rho\, d\tau$ is $\rho\, d\tau[\mathbf{E} + (\mathbf{v} \times \mathbf{B}]$ and the rate at which the field does work on this element is $d\mathbf{F} \cdot \mathbf{v} = (\rho \mathbf{v}\, d\tau) \cdot \mathbf{E} = \mathbf{J} \cdot \mathbf{E}\, d\tau$. So $\mathbf{J} \cdot \mathbf{E}$ is the work done by the field per unit volume.

If the first term of the equation is now interpreted as the energy flux out of $\mathcal{V}$, the meaning of the equation is clear. It expresses the conservation of energy: the radiant flux of energy through the boundary of $\mathcal{V}$, plus the rate of change of electromagnetic energy within $\mathcal{V}$, plus the rate at which the field is doing work on the charges inside $\mathcal{V}$ must all total zero. It is important to notice that this equation involves integrals and is applicable to a region, not a point. Suppose we wish to find the vector which, at all points of space, points in the direction of energy flux and gives the magnitude of the flux there. From the above equation one can only conclude that this vector is $\mathbf{S}' = (1/\mu_0)(\mathbf{E} \times \mathbf{B}) + \mathrm{s}$, where $\nabla \cdot \mathrm{s} = 0$.

We leave it to Prob. 1 to work out the details of one case where $\mathrm{s} \neq 0$. This involves the superposition of a static field and an alternating field. For most of the cases we will be interested in, however, it will be found that $\mathrm{s} = 0$. In that case $\mathbf{S}'$ will be designated as $\mathbf{S}$; i.e., $\mathbf{S}' = \mathbf{S} + \mathrm{s}$. Then

$$\mathbf{S} = \left(\frac{1}{\mu_0}\right) \mathbf{E} \times \mathbf{B}$$

will be the vector which points in the direction of energy flux and gives its magnitude at all points of space. Named after J. H. Poynting, who first discovered it, this vector is called, by pure serendipity, the Poynting vector.

Until very recently people were extremely loathe to attribute this meaning for $\mathbf{S}$ to static fields. For, consider the configuration of Fig. 11-22, which shows two completely independent static fields, one for $\mathbf{E}$ and one for $\mathbf{B}$. A permanent magnet in the shape of a thin needle produces a dipolelike magnetic field which is symmetrical about the needle axis, the figure showing only two lines of **B** in the plane of the page. Suppose a narrow strip of insulation is painted on the midsection of the magnet and a static electric charge Q is then sprayed on the paint. The lines of electric field **E** are then essentially radial, and the Poynting vector will be azimuthal. The figure shows two such circulating vectors in cross section, $\mathbf{S}_1$-$\mathbf{S}_1$ and $\mathbf{S}_2$-$\mathbf{S}_2$; this is only typical—there is an infinity of such rings.

It would seem that, though the fields are static, a dynamic flux of energy is continually being transferred in rings about the magnet axis. In any given volume, however, as much energy is being brought in as is being taken out, so there is no net transfer of energy anywhere. This idea of circulating energy in a static situation is hard to accept. However, a number of authors have recently shown that, because energy flux is intimately connected with linear momentum density, and because linear momentum is intimately connected with angular momentum, it follows that, even for the static case, if angular momentum is to be conserved it is absolutely necessary that the energy flux exist. In Example 1 we present in detail the argument given by one pair of authors. The following is a list of several articles by other authors:

1. R. P. Feynman, R. B. Leighton, and M. L. Sands, "The Feynman Lectures on Physics," vol. II, pp. 17-6 and 27-11, Addison-Wesley, Reading, Mass., 1964.
2. W. Shockley and R. P. James, *Phys. Rev. Lett.*, **18**:876 (1967).
3. W. Shockley, *Phys. Rev. Lett.*, **20**:343 (1968).
4. S. Coleman and J. H. Van Vleck, *Phys. Rev.*, **171**:1370 (1968).
5. W. H. Furry, *Amer. J. Phys.*, **37**:621 (1969).

Figure 11-23 summarizes the relations that exist among the four quantities:

u energy density
S energy flux
g momentum density
f momentum flux

in a system composed of particles, all moving with the same velocity **v** with respect to an observer. c is the speed of an electromagnetic wave in vacuum for this observer: its value will be calculated later in this section. Since it is known that light is an electromagnetic wave, c is usually called the speed of light (the vacuum medium being implicit). It is found experimentally that for this speed, and only for this speed, the magnitude of c is independent of the speed of the source. This non-common-sense result is a well-established fact; it will be discussed further in the next chapter.

The relation between S and u is apparent if we consider a cylinder of unit cross-sectional area and length v meters perpendicular to a plane. The number of particles in the cylinder is $N = nv$, where n is the particle density. Let each particle have energy ϵ

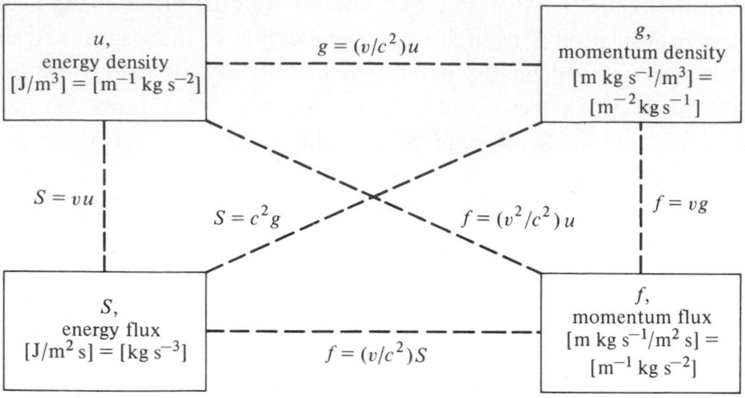

FIGURE 11-23

and momentum p. The energy density is then $u = n\epsilon$. All the particles in this cylinder, and only these particles, cross the given unit area in unit time. So the particle flux is N and the energy flux is $S = N\epsilon = (nv)\epsilon = v(n\epsilon) = vu$.

If one substitutes p for ϵ in these last equations one obtains, similarly, $f = Np = (nv)p = v(np) = vg$.

The relation between u and g is one which is derived relativistically and, therefore, should properly be left for the next chapter. It depends, however, on such a well-known equation, and the derivation is so direct, that we will give it here. From $E = mc^2$, if the energy density is u then the mass density is u/c^2. From $p = mv$, then, the momentum density is

$$g = \frac{u}{c^2}\, v$$

To find the relation between S and g is now simple:

$$S = vu = v\left(\frac{c^2}{v}\, g\right) = c^2 g$$

Since the velocity of the particles no longer appears, it should not be surprising that this relation is quite general and applies to any system containing energy and momentum, in whatever form. It does *not* involve v and is thus a more basic relation than the others. This equation applies to systems having particles of different speeds; or having quasiparticles, e.g., photons, or having no particles, such as a continuum.

The other relations in Fig. 11-23 are derived in a similar fashion.

It is clear from this diagram that momentum density and energy flux stand on an equal footing for any system—one is the concomitant of the other, and they are both of equal importance. So, if $\mathbf{S} = (1/\mu_0)\mathbf{E} \times \mathbf{B}$, it follows from $\mathbf{g} = (1/c^2)\mathbf{S}$ that

$$\mathbf{g} = \left(\frac{1}{\mu_0 c^2}\right)\mathbf{E} \times \mathbf{B}$$

Making use of the relation $\epsilon_0 \mu_0 = 1/c^2$, which we have accepted on faith hitherto but which we will derive later in this section, it follows that

$$\mathbf{g} = \epsilon_0 \mathbf{E} \times \mathbf{B}$$

Note the symmetry between the expressions for **S** and g. This symmetry vanishes if the equations are written $\mathbf{S} = \mathbf{E} \times \mathbf{H}$, $\mathbf{g} = (1/c^2)\mathbf{E} \times \mathbf{H}$.

Incidentally when the system under consideration is the electromagnetic field—regardless of whether the field is static or a wave—then $v \to c$ in Fig. 11-23. The reason for this is quantum mechanical: in a static field the forces (more important, the energy and momentum) are transmitted by virtual quanta—quasiparticles which, like ordinary photons, only travel at the speed of light. So, e.g., for an electromagnetic field, $\mathbf{f} = (1/c)\mathbf{S}$.

The Poynting vector **S** shows that there is a flux of energy associated with the simultaneous existence of an **E** and a **B** at any point. If the energy flows, with what speed does it flow? Maxwell answered this question by a simple manipulation of his equations. Suppose we first consider only free space (a region without matter, charges, or currents, which is far from any boundaries). This is the simplest case to take. Here Maxwell's equations become

$$\begin{cases} \nabla \cdot \mathbf{E} = 0 \\ \nabla \times \mathbf{E} = -\dfrac{\partial \mathbf{B}}{\partial t} \end{cases} \qquad \begin{cases} \nabla \cdot \mathbf{B} = 0 \\ \nabla \times \mathbf{B} = +\epsilon_0 \mu_0 \dfrac{\partial \mathbf{E}}{\partial t} \end{cases}$$

Then

$$\nabla \times \nabla \times \mathbf{E} = -\frac{\partial}{\partial t}(\nabla \times \mathbf{B})$$

$$\nabla(\nabla \cdot \mathbf{E}) - \nabla^2 \mathbf{E} = -\frac{\partial}{\partial t}\left(\epsilon_0 \mu_0 \frac{\partial \mathbf{E}}{\partial t}\right)$$

$$\nabla^2 \mathbf{E} - \epsilon_0 \mu_0 \frac{\partial^2 \mathbf{E}}{\partial t^2} = 0$$

This differential equation is the wave equation,

$$\nabla^2 \psi - \frac{1}{c^2}\frac{\partial^2 \psi}{\partial t^2} = 0$$

with ψ representing any one of the three components of **E**, if the velocity of transport of the vector **E** in free space c is given by

$$c = \frac{1}{\sqrt{\epsilon_0 \mu_0}}$$

Substituting the defined value of μ_0 and the measured value of ϵ_0 gives $c \approx 3 \times 10^8$ m/s. A similar derivation shows that the same wave equation holds true for the three components of **B**.

The wave equation is a second-order partial differential equation, so its solution in any one particular case can be quite complicated, depending either on the boundary conditions or on the initial conditions. But no matter how complicated, because the equation is linear, the superposition theorem holds true and the solution can be built up as a sum of simple solutions. In cartesian coordinates the simplest solution is a plane-wave solution. We may take $\psi = \psi(z)$ only, for example, and make $\partial\psi/\partial x = \partial\psi/\partial y = 0$. For any value of z, then, the solution is ψ in an xy plane in which ψ has everywhere the same value. The term wave enters since there are two linearly independent general solutions, $f(z - ct)$ and $g(z + ct)$, where f and g are any functions of the argument. $f(z - ct)$ represents a plane wave moving toward more positive z with speed c; to see this, note that the value of the function (e.g., a particular peak) is determined by the value of its argument, which depends on two independent variables. At $t = 0$ the value of the argument of $f(z - ct)$, i.e., $(z - ct)$, at which the particular peak occurs is z. Say the peak occurs when the argument is 2; at $t = 0$ the peak occurs at $z = 2$. At a later time t, the peak will still occur when the argument has the value 2, but this now occurs at $z = 2 + ct$. The peak has moved the distance ct toward greater z in the time t, so its speed is c. Similarly, $g(z + ct)$ represents a plane wave moving toward $-z$ with speed c. The general solution along the z direction is then $f(z - ct) + g(z + ct)$. In Example 2 we show that this result need not necessarily be a traveling wave, though it does have some wave qualities. If we take similar solutions for waves along the x and y directions and add them all together with the proper multiplying factors we can synthesize any solution in the three-dimensional volume. This may be a plane wave along some direction making arbitrary angles with the x, y, and z axes; or it may be a spherical wave rather than a plane wave, where ψ is independent of θ and ϕ and depends only on r; or it may not be a wave at all. In the problems we consider some other properties of the plane-wave solutions pertinent to free space. In Chap. 19 we will consider radiation in free space in much greater detail.

Examples

1. The necessity for circulating energy with static fields This example is based on "Physical Significance of the Poynting Vector in Static Fields," by Emerson M. Pugh and George E. Pugh, *Amer. J. Phys.*, **35**:153 (1967); for other examples, see the bibliography cited earlier in the section.

In Fig. 11-24 the geometry is that of a spherical capacitor, but there are several modifications. The inner sphere, of radius a, is a permanent magnet having magnetization $\mathbf{M} = \hat{\mathbf{z}}M$. It produces a field outside this sphere identical with that which would be

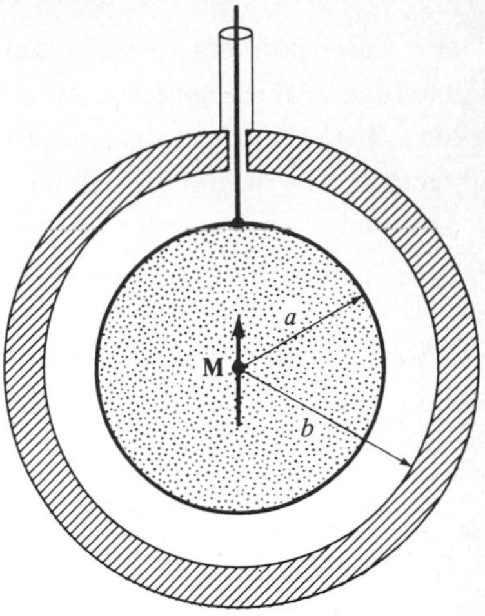

FIGURE 11-24

produced by a point magnetic dipole having $\mathbf{m} = \mathbf{M}(\frac{4}{3}\pi a^3)$ at the center of the sphere (the sphere itself being removed). So

$$\mathbf{B} = \frac{\mu_0}{4\pi}(M\tfrac{4}{3}\pi a^3)\,\frac{\hat{\mathbf{r}}2\cos\theta + \hat{\boldsymbol{\theta}}\sin\theta}{r^3}$$

The sphere is made of a ferrite (a nonconducting material) but its surface is covered with a conducting film, connected at the point $\theta = 0$ by a thin wire along the z axis to a battery. A charge $+Q$ coming in along this wire will flow only on the surface of the sphere and distribute itself uniformly there.

The outer shell, of inner radius b, is made of copper. It is connected to the negative terminal of the battery by a cylindrical lead, coaxial with the wire on the z axis to the inner sphere. A charge $-Q$ coming in to the shell will also distribute itself uniformly on the inner surface. The electric field established in the gap between sphere and shell will be

$$\mathbf{E} = \hat{\mathbf{r}}\,\frac{Q}{4\pi\epsilon_0}\,\frac{1}{r^2}$$

In the gap region of the spherical capacitor the value of the Poynting vector will be

$$\mathbf{S} = \frac{1}{\mu_0}\,\mathbf{E} \times \mathbf{B} = \hat{\boldsymbol{\phi}}\,\frac{QMa^3\sin\theta}{12\pi\epsilon_0}\,\frac{1}{r^5}$$

The energy flux circulates about the z axis azimuthally in the positive sense. The momentum density in this region is, making use of $\mu_0 = 1/\epsilon_0 c^2$,

$$\mathbf{g} = \frac{1}{c^2}\,\mathbf{S} = \hat{\boldsymbol{\phi}}\,\frac{Q\mu_0 Ma^3\sin\theta}{12\pi}\,\frac{1}{r^5}$$

Corresponding to this momentum density there is an angular momentum density. Switching from a spherical to a cylindrical coordinate system momentarily, $\mathbf{l} = \mathbf{R} \times \mathbf{g}$ (with R a cylindrical distance to the z axis) will give an angular momentum $\mathbf{l}$ in the z direction. The magnitude will be

$$\ell = r \sin\theta\,(1)\,\frac{Q\mu_0 M a^3 \sin\theta}{12\pi}\,\frac{1}{r^5}$$

with r back in the spherical system again. The total angular momentum of the field is then

$$\mathbf{L} = \hat{z}\int_{\theta=0}^{\pi}\int_{\phi=0}^{2\pi}\int_{r=a}^{b}\frac{Q\mu_0 M a^3 \sin^2\theta}{12\pi r^4}\cdot r^2 \sin\theta\, d\theta\, d\phi\, dr$$

$$= \hat{z}\,\frac{2Q\mu_0 M a^3}{9}\left(\frac{1}{a} - \frac{1}{b}\right)$$

Next we calculate the mechanical forces and torques exerted on the two parts of the spherical capacitor in the process of charging them from charge zero to charges $+Q$ and $-Q$. Let the entire charges on the two surfaces at any time in the charging process be $\pm q$. Below any line of latitude (colatitude $\theta°$) on the inner surface of the outer sphere there is then a charge at any instant of

$$-q\,\frac{2\pi b \int_\theta^\pi \sin\theta\, d\theta}{2\pi b \int_0^\pi \sin\theta\, d\theta} = -q\,\frac{1+\cos\theta}{2}$$

The surface current flowing downward past this line of latitude at that instant is

$$\mathbf{I} = -\hat{\boldsymbol{\theta}}\,\frac{1+\cos\theta}{2}\,\frac{dq}{dt}$$

The surface azimuthal force on a zone of width $b(d\theta)$ between colatitudes θ and $\theta + d\theta$ is $b(d\theta)$ times the force on a zone of unit width:

$$d\mathbf{F} = \mathbf{I} \times \mathbf{B}(b\, d\theta)$$

$$= \hat{\boldsymbol{\phi}}\,\frac{1+\cos\theta}{2}\,\frac{dq}{dt}\,\frac{\mu_0 M a^3\, 2\cos\theta}{3b^3}\, b\, d\theta$$

$$= \hat{\boldsymbol{\phi}}\,\frac{\mu_0 M a^3}{3b^2}\,(1+\cos\theta)\cos\theta\, d\theta\,\frac{dq}{dt}$$

The torque about the $\hat{\mathbf{z}}$ axis is in the $\hat{\mathbf{z}}$ direction with magnitude $(dF)b\sin\theta$:

$$d\mathbf{T} = \hat{\mathbf{z}}\,\frac{\mu_0 M a^3}{3b}\,(1+\cos\theta)\cos\theta\sin\theta\, d\theta\,\frac{dq}{dt}$$

Integrating this expression over a time ranging from zero to infinity gives the angular impulse exerted on this zone of the outer surface:

$$d\mathcal{I}_{\text{outer}} = \int_{t=0}^{\infty} d\mathbf{T}\, dt = \hat{\mathbf{z}} \frac{Q\mu_0 M a^3}{3b} (1 + \cos\theta) \cos\theta \sin\theta\, d\theta$$

The angular impulse on the entire outer sphere is the integral of this over θ from 0 to π:

$$\mathcal{I}_{\text{outer}} = \hat{\mathbf{z}} \frac{2Q\mu_0 M a^3}{9b}$$

This calculation must now be repeated for the inner sphere, but the result may be written down at once from the expression above:

$$\mathcal{I}_{\text{inner}} = -\hat{\mathbf{z}} \frac{2Q\mu_0 M a^3}{9a}$$

The net mechanical angular impulse delivered to the system is, then,

$$\mathcal{I} = -\hat{\mathbf{z}} \frac{2Q\mu_0 M a^3}{9} \left(\frac{1}{a} - \frac{1}{b}\right)$$

If the two spheres are free to rotate freely without friction this angular impulse gives the net change in the mechanical angular momentum of the system. It is seen to be exactly the negative of the angular momentum of the field. The angular momentum of the total isolated system, mechanical and electromagnetic, is therefore the same after the charging process as it was before. The angular momentum of the system is conserved, although the angular momenta of the individual constituents vary.

2. **Waves: traveling, standing, and pulse** **a. Traveling** The function $\cos(z - ct)$ represents a sinusoidal wave traveling toward greater z with speed c. At a given time it looks like a sinusoidal curve along the z axis; at a given point on the z axis it looks like a sinusoidal curve in time.

The function $\cosh(z - ct)$ also represents a wave traveling toward greater z with speed c. At a given time it looks like a hyperbolic cosine curve with its minimum point somewhere along the z axis; at a given point on the z axis it looks like a hyperbolic cosine curve in time, with its minimum at $t = z/c$. It could be described as a traveling wave.

b. Standing

$$\cos(z - ct) + \cos(z + ct) = 2 \cos z \cos ct$$

This function, the sum of two sinusoidal traveling waves, represents a standing (sinusoidal) wave. This is a wave in the sense that there is a sinusoidal oscillation with time at a given point, and there is a sinusoidal oscillation in space at a given time. This is not a wave in the sense of a traveling wave. The latter maintains its appearance, more or less, as time goes by but is continually displaced in position. A standing wave changes its appearance, however; e.g., at some instants the amplitude disappears altogether. Also, its maximum, e.g., does not progress in space with some given speed.

c. Pulse

$$\cosh (z - ct) + \cosh (z + ct) = 2 \cosh z \cosh ct$$

This function is the sum of two nonsinusoidal traveling waves, but it is not a standing wave in any sense. It is a stationary pulse in which values at all points change nonsinusoidally by a given factor with time. For $t < 0$ this factor is a decreasing one, while for $t > 0$ it is an increasing one. In this text we will not be concerned with such waves.

PROBLEMS

1 (*a*) Given a plane wave in free space with $\mathbf{E}_1 = \hat{\mathbf{x}}E_1 \sin(\omega t - kz)$, $E_1 = cB_1$, and $\mathbf{B}_1 = \hat{\mathbf{y}}B_1 \sin(\omega t - kz)$. Here $\omega = 2\pi f$, $k = 2\pi/\lambda$, and $c = \omega/k$ so $\sin(\omega t - kz) = -\sin(2\pi/\lambda)(z - ct)$. Find the energy density u_1 and the Poynting vector $\mathbf{S}_1$. Do they obey $S_1 = cu_1$?

(*b*) Add the static field $\mathbf{E}_2 = \hat{\mathbf{x}}E_2$ to the plane wave above. Find u_2 and S_2. Do they obey $S_2 = cu_2$?

(*c*) Find s such that $\mathbf{S}' = \mathbf{S}_2 + \mathbf{s}$ with $\nabla \cdot \mathbf{s} = 0$ and $S' = cu_2$.

2 (*a*) Modify the derivation in the text for the Poynting vector so that it is applicable to a region with homogeneous ϵ and μ instead of ϵ_0 and μ_0, though ρ and **J** are still equal to zero in this region.

(*b*) Modify Fig. 11-23 so that it is applicable to this case.

(*c*) Find **g**.

3 Modify Example 1 so that it applies to the earth, assumed to have a uniform magnetization $\hat{\mathbf{z}}M$ and a total charge Q. What is the total angular momentum of the earth's electromagnetic field if the earth's radius is R?

4 Given a region containing an electromagnetic field such that Fig. 11-23 applies. What pressure would this field exert on a material object in this region?

5 (This problem is preparatory for Prob. 7.) A parallel-plane capacitor with circular plates of radius R and a plate separation h is being charged by a battery. Assume E to be uniform within the plates at any instant of time.

(*a*) What is the electric energy stored in the region between the plates?

(*b*) What is the magnetic energy stored between the plates?

(*c*) What is the ratio $U_{\text{mag}}/U_{\text{elec}}$?

(*d*) How long after charging commences need one wait to have this ratio become extremely small if E is taken to be $E = E_0[1 - e^{-t/\tau}]$ (E_0 will cancel out; τ is the so-called time factor of the charging circuit).

(*e*) Find a value for this time, assuming $R = 10$ cm.

6 Derive the wave equation for the components of **B** from Maxwell's equations.

7 (*a*) Assuming, in Prob. 5, that the energy in the region between the capacitor plates when it is at all appreciable is practically all electrical energy, find an expression for the rate of change of this energy with time at any instant.

(*b*) Find an expression for the Poynting vector in the region between the plates for $r = R$.

(*c*) Find the energy flux into the region between the plates.

(*d*) Compare the answers of (*a*) and (*c*).

(*e*) In what direction does energy flux enter the volume in question

8 A cylindrical conductor of radius a carries a current I. A length L of this conductor has a resistance of R ohms and dissipates I^2R watts in heat. Find the energy flux radial into the conductor supplied by the Poynting vector. (There is a uniform longitudinal **E** in the wire and at the surface given by $E = V/L$, where V can be obtained from Ohm's law.)

9 A certain plane wave is represented by $\mathbf{E} = \hat{\mathbf{x}}E \sin (2\pi/\lambda)(z - ct)$, $E = cB$, $\mathbf{B} = \hat{\mathbf{y}}B \sin (2\pi/\lambda)(z - ct)$. Find (*a*) the instantaneous **S**, (*b*) the value of **S** time-averaged over one cycle.

10 Convert the expression for the field's momentum $\int \epsilon_0 \mathbf{E} \times \mathbf{B}\, d\tau$ to the following equation, involving only potentials, which is valid only under static conditions: $(1/c^2) \int \phi \mathbf{J}\, d\tau$.

11 Let the plane wave of Prob. 9 represent sunlight. (The distance from the sun is so large that the sun's spherical wave is approximately a plane at the earth's surface.) The average solar flux on the earth is 1.4 kW/m^2. Find the amplitudes of **E** and **B**.

12 Show that the relation in Prob. 10 is equal to $\int \rho \mathbf{A}\, d\tau$.

13 Find **S** and **g**, for the geometry of Fig. 11-22, at points far from the magnet and charge, so that they can be considered as a superimposed point magnet, of dipole moment $\hat{\mathbf{z}}m$, and point charge Q.

14 R. P. Feynman has proposed the following problem in his book: the insulated disk of Fig. 11-25, supported by virtually frictionless bearings, carries a coaxial turn of superconductor having a current I. Around the disk circumference there are a number of insulated metal cylinders, each of which carries an electric charge Q. Suppose the ambient temperature is raised, while the disk is stationary, above the critical temperature of the superconducting material; then the wire will become nonsuperconducting and the current will become zero. During the transient period the azimuthal **A** at the rim will decay to zero, accompanied by an induced electric field which is also azimuthal. Thus an azimuthal force will be exerted on each of the metal cylinders, all in the same sense. Will the disk start to rotate without the intervention of any external torque?

15 (*a*) If **E** and **B** are plane waves traveling in free space in the $+z$ direction, prove from Maxwell's equations that they must be transverse waves.

(*b*) Is this true also of **A**?

16 The differential equation satisfied by the six components of **E** and **B** in free space is $\nabla^2 \psi - (1/c^2)(\partial^2 \psi/\partial t^2) = 0$.

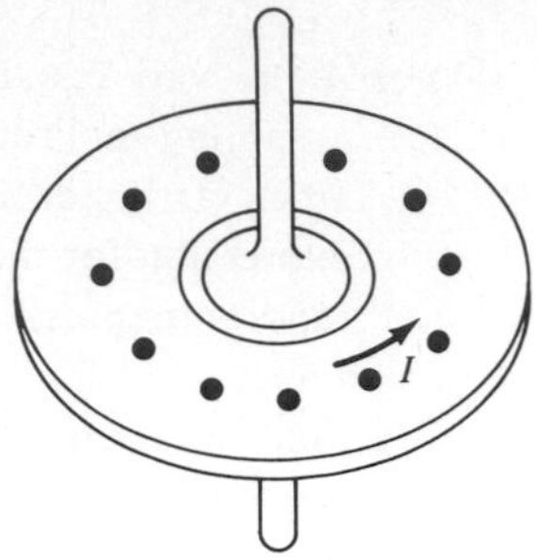

FIGURE 11-25

(*a*) Find the differential equation satisfied by the three components of **A**.
(*b*) Find the differential equation satisfied by ϕ.

17 In Prob. 15 **E** must have only x and y components. If **E** oscillates along a straight line (e.g., $\mathbf{E} = \hat{\mathbf{x}}E$), the wave is said to be linearly polarized.
(*a*) What do Maxwell's equations become here?
(*b*) Prove from Maxwell's equations that the **B** vector will also be linearly polarized and will lie in a plane perpendicular to that in which **E** lies, i.e., **B** will lie in the yz plane ($\mathbf{B} = \hat{\mathbf{y}}B$).

18 Derive the relations $f = (v/c^2)S$ and $f = (v^2/c^2)u$ shown in Fig. 11-23.

19 Given a plane electromagnetic wave moving in the $+z$ direction with speed c. The wave is linearly polarized with **E** oscillating sinusoidally along the x axis.

$$\mathbf{E} = \hat{\mathbf{x}}E_0 \sin (2\pi/\lambda)(z - ct)$$

(*a*) Find, by the use of Maxwell's equation for $\nabla \times \mathbf{E}$, the expression for **B**.
(*b*) What is the value of E_0/B_0?

20 Considering that the solution of the wave equation is one which travels with a finite velocity, explain why $\frac{1}{2}\int [\epsilon_0 E^2 + (1/\mu_0)B^2]\,d\tau$ can still give the energy in dynamic as well as static conditions while $\frac{1}{2}\int (\rho\phi + \mathbf{J}\cdot\mathbf{A})\,d\tau$ cannot.

21 In the wave of Prob. 9 the instantaneous Poynting vector amplitude represents power. An expression for the power loss in a resistor is V^2/R. Comparing the two expressions, what is R in the case of the wave in free space? (This is generally called Z_0, the impedance of free space, though it is resistive.) What is its value in ohms?

22 Maxwell's equations in free space are asymmetric because of the c^2 in the fourth equation:

$$\nabla \cdot \mathbf{E} = 0 \qquad \nabla \cdot \mathbf{B} = 0$$

$$\nabla \times \mathbf{E} = -\dot{\mathbf{B}} \qquad \nabla \times \mathbf{B} = \frac{1}{c^2}\dot{\mathbf{E}}$$

Suggest a simple procedure for making them symmetric by letting

$$B \to ? \qquad t \to ?$$

The significance of this will only become apparent in Chap. 14, on relativity.

23 The wave $\mathbf{E} = \hat{\mathbf{x}}E_0 \sin(kz - \omega t)$ is linearly polarized along the x direction. Without altering either the amplitude E_0, the propagation constant $k = 1/\lambdabar$ ($\lambdabar \equiv \lambda/2\pi$, where λ is the wavelength), or the angular frequency ω($\omega = 2\pi f$, where f is the frequency), convert this wave so it is

(*a*) linearly polarized in the y direction

(*b*) linearly polarized along a line making 30° with $+\hat{\mathbf{x}}$

(*c*) circularly polarized (the amplitude is in the xy plane, but executes a circle there)

24 Hazard a guess on the significance of the impedance of free space in connection with a television transmitter antenna. Let the current to the antenna be sinusoidal with an amplitude of 1 A, say. What would be the average transmitted power?

25 The equation $c = f\lambda$ is equivalent to $c = \omega/k$. From $\mathbf{E} = \hat{\mathbf{x}}E_0 \sin(\omega t - kz)$ with $E_0 = 10$ V/m and $f = 10^6$ Hz, find (*a*) the wavelength, (*b*) the propagation constant, (*c*) the polarization, and (*d*) the amplitude of the **B** wave.

26 A radio-frequency transmitting antenna might consist of two metal bars, lying along one line, with their nearer points connected to lead-in wires. A parabolic reflector has this assembly at its focus. Would you expect the transmitted wave to be polarized? Compare this with a searchlight having a bulb at the focus of a parabolic reflector.

27 (*a*) In the case of sinusoidal charges and currents, derive expressions for **E** and **B** in terms of **A** alone (i.e., without ϕ).

(*b*) Is it possible to do this, similarly, for nonsinusoidal time variations?

(*c*) Explain why ϕ can be excluded.

28 From $c = 1/\sqrt{\epsilon_0 \mu_0}$, show that all magnetic fields would vanish if the speed of electromagnetic waves in vacuum were infinite.

12
ELECTRIC CIRCUITS

It is not necessary to wait for the introduction of time variation before beginning the study of electric circuits, for circuits exist in which only steady currents flow—that is, currents that are constant in time and flowing in a given direction—but such circuits are primitive in nature and of minor importance, technically and economically, compared to circuits that carry time-varying currents. We will here undertake an introductory treatment of circuits, only for the sinusoidal time-varying case, in order to show how the circuit concepts arise from those of the field.

What one means by an electric circuit is a collection of a number of closed loops of filamentary wires (ideally taken to be made of resistanceless conductors) connecting various combinations of sources of generated or induced voltage with the passive elements—resistors, capacitors, and inductors. In such a situation it is not practical to deal with the fields in the region bounded by the metallic conductors, or even only in the immediate vicinity: the mathematics would be much too complicated. Instead, Maxwell's equations yield a number of results in connection with the voltages and the resultant currents in the circuit itself. The study of these relations is the subject matter of circuit theory. Because of its great importance, this part of electricity and magnetism has been very highly developed and a tremendous amount of literature exists on the subject.

Usually the study of electric circuits is undertaken as a subject in itself, so we will only touch on it briefly here.

It is convenient to divide the study of time-varying currents into two classes: repetitive and nonrepetitive. If a current is repetitive there is some minimum time T during which it executes a complete cycle of its pattern and $f = 1/T$ is the frequency of a sinusoidal wave that could complete one whole cycle in this period. It can be shown that by properly combining such a sinusoidal wave at this base frequency with other sinusoidal waves having proper values of amplitude, frequency, and phase, one may synthesize any recurring wave, no matter what its shape may be, provided it is not psychopathic but is moderately well behaved. Sometimes the number of such other waves is finite in number, sometimes infinite, but always it turns out that the frequencies of the other sinusoidal waves are integral multiples of the base frequency, so-called harmonic frequencies. Actually, this is nothing but a practical example of the application of that branch of mathematics called Fourier analysis.

If a current is nonrepetitive it may again be synthesized, in similar fashion, by component sinusoidal waves; this time, however, the frequencies are not integrally related but, instead, form a continuum; so there is always an infinite number of them. In any event, the study of any wave shape can be broken down into the study of the sinusoidal wave shape. When one adds to this the fact that the sinusoidal wave shape is a particularly easy one to generate, one has an explanation for the overwhelming importance of the sinusoidal wave in circuit theory.

A typical form of the sinusoidal wave $A \sin(\omega t + \phi)$ is the sine wave $A \sin \omega t$, for which $\phi = 0$. Such a function has the disadvantage, from the viewpoint of mathematical manipulation, of turning into the wave $\omega A \sin(\omega t + \pi/2) = \omega A \cos \omega t$ on differentiation and into $(1/\omega)A \sin(\omega t - \pi/2) = -(1/\omega)A \cos \omega t$ on integration. So it was useful to employ the complex exponential form $Ae^{(j\omega t)}$, where $j = \sqrt{-1}$; this maintains its form after either differentiation or integration. Since this function has both a real and an imaginary part, while the circuit current and potential differences are all real, it is necessary to restrict oneself either to the real or imaginary part of the function. The actual current, e.g., may be written $\mathcal{R}(Ie^{j\omega t})$ where I is independent of time but may be complex; usually the symbol $\mathcal{R}$ is not written but is implicit.

Suppose a sinusoidal voltage is suddenly applied to the terminals of a circuit, say by the closing of a switch. The current that flows initially is noncyclic but, since the applied voltage is cyclic, as time goes by the current will more and more approach a periodic function. The asymptotic cyclic function is called the steady-state condition (it is the state that is steady, though the current changes), while the initial aperiodic function is called a transient (even though it lasts, theoretically, for an infinitely long time). This twofold behavior stems from the fact that the differential equations which are obeyed by the current are second-order and linear but inhomogeneous; the inhomogeneous term is given by the voltage. The general solution to such an equation is the sum of two parts:

1 The complementary function (the transient)—the general solution of the homogeneous equation containing two arbitrary constants. This is characteristic of the circuit itself, without any forcing voltage.

2 A particular integral (the steady state)—any solution, whatever, of the inhomogeneous equation with no arbitrary constants. This part is specified by the forcing voltage as well.

In a particular solution the two arbitrary constants of the general solution assume definite values which make the solution fit the two initial conditions.

In the case of the steady-state sinusoidal operation of a circuit, the fields in space created by the generator will be waves of the form $A(r)e^{j(\omega t-kr)}$. The argument of the complex exponentials is similar to those of the plane waves of the previous section but with $z \to r$: they are waves propagating radially outward. But $e^{-jkr} = e^{-j2\pi r/\lambda}$, so if $r \ll \lambda$ the field becomes $\approx A(r)e^{j\omega t}$. If the circuit dimensions are sufficiently small compared to the wavelength then the arguments have the same values at all parts of the circuit; i.e., if the frequency is small enough then all parts of the circuit behave as if the time of propagation of the waves is negligibly small. Under these conditions, for every current element there is another one which is practically the negative of the first and not far away, so that at some distance from the circuit the effects of these two elements practically cancel and there is, then, negligible energy loss by radiation. It is necessary to impose this restriction for the circuit concept to have validity. It is called the quasistatic condition. In this chapter we will assume the frequency of operation to be so low that the quasistatic condition holds true and there is negligible radiation.

When the sinusoidal functions on which the operators act are written in complex exponential form, then a very useful simplification with respect to integrating and taking derivatives occurs.

$$\frac{d}{dt} e^{j(\omega t-kr)} = j\omega e^{j(\omega t-kr)}$$

$$\int e^{j(\omega t-kr)} dt = -\frac{j}{\omega} e^{j(\omega t-kr)}$$

If we have either integral or differential equations involving only such functions then these equations become much simpler—algebraic equations with j multipliers—than the original differential or integral equations. Using this complex form for the sinusoidal voltage source in a circuit, we will now derive the relations between potential difference and current pertaining to the various passive circuit elements.

Suppose we first consider a sinusoidal current $I = I_0 e^{j\omega t}$ going through a resistor R as in Fig. 12-1*a*. The arrow shows an arbitrarily assumed positive direction for the current. At any instant when the current flows in this direction it is taken as positive; in

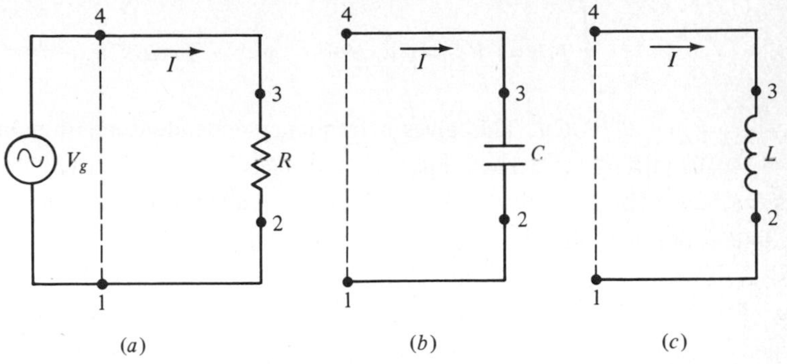

FIGURE 12-1

the reverse direction it will be negative. When the generator voltage V_g (the real part of $V_0 e^{j\omega t}$) is positive, the point 4 has a positive voltage relative to point 1. As a result of the generator voltage V_g, the current I will flow through the resistor, where $I = V_g/R$. There will then be a potential difference, $V_R = \phi_3 - \phi_2$, across the resistor terminals such that

$$V_R = RI$$

When the current flows downward through R the potential of point 4 is positive with respect to point 1. The potential difference between 3 and 2 equals the generated voltage between 4 and 1.

In Fig. 12-1b a capacitor has been substituted for the resistor. If we take the path of integration 12341, where 4-1 is in the region between the resistanceless conductors, the voltage around the path is

$$V = \oint \mathbf{E} \cdot d\mathbf{r} = \int_2^3 \mathbf{E} \cdot d\mathbf{r} + \int_4^1 \mathbf{E} \cdot d\mathbf{r} = 0$$

since there is no induced voltage in the path we have selected. Here both integrals go through the region intervening between the two conductors, so $\mathbf{E} = -\nabla\phi$ and we can write

$$\int_4^1 (-\nabla\phi) \cdot d\mathbf{r} = -\int_2^3 (-\nabla\phi) \cdot d\mathbf{r}$$

The right side is $+Q/C$. Letting $V_C = \phi_4 - \phi_1$ be the potential of point 4 relative to point 1, the left side gives V_C, so this merely affirms $V_C = Q/C$. Take $V_C = V_0 e^{j\omega t}$. Since $Q = \int I\, dt$,

$$\int I\,dt = CV_C = CV_0 e^{j\omega t} \quad \text{and} \quad I = j\omega CV_0 e^{j\omega t} = (j\omega C)V_C$$

Thus $V_C = (1/j\omega C)I$. This gives a frequency-dependent relation between V_C, the rms potential difference across capacitor terminals, and I, the rms current flowing into the capacitor. The quantity $-jX_C \equiv 1/j\omega C$ is called the reactance of the capacitor; its dimensions are [ohms].

Next, consider an inductor, as in Fig. 12-1*c*. Then, again, $V = \oint \mathbf{E} \cdot d\mathbf{r}$ equals $\int_2^3 \mathbf{E} \cdot d\mathbf{r} + \int_4^1 \mathbf{E} \cdot d\mathbf{r}$. But here $\int_2^3 \mathbf{E} \cdot d\mathbf{r} = 0$ since we are assuming an ideal inductor having zero resistance, and there can be no $\mathbf{E}$ within the wire of such a coil without giving an infinite $\mathbf{J}$. $\int_4^1 \mathbf{E} \cdot d\mathbf{r} = \int_4^1 (-\nabla\phi) \cdot d\mathbf{r} = -\int_4^1 d\phi = \phi_4 - \phi_1$ as before. Designate this now by V_L; this is the potential difference across the inductor terminals. In this case there *is* an induced voltage in 12341: if Φ is the flux through the area bounded by 12341, then

$$V = -\frac{d\Phi}{dt} = -\frac{d}{dt}(LI) = -L\frac{dI}{dt}$$

Note that we have written V, not V_L. The sign for V in this equation is given with reference to a circumnavigation in the positive direction, i.e., that in which I is positive (dI/dt is also positive in this sense); but here we have taken a negative direction of navigation so the sign must be reversed. Taking $V = -V_L$ gives $V_L = L(dI/dt)$. With $V_L = V_g = V_0 e^{j\omega t}$,

$$\frac{dI}{dt} = \left(\frac{V_0}{L}\right) e^{j\omega t} \quad \text{or} \quad I = \left(\frac{V_0}{j\omega L}\right) e^{j\omega t} = \frac{V_L}{j\omega L}$$

So

$$V_L = (j\omega L)I$$

The inductive reactance $jX_L = j\omega L$ again gives a frequency-dependent connection between V_L and I.

The three quantities R, C, and L are the basic passive elements from which all circuits, no matter how complicated, are designed. (If we allow magnetic interaction between the elements it will be necessary to take a fourth parameter into account—M, the mutual inductance. Electric interaction can be provided for by extra capacitors.) A typical network problem is one containing several meshes in which all the passive parameters are given, as well as some driving voltages somewhere in the system; it is desired to find the current in a particular branch. To solve such problems in a systematic manner it is only necessary to employ two procedures, Kirchhoff's laws.

1 The law of the loop. At any instant the sum of the generated or induced voltages around a closed loop minus the sum of the individual potential differences equals zero. This law is only a consequence of the definition of voltage as the work per unit charge in going around any closed path. It is usually convenient to take

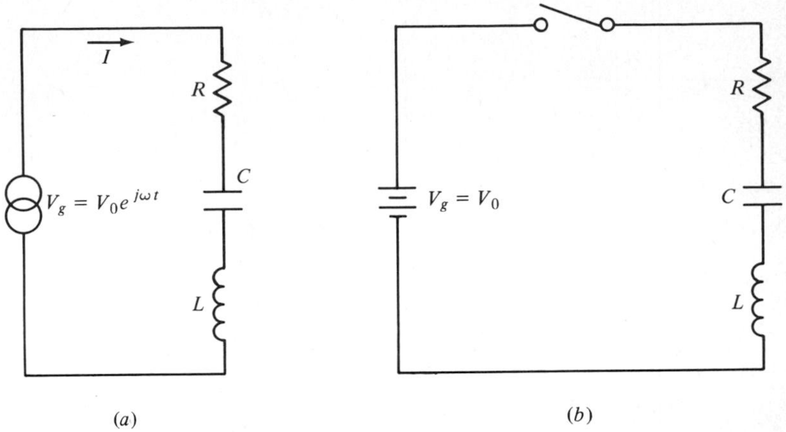

FIGURE 12-2

$L(dI/dt) = j\omega LI$ as the potential difference across an inductor instead of taking $-L(dI/dt) = -j\omega LI$ as an induced voltage in a loop.

2 The law of the node. At any junction the algebraic sum of the individual currents is zero. This law merely states that at any point as much current flows toward the point as away from it, so that it is neither a source nor a sink for charge.

Examples

1. The simple series circuit a. Steady state Figure 12-2*a* shows a sinusoidal voltage generator connected to a series combination of R, L, and C. This circuit, assuming the transient current has decayed to a negligible value, is appropriate for finding the steady-state current. The arrow in the diagram does not show this current but only the direction in which the current is positive.

By Kirchhoff's law of the loop

$$V = RI + \frac{1}{j\omega C} I + j\omega LI$$

$$= \left[R + j\left(\omega L - \frac{1}{\omega C}\right)\right] I$$

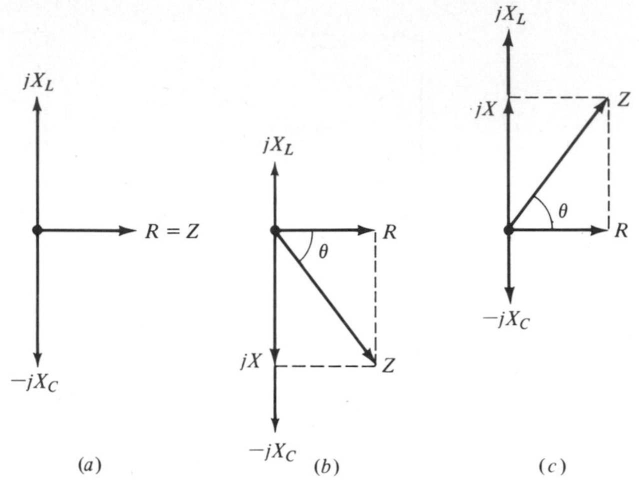

FIGURE 12-3

The quantity

$$Z = R + j\left(\omega L - \frac{1}{\omega C}\right)$$
$$= R + j(X_L - X_C) = R + jX$$

is called the impedance of the circuit. Its dimensions and units are the same as those of resistance: [ohms].

At some sinusoidal generator frequency the capacitive and inductive reactances will be equal, so $X = 0$ and $Z = R$. The circuit impedance is then a real quantity: it is resistive. This is shown in Fig. 12-3*a*. At a lower frequency than this, $X_C > X_L$, so X is negative. The circuit impedance is here complex, with X capacitive. See Fig. 12-3*b*. When the frequency is higher than that in Fig. 12-3*a* the reactance becomes inductive, as shown in Fig. 12-3*c*. In any case

$$|Z| = \sqrt{R^2 + X^2} = \sqrt{R^2 + (X_L - X_C)^2}$$
$$\theta = \tan^{-1}\left(\frac{X_L - X_C}{R}\right)$$
$$Z = |Z|e^{j\theta}$$

The Kirchhoff equation now becomes

$$I = \frac{V}{Z} = \frac{V_0 e^{j\omega t}}{|Z| e^{j\theta}} = \frac{V_0}{|Z|} e^{j(\omega t - \theta)}$$

By our convention, this means that the current in the circuit is actually $I_0 \cos(\omega t - \theta)$, where $I_0 = V_0/|Z|$. The steady-state current is sinusoidal with the same frequency as that of the generator voltage; but the two are not in phase, generally. The current and voltage amplitudes obey a revised form of Ohm's law, with $|Z|$ replacing R.

b. Transient Figure 12-2*b* shows an ideal (constant voltage) battery which is connected to the R, L, C series circuit at $t = 0$, when the switch is closed. Kirchhoff's loop law gives, at any time $t > 0$,

$$V_0 - \left(RI + \frac{1}{C} \int I \, dt + L \frac{dI}{dt} \right) = 0$$

This integrodifferential equation can be converted to an ordinary differential equation by taking the derivative of both sides with respect to time:

$$L \frac{d^2 I}{dt^2} + R \frac{dI}{dt} + \frac{1}{C} I = 0$$

The solution to this differential equation, the complementary function referred to previously, is obtained by substituting $I = Ae^{\gamma t}$ into the equation. This gives two important cases.

Case 1: If

$$\frac{R}{2L} > \frac{1}{\sqrt{LC}}$$

$$\begin{cases} \gamma_1 = -\alpha + \beta \\ \gamma_2 = -\alpha - \beta \end{cases}$$

where

$$\begin{cases} \alpha = \dfrac{R}{2L} \\ \beta = \sqrt{\left(\dfrac{R}{2L}\right)^2 - \dfrac{1}{LC}} \end{cases}$$

Case 2: If

$$\frac{R}{2L} < \frac{1}{\sqrt{LC}}$$

then it gives

$$\begin{cases} \gamma_3 = -\alpha + j\sqrt{-\beta^2} \\ \gamma_4 = -\alpha - j\sqrt{-\beta^2} \end{cases}$$

CASE 1 The solution here, when the resistance is large enough so that $R/2L > 1/\sqrt{LC}$, is

$$I = A_1 \exp \left[-\frac{R}{2L} \left(1 - \sqrt{1 - \frac{1/LC}{R^2/4L^2}} \right) \right] t$$

$$+ A_2 \exp \left[-\frac{R}{2L} \left(1 + \sqrt{1 - \frac{1/LC}{R^2/4L^2}} \right) \right] t$$

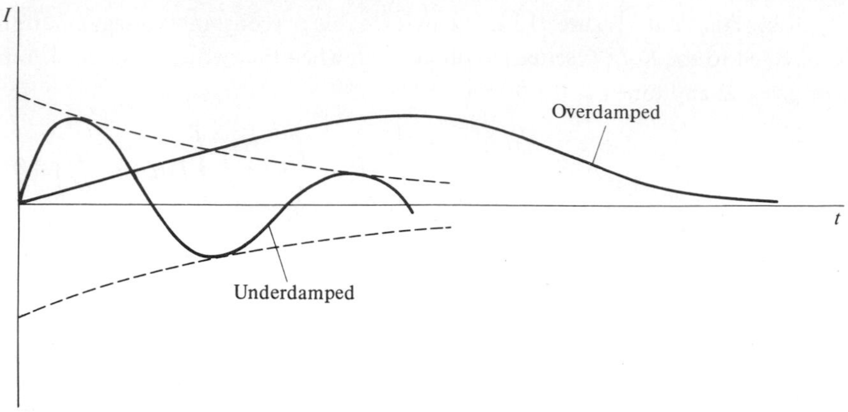

FIGURE 12-4

This is called the overdamped case and each of the terms in the solution decays monotonically. Because $I = 0$ when $t = 0$ it is necessary that $A_2 = -A_1$. Then

$$I = 2A_1 \exp\left(-\frac{R}{2L}t\right) \sinh\left(\sqrt{\frac{R^2}{4L^2} - \frac{1}{LC}}\, t\right)$$

The value of A_1 may be found from the facts that (1) at $t = 0_+$ (immediately after closing the switch) the charge of the capacitor is still zero if it was initially zero (because $I = dq/dt$, and a discontinuity in q would imply infinite current); and (2) at $t = 0_+$ the current is still zero (because $V_L = L(dI/dt)$, and a discontinuity in I would imply infinite voltage). Then at $t = 0_+$, with $V_C = 0$ and $V_R = 0$, we must have $V_L = V_0$. Figure 12-4 is a typical graph of this equation.

CASE 2 If the resistance is small enough so that $(R/2L) < (1/\sqrt{LC})$ the solution is

$$I = A_1 \exp\left(-\frac{R}{2L}t\right) \exp\left[j\sqrt{\frac{1}{LC} - \frac{R^2}{4L^2}}\, t\right] + A_2 \exp\left(-\frac{R}{2L}t\right) \exp\left[-j\sqrt{\frac{1}{LC} - \frac{R^2}{4L^2}}\, t\right]$$

This is the underdamped case and represents the damped oscillation shown in Fig. 12-4. Since $I_{t=0} = 0$ it follows that $A_2 = -A_1$ so

$$I = j2A_1 \exp\left(-\frac{R}{2L}t\right) \sin\left(\sqrt{\frac{1}{LC} - \frac{R^2}{4L^2}}\, t\right)$$

From the facts that the current must be real and that the capacitor charge is initially zero, we can determine A_1. The final result is

$$I = \frac{V_0/L}{\sqrt{(1/LC) - (R^2/4L^2)}} e^{-(R/2L)t} \sin\left(\sqrt{\frac{1}{LC} - \frac{R^2}{4L^2}}\, t\right)$$

The transient found here, for Fig. 12-2*b*, cannot simply be added to the steady-state result of Fig. 12-2*a* to find the total current when the switch of Fig. 12-2*a* is closed. One must take into account the phase of the cycle at the moment when the switch was closed. Under certain conditions the resultant peak values may be dangerously high.

2. Ideal current generators Any real, common source of voltage, such as an ordinary dry cell, is not an ideal voltage generator. This is so, unfortunately, because the potential difference between its terminals, when it is connected to a load, depends on the current which the generator supplies to the load. The real situation may be represented by asserting that the actual generator consists of an ideal voltage generator V in series with an internal resistance r as shown in Fig. 12-5*a*. The ideal voltage generator possesses the property of having a given voltage between its terminals, regardless of the load impedance or the current. Although the voltage is fixed, the current supplied by the generator is determined by the load impedance. The voltage V_g need not necessarily be constant, as in the figure, but may have any time variation we assign to it. For example, it is common to have $V_g = V_0 e^{j\omega t}$ as in Fig. 12-2*a*, where we set $r = 0$. Nevertheless, the ideal ac voltage generator is often referred to as a constant-voltage generator: the rms value is constant.

The representation of an actual voltage generator by an ideal voltage generator in series with a resistor is a natural one. It is a very useful idealization for the analysis of series circuits such as that in Fig. 12-2*a* or *b*. Suppose we are interested in a parallel circuit, however. An analysis of such a circuit by an ideal voltage generator, as in Fig. 12-6*a*, will only give information about the currents in the separate branches of the circuit, for the potential differences across the R, C, and L are fixed to have the same value as the generator voltage. This is not too helpful if we are interested in the behavior of the parallel circuit when it appears in conjunction with another circuit element such as r in Fig. 12-6*b*. In such a case it may be more useful to have a different representation of the generator.

Suppose, in Fig. 12-6*b*, that the magnitude of the impedance of the RCL circuit is $|Z|$; and suppose we imagine that we let r get larger and larger compared to $|Z|$. Then the current supplied by the generator will depend primarily on r and will be more or less independent of R, C, or L: between points a and b, for large r, this source is more a constant-current generator than a constant-voltage generator. If the amplitude V_0 is kept fixed while we increase r in our mind's eye, then, of course, the magnitude of the

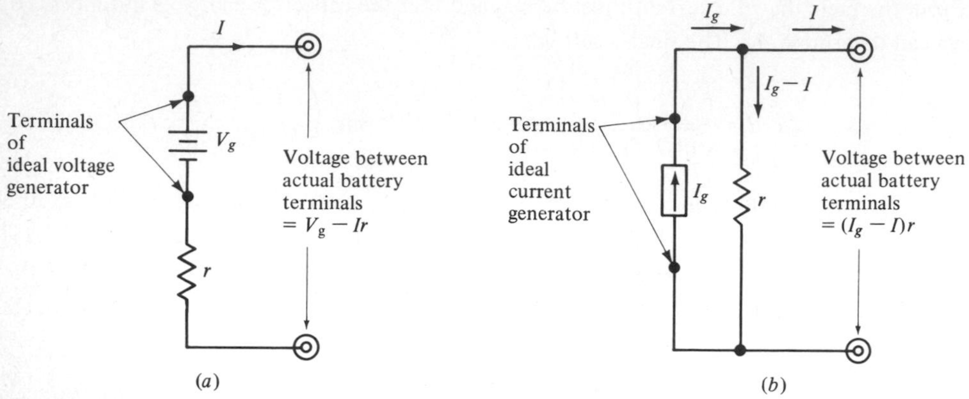

FIGURE 12-5

resultant current will decrease. But it is just as easy to imagine that V_0 is increased in step with r to keep the current amplitude at a fixed value. In the limit, when $r \to \infty$ and $V_g \to \infty$, we arrive at the concept of an ideal current generator: one which delivers a predetermined current to the load impedance regardless of the load impedance or the terminal voltage.

The representation of a current generator is awkward in terms of an ideal voltage generator in series with a very large internal resistor. It is more conveniently done in terms of an ideal current generator in parallel with an internal resistor, as shown in Fig. 12-5*b*. Comparing this with Fig. 12-5*a* we see that if $I_g r = V_g$ the two representations are equivalent. If $r \to \infty$ this becomes the ideal current generator. Such a representation is used in Fig. 12-6*c* for the analysis of the parallel network.

3. Kirchhoff's laws Suppose we are given the two-mesh network of Fig. 12-7*a* and we wish to find the current supplied by each battery as well as the potential difference across the 10-Ω load resistor. We will solve the problem by two methods: in Fig. 12-7*b* by the use of the three branch currents I_1, I_2, and I_3; in Fig. 12-7*c* by the use of the two loop currents I_1 and I_2.

a. Branch currents Each branch is assigned a current and the positive direction of each is arbitrarily designated. By the law of the loop,

$$12 - (1)I_1 + (10)I_2 = 0$$

$$-(10)I_2 + (2)I_3 - 13 = 0$$

(The third loop, around the outside, gives an equation which is only the sum of the two above.) Then, by the law of the node,

$$I_1 + I_2 + I_3 = 0$$

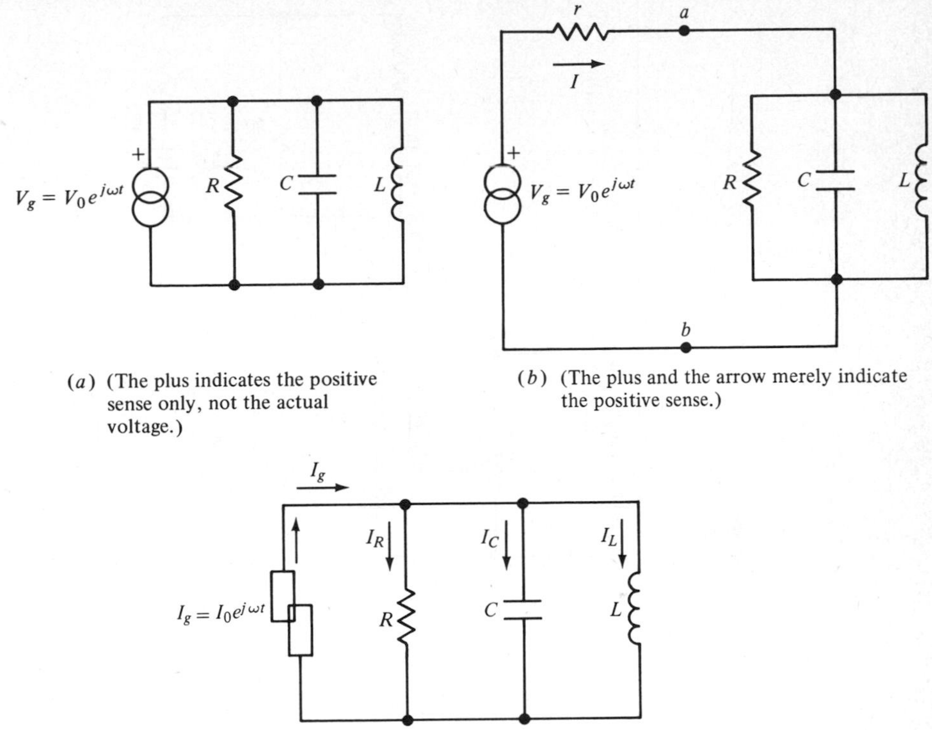

(*a*) (The plus indicates the positive sense only, not the actual voltage.)

(*b*) (The plus and the arrow merely indicate the positive sense.)

(*c*) (The arrows only indicate the + sense not the actual instantaneous directions.)

FIGURE 12-6

These three linear equations in three unknowns give

$$I_1 = \frac{\begin{vmatrix} -12 & 10 & 0 \\ 13 & -10 & 2 \\ 0 & 1 & 1 \end{vmatrix}}{\begin{vmatrix} -1 & 10 & 0 \\ 0 & -10 & 2 \\ 1 & 1 & 1 \end{vmatrix}} = \frac{14}{32}$$

$$I_2 = \frac{\begin{vmatrix} -1 & -12 & 0 \\ 0 & 13 & 2 \\ 1 & 0 & 1 \end{vmatrix}}{32} = -\frac{37}{32}$$

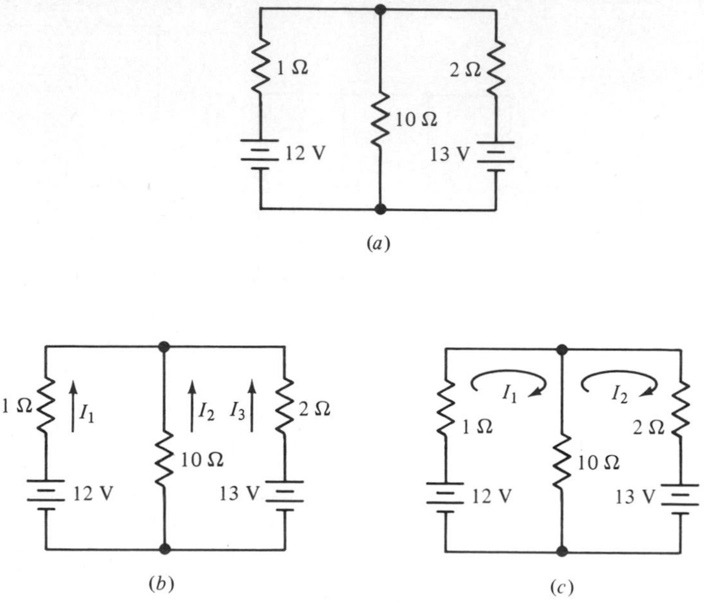

FIGURE 12-7

$$I_3 = \frac{\begin{vmatrix} -1 & 10 & -12 \\ 0 & -10 & 13 \\ 1 & 1 & 0 \end{vmatrix}}{32} = \frac{23}{32}$$

The 12-V battery supplies a considerably smaller portion of the load current. The potential difference across the load resistor will be $\left(\frac{37}{32}\right) 10 = 11.6$ V, with the top part positive.

b. Loop currents Assign individual currents in each loop as shown in Fig. 12-7*c*. The current in the 10-Ω resistor is then either $(I_1 - I_2)$ downward or $(I_2 - I_1)$ upward, as one wishes. Here we need only use the law of the loop:

$$12 - (1)I_1 - (10)(I_1 - I_2) = 0$$

$$10(I_1 - I_2) - (2)I_2 - 13 = 0$$

(The third loop is redundant.) So

$$\begin{cases} 11I_1 - 10I_2 = 12 \\ 10I_1 - 12I_2 = 13 \end{cases}$$

giving
$$I_1 = \frac{14}{32}\text{A} \qquad I_2 = -\frac{23}{32}\text{A} \qquad I_1 - I_2 = \frac{37}{32}\text{A}$$

Then
$$V_{\text{load}} = \left(\frac{37}{32}\right)10\text{ V} = 11.6\text{ V}$$

PROBLEMS

1 In Example 1 find the value of A_1 to give I in the case of the transient when $R/2L > 1/\sqrt{LC}$.

2 How is it possible for the transient solution in Example 1 to consist of two terms in the overdamped case, each of which is monotonically decreasing, yet have the transient solution itself not necessarily monotonically decreasing?

3 Suppose there is no inductor in Example 1. Find (*a*) the steady-state and (*b*) the transient currents, for a sinusoidal and a constant voltage, respectively.

4 (*a*) Is it permissible for the potential difference across a capacitor to change discontinuously?
(*b*) Is it permissible for the current into a capacitor to change discontinuously?
(*c*) Is it permissible for the potential difference across an inductor to change discontinuously?
(*d*) Is it permissible for the current into an inductor to change discontinuously?

5 In Prob. 3, what is the potential difference across the capacitor as a function of time in the case of the battery and the transient conditions?

6 Suppose the capacitor of Example 1 is to be omitted. Should one set $C = 0$? To what does $C = 0$ correspond? What should C be set to in order to omit it?

7 Figure 12-8 shows a widely used circuit called the Wheatstone bridge. When $R_1/R_2 = R_3/R_4$ the bridge is called balanced and there is no current through the milliammeter, whose resistance is R_A. Thus, if three of the resistors are known, the fourth can be calculated. Find the current expression for I, in general.

8 In Prob. 3*b* define a time constant τ in terms of circuit parameters. What is dI/dt at $t = 0$ in terms of τ? From this give a physical meaning of τ in terms of initial behavior.

9 Suppose there is no capacitor in Example 1. Find (*a*) the steady-state current with sinusoidal excitation and (*b*) the transient current with a battery.

10 Let $R = 0$ in Example 1. Find the impedance and the current as functions of the frequency.

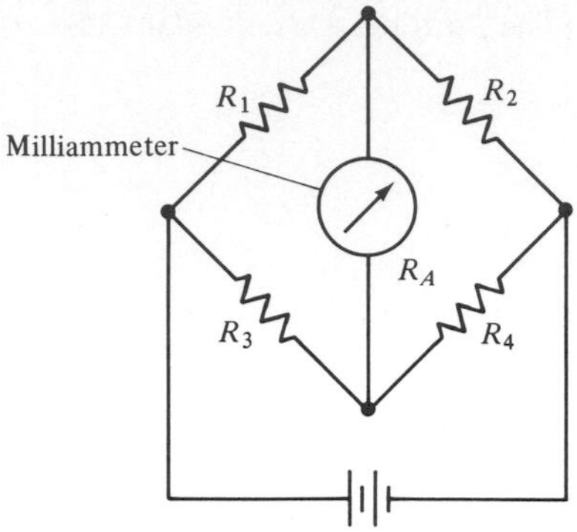

FIGURE 12-8

11 The resonant frequency of a series circuit is that frequency for which the circuit impedance is a minimum. An important parameter in describing the frequency response is the quantity $Q = X_0/R$, where $X_0 = \omega_0 L = 1/\omega_0 C$ is the reactance of either reactive element at resonance.

(*a*) Let $\gamma = \omega/\omega_0$. Find $|I|/(V_0/R)$ for the steady-state current in terms of Q and γ.

(*b*) Let $\gamma = 1 + \delta$; i.e., $\delta = (\Delta\omega)/\omega_0$. Repeat (*a*) in terms of Q and δ when $\delta \ll 1$.

(*c*) When I falls to $\sqrt{2}/2$ times its value at maximum current then the power $I^2 R$ falls to one-half its maximum value. This provides a convenient measure of the sharpness of the peak in current near the circuit resonance frequency. At what values of δ does I fall to $0.707\, I_{max}$?

(*d*) From (*c*) find an expression for Q in terms of f_0 and the difference in frequency between the two half-power points, $f_+ - f_-$.

12 In the transient case of Example 1 suppose $(R/2L) = (1/\sqrt{LC})$. This is known as the critically damped case. Find the current as a function of time.

13 Let the impressed voltage on a series RLC circuit be $V = V_0 e^{j\omega t}$. The current is then $I = I_0 e^{j(\omega t - \theta)}$. Find the average power, delivered to the circuit by the generator over a period of one cycle, from $(1/2\pi)\int_0^{2\pi} VI\, d(\omega t)$ by using (*a*) the sinusoidal functions for V and I and (*b*) the complex exponential functions.

(*c*) When dealing with the multiplication of two sinusoids in exponential form what extra step must be taken? Consider the power with sinusoidal voltage and current, for example.

14 In Prob. 13 find the average power delivered to the inductor in one cycle, from $(1/2\pi)\int_0^{2\pi} V_L I\, d(\omega t)$. Similarly, find the average power delivered to the resistor and to the capacitor.

15 If a steady current were to deliver the same power to a resistor R as is delivered on the average over one cycle by a sinusoidal current of amplitude I_0, what is the relation of the steady current to I_0? This value of the steady current is called the rms value (from root mean square):

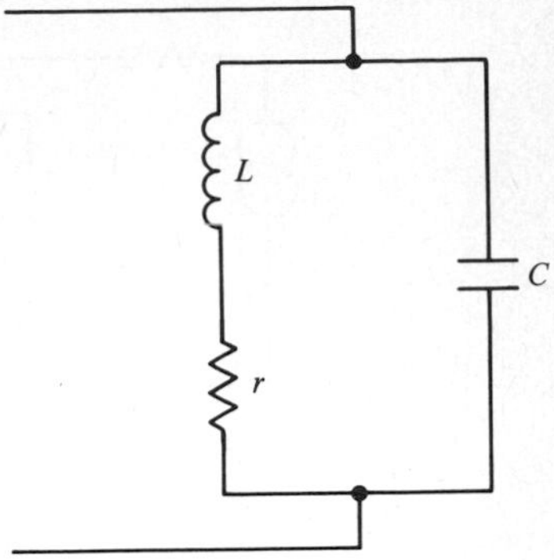

FIGURE 12-9

$$I_{\text{rms}} = \sqrt{\frac{1}{2\pi}\int_0^{2\pi} I_0^2 \cos^2(\omega t)\,d\omega t}$$

since

$$I_{\text{rms}}^2 R = \frac{1}{2\pi}\int_0^{2\pi} (I_0 \cos \omega t)^2 R \, d\omega t$$

16 If the cyclical current to a resistor is not sinusoidal, is the relation between I_0 and I_{rms} the same as it is for a sinusoidal current?

17 (*a*) Find the impedance of the parallel resonant circuit of Fig. 12-6 in terms of $\gamma = \omega/\omega_0$ [where $\omega_0 L = 1/(\omega_0 C)$] and $Q = R/X_0$ (note the difference between this and the definition of Q for the series circuit).

(*b*) Set $\gamma = 1 + \delta$ where $\delta \ll 1$ and find $|Z|/R$.

(*c*) What is $|I|/(V_0/R)$?

(*d*) Is $|I|$ a maximum or minimum at resonance?

(*e*) Does $\omega_0 L = 1/(\omega_0 C)$ give the frequency at which the impedance is real (as well as where its magnitude is a maximum)?

18 Consider the circuit of Fig. 12-9. This represents a parallel combination of an actual (i.e., lossy) inductor and a capacitor. (An actual capacitor also is lossy, but much less so than an inductor. Its power loss is represented by a parallel resistor of very large value, here neglected.)

(*a*) Find the impedance as a function of frequency.

(*b*) Is that frequency given by $\omega L = 1/(\omega C)$ the same as the frequency for which the impedance is a maximum?

(*c*) Are these the same as the frequency where the impedance is real?

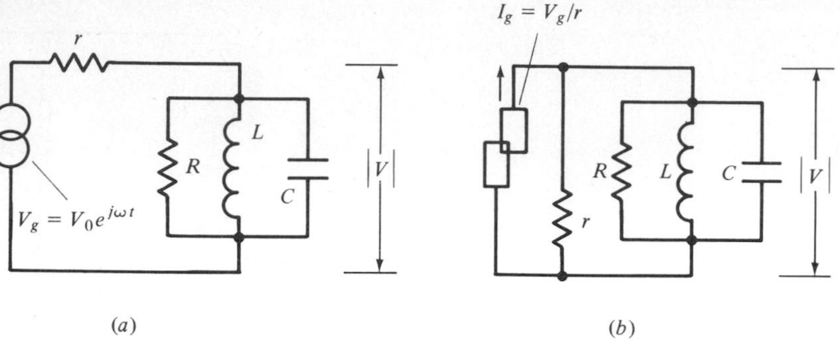

FIGURE 12-10

(d) What is the resonant frequency?

(e) Repeat (d) when $Q \gg 1$, where $Q = X_0/r$ as in the series case (not as in Prob. 17) and $\omega_0 = 1/\sqrt{LC}$. (The circuits of Probs. 17 and 18, especially the latter, are much more important in practice than those of Example 1.)

19 Modify the parallel resonant circuit of Fig. 12-6 by adding a resistor r in series with the parallel combination of the three elements. Find the magnitude of the potential difference across the parallel combination near the resonant frequency: (a) when the generator is treated as an ideal voltage generator $V_g = V_0 e^{j\omega t}$ in series with r and the RLC combination and (b) when the generator is treated as an ideal current generator $I_g = (V_g/r)e^{j\omega t}$ supplying r in parallel with the RLC combination. See Fig. 12-10. It is seen immediately from (b) that we must set $Q = [rR/(r + R)]/X_0$ in this problem. Note how much easier it is to employ the constant-current generator here.

20 In the circuit of Fig. 12-9 let $L = 100\ \mu$H and $C = 100$ pF (μH = microhenry, pF = picofarad).

(a) Find Q when $r = 10\ \Omega$, $100\ \Omega$, and $1\ \text{k}\Omega$, respectively (kΩ = kilohm), using $\omega_0 = 1/\sqrt{LC}$.

(b) For these values of r find the frequencies where the circuit impedance is real.

(c) Repeat for the frequencies where the circuit impedance is a maximum.

21 A dry cell ages primarily by increasing its internal resistance rather than by decreasing its open-circuit generated voltage. If the open-circuit voltage is 1.5 V, how much short-circuit current will flow if the internal resistance is (a) $0.05\ \Omega$, (b) 0.5?

(c) How would case (a) be represented using an ideal current generator?

22 Reverse the polarity of the 12-V battery in Fig. 12-7a; what is then the voltage across the 10-Ω resistor? Why is such a connection of two batteries to a load resistor to be avoided?

13

SPECIAL RELATIVITY

13-1 SPACE–TIME

Why does the theory of the electromagnetic field involve two fields—ϕ and **A**, say, or **E** and **B**? Why isn't one field sufficient? The theory of relativity addresses itself to this problem, among others. In most cases, it is not necessary to know relativity theory in order to solve problems in electricity and magnetism; but special relativity throws a new light on the subject and is, in addition, simple, elegant, and useful. So there is a growing tendency to use the course in the electromagnetic field as the place to expose the student to relativity. This chapter will be devoted to a treatment of relativistic kinematics and dynamics; the next one deals with the applications to electricity and magnetism. The treatment here is, of necessity, greatly abbreviated and the student who has never been exposed to relativity would be well advised to consult a textbook devoted to this subject. There are numerous texts devoted to this topic. "Special Relativity," by A. P. French (W. W. Norton, New York, 1968), is one which uses an analytic approach. "Special Relativity," by A. Shadowitz (W. B. Saunders, Philadelphia, 1968), employs a geometric approach, that of the Loedel diagram.

Consider two distinct events, not necessarily having any relation to each other, that occur at different places and times. The event E_1 occurs at (x_1,y_1,z_1) at time t_1, while the event E_2 occurs at (x_2,y_2,z_2) at time t_2; the positions and the times are both measured by one observer A. Suppose we simplify matters by taking E_1 to occur at the origin of A's coordinate system at time $t_1 = 0$ and, further, take E_2 (and all events) to occur on the x axis, so that $y_2 = z_2 = 0$. One event is, therefore, specified for A by the numbers (0,0), where the first number gives the x coordinate and the second gives the t coordinate, while the other event is specified for A by (x,t).

How should one measure the separation between two events? For points in three-dimensional space the spatial separation between them is called the distance and is given by $[(x_2 - x_1)^2 + (y_2 - y_1)^2 + (z_2 - z_1)^2]^{1/2}$, and in one-dimensional space this becomes $x_2 - x_1$; for events in one-dimensional time which occur at the same place in space, the separation between the events is called the time interval and is given by $t_2 - t_1$. How can we combine these concepts to obtain separation between events?

It may ease the transition to the proper generalization if we first arrange matters such that the two measures which characterize an event have the same dimensions. To do this we can multiply the time by a velocity; then the dimension, instead of [seconds], becomes [meters]. The most logical velocity to use might seem to be 1 m/s; the time interval between two events, t seconds, would be transformed into a distance of t meters—the distance covered by a body moving at 1 m/s for t seconds. This has a difficulty, however: a body moving at 1 m/s for observer A does not have a speed of 1 m/s for another observer B, who is moving relative to A. So a time interval of t seconds between two events, presumably having the same value for A and B, would correspond to t meters for A but not for B.

Measurements have shown, however, that there *is* a speed which has the same value for all observers, regardless of their velocities relative to each other—the speed of any electromagnetic radiation in vacuum, c. This is a non-common-sense result, but the experimental evidence, accumulated now for almost a century, is firm and convincing. So we will use c meters per second as the multiplying speed, to change a time interval of t seconds to ct meters.

This is an awkward result that is due to the very large magnitude of c, 2.998×10^8 m/s. Small, easily remembered time intervals, such as 1 or 2 s, will now be represented by large numbers with many significant figures. To circumvent this, an artifice has been universally adopted. Instead of using the meter as the unit of distance we will adopt the distance traveled by light in one unit of time as the new unit of distance. If the unit of time is the second, then the unit of distance will be the so-called light-second ($\approx 3 \times 10^8$ m); if the unit of time is the year, then the unit of distance will be the light-year ($\approx 5.87 \times 10^{12}$ mi); if the unit of time is the microsecond, then the unit of distance will be the light-microsecond (≈ 300 m). By this scheme, whatever the unit of time, the large numbers representing the time interval are not only avoided but the small numbers have the same numerical value as the time intervals themselves. Thus:

A time interval of 3 seconds → 3 light-seconds (9×10^8 m) for the distance interval.
A time interval of 2.4 years → 2.4 light-years (1.4×10^{13} mi) for the distance interval.
A time interval of 1.7 seconds → 1.7 light-seconds (510 m) for the distance interval.

In effect, we are simply choosing such a unit of distance that the speed of light equals unity: $c = 1$ light-second per second and

$$t \text{ seconds} \rightarrow ct \text{ light-seconds} = \left(1 \, \frac{\text{light-second}}{\text{second}}\right)(t \text{ seconds}) = t \text{ light-seconds}$$

The numeric for the time interval is identical to the numeric for the distance interval.

We now return to the problem of measuring the separation between two events which occur, in general, at different places at different times. The speed of light ("light" is shorthand here for "electromagnetic wave in vacuum") has one other unique characteristic besides that of having the same magnitude for all observers: it is the fastest speed, to our knowledge, that occurs in nature. The speed of light, as we shall see below, acts for material particles like an ultimate limit that can only be approached but never reached—something like 0°K in the temperature scale. Physically speaking, for material particles c acts like an infinite speed; i.e., as if an infinitesimal time interval elapses between two events connected by such a light wave. This is a clue for the definition of the "interval" between the two events, E_1 and E_2.

Let us define the interval (*note*: interval, not space interval or time interval) as $s = [(ct_2 - ct_1)^2 - (x_2 - x_1)^2]^{1/2}$. In our case this becomes $s = \sqrt{c^2t^2 - x^2}$; and if a light wave emitted at the origin at $t = 0$ (coincident with E_1) reaches x_2 at $t = t$ (coincident with E_2) then the interval between E_1 and E_2 for this fastest of all signals becomes zero. If E_1 and E_2 occur at the same place but not at the same time then $x = 0$ and $s = ct$: the interval between E_1 and E_2 will be numerically equal to the time interval between them. But if E_1 and E_2 occur simultaneously but not at the same place then this definition gives $s = ix$ in our case: the distance between the points where E_1 and E_2 occurred, but multiplied by $\sqrt{-1}$.

We now have two events, $E_1(0,0)$ and $E_2(x,ct)$; one observer of the two events, A; and the interval between these two events for this observer, $s = \sqrt{c^2t^2 - x^2}$. Question: suppose we have a second observer B, moving relative to A along A's x axis toward $+x$. What would the interval between E_1 and E_2 be for B?

Suppose we let B's origin coincide with A's origin, and also have B start counting time, when the event E_1 occurs; then, for B too, E_1 is characterized by (0,0). The other event E_2, let us say, will be characterized by (X,cT). So the interval between E_1 and E_2, for B, will be $S = \sqrt{c^2T^2 - X^2}$. Common sense would say $T = t$ and $X = x - vt$, as shown in Fig. 13-1, but if we use common sense we cannot possibly obtain the fact that c, the

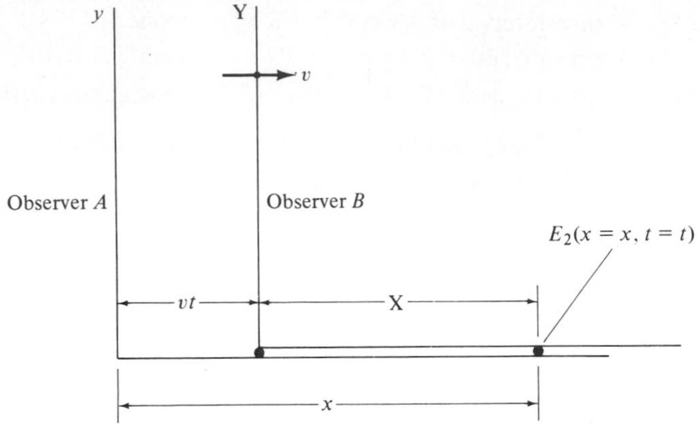

FIGURE 13-1

speed of a light ray, has the same value for A and B. So we leave open the question: what are X and T, in terms of x and t? Instead, we will be guided by the fact that in three-dimensional space the distance between two points is an invariant; its value is independent of the observer. Thus, here let us demand that the value of the interval between two events is an invariant: $s = S$, or, what is simpler mathematically, $s^2 = S^2$. Our problem now is to find the connection between (X,cT) and (x,ct) that will make $s^2 = S^2$. We will solve this problem graphically, employing the Loedel space-time diagram and the Minkowski space-time diagram.

Figure 13-2*a* shows a set of axes in which the abscissa gives the x position of an event while the ordinate gives the time-related value ct for the event. Every point on such a graph corresponds to an event rather than to merely a position. In the illustration, E_1 occurs at $x = 0$, $ct = 0$ while E_2 occurs at $x = 1$ light-second and $ct = 2$ light-seconds; i.e., E_2 occurs $t = 2$ s after E_1, and at a point 3×10^8 m to the right of the place where E_1 occurred. In Fig. 13-2*b* a dotted line is shown bisecting the angle between the two axes, which have a common scale. For every event E along this line the x coordinate and the ct coordinate are related by $x = ct$. Both x and ct increase as one goes away from the origin, so this line, each of whose points is an event, represents the sequence of events of a ray of light emitted at the origin at $t = 0$ and going toward $+x$. Such a line is called a world line of light.

Figure 13-2*c* shows a similar line from the origin but making the angle $\delta < 45°$ with the ct axis. For any event E' along this line the place x where the event occurred will be less than ct, where t is the time at which it occurred. This line represents an object moving toward higher x with the speed $v = c \tan \delta$. It could, in particular, represent the origin of observer B's coordinate system, with B moving toward $+x$ at velocity v. But the position, instant by instant, of an observer's origin is the time axis for that observer. So the solid world line of Fig. 13-2*c* is the time axis ct' of observer B.

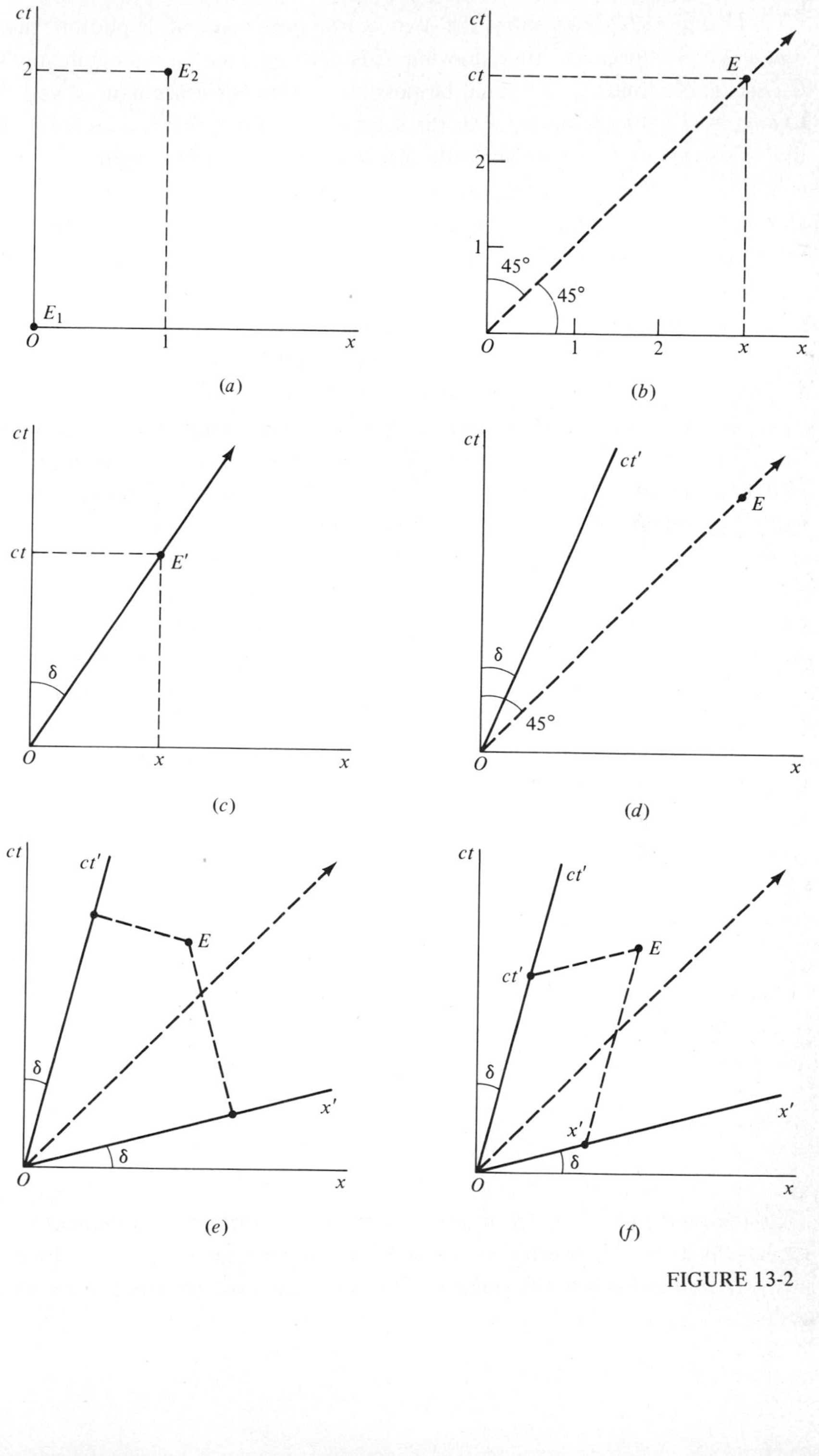

ct
2
E_2
E_1
O
1
x
(a)
ct
ct
E
2
1
45°
45°
O
1
2
x
x
(b)
ct
ct
E′
δ
O
x
x
(c)
ct
ct′
E
δ
45°
O
x
(d)
ct
ct′
E
δ
x′
δ
O
x
(e)
ct
ct′
ct′
E
δ
x′
x′
δ
O
x
(f)

FIGURE 13-2

In Fig. 13-2*d* we show the two world lines—one of a photon and one of *B*'s origin—on *A*'s space and time drawing. *E* is one event on the right-going ray's world line. If O-ct' is *B*'s time axis it should be possible to find *B*'s space axis as well, for we must have $x' = ct'$: the ray moves with the same speed c for *B* that it does for *A*. This requires that *B*'s space axis be symmetrically placed with respect to the light world line, as shown in Fig. 13-2*e*. Now it becomes necessary to specify how the space and time coordinates of an event are to be measured, for the x'-ct' axes of observer *B* are not orthogonal to each other, and two methods for finding coordinate values are apparent. This figure shows one of the methods for finding the coordinates of an arbitrary event, neither on an axis nor on the world line of the light ray: drop perpendiculars to the axes. This is a perfectly valid scheme and when this method of determining the coordinates of a vector is employed, the vector is called a covariant vector or a covector.

The Minkowski diagram and the Loedel diagram employ the second alternative for finding the coordinate of an event. The Minkowski diagram is shown in Fig. 13-2*f*. To find the ct' (time) coordinate for observer *B* we draw a line E-ct' from *E*, parallel to x', the mating space axis, until it hits the ct' axis. Similarly, to find the x' (space) coordinate for *B* we draw E-x' parallel to ct', the mating time axis. Notice that in this method O-x' and O-ct' combine according to the usual parallelogram law of addition to give *O-E*. This is not true of the method used for covariant vectors (covectors). A vector whose components are determined in this fashion is called a contravariant vector, or a contravector. When the axes are orthogonal both methods give the same result and there is no distinction, then, between covariant and contravariant vectors. The x coordinates of *E* are the same in Figs. 13-2*e* and *f*; the ct coordinates of *E* are also the same in the two figures.

We still require one more step to obtain the Minkowski space-time diagram. It is necessary to show that the interval between events *O* and *E* has the same values for observers *A* and *B*. In Fig. 13-3*a*, where the scales on all four axes are identical, we obtain from the drawing

$$x = x' \cos \delta + ct' \sin \delta$$

$$ct = x' \sin \delta + ct' \cos \delta$$

So $s^2 = c^2 t^2 - x^2 = (\cos^2 \delta - \sin^2 \delta)[c^2(t')^2 - (x')^2]$. But $\tan \delta = v/c$. Call this β. Then

$$s^2 = \frac{1-\beta^2}{1+\beta^2}\,(s')^2$$

Thus the diagram of Fig. 13-3*a* does not satisfy the original requirement we imposed on it, that the interval between events *O* and *E* be an invariant for the two observers.

The situation is easily remedied, however. Suppose we employ a scale factor such that

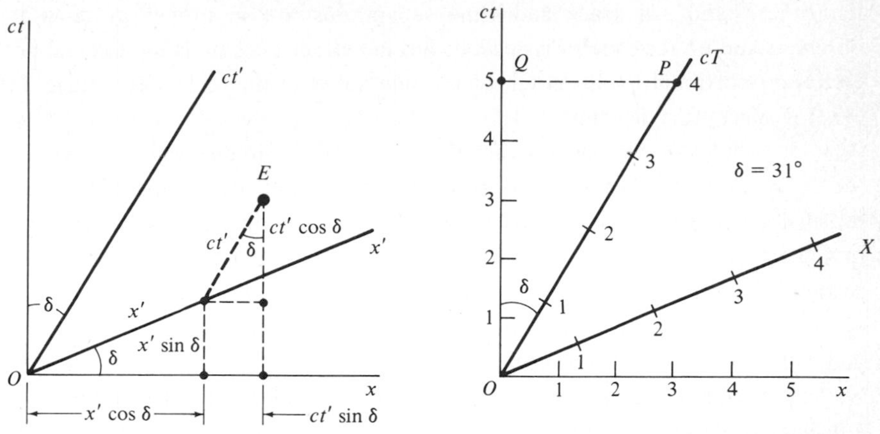

(*a*) (*b*)

FIGURE 13-3

$$ct' = \sqrt{\frac{1+\beta^2}{1-\beta^2}}\, cT$$

$$x' = \sqrt{\frac{1+\beta^2}{1-\beta^2}}\, X$$

thereby defining axes for B whose units, while identical with each other, differ from those of A. The units along the (X,cT) axes are spaced farther apart than those along the (x,ct) axes, so the measure of a distance along X will be smaller than one of equivalent length along x. Figure 13-3*b* shows the situation when $\delta = 31°$, which corresponds to a relative velocity between the two observers of $v = 0.6c$. The scale factor here is 1.46. (The scale factor changes if the velocity between A and B changes.)

It is important to avoid a likely source of error in using this diagram. In Fig. 13-3*b* $OQ \neq OP \cos 31°$, for $5 \neq 4(0.857)$. Whenever we switch from one set of axes to the other it is necessary to take the scale factor into account. To obtain an (x,ct) value, first multiply the (X,cT) value by the scale factor (1.46 in the above example). To obtain an (X,cT) value, first divide the (x,ct) value by the scale factor.

Any diagram having space and time axes is called a space-time diagram *if it leaves the value of the interval between two events invariant* for different observers moving at constant speed relative to each other. The Minkowski diagram, only one of several different types of such diagrams is, by far, the one most widely used. When comparing the measurements of the two different observers by this diagram it is necessary to keep in mind the fact that distortion exists: a scale factor is needed. In the Loedel diagram, on the other hand, all space and time separations are in proper proportion for both observers and no scale factor is needed. But in neither diagram is the interval between two events proportional to the length of the line between them. This is a grave defect of all existing space-time diagrams.

It should be emphasized that there is a very basic distinction between space-time (or a space-time diagram) and ordinary space. Let us compare an ordinary two-dimensional diagram with a space-time diagram having one spatial and one temporal dimension. In the ordinary diagram if we rotate the (x,y) axes to a new set (X,Y) for a second observer we obtain for the distance between two points $x^2 + y^2 = X^2 + Y^2$. In a space-time diagram the interval between two events is the quantity that is unchanged in going from one observer to another: $(ct)^2 - x^2 = (cT)^2 - X^2$. Note that in ordinary space we are comparing observers whose attitudes are different (one is turned relative to another) while in space-time we are comparing observers who are moving with constant velocity relative to each other. But that is not the important distinction, nor is it that time, rather than space, is measured in space-time. The crucial point is that in a space-time diagram the difference between squares is left invariant while in an ordinary diagram the sum of squares is invariant. Whenever a Minkowski diagram is used, whether its axes refer to x and ct or to any other quantities (the difference of whose squares is to remain constant), the properties of those quantities will be the properties of space-time.

We may now obtain the connection between (x,ct) and (X,cT). Directly from Fig. 13-4, which is Fig. 13-3*a* modified by the scale factor (SF),

$$x = X\left(\sqrt{\frac{1+\beta^2}{1-\beta^2}}\right)\frac{1}{\sqrt{1+\beta^2}} + cT\left(\sqrt{\frac{1+\beta^2}{1-\beta^2}}\right)\frac{\beta}{\sqrt{1+\beta^2}}$$

$$ct = X\left(\sqrt{\frac{1+\beta^2}{1-\beta^2}}\right)\frac{\beta}{\sqrt{1+\beta^2}} + cT\left(\sqrt{\frac{1+\beta^2}{1-\beta^2}}\right)\frac{1}{\sqrt{1+\beta^2}}$$

or

$$x = \frac{X + \beta(cT)}{\sqrt{1-\beta^2}}$$

$$ct = \frac{\beta X + (cT)}{\sqrt{1-\beta^2}}$$

These equations, which connect the space-time coordinates of an event for one observer with those of the same event for the second observer, are called the Lorentz transformation after the Dutch physicist H. A. Lorentz who first discovered them in 1904; he based them on an erroneous idea—the ether—that was first proven wrong much later, by the Kennedy-Thorndike experiment in 1932. The equations were independently discovered in 1905 by Albert Einstein in his paper introducing special relativity.

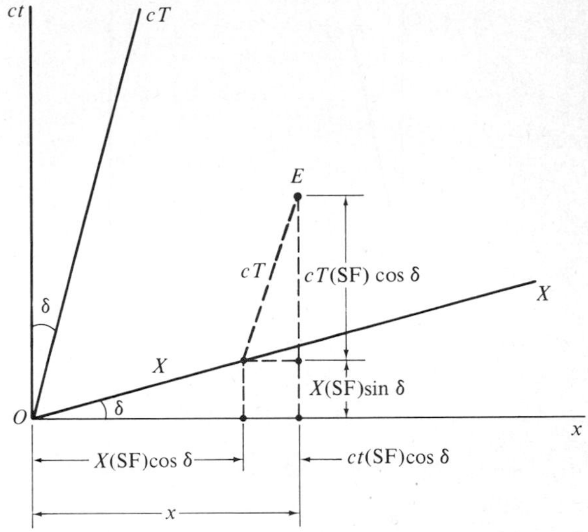

FIGURE 13-4

The velocity v which appears in these equations is the speed with which the (X,cT) observer moves toward $+x$ relative to the (x,ct) observer. In the directions transverse to the relative velocity we still have the usual relations

$$y = Y$$
$$z = Z$$

When the quantity $v/c \to 0$, the Lorentz transformation becomes the ordinary, common-sense, galilean transformation

$$x = X + vT \qquad y = Y \qquad z = Z \qquad t = T$$

In fact, all the non-common-sense consequences of the Lorentz transformation occur to an appreciable extent only when v is comparable to c. When v is smaller than $0.001c$, say, then the results of the two transformations are very nearly the same. The usual speeds of ordinary macroscopic bodies are very much smaller than this, so any relativistic effects are scarcely noticeable.

The expressions for x and ct in the Lorentz transformation become infinite for $v = c$; and for $v > c$ they become imaginary. So c is an upper limit to the velocity with which a real observer may move relative to a second. Although, there has recently been considerable speculation about the possibility that bodies with imaginary mass exist, called tachyons, which are born at superluminal speeds. No experimental evidence for these has yet been found. Everyone agrees that it is impossible to push bodies moving at subluminal speeds into the superluminal range. In any event, since c appears explicitly in

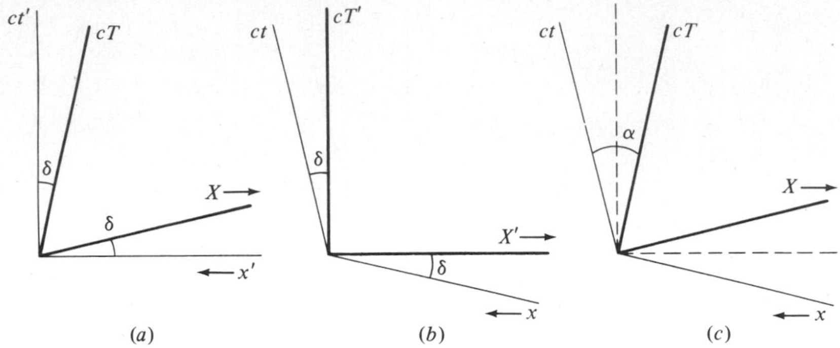

FIGURE 13-5

the Lorentz equations, the velocity of light in vacuum plays a very fundamental role in kinematics as well as dynamics.

The scale factor distortion of the Minkowski diagram may be eliminated by altering it to make the two observers symmetric. In Fig. 13-5*a* the (x',ct') observer has the orthogonal axes, in contrast to Fig. 13-3*a*. The (X,cT) observer, with acute axes, moves toward more positive x in real space. In Fig. 13-5*b* an (x,ct) observer has been assigned the nonorthogonal axes, which are now obtuse, while the (X',cT') observer (here he moves toward more positive x) has the orthogonal axes. Let us make the (x',ct') observer of Fig. 13-5*a* the same as the (X',cT') observer of Fig. 13-5*b*. In part (*c*) this observer is omitted, and the diagram shows the result of combining the (X,cT) observer from part (*a*) with the (x,ct) observer from part (*b*). This gives the Loedel space-time diagram.

The angle α in the Loedel diagram is equal to 2δ, where $\tan\delta = v/c$ in the Minkowski diagram. But this does *not* mean that the relative velocity between the (X,cT) and (x,ct) observers is twice that between each of the two sets of observers in the Minkowski diagram. Instead, if (X,cT) has the velocity $+u$ relative to the orthogonal observer while the latter has the velocity $+v$ relative to (x,ct), the velocity of (X,cT) relative to (x,ct) is obtained (see Probs. 15 and 18) as

$$w = \frac{u + v}{1 + uv/c^2}$$

or here, with $u = v$,

$$\frac{w}{c} = \frac{(v/c) + (v/c)}{1 + (v/c)(v/c)} = \frac{2\beta}{1 + \beta^2}$$

But $$\sin\alpha = \sin 2\delta = 2\sin\delta\cos\delta = 2\left(\frac{v}{\sqrt{v^2 + c^2}}\right)\left(\frac{c}{\sqrt{v^2 + c^2}}\right) = \frac{2\beta}{1 + \beta^2}$$

Thus, the angle α in the Loedel diagram is given by $\sin\alpha = w/c$. When we do not use both diagrams we will use v alone, and write $\sin\alpha = v/c$.

The (X,cT) observer in Fig. 13-5*a* has the same scale factor as the (x,ct) observer in Fig. 13-5*b*. Consequently the scale factor in part (*c*) between these two observers is unity. No conversions are necessary and all space or time separations are proportional to the line lengths for either observer. The price that has to be paid for this is the fact that now neither observer has an orthogonal set of axes. To find the X coordinate of an event E, we would draw a line from E, parallel to the cT axis, until it intersected the X axis; to find the x coordinate we would draw a line parallel to the ct axis to the x axis; and similarly for cT and ct.

Examples

1. Time dilation One of the famous, non-common-sense, consequences of the Lorentz transformation is the fact that time does not flow at the same rate for all observers. Our ingrained, intuitive, feelings to the contrary are based exclusively on experience with low velocities and it is, simply, not correct to extrapolate these results to high velocities.

The Loedel diagram of Fig. 13-6*a* and *b* shows the world line *w-w* of an object which is at rest relative to the (X,cT) observer: the line is parallel to the cT axis, and all events on this world line occur at one and the same point X_w for this observer. Consider the two events E_1 and E_2 on w's world line. We will characterize a time interval between two events that occur at the same point for a given observer as a proper time interval and label it with a zero subscript. The time interval between E_1 and E_2 is the proper time interval cT_0 for the (X,cT) observer; it is formed by drawing lines from E_1 and E_2 parallel to the X axis.

The (X,cT) observer is moving to the right with velocity $v = c \sin \alpha$ relative to the (x,ct) observer. An object at rest relative to the (X,cT) observer is, therefore, also moving to the right relative to the (x,ct) observer; so the time interval between the events E_1 and E_2 for the (x,ct) observer is a nonproper time interval. The adjectives here are not pejorative but merely descriptive; they are definitions. The nonproper time interval between E_1 and E_2 for (x,ct) is obtained by drawing lines from E_1 and E_2, parallel to the x axis, to the ct axis. But this nonproper time interval ct is the same as the hypotenuse of the right triangle shown, of which cT_0 is one side. So

$$cT_0 = ct \cos \alpha = ct\sqrt{1-\beta^2}$$

or

$$t = \frac{T_0}{\sqrt{1-\beta^2}}$$

The nonproper time interval between the two events E_1 and E_2, here measured by (x,ct), is larger by the factor $1/\sqrt{1-\beta^2}$ than the proper time interval between those events as measured by (X,cT).

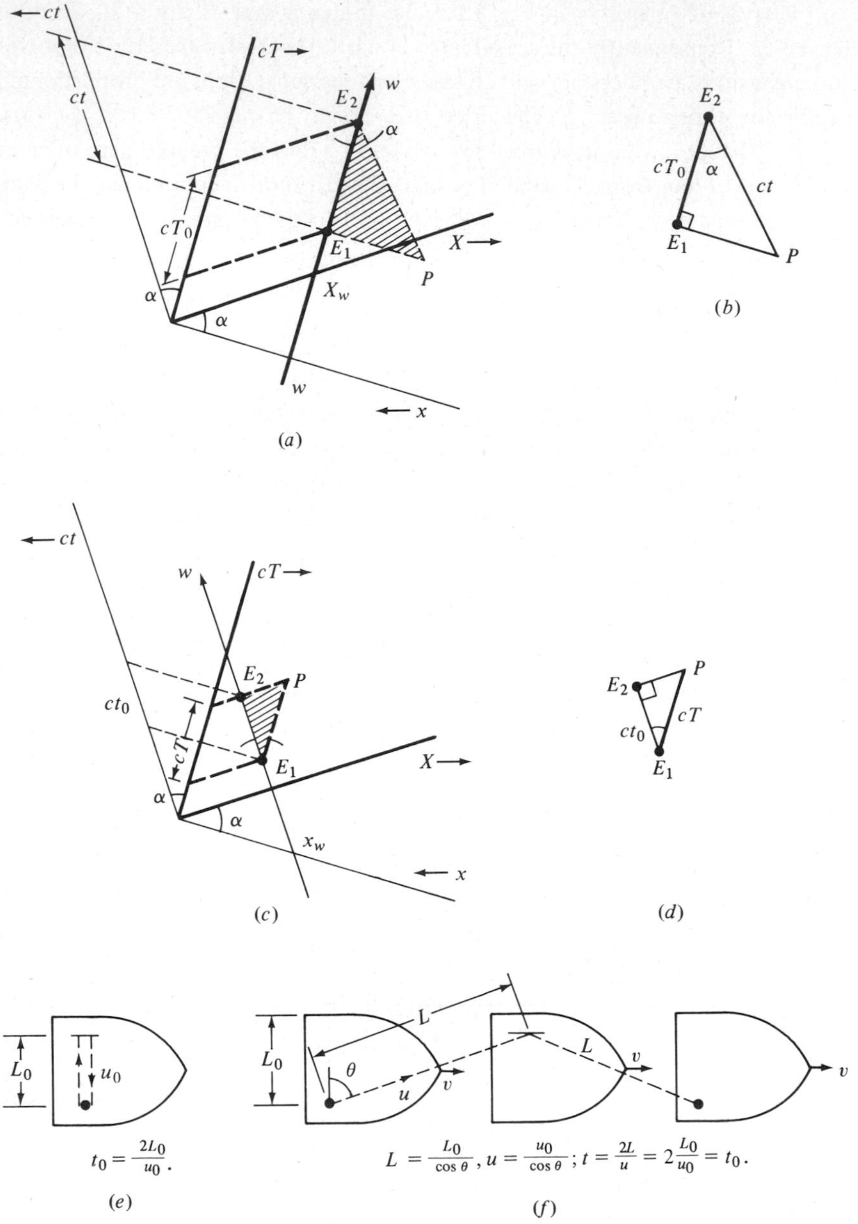

ct
cT
w
E2
α
ct
cT0
E1
X
P
Xw
x
(a)
E2
cT0
α
ct
E1
P
(b)
ct
w
cT
E2
P
ct0
cT
E1
X
α
xw
x
(c)
E2
P
cT
ct0
E1
(d)
L0
u0
t0 = 2L0/u0.
(e)
L
L0
θ
u
v
L
v
v
L = L0/cos θ, u = u0/cos θ; t = 2L/u = 2 L0/u0 = t0.
(f)

FIGURE 13-6

Suppose we had taken, instead of the above, two events which occurred at the same place for (x,ct). Figure 13-6*c* and *d* illustrates this case. The acute (X,cT) axes again belong to the observer going toward more positive x; the obtuse (x,ct) axes are those of the observer going toward more negative X. The world line *w-w* is here parallel to the ct axis and the time interval between E_1 and E_2 for (x,ct) is now a proper time ct_0. The nonproper time cT between these events for the (X,cT) observer is obtained on the cT axis by drawing lines from E_1 and E_2 parallel to the X axis, as indicated. The triangle at the right now gives

$$ct_0 = cT \cos \alpha$$

or

$$T = \frac{t_0}{\sqrt{1-\beta^2}}$$

Here it is T which is larger than t: the opposite of the previous case, apparently. Actually the result is the same as in the previous case, for in both cases the nonproper time interval between two events is greater than the proper time interval between them.

Figure 13-6*a* and *c* may be made simpler by taking the world line *w-w* to coincide with a time axis, the cT axis in the first case and the ct axis in the second. The result is found at once by choosing E_1 and E_2 and completing a triangle to P. We drew *w-w* off the time axis to illustrate the general process of finding coordinates.

The reason for this implausible behavior—the unequal flow of time for different observers—is not hard to find. It is the non-common-sense fact that c has the same value for A and B even when there is relative motion between them. In Fig. 13-6*e* we have shown a proper time interval in a rocket ship for the motion of a *particle*, to and from a reflector, with speed u_0. Suppose this to-and-fro time is measured from the ground, relative to which the rocket ship has speed v, as shown in Fig. 13-6*f*. The distance is larger than before, $2L = 2L_0/\cos\theta$; the speed of the particle is also larger, and in the same proportion, $u = u_0/\cos\theta$; so the total elapsed time is the same as before, $t = t_0$.

But this argument is completely demolished if we insist that in case (*f*) the speed $u = c$ shall be the same as in case (*e*), $u_0 = c$; this is what we do for a *light ray* rather than a particle. The distance in case (*e*) is larger than in case (*f*); if the speed of the photon relative to the observer is the same in the two cases, then it is necessary that the nonproper total elapsed time interval be larger than the proper time interval. It is left to Prob. 4 to show that when $u = u_0 = c$ then $t = t_0/\sqrt{1-\beta^2}$.

2. Lorentz-Fitzgerald contraction As another consequence of the Lorentz transformation, we consider a comparison of the measurements of a length, made by two observers, when the length lies along their direction of relative velocity. Let the object being measured be a rod lying along the x axis, stationary relative to the (x,ct) observer who is moving toward the left. In Fig. 13-7 the world line *f-f* gives the history of events associated with the front of the rod; they all take place at the one point x_f of the

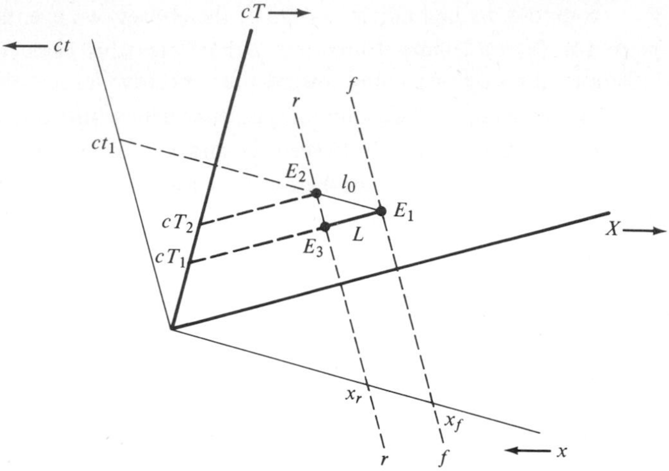

FIGURE 13-7

left-moving observer. Similarly, the world line of the rear of the rod is *r-r*. The length of the rod is $\ell_0 = x_f - x_r$, where the zero subscript designates this as a proper length measurement since the rod is stationary relative to this observer. The same length ℓ_0 would be obtained if the measurement were made at any given time t_1 associated with the events E_1 and E_2.

This rod is moving to the left, along with the (x,ct) observer, compared to the (X,cT) observer; but for the latter, also, the world lines *f-f* and *r-r* still give the events associated with the ends of the rod. Of course, if (X,cT) wishes to measure the length of a moving object he must note the simultaneous position of the two ends; then, by subtracting one position from the other, he obtains the length. But for the (X,cT) observer the events E_1 and E_2 are not simultaneous, though they are simultaneous for (x,ct), since E_1 occurs at cT_1 while E_2 occurs at cT_2. So the (X,cT) observer does not consider it correct to take a length measurement by comparing the positions at those events.

To (X,cT) the events E_1 and E_3 occur simultaneously at T_1, so he could obtain a length measurement that was correct for him by noting and subtracting the two positions at that time. Since this is a nonproper length measurement made by (X,cT) we will call its value L.

We now compare the value of L and ℓ_0 in the Loedel diagram:

$$L = \ell_0 \cos \alpha$$

or

$$L = \ell_0 \sqrt{1 - \beta^2}$$

The nonproper length measured by the observer relative to whom the longitudinal rod is moving will be less than the proper length measured by an observer relative to whom the rod is stationary. This is called the Lorentz-Fitzgerald contraction.

The basic reason for this strange behavior of a longitudinal length lies in the realm of time: the simultaneity of two events is not an immutable, absolute fact, but is only a relative fact. Two events can be simultaneous for one observer and yet be nonsimultaneous for another. Note, however, that when α, the angle between the two time axes, is very small, the difference in time is also very small.

Suppose the length being compared lies in a direction transverse to the direction of relative motion; how do the measurements compare then? Directly from the Lorentz transformation as we have written it, with relative motion along the X direction, it is seen that y and z are not intermingled with t–only x is involved with such behavior. So y and z lengths have the same properties as in the galilean transformation–a transverse length has the same value for all observers. One may modify the Loedel diagram by assigning the direction perpendicular to the plane of the paper to the y axis. One world line might be as shown in Fig. 13-7; another would be parallel to it, say one inch above the page. The time for any event not in the plane of the page is found, in such a diagram, by first dropping a perpendicular to the page and then proceeding as in Fig. 13-7. So two events, one on each world line, which are simultaneous for (y,ct) are also simultaneous for (Y,cT) and there is no distinction here between the measurement of a proper length and a nonproper length when this length is perpendicular to the direction of relative motion.

Instead of choosing y as the added axis we could have taken z. But suppose we had wished to take both y and z. We would have four independent space-time axes–x,y,z,ct– but only three independent directions in real space with which to represent them. This can be done either by showing four separate three-dimensional slices of four-dimensional space-time (x,y,z; x,y,ct; x,z,ct; y,z,ct) or by imagining the fourth dimension. It is too cumbersome for practical work, in either case, and an analytic approach is then preferable.

3. Presumed time Only one clock C is needed to record the time interval between event 1 and event 2 if both events occur at the same place, the place where the clock is. C measures a proper time interval. To measure a nonproper time interval between these two events we require two clocks, A and B, at two different points. A and B are stationary relative to each other, but they move with a common velocity relative to C. The nonproper time interval is always greater than the proper time interval.

What time interval does the moving clock A, itself, record between the two events according to C? It might seem that this would be the nonproper time above, since A and B run at the same rate; but this turns out to be incorrect. We will now show that according to C, which is at the place where events 1 and 2 occur, the time interval that A assigns between the two events is *less* than the proper time interval between those events.

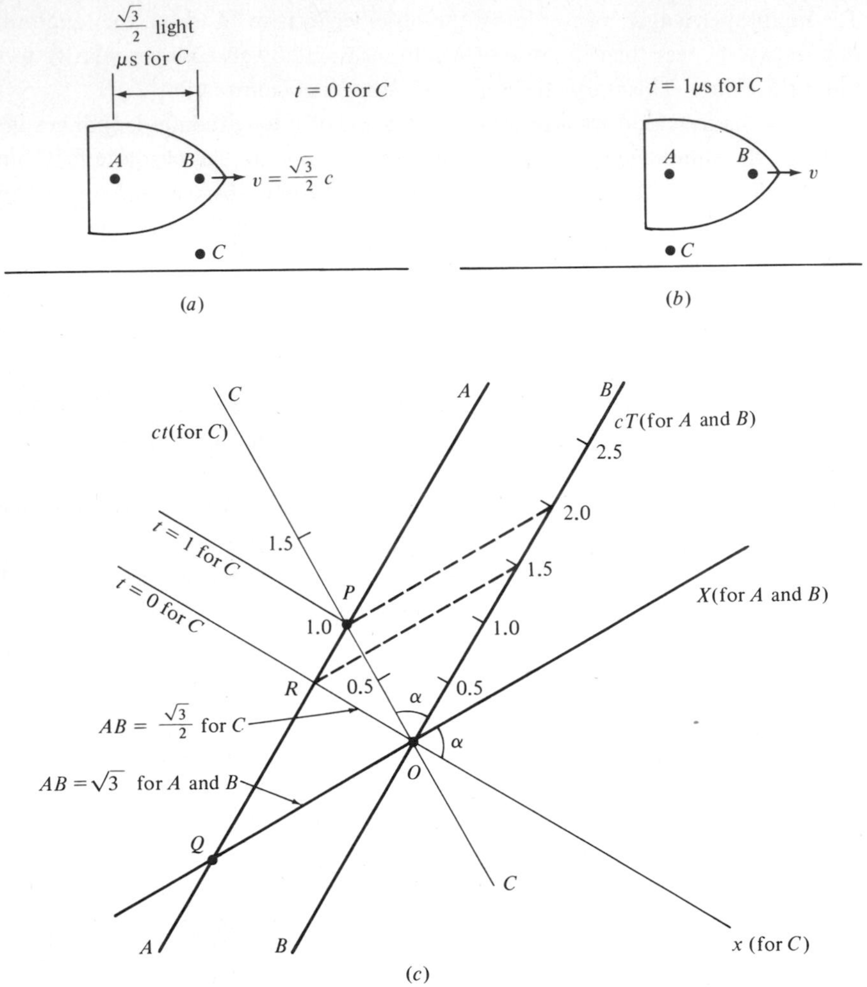

FIGURE 13-8

A, the moving clock, is presumed to run slowly by C. We will designate the stationary clock's version of the time measured by this moving clock as the presumed time. Note that the presumed time differs in principle from the actual time measured by A.

Consider Fig. 13-8a, which shows a spaceship carrying two clocks A and B, such that B is, at event 1, opposite clock C on the ground. Let the speed of the spaceship to the right be $v = (\sqrt{3}/2)c$, so $\sqrt{1-\beta^2} = 0.5$; and suppose the proper length between A and B to be $\sqrt{3}$ light-microseconds, i.e., $(1.73)300 = 520$ m. The nonproper distance between A and B is then $\frac{1}{2}\sqrt{3}$ light-microsecond for C. In the Loedel space-time diagram of Fig. 13-8c the conditions at this instant, event 1, are shown at O. The two clocks B and C are synchronized to start at $t = T = 0$ at this moment. The angle α is 60°. The world line BB along the cT axis gives the history of events for the clock B; AA, the world line of clock A, lies 1.732 units to the left of BB along the X axis or 0.866 unit to the left along the x axis. AA and BB are parallel to each other: they move at the same velocity relative to C.

CC is the world line of clock C: it lies along C's ct axis. It crosses A's world line at P, the event 2 shown in Fig. 13-8b. Event 1 occurred at $t = 0$ for clock C; event 2 occurred at $t = 1$ μs for C. On the Loedel diagram O and P are at $t = 0$ and $t = 1$, respectively.

What was clock A's time at event 1? A and C will differ on this. According to A this occurred at $T = 0$, the event Q on A's world line, where $T = 0$; but according to C this occurred at $t = 0$, the event R on A's world line. A's time at R, obtained by the dotted lines, is 1.5 μs.

A's time at event 2 is agreed to by all the clocks: it is 2 μs. The elapsed time between the two events is thus $2 - 0 = 2$ μs using the two clocks A and B which, according to them, are synchronized with each other. But using clock A alone, the elapsed time is $2 - 1.5 = 0.5$ μs, according to C. Thus, according to C the clock A is running slower than it should. From the diagram one can show that, in general, when C measures t_0 then C will say that A and B measure

$$\frac{t_0}{\sqrt{1-\beta^2}} - \frac{vx/c^2}{\sqrt{1-\beta^2}} \qquad \text{or} \qquad t_0\sqrt{1-\beta^2}$$

Note the distinction between (1) $t_0/\sqrt{1-\beta^2}$, the nonproper time interval between two events actually measured by an observer for whom the events occurred at two different places, and (2) $t_0\sqrt{1-\beta^2}$, the presumed time interval between two events ascribed to a moving observer by an observer for whom the two events occurred at the same place.

4. Electric charge of a conducting wire It is found experimentally that electric charge q is an invariant for all observers. If there is a total charge Q within a box as measured by observer A then—whatever the dimensions of the box may be for another observer B moving with respect to A—the same charge Q will be measured within the box by B. This means, of course, that charge density is not an invariant. If the charge is distributed uniformly along a straight line, with linear charge density λ_0 coulombs per meter for an observer at rest relative to the line, then the linear charge density will be $\lambda = \lambda_0/\sqrt{1-\beta^2}$ for an observer moving with velocity $v = \beta c$ in a direction parallel to the line; for an observer moving perpendicular to the line the charge density remains λ_0. These results follow from the Lorentz-Fitzgerald contraction and the invariance of charge. Similarly, a uniform surface charge density σ_0 will either become $\sigma = \sigma_0/\sqrt{1-\beta^2}$ or remain σ_0 for a

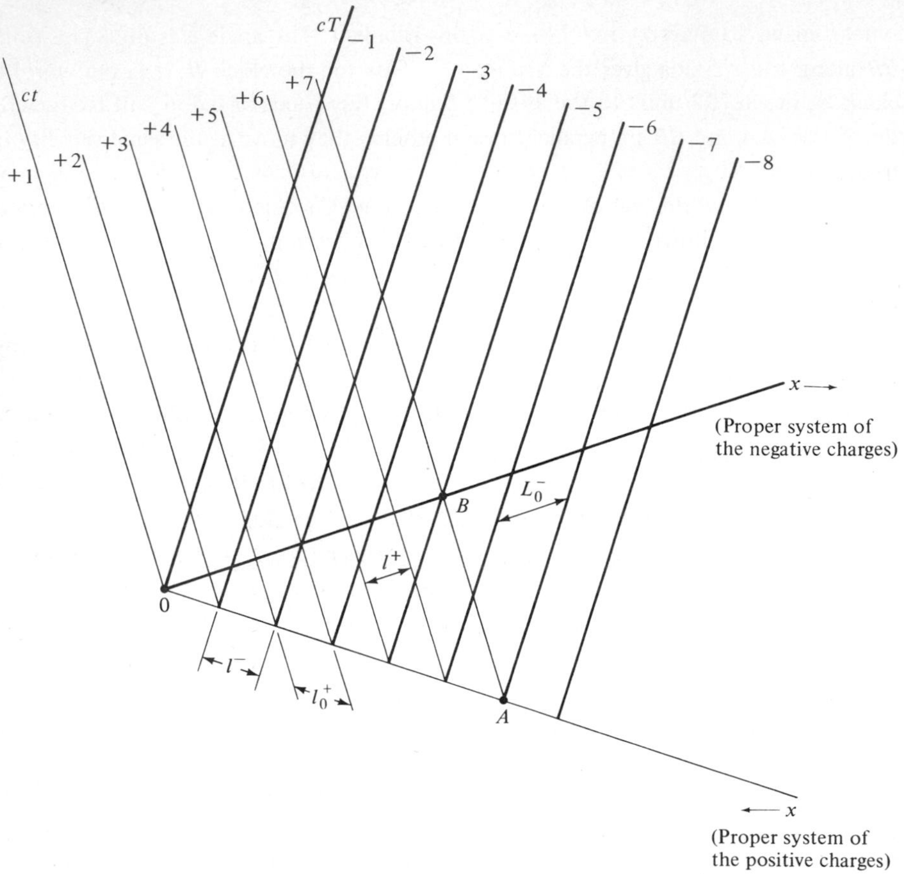

FIGURE 13-9

moving observer, depending on whether the velocity **v** is in one of the two independent directions parallel to the surface or is perpendicular to it. A uniform volume charge density ρ_0, however, becomes $\rho = \rho_0/\sqrt{1-\beta^2}$ whatever the direction of motion.

Suppose we consider a long, straight conductor carrying a current to the left—that is, electrons to the right. We will endow this wire with some simplifying properties which are not actually realistic. The wire will be assumed to consist of two linear arrays of charges, one positive and one negative; both have the same separation between charges, but the positive charges are stationary with respect to the conductor while the negative charges all move to the right with velocity **v**.

In Fig. 13-9 a number of world lines, labeled +1, +2, . . ., represent the life history of some positive charges. They are parallel to the ct axis. The world lines of the negative charges are parallel to the cT axis and some are shown in the figure, labeled −1, −2, ... The separation between the positive charges ℓ_0^+ is a proper distance for (x,ct). The separation between the negative charges ℓ^- is a nonproper distance for (x,ct). Between the points O and A there are shown, for (x,ct), seven positive charges and seven negative charges. So the wire is electrically neutral for (x,ct). The negative charges flowing to the right constitute an electrical current to the left, and the picture expresses the assumption that the flow of current in this wire is such that the wire is electrically neutral when carrying this current. With $\ell_0^+ = \ell^-$ there are equal numbers of positive and negative world lines between O and A.

Let the (X,cT) observer move to the right with the same speed as the negative charges. For the observer the separation between the negative charges is a proper length, so $L_0^- = \ell^-/\sqrt{1-\beta^2}$, but the separation between the positive charges is now a nonproper length, $L^+ = \ell_0^+\sqrt{1-\beta^2}$. So for the (X,cT) observer, since $\ell_0^+ = \ell^-$, we have $L^+ = L_0^-(1-\beta^2)$. In the figure there are shown the world line of seven positive charges lying between the points O and B but, since the separation between the negative charges is larger, there are fewer negative charges between O and B—in the figure there are five. So for the moving observer (despite the facts that the charge is conserved and charge is invariant) the wire will be positively charged, although it is neutral for the observer who is stationary relative to the wire. This paradoxical behavior has its explanation in the relative nature of simultaneity.

The simplified version of a conductor carrying current presented above is useful for showing how simultaneity enters the domain of electricity to affect the existence or nonexistence of a net electric charge. Despite the widespread use of this illustration in textbooks, however, it does not accurately reflect the actual facts. We have previously discussed the case of a properly simplified version of a conductor carrying current—a superconducting transmission line—in Chap. 3, Sec. 3-3, Example 3. A fuller treatment is given in the reference cited at the end of that section. The assumption made there is that the positive and negative linear *charge densities are equal when no current is flowing.* When current does flow the surface of the conductor (not the interior) develops a charge, one wire developing a positive surface charge and the other developing a negative surface charge. The positive wire develops this charge because the mean separation between the moving negative charges is increased when current flows, while the separation between the stationary positive charges remains the same as before. The net charge on this wire, therefore, changes from zero to positive. The other wire develops a negative charge when current flows because the mean separation between the negative charges is decreased below its static value. So, in actuality, the two wires, which were originally neutral for a stationary observer when there was no current, do become charged, for this observer, when current flows.

PROBLEMS

1 The proper time interval between two events is 1 s. What is the correction to this for the nonproper observer who is moving relative to the first one with a speed of (*a*) 6.7 mi/h, (*b*) 670 mi/h, (*c*) 67,000 mi/h, (*d*) 6,700,000 mi/h?

2 In the Minkowski diagram with three spatial dimensions let the observer with the orthogonal coordinate system use the scale 1 cm = 1 unit for each of his four axes; and suppose the observer with the nonorthogonal axes uses the scale factor to give him the scale 2 cm = 1 unit for his X and cT axes, with the relative velocity $\mathbf{v}$ between the two observers. Can the nonorthogonal observer also use the 2-cm-per-unit scale for his Y and Z axes? Or must he use a 1-cm-per-unit scale for his Y and Z axes?

3 Two events occur simultaneously for an observer but at two different points in space, separated by a distance of 1 light-second. Find the time separation between these events, as measured by two clocks, for an observer moving along the direction of the line connecting the two points (*a*) with a velocity $\beta = \sqrt{3}/3$, (*b*) with a velocity of 67,000 mi/h.

(*c*) Let the velocity be 67,000 mi/h, but suppose the points are separated by 1 light-microsecond.

4 Example 1 employed an imagined experiment, illustrated in Fig. 13-6*e* and *f*, to show the necessity for time dilation. Complete the derivation to show that when $u_0 = c = u$ then $t = t_0/\sqrt{1-\beta^2}$.

5 The mean life of a meson is 2.2 μs when at rest. The particle moves with $v/c = 0.8$ relative to the earth. How far does the meson travel between birth and death as measured by the meson? By the earth?

6 A room having a proper length of 1 m has an opening in one wall which can be closed in negligible time. An arrow, of proper length 1 m, is aimed directly at the opening with a velocity $\beta = 0.6$.

(*a*) From the room's viewpoint the arrow's length is contracted to 0.8 m, so it can fit into the room.

(*b*) From the arrow's viewpoint the room's length is contracted to 0.8 m, so the arrow cannot fit into the room. Which is correct?

7 In Fig. 13-4 when B passes A one clock reads $t = 0$ at $x = 0$ and a second clock gives $T = 0$ at $X = 0$. Suppose, instead, the first clock reads t_1 at x_1 when it is passed by the second clock reading T_2 at X_2. What does the Lorentz transformation become for this case? What does it give for x and t at the event (X, T)?

8 The mean earth-sun distance is 93,000,000 mi and one orbit is completed in 365 days. What is the Lorentz-Fitzgerald contraction, in miles, of the earth's 8000-mi diameter that is caused by this motion?

9 Let B be in a rocket ship moving with speed v to the right relative to D on earth and let A and C be in the rocket ship also, the same proper distance ℓ_0 to the left and right, respectively, relative to B. The clocks of A,B,C are synchronized with each other at the start of a journey and remain synchronized; B's clock and D's clock both read zero when B and D pass each other. D also has two extra clocks, A' and C', synchronized with himself, one at each of the points where A and C are when B and D pass each other. Do A and A' read the same at the instant when B and D are opposite each other? C and C'? Derive the formulas.

10 When two rockets are at rest on earth a string is stretched taut between them, to just below the breaking point. The rockets then are accelerated simultaneously, as measured by the earth, to the velocity v. Does the string break because of the Lorentz-Fitzgerald contraction of the string?

11 In Prob. 9 the clock A' does not agree with the clock A when D and B are passing each other. At some subsequent instant D will be adjacent to A. Will the two clocks A and D agree? Explain with a Loedel diagram.

12 A rocket of proper length 300 m is moving away from the earth, to the right, with constant speed. A light ray sent out from the earth is reflected by mirrors at the front and back of the rocket. The first pulse is received 100 s after it was transmitted; the second one arrives 10 μs later. Draw a Loedel diagram showing the emission event O, the reflection events A and B, and the reception events C and D. (The world line of a light ray going left makes a 45° angle with the $-x$ direction. Make the world lines of the rocket front and rear parallel to the cT axis; convert 30 m to light-seconds.) Show that the speed of the rocket is $\beta = 0.98$ (*Hint*: see answers to Prob. 23.)

13 Proxima Centauri, 4.3 light-years from earth, is the nearest of all stars to us. The farthest star in our galaxy is 10^5 light-years away. With what speed would a rocket ship have to go in order to get to each of them in 50 years elapsed time, as measured by the inhabitants of the rocket ship?

14 An antenna on a rocket ship moving with velocity v relative to the earth makes contact with an antenna on earth for a moment and a radio pulse wave is emitted. On the rocket ship there are two receivers; one is 100 m in front and the other is 100 m behind the flying transmitter antenna. Each receives the radio signal $(100/c)$ seconds after the pulse is transmitted, so the rocket antenna is considered the center of a wave moving outward from the wire with speed c. But on the ground there are also two receivers, one 100 m in front of the antenna and the other 100 m behind; and these receive the radio signal $(100/c)$ seconds after transmission. So the earth antenna is considered the center of an outgoing wave moving with speed c. Which is the true center? On a Loedel diagram draw a right-going world line for a light ray, and also a left-going one, both starting from O. Show how simultaneity explains the paradox.

15 From the Lorentz transformation obtain an expression for $\Delta x = x_2 - x_1$, the distance between two events, and $\Delta t = t_2 - t_1$, the time interval between them, both in terms of ΔX and ΔT. Setting $u_x = \Delta x/\Delta t$ and $U_x = \Delta X/\Delta T$, find the relation between u_x and U_x.

16 From the transformation formulas for velocity find u_x if (*a*) $U_X = 0.7c$, $v = 0.8c$; (*b*) $U_X = c$, $v = c$; (*c*) $U_X = c$, $v = 0.1c$; (*d*) $U_X = 0.1c$, $v = c$.

17 What are the transformation formulas, obtained in a manner similar to that in Prob. 15, for the transverse velocity components?

18 Show that the longitudinal velocity transformation formula may be written

$$\tanh^{-1}\left(\frac{u_x}{c}\right) = \tanh^{-1}\left(\frac{U_X}{c}\right) + \tanh^{-1}\left(\frac{v}{c}\right)$$

19 Suppose the velocity of light in water at rest relative to an observer is $u = c/n$, where n is the index of refraction. Let the water move to the right with speed v. What is the speed of the light with respect to the observer now? (This, the Fizeau experiment, was important in establishing the validity of special relativity.)

20 Find the locus of all events, in a Minkowski diagram and in a Loedel diagram, which have a given interval s between each such event and the event O at the origin. What is the locus of the analog of this in real (i.e., three-dimensional, ordinary) space?

21 Let an electromagnetic wave be emitted by a transmitter moving toward $+x$ to the right, with velocity $\mathbf{v}$ relative to a receiver. The wave comes to the receiver making an angle θ_r with the $+x$ direction, but for an observer at the point of arrival who is moving with the velocity of the transmitter the angle of arrival with the $+x$ direction is θ_t. Then θ_t is also the angle the emitted wave makes at the transmitter with the $-x$ direction. Find the aberration formula, i.e., the relation between θ_r and θ_t from the equations for velocity transformation.

22 The Lorentz transformation refers to the coordinates of events that are obtained by measurement. In the measurement of the position of an extended body moving relative to an observer, the latter must note the positions of different points, simultaneously for him. If photons were to travel from these many individual points to one central collection point they would not arrive simultaneously if the distances differed. Compare this measurement process with the process of one-eyed seeing (or of taking a photograph with a camera). Will the Lorentz transformation apply to seeing? Would a cube seem foreshortened by the Lorentz-Fitzgerald contraction factor?

23 Figure 13-10 is a Minkowski diagram explanation of the ordinary Doppler effect, i.e., the apparent change in frequency of a wave obtained by a receiver when it is moving with respect to the transmitter along the line connecting the two. The X axis is unnecessary for this effect and has been omitted for simplicity. (Actually, the x axis is superfluous.) The events $ct = \ldots, -3, -2, -1, 0$ are cyclical events, such as the crests of sine waves, which are emitted by the ct observer. Radio waves are

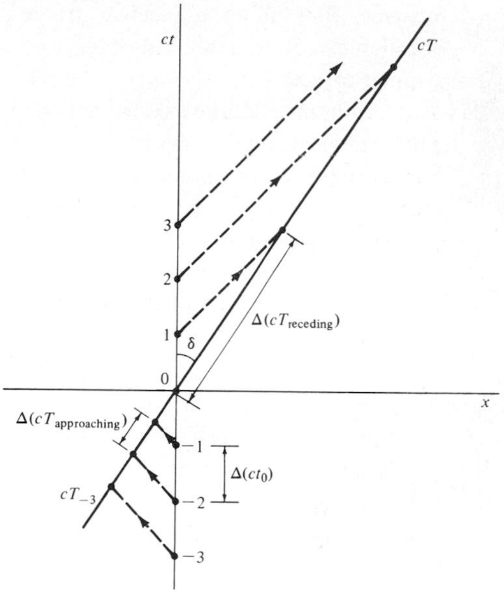

FIGURE 13-10

directed from the emitter toward the cT observer who is approaching from the left; these are shown by the world lines at 45° angles to the $-x$ direction. The signals are received by the cT observer at cT_{-3}, cT_{-2}, ...

(*a*) From the diagram find $\Delta(cT_{\text{approaching}})$, the time between received events, in terms of $\Delta(ct_0)$, the proper time between the transmitted events (when the observers are approaching each other).

(*b*) From this find the relation between the apparent frequency and the proper frequency on approaching.

(*c*) At event O the two observers pass each other and thereafter they recede from each other. The signals from the ct observer must now go to the right to reach the cT observer, as shown. Repeat (*a*) for this case.

(*d*) Repeat (*b*) for this case.

24 Add one spatial dimension to the Loedel diagram. Let the relative motion between the two observers (receiver and transmitter) be parallel to the x-X direction but with a $y(=Y)$ separation between them, so that a light signal arriving at the origin has x and y components. Suppose the ray is perpendicular ($X = 0$) for (X, Y, cT) with a frequency f_{per}. For (x,y,ct) the ray is not perpendicular and the frequency is f. The problem can be reduced to a Loedel diagram in the plane of the page by considering the projection of the ray in that plane. Show that f_{per} equals $f/\sqrt{1-\beta^2}$. This is the transverse doppler effect, which has no classical counterpart. Suppose (X, Y, cT) is the receiver and (x,y,ct) the transmitter. The case above corresponds to a ray that is perpendicular for the receiver but not for the transmitter ($\theta < 90°$). The receiver frequency will then be slightly higher. On the other hand, if the ray is perpendicular to the transmitter (x,y,ct) but not for the receiver ($\theta > 90^\circ$) the receiver frequency will be slightly lower. The transverse doppler effect is a second-order one and is much smaller than the longitudinal doppler effect.

25 Any event in a Minkowski or a Loedel space-time diagram has a unique position. The interval between two events also has a unique value for the straight world line connecting the events, but it has a different value for any other line between these events. In fact, it is a property of space-time that a straight line has the longest interval between two events. This leads to the famous paradox of the twins.

(*a*) Find the relation between interval and proper time.

(*b*) Consider two identical twins. One stays on earth while the other journeys off into space at $v/c = 0.6$. After a period of 5 years for the earth twin, the rocket twin turns about in negligible time and returns to earth, 10 years after he left, according to the earth twin. Using a Loedel diagram, find the interval between the start and finish events for the earth twin.

(*c*) What is his proper time?

(*d*) What is the proper time of the journeying twin?

(*e*) What is the interval between those events for him?

26 Consider a set of triplets. One goes off toward $+x$ with velocity $v/c = 0.6$ for 5 years, earth time, then turns about and returns to earth at the same speed; 10 years, earth time, elapse between his going and coming back. The second triplet goes off toward $-x$ with velocity $v/c = 0.6$ for 5 years, earth time; then he returns to earth at the same speed. The third stays on earth, and is 10 years older after the second event. How much older are the other two? Is one older than the other? What distinguishes them?

13-2 MOMENTUM, MASS, ENERGY

The relations among mass, momentum, and energy constitute the subject matter of dynamics, as distinct from kinematics. The latter just deals with space and time. In this section we will show how the kinematic transition from the galilean transformation to the Lorentz transformation produces profound modifications in dynamics. We will consider, in order, the connections between (1) momentum and mass, (2) mass and energy, and (3) energy and momentum.

1. Consider a head-on inelastic collision between two identical particles. The term inelastic means that the kinetic energy of the particles is not conserved in the collision; after the collision of two lumps of putty, e.g., there would be less total kinetic energy than before the collision but there would be a corresponding increase in the internal energy of the system—binding forces, surface tension, or something similar. In the table below, line 1 shows the two particles before and after the completely inelastic collision in the center-of-momentum system of observer A (i.e., the system in which the total

Observer	Before		After
A	m $\xrightarrow{v}$	$\xleftarrow{v}$ m	$2m$ •
B_{galilean}	m $\xrightarrow{2v}$	m •	$2m$ $\xrightarrow{v}$
B_{Lorentz}	m $\xrightarrow{\quad}$ $w = \dfrac{2v}{1 + v^2/c^2}$	m •	$2m$ $\xrightarrow{v}$
B'_{Lorentz}	m $\xrightarrow{\quad}$ $w = \dfrac{2v}{1 + v^2/c^2}$	m •	M_v $\xrightarrow{v}$

momentum is zero). The mass is conserved in the collision—the mass of the system is $2m$ before and $2m$ after; so is the momentum, but there is a loss in kinetic energy of mv^2.

A galilean observer B, who is moving to the left relative to A with velocity v, would describe the collision according to line 2 of the table. The masses are invariant, while the velocities are obtained from the first line by adding $+v$ to each one. Momentum is again conserved and there is still a loss of mv^2 in kinetic energy.

Suppose observer B, moving from the left with velocity v, uses the Lorentz transformation instead of the galilean transformation to compute the velocities. These are given in line 3 of the table. Mass is conserved, since the masses are still invariant, but the momentum is no longer conserved. Since the conservation laws of physics are probably among the most fundamental laws there are, it is necessary to modify the definitions of mass and/or momentum. To do this, Einstein assumed that the mass of a particle at rest would be one thing—the proper mass m_0—while its mass when moving relative to an observer would be a function of the velocity—m_v, the moving mass. We now find the dependence of m_v on v.

Line 4 of the table shows that the Lorentz B' observer, who is identical with Lorentz B observer in velocity calculations but who differs from him in using a variable mass, will express conservation of mass and conservation of momentum by

$$m_w + m_0 = M_v$$

$$m_w w = M_v v$$

Putting $m_w = (v/w)M_v$ from the second equation into the first gives

$$M_v = \frac{m_0}{1 - v/w} = \frac{2m_0}{1 - v^2/c^2}$$

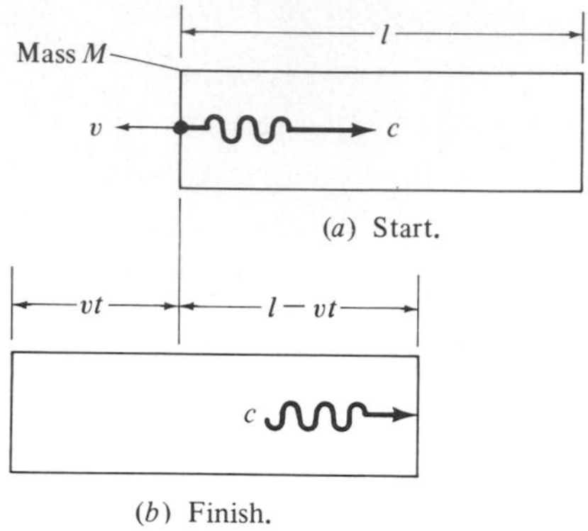

FIGURE 13-11

Then

$$m_w = \frac{v}{w}\frac{2m_0}{1-\beta^2} = \frac{1+\beta^2}{2}\frac{2m_0}{1-\beta^2} = \frac{m_0}{[(1-\beta^2)/(1+\beta^2)]}$$

$$= \frac{m_0}{\sqrt{1-(w/c)^2}}$$

Since v is arbitrary, w is also arbitrary. We can, then, simply write

$$m_v = \frac{m_0}{\sqrt{1-\beta^2}}$$

The momentum now becomes

$$p = m_v v = \frac{m_0 v}{\sqrt{1-\beta^2}} = m_0 c \frac{\beta}{\sqrt{1-\beta^2}}$$

Material particles are those with $m_0 \neq 0$; for them it is necessary that $\beta < 1$ if p is to be noninfinite. Nonmaterial particles are those having $m_0 = 0$; for them it is necessary that $\beta = 1$ if p is to be nonvanishing. The only nonmaterial particle we know for certain is the photon; neutrinos are probably nonmaterial (there are four types: ν_e and ν_μ, the electron neutrino and the muon neutrino; and their antiparticles $\tilde{\nu}_e$ and $\tilde{\nu}_\mu$); gravitons are possibly nonmaterial, if they exist.

The connection between momentum and mass was derived by considering an inelastic collision, but the result is general—the same result would be obtained from an elastic collision. In fact, it does not depend on collisions but applies whenever a mass is moving relative to an observer. The formula is a consequence of only two assumptions: (1) the constant velocity of electromagnetic waves in vacuum for all observers and (2) the principle of relativity, which here states that if momentum is conserved for one observer it is conserved for all observers moving at a constant rate relative to the first. These two assumptions are both combined in the statement of the invariance of the interval.

2. Having obtained the relation between momentum and mass we turn next to that between mass and energy. Einstein derived this (the result that is most famous with the

general public) from a consideration of the imagined experiment shown in Fig. 13-11. In the closed box a transmitter at the left emits a short pulse of radiation to the receiver at the right. During the transmission pulse there is a mechanical recoil to the left exerted on the transmitter and the box to which it is attached; this is caused by the radiation pressure of the emitted pulse. So, during the time it takes the pulse to travel to the right end of the box, the box would move to the left with some velocity **v**. The box would stop moving when the pulse reached the right end of the box, for the radiated pulse would give it an impulse to the right equal to the initial impulse to the left.

Although the box did not interact with the environment it would be displaced to the left of its original position. This is a violation of Newton's first law. To avoid this it is necessary to consider that the displacement of the electromagnetic pulse from the left end of the box to the right end is equivalent to the transfer of some mass. From quantum theory $E = hf$ and $p = h/\lambda$. Here f and λ, the frequency and wavelength, are the field aspects of radiation while E and p, the energy and momentum, are the particle aspects. h is Planck's constant. But $\lambda f = c$ so $p = E/c$. If M is the mass of the box and v is the velocity imparted to it by the impulse,

$$Mv = \frac{E}{c}$$

The time of travel of the pulse is $t = (\ell - vt)/c$ if ℓ is the length of the box. Solving for t:

$$t = \frac{\ell/c}{1 + \beta}$$

Then $\ell - vt = 1/(1 + \beta)$. Meanwhile, the box has moved the distance vt to the left. So, if m is the equivalent mass of the energy E,

$$M(vt) = m(\ell - vt) = (Mv)t$$

Therefore

$$\frac{E}{c}\frac{\ell/c}{1 + \beta} = m\frac{\ell}{1 + \beta}$$

$$E = mc^2$$

The mass and the energy are two different, but concomitant, aspects of the same quantity: mass-energy. When we say a certain amount of electromagnetic energy has been transferred from the left end to the right end of the box we must also say a certain equivalent amount of electromagnetic mass has been transferred. Because c^2 is such a large number, the amount of mass transferred for a given amount of energy is very small. On the other hand, if we transfer a given amount of mass then an equivalent energy is also transferred. Because c^2 is so large, the amount of energy transferred for a given amount of mass is very large. In chemical reactions the binding energies are of the order of 10 eV, corresponding to a mass of the order of 10^{-35} kg. This is at most 10^{-8} of the rest-masses of the atoms, so it may be neglected. The law of conservation of mass-energy then becomes the law of conservation of mass. But in nuclear reactions the binding energies are millions, even billions, of times as large as in chemical reactions. The law of conservation of mass is then simply not true but is replaced by the law of conservation of mass-energy.

3. The energy-momentum relation is the third of our trilogy. From

$$p = mv = \frac{m_0 v}{\sqrt{1-\beta^2}} \qquad \text{and} \qquad E = mc^2 = \frac{m_0 c^2}{\sqrt{1-\beta^2}}$$

we see that

$$E^2 - (cp)^2 = (m_0 c^2)^2$$

The quantities E and p for a particle vary from one observer to another, but $(m_0 c^2)$ is an invariant for all observers—for a given particle. This is just the kind of relation we had in discussing the interval: the difference of two squares is an invariant. Here E plays the role of ct; cp takes the part of r, i.e., $cp_x \leftrightarrow x$, $cp_y \leftrightarrow y$, $cp_z \leftrightarrow z$; the rest-energy corresponds to the interval. Using these analogs we may immediately write the transformation equations for momentum and energy from the Lorentz transformation equations for space and time:

$$cp'_x = \frac{cp_x - \beta E}{\sqrt{1-\beta^2}}$$

$$cp'_y = cp_y$$

$$cp'_z = cp_z$$

$$E' = \frac{-\beta(cp_x) + E}{\sqrt{1-\beta^2}}$$

Just as space and time are really the components of one entity, space-time, so momentum and energy are really the components of one entity, momentum-energy. While only one component of momentum enters it is both convenient and enlightening to use a Loedel-type diagram, as in Fig. 13-12. This momentum-energy diagram gives p_x and E for a specific particle as measured by two observers who are moving relative to each other with velocity v along the x and X direction.

Examples

1. The kinetic energy We will find the relativistic formulas for K, the kinetic energy of a particle, in terms of (*a*) its velocity and (*b*) its momentum.

Definition: the total energy of a free particle equals rest-energy plus the kinetic energy. Thus $mc^2 = m_0c^2 + K$.

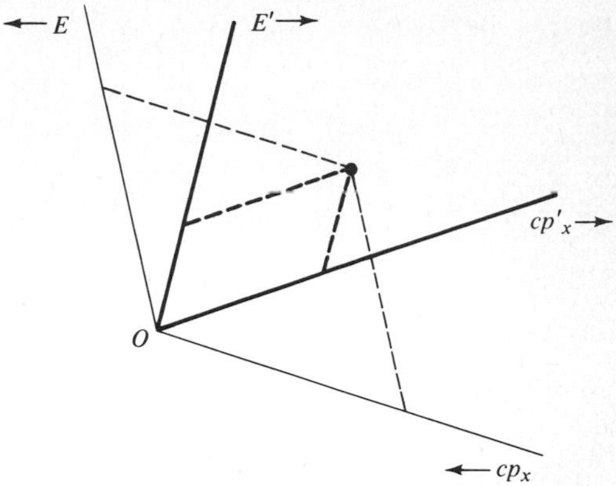

FIGURE 13-12

a. $K = mc^2 - m_0c^2 = m_0c^2 [(1/\sqrt{1-\beta^2}) - 1]$. This is the answer to (*a*) in general. When $\beta \ll 1$,

$$K = m_0c^2 [(1 + \tfrac{1}{2}\beta^2 + \cdots) - 1] = \tfrac{1}{2}m_0v^2 \left(1 + \frac{3}{4}\frac{v^2}{c^2} + \cdots\right) \approx \tfrac{1}{2}m_0v^2$$

It is incorrect to insert the equation $m = m_0/\sqrt{1-\beta^2}$ into the classical formula for the kinetic energy:

$$\tfrac{1}{2}mv^2 = \frac{1}{2}\frac{m_0}{\sqrt{1-\beta^2}}v^2 = \tfrac{1}{2}m_0v^2 \left(1 + \frac{1}{2}\frac{v^2}{c^2} + \cdots\right)$$

This differs from the correct formula by $\frac{1}{2}m_0v^2(v^2/4c^2)$ to the second order.

b. $K = mc^2 - m_0c^2$. But $mc^2 = E$ and, from $E^2 - c^2p^2 = (m_0c_2)^2$, we have

$$mc^2 = +\sqrt{m_0{}^2c^4 + c^2p^2}$$

So

$$K = \sqrt{m_0{}^2c^4 + c^2p^2} - m_0c^2$$

For $p \ll m_0c$ this becomes $K = p^2/2m_0$, the classical result. In the relativistic domain $p \gg m_0c$ and $K \approx cp$.

2. Spontaneous disintegration of particles Suppose a particle of rest-mass M_0 decays spontaneously into two particles, one with rest-mass m_0 and one with rest-mass μ_0. Then conservation of mass-energy gives, if v_1 and v_2 are the velocities of the two decay particles,

$$M_0c^2 = \frac{m_0c^2}{\sqrt{1-\beta_1{}^2}} + \frac{\mu_0c^2}{\sqrt{1-\beta_2{}^2}}$$

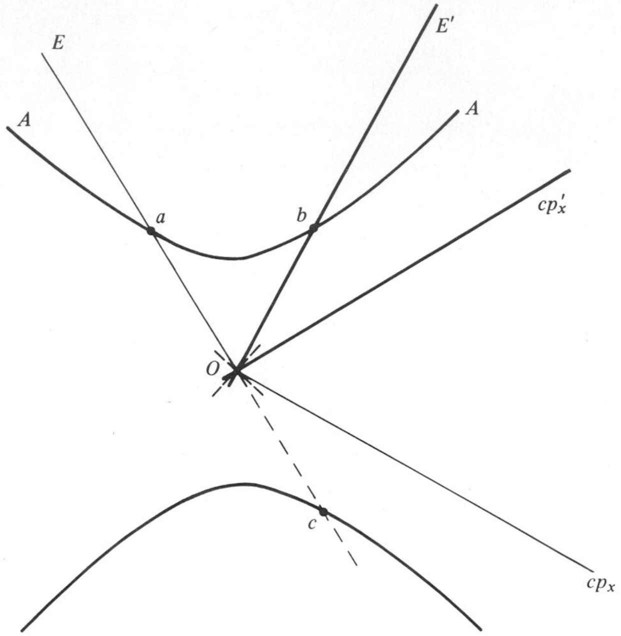

FIGURE 13-13

From conservation of momentum

$$0 = \frac{m_0 v_{1x}}{\sqrt{1-\beta_1^2}} + \frac{\mu_0 v_{2x}}{\sqrt{1-\beta_2^2}}$$

where the x direction is arbitrary. If $\beta_1 = 0$ then $v_{1x} = 0$, so $v_{2x} = 0$ and $\beta_2 = 0$. Thus, if one decay product has zero velocity, both have zero velocity. In this case the two particles formed occupy exactly the same position as the original particle; the two systems are, in fact, indistinguishable. So we have the result that neither β_1 nor β_2 vanishes.

Coming back to the equation for conservation of mass-energy, this now gives $M_0 > m_0 + \mu_0$. This is a necessary condition for spontaneous disintegration. If this condition is not fulfilled, the particle may still disintegrate in this fashion if external energy E is given to it:

$$M_0 + \frac{E}{c^2} = \frac{m_0}{\sqrt{1-\beta_1^2}} + \frac{\mu_0}{\sqrt{1-\beta_1^2}}$$

The greater E, the greater the velocities v_1 and v_2. The smallest E required to induce the disintegration is called the binding energy E_b and it corresponds to infinitesimal velocities for the product particles. Thus,

$$M_0 + \frac{E_b}{c^2} = m_0 + \mu_0$$

3. Energy-momentum curves of electrons and positrons In Fig. 13-13 a space-time-like diagram has been drawn for the momentum-energy of an electron, $m_0 = 9.1 \times 10^{-31}$ kg, $m_0c^2 = 0.82 \times 10^{-13}$ J $= 0.51 \times 10^6$ eV. The point a represents an electron for which $cp_x = 0$ for one observer; E is then the rest-energy for this observer. For a second observer, moving right relative to the first with velocity v, the electron at a is moving to the left; its momentum p_x' will be negative and its energy E' will be greater than the rest-energy. For this observer the electron indicated by b would be the one having zero momentum and only rest-energy. In this way, for a particle of a given type, with a given rest-mass m_0 (or rest-energy $E_0 = m_0c^2$), there is an infinity of observers, moving at all the different speeds along the x axis relative to each other, such that one point on the curve A-A corresponds to the rest-particle of this type, one for each observer. The curve $E^2 - (cp_x)^2 = E_0{}^2$ is a hyperbola, of which A-A is one branch, with the two light rays through O as asymptotes.

Since E_0 is real the right-hand side of this equation is positive, so cp_x can be zero, in which case $E = \pm E_0$ becomes the rest-energy. One branch of the hyperbola must, therefore, lie above or below the origin between the two light rays. Such a curve is called energylike (in the case of the space-time diagram: timelike). It cannot lie (again, as two branches of a hyperbola) to the left and right of the origin between the light rays. There it would be called momentumlike (spacelike); and, for a given point, it would be possible to find an observer for whom the energy of the particles was zero but for whom the momentum was not zero. Particles of imaginary rest-mass, having speeds *always* greater than c—the so-called tachyons, which some investigators are presently seeking—would have precisely such behavior, but such particles have so far not been discovered.

What about the hyperbola that lies below O? It is still given by $E^2 - c^2p_x{}^2 = (E_0)^2$ with E_0 real; but now E_0 is negative. An electron with negative rest-mass (equal in magnitude to that of the ordinary, positive rest-mass, electron) does exist; it has an electric charge equal in magnitude but opposite in sign to that of the ordinary, negatively charged electron; this antiparticle of the electron is the positron. The positron, represented by point c in the figure, is one having zero momentum for the (cp_x,E) observer. Any other positron on the lower branch of the hyperbola would have nonvanishing momentum for this observer with an energy $E = -\sqrt{E_0{}^2 + c^2p_x{}^2}$, which is more negative than that represented by point c.

A particle with negative rest-mass has a momentum in the direction opposite to its velocity and a rate of change of momentum in the direction opposite to that of the force producing it. If a positron were given sufficient energy it could presumably be lifted from the lower branch of the hyperbola to the upper one. For the (cp_x,E) observer the least amount of energy required would be twice the rest-energy of the electron: 1.02 MeV. But this is not observed in nature. The reason we assign to this fact is that the process would violate the conservation of charge. We do, however, observe something similar, but more complicated: pair production, the creation of one electron and one positron by a photon of electromagnetic energy: $hf = m_+c^2 + m_-c^2 + K_+ + K_-$. If the pair were created *from* the photon we would need to put $m_- = -m_+$, so if the kinetic energies were zero this process would require only infinitesimal energy to initiate it. What is observered, however, is that a minimum photon energy of 1.02 MeV is required.

P. A. M. Dirac postulated that the pair is created *by* the photon in the following manner. An infinite sea of ordinary electrons of positive rest-mass m_0 is presumed to fill *all* the negative energy levels $E \leqslant -m_0c^2$. Any electron in this sea, however, is unobservable; when *all* the levels are filled, the presence of one such electron cannot be detected, so there is no contradiction between positive rest-mass and negative rest-energy. A photon hf comes along and lifts one such electron, say the one at c, out of the sea of negative levels to one of the positive energy level $E = (+m_0)c^2$ where, because this sea is *not* completely filled, it becomes an observable ordinary electron. Conservation of mass-energy requires the photon to have a minimum energy of $hf = (+m_0)c^2 - (-m_0)c^2$ or $2m_0c^2$. The hole, or bubble, in the infinite sea of unobservable (positive rest-mass but negative rest-energy) ordinary electrons now becomes an observable (negative rest-mass with negative rest-energy) positron. Pair production! It can be shown that pair production is not possible in free space, i.e., at a place remote from any other real particle. Because of the need to satisfy both conservation of mass-energy and conservation of momentum this process can only occur in the vicinity of another particle, such as a nucleus. The latter acquires negligible energy in the process, though it may acquire appreciable momentum. The proof of this necessity is left to one of the problems.

PROBLEMS

1 The rest-mass of a μ meson is 206.8 electron rest-masses. If ν is a neutrino and $\bar{\nu}$ an antineutrino, both with zero rest-mass, find the maximum kinetic energy for the electron in the disintegration $\mu \rightarrow e + \nu + \bar{\nu}$.

2 Show, through the laws of conservation of mass-energy and momentum, that it is not possible for a single isolated photon to create an electron-positron pair.

3 A particle of charge q and rest-mass m_0 is accelerated through a potential difference of ϕ volts starting from rest. What is its velocity?

4 The derivation of $E = mc^2$ by means of the isolated box has been criticized on the ground that, since it assumes the box starts and stops as a unit, it implies the existence of an ideal rigid body. But a rigid body is impossible relativistically, since this requires an infinite signal velocity. Modify the derivation to meet the objection.

5 In the derivation of $p = m_v v$ an inelastic collision was considered. What is the rest-mass of the coalesced particle there?

6 The formula for moving mass, $m = m_0/\sqrt{1-\beta^2}$, becomes 0/0 when $\beta = 1$, $m_0 = 0$. How is this indeterminate quantity evaluated?

7 What is the minimum frequency of a photon that can give pair production? Compare with the frequency of an optical photon with $\lambda = 5000$ Å.

8 In the photoelectric effect a photon is completely absorbed by an atom and an electron is released. Show, by conservation of mass-energy and conservation of momentum, that it is not possible for a free electron to absorb a photon completely.

9 In the Compton effect a photon is partially absorbed by a free electron, the difference in energy and momentum going off as a scattered photon in a direction making the angle θ with the direction of the impinging photon. From conservation of mass-energy and conservation of momentum find $\Delta\lambda$, the change in photon wavelength.

10 Find the rest-energies, in million electron volts, of the following particles whose rest-masses are given in terms of that of the electron: μ meson, 207; π° pion, 264; $\pi^\pm$ pion, 273; proton, 1836; Λ hyperon, 2181.

11 (*a*) A particle of rest-mass m_0 has momentum p_x for observer 1; observer 1 moves to the right with velocity v relative to observer 2. If the particle were at rest relative to observer 1 its momentum for observer 2 would be p_v. Find its momentum, p_X, for observer 2 in terms of p_x and p_v.

(*b*) Repeat (*a*) with E_x, E_v, and E_X—as the total energies—substituted for the momenta.

(*c*) Repeat (*a*) with K_x, K_v, and K_X as the kinetic energies.

12 Consider the completely inelastic collision between a particle of rest-mass m_0 that moves with velocity v and a stationary particle of rest-mass μ_0. Show that the velocity of the resultant mass is

$$V = \frac{v}{1 + (\mu_0/m_0)\sqrt{1-\beta^2}}$$

13 Using $\mathbf{p} = (m_0/\sqrt{1-\beta^2})\mathbf{v}$ and $\mathbf{F} = d\mathbf{p}/dt$ find $\mathbf{F}$. Let $\mathbf{a}$ be the acceleration and a_v its component along $\mathbf{v}$.

14 Instead of $\mathbf{F} = \mathbf{F}_a + \mathbf{F}_v$, as in Prob. 13, decompose $\mathbf{F}$ into components parallel and perpendicular to the velocity. These components will involve the longitudinal and transverse components of the acceleration, respectively. The remaining factors, called the longitudinal mass and the transverse mass, are not equal to each other. This complication is the price it is necessary to pay if one wishes to use $\mathbf{F} = m\mathbf{a}$ instead of $\mathbf{F} = d\mathbf{p}/dt$.

15 A particle has the normalized velocity components $\delta_x = u_x/c$, $\delta_y = u_y/c$, $\delta_z = u_z/c$ relative to the (x,ct) observer; and also the acceleration components a_x, a_y, a_z. An (X,cT) observer moves left with normalized velocity $\beta = v/c$ relative to the first one. Find the acceleration components of the particle A_x, A_y, A_z relative to him.

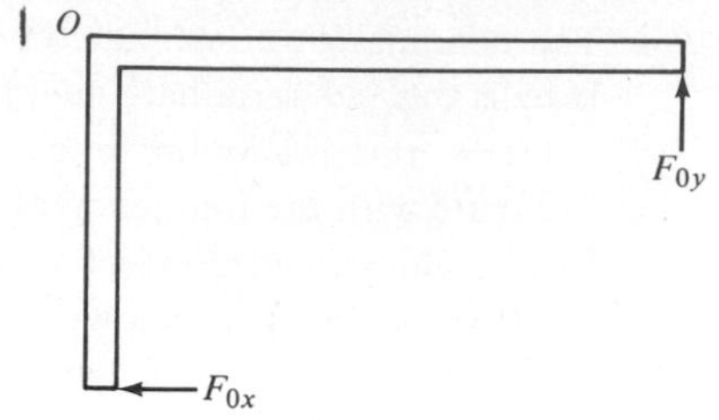

FIGURE 13-14

16 (a) From Prob. 13 find the force components in terms of mass and acceleration for a particle in the system where it is instantaneously at rest. This is a proper force.

(b) Let a particle move relative to an observer along the x axis. Find the nonproper force components in terms of mass and acceleration from Prob. 14.

(c) Modify the general results of Prob. 15 to make them applicable to the case of a particle instantaneously at rest.

(d) Insert the results of (c) into those of (b), and by comparing them with (a) show that when a particle moves relative to an observer only along the x direction, the relations between the proper and the nonproper force components are

$$F_x = F_{0x}$$
$$F_y = \sqrt{1-\beta^2}\, F_{0y}$$
$$F_z = \sqrt{1-\beta^2}\, F_{0z}$$

17 Figure 13-14 shows a right-angled lever with two applied forces such that the net torque is zero: $F_{0y}L_{0x} - F_{0x}L_{0y} = 0$. The lever is then in equilibrium, and if the system was originally at rest it continues to stay at rest. The distances are proper lengths and the forces are proper forces. What is the torque for an observer moving toward $+x$ with velocity v? Is there any rate of change of angular momentum for this observer? Explain.

18 Prove that for the two observers in a momentum-energy Minkowski diagram $\tan\delta = \beta$ just as in a space-time diagram; i.e., the observers in the two diagrams can be taken to be identically the same.

19 Use the laws of conservation of mass-energy and conservation of momentum, in conjunction with invariance of the rest-energy for different observers, to find the energies E_m and E_μ of the decay products of Example 2.

20 Interpret the Heisenberg indeterminacy principle graphically in connection with a space-time diagram and a momentum-energy diagram.

21 The equations $E = mc^2$ and $m = m_0/\sqrt{1-\beta^2}$ give $E = E_0/\sqrt{1-\beta^2}$ for the total energy of a particle or system in terms of the total energy for a rest, or proper, observer. Find the corresponding relation between the work W performed by a force in displacing a body a finite distance for a nonproper observer and W_0, the

work produced by this force for an observer for whom the displacement is infinitesimal, i.e., a proper observer.

22 What effect will the answer to Prob. 21 have on an attempt to write the first law of thermodynamics in such a way that it holds true for A and also for B, moving relative to A?

23 The earth receives approximately 1.35 kW/m^2 of radiant energy from the sun. Taking the distance from earth to sun as 1.5×10^8 km and assuming the sun's loss of energy as isotropic, what is the rate at which the sun loses mass?

14

THE CONNECTION BETWEEN ELECTRICITY AND MAGNETISM

14-1 FIELD TRANSFORMATIONS

In the previous chapter we saw that $\Box^2 \phi = -(1/\epsilon_0)\rho$ and $\Box^2 \mathbf{A} = -\mu_0 \mathbf{J}$ in vacuum. $\Box^2$ is the d'alembertian operator,

$$\Box^2 \equiv \frac{\partial^2}{\partial x^2} + \frac{\partial^2}{\partial y^2} + \frac{\partial^2}{\partial z^2} - \frac{1}{c^2}\frac{\partial^2}{\partial t^2}$$

Using $\epsilon_0 \mu_0 = 1/c^2$, the vector equation may be written

$$\Box^2 (c\mathbf{A}) = -\frac{1}{\epsilon_0}\left(\frac{1}{c}\mathbf{J}\right)$$

The reason for doing this will become apparent shortly.

The d'alembertian has an important property, the proof of which is left to one of the problems. Let there be two observers, $O(x,y,z,t)$ and $O'(x',y',z',t')$, with O moving toward $+x'$ at the constant speed v. Then, using the Lorentz transformation between the two observers for the coordinates of an event, and employing the chain rule of calculus,

$$\frac{\partial^2}{\partial x^2} + \frac{\partial^2}{\partial y^2} + \frac{\partial^2}{\partial z^2} - \frac{1}{c^2}\frac{\partial^2}{\partial t^2} = \frac{\partial^2}{\partial x'^2} + \frac{\partial^2}{\partial y'^2} + \frac{\partial^2}{\partial z'^2} - \frac{1}{c^2}\frac{\partial^2}{\partial t'^2}$$

This does *not* imply that $\Box^2\phi$, say, the d'alembertian of the electric potential at some point, has an invariant value for all inertial observers. (These are observers, moving relative to each other with constant velocity, for whom no inertial forces need be introduced to make Newton's first law applicable. This requirement thereby excludes accelerating observers and rotating observers. We will carefully refrain from specifying what that is, relative to which they are not accelerating.) The $\Box^2$ operator has an invariant form, but the value of a function $\Box^2\psi$ is not necessarily an invariant for different inertial observers. In the case of the electric potential, in fact, $\Box^2\phi$ at any point is proportional to the value of ρ there, and we saw in the preceding section that the proper charge density has a value ρ_0, while other observers have the nonproper value $\rho = \rho_0/\sqrt{1-\beta^2}$. So $\Box^2\phi$ is not an invariant quantity though $\Box^2$ itself is form-invariant.

The proper observer, i.e., the one for whom the charge is at rest, measures ρ_0; for him there is no current and no current density. A nonproper observer does have a current density: $\mathbf{J} = \rho\mathbf{v} = \rho_0\mathbf{v}/\sqrt{1-\beta^2}$. This is a convection current density, with the components

$$J_x = \frac{\rho_0 v_x}{\sqrt{1-\beta^2}} \qquad J_y = \frac{\rho_0 v_y}{\sqrt{1-\beta^2}} \qquad J_z = \frac{\rho_0 v_z}{\sqrt{1-\beta^2}}$$

The quantity $c^2\rho^2 - (J_x{}^2 + J_y{}^2 + J_z{}^2)$ is, consequently, equal to $c^2\rho_0{}^2$: the charge density and current density at a point are not unrelated quantities but are connected as shown. However, the right-hand side is an invariant quantity: every inertial observer, regardless of his velocity, will assign the same value to it in any given case. So the quantity $(J_x, J_y, J_z, c\rho)$ behaves in a manner analogous to that of the quantity (x,y,z,ct). We can draw a "space-time" diagram, actually a current density–charge density diagram, that portrays the invariant difference of the squares of two quantities and thereby we can obtain the transformation equations for ρ and J between different inertial observers. But this is not actually necessary, for we can write the answer down at once by analogy. For relative motion only in the x direction the space-time diagram can leave out the y and z dimensions and gives, with $\mathbf{r} \to \mathbf{J}$ and $ct \to c\rho$,

$$\begin{aligned} J'_x &= \frac{(J_x) + (v/c)(c\rho)}{\sqrt{1-\beta^2}} \\ J'_y &= J_y \\ J'_z &= J_z \\ c\rho' &= \frac{(v/c)(J_x) + (c\rho)}{\sqrt{1-\beta^2}} \end{aligned} \qquad \text{or} \qquad \begin{aligned} J'_x &= \frac{J_x + \rho v}{\sqrt{1-\beta^2}} \\ J'_y &= J_y \\ J'_z &= J_z \\ \rho' &= \frac{\rho + (v/c^2)J_x}{\sqrt{1-\beta^2}} \end{aligned}$$

If λ coulombs per meter is the linear charge density on a filamentary circuit and I is the current in it, then $I = \lambda v$. The quantity $c^2\lambda^2 - I^2$ is an invariant, just as is $c^2\rho^2 - J^2$. The sign of this invariant has absolute significance. It also determines the sign of the force between two such parallel filaments. This is the basis of the distinction made in Chap. 3 between conduction and convection currents. See also Prob. 21 following this section.

Any quantity which transforms this way between O and O' is called a four-vector. So far we have $(\mathbf{r},ct) = (x,y,z,ct)$, the space-time four-vector; $(c\mathbf{p},E) = (cp_x,cp_y,cp_z,E)$ the momentum-energy four-vector; and $(\mathbf{J},c\rho)$, the current density–charge density four-vector. If the components of a four-vector are all divided by an invariant then the new components will still transform as above, so $[(1/c)\mathbf{J},\rho]$ is also a four-vector. The choice of *the* current density–charge density four-vector or, indeed, any particular four-vector is arbitrary to within a multiplicative constant.

GENERAL POTENTIALS The equations

$$\Box^2 \phi = -\frac{1}{\epsilon_0}\rho \qquad \text{and} \qquad \Box^2 (c\mathbf{A}) = -\frac{1}{\epsilon_0}\left(\frac{1}{c}\mathbf{J}\right)$$

now show that $\Box^2(c\mathbf{A},\phi)$ is a four-vector. But we have seen that the $\Box^2$ operator is form-invariant, so $(c\mathbf{A},\phi) = (cA_x,cA_y,cA_z,\phi)$ is a four-vector too: the potential four-vector. If A_x,A_y,A_z, and ϕ are the components of this four-vector at any point for the observer O then the components for O', moving toward O's left with speed v, are

$$cA'_x = \frac{(cA_x) + (v/c)\phi}{\sqrt{1-\beta^2}} \qquad\qquad A'_x = \frac{A_x + (v/c^2)\phi}{\sqrt{1-\beta^2}}$$

$$cA'_y = cA_y \qquad\qquad A'_y = A_y$$

$$cA'_z = cA_z \qquad \text{or} \qquad A'_z = A_z$$

$$\phi' = \frac{(v/c)(cA_x) + \phi}{\sqrt{1-\beta^2}} \qquad\qquad \phi' = \frac{vA_x + \phi}{\sqrt{1-\beta^2}}$$

These are the general transformation equations for the scalar and vector potentials (for observers in relative motion along the x direction); they are basic. Among other things, they show that if one observer has a ϕ but no $\mathbf{A}$ then another observer will have both a ϕ' and an $\mathbf{A}'$; similarly, one can have an $\mathbf{A}$ alone while another will have, again, a ϕ' and an $\mathbf{A}'$.

POINT CHARGE POTENTIALS Suppose, now, that the potential ϕ is caused exclusively by one charge Q located at the origin of O. Figure 14-1 indicates this. For O the source charge Q is permanently at rest; so, too, any observation point P is permanently at rest; then the potentials at the point P are given by

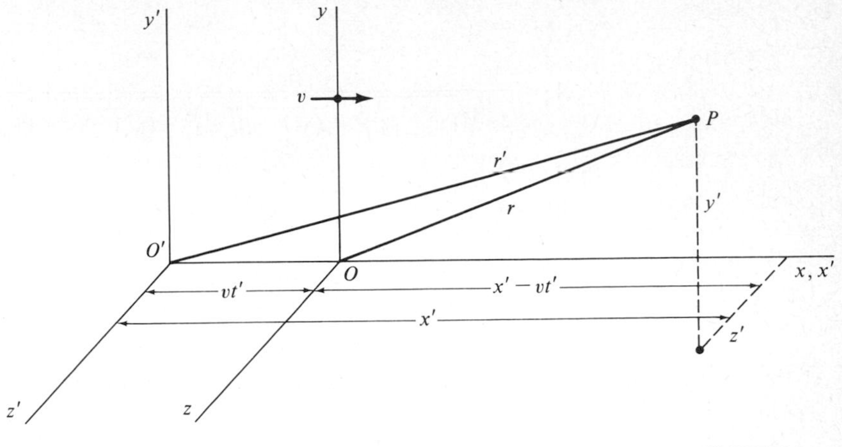

FIGURE 14-1

$$A_x = 0$$
$$A_y = 0$$
$$A_z = 0$$
$$\phi = \frac{1}{4\pi\epsilon_0}\left(\frac{Q}{r}\right)$$

Putting this into the set of transformation equations above, we obtain the potentials at P as measured by O'. The point P is, therefore, moving with respect to O'. The potentials at P for O' are

$$A'_x = \frac{1}{\sqrt{1-\beta^2}}\left(\frac{v}{c^2}\right)\left(\frac{Q}{4\pi\epsilon_0 r}\right)$$
$$A'_y = 0$$
$$A'_z = 0$$
$$\phi' = \frac{1}{\sqrt{1-\beta^2}}\left(\frac{Q}{4\pi\epsilon_0 r}\right)$$

To eliminate r and express the results, instead, in terms of r', we use the Lorentz transformation:

$$r = \sqrt{x^2 + y^2 + z^2} = \sqrt{\frac{(x' - vt')^2}{1-\beta^2} + (y')^2 + (z')^2}$$

So

$$A'_x = \frac{\mu_0 Qv}{4\pi\sqrt{(x'-vt')^2 + (\sqrt{1-\beta^2}\,y')^2 + (\sqrt{1-\beta^2}\,z')^2}}$$

$$A'_y = 0$$

$$A'_z = 0$$

$$\phi' = \frac{Q}{4\pi\epsilon_0\sqrt{(x'-vt')^2 + (\sqrt{1-\beta^2}\,y')^2 + (\sqrt{1-\beta^2}\,z')^2}}$$

At the time $t' = 0$, the instant when the origins O and O' coincide and Q is at O', these become

$$A'_x = \frac{\mu_0 Qv}{4\pi\sqrt{(x')^2 + (\sqrt{1-\beta^2}\,y')^2 + (\sqrt{1-\beta^2}\,z')^2}}$$

$$A'_y = A_y$$

$$A'_z = A_z$$

$$\phi' = \frac{Q}{4\pi\epsilon_0\sqrt{(x')^2 + (\sqrt{1-\beta^2}\,y')^2 + (\sqrt{1-\beta^2}\,z')^2}}$$

These four equations above give the result we seek: the potentials produced by a point charge in uniform motion relative to an observer or, interchangeably, the potentials measured by an observer moving relative to a fixed charge. Two features are worth pointing out. Consider $(x' - vt', y', z')$ for the first set of equations or (x', y', z') for the second set: these coordinates give the spatial separation between P and Q at the instant when A' and ϕ' are being evaluated at P. The potentials at any instant are, therefore, being evaluated in terms of the position of the source charge *at that same instant*. In Chap. 19, where we consider radiation, it will be found that the solution for **A** and ϕ, in general, is given in terms of the position of Q at a *retarded time*, thereby allowing for the finite time of transmission of the signal informing P of Q's value and position. The general solution simplifies, however, for the particular case of uniform velocity of the source and one obtains the remarkable result that only the present, instantaneous, position need be known. This is in no way to be interpreted to mean infinite speed of propagation of signal.

The second feature worth noting is the presence of the $\sqrt{1-\beta^2}$ factor in a seemingly unusual arrangement in connection with the distance. The coordinates transverse to the direction of relative motion are decreased by the factor $\sqrt{1-\beta^2}$ as if these coordinate units themselves were increased in separation by the factor $1/\sqrt{1-\beta^2}$. By contrast, in the Lorentz-Fitzgerald contraction it is the longitudinal axis units which

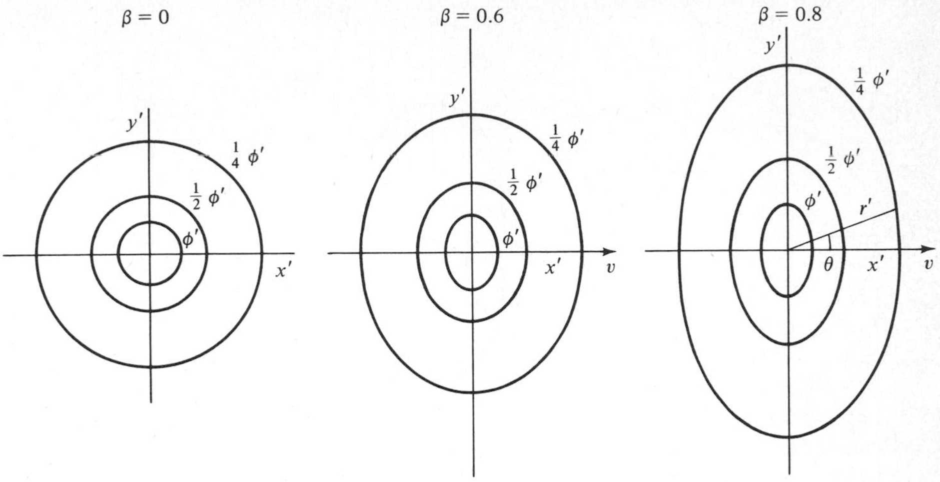

FIGURE 14-2

appear to spread apart. In the present instance we are measuring a quantity which depends on the distance in an inverse fashion. The distinction between the two, expressed in terms of quantities which will only be first discussed later in this chapter, is that between contravariant and covariant four-vectors.

Figure 14-2 shows several equipotentials for $\beta = 0, 0.6, 0.8$ with $z' = 0$. (The curves for constant A'_x would have a similar appearance.)

POINT CHARGE FIELDS Having determined the transformation equations for the potentials, we next seek the corresponding transformations for the quantities commonly called *the* fields, **E** and **B**. (The potentials, ϕ and **A**, are actually also fields.) We will reverse the previous order, first obtaining the result for the point charge and then finding the equations for the general case. To treat the case of the point charge we apply $\mathbf{E}' = -\nabla'\phi' - \partial\mathbf{A}'/\partial t'$ and $\mathbf{B}' = \nabla' \times \mathbf{A}'$ to the $t' = t'$ set of equations above.

$$E'_x = -\frac{\partial\phi'}{\partial x'} - \frac{\partial A'_x}{\partial t'} = +\frac{Q}{4\pi\epsilon_0}\frac{(x' - vt') - (v^2/c^2)(x' - vt')}{[(x' - vt')^2 + (\sqrt{1-\beta^2}\,y')^2 + (\sqrt{1-\beta^2}\,z')^2]^{3/2}}$$

$$= \frac{Q}{4\pi\epsilon_0}\frac{(1-\beta^2)(x' - vt')}{[(x' - vt')^2 + (\sqrt{1-\beta^2}\,y')^2 + (\sqrt{1-\beta^2}\,z')^2]^{3/2}}$$

$$E'_y = -\frac{\partial\phi'}{\partial y'} = \frac{Q}{4\pi\epsilon_0}\frac{(1-\beta^2)y'}{[(x' - vt')^2 + (\sqrt{1-\beta^2}\,y')^2 + (\sqrt{1-\beta^2}\,z')^2]^{3/2}}$$

$$E'_z = -\frac{\partial\phi'}{\partial z'} = \frac{Q}{4\pi\epsilon_0}\frac{(1-\beta^2)z'}{[(x' - vt')^2 + (\sqrt{1-\beta^2}\,y')^2 + (\sqrt{1-\beta^2}\,z')^2]^{3/2}}$$

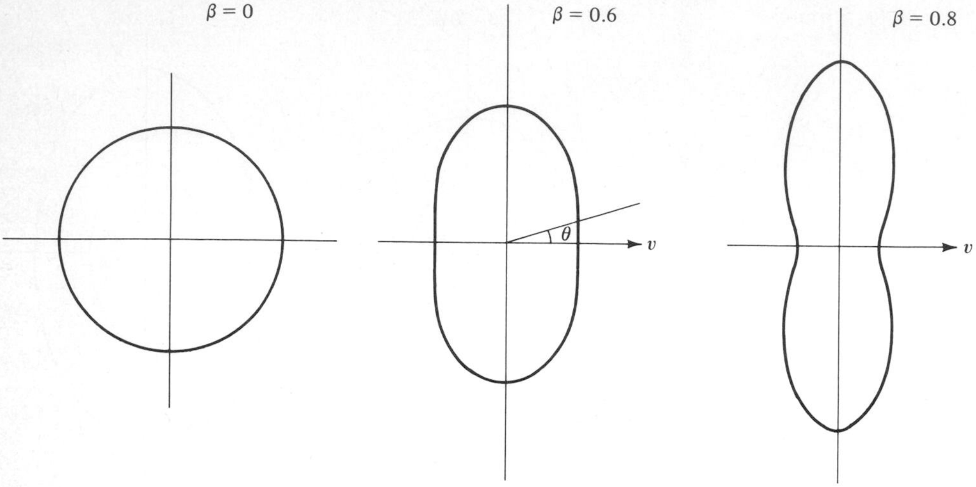

FIGURE 14-3

At the time $t' = 0$, when Q is at O' as well as at O, the equations for $\mathbf{E}'$ reduce to

$$E'_x = \frac{Q}{4\pi\epsilon_0} \frac{(1-\beta^2)x'}{[(x')^2 + (\sqrt{1-\beta^2}\,y')^2 + (\sqrt{1-\beta^2}\,z')^2]^{3/2}}$$

$$E'_y = \frac{Q}{4\pi\epsilon_0} \frac{(1-\beta^2)y'}{[(x')^2 + (\sqrt{1-\beta^2}\,y')^2 + (\sqrt{1-\beta^2}\,z')^2]^{3/2}}$$

$$E'_z = \frac{Q}{4\pi\epsilon_0} \frac{(1-\beta^2)z'}{[(x')^2 + (\sqrt{1-\beta^2}\,y')^2 + (\sqrt{1-\beta^2}\,z')^2]^{3/2}}$$

We see $E'_x/E'_y = x'/y'$, $E'_y/E'_z = y'/z'$, $E'_z/E'_x = z'/x'$; so the vector $\mathbf{E}'$ is radial, just as $\mathbf{E}$ is—Q would produce a central force, in any inertial reference frame, on a charge at P. But unlike the Coulomb intensity, $\mathbf{E}'$ is not spherically symmetric (because of the $\sqrt{1-\beta^2}$ before y' and z', but not x', in the denominators). Instead, the lines of the electric field are shifted away from the regions before and behind the charge to the regions at the side giving a pancakelike effect, as indicated in Fig. 14-3. This effect differs appreciably from that of the Coulomb law only when $\beta \to 1$. The actual working out of the result for E' is left for one of the problems. If θ is the angle shown in Fig. 14-2 this result is

$$E' = \frac{Q}{4\pi\epsilon_0 (r')^2} \frac{1-\beta^2}{(1-\beta^2 \sin^2\theta)^{3/2}}$$

The $(1-\beta^2)$ factor for E'_x, e.g., behaves in an even more complicated way than it does for A'_x. This is an indication of the fact that E'_x, E'_y, E'_z are not the components of a four-vector but of some other, more complicated, quantity. See the next section.

The **B** components may be calculated in similar fashion:

$$B'_x = \frac{\partial A'_z}{\partial y'} - \frac{\partial A'_y}{\partial z'} = 0 - 0 = 0$$

$$B'_y = \frac{\partial A'_x}{\partial z'} - \frac{\partial A'_z}{\partial x'} = -\frac{\mu_0 Qv}{4\pi} \frac{(1-\beta^2)z'}{[(x')^2 + (\sqrt{1-\beta^2}y')^2 + (\sqrt{1-\beta^2}z')^2]^{3/2}}$$

$$B'_z = \frac{\partial A'_y}{\partial x'} - \frac{\partial A'_x}{\partial y'} = +\frac{\mu_0 Qv}{4\pi} \frac{(1-\beta^2)y'}{[(x')^2 + (\sqrt{1-\beta^2}y') + (\sqrt{1-\beta^2}z')^2]^{3/2}}$$

The $\mathbf{B}'$ vector is circular about the x axis since $B'_y/B'_z = -z'/y'$. Also, it is not difficult to show that

$$\mathbf{B}' = \left(\frac{1}{c^2}\right) \mathbf{v} \times \mathbf{E}'$$

Note that $\mathbf{B}'$ is circular even when $\beta \to 1$.

GENERAL FIELDS Now we come to the general transformation equations between two observers moving relative to each other along x and x'. These give $\mathbf{E}'$ and $\mathbf{B}'$ in terms of $\mathbf{E}$ and $\mathbf{B}$ for all cases, i.e., regardless of the particular ρ and $\mathbf{J}$ which produce the fields. We already have the following equations (note that $\hat{\mathbf{x}} = \hat{\mathbf{x}}'$, $\hat{\mathbf{y}} = \hat{\mathbf{y}}'$, $\hat{\mathbf{z}} = \hat{\mathbf{z}}'$):

1 $\quad \gamma \equiv 1/\sqrt{1-\beta^2}; \quad x = \gamma x' - \gamma v t', \quad y = y', \quad z = z', \quad t = -\frac{\gamma v}{c^2} x' + \gamma t'$

2 $\quad A'_x = \frac{\gamma v}{c^2}\phi + \gamma A_x, \quad A'_y = Ay, \quad A'_z = A_z, \quad \phi' = \gamma v A_x + \gamma \phi$

3 $\quad \nabla' \equiv \hat{\mathbf{x}}\frac{\partial}{\partial x'} + \hat{\mathbf{y}}\frac{\partial}{\partial y'} + \hat{\mathbf{z}}\frac{\partial}{\partial z'}; \quad \mathbf{E}' = -\nabla'\phi' - \frac{\partial \mathbf{A}'}{\partial t'}, \quad \mathbf{B}' = \nabla' \times \mathbf{A}'$

Therefore

$$E'_x = -\frac{\partial \phi'}{\partial x'} - \frac{\partial A'_x}{\partial t'}$$

$$= -\left(\frac{\partial \phi'}{\partial x}\frac{\partial x}{\partial x'} + \frac{\partial \phi'}{\partial y}\frac{\partial y}{\partial x'} + \frac{\partial \phi'}{\partial z}\frac{\partial z}{\partial x'} + \frac{\partial \phi'}{\partial t}\frac{\partial t}{\partial x'}\right)$$

$$-\left(\frac{\partial A'_x}{\partial x}\frac{\partial x}{\partial t'} + \frac{\partial A'_x}{\partial y}\frac{\partial y}{\partial t'} + \frac{\partial A'_x}{\partial z}\frac{\partial z}{\partial t'} + \frac{\partial A'_x}{\partial t}\frac{\partial t}{\partial t'}\right)$$

$$= -\left(\gamma v \frac{\partial A_x}{\partial x} + \gamma \frac{\partial \phi}{\partial x}\right)(\gamma) - 0 - 0 - \left(\gamma v \frac{\partial A_x}{\partial t} + \gamma \frac{\partial \phi}{\partial t}\right)\left(-\frac{\gamma v}{c^2}\right)$$

$$-\left(\frac{\gamma v}{c^2}\frac{\partial \phi}{\partial x} + \gamma \frac{\partial A_x}{\partial x}\right)(-\gamma v) - 0 - 0 - \left(\frac{\gamma v}{c^2}\frac{\partial \phi}{\partial t} + \gamma \frac{\partial A_x}{\partial t}\right)(\gamma)$$

$$= -\frac{\partial \phi}{\partial x} - \frac{\partial A_x}{\partial t}$$

$$= E_x$$

By a similar process the other components are found, yielding

$$E'_x = E_x \qquad B'_x = B_x$$

$$E'_y = \frac{E_y + vB_z}{\sqrt{1-\beta^2}} \qquad B'_y = \frac{B_y - (v/c^2)E_z}{\sqrt{1-\beta^2}}$$

$$E'_z = \frac{E_z - vB_y}{\sqrt{1-\beta^2}} \qquad B'_z = \frac{B_z + (v/c^2)E_y}{\sqrt{1-\beta^2}}$$

These are the transformation equations when the O' observer is moving toward $-x$ with speed v relative to the O observer. To find the transformation in the reverse direction simply interchange the primed and unprimed letters and let $v \rightarrow -v$.

Suppose we rewrite these equations as follows:

$$E'_x = E_x \qquad E'_y = \frac{E_y + \beta(cB_z)}{\sqrt{1-\beta^2}} \qquad E'_z = \frac{E_z - \beta(cB_y)}{\sqrt{1-\beta^2}}$$

$$cB'_x = cB_x \qquad cB'_z = \frac{\beta E_y + (cB_z)}{\sqrt{1-\beta^2}} \qquad cB'_y = \frac{-\beta E_z + (cB_y)}{\sqrt{1-\beta^2}}$$

Then (E_y,cB_z) transforms like (x,ct):

$$x' = \frac{x + \beta(ct)}{\sqrt{1-\beta^2}}$$

$$ct' = \frac{\beta x + (ct)}{\sqrt{1-\beta^2}}$$

with O' moving left relative to O. But (E_z,cB_y) transforms like (x,ct) with O' moving right relative to O:

$$x' = \frac{x - \beta(ct)}{\sqrt{1-\beta^2}}$$

$$ct' = \frac{-\beta x + (ct)}{\sqrt{1-\beta^2}}$$

In a very limited sense we could consider E_x and B_x as scalars; $(E_y,0,0,cB_z)$ as a four-vector; and $(E_z,0,0,-cB_y)$ as a second four-vector. But actually this is not so, since the x component of these four-vectors is a transverse component rather than a longitudinal one. It will be seen shortly that the components of **E** and c**B** are not really the components of a four-vector at all, even though certain combinations of some of the components do transform as if they were. They are the components of a more complicated quantity, a four-tensor of the second rank; this will be considered further in the next section.

A more useful rewriting of the transformation equations is the following, in which **v** is the velocity of O' relative to O. (In the cases worked above we had $\mathbf{v} = -\hat{\mathbf{x}}v$.)

$$E'_x = E_x \qquad B'_x = B_x$$

$$E'_y = \frac{[\mathbf{E} + (\mathbf{v} \times \mathbf{B})]_y}{\sqrt{1-\beta^2}} \qquad B'_y = \frac{[\mathbf{B} - (1/c^2)(\mathbf{v} \times \mathbf{E})]_y}{\sqrt{1-\beta^2}}$$

$$E'_z = \frac{[\mathbf{E} + (\mathbf{v} \times \mathbf{B})]_z}{\sqrt{1-\beta^2}} \qquad B'_z = \frac{[\mathbf{B} - (1/c^2)(\mathbf{v} \times \mathbf{E})]_z}{\sqrt{1-\beta^2}}$$

The shadow of the Lorentz force is clearly shown in the first group here.

Examples

1. The force on a moving test charge produced by stationary charges

a. Source charges which are at rest relative to an observer produce an **E** for that observer—a superposition of the individual Coulomb fields of the various sources. But there will be no **B**—the force on a stationary test charge is simply $q_t\mathbf{E}$.

b. When the source charges are moving with constant velocity relative to an observer then the **E** for the observer will be radial but non-Coulombic for the individual source charges; and there will also be a circular **B**. But only the **E** will give a force on this observer's stationary test charge. For a collection of moving source charges the net **E** will be the superposition of the individual **E**'s (and similarly for the **B** field).

c. We will now use the equations of relativity to find the force on a test charge which is moving in the neighborhood of a source charge at rest relative to an observer. We want to know if the same **E** which gives the force on a stationary test charge also gives the force on a moving test charge.

In Fig. 14-4*a* O is the source charge q_s, moving with velocity **v** relative to the test charge q_t, which is at rest relative to the observer O'. Let $\mathbf{E}'$ be the electric field and $\mathbf{B}'$ be the magnetic field as measured by O'; the force on q_t, as measured by O', is then a proper force with the components

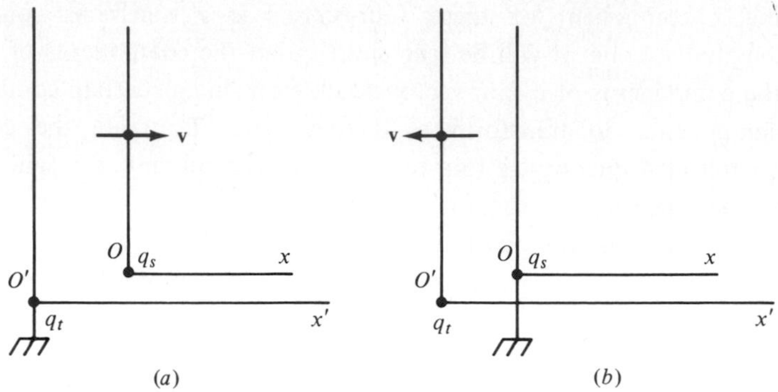

FIGURE 14-4

$$F'_{0x} = q_t E'_x$$
$$F'_{0y} = q_t E'_y$$
$$F'_{0z} = q_t E'_z$$

The $\mathbf{B}'$ produced by q_s does not contribute to F'_0, as q_t is at rest for O'. (Since the $\mathbf{E}'$ acting on q_t produces a force and an acceleration, q_t does not remain at rest permanently relative to O' unless it is constrained to do so; otherwise we are merely considering it in an instantaneous rest system.) Now, as measured by O, the force on q_t is a nonproper force $\mathbf{F}$. See Fig. 14-4*b*. This force is given by the transformation equations of the previous section:

$$F_x = F'_{0x}$$
$$F_y = \sqrt{1-\beta^2}\, F'_{0y}$$
$$F_z = \sqrt{1-\beta^2}\, F'_{0z}$$

Then, since

$$E'_x = E_x$$
$$E'_y = \frac{E_y + vB_z}{\sqrt{1-\beta^2}}$$
$$E'_z = \frac{E_z - vB_y}{\sqrt{1+\beta^2}}$$

we have

$$F_x = q_t E_x$$
$$F_y = q_t(E_y + vB_z)$$
$$F_z = q_t(E_z - vB_y)$$

or

$$\mathbf{F} = q_t[\mathbf{E} + (\mathbf{v} \times \mathbf{B})]$$

where $\mathbf{v} = -\hat{\mathbf{x}}v$ for O.

The force on q_t measured by O is the Lorentz force, which has here been obtained relativistically by the simple and direct transformation from a proper to a nonproper observer. In the present case, however, $\mathbf{B} = 0$ since the only source charge present is at rest relative to O'. So

$$F_x = q_t E_x$$
$$F_y = q_t E_y$$
$$F_z = q_t E_z$$

Since $\mathbf{E}$ is the electric field produced for O by the one stationary source charge q_s, it is given by the Coulomb field.

When more than one stationary source charge is present we employ the superposition theorem. We obtain the result that when all the sources are stationary relative to an observer then the force on a moving test charge is identically the same as that on a stationary test charge at that point.

2. **The Lorentz force**

d. The remaining case is that of the force measured by an observer on a test charge which is moving with respect to him when the source charges are also moving relative to him. This result has already been obtained implicitly in the previous example: it is the Lorentz force expression with $B \neq 0$:

$$\mathbf{F} = q_t\ [\mathbf{E} + (\mathbf{v} \times \mathbf{B})]$$

Table 14-1 summarizes the results for the various possibilities. Actually, these four possibilities are all special cases of the Lorentz force. In the first row $\mathbf{B} = 0$; in the first column $\mathbf{v} = 0$.

Table 14-1 THE FORCE ON A TEST CHARGE

		Relative to the observer O, the *test* charge is	
		Stationary	Moving
Relative to the observer O, the *source* charge is	Stationary	(*a*) $q_t\mathbf{E}$ (Coulombic)	(*c*) $q_t\mathbf{E}$ (Coulombic)
	Moving	(*b*) $q_t\mathbf{E}$ (non-Coulombic for $\beta \to 1$)	(*d*) $q_t[\mathbf{E} + (\mathbf{v} \times \mathbf{B})]$ (non-Coulombic $\mathbf{E}$ for $\beta \to 1$)

3. Motion through uniform static fields Suppose there is relative transverse motion at constant velocity between an observer and a uniform magnetic field. (It does not matter whether the field is considered stationary, or the observer is considered stationary.) This is the case of the linear homopolar generator. Let O be the observer relative to whom the fields are defined as $\mathbf{B} = \hat{\mathbf{z}}B$, $\mathbf{E} = 0$. If the observer O' moves toward $+x$ with nonrelativistic speed v, then the fields for him are

$$E'_x = 0 \qquad B'_x = 0$$

$$E'_y = \frac{0 - vB}{\sqrt{1-\beta^2}} \approx -vB \qquad B'_y = \frac{0-0}{\sqrt{1-\beta^2}} = 0$$

$$E'_z = \frac{0+0}{\sqrt{1-\beta^2}} = 0 \qquad B'_z = \frac{B+0}{\sqrt{1-\beta^2}} \approx B$$

So, for the ordinary slow speeds, the moving observer (the long metal sheet in the Faraday linear analog) will encounter the same magnet field as that measured by an observer at rest relative to the field; but he will also experience an electric field transverse to both $\mathbf{v}$ and $\mathbf{B}$. One observer has only a $\mathbf{B}$ but another has an $\mathbf{E}$ as well as a $\mathbf{B}$. Note that E'_y leads to a motional induced voltage when it is integrated over the width of the sheet. There is no reference to cutting lines of flux. Only the transformation equations are involved.

Next, consider the other case: relative transverse motion at constant velocity between an observer and a uniform electric field. Say an electron is the observer O' moving toward $+x$ through the region between the plates of a parallel-plane capacitor. The capacitor gives the field $\mathbf{E} = \hat{\mathbf{z}}E$ for the stationary observer O, but there is no $\mathbf{B}$. Then

$$E'_x = 0 \qquad B'_x = 0$$

$$E'_y = \frac{0-0}{\sqrt{1-\beta^2}} = 0 \qquad B'_y = \frac{0 + (v/c^2)E}{\sqrt{1-\beta^2}} \approx \frac{v}{c^2} E$$

$$E'_z = \frac{E+0}{\sqrt{1-\beta^2}} \approx E \qquad B'_z = \frac{0-0}{\sqrt{1-\beta^2}} = 0$$

For ordinary, slow speeds the moving observer (the electron) encounters, in its own reference frame, essentially the same electric intensity as that experienced by O, the observer stationary relative to the capacitor; but the electron also experiences a magnetic field.

Conclusion: the terms *electric* and *magnetic* have relative, not absolute, significance.

PROBLEMS

1 The Coulomb field of an electron fixed at the origin of O is $\mathbf{E} = (-e/4\pi\epsilon_0 r^2)\hat{\mathbf{r}}$. Suppose O moves, relative to O', to the right (toward $+x'$) with normalized speed β. Find $\mathbf{E}'$, if θ is given by Fig. 14-2.

2 In finding the fields from the potentials in the case of the moving point charge the expressions used for the potentials were the first set of the two sets given. This set is more complicated than the second set, as it involves t'. Since the desired expressions for the fields did not involve the time, why was not the simpler second set employed?

3 Two infinitely long parallel wires, separated by the distance D, are each charged with λ_0 coulombs per meter, stationary with respect to the wires.

(*a*) Using the Coulomb law, find the magnitude of the electric field at wire 2 produced by wire 1.

(*b*) Suppose an observer moves relative to the wires with velocity $\mathbf{v}$ parallel to their lengths. Using the results of Prob. 1, find the electric field magnitude at wire 2 produced by wire 1 for this observer.

4 In Prob. 3*b*, what would be the electric field at wire 2 if Coulomb's law were employed instead of the correct expression from Prob. 1? Explain.

5 Express the results of Prob. 1 in terms of $\mathbf{r}'$ and β instead of $\mathbf{r}'$, β, and θ.

6 Show that the d'alembertian operator is form-invariant between two observers moving with constant speed relative to each other along the common x,x' axes.

7 In Prob. 3 find the electrical force of repulsion, per unit length of wire 2, produced by all of wire 1: (*a*) in the frame of the wires and (*b*) in the frame of the moving observer.

(*c*) Find the magnetic force of attraction, per unit length of wire 2, caused by all of wire 1.

(*d*) What is the net repulsive force, per unit length of wire 2, in the observer's frame?

(*e*) Compare (*d*) with (*a*).

8 Reconcile the answers to Prob. 7*a* and *d*. Transverse force transforms with a $\sqrt{1-\beta^2}$ factor, while here the transverse forces seem to be equal.

9 Given two infinitely long parallel wires, let O measure: (*a*) a linear proper charge density λ_{0p} of positive charges in each conductor and (*b*) a linear nonproper charge density λ_n of negative charges in each. The negative charges all move with the same constant velocity $\mathbf{v}$ in each wire. Find the force, per unit length of wire 2, that is due to all of wire 1 considering only the force of

(*a*) stationary charges on stationary charges

(*b*) moving charges on stationary charges

(*c*) stationary charges on moving charges

(*d*) moving charges on moving charges

10 Prove that $\mathbf{E} \cdot \mathbf{B}$ and $E^2 - c^2B^2$ are invariant quantities between two Lorentz observers (i.e., inertial observers moving relative to each other with constant velocity). Repeat for $E_y^2 - (cB_z)^2$ and $E_z^2 - (cB_y)^2$.

11 In Prob. 9 assume $\lambda_n = \lambda_{0p}$. Find the net force, per unit length of wire 2, exerted on it by all of wire 1.

12 In Example 3 above the relative motion between the observer and the magnetic field was transverse. Obtain the results when the motion is along **B**. Repeat for the case of the electric field.

13 Take the earth's magnetic field to be uniform, with $\mathbf{B} = \hat{\mathbf{z}}(0.03)$ teslas and also take the earth flat. A plane flies horizontally at a speed of 600 mi/h. What is the motionally induced electric field in a transverse wire inside the plane E'_y?

14 A uniform $\mathbf{B} = \hat{\mathbf{z}}B$ is created by two parallel magnet poles, one in the $z = -h/2$ plane and one in the $z = +h/2$ plane. A man stands on the bottom pole face holding a device for measuring the magnetic field. Both pole faces (and the man with instrument) are then made to move toward $+x$ with a constant velocity **v** that has a large magnitude, approaching c. At all passing points on the magnet pole the stationary observer continues to read the same B. What magnetic field does the moving man's instrument record? Does it differ from his previous reading, when the magnet and he were at rest, although he continues to stand on the pole face? What electric field does the moving man have? Is $\mathbf{E}'$ still equal to $\mathbf{v} \times \mathbf{B}'$? Then repeat this for nonrelativistic speeds.

15 A charged particle moves in a uniform magnetic field. Does the relativistic variation of mass with velocity have any effect on the motion? Find the radius of curvature and the angular frequency.

16 Is it possible to transform an **E** field (with $\mathbf{B} = 0$) for one observer completely into a $\mathbf{B}'$ field for another observer? Can a **B** field (with $\mathbf{E} = 0$) for one observer be completely transformed into an $\mathbf{E}'$ field?

17 Given $\mathbf{E} = \hat{\mathbf{y}}E$, $\mathbf{B} = \hat{\mathbf{z}}B$ for O. An observer O' moves with velocity $\mathbf{v} = -\hat{\mathbf{x}}v$ relative to O.

(*a*) Is there a velocity such that $\mathbf{B}' = 0$?
(*b*) What is $\mathbf{E}'$ then?
(*c*) Is there a velocity such that $\mathbf{E}' = 0$?
(*d*) What is $\mathbf{B}'$ then?

18 From the relation $\mathbf{B} = (1/c^2)\mathbf{v} \times \mathbf{E}$ show that when $v \ll c$ this becomes the law of Biot-Savart. It holds true, also, for $d\mathbf{B}$ and $d\mathbf{E}$.

19 Given $\mathbf{E} = \hat{\mathbf{y}}E_y + \hat{\mathbf{z}}E_z$, $\mathbf{B} = \hat{\mathbf{y}}B_y + \hat{\mathbf{z}}B_z$ for O. An observer O' moves with velocity $\mathbf{v} = -\hat{\mathbf{x}}v$ relative to O. Find the speed such that $\mathbf{E}'$ and $\mathbf{B}'$ are parallel to each other.

20 Derive the equations for E'_y, E'_z and for B'_x, B'_y, B'_z from ϕ' and A' as was done for E'_x earlier in this section.

21 A filamentary conductor has the nonproper linear charge density λ coulomb per meter and the current $\hat{\mathbf{x}}I$ amperes for O, the observer stationary relative to the wire.

(*a*) What velocity $\mathbf{v}'$ must an observer O' have such that $\mathbf{B}' = 0$?
(*b*) What is $\mathbf{E}'$?
(*c*) What velocity $\mathbf{v}''$ must an observer O'' have such that $\mathbf{E}'' = 0$?
(*d*) What is $\mathbf{B}''$?
(*e*) When are these observers possible?

(Use the formulas for **E** and **B** for an infinitely long line. Pick a point on the z axis and employ the transformation equations. Then generalize to cylindrical coordinates.)

22 An electric dipole having the rest momentum **p** moves with the velocity $\mathbf{v} = \hat{\mathbf{x}}'v$ relative to the observer O'. Find $\phi', \mathbf{A}', \mathbf{E}', \mathbf{B}'$ for O'.

23 Find the total energy of a charged particle moving in an electromagnetic field in terms of the potentials.

24 A magnetic dipole having the rest moment **m** moves with the velocity $\mathbf{v} = \hat{\mathbf{x}}v$ relative to the observer O'. Find $\phi', \mathbf{A}', \mathbf{E}', \mathbf{B}'$ for O'.

25 If the field components for O are **E** and **B**, find the field components $\mathbf{E}'$ and $\mathbf{B}'$ for O' in the general case, when O's velocity relative to O is **v**. Take $\gamma \equiv 1/\sqrt{1-\beta^2}$ and $\boldsymbol{\beta} = (1/c)\mathbf{v}$.

26 Prove that the statement $\mathbf{E} \perp \mathbf{B}$ has absolute significance.

14-2 SCALARS, VECTORS, AND TENSORS

The theory of relativity is a body of knowledge which is the logical consequence of two basic axioms. The second of these we have already used many times in this chapter, the assumption that the velocity of electromagnetic waves in vacuum has the same value c, for all inertial observers; the invariance of the interval and the possibility of employing space-time diagrams are merely expressions of this premise. We now turn to the first axiom: the principle of relativity. This asserts that the fundamental laws of physics must be expressible in a similar fashion for all inertial observers. In other words, the equations that express such a law for A must be form-invariant, so that a similar equation expresses the law for B (if B and A are inertial observers moving relative to each other with constant velocity). Thus if $F = dp/dt$ for A then $F' = dp'/dt'$ for B, where F', p', and t' are the transforms of F, p, and t in going from A to B.

Not all the laws of nature obey this principle. The propagation of sound in air, for example, depends on a medium and there is a preferred observer: one who is stationary relative to the air. The differential equation which expresses the motion of the sound wave will be different for this observer than for others moving relative to the air. But this may be taken to express the fact that the propagation of sound is not a truly fundamental phenomenon of nature; fundamental, by definition then, becomes synonymous with form-invariant. Another way of stating this is to assert that for fundamental laws all the different inertial observers, moving relative to each other with any constant speeds whatever, are the members of a mathematical group, any one of which is no more distinguished or important than any other.

We now proceed to write the equations for these fundamental laws in such a way that they are automatically form-invariant when transformed from one inertial observer to another. Satisfying the four properties which define a mathematical group, the various inertial observers are then all on an equal footing. (The question of how to frame the equations so that they apply also to rotating or accelerating observers–in fact, to all observers–belongs to the province of general relativity.) The language we will employ is that of tensor analysis. It turns out that if we can express all physical quantities by tensors then this guarantees that the equations connecting them will be form-invariant. The use of tensors, however, is not a necessary condition for form invariance; e.g., spinors–quantities which are closely related to the quaternions invented by Hamilton–accomplish the same result in a quite different fashion.

The most elementary form of a tensor is the scalar, a tensor of zero order. A scalar is a quantity, characterized by one number, whose magnitude is an invariant for all inertial observers. The charge of an electron is a scalar; its mass, though a number, is not a scalar since its magnitude depends on the velocity of the electron. The electron's proper mass, however, is a scalar. Clearly, if an equation were written in terms of quantities which were all scalars then the equation would be form-invariant; indeed, not only would it be form-invariant but each factor, each term, and each side would be completely invariant. This is, however, much too tight a restriction since there are few physical quantities which are scalars. Only a very, very small percent of the equations of interest could be written exclusively in terms of scalars. The chief utility of scalars lies in their use as factors in equations involving tensors of higher order. The speed of light in vacuum c is an outstanding example.

Next, in order of complexity, are vectors–tensors of the first order. Consider the point P in ordinary three-dimensional space. Suppose we write its coordinates as (x^1,x^2,x^3), instead of (x,y,z), relative to one inertial observer. As measured by another inertial observer its coordinates might be (X^1,X^2,X^3). The rules of calculus then give, for a differential distance in the neighborhood of P,

$$dx^1 = \frac{\partial x^1}{\partial X^1}\,dX^1 + \frac{\partial x^1}{\partial X^2}\,dX^2 + \frac{\partial x^1}{\partial X^3}\,dX^3$$

where the numerics are superscripts, not exponents or powers. This equation, together with the similar ones which hold for dx^2 and dx^3, can be summed up by

$$dx^i = \sum_{J=1}^{3} \frac{\partial x^i}{\partial X^J}\,dX^J \qquad (i = 1, 2, 3)$$

The nomenclature may be made even more succinct by adopting the Einstein convention: whenever the same superscript appears in a term both in the numerator and in the denominator it is intended that the superscript shall be a dummy index and that the sum over all its values be taken. Thus

$$dx^i = \frac{\partial x^i}{\partial X^J}\,dX^J \qquad (i = 1, 2, 3)$$

(It is customary to employ $(x')^i$ instead of X^J, but our terminology is simpler both for the typographer and for the student. For a lower-case letter the index will be lower-case, for a capital letter the index will be capital. A mnemonic device for the transformation rules is then simply obtained as a by-product.) Any collection of quantities in one coordinate system w^i which behaves, when transformed to another coordinate system, in the similar fashion

$$w^i = \frac{\partial x^i}{\partial X^J} W^J$$

is defined as a contravariant vector, or a contravector.

The differential position vector at a point in ordinary space is thus a contravector in three-dimensional space; we may call it a three-contravector. The generalization to a space of two dimensions, or of n dimensions, is obvious, but we will deal only with real-space three-vectors. Real-space vectors in four dimensions, though conceivable, are scarcely ever used and the term four-vectors will be reserved for the four-dimensional space-time vectors. Thus, the event O will be assigned the space-time coordinates

$$\{x^1, x^2, x^3, x^4\} = \{x, y, z, ct\}$$

Much more than the extension of the number of dimensions from three to four is involved in going from a three-vector to a four-vector. The magnitude of a three-vector is an invariant if we compare the measurements of two observers, one of whom is turned at some angle relative to the other:

$$\sum_{i=1}^{3} (x^i)^2 = \sum_{J=1}^{3} (X^J)^2$$

But the magnitude of a four-vector $\{u\}$ is not an invariant

$$(u^1)^2 + (u^2)^2 + (u^3)^2 + (u^4)^2 \neq (U^1)^2 + (U^2)^2 + (U^3)^2 + (U^4)^2$$

Instead,

$$-(u^1)^2 - (u^2)^2 - (u^3)^2 + (u^4)^2 = -(U^1)^2 - (U^2)^2 - (U^3)^2 + (U^4)^2$$

If $\{u\}$ is the space-time four-vector—$\{x,y,z,ct\}$, for example—this equality corresponds to the invariance of the interval for inertial observers.

CONTRAVARIANT VECTORS Contravectors can be given a geometric significance. In Fig. 14-5 we consider a three-dimensional vector for simplicity, and two axes have been drawn as straight lines which are not orthogonal to each other. This, of course, excludes the polar, cylindrical, and spherical coordinate systems of common everyday use; it also leaves out the systems, common to general relativity, which employ nonlinear

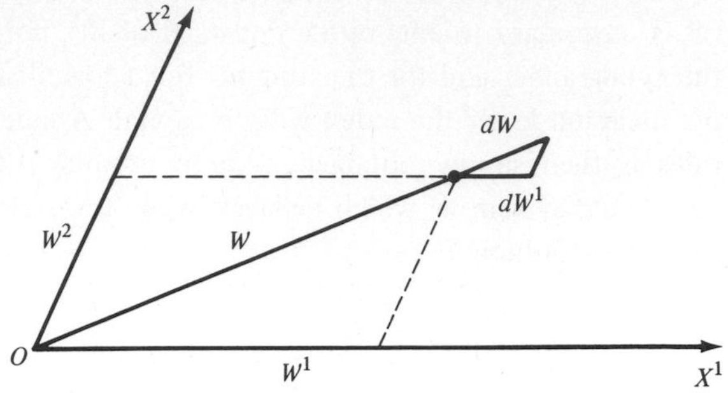

FIGURE 14-5

transformations in going from one set of coordinates to another. The systems we are concerned with have transformations which may be written $w^i = A_J{}^i W^J$ (summation convention) with constant coefficients $A_J{}^i$. Such linear transformations are called affine; they correspond to linear axes.

In Fig. 14-5 the components of the vector are determined by lines parallel to the mating axes, as in the Loedel diagram and the Minkowski diagram. A change dW^1 in the component W^1 then produces a change dW in the magnitude of the vector itself. Since the two triangles are similar

$$\frac{W^1}{W} = \frac{dW^1}{dW} = \frac{\partial X^1}{\partial W}$$

so

$$W^1 = \frac{\partial X^1}{\partial W} W$$

Similarly

$$W^2 = \frac{\partial X^2}{\partial W} W \qquad W^3 = \frac{\partial X^3}{\partial W} W$$

Suppose the same vector **W** is now referred to a different set of oblique cartesian axes, with the coordinates determined in a similar fashion. Then its components—w^1, w^2, w^3—are

$$w^1 = \frac{\partial x^1}{\partial W} W \qquad w^2 = \frac{\partial x^2}{\partial W} W \qquad w^3 = \frac{\partial x^3}{\partial W} W$$

The first gives

$$w^1 = \left(\frac{\partial x^1}{\partial X^1}\frac{\partial X^1}{\partial W} + \frac{\partial x^1}{\partial X^2}\frac{\partial X^2}{\partial W} + \frac{\partial x^1}{\partial X^3}\frac{\partial X^3}{\partial W}\right) W$$

so, using the previous results,

$$w^1 = \frac{\partial x^1}{\partial X^1} W^1 + \frac{\partial x^1}{\partial X^2} W^2 + \frac{\partial x^1}{\partial X^3} W^3 = \frac{\partial x^1}{\partial X^J} W^J$$

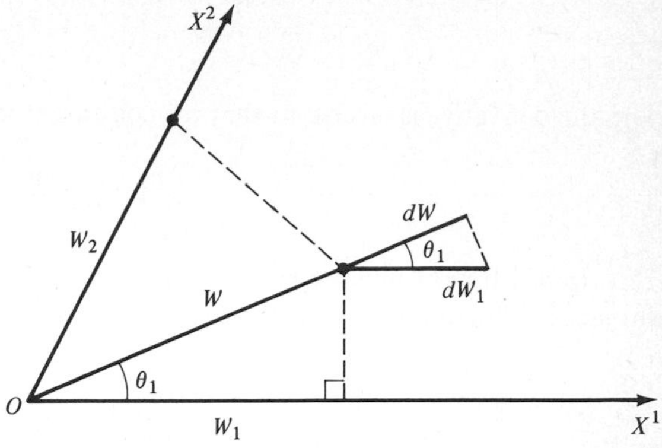

FIGURE 14-6

The other two equations give similar results. These are precisely the transformations which we have used to define contravectors. The asymmetric (half orthogonal, half nonorthogonal) Minkowski diagram and the symmetric (completely nonorthogonal) Loedel diagram are two different representations of a contravector.

Let us assume we are given the equation $\mathbf{W} = a\mathbf{U} + b\mathbf{V}$ connecting three such vectors for some observer. Then $W^1 = aU^1 + bV^1$ and $A_1{}^1 W^1 = A_1{}^1 aU^1 + A_1{}^1 bV^1$. Similarly $A_2{}^1 W^2 = A_2{}^1 aU^2 + A_2{}^1 bV^2$ and $A_3{}^1 W^3 = A_3{}^1 aU^3 + A_3{}^1 bV^3$. Adding all three gives $A_J{}^1 W^J = aA_J{}^1 U^J + bA_J{}^1 V^J$ or $w^1 = au^1 + bv^1$. We can do the same for the other components. Thus $\mathbf{w} = a\mathbf{u} + b\mathbf{v}$: the original equation is automatically form-invariant.

COVARIANT VECTORS The geometric interpretation of contravectors suggests an alternative way for obtaining coordinates in a nonorthogonal cartesian system, that shown in Fig. 14-6. Here subscripts rather than superscripts are employed, and the coordinates are obtained by dropping perpendiculars to the axes. There is no counterpart diagram in use which employs this method for the analysis or synthesis of vectors on a Minkowski-like combination of one orthogonal set of axes with one nonorthogonal set, but the symmetric Loedel system of two nonorthogonal sets of axes does have such an analog in the Brehme diagram. The latter, like any diagram which uses the decomposition system of Fig. 14-6, is employed for covariant vectors or, for short, covectors.

As in the previous figure, suppose the covariant component W_1 of the vector $\mathbf{W}$ receives an infinitesimal change dW_1. This produces a change dW, obtained by dropping a perpendicular to W. Note that the right angle is on dW, not on dW_1, while in the other triangle in this figure it is on W_1. Then

$$\cos\theta_1 = \frac{W_1}{W} = \frac{dW}{dW_1} = \frac{\partial W}{\partial X^1}$$

so
$$W_1 = \frac{\partial W}{\partial X^1} W$$

Here the derivative is inverse to that for contravectors. Similarly,

$$W_2 = \frac{\partial W}{\partial X^2} W \quad \text{and} \quad W_3 = \frac{\partial W}{\partial X^3} W$$

In a different oblique system, using the same method for determining components, the vector **W**'s components would be

$$w_1 = \frac{\partial W}{\partial x^1} W \qquad w_2 = \frac{\partial W}{\partial x^2} W \qquad w_3 = \frac{\partial W}{\partial x^3} W$$

The first gives

$$w_1 = \left(\frac{\partial W}{\partial X^1}\frac{\partial X^1}{\partial x^1} + \frac{\partial W}{\partial X^2}\frac{\partial X^2}{\partial x^1} + \frac{\partial W}{\partial X^3}\frac{\partial X^3}{\partial x^1}\right) W$$
$$= \frac{\partial X^1}{\partial x^1} W_1 + \frac{\partial X^2}{\partial x^1} W_2 + \frac{\partial X^3}{\partial x^1} W_3 = \frac{\partial X^J}{\partial x^1} W_J$$

The other two equations give similar results; so

$$w_i = \frac{\partial X^J}{\partial x^i} W_J$$

This transformation is then the defining equation for covectors. The derivatives are the inverse of those in the definition for contravectors.

The gradient of a scalar function is an example of a covector. Thus, suppose for the capital observer we have

$$\nabla\phi = \hat{\mathbf{X}}\frac{\partial\phi}{\partial X} + \hat{\mathbf{Y}}\frac{\partial\phi}{\partial Y} + \hat{\mathbf{Z}}\frac{\partial\phi}{\partial Z} \quad \text{or} \quad (\nabla\phi)_I = \frac{\partial\phi}{\partial X^I}$$

For a different, lower-case, observer

$$\frac{\partial\phi}{\partial x} = \frac{\partial\phi}{\partial X}\frac{\partial X}{\partial x} + \frac{\partial\phi}{\partial Y}\frac{\partial Y}{\partial x} + \frac{\partial\phi}{\partial Z}\frac{\partial Z}{\partial x}$$

or
$$\frac{\partial\phi}{\partial x} = \frac{\partial\phi}{\partial X^J}\frac{\partial X^J}{\partial x}$$

Indeed, for any component

$$\frac{\partial\phi}{\partial x^i} = \frac{\partial X^J}{\partial x^i}\frac{\partial\phi}{\partial X^J}$$

So if

$$W_J \equiv \frac{\partial \phi}{\partial X^J}$$

then

$$w_i = \frac{\partial \phi}{\partial x^i}$$

A contravariant index in the denominator on the right side of this relation is seen to be associated with a covariant index in the numerator on the left side. This, too, is the reason why, in the example of the Coulomb potential ϕ in the previous section, the units on the transverse axes were effectively more widely separated. Note that **E** and **B** in that example do *not* behave so simply. This is a reflection of the fact that, unlike $\{c\mathbf{A}, \phi\}$, they are not components of four-vectors but are more complicated quantities.

Just as an equation written in terms of scalars and contravectors is automatically form-invariant so, too, is an equation written with scalars, contravectors, and covectors. Each term of such an equation must have the same number of superscripts and subscripts, i.e., all terms must be contravariant to the same extent and covariant to the same extent. The reason for this is the same as that which requires all terms to have the same dimensions: the need for similar behavior under transformations. Just as a contravector in a denominator is equivalent to a covector, a covector in a denominator is equivalent to a contravector. Dummy indices do not have to be considered in determining the extent of covariance or contravariance since they always occur in pairs having one superindex and one subindex, or their equivalents. The process of summation implied by a dummy index is called contraction. For an orthogonal system there is no distinction between contravectors and covectors.

Suppose we take a three-vector $\mathbf{A} = (A^J) = (A^1, A^2, A^3)$, writing it as a contravector, and contract it with another three-vector written as a covector, $\mathbf{B} = (B_J) = (B_1, B_2, B_3)$:

$$A^J B_J \equiv \sum_{J=1}^{3} A^J B_J = A^1 B_1 + A^2 B_2 + A^3 B_3$$

This result is the scalar product of the two vectors. We leave it to the problems to show that the result of such a contraction is a scalar rather than just a number.

Figure 14-7*a* shows one observer O (lower-case, moving right) having a velocity v relative to a second observer A (capitals, moving toward the left in real space). The superscripts indicate contravariant unit vectors. Both observers have right-handed systems. Figure 14-7*b* shows a Minkowski space-time diagram (only two of the four dimensions of each observer are drawn). Observer A has the orthogonal axes in this diagram, moving toward $-x^1$, the vertical line being his world line. Figure 14-7*c* is a Loedel space-time diagram which, like the Minkowski diagram, is contravariant; $\tan \delta = v/c$ for the Minkowski diagram, but $\sin \alpha = v/c$ for the Loedel diagram. Here, again, observer A

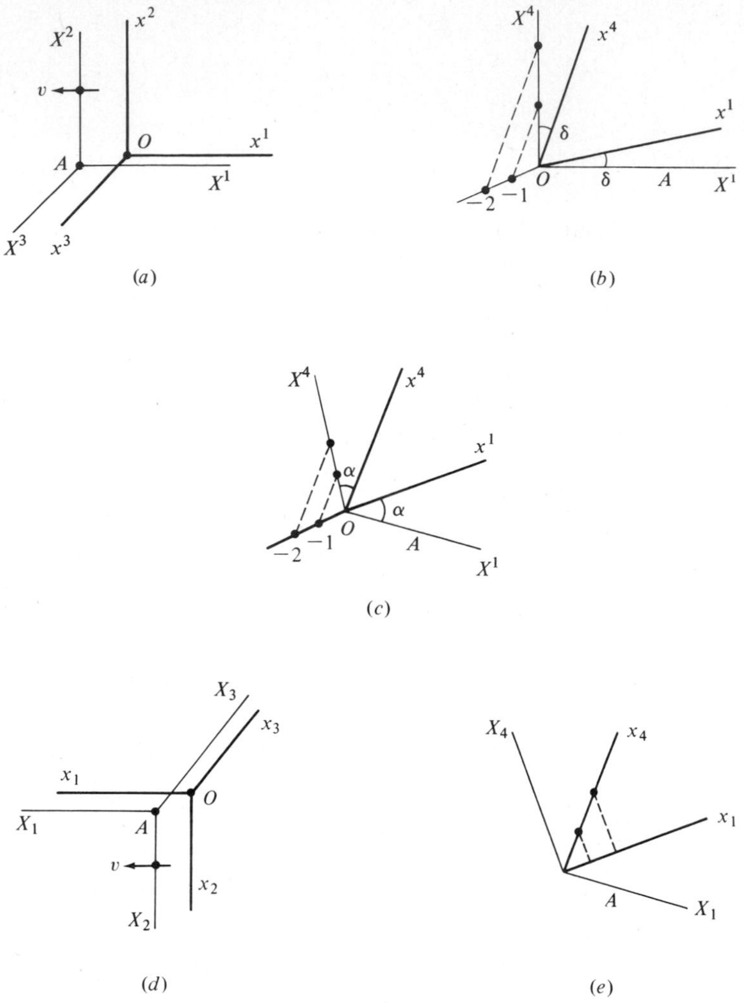

FIGURE 14-7

has the outside axes. A's world line is toward the northwest; as time goes by A assumes increasingly negative values of x^1.

In Fig. 14-7*d* the axes of (*a*) have each been reversed, thereby converting the two right-handed systems to left-handed systems. Observer A still moves to the left in real space, but now it is toward increasingly positive values of x_1. The space-time diagram corresponding to this is a Brehme diagram, shown in Fig. 14-7*e*.

At this point we cannot employ the Minkowski diagram unless we introduce a similar one with covariant coordinates; instead, we will use the Loedel and Brehme diagrams. An event E in the contravariant Loedel diagram is characterized as $\{x^1,x^2,x^3,x^4\}$ by O and as $\{X^1,X^2,X^3,X^4\}$ by A. In the covariant Brehme diagram the same event has the coordinates $\{x_1,x_2,x_3,x_4\}$ for O and $\{X_1,X_2,X_3,X_4\}$ for A. Let all scales be identical in the Loedel and Brehme diagrams, including the orthogonal y and z axes. Then,

$$\begin{aligned} x_1 &= -x^1 \\ x_2 &= -x^2 \\ x_3 &= -x^3 \\ x_4 &= x^4 = ct \end{aligned} \qquad \text{and} \qquad \begin{aligned} X_1 &= -X^1 \\ X_2 &= -X^2 \\ X_3 &= -X^3 \\ X_4 &= X^4 = cT \end{aligned}$$

This reversal of signs for all space coordinates, but not for the time coordinates, is useful in writing quadratic forms. For example,

$$\begin{aligned} s^2 &= (ct)^2 - (x^2+y^2+z^2) = x^4x^4 - (x^1x^1 + x^2x^2 + x^3x^3) \\ &= -[(-x_1)x^1 + (-x_2)x^2 + (-x_3)x^3] + x_4x^4 \\ &= +x_1x^1 + x_2x^2 + x_3x^3 + x_4x^4 \\ &= x_ix^i \qquad\qquad (i = 1, 2, 3, 4) \end{aligned}$$

In this equation we follow the somewhat common procedure of letting a Latin index run through the values 1,2,3,4 while a Greek index has the range 1,2,3. (An equally common rule is the reverse.) When the intent is clear we will sometimes not follow any rule. We will return to covectors and contravectors later in order to show how to convert one to the other mathematically. But first we must discuss tensors.

TENSORS The scalar product is not the only means one has for multiplying two vectors. In three-space, e.g., we may write

$$\mathbf{UV} = (U_J)(V_K) = \begin{bmatrix} U_1V_1 & U_1V_2 & U_1V_3 \\ U_2V_1 & U_2V_2 & U_2V_3 \\ U_3V_1 & U_3V_2 & U_3V_3 \end{bmatrix}$$

The quantity on the right has nine independent components and is obviously more complicated than either of the two original vectors. These components are given in the same coordinate system as that of (U_J) and (V_K). (We use Latin indices here since the equations we are writing apply also to the four-space in which we are really interested.) In a different system $U_J = (\partial x^i/\partial X^J)u_i$ and $V_K = (\partial x^\ell/\partial X^K)v_\ell$, so in this system the JK element or component is

$$U_JV_K = \frac{\partial x^i}{\partial X^J}\frac{\partial x^\ell}{\partial X^K}u_iv_\ell$$

If we label the quantity on the left T_{JK} we have the definition of a second-order covariant tensor, or cotensor:

$$T_{JK} = \frac{\partial x^i}{\partial X^J} \frac{\partial x^\ell}{\partial X^K} t_{i\ell}$$

Note, however, that any T_{JK} must not necessarily be capable of being factored into this tensor product of two vectors; for it may be the sum of several such products. The inverse transformation (the method used for finding this is similar to that for finding the inverse of a contravector–Prob. 11) is

$$t_{jk} = \frac{\partial X^I}{\partial x^j} \frac{\partial X^L}{\partial x^k} T_{IL}$$

The type of multiplication of two vectors which produces a tensor is called the tensor product. The type of tensor above is only one type. A contravariant tensor (or contratensor) of the second rank is given, similarly, by

$$T^{JK} = \frac{\partial X^J}{\partial x^i} \frac{\partial X^K}{\partial x^\ell} t^{i\ell}$$

Its prototype is the tensor product of two contravectors. Finally, a mixed tensor of the second rank follows the pattern set by the tensor product of a covector and a contravector. The rank refers to the number of indices.

This method may be continued to define tensors of the third rank, fourth rank, etc. Some such are, indeed, used in general relativity but for our purposes we will only require tensors of rank zero (scalars), rank one (vectors), and rank two. The same equations we have used for defining tensors in three-space ($i = 1,2,3$) may also be used for tensors in the four-dimensional space of special relativity ($i = 1,2,3,4$). The importance of tensors rests on the fact that an equation written exclusively with tensors is automatically form-invariant.

COVARIANT INDICES ↔ CONTRAVARIANT INDICES We now consider a transfer between covariant indices and contravariant indices. The method of contraction, summing over a pair of dummy indices (one contra- and one co-), supplies us with such a device. For, suppose we had a contratensor of the second rank T^{ij}. (When writing a transformation we will use capital indices with capital tensor symbols and lower-case indices with lower-case tensors; but otherwise we will conform with the general custom of lower-case indices and capital tensor symbols.) We could lower the first index by first multiplying with an arbitrary second-rank cotensor and then summing over the dummy first index:

$$a_{ik} T^{ij} = S_k{}^j$$

Or, we could lower the second index in a similar manner:

$$a_{kj} T^{ij} = R_k{}^i$$

These operations have the defect that $S_k{}^j \neq R_k{}^i$ in general; in the above we have arbitrarily put the subscript before the superscript, but there is no justification for this.

The difficulty vanishes if the tensor a_{ik} is symmetric, i.e., if $a_{ik} = a_{ki}$. If we lower the first index we then obtain $a_{ik}T^{ij} = a_{ki}T^{ij} = T_k{}^j$; if we lower the second, $a_{kj}T^{ij} = a_{jk}T^{ij} = T^{ij}a_{jk} = T_k{}^i$. In the first case, it is clear that the k comes before the j. Considered as a matrix, the k gives the row and the j gives the column. In the case of the lowered second index, the i clearly precedes the k. Note that $T_i{}^j$ and $T^i{}_j$ are not, in general, equal to one another.

Which symmetric matrix shall be arbitrarily selected for this contraction process? There is universal agreement to let the so-called metric tensor $\{g_{ij}\}$ play this role. In gaussian differential geometry the square of the distance between two points on a surface in three-space is

$$(ds)^2 = \sum_{i,j=1}^{2} g_{ij}du^i du^j$$

where u^i and u^j are two parameters which fix the coordinates of any point (x,y,z) on the surface. The g_{ij} are coefficients given by certain rules which we need not go into. The g_{ij} coefficients contain full information on the nature of the surface, independent of the particular u's. Riemann extended Gauss' formulation for a two-dimensional surface to an n-dimensional surface. One would think that a form employing $(n-1)$ independent variables would be the suitable one for a surface in n-space, but it was shown that a bilinear form was not only sufficient but also necessary. Einstein, accordingly, adopted this form for relativity.

In the four-dimensional space-time of special relativity the interval is given by

$$(ds)^2 = -(dx^1)^2 - (dx^2)^2 - (dx^3)^2 + (dx^4)^2$$

Comparing this with the Riemann bilinear form we see that the metric coefficients are given by

$$(g_{ij}) = \begin{bmatrix} -1 & 0 & 0 & 0 \\ 0 & -1 & 0 & 0 \\ 0 & 0 & -1 & 0 \\ 0 & 0 & 0 & +1 \end{bmatrix}$$

$$g_{11} = g_{22} = g_{33} = -1, \quad g_{ij} = 0 (i \neq j), \quad g_{44} = +1$$

Do the coefficients g_{ij} constitute a tensor $\{g_{ij}\}$, or are they merely a matrix (g_{ij})? The question is answered by the so-called quotient theorem: a test matrix (T) is also a tensor $\{T\}$ if, in $(R) = (S)(T)$, the resultant product matrix (R) is a tensor $\{R\}$ for an arbitrary tensor $\{S\}$. Applying this test to $(ds)^2 = g_{ij}dx^i dx^j$ with arbitrary $dx^i dx^j$ then shows, since $(ds)^2$ is a scalar, that the g_{ij} form a tensor. Suppose that we start with a contravariant right-hand coordinate system

$$\{x^i\} = \{x^1, x^2, x^3, x^4\} = \{x, y, z, ct\}$$

To find the covariant system we would write $x_i = g_{ij}x^j$. Thus, in space-time $x_4 = g_{41}x^1 + g_{42}x^2 + g_{43}x^3 + g_{44}x^4 = x^4$, while $x_1 = g_{11}x^1 + g_{12}x^2 + g_{13}x^3 + g_{14}x^4 = -x^1$. Similarly $x_2 = -x^2$, $x_3 = -x^3$. As seen before, geometrically, the space axes of the right-hand contravariant Loedel diagram transform into a left-hand covariant Brehme diagram, while the time axis is unaltered. But in three-dimensional real space,

$$(ds)^2 = \sum_{\alpha,\beta=1}^{3} g_{\alpha\beta}dx^\alpha dx^\beta$$

with $g_{11} = g_{22} = g_{33} = 1$ and $g_{\alpha\beta} = 0$ when $\alpha \neq \beta$, so there is then no distinction in handedness here between a contravariant and a covariant system: $x_1 = x^1$, $x_2 = x^2$, $x_3 = x^3$.

Having g_{ij} we can now define g^{ij}:

$$g^{ij} = \frac{C^{ij}}{|g_{ij}|}$$

where C^{ij} is the cofactor of g_{ij}. It is found that the g^{ij} form a tensor such that $\{g^{ij}\} = \{g_{ij}\}$ both for the space-time of special relativity and for ordinary three-space. This gives us the following method for raising an index:

$$g^{ik}T_{ij} = g^{ki}T_{ij} = T^k{}_j$$

$$g^{kj}T_{ij} = g^{jk}T_{ij} = T_{ij}g^{jk} = T_i{}^k$$

$$g^{ki}g^{\ell j}T_{ij} = g^{ki}(T_i{}^\ell) = T^{k\ell}$$

It is seen that the extent to which a tensor is contravariant or covariant is arbitrary, simply a matter of convenience. There is no fundamental distinction such as exists with the rank of a tensor. See, also, Example 1 of the following section for proof that the symmetric or antisymmetric properties are not absolute under interchange of index position. But the arbitrariness has limits: all the terms in an equation must be covariant to the same extent and contravariant to the same extent in order that the principle of index balance can apply to the equation.

There is one other type of multiplication of vectors, besides the scalar product and the tensor product, which deserves some mention. This type, the vector (or cross) product, is unique to three-dimensional ordinary space. Consider an antisymmetric tensor of the second rank in such a space. Here antisymmetric means $a_{\alpha\beta} = -a_{\beta\alpha}$:

$$\{a_{\alpha\beta}\} = \begin{bmatrix} 0 & a_{12} & a_{13} \\ -a_{12} & 0 & a_{23} \\ -a_{13} & -a_{23} & 0 \end{bmatrix}$$

Instead of nine independent elements, such a tensor has only three. We will compare this with the cross product of two vectors,

$$\mathbf{B} = (B) = (B_1, B_2, B_3) \quad \text{and} \quad \mathbf{C} = (C) = (C_1, C_2, C_3)$$

which seemingly forms another vector

$$\mathbf{B} \times \mathbf{C} = (B)(C) = (B_2C_3 - B_3C_2,\ B_3C_1 - B_1C_3,\ B_1C_2 - B_2C_1)$$

Now, forming the tensor product $(B)(C)$ gives

$$(B)(C) = \begin{bmatrix} B_1C_1 & B_1C_2 & B_1C_3 \\ B_2C_1 & B_2C_2 & B_2C_3 \\ B_3C_1 & B_3C_2 & B_3C_3 \end{bmatrix}$$

while the tensor product $(C)(B)$ gives

$$(C)(B) = \begin{bmatrix} C_1B_1 & C_1B_2 & C_1B_3 \\ C_2B_1 & C_2B_2 & C_2B_3 \\ C_3B_1 & C_3B_2 & C_3B_3 \end{bmatrix}$$

So, taking the difference of the two,

$$(B)(C) - (C)(B) = \begin{bmatrix} 0 & (B_1C_2 - B_2C_1) & -(B_3C_1 - B_1C_3) \\ -(B_1C_2 - B_2C_1) & 0 & (B_2C_3 - B_3C_2) \\ (B_3C_1 - B_1C_3) & -(B_2C_3 - B_3C_2) & 0 \end{bmatrix}$$

We see the correspondence

$$a_{23} = B_2C_3 - B_3C_2$$
$$a_{31} = B_3C_1 - B_1C_3$$
$$a_{12} = B_1C_2 - B_2C_1$$

The "vector" that is formed as a result of the vector product of two vectors is really the antisymmetric combination of two tensor products of these vectors. An ordinary three-vector is called a polar vector. The kind of three-vector which is really an antisymmetric three-tensor is called an axial vector. It turns out that the transformation properties of the two types are the same in all respects but one: on reflection in a plane, the longitudinal component of a polar vector is reversed while the transverse components of an axial vector are reversed. So an axial vector is also called a pseudovector.

Examples

1. **Transformation of vector functions** What is the vector, or nonvector, behavior of $\nabla\phi$ and of $\hat{\mathbf{x}}(\partial\phi/\partial x) + \hat{\mathbf{y}}3(\partial\phi/\partial y) + \hat{\mathbf{z}}(\partial\phi/\partial z)$ with respect to the rotation of axes ($X = x\cos\alpha - y\sin\alpha,\ \ Y = x\sin\alpha + y\cos\alpha,\ \ Z = z$)? Consider

$$\frac{\partial\phi}{\partial x} = \frac{\partial\phi}{\partial X}\frac{\partial X}{\partial x} + \frac{\partial\phi}{\partial Y}\frac{\partial Y}{\partial x} + \frac{\partial\phi}{\partial Z}\frac{\partial Z}{\partial x} = \frac{\partial\phi}{\partial X}\cos\alpha + \frac{\partial\phi}{\partial Y}\sin\alpha$$

Similarly,

$$\frac{\partial\phi}{\partial y} = \frac{\partial\phi}{\partial X}(-\sin\alpha) + \frac{\partial\phi}{\partial Y}\cos\alpha$$

and

$$\frac{\partial\phi}{\partial z} = \frac{\partial\phi}{\partial Z}(1)$$

So

$$\nabla\phi = \hat{\mathbf{x}}\left(\frac{\partial\phi}{\partial X}\cos\alpha + \frac{\partial\phi}{\partial Y}\sin\alpha\right) + \hat{\mathbf{y}}\left(-\frac{\partial\phi}{\partial X}\sin\alpha + \frac{\partial\phi}{\partial Y}\cos\alpha\right) + \hat{\mathbf{z}}\left(\frac{\partial\phi}{\partial Z}\right)$$

$$= \frac{\partial\phi}{\partial X}(\hat{\mathbf{x}}\cos\alpha - \hat{\mathbf{y}}\sin\alpha) + \frac{\partial\phi}{\partial Y}(\hat{\mathbf{x}}\sin\alpha + \hat{\mathbf{y}}\cos\alpha) + \frac{\partial\phi}{\partial Z}(\hat{\mathbf{z}})$$

$$= \frac{\partial\phi}{\partial X}\hat{\mathbf{X}} + \frac{\partial\phi}{\partial Y}\hat{\mathbf{Y}} + \frac{\partial\phi}{\partial Z}\hat{\mathbf{Z}}$$

Thus $\nabla\phi$ is form-invariant under this rotation.

But

$$\hat{\mathbf{x}}\frac{\partial\phi}{\partial x} + \hat{\mathbf{y}}3\frac{\partial\phi}{\partial y} + \hat{\mathbf{z}}\frac{\partial\phi}{\partial z}$$

becomes

$$\hat{\mathbf{x}}\left(\frac{\partial\phi}{\partial X}\cos\alpha + \frac{\partial\phi}{\partial Y}\sin\alpha\right) + \hat{\mathbf{y}}3\left(-\frac{\partial\phi}{\partial X}\sin\alpha + \frac{\partial\phi}{\partial Y}\cos\alpha\right) + \hat{\mathbf{z}}\frac{\partial\phi}{\partial Z}$$

$$= \frac{\partial\phi}{\partial X}(\hat{\mathbf{x}}\cos\alpha - 3\hat{\mathbf{y}}\sin\alpha) + \frac{\partial\phi}{\partial Y}(\hat{\mathbf{x}}\sin\alpha + 3\hat{\mathbf{y}}\cos\alpha) + \hat{\mathbf{z}}\frac{\partial\phi}{\partial Z}$$

$$= \hat{\mathbf{X}}\frac{\partial\phi}{\partial X} - 2\hat{\mathbf{y}}\frac{\partial\phi}{\partial X}\sin\alpha + 3\hat{\mathbf{Y}}\frac{\partial\phi}{\partial Y} - 2\hat{\mathbf{x}}\frac{\partial\phi}{\partial Y}\sin\alpha + \hat{\mathbf{Z}}\frac{\partial\phi}{\partial Z}$$

$$= \left(\hat{\mathbf{X}}\frac{\partial\phi}{\partial X} + \hat{\mathbf{Y}}3\frac{\partial\phi}{\partial Y} + \hat{\mathbf{Z}}\frac{\partial\phi}{\partial Z}\right) - 2\sin\alpha\left(\hat{\mathbf{x}}\frac{\partial\phi}{\partial Y} + \hat{\mathbf{y}}\frac{\partial\phi}{\partial X}\right)$$

This is not form-invariant under the rotation, so it is not a vector;

$$\hat{\mathbf{X}}\frac{\partial\phi}{\partial X} + \hat{\mathbf{Y}}3\frac{\partial\phi}{\partial Y} + \hat{\mathbf{Z}}\frac{\partial\phi}{\partial Z}$$

is a very different function from the transform of $\hat{\mathbf{x}}(\partial\phi/\partial x) + \hat{\mathbf{y}}3(\partial\phi/\partial y) + \hat{\mathbf{z}}(\partial\phi/\partial z)$.

2. The four-dimensional del operator The d'alembertian operator

$$\Box^2 \equiv \frac{\partial^2}{\partial x^2} + \frac{\partial^2}{\partial y^2} + \frac{\partial^2}{\partial z^2} - \frac{1}{c^2}\frac{\partial^2}{\partial t^2}$$

becomes the laplacian operator when variations in time are absent. Is there a four-dimensional operator which becomes ∇ in similar conditions?

∇^2 can be written

$$\begin{aligned}\nabla^\alpha \nabla_\alpha &= (\nabla^\alpha) \cdot (\nabla_\alpha) \\ &= (\hat{\mathbf{x}}\,\partial_x + \hat{\mathbf{y}}\,\partial_y + \hat{\mathbf{z}}\,\partial_z) \cdot (\hat{\mathbf{x}}\,\partial_x + \hat{\mathbf{y}}\,\partial_y + \hat{\mathbf{z}}\,\partial_z) \\ &= \partial_x^{\,2} + \partial_y^{\,2} + \partial_z^{\,2}\end{aligned}$$

in three-space. No reversal of axes occurs in switching between contravariant and covariant operators in three-space, just as with three-vectors. But in four-space the spatial components of such operators change sign in going from contra- to co- or vice versa. Thus

$$\begin{aligned}\nabla^i \nabla_i &= \{\nabla^i\} \cdot \{\nabla_i\} \\ &= \{\hat{\mathbf{x}}\,\partial_x + \hat{\mathbf{y}}\,\partial_y + \hat{\mathbf{z}}\,\partial_z + \hat{\mathbf{ct}}u_t\} \cdot \{-\hat{\mathbf{x}}\,\partial_x - \hat{\mathbf{y}}\,\partial_y - \hat{\mathbf{z}}\,\partial_z + \hat{\mathbf{ct}}u_t\} \\ &= -\partial_x^{\,2} - \partial_y^{\,2} - \partial_z^{\,2} + u_t^2 \qquad (i = 1, 2, 3, 4)\end{aligned}$$

Here u_t is an unknown operator on the time, to be determined. Very simply, if $u_t = \partial/[\partial(ct)]$ we have, with $\alpha = 1,2,3$,

$$\Box^2 = -\nabla^\alpha \nabla_\alpha + \frac{1}{c^2}\frac{\partial^2}{\partial t^2}$$

When time does not enter the picture, $\Box^2 = -\nabla^\alpha \nabla_\alpha$ where $\alpha = 1,2,3$. Thus the del four-operators are

$$\{\nabla^i\} \equiv \hat{\mathbf{x}}\frac{\partial}{\partial x} + \hat{\mathbf{y}}\frac{\partial}{\partial y} + \hat{\mathbf{z}}\frac{\partial}{\partial z} + \hat{\mathbf{ct}}\frac{\partial}{\partial(ct)}$$

and

$$\{\nabla_i\} \equiv -\hat{\mathbf{x}}\frac{\partial}{\partial x} - \hat{\mathbf{y}}\frac{\partial}{\partial y} - \hat{\mathbf{z}}\frac{\partial}{\partial z} + \hat{\mathbf{ct}}\frac{\partial}{\partial(ct)}$$

The equations $\Box^2 \mathbf{A} = -\mu_0 \mathbf{J}$, $\Box^2 \phi = -(1/\epsilon_0)\rho$ which were obtained individually above, in separate chapters, now become the components of *one* relativistic equation. For if

$$\{J^i\} = \{J_x, J_y, J_z, c\rho\} \quad \text{and} \quad \{A^i\} = \{cA_x, cA_y, cA_z, \phi\}$$

we can write

$$\nabla^i \nabla_i A^k = \mu_0 c J^k \qquad (i = 1, 2, 3, 4;\; k = 1, 2, 3, 4)$$

The two $\Box^2$ equations are not independent. If ρ, for example, is given then it is not possible to pick any $\mathbf{J}$ whatever. But in electrostatic and magnetostatic conditions, when time does not enter the picture, $\nabla^i \nabla_i A^k = \mu_0 c J^k$ becomes two separate equations:

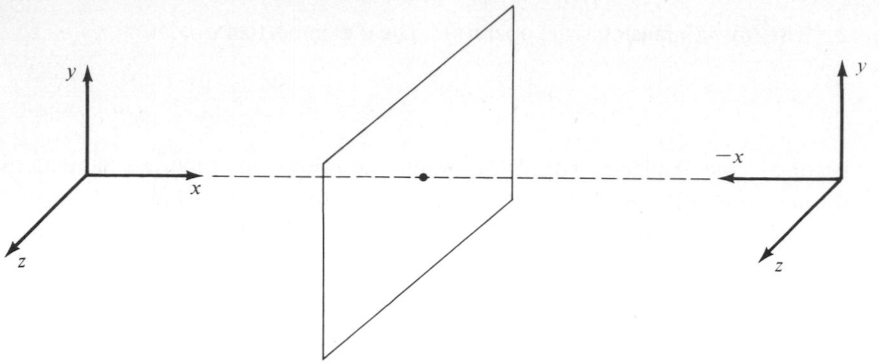

(*a*) *Polar* vectors reverse the *longitudinal* component.

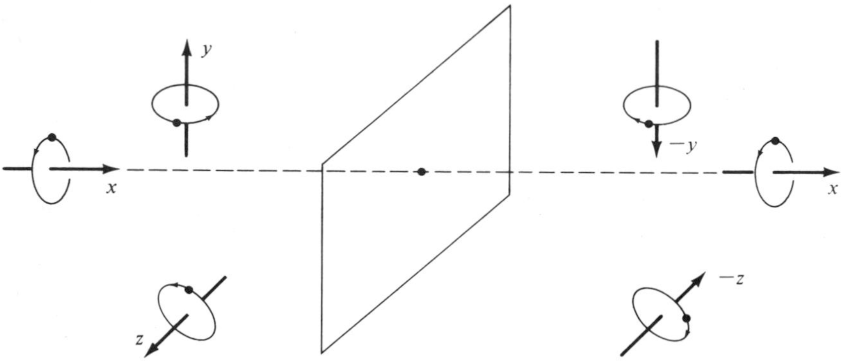

(*b*) *Axial* vectors reverse the *tranverse* components.

FIGURE 14-8

$-\nabla^2(c\mathbf{A}) = \mu_0 c\mathbf{J}$ and $-\nabla^2(\phi) = \mu_0 c(c\rho)$. These are now unrelated to each other and it *is* possible to pick **J** and ρ independently.

3. Reflection of polar and axial vectors in a plane A displacement is a polar vector. Consider an arrow directed toward $+x$ to represent such a displacement and let there be a mirror in the plane $x = 0$. The image of the arrow will point toward $-x$. But an arrow pointing toward $+y$ will have an image which points toward $+y$; and an arrow pointing toward $+z$ will also have an image which points toward $+z$. See Fig. 14-8.

The arrow can also represent an axial vector as, for example, the angular velocity of a ball revolving in a circle. If the ball revolves CW, as seen by an observer looking toward $+x$, then the arrow points to $+x$ by convention. Consider the image of the ball at various instants in its cycle, making use of the fact that the distance from the ball to the mirror is always equal to the distance from the image to the mirror. It will be found that the ball rotates in the same sense when the axis of rotation is perpendicular to the mirror, but the sense of rotation of the image is opposite to that of the ball when the axis of rotation is parallel to the plane of the mirror.

4. Aharanov-Bohm experiment The Aharanov-Bohm effect is quantum mechanical in essence, since it involves Planck's constant h. As such, a description of this effect does not properly belong in this text which is limited only to the classical electromagnetic field; we place it here nevertheless, in the chapter on relativity, as an example of the way four-vectors may be combined to give invariant quantities which have significance, both classically and quantum mechanically.

Consider the inner product of two four-vectors, the momentum-energy and the differential space-time:

$$\{c\mathbf{p}, E\} \cdot \{d\mathbf{r}, c\,dt\} = c[(\mathbf{p} \cdot d\mathbf{r}) - E\,dt]$$

This quantity is an invariant for all inertial observers. As it stands it may be applied directly to a stream of particles along a differential element of their trajectory. By making use of the de Broglie relations, $\mathbf{p} = \hbar\mathbf{k}$ and $E = \hbar\omega$, we may also apply this invariant to a wave, obtaining $c\hbar[\mathbf{k} \cdot d\mathbf{r} - \omega\,dt]$. For plane waves we then have the invariant expression $(\mathbf{k} \cdot \mathbf{r} - \omega t)$. This appears continually throughout the following chapter as the phase Φ in $e^{i\Phi}$, but in general, for points separated by finite distances and times, the invariant will be an integral over the trajectory:

$$\exp\left[i \int_{\text{traj}} (\mathbf{k} \cdot d\mathbf{r} - \omega\,dt)\right]$$

There is another four-vector, characteristic of the electromagnetic field, which has the same dimensions as the momentum-energy four-vector of the particles or waves—the electric charge times the potential four-vector, $q\{c\mathbf{A},\phi\}$. We may, thus, add these two four-vectors and form an invariant quantity by taking the inner product of the resultant with the space-time:

$$\{c\mathbf{p} + qc\mathbf{A}, E + q\phi\} \cdot \{d\mathbf{r}, c\,dt\} = c[(\mathbf{p} + q\mathbf{A}) \cdot d\mathbf{r} - (E + q\phi)dt]$$

The quantity $(\mathbf{p} + q\mathbf{A})$, in fact, is the canonical momentum discussed in Appendix 3; it is a generalization of the ordinary momentum, $\mathbf{p} = m\mathbf{v}$, and turns out to be more useful than its simpler prototype when there is an electromagnetic field (with its velocity-dependent forces), either classically or quantum mechanically.

In the Aharanov-Bohm experiment there is no electric field, so we put $\phi = 0$. Setting $\mathbf{p} = \hbar\mathbf{k}$ and $E = \hbar\omega$, then dividing the invariant above by $c\hbar$, we obtain the phase difference for one particular electron trajectory in covering the distance between P and Q:

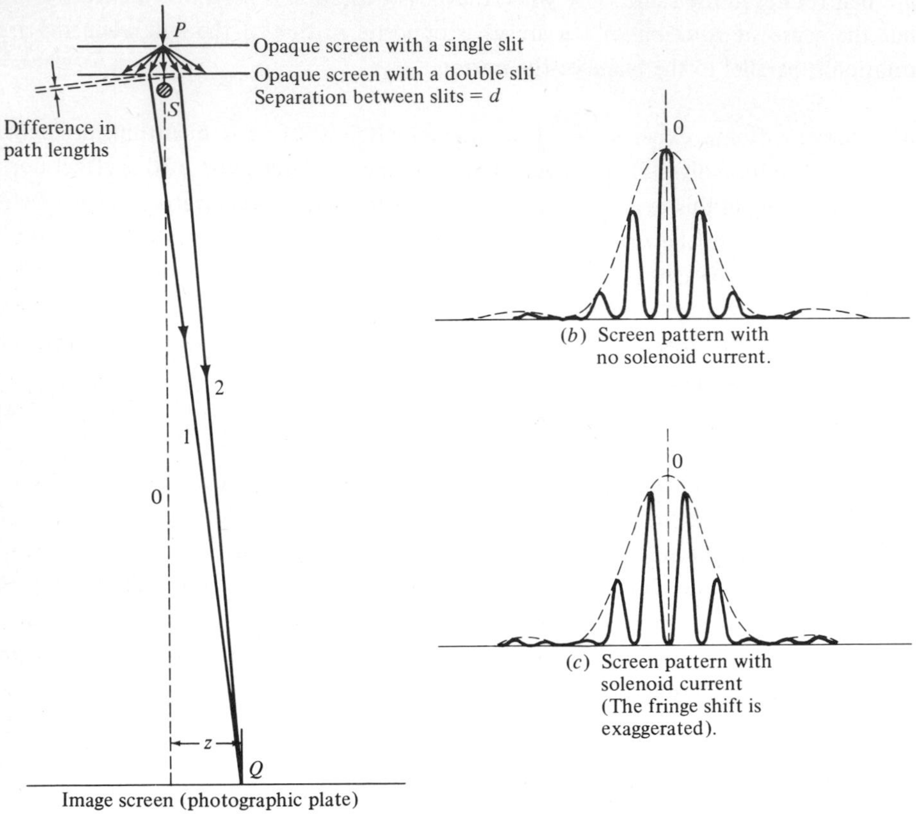

(a) Geometry.

(b) Screen pattern with no solenoid current.

(c) Screen pattern with solenoid current (The fringe shift is exaggerated).

FIGURE 14-9

$$\Phi_1 = \int_1 (\mathbf{k} \cdot d\mathbf{r} - \omega\, dt) + \frac{q}{\hbar} \int_1 \mathbf{A} \cdot d\mathbf{r}$$

See Fig. 14-9. Let there be a second trajectory between the two points P and Q; then the phase difference that results on going from P to Q along this second trajectory is

$$\Phi_2 = \int_2 (\mathbf{k} \cdot d\mathbf{r} - \omega\, dt) + \frac{q}{\hbar} \int_2 \mathbf{A} \cdot d\mathbf{r}$$

The difference in these two phase differences, one that results from going between the two points P and Q by the two different paths 1 and 2, is

$$\Phi_1 - \Phi_2 = \left[\int_1 (\mathbf{k} \cdot d\mathbf{r} - \omega\, dt) - \int_2 (\mathbf{k} \cdot d\mathbf{r} - \omega\, dt) \right]$$

$$+ \frac{q}{\hbar} \left[\int_1 \mathbf{A} \cdot d\mathbf{r} - \int_2 \mathbf{A} \cdot d\mathbf{r} \right]$$

In Fig. 14-9*a*, *S* symbolizes a very long, very thin, solenoid–entirely contained within the narrow region between the two trajectories 1 and 2. When no current flows through the solenoid winding there is no magnetic field anywhere and the difference in phases $(\Phi_1 - \Phi_2)_0$ between the two paths is then simply the first square bracketed expression in the equation above. This results in a variation of the electron intensity on the distant screen that is similar to the variation of light intensity in the corresponding double-slit experiment in optics. A series of fringes is formed (alternating between maximum intensity and zero intensity) by the interference between the two slits. The maximum intensity of these fringes follows an envelope that is due to the diffraction of waves from a single slit. Figure 14-9*b* shows that in this case the maximum of the zero-order fringe occurs on the central axis of the diagram, where $(\Phi_1 - \Phi_2)_0 = 0$. For each other point on the screen there will be some value of $(\Phi_1 - \Phi_2)_0$ for each z, resulting in the pattern shown in Fig. 14-9*b*.

When current does pass through the solenoid winding the **B** field is entirely restricted to the region interior to the solenoid. In the region of the electron trajectories the value of **B** remains zero. Nevertheless, the value of **A** is no longer zero here, so at each z the value of $\Phi_1 - \Phi_2$ is altered from its previous value by the second square bracketed expression in the equation above. The entire pattern of double-slit interference fringes is then simply shifted inside the single-slit diffraction envelope, as shown in Fig. 14-9*c*. This can be seen directly, since the additional phase shift at each point is a constant, independent of z. For its value is

$$\frac{q}{\hbar} \oint \mathbf{A} \cdot d\mathbf{r}$$

where the integral goes from P to Q along 1 and from Q to P along 2. Now this is nothing but $(q/\hbar)$ times the magnetic flux produced within the solenoid.

The experiment is a difficult one to perform, for the electron wavelength is very small, and this results in an extremely small spacing between the fringes. In actuality a magnetized metal whisker, only microns thick, was used instead of a solenoid. The fringe system had to be examined with a microscope. But the experimental results were in excellent agreement with the theory. Despite the fact there was no **B**, and consequently no force, acting on the electron beams when the solenoid (or whisker) was added, there was a physical displacement of the electron distribution in space. This was caused by the nonvanishing vector potential **A**, which existed in the region external to the solenoid (or whisker). The conclusion must then be reached that **A** is not just a mathematical aid in calculating **B**; it is a real field, with the possibility of producing physical effects even in a region where there is no **B**. The fact that **A** is not unique, for a given **B**, but may be modified by adding a gradient of an arbitrary function does not alter this conclusion.

Note that while the passage of current through the solenoid winding shifted the fringe system on the screen, this current produced no additional force on the electrons themselves. The Lorentz force $q\mathbf{v} \times \mathbf{B}$ remains zero throughout the experiment. Nevertheless, the probability distribution of the electrons is altered by the **A**.

PROBLEMS

1 $\Sigma_{i=1}^{4} A^i A_i$ is a quantity which is an invariant for the four-vector $\{A^i\}$. (For the space-time four-vector this is the interval squared.) If this quantity is positive then $\{A^i\}$ is called spacelike; if it is negative then $\{A^i\}$ is called timelike; if it is zero then $\{A^i\}$ is called a null vector. Do these names have absolute significance?

2 Show that every four-tensor can be written as the sum of a symmetric tensor and an antisymmetric tensor.

3 (*a*) Write a four-vector for the velocity.
(*b*) How did you find it?
(*c*) Is it unique? If not, write another one; if yes, show why.
(*d*) What is $v^i v_i$?

4 Show that the g^{ij} form a tensor if $g^{ij} = C^{ij}/|g_{ij}|$, where C^{ij} is the cofactor of g_{ij}. Why is C^{ij} doubly contravariant if g_{ij} is doubly covariant?

5 (*a*) Write a force four-vector.
(*b*) Is it possible to state whether this is contravariant or covariant?

6 Show that the definitions for a covector and a contravector give the same result when the axes are orthogonal.

7 A tensor that is contracted by a rank of two (in letting one superindex and one subindex become a dummy) produces a tensor. Suppose two superindices are turned into a dummy index and added through all values. Is the result a tensor? Two subindices?

8 (*a*) If the temperature of a body is defined by a number at all points of space we have a temperature field. Is this a scalar field?
(*b*) Is a pressure field a scalar field?
(*c*) Is a density field a scalar field?

9 Consider the antisymmetric tensor which has the following properties in some system (not in all: the values change in a transformation). If any two indices are alike the value is zero; otherwise the value is +1 if the indices can be brought to the order 1, 2, 3, . . . , by an even number of transpositions, or −1 if an odd number of transpositions are required, In two-space ϵ^{jk} has four elements: $\epsilon^{11}, \epsilon^{12}, \epsilon^{21}, \epsilon^{22}$ with the values 0, 1, −1, 0. How many elements are there to such a three-dimensional tensor and how many are positive and how many negative? How many elements are there to such a four-dimensional tensor and how many are zero?

10 What are the four properties satisfied by a group?

11 Given $w^i = (\partial x^i/\partial X^J)W^J$. Find W^J in terms of the w^i.

12 Show that the scalar product of two vectors is a scalar.

13 (*a*) Find the value of the determinant of the coefficients in the transformation formula for vectors in three-space, and attach physical significance to the results.

(*b*) Repeat for contravectors in four-space.

(*c*) Covectors in four-space.

14 Write the nine components of T^{ij} in three-space after the following reversal of axes:

(*a*) The x axis (*b*) The y axis (*c*) The z axis

(*d*) The x and y axes (*e*) The y and z axes (*f*) The z and x axes

(*g*) The x, y, and z axes

15 By taking the derivative of the velocity four-vector with respect to the proper time ($s = ct_0$ so t_0 is a scalar, like s and c) find a four-vector for acceleration. Let $\mathbf{a} \equiv d\mathbf{v}/dt$.

16 Show that a covector in a denominator is equivalent to a contravector in the numerator.

17 Let

$$\gamma_1 = \frac{1}{\sqrt{1-(v_1/c)^2}} \qquad \text{and} \qquad \gamma_2 = \frac{1}{\sqrt{1-(v_2/c)^2}}$$

Add one velocity four-vector $\{\gamma_1 v_1, 0, 0, \gamma_1 c\}$ to another $\{\gamma_2 v_2, 0, 0, \gamma_2 c\}$ to obtain the resultant four-vector $\{\gamma v, 0, 0, \gamma c\}$. Determine v in terms of v_1 and v_2. This is an example of the usefulness of dealing with four-vector equations.

18 The Kronecker delta symbol $\delta_{\alpha\beta}$ is defined as $\delta_{\alpha\beta} = 0\ (\alpha \neq \beta)$, $\delta_{\alpha\beta} = 1\ (\alpha = \beta)$.

(*a*) If one makes a mixed tensor $\{\delta_\beta{}^\alpha\}$ from this, one whose matrix is the unit matrix for one particular observer, will its matrix also be the unit matrix for other observers?

(*b*) Repeat for $\{\delta_{\alpha\beta}\}$.

(*c*) Repeat for $\{\delta^{\alpha\beta}\}$.

(*d*) Given: $\{\delta_\beta{}^\alpha\}$, $\{\delta_{\alpha\beta}\}$, $\{\delta^{\alpha\beta}\}$. Which is an accidental unit tensor in one system only? Which is a basically important unit tensor?

19 The transformation equations

$$w^i = \frac{\partial x^i}{\partial X^J} W^J \qquad \text{and} \qquad w_i = \frac{\partial X^J}{\partial x^i} W_J$$

give
$$w^i w_i = \frac{\partial x^i}{\partial X^J}\frac{\partial X^K}{\partial x^i} W^J W_K$$

When this quantity is an invariant, i.e., when $w^i w_i = W^I W_I$, the transformation is called orthogonal and the transformation may be looked on as a rotation. (If the indices go from 1 to 4 it is a rotation in four-space.) Find the condition for the transformation to be orthogonal.

20 Show that $g \equiv |g_{ij}|$, the determinant of the metric tensor, is not a scalar.

21 (*a*) Is the symmetry or antisymmetry with respect to the interchange of a pair of superscripts an invariant property of a tensor?

(*b*) Of a pair of subscripts?

(*c*) Of the interchange of a subscript and superscript?

14-3 ELECTRODYNAMICS

Let the (X,Y,Z,cT) observer move toward $+x$ with velocity v relative to another inertial observer (x,y,z,ct). Then, if $\gamma \equiv 1/\sqrt{1-\beta^2}$,

$$\begin{aligned}
X^1 &= \gamma x^1 - \beta\gamma x^4 \\
X^2 &= x^2 \\
X^3 &= x^3 \\
X^4 &= -\beta\gamma x^1 + \gamma x^4
\end{aligned}$$

So $W^J = [(\partial X^J/\partial x^k)]\, w^k$ gives

$$\begin{aligned}
W^1 &= \gamma w^1 + 0w^2 + 0w^3 - \beta\gamma w^4 \\
W^2 &= 0w^1 + 1w^2 + 0w^3 + 0w^4 \\
W^3 &= 0w^1 + 0w^2 + 1w^3 + 0w^4 \\
W^4 &= -\beta\gamma w^1 + 0w^2 + 0w^3 + \gamma w^4
\end{aligned}$$

The matrix of the coefficients, representing the Lorentz transformation for relative motion along the x,X axes, is

$$(L^J{}_k) = \begin{bmatrix} \gamma & 0 & 0 & -\beta\gamma \\ 0 & 1 & 0 & 0 \\ 0 & 0 & 1 & 0 \\ -\beta\gamma & 0 & 0 & \gamma \end{bmatrix}$$

As usual, the first index—here a superscript—stands for the row and the second—here a subscript—designates a column. For a symmetric matrix, this is reversible. The Lorentz transformation, itself, may be written $W^J = L^J{}_k w^k$, which is tensor shorthand for four separate equations, each having four terms on the right. Since this is true for an arbitrary four-vector $\{w^k\}$ and the resultant four-vector $\{W^J\}$, where

$$\{W^J\} = \begin{bmatrix} W^1 \\ W^2 \\ W^3 \\ W^4 \end{bmatrix} \quad \text{and} \quad \{w^k\} = \begin{bmatrix} w^1 \\ w^2 \\ w^3 \\ w^4 \end{bmatrix}$$

(both written as 4×1 matrices), the quotient theorem mentioned above tells us that $\{L^J{}_k\}$ is not only a matrix but also a tensor. The tensor equation is $W^J = L^J{}_k w^k$, or $\{W^J\} = \{L^J{}_k\}\{w^k\}$. It may also be written as the matrix equation $(W^J) = (L^J{}_k)(w^k)$, with the usual rules for multiplication of matrices.

If $\{L_K{}^J\}$ gives the components for the (X,Y,Z,cT) observer in terms of those for (x,y,z,ct), then the reserve transformation is given by

$$\{L^J{}_k\}^{-1} = \{L^j{}_K\} = \begin{bmatrix} \gamma & 0 & 0 & \beta\gamma \\ 0 & 1 & 1 & 0 \\ 0 & 0 & 0 & 0 \\ \beta\gamma & 0 & 0 & \gamma \end{bmatrix}$$

In the above, the capital observer is moving to the right in real space relative to the lower-case observer; for contravectors both observers have right-hand systems, so the capital observer is moving toward $+x$. Suppose we were dealing with covectors. Then both observers would have left-hand space systems. If we still consider that the capital observer is moving to the right in real space then the capital observer is now moving toward $-x$. From

$$W_J = \frac{\partial x^k}{\partial X^J} w_k = L_J{}^k w_k$$

we obtain an $L_J{}^k$ here that is identical with the $L^j{}_K$ above. The matrix or tensor with all coefficients positive gives the transformation to the system which is going toward more negative values on the axis of the first system (regardless of whether more positive values means to the left or to the right in real space).

The Lorentz transformation can be applied to other quantities than vectors; for example, it may be applied to a tensor of the second rank. Thus, applying it to a doubly covariant tensor,

$$T_{KM} = L_K{}^i L_M{}^j t_{ij}$$

This holds for any one of the 16 components of the resultant tensor T_{KM}; each equation contains 16 terms on the right-hand side, one for each term of the original tensor t_{ij}.

Using the expressions for $L_K{}^i$ above it is not difficult, though it is tedious, to show that $\{T_{KM}\}$ is given by

$$\begin{bmatrix} \gamma^2\left[t_{11}+\beta(t_{14}+t_{41})+\beta^2 t_{44}\right] & \gamma(t_{12}+\beta t_{42}) & \gamma(t_{13}+\beta t_{43}) & \gamma^2\left[t_{14}+\beta(t_{11}+t_{44})+\beta^2 t_{41}\right] \\ \gamma(t_{21}+\beta t_{24}) & t_{22} & t_{23} & \gamma(t_{24}+\beta t_{21}) \\ \gamma(t_{31}+\beta t_{34}) & t_{32} & t_{33} & \gamma(t_{34}+\beta t_{31}) \\ \gamma^2\left[t_{41}+\beta(t_{11}+t_{44})+\beta^2 t_{14}\right] & \gamma(t_{42}+\beta t_{12}) & \gamma(t_{43}+\beta t_{13}) & \gamma^2\left[t_{44}+\beta(t_{14}+t_{41})+\beta^2 t_{11}\right] \end{bmatrix}$$

These results become very much simpler if we consider symmetric tensors (10 independent components in four-space) or antisymmetric tensors (six independent components in four-space). In the latter case $\{T_{KM}\}_{\text{anti}}$ is given by

$$\begin{bmatrix} 0 & \gamma t_{12}-\gamma\beta t_{24} & \gamma t_{13}-\gamma\beta t_{34} & t_{14} \\ -\gamma t_{12}+\gamma\beta t_{24} & 0 & t_{23} & \gamma t_{24}-\gamma\beta t_{12} \\ -\gamma t_{13}+\gamma\beta t_{34} & -t_{23} & 0 & \gamma t_{34}-\gamma\beta t_{13} \\ -t_{14} & -\gamma t_{24}+\gamma\beta t_{12} & -\gamma t_{34}+\gamma\beta t_{13} & 0 \end{bmatrix}$$

This case becomes especially interesting when we consider that

1 **E** and **B** have six components between them
2 **E** and **B** are given by combinations of space and time derivatives of ϕ and **A**
3 $\{c\mathbf{A}, \phi\}$ is a four-vector

It then seems plausible that the components of **E** and **B** may be associated with a second-rank tensor. In fact, by comparing the components of $\{T_{KM}\}_{\text{anti}}$ with the transformation equations for **E** and **B** in Sec. 3, we see that **E** and **B** are actually the components of an antisymmetric second-rank tensor. When written in doubly covariant form this tensor is

$$\{F_{KM}\} = \begin{bmatrix} 0 & cB_z & -cB_y & E_x \\ -cB_z & 0 & cB_x & E_y \\ cB_y & -cB_x & 0 & E_z \\ -E_x & -E_x & -E_z & 0 \end{bmatrix}$$

Note that the general rule for the transformation of an arbitrary second-rank doubly covariant tensor $t_{i\ell}$ with 16 components is

$$T_{JK} = \frac{\partial x^i}{\partial X^J}\frac{\partial x^\ell}{\partial X^K}\, t_{i\ell}$$

But when $t_{i\ell}$ is antisymmetric then, as we can see from the results of Sec. 3 with the relative motion along the x,X direction, the following simplifications occur.

1 The four components along the main diagonal—$F_{11}, F_{22}, F_{33}, F_{44}$—vanish: $t_{ii} = 0$.
2 The four components along the main antidiagonal—$F_{41} = -F_{14} = E_x$ and $F_{23} = -F_{32} = cB_x$—behave like scalars: $t_{i,(5-i)} = T_{I,(5-I)}$.
3 The other two components on the second column, $F_{12} = cB_z$, $F_{42} = -E_y$ (as well as the other two components on the second row $F_{21} = -cB_z, F_{24} = E_y$), behave like the components $\{t_1, 0, 0, t_4\}$ of a covector, $T_J = (\partial x^i/\partial X^J)t_i$.
4 The other two components on the third column $F_{13} = -cB_y$, $F_{43} = -E_z$ (as well as the other two components on the third row $F_{31} = cB_y, F_{34} = E_z$) behave like the components $\{t_1, 0, 0, t_4\}$ of a covector, also.

THE TENSOR EQUATIONS We are now in a position to write the important equations of the previous chapters in form-invariant fashion as tensor equations. Thus, let

$$\{x^i\} = \{\mathbf{r}, ct\}$$
$$\{A^i\} = \{c\mathbf{A}, \phi\}$$
$$\{J^i\} = \{\mathbf{J}, c\rho\}$$
$$\{v^i\} = \{\gamma\mathbf{v}, \gamma c\}$$
$$\{F^i\} = \{\gamma\mathbf{F}, \gamma\mathbf{F}\cdot\boldsymbol{\beta}\}$$

Then the following equations may be written.

1. The Lorentz force expression $\mathbf{F} = q[\mathbf{E} + (\mathbf{v} \times \mathbf{B})]$ comes from the first three components of

$$f_i = -\frac{q}{c}F_{ik}v^k = -\frac{1}{c}F_{ik}J^k$$

This Lorentz force four-vector depends explicitly on the velocity; $\{f_i\}$ (or $\{f^i\}$) is quite different from the generic Minkowski force four-vector $\{F^i\}$ above.

2. The equation of continuity $\nabla \cdot \mathbf{J} + \partial\rho/\partial t = 0$ is

$$\frac{\partial J^i}{\partial x^i} = 0$$

This says that the four-dimensional divergence (or four-divergence) of the current four-vector vanishes when the three-dimensional equation of continuity is satisfied. All electricity and magnetism is restricted to a solenoidal current four-vector.

3. The Lorentz condition $\nabla \cdot \mathbf{A} + (1/c^2)(\partial\phi/\partial t) = 0$ becomes

$$\frac{\partial A^i}{\partial x^i} = 0$$

One sees at once that this condition is not an ad hoc choice but a basic statement that the four-divergence of the potential four-vector vanishes.

4. The gage transformation ($\phi' = \phi + \partial f/\partial t$, $\mathbf{A}' = \mathbf{A} - \nabla f$) can be written $[\phi' = \phi + \partial(cf)/\partial(ct)$, $c\mathbf{A}' = c\mathbf{A} - \nabla(cf)]$. Setting $h = cf$, the two combine to give $A_i' = A_i + \partial h/\partial x^i$. It seems that one cannot attach physical significance to the invariant $A^i A_i$: the four-gradient of an arbitrary function may be added to A_i without affecting F_{ik}. Most other such invariants do have physical significance: usually there is no four-curl function, similar to $F_{ik} = (\partial A_i/\partial x^k) - (\partial A_k/\partial x^i)$, to consider.

5. The two inhomogeneous Maxwell equations

$$\nabla \cdot \mathbf{E} = \frac{1}{\epsilon_0}\rho \qquad \text{and} \qquad \nabla \times \mathbf{B} = \mu_0 \mathbf{J} + \epsilon_0 \mu_0 \left(\frac{\partial \mathbf{E}}{\partial t}\right)$$

may be written

$$\frac{\partial F_{ij}}{\partial x_j} = \sqrt{\frac{\mu_0}{\epsilon_0}}\, J_i$$

The two homogeneous Maxwell equations come from

$$\frac{\partial F_{ij}}{\partial x^k} + \frac{\partial F_{jk}}{\partial x^i} + \frac{\partial F_{ki}}{\partial x^j} = 0$$

when $(i,j,k) = (1,2,3)$, $(1,2,4)$, $(1,3,4)$, and $(2,3,4)$. These two equations may be taken as the central equations for the study of electricity and magnetism, replacing the four usual Maxwell's equations, if we wish to concentrate our attention on $\mathbf{E}$ and $c\mathbf{B}$, or F_{ij}. But if we wish to stress $\mathbf{A}$ and ϕ, or A^i, then the central equation of electricity and magnetism is

$$\nabla^i \nabla_i A^k = \sqrt{\frac{\mu_0}{\epsilon_0}}\, J^k$$

Here $\{\nabla^i\}$ is $\hat{\mathbf{x}}(\partial/\partial x) + \hat{\mathbf{y}}(\partial/\partial y) + \hat{\mathbf{z}}(\partial/\partial z) + \hat{\mathbf{ct}}[\partial/\partial(ct)]$ and the equation is of the first rank. When using the tensor $\{F_{ij}\}$, however, while one equation has rank one also, the other has rank three. The potential equations are simpler.

Examples

1. The forms $\{F_{JK}\}$, $\{F_K{}^J\}$, $\{F^{JK}\}$ We start with the doubly covariant form of the electromagnetic field tensor:

$$\{F_{JK}\} = \begin{bmatrix} 0 & cB_z & -cB_y & E_x \\ -cB_z & 0 & cB_x & E_y \\ cB_y & -cB_x & 0 & E_z \\ -E_x & -E_y & -E_z & 0 \end{bmatrix}$$

where the row in F_{JK} is given by the J, the column by the K. We proceed to find the other two forms by raising the indices, one at a time.

$F^L{}_K = g^{LJ}F_{JK}$ (sum convention). Since only the four metric components $g^{11} = g^{22} = g^{33} = -1$ and $g^{44} = +1$ do not vanish, this gives $F^L{}_K = g^{LL}F_{LK}$(no sum). Thus $F^1{}_3 = g^{11}F_{13} = (-1)(-cB_y) = +cB_y$. In this manncr

$$\{F^J{}_K\} = \begin{bmatrix} 0 & -cB_z & cB_y & -E_x \\ cB_z & 0 & -cB_x & -E_y \\ -cB_y & cB_x & 0 & -E_z \\ -E_x & -E_y & -E_z & 0 \end{bmatrix}$$

Note that, while $\{F_{JK}\}$ is antisymmetric, $\{F^J{}_K\}$ is neither symmetric nor antisymmetric. The "space" part of $\{F^J{}_K\}$ is antisymmetric, though each half is reversed from that in $\{F_{JK}\}$; the "space-time" part is symmetric; the "time" part vanishes. If we had raised the second index instead of the first, $F_J{}^L F_{JK}g^{KL}$ (sum convention) = $F_{JL}g^{LL}$ (no sum), the resulting matrix would differ from $F^J{}_K$ in the signs before the six "space-time" E terms: they would all be positive.

Next we raise the second index.

$$\begin{aligned} F^{LM} &= F^L{}_K\, g^{KM} \text{ (sum convention)} \\ &= F^L{}_M\, g^{MM} \text{ (no sum)} \end{aligned}$$

For example, $F^{13} = F^1{}_3\, g^{33} = (-1)(cB_y) = -cB_y$. Thus,

$$\{F^{JK}\} = \begin{bmatrix} 0 & cB_z & -cB_y & -E_x \\ -cB_z & 0 & cB_x & -E_y \\ cB_y & -cB_x & 0 & -E_z \\ E_x & E_y & E_z & 0 \end{bmatrix}$$

We see that $\{F^{JK}\}$, like $\{F_{JK}\}$, is antisymmetric; but the fourth row and column in the two are interchanged.

The lowering (or raising) of the second, or column, index is equivalent to multiplying columns 1, 2, and 3 by -1. Consequently the lowering (or raising) of both indices results only in a change of sign in the "space-time" positions, the "space" and "time" positions being unaffected.

Figure 14-10 shows how the various parts of $\{F_{JK}\}$, $\{F_K{}^J\}$, and $\{F^{JK}\}$ appear, starting with a doubly covariant tensor that is either antisymmetric, symmetric, or neither. The symmetry or antisymmetry of two indices always has significance when both are contravariant or both are covariant, i.e., when the same method of obtaining

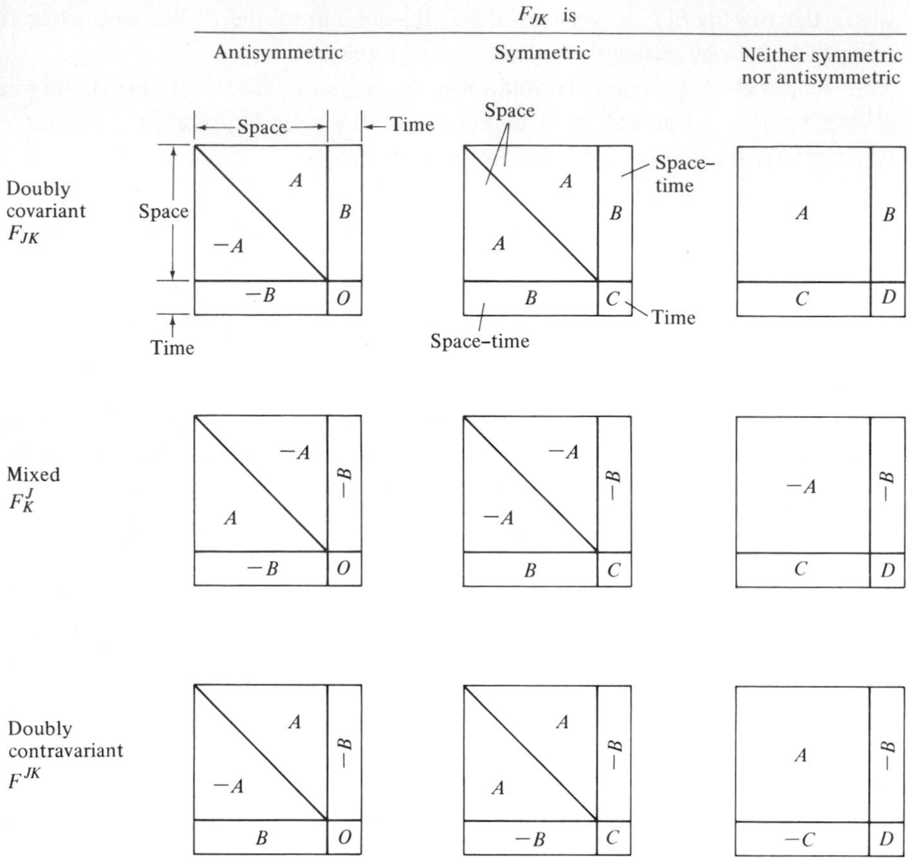

FIGURE 14-10

coordinates is used for both indices. When one index is up and the other is down then the symmetry is, in general, not an invariant property. But special cases exist, such as $\delta_k{}^j$, where the symmetry is invariant.

2. The electromagnetic momentum-energy tensor At the end of Chap. 4, without explaining what we mean by "tensor," we introduced the electrical part of a 3 × 3 matrix which we called the Maxwell stress tensor, and at the end of Chap. 5 the magnetic part of the 3 × 3 matrix was added to this tensor, which we will henceforth call $\{M\}$. Now that the term "tensor" has been defined, it is obvious that, while $\{M\}$ may be a three-tensor, it cannot possibly be a four-tensor. It turns out that $\{M\}$ becomes the "space" part of a four-tensor $\{T\}$ called the electromagnetic momentum-energy tensor.

The 1 X 1 "time" component of this tensor is found to be the *energy density* of the electromagnetic field, while 1 X 3 and 3 X 1 "space-time" components of the tensor are found to be the *energy flux* (Poynting) vector and the *field momentum density* vector, respectively. So $\{T\}$, combining so many fundamental quantities, is a basic tensor.

Given the 3 X 3 Maxwell stress tensor:

$$\{M_{\alpha\beta}\} = \begin{bmatrix} \epsilon_0 E_x^2 + \frac{1}{\mu_0} B_x^2 - u & \epsilon_0 E_x E_y + \frac{1}{\mu_0} B_x B_y & \epsilon_0 E_x E_z + \frac{1}{\mu_0} B_x B_z \\ \epsilon_0 E_y E_x + \frac{1}{\mu_0} B_y B_x & \epsilon_0 E_y^2 + \frac{1}{\mu_0} B_y^2 - u & \epsilon_0 E_y E_z + \frac{1}{\mu_0} B_y B_z \\ \epsilon_0 E_z E_x + \frac{1}{\mu_0} B_z B_x & \epsilon_0 E_z E_y + \frac{1}{\mu_0} B_z B_y & \epsilon_0 E_z^2 + \frac{1}{\mu_0} B_z^2 - u \end{bmatrix}$$

where u, the energy density of the field, is

$$u = \frac{\epsilon_0}{2} (E_x^2 + E_y^2 + E_z^2) + \frac{1}{2\mu_0} (B_x^2 + B_y^2 + B_z^2) = \frac{\epsilon_0}{2} E^2 + \frac{1}{2\mu_0} B^2$$

The Poynting vector is $\mathbf{S} = (1/\mu_0)\mathbf{E} \times \mathbf{B}$; so $(1/c)\mathbf{S}$ has the components

$$\sqrt{\frac{\epsilon_0}{\mu_0}} (E_y B_z - E_z B_y \quad E_z B_x - E_x B_z \quad E_x B_y - E_y B_x)$$

The momentum density is $\mathbf{g} = (1/c^2)\mathbf{S} = \epsilon_0 \mathbf{E} \times \mathbf{B}$. The components of $c\mathbf{g}$ are seen to be the same as those of $(1/c)\mathbf{S}$.

Suppose we define the 4 X 4 tensor consisting of the separate tensor components shown.

$$\{T_{ik}\} = \left[\begin{array}{c|c} \{T_{\alpha\beta}\} = \{M_{\alpha\beta}\} & T_{\alpha 4} = c g_\alpha \\ \hline T_{4\beta} = \frac{1}{c} S_\beta & -u \end{array}\right]$$

Although $\{T_{ik}\}$ is symmetric, the matrix has been written with $(1/c)\mathbf{S}$ in the fourth row and $c\mathbf{g}$ in the fourth column. The reason for this is that the following equation for the four-divergence of $\{T_{ik}\}$ then holds true:

$$\frac{\partial T_{ik}}{\partial x_k} = f_i = -\frac{1}{c} F_{i\ell} J^\ell$$

For example, take $i = 1$. Then

$$\frac{\partial T_{1k}}{\partial x_k} = \frac{\partial T_{11}}{\partial x_1} + \frac{\partial T_{12}}{\partial x_2} + \frac{\partial T_{13}}{\partial x_3} + \frac{\partial T_{14}}{\partial x_4}$$

$$= \frac{\partial M_{11}}{\partial(-x)} + \frac{\partial M_{12}}{\partial(-y)} + \frac{\partial M_{13}}{\partial(-z)} + \frac{\partial}{\partial(ct)} \left[\sqrt{\frac{\epsilon_0}{\mu_0}} (E_y B_z - E_z B_y)\right]$$

$$= -\frac{\epsilon_0}{2}\left(2E_x \frac{\partial E_x}{\partial x} - 2E_y \frac{\partial E_y}{\partial x} - 2E_z \frac{\partial E_z}{\partial x}\right) - \epsilon_0 \left(E_x \frac{\partial E_y}{\partial y} + E_y \frac{\partial E_x}{\partial y}\right)$$

$$- \epsilon_0 \left(E_x \frac{\partial E_z}{\partial z} + E_z \frac{\partial E_x}{\partial z}\right)$$

$$+ \epsilon_0 (E_y \dot{B}_z + \dot{E}_y B_z - E_z \dot{B}_y - \dot{E}_z B_y)$$

$$+ 3 \text{ similar groups of terms with } \epsilon_0 \to \frac{1}{\mu_0}, \mathbf{E} \to \mathbf{B}$$

$$= -\epsilon_0 E_x (\nabla \cdot \mathbf{E}) + \epsilon_0 [E_y (\nabla \times \mathbf{E})_z - E_z (\nabla \times \mathbf{E})_y] - \frac{1}{\mu_0} B_x (\nabla \cdot \mathbf{B})$$

$$+ \frac{1}{\mu_0} [B_y (\nabla \times \mathbf{B})_z - B_z (\nabla \times \mathbf{B})_y]$$

$$+ \epsilon_0 (E_y \dot{B}_z + \dot{E}_y B_z - E_z \dot{B}_y - \dot{E}_z B_y)$$

$$= -\rho E_x - \epsilon_0 E_y \dot{B}_z + \epsilon_0 E_z \dot{B}_y - 0 + \frac{1}{\mu_0} [B_y (\mu_0 J_z + \mu_0 \epsilon_0 \dot{E}_z)$$

$$- B_z (\mu_0 J_y + \mu_0 \epsilon_0 \dot{E}_y)] + \epsilon_0 (E_y \dot{B}_z + \dot{E}_y B_z - E_z \dot{B}_y - \dot{E}_z B_y)$$

$$= -\rho E_x + B_y J_z - B_z J_y$$

$$= -[\rho \mathbf{E} + (\mathbf{J} \times \mathbf{B})]_x$$

Next we find f_1:

$$f_1 = -\frac{1}{c} F_{1k} J^k = -\frac{1}{c} (F_{11} J^1 + F_{12} J^2 + F_{13} J^3 + F_{14} J^4)$$

$$= -\frac{1}{c} (0)(J_x) + (cB_z)(J_y) + (-cB_y)(J_z) + (E_x)(c\rho)$$

$$= -J_y B_z + J_z B_y - \rho E_x$$

$$= -[\rho \mathbf{E} + (\mathbf{J} \times \mathbf{B})]_x$$

So $f_i = \partial T_{ik}/\partial x_k$ when $i = 1$.

The components with $i = 2$ and $i = 3$ give similar results showing that $-f_\alpha = -\partial T_{\alpha k}/\partial x_k$ are the Lorentz force density components $(\text{LFD})_\alpha$, exerted on the charges and currents, per unit volume, by the field. For the first three rows of $\{T_{ik}\}$, then,

$$-\frac{\partial T_{\alpha k}}{\partial x_k} = -\frac{\partial M_{\alpha\beta}}{\partial x_\beta} - \frac{\partial (c g_\alpha)}{\partial (ct)}$$

But $$\frac{\partial M_{\alpha\beta}}{\partial x_\beta} = \frac{\partial M_{\alpha 1}}{\partial(-x)} + \frac{\partial M_{\alpha 2}}{\partial(-y)} + \frac{\partial M_{\alpha 3}}{\partial(-z)} \qquad \text{or} \qquad \frac{\partial M_{\alpha\beta}}{\partial x_\beta} = -\nabla \cdot \mathbf{M}_\alpha$$

where $\mathbf{M}_\alpha = (M_{\alpha 1}, M_{\alpha 2}, M_{\alpha 3})$. So

$$(\text{LFD})_\alpha = -\nabla \cdot \mathbf{M}_\alpha - \frac{\partial g_\alpha}{\partial t} \qquad \text{or} \qquad -\frac{\partial g_\alpha}{\partial t} = \nabla \cdot \mathbf{M}_\alpha + (\text{LFD})_\alpha$$

The first term on the right side may be related, by Gauss' divergence theorem, to the stress transmitted outward through the borders of a given volume. The second term on the right gives the force density exerted on the charges and currents within this volume. The left side represents the rate of loss of momentum density of the field (so it is the force density it exerts) within the volume. By letting $T_{\alpha 4} = cg_\alpha$, therefore, we obtain an equation for the equilibrium of forces which involves both the field and the particles.

We leave it to the problems to show that $(\partial T_{4k}/\partial x_k) = f_4$. That is, the $i = 4$ component of the four-divergence of $\{T_{ik}\}$ gives $-\partial U/\partial t = (\mathbf{E} \cdot \mathbf{J}) + \nabla \cdot \mathbf{S}$ if we identify $T_{4\beta}$ with $(1/c)S_\beta$ (instead of calling it cg_β, which has the same magnitude and dimensions). This equation, representing an equilibrium of energies, is the same as that found in the previous chapter in dealing with the Poynting vector. We see, again, that $\mathbf{S} = (1/\mu_0)\mathbf{E} \times \mathbf{B}$ and $\mathbf{g} = \epsilon_0 \mathbf{E} \times \mathbf{B}$ play equally important roles.

3. The polarization-magnetization tensor The equations for **E** and **B** in the presence of matter,

$$\mathbf{D} = \epsilon_0(\mathbf{E}) + \mathbf{P}$$

$$\left(\frac{1}{c}\right)\mathbf{H} = \epsilon_0(c\mathbf{B}) - \left(\frac{1}{c}\right)\mathbf{M}$$

may be made form-invariant by defining a tensor H_{ij} for $[\mathbf{D},(1/c)\mathbf{H}]$ similar to the tensor F_{ij} for $(\mathbf{E},c\mathbf{B})$. The factor c is needed, in the combinations shown, in order to give the different terms in an equation the same dimensions. The previous equation, $\partial F_{ij}/\partial x_j = c\mu_0 J_i$, is then interpreted to mean that J_i refers to all current densities—conventional (conduction and convection) and bound (magnetization and polarization). A corresponding equation for the new tensor will give fields produced only by free charge densities and conventional current densities. Another tensor, the polarization-magnetization tensor P_{ij}, will also be defined for $[\mathbf{P},(1/c)\mathbf{M}]$.

The tensors H_{ij} and P_{ij} are not unique, but the fashion in which one H_{ij} differs from another is essentially trivial, depending on, say, whether a c is put into a denominator in one part or into a numerator in another part. The pair we define is no better and no worse than any other choice. Thus, with

$$\{H_{ij}\} = \begin{bmatrix} 0 & \frac{1}{c}H_z & -\frac{1}{c}H_y & D_x \\ & 0 & \frac{1}{c}H_x & D_y \\ & & 0 & D_z \\ & & & 0 \end{bmatrix} \qquad \{F_{ij}\} = \begin{bmatrix} 0 & cB_z & -cB_y & E_x \\ & 0 & cB_x & E_y \\ & & 0 & E_z \\ & & & 0 \end{bmatrix}$$

and $$\{P_{ij}\} = \begin{bmatrix} 0 & -\frac{1}{c}M_z & \frac{1}{c}M_y & P_x \\ & 0 & -\frac{1}{c}M_x & P_y \\ & & 0 & P_z \\ & & & 0 \end{bmatrix}$$

we have

$$H_{ij} = \epsilon_0 F_{ij} + P_{ij}$$

Then $\partial H_{ij}/\partial x_j = (1/c)J_i^{(c)}$, where the superscript (c) stands for conventional current densities and free charge density.

PROBLEMS

1 Find the time component (i.e., the $i = 4$ component) of the Lorentz force density four-vector.

2 Show explicitly that $T_{KM} = L_k{}^i L_m{}^j t_{ij}$ gives the $\{T_{KM}\}$ shown in the text.

3 Write Maxwell's equations in terms of the contravariant field tensor $\{F_{ij}\}$.

4 Write the expression for the Coulomb potential in a form-invariant fashion.

5 Find the tensor equation giving **E** and **B** in terms of ϕ and **A**.

6 Write the four-dimensional analogs of the three-dimensional del operators—the gradient, divergence, curl, and laplacian.

7 The completely antisymmetric tensor ϵ^{ijkl} introduced in Prob. 9 of the previous section can be used to define a quantity called the dual: $D^{ij} \equiv \frac{1}{2}\epsilon^{ijkl}F_{kl}$. Find $\{D^{ij}\}$.

8 Show that the four-gradient of a scalar is a four-vector.

9 Write the tensor expression for the homogeneous Maxwell equations in terms of the dual of F_{kl} defined in Prob. 7.

10 The equation

$$\frac{\partial F_{ij}}{\partial x^k} + \frac{\partial F_{jk}}{\partial x^i} + \frac{\partial F_{ki}}{\partial x^j} = 0$$

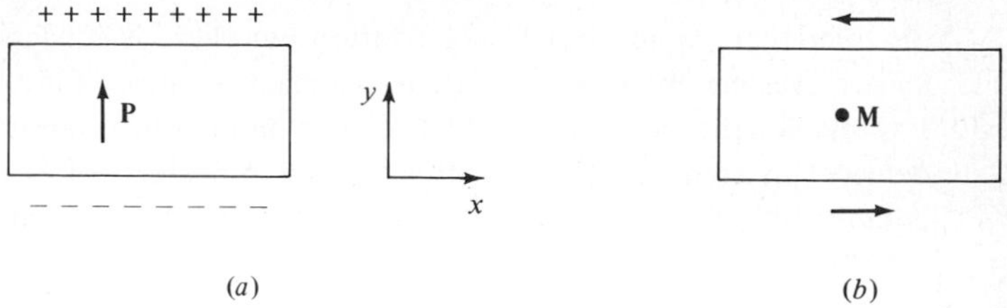

FIGURE 14-11

gives the two homogeneous Maxwell equations for four specific combinations of the indices: (1,2,3), (1,2,4), (1,3,4), and (2,3,4). There are 60 other combinations. What do they yield?

11 Find the components of the bound current density four-vector $J_i^{(b)}$ and the equation relating it to the tensor P_{ij}.

12 Write the electromagnetic momentum-energy tensor, explicitly, in contravariant form and give the equation connecting it to f^i, the contravariant Lorentz force four-vector.

13 From the answers to Prob. 11, find a bound current density different from the magnetization current density $\nabla \times \mathbf{M}$, previously discussed in Chap. 7. Give a physical meaning to this bound current density.

14 Show that $\partial T_{4k}/\partial x_k = f_4$ gives $-\partial U/\partial t = (\mathbf{E} \cdot \mathbf{J}) + \nabla \cdot \mathbf{S}$.

15 A particle with charge q and mass m_0 moves through an electromagnetic field $(\mathbf{E},\mathbf{B})$ with velocity $\mathbf{v}$. Write its equation of motion (*a*) in vector form and (*b*) in tensor form.

16 Verify that, for the H_{ij} and P_{ij} given, the equations

$$\frac{\partial H_{ij}}{\partial x_j} = \frac{1}{c} J_i^{(c)} \qquad \frac{\partial P_{ij}}{\partial x_j} = -\frac{1}{c} J_i^{(b)}$$

hold true.

17 A substance has the polarization **P** and magnetization **M** for one observer. Use $p_{KM} = L_K{}^i L_M{}^j p_{ij}$ to find $\mathbf{P}'$ and $\mathbf{M}'$ for another observer whose relative motion is toward $+x$.

18 (*a*) Figure 14-11*a* shows a plate of polarized material polarized along $+\hat{\mathbf{y}}$. To an observer moving with velocity $\hat{\mathbf{x}}v$ the plate is moving to the left. Explain the answer for M_z' for this observer, in the previous problem, in terms of this figure. $M_z = 0$.

(*b*) Figure 14-11*b* shows a plate with bound magnetization current giving a magnetization along $+\hat{\mathbf{z}}$. Explain the answer for P_z' in the previous problem in terms of the different charge densities on the upper and lower surfaces.

19 Let a filamentary circuit be rectangular, with two sides of length a parallel to x and two sides of length b parallel to y, and let it carry a CCW current I. For a stationary observer there is no electric polarization but there is a magnetization, $m = abI$. For an observer moving toward $+x$ there is a magnetization. Find its value.

20 Express **S** (the Poynting vector), g (the momentum density), and u (the energy density) in terms of the potentials ϕ and **A** instead of in terms of **E** and **B**. The complexity of these expressions is a counterargument against the view that ϕ, **A** should have preference over **E**, **B**.

15

PLANE WAVES

15-1 FREE SPACE

We first consider plane waves in free space—a region, devoid of charges or currents, which is so far from any physical boundaries that we are dealing, in effect, with an infinite domain. In the last chapter we saw that **E** and $c\mathbf{B}$ are actually the components of one tensor. We may utilize this fact here by writing Maxwell's equations in terms of $c\mathbf{B}$ rather than **B**. This yields the following elegant equations:

$$\begin{cases} \nabla \cdot \mathbf{E} = 0 & \nabla \cdot (c\mathbf{B}) = 0 \\ \nabla \times \mathbf{E} = -\dfrac{1}{c}\dfrac{\partial}{\partial t}(c\mathbf{B}) & \nabla \times (c\mathbf{B}) = +\dfrac{1}{c}\dfrac{\partial}{\partial t}\mathbf{E} \end{cases}$$

Usually, however, the two curl equations are written, for simplicity,

$$\nabla \times \mathbf{E} = -\frac{\partial \mathbf{B}}{\partial t} \qquad \nabla \times \mathbf{B} = +\frac{1}{c^2}\frac{\partial \mathbf{E}}{\partial t}$$

The symmetry is then lost; but it is incorrect to attribute this to the use of the mksa system.

By taking the curl of the $\nabla \times \mathbf{E}$ equation we obtain, using the vector identity $\nabla \times \nabla \times \mathbf{E} = \nabla\nabla \cdot \mathbf{E} - \nabla^2 \mathbf{E}$,

$$\nabla^2 \mathbf{E} - \frac{1}{c^2}\frac{\partial^2 \mathbf{E}}{\partial t^2} = 0$$

and by taking the curl of the $\nabla \times \mathbf{B}$ equation we obtain

$$\nabla^2 \mathbf{B} - \frac{1}{c^2}\frac{\partial^2 \mathbf{B}}{\partial t^2} = 0$$

Letting ψ represent any one of the six components of $\mathbf{E}$ or $\mathbf{B}$ (or of $c\mathbf{B}$, if we wish) we then need to solve $\square^2 \psi = 0$. There are very many different kinds of solutions to this equation. Suppose we seek sinusoidal plane-wave solutions which are functions only of x, the distance of the plane from the origin. The values of the components of $\mathbf{E}$ are then identically the same everywhere in any yz plane, and the components of $\mathbf{B}$ also have a fixed value in such a plane, though of course the values vary from one plane to another.

A sinusoidal plane wave moving along the x axis toward $+x$ may be represented by

$$\psi = \psi_0 \sin\left[2\pi\left(ft - \frac{x}{\lambda}\right) + \alpha\right] = \psi_0 \sin\left[(\omega t - kx) + \alpha\right]$$

where

$$\omega = 2\pi f = \frac{2\pi}{T}$$

is the angular frequency, $k = 2\pi/\lambda$ is the propagation constant, α is a phase angle, and the amplitude ψ_0 is a constant. We simplify matters by taking $\alpha = 90°$ and writing the sinusoid in complex exponential form,

$$\psi = \psi_0 \, e^{i(\omega t - kx)}$$

implicitly understanding that the physical wave with which we are concerned is actually given only by the real part of ψ.

The del operator and the time derivative assume special forms in this case.

1. $\nabla \psi \rightarrow \hat{\mathbf{x}} \dfrac{\partial \psi}{\partial x} = \hat{\mathbf{x}}(-ik\psi)$

2. $\nabla \cdot \mathbf{E} \rightarrow \dfrac{\partial E_x}{\partial x} = \dfrac{\partial E_n}{\partial x}$

where E_n is the normal component of $\mathbf{E}$. Treating E_n like ψ above gives

$$\frac{\partial E_n}{\partial x} = -ikE_n$$

so

$$\nabla \cdot \mathbf{E} \rightarrow -ikE_n$$

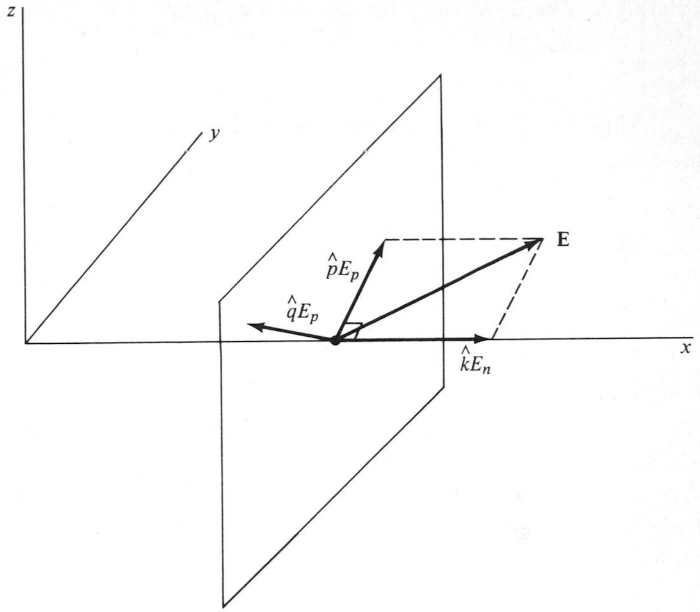

FIGURE 15-1

Similarly

$$\nabla \cdot \mathbf{B} \to -ikB_n$$

for a sinusoidal plane wave.

3. Figure 15-1 shows the vector **E** decomposed into two components, $\hat{\mathbf{k}}E_n$ normal to the plane and $\hat{\mathbf{p}}E_p$ parallel to the plane. We first assume the wave is linearly polarized: **E** at different points along the x axis is a vector which varies in length and sense along some straight line. This case is often called plane polarized, rather than linear polarized; the plane is that which includes this line and the x axis. Then

$$\nabla \times \mathbf{E} \to -\hat{\mathbf{y}}\frac{\partial E_z}{\partial x} + \hat{\mathbf{z}}\frac{\partial E_y}{\partial x}$$

But $\hat{\mathbf{k}} \times \mathbf{E} = -\hat{\mathbf{y}}E_z + \hat{\mathbf{z}}E_y$, so

$$\nabla \times \mathbf{E} \to \frac{\partial}{\partial x}(\hat{\mathbf{k}} \times \mathbf{E})$$

Further, $\hat{\mathbf{k}} \times \mathbf{E} = \hat{\mathbf{k}} \times (\hat{\mathbf{k}}E_n + \hat{\mathbf{p}}E_p) = (\hat{\mathbf{k}} \times \hat{\mathbf{p}})E_p$. We will set this equal to $\hat{\mathbf{q}}E_p$ where $\hat{\mathbf{q}}$ is, like $\hat{\mathbf{p}}$, in the plane of the wave but is turned 90° CW if one looks in the $\hat{\mathbf{k}}$ direction. Since $\hat{\mathbf{q}}$ is perpendicular to the fixed $\hat{\mathbf{p}}$ direction, $\hat{\mathbf{q}}$ does not vary with x and

$$\nabla \times \mathbf{E} \to \hat{\mathbf{q}}\frac{\partial E_p}{\partial x}$$

Treating E_p like ψ, above, gives $\partial E_p/\partial x = ikE_p$. So

$$\nabla \times \mathbf{E} \to -\,\hat{\mathbf{q}}\, ikE_p$$

Similarly, let $\mathbf{B} = \hat{\mathbf{k}}B_n + \hat{\mathbf{P}}B_P$, where $\hat{\mathbf{P}}$ is parallel to the plane but is not necessarily the same as $\hat{\mathbf{p}}$. Then

$$\nabla \times \mathbf{B} \to -\hat{\mathbf{Q}}ikB_P$$

where $\hat{\mathbf{Q}}$ is 90° from $\hat{\mathbf{P}}$, CW looking in the $\hat{\mathbf{k}}$ direction.

4. $\partial\psi/\partial t \to +\, i\omega\psi$. For a sinusoidal plane wave, then, after the i's are canceled, Maxwell's equations become

$$\begin{cases} kE_n = 0 & kB_n = 0 \\ \hat{\mathbf{q}}kE_p = \omega\mathbf{B} & -\hat{\mathbf{Q}}kB_P = \dfrac{1}{c^2}\,\omega\mathbf{E} \end{cases}$$

Consider, first the top two equations: the divergence expressions. If $k = 0$ then $\lambda = \infty$. From $v = f\lambda$, for a finite velocity, this corresponds to $f = 0$, i.e., to static conditions. So, for the nonstatic conditions here, $k \neq 0$. Then the top two equations show that $\mathbf{E} = 0$ and $\mathbf{B} = 0$.

We have thus arrived at the result that a sinusoidal plane electromagnetic wave in free space is a transverse wave. By itself this possibly might not be a particularly important result, for a sinusoidal wave is a monochromatic wave, one having a single frequency and wavelength, and such a wave must exist from $t = -\infty$ to $t = +\infty$. No such waves actually ever occur. What makes the result important is the fact, mentioned earlier in Chap. 7, that any plane electromagnetic wave—periodic or not—may be decomposed into constituent monochromatic sinusoidal waves. The process by which this is done is Fourier analysis, yielding either a series for a periodic wave or a Fourier integral for a nonperiodic wave. Since each constituent of any plane electromagnetic wave is a transverse wave, it follows that *any* plane electromagnetic wave in free space is a transverse wave.

The third equation tells us that $\mathbf{B}$ must lie along $\hat{\mathbf{q}}$, in the plane of the wave. But $\hat{\mathbf{q}}$ is 90° CW, looking in the direction of propagation, from $\hat{\mathbf{p}}$; and $\hat{\mathbf{p}}$, originally taken to be the direction of only the tangential component of $\mathbf{E}$, is now seen to be actually the direction of $\mathbf{E}$. So $\mathbf{B}$ is 90° CW, looking toward $+\hat{\mathbf{k}}$, from $\mathbf{E}$. Figure 15-2 shows this situation at some fixed time instant, $t = nT$; say $t = 4T$. Further, this equation gives a relation between the magnitudes of $\mathbf{E}$ and $\mathbf{B}$. Dropping the subscript on E_p we have $kE = \omega B$. But $\omega/k = 2\pi f/(2\pi/\lambda) = f\lambda = c$, where c is the velocity of the plane wave in free space; so $E = cB$. Also, the vectors $\mathbf{E}$ and $\mathbf{B}$ are seen to be in time phase with each other at any value of x; and in space phase with each other at any value of t.

The fourth equation corroborates these results: $\mathbf{E}$ must lie in the $-\hat{\mathbf{Q}}$ direction, where $-\hat{\mathbf{Q}}$ is 90° CCW from $\hat{\mathbf{P}}$. Since $\mathbf{E}$ is known to lie along $\hat{\mathbf{p}}$, this makes $\hat{\mathbf{p}} = \hat{\mathbf{q}}$. Again,

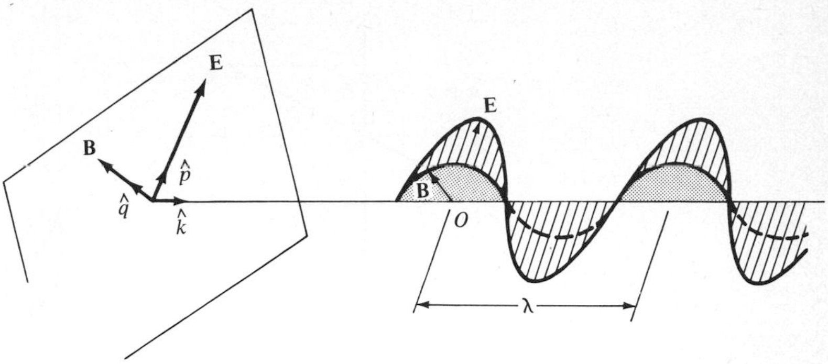

FIGURE 15-2

dropping the subscript on B_P we obtain $E = cB$. The three unit vectors $\hat{\mathbf{k}}, \hat{\mathbf{p}}, \hat{\mathbf{P}}$ are thus mutually orthogonal to each other, in the sense

$$\hat{\mathbf{k}} \times \hat{\mathbf{p}} = \hat{\mathbf{P}} \qquad \hat{\mathbf{p}} \times \hat{\mathbf{P}} = \hat{\mathbf{k}} \qquad \hat{\mathbf{P}} \times \hat{\mathbf{k}} = \hat{\mathbf{p}}$$

The linearly polarized wave is the simplest type to treat. If, for instance, we let **E** vary along the y axis then **B** lies on the z axis.

$$\mathbf{E} = \hat{\mathbf{y}} E_0 \, e^{i(\omega t - kx)} \qquad \mathbf{B} = \hat{\mathbf{z}} B_0 \, e^{i(\omega t - kx)}$$

(**E** is plane polarized in the xy plane, while **B** is plane polarized in the zx plane.) Here E_0 and B_0 are real constants with $E_0 = cB_0$. We may generalize this result by making **E** the sum of two linearly independent waves, one linearly polarized with $\mathbf{E}_1$ along $\hat{\mathbf{y}}$ and the other linearly polarized with $\mathbf{E}_2$ along $\hat{\mathbf{z}}$:

$$\mathbf{E} = (\hat{\mathbf{y}} E_1 + \hat{\mathbf{z}} E_2) e^{i(\omega t - kx)} \qquad \mathbf{B} = (-\hat{\mathbf{y}} B_2 + \hat{\mathbf{z}} B_1) e^{i(\omega t - kx)}$$

To allow for a possible phase difference between the two E constituents we will take

$$E_1 = E_{01} e^{-i\alpha} \qquad \text{and} \qquad E_2 = E_{02} e^{-i\beta}$$

Then

$$B_1 = B_{01} e^{-i\alpha} = \frac{1}{c} E_{01} e^{-i\alpha}$$

$$B_2 = B_{02} e^{-i\beta} = \frac{1}{c} E_{02} e^{-i\beta}$$

We now consider various special cases.

CASE 1: $\alpha = \beta$ The case we took at the beginning is an example of this, with $\alpha = \beta = 0$. But, in general, we would have

$$\alpha = \beta \neq 0 \qquad \text{and} \qquad \mathbf{E} = (\hat{\mathbf{y}} E_{01} + \hat{\mathbf{z}} E_{02}) e^{i(\omega t - kx - \beta)}$$

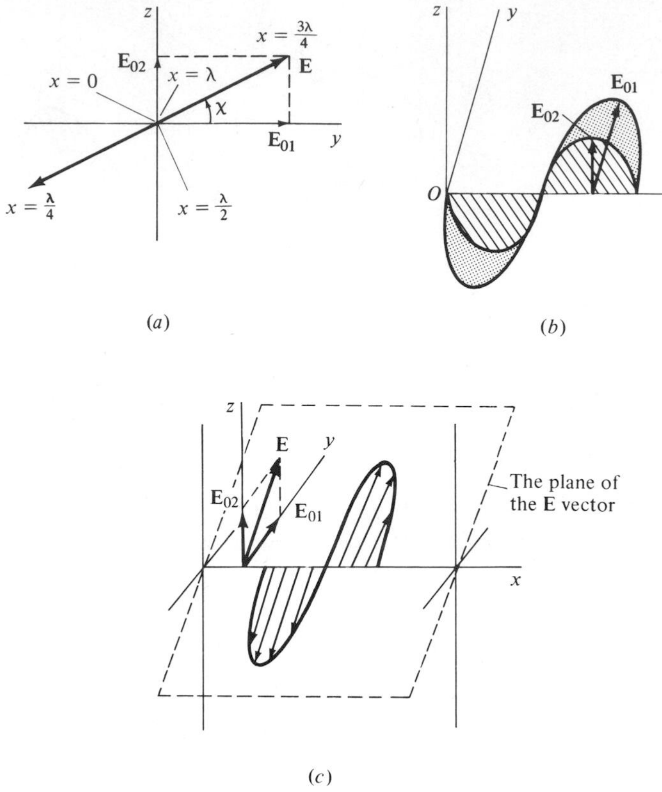

FIGURE 15-3

The real part of this case, illustrated in Fig. 15-3*a*, yields a linearly polarized wave. The value of β serves only to fix the phase of the waves at the origin O in Fig. 15-3*b*. This shows the situation at $t = 4T$, say, with $\alpha = \beta = \pi/2$ and $E_{01} = 2E_{02}$. The linearly polarized vector **E** has a magnitude of $\sqrt{E_{01}^2 + E_{02}^2}$ and its direction of polarization makes an angle $\chi = \tan^{-1}(E_{02}/E_{01})$ with the y direction. Figure 15-3*c* shows the variation of the resultant E with x under the same conditions as Fig. 15-3*b*. The straight line of Fig. 15-3*a* is obtained by moving the yz plane along the x axis of Fig. 15-3*c* through a distance of $kx = 2\pi$. At each x the position of the tip of **E** puts a dot on the straight line. Figure 15-3*a* shows why this wave is called linearly polarized; Figure 15-3*c* shows why it is called plane polarized.

CASE 2: $\alpha = \beta + \pi/4$ Here

$$\mathbf{E} = \hat{\mathbf{y}} E_{01} e^{-i(\beta+\pi/4)} e^{i(\omega t - kx)} + \hat{\mathbf{z}} E_{02} e^{-i\beta} e^{i(\omega t - kx)}$$

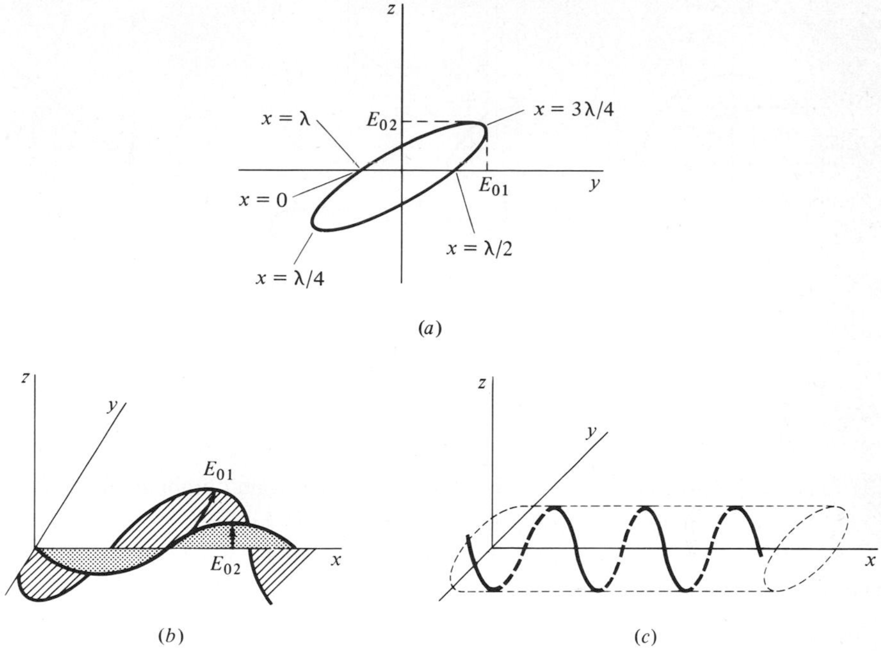

FIGURE 15-4

The real part of this is

$$\mathbf{E} = \hat{\mathbf{y}}E_{01} \cos\left[\omega t - kx - \left(\beta + \frac{\pi}{4}\right)\right] + \hat{\mathbf{z}}E_{02} \cos\left[\omega t - kx - \beta\right]$$

Take $\beta = \pi/2$ and $t = 4T$ again, and let $E_{01} = E_{02}$. Then

$$\mathbf{E} = \hat{\mathbf{y}}E_{01} \cos\left(kx + \frac{3\pi}{4}\right) + \hat{\mathbf{z}}E_{02} \cos\left(kx + \frac{\pi}{2}\right)$$

$$= -\hat{\mathbf{y}}E_{01} \sin\left(kx + \frac{\pi}{4}\right) - \hat{\mathbf{z}}E_{02} \sin kx$$

Figure 15-4 shows the locus of $\mathbf{E}$ for all possible values of kx. The result is an ellipse. Looking toward $\hat{\mathbf{k}}$ the ellipse is swept out CW. The wave is then said to be elliptically polarized with positive helicity: a screw turned in the sense of the arrow of Fig. 15-4*a* advances along $+\hat{\mathbf{k}}$. (The convention in optics is to call this a left elliptically polarized wave.) Figure 15-4*b* shows the individual y and z components of the wave plotted against x at a given instant, while Fig. 15-4*c* gives the resultant.

Case 3: $\alpha = \beta + \dfrac{\pi}{2} \qquad E_{01} = E_{02}$

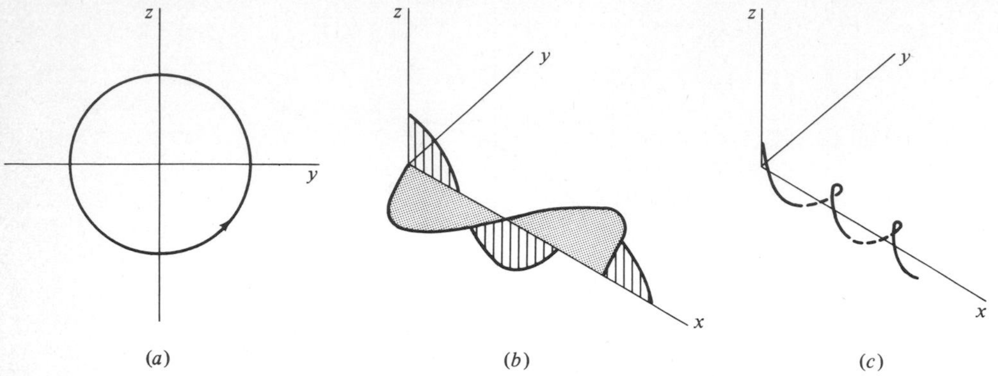

FIGURE 15-5

If only the condition $\alpha = \beta + \pi/2$ is imposed then we again obtain an ellipse; this time the major and minor axes lie along the y and z axes. (If $E_{01} > E_{02}$, then the major axis is along the y axis; if $E_{02} > E_{01}$ the major axis is along the z axis.) But if we also impose the additional condition $E_{01} = E_{02}$ then the ellipse becomes a circle: the wave is said to be circularly polarized. Figure 15-5 shows the conditions for this case.

It is worth noting that in the case of circular polarization we may write

$$\begin{aligned}\mathbf{E}_+ &= \hat{\mathbf{y}}E_0e^{-i\alpha}e^{i(\omega t-kx)} + \hat{\mathbf{z}}E_0e^{-i(\alpha-\pi/2)}e^{i(\omega t-kx)}\\ &= E_0\,[\hat{\mathbf{y}} + \hat{\mathbf{z}}e^{i(\pi/2)}]\,e^{i(\omega t-kx-\alpha)}\\ &= E_0(\hat{\mathbf{y}} + i\hat{\mathbf{z}})\,e^{i(\omega t-kx-\alpha)}\end{aligned}$$

This represents a wave with positive helicity. One with negative helicity would be given by $\alpha = \beta - \pi/2$ and is represented by

$$\mathbf{E}_- = E_0(\hat{\mathbf{y}} - i\hat{\mathbf{z}})\,e^{i(\omega t-kx-\alpha)}$$

Instead of using the sum of the fundamental pair of $\mathbf{E}$ vectors, $\hat{\mathbf{y}}E_1e^{i(\omega t-kx)}$ and $\hat{\mathbf{z}}E_2e^{i(\omega t-kx)}$ to give the sum $(\hat{\mathbf{y}}E_1 + \hat{\mathbf{z}}E_2)e^{i(\omega t-kx)}$, we could then use, just as well, the fundamental pair combination

$$\mathbf{E} = (\hat{\boldsymbol{\epsilon}}_+E_+ + \hat{\boldsymbol{\epsilon}}_-E_-)e^{i(\omega t-kx)}$$

where $$\hat{\boldsymbol{\epsilon}}_+ = \frac{\sqrt{2}}{2}(\hat{\mathbf{y}} + i\hat{\mathbf{z}}) \qquad \text{and} \qquad \hat{\boldsymbol{\epsilon}}_- = \frac{\sqrt{2}}{2}(\hat{\mathbf{y}} - i\hat{\mathbf{z}})$$

Any kind of plane-polarized wave may be represented by various proportions (amplitudes) of the two components of the fundamental pair. It is natural to take the fundamental pair $\hat{\mathbf{y}}E_1 + \hat{\mathbf{z}}E_2$ for a linearly polarized wave, and the fundamental pair $\hat{\boldsymbol{\epsilon}}_+E_+ + \hat{\boldsymbol{\epsilon}}_-E_-$ for a circularly polarized wave. For an elliptically polarized wave either pair is just as simple as the other.

15-2 UNCHARGED, NONCONDUCTING MEDIUM

The results above are applicable to plane waves in free space. Suppose we substitute an uncharged, nonconducting, medium; what changes need we make? If we assume that the medium is

1 homogeneous (no lumps)
2 isotropic (this excludes most single crystals)
3 nondispersive (the velocity is the same for all the monochromatic plane-wave components)
4 nondissipative (no attenuation by energy loss in the medium)

then the answer is simple: we must let $\epsilon_0 \to \epsilon$ and $\mu_0 \to \mu$ in Maxwell's equations. Thus $1/\sqrt{\epsilon_0 \mu_0} \to 1/\sqrt{\epsilon\mu}$, i.e., $c \to v$. The equation $c = \lambda f$ in vacuum becomes $v = \lambda_m f_m$, where the subscript refers to the quantities measured in the medium.

Consider a plane interface, now, between a vacuum and the medium, as shown in Fig. 15-6, and let a plane wave be incident on the surface at the point P. Suppose an observer A, on the vacuum side but very close to P, counts the number of cycles that impinge on the surface at P in unit time. Let this number be f. Then an observer B, on the medium side but very close to P, i.e., essentially at the same point as A, must also count f cycles per unit time. Nothing is changed if the incident wave is not plane, or if it is not incident normally. The general result is that the frequency of the wave is the same in the medium as in vacuum. Since $v < c$, however, it follows that $\lambda_m < \lambda$. We will defer further consideration of what happens at the interface until a subsequent section.

15-3 UNCHARGED, CONDUCTING MEDIUM

Suppose that we again consider an uncharged medium, but this time let there be a finite conductivity, for example a metal, or sea water. With $\rho = 0$ and $\mathbf{J} = g\mathbf{E}$, Maxwell's equations now give (the derivation is left for one of the problems):

$$\nabla^2 \mathbf{E} - \epsilon\mu \frac{\partial^2 \mathbf{E}}{\partial t^2} - g\mu \frac{\partial \mathbf{E}}{\partial t} = 0$$

$$\nabla^2 \mathbf{B} - \epsilon\mu \frac{\partial^2 \mathbf{B}}{\partial t^2} - g\mu \frac{\partial \mathbf{B}}{\partial t} = 0$$

Here ϵ and μ are assumed to be uniform in space and constant in time. These equations, which are generalizations of the wave equation of Sec. 15-1 by the addition of a term involving σ, are examples of the "telegrapher's equation," a name which arose from the role it plays in transmission-line theory. For a plane wave traveling toward $+x$, we may now take ψ to be any one of the six components of $\mathbf{E}$ or $\mathbf{B}$. Substituting $\psi = \psi_0 e^{i(\omega t - kx)}$ in Maxwell's equation gives $-k^2 + \epsilon\mu\omega^2 - ig\mu\omega = 0$.

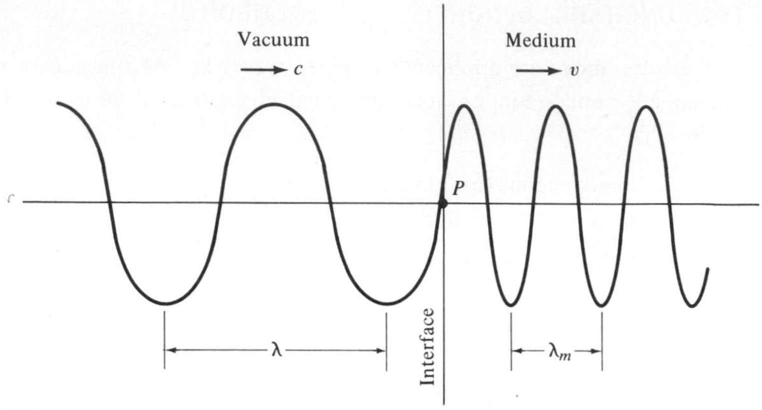

FIGURE 15-6

To solve this complex equation (it is a condition that must be satisfied if the differential equations for **E** and **B** are to hold for a wave solution) we make k complex. We will temporarily set $k = k_r + ik_i'$. Then

$$\psi = [\psi_0 e^{k_i' x}] e^{i(\omega t - k_r x)}$$

For this to represent a wave traveling to the right we must have $k_r > 0$ and for the amplitude to attenuate, rather than increase, with distance it is necessary that $k_i' < 0$. So we set $k_i = -k_i'$ with $k_i > 0$:

$$k = k_r - ik_i \qquad (k_r > 0, k_i > 0)$$

giving

$$\psi = (\psi_0 e^{-k_i x}) e^{i(\omega t - k_r x)}$$

We now adopt the conventional notation setting

$$\psi = (\psi_0 e^{-\alpha x}) e^{i(\omega t - \beta x)}$$

so that in effect we have put $k = \beta - i\alpha$. Then

$$k^2 = (\beta^2 - \alpha^2) - i(2\alpha\beta)$$

[Note: $k^2 \neq |k|^2$.] So $\beta^2 - \alpha^2 = \epsilon\mu\omega^2$, $2\alpha\beta = g\mu\omega$. Squaring the last two equations and adding:

$$\beta^4 + 2\alpha^2\beta^2 + \alpha^4 = \mu^2\omega^2(\epsilon^2\omega^2 + g^2)$$

Taking the square root of this gives $\beta^2 + \alpha^2 = +\mu\omega\sqrt{\epsilon^2\omega^2 + g^2}$, where the positive root must be taken since the left side is positive. We now have

$$\begin{cases} \beta^2 - \alpha^2 = \epsilon\mu\omega^2 \\ \beta^2 + \alpha^2 = \mu\omega\sqrt{\epsilon^2\omega^2 + g^2} \end{cases}$$

Adding these two gives

$$\beta^2 = \left(\frac{\epsilon\mu}{2}\right)\omega^2 \left[\sqrt{1 + \left(\frac{g}{\epsilon\omega}\right)^2} + 1\right]$$

We define

$$Q \equiv \frac{\epsilon\omega}{g}$$

Then

$$\beta^2 = \left(\frac{\epsilon\mu}{2}\right)\omega^2 [\sqrt{1 + Q^{-2}} + 1]$$

or

$$\beta = \sqrt{\frac{\epsilon\mu}{2}} \left[\sqrt{1 + Q^{-2}} + 1\right]^{1/2} \omega$$

Similarly, subtracting the two equations above gives

$$\alpha^2 = \left(\frac{\epsilon\mu}{2}\right)\omega^2 \left[\sqrt{1 + Q^{-2}} - 1\right]$$

$$\alpha = \sqrt{\frac{\epsilon\mu}{2}} \left[\sqrt{1 + Q^{-2}} - 1\right]^{1/2} \omega$$

The expression

$$\psi = [\psi_0 e^{-\alpha x}] e^{i(\omega t - \beta x)}$$

gives the wave velocity $v = \omega/\beta$. So

$$v = \frac{1}{\sqrt{\epsilon\mu}} \left[\frac{2}{\sqrt{1 + Q^{-2}} + 1}\right]^{1/2}$$

If $Q > 10$, say, then $v \approx 1/\sqrt{\epsilon\mu}$; on the other hand, when $Q < 0.1$ then $v \approx \sqrt{2Q}\,(1/\sqrt{\epsilon\mu})$. The parameter Q, which is important in determining the behavior of the wave, may be easily related to other, well-known concepts:

$$Q \equiv \frac{\epsilon\omega}{g} = \frac{\epsilon(\omega E)}{gE} = \frac{\epsilon|\dot{E}|}{J} = \left|\frac{\dot{D}}{J}\right|$$

This shows that Q is the ratio of the "displacement current density" to the conduction current density.

CASE 1: $Q \gg 1$ This case occurs when we deal either (1) with very high frequencies, or (2) with insulators or poor conductors. Now $v = \omega/\beta$ gives

$$v = \sqrt{\frac{2}{\epsilon\mu}}\,[\sqrt{1+Q^{-2}}+1]^{-1/2} \qquad \text{so} \qquad \sqrt{\frac{\epsilon\mu}{2}} = \frac{1}{v}\,[\sqrt{1+Q^{-2}}+1]^{-1/2}$$

Therefore

$$\alpha = \frac{1}{v}[\sqrt{1+Q^{-2}}+1]^{-1/2}\,[\sqrt{1+Q^{-2}}-1]^{-1/2}\,\omega \approx \frac{g}{2\epsilon v}$$

Here, then,

$$\psi = \left[\psi_0 \exp\left(-\frac{x}{2\epsilon v/g}\right)\right] \exp i\omega\left(t - \frac{x}{v}\right)$$

gives the behavior of any of the components of **E** or **B**. A light beam going through glass would behave in this fashion.

The value of the magnitude of the complex propagation constant here is

$$|k| = \sqrt{\beta^2 + \alpha^2} = \omega\sqrt{\epsilon\mu}\,(1+Q^{-2})^{1/4} \approx \omega\sqrt{\epsilon\mu}$$

The angle ζ between the imaginary and real parts of the complex k is (see Fig. 15-7)

$$\zeta = \tan^{-1}\left(\frac{\alpha}{\beta}\right) = \tan^{-1}\left[\frac{\sqrt{1+Q^{-2}}-1}{\sqrt{1+Q^{-2}}+1}\right]^{1/2}$$

$$\approx \tan^{-1}\left[\frac{\frac{1}{2}Q^{-2}}{2}\right]^{1/2} = \tan^{-1}\left(\frac{1}{2Q}\right) \approx \frac{1}{2Q}$$

The distance $2\epsilon v/g$ is called the attenuation distance δ, or the skin depth. It is that distance in which the amplitude drops off by the factor $1/e$.

The phase angle between **E** and **B** in this case of an attenuated wave is not exactly zero, as it is in vacuum, but it is very small. To see this, we again consider Maxwell's equation $\nabla \times \mathbf{E} = -\dot{\mathbf{B}}$ for a plane wave. Just as before, we have $-\hat{\mathbf{q}}ikE = -i\omega\mathbf{B}$ where $\hat{\mathbf{q}}$ is in the transverse plane of the wave, turned 90° CW from $\hat{\mathbf{p}}$ looking toward $+x$; but now k is complex instead of real. **B** lies along $\hat{\mathbf{q}}$, while **E** lies along $\hat{\mathbf{p}}$. Thus $kE = \omega B$ or

$$\frac{E}{B} = \frac{\omega}{k} = \frac{\omega}{|k|^{e^{-i\zeta}}} \approx \frac{\omega}{\beta}\,e^{+i\zeta} = ve^{i\zeta}$$

This means that, while **E** and **B** are each linearly polarized, **E** leads **B** slightly in phase. Note, however, that this does not give a slightly elliptically polarized wave. The **E** and **B** are each waves linearly polarized. **E** and **B** are separate quantities; they are actually different components of a four-tensor, as we saw in the previous chapter. They cannot be added to each other, vectorially or otherwise, even if **B** is multiplied by c to make the two quantities have the same dimensions.

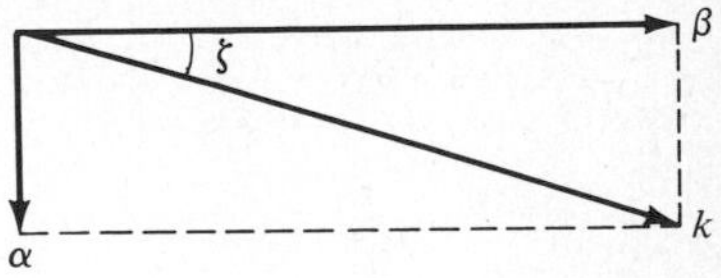

FIGURE 15-7

CASE 2: $Q \ll 1$ This case occurs when we deal either with low frequencies or with good conductors. $g \approx 10^7$ mho/m for the ordinary metals used in circuits, while $g/\epsilon \approx 10^{18}$. Then, for any frequencies below, say, 10^{17} Hz, we have $\omega/10^{18} \ll 1$. We see, therefore, that this condition is satisfied by metals for all parts of the electromagnetic spectrum with frequencies below the ultraviolet. Here

$$\beta \approx \sqrt{\frac{\epsilon\mu}{2}\frac{\omega}{\sqrt{Q}}} = \sqrt{\tfrac{1}{2}\mu g\omega} \qquad \alpha \approx \sqrt{\tfrac{1}{2}\mu g\omega}$$

and

$$\psi \approx \psi_0 \exp\left(-\sqrt{\tfrac{1}{2}\mu g\omega}\,x\right) \exp i\omega\left(t - \frac{x}{v}\right)$$

The magnitude of the complex propagation constant is

$$|k| = \sqrt{\beta^2 + \alpha^2} = \sqrt{\mu g\omega}$$

here, while ζ, the angle between the real and imaginary parts of k, is now 45°. So

$$\frac{E}{B} = \frac{\omega}{k} = \frac{\omega}{|k|e^{-i\zeta}} = \frac{\omega}{\sqrt{2}\,\beta}\, e^{i\zeta} = \frac{v}{\sqrt{2}}\, e^{i(\pi/4)}$$

The velocity of the wave now is

$$v = \frac{\omega}{\beta} = \sqrt{\frac{2\omega}{\mu g}} = \sqrt{\frac{2Q}{\epsilon\mu}} = \sqrt{2Q}\, v'$$

where v' is the wave velocity in a nonconducting medium with the same values of ϵ and μ. The v in a good conductor, having $Q \ll 1$, is therefore very much lower than in a good insulator. Also, since $|E/B| = v/\sqrt{2}$ here, while for a plane wave in vacuum $E = cB$, the electric field vector is much smaller in relation to the magnetic field in a metal than it is in a good insulator.

The attenuation distance in a metal is $\delta = 1/\alpha = \sqrt{2/\mu g\omega}$. At very low frequencies δ is large and the wave penetrates the metal with very little attenuation. But at microwave frequencies, such as 3×10^9 Hz for example, δ is only 1.2×10^{-6} m for gold. So the waveguides for microwaves are made with a very thin coating of silver or gold on the surface of a cheap material with poor conductivity—the wave never reaches the base material. Figure 15-8 shows the relation between the **E** and **B** waves at a given instant of time for a wave penetrating normally into a metal. The attenuation is so great that it cannot be conveniently shown in the diagram. The attenuation is given by $e^{-k_i x}$; here $\alpha \approx$

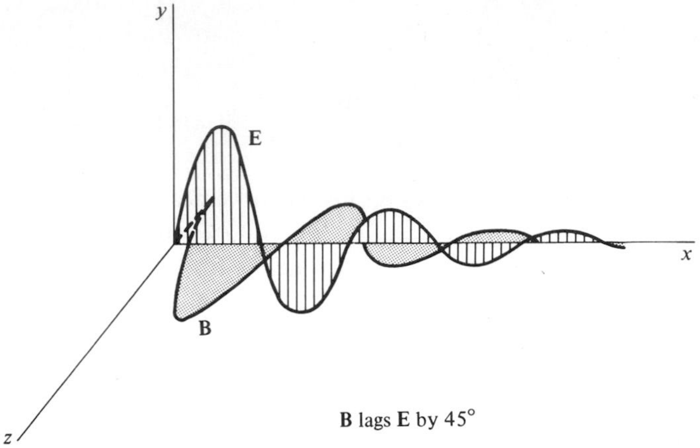

FIGURE 15-8

β; also, from $(\omega t - k_r x)$ we have $k_r = 1/\lambdabar$; so the attenuation is given by $e^{-2\pi x/\lambda}$. In a distance of penetration of one wavelength (much smaller than one wavelength in vacuum) the amplitude drops to $e - 2\pi$, i.e., the attenuation factor is $\frac{1}{400}$ in the minute distance of one wavelength; and here $\lambda \ll \lambda_0$!

15-4 CHARGED, NONCONDUCTING MEDIUM

This is not a practical case and we mention it only because it fits, logically, into the sequence we are considering. Such a medium could be artificially produced, say, by disposing a three-dimensional array of ions which could vibrate about fixed positions but could not move away from them (no electrons being present).

Maxwell's equations here, with $\rho \neq 0$ but $\mathbf{J} = 0$, lead to

$$\nabla^2 \mathbf{E} - \epsilon\mu \frac{\partial^2 \mathbf{E}}{\partial t^2} = \frac{1}{\epsilon} \nabla \rho$$

$$\nabla^2 \mathbf{B} - \epsilon\mu \frac{\partial^2 \mathbf{B}}{\partial t^2} = 0$$

The derivation of these equations is left to one of the problems. For the special case of a plane wave propagating along the $+x$ direction these equations divide into two groups, one group containing just one equation, the other group containing the other five. For the first equation, i.e., group, we first replace the right side by $\nabla(\nabla \cdot \mathbf{E})$; then, writing out the vector equation in full,

$$\nabla^2(\hat{\mathbf{x}}E_x + \hat{\mathbf{y}}E_y + \hat{\mathbf{z}}E_z) - \epsilon\mu(\hat{\mathbf{x}}\ddot{E}_x + \hat{\mathbf{y}}\ddot{E}_y + \hat{\mathbf{z}}\ddot{E}_z) = \hat{\mathbf{x}}\frac{\partial}{\partial x}\left(\frac{\partial E_x}{\partial x}\right)$$

Since

$$\nabla^2(\hat{\mathbf{x}}E_x) = \hat{\mathbf{x}}\frac{\partial^2 E_x}{\partial x^2}$$

here, the x component of this equation gives $-\epsilon\mu\ddot{E}_x = 0$. This is the one equation that is different. Its solution is $E_x = a(x)t + b(x)$. Any solution with nonvanishing $a(x)$ is physically unacceptable since it would grow without bound in time, either positively or negatively. So we set $a(x) = 0$. $E_x = b(x)$ is acceptable both physically and mathematically, but it represents a fixed field rather than a wave, so it is convenient to set $b(x) = 0$; i.e., the **E** wave is a transverse wave here also.

The $\hat{\mathbf{y}}$ and $\hat{\mathbf{z}}$ components of **E**, as well as all three components of **B**, i.e., the second group of five equations, obey the differential equation $\nabla^2\psi - \epsilon\mu\ddot{\psi} = 0$, so their solutions all fall into the second group:

$$-k^2\psi + \epsilon\mu\omega^2\psi = 0 \qquad \text{i.e.,} \qquad v = \frac{\omega}{k} = \frac{1}{\sqrt{\epsilon\mu}}$$

The solutions for E_y, E_z, B_x, B_y, B_z are similar to those in vacuum. Thus, the wave is transverse here also.

15-5 CHARGED, CONDUCTING MEDIUM

It is now only a short step to the final, most general, case—a medium having both finite charge density and finite current density. An ionized gas with a very low particle density, called a plasma, is an example of such a medium. A plasma is neutral over a sufficiently large region, but it can have a positive or negative charge density in a very small region. The charges may oscillate about their mean positions; the electrons, however, do so much more than the ions. Maxwell's equations now lead to the following (if we first put, as in Sec. 15-3, $\mathbf{J} = g\mathbf{E}$)

$$\begin{cases} \nabla^2\mathbf{E} - \epsilon\mu\ddot{\mathbf{E}} - g\mu\dot{\mathbf{E}} = \dfrac{1}{\epsilon}\nabla\rho \\ \nabla^2\mathbf{B} - \epsilon\mu\ddot{\mathbf{B}} - g\mu\mathbf{B} = 0 \end{cases}$$

The $\hat{\mathbf{x}}$ component of **E** again satisfies one differential equation while all the other components of **E** and **B** satisfy another. The equation for E_x is, substituting $\partial E_x/\partial x$ for $(1/\epsilon)\rho$,

$$\nabla^2 E_x - \epsilon\mu\ddot{E}_x - g\mu\dot{E}_x = \frac{\partial^2 E_x}{\partial x^2}$$

or $\epsilon\ddot{E}_x + g\dot{E}_x = 0$. As before, this is a second-order partial differential equation though it differs from the previous one. Its solution has two arbitrary functions of x:

$$E_x = a(x)e^{-gt/\epsilon} + b(x)$$

This is not interesting physically, for it simply represents an exponential decay rather than a wave, so we will ignore it. For simplicity we take $b(x) = 0$. Just as before, if an **E** wave exists it must be a transverse wave. The proof that the **B** wave is also transverse here is identically the same as the proof in the previous case.

We now make a slight digression to discuss the nature of the conductivity in such a medium. The force experienced by an electron subjected to a sinusoidal wave, $(-e)\mathbf{E}$ or $(-e)\mathbf{E}_0 \exp i(\omega t - kx)$, may be set equal to the sum of two terms: $m(d\mathbf{v}/dt) + \xi\,(m\mathbf{v})$. The first of these is the usual inertial force connected with the electron's acceleration; the second is related to the electron's transfer of momentum by collisions–ξ is a mean rate of collisions with various particles. The resulting equation, when solved for $\mathbf{v}$, gives

$$\mathbf{v} = \frac{(-e)}{m(\xi + i\omega)}\,\mathbf{E}_0 \exp i(\omega t - kx) = \frac{-e}{m(\xi + i\omega)}\,\mathbf{E}$$

Then, if N is the electron number density,

$$\mathbf{J} = N(-e)\mathbf{v} = \frac{Ne^2}{m(\xi + i\omega)}\,\mathbf{E}$$

Here we have used N rather than n to avoid confusion with the index of refraction. From $\mathbf{J} = g\mathbf{E}$, this yields

$$g = \frac{Ne^2}{m(\xi + i\omega)} = \frac{(Ne^2/m\xi)}{1 + i(\omega/\xi)}$$

For metals $\xi \sim 3 \times 10^{13}\ \text{s}^{-1}$, so for all frequencies up to about $f = 10^{11}$ Hz, $\xi \gg \omega$ and g is real and equal to $Ne^2/m\xi$. Thus, up to and including the microwave part of the electromagnetic spectrum, the real conductivities of copper and the other metals used for conductors give a **J** which is in phase with **E**. For the far infrared part of the spectrum, however, g is complex and in the optical part of the spectrum g becomes imaginary.

In plasmas the collision rate is extremely small compared to that in metals, primarily because of the much lower number density, and for all but the very lowest frequencies $\xi \ll \omega$. Then $g = -i(Ne^2/m\omega)$, i.e., the current density sinusoid lags the electric field sinusoid by 90°; thus there is no loss in the energy of the wave. This completes our digression.

We again turn to Maxwell's equations for this case, just as we did in the previous cases. As we saw at the beginning of Sec. 15-5, plane waves going through such a medium are transverse; there can be a longitudinal component in **E**, but this is a damped exponential rather than a wave. For our purposes, then, we may as well return to the

Maxwell's equations without charge in Sec. 15-3. The condition that results is then, again, $-k^2 + \epsilon\mu\omega^2 - ig\mu\omega = 0$, but this time g is imaginary. So $k^2 = \epsilon\mu\omega^2 - Ne^2\mu/m$. Setting $\omega_p = \sqrt{Ne^2/\epsilon m}$ this gives $k^2 = \epsilon\mu\omega^2\,[1-(\omega_p/\omega)^2]$; if we take $\epsilon \approx \epsilon_0$ and $\mu \approx \mu_0$ for this tenuous medium

$$k = \pm\frac{2\pi f}{c}\sqrt{1-\left(\frac{f_p}{f}\right)^2}$$

$f_p \equiv \omega_p/2\pi$ is called the plasma frequency. It is clear that for $f > f_p$ this gives a real k, while for $f < f_p$ the value of k is imaginary.

We first consider $f > f_p$. Then $f\lambda_0 = c$, $k/2\pi = 1/\lambda$, and the positive root corresponds to a wave going toward $+x$. So

$$\lambda = \frac{\lambda_0}{\sqrt{1-(f_p/f)^2}}$$

For a given frequency the wavelength here is larger than the corresponding wavelength in vacuum. This is just the opposite of the situation in an uncharged dielectric (Sec. 15-2 above). In fact, here,

$$v = f\lambda = \frac{c}{\sqrt{1-(f_p/f)^2}}$$

gives a value for the velocity of the monochromatic wave which is greater than the speed of an electromagnetic wave in vacuum. This, however, is not in violation of any of the results of special relativity, for a monochromatic wave is a sinusoidal wave that extends from $t = -\infty$ to $t = +\infty$ and cannot be used for transmitting signals or information. The group velocity, either for a series of discrete monochromatic waves or for a part of the spectrum with a continuum of such waves, is the velocity with which a signal may propagate. This is not always a quantity which is well defined for all possible dependences of phase velocity on frequency, since in some cases so much distortion results that it is difficult to pinpoint the signal. When it does have meaning, the group velocity always turns out to be less than c but a discussion of this is lengthy and leads us too far astray. Note, however, that when the phase velocity depends on frequency (as it does here) then different parts of the group of component signals travel at different rates, so that eventually the group must be distorted beyond recognition. The greater the dispersion, i.e., the dependence of phase velocity on frequency, the shorter the time during which the signal is recognizable.

Finally we consider the case $f < f_p$. Here k is a pure imaginary so

$$\psi = \psi_0 e^{i(\omega t - kx)}$$

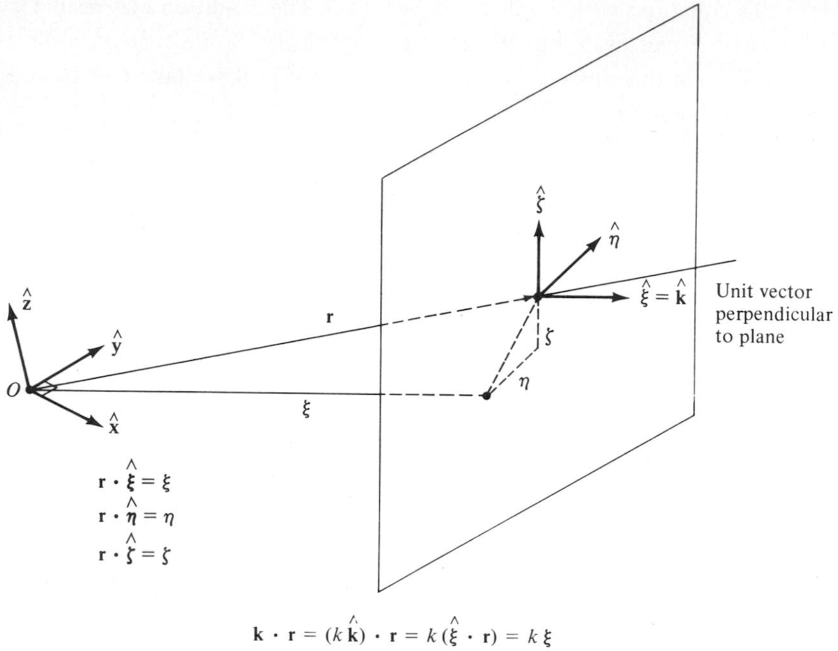

FIGURE 15-9

becomes either

$$\psi = [\psi_0 e^{kx}]e^{i\omega t} \qquad \text{or} \qquad \psi = [\psi_0 e^{-kx}]e^{i\omega t}$$

Neither of these expressions represents a traveling wave. f_p therefore represents a minimum cutoff frequency below which it is not possible for electromagnetic waves to propagate through a plasma.

Examples

1. Plane wave propagating in an arbitrary direction In the text, above, the wave was assumed to move along the x axis. Suppose we wish to consider a plane wave moving in an arbitrary direction characterized by the propagation vector $\mathbf{k} = (2\pi/\lambda)\hat{\mathbf{k}}$. The original cartesian coordinate system—$\hat{\mathbf{x}},\hat{\mathbf{y}},\hat{\mathbf{z}}$—may be rotated about the origin to give a new cartesian coordinate system—$\hat{\boldsymbol{\xi}}, \hat{\boldsymbol{\eta}}, \hat{\boldsymbol{\zeta}}$—with $\hat{\boldsymbol{\xi}}$ longitudinal and pointing in the direction of propagation ($\hat{\mathbf{k}} = \hat{\boldsymbol{\xi}}$); $\hat{\boldsymbol{\eta}}$ and $\hat{\boldsymbol{\zeta}}$ are then in a transverse plane. See Fig. 15-9.

The various forms of the del operator are form-invariant expressions. Under the rotation of axes $\nabla^2\psi$ transforms, e.g., from

$$\frac{\partial^2 \psi}{\partial x^2} + \frac{\partial^2 \psi}{\partial y^2} + \frac{\partial^2 \psi}{\partial z^2} \quad \text{to} \quad \frac{\partial^2 \psi}{\partial \xi^2} + \frac{\partial^2 \psi}{\partial \eta^2} + \frac{\partial^2 \psi}{\partial \zeta^2}$$

where ξ, η, and ζ are the coordinates of an arbitrary point P relative to the new axes. So the wave equation in the new coordinate system is obtained from that in the old system by merely letting

$$x \to \xi \qquad y \to \eta \qquad z \to \zeta$$

In the new system, let ψ be any one of the six components of **E** or **B** of a plane wave propagating along $\hat{\mathbf{k}}$; let $\boldsymbol{\Psi}$ be either **E** or **B**; and suppose **E** is linearly polarized along the $\hat{\boldsymbol{\eta}}$ direction. Then, as in the previous case,

$$\nabla\psi = \hat{\boldsymbol{\xi}}\frac{\partial \psi}{\partial \xi} + \hat{\boldsymbol{\eta}}\frac{\partial \psi}{\partial \eta} + \hat{\boldsymbol{\zeta}}\frac{\partial \psi}{\partial \zeta} = \hat{\boldsymbol{\xi}}\frac{\partial \psi}{\partial \xi} = \hat{\mathbf{k}}\frac{\partial}{\partial \xi}\left[\psi_0 e^{i(\omega t - \mathbf{k}\cdot\mathbf{r})}\right] = \hat{\mathbf{k}}(-ik\psi)$$

$$\nabla\cdot\boldsymbol{\Psi} = \frac{\partial \psi_\xi}{\partial \xi} + \frac{\partial \psi_\eta}{\partial \eta} + \frac{\partial \psi_\zeta}{\partial \zeta} = \frac{\partial \psi_\xi}{\partial \xi} = \frac{\partial}{\partial \xi}\left[\psi_{0\xi} e^{i(\omega t - k\xi)}\right] = -ik\psi_\xi = -ik\psi_n$$

$$\nabla\times\boldsymbol{\Psi} = \hat{\boldsymbol{\xi}}\left(\frac{\partial \psi_\zeta}{\partial \eta} - \frac{\partial \psi_\eta}{\partial \zeta}\right) + \hat{\boldsymbol{\eta}}\left(\frac{\partial \psi_\xi}{\partial \zeta} - \frac{\partial \psi_\zeta}{\partial \xi}\right) + \hat{\boldsymbol{\zeta}}\left(\frac{\partial \psi_\eta}{\partial \xi} - \frac{\partial \psi_\xi}{\partial \eta}\right) = -\hat{\boldsymbol{\eta}}\frac{\partial \psi_\zeta}{\partial \xi} + \hat{\boldsymbol{\zeta}}\frac{\partial \psi_\eta}{\partial \xi}$$

$$= \frac{\partial}{\partial \xi}(\hat{\boldsymbol{\xi}} \times \boldsymbol{\Psi})$$

$$\left\{\nabla\times\mathbf{E} = \frac{\partial}{\partial \xi}(\hat{\boldsymbol{\xi}} \times \hat{\boldsymbol{\eta}}E) = \hat{\boldsymbol{\zeta}}\frac{\partial E}{\partial \xi} = -\hat{\boldsymbol{\zeta}}ikE_\eta \qquad \nabla\times\mathbf{B} = \frac{\partial}{\partial \xi}(\hat{\boldsymbol{\xi}} \times \hat{\boldsymbol{\zeta}}B) = \hat{\boldsymbol{\eta}}ikB_\zeta\right\}$$

$$\nabla^2\psi = \frac{\partial^2 \psi}{\partial \xi^2} + \frac{\partial^2 \psi}{\partial \eta^2} + \frac{\partial^2 \psi}{\partial \zeta^2} = \frac{\partial^2 \psi}{\partial \xi^2} = \frac{\partial}{\partial \xi}\left[\frac{\partial}{\partial \xi}\psi_0 e^{i(\omega t - k\xi)}\right] = -k^2\psi$$

$$\frac{\partial \psi}{\partial t} = \frac{\partial}{\partial t}\left[\psi_0 e^{i(\omega t - \mathbf{k}\cdot\mathbf{r})}\right] = i\omega\left[\psi_0 e^{i(\omega t - \mathbf{k}\cdot\mathbf{r})}\right] = i\omega\psi$$

$$\frac{\partial^2 \psi}{\partial t^2} = \frac{\partial}{\partial t}\left(\frac{\partial \psi}{\partial t}\right) = \frac{\partial}{\partial t}\left[i\omega\psi_0 e^{i(\omega t - \mathbf{k}\cdot\mathbf{r})}\right] = (i\omega)(i\omega)\psi_0 e^{i(\omega t - \mathbf{k}\cdot\mathbf{r})}] = -\omega^2\psi$$

$$\Box^2\psi = \left(-k^2 - \frac{\omega^2}{c^2}\right)\psi$$

Maxwell's equations for a plane wave in vacuum become

$$-ikE_\xi = 0 \qquad\qquad \left\{\begin{aligned} &E_\xi = 0 \\ &E_\eta = \frac{\omega}{k}B_\zeta \end{aligned}\right.$$

$$-\hat{\boldsymbol{\zeta}}ikE_\eta = -i\omega\hat{\boldsymbol{\zeta}}B_\zeta \qquad \text{or}$$

$$-ikB_\xi = 0 \qquad\qquad \left\{\begin{aligned} &B_\xi = 0 \\ &E_\eta = \frac{1}{\epsilon_0\mu_0(\omega/k)}B_\zeta \end{aligned}\right.$$

$$\hat{\boldsymbol{\eta}}ikB_\zeta = \epsilon_0\mu_0 i\omega\hat{\boldsymbol{\eta}}E_\eta \qquad \text{or}$$

2. The impedance of free space The wave equation for free space, derived from Maxwell's equations, gives the velocity $c = (\epsilon_0 \mu_0)^{-1/2}$. Another relation between the two constants ϵ_0 and μ_0 is obtained from the fact that $E = cB$ for a plane wave. Thus

$$E = \frac{1}{\sqrt{\epsilon_0 \mu_0}} (\mu_0 H) \qquad \text{or} \qquad \frac{E}{H} = \sqrt{\frac{\mu_0}{\epsilon_0}} = 377\ \Omega$$

This is known as the impedance of free space, a name which has significance beyond the fact that the dimensions of E/H are the same as those of ohms.

When an antenna is designed to radiate power into space with maximum efficiency, one is faced with the same type of problem as a circuit designer encounters when it is desired to transfer the maximum power from a generator to a load resistor. In Fig. 15-10 a generator with internal impedance r is to deliver the maximum possible power to a fixed load resistor R. The instantaneous current is

$$I = \frac{V}{R + r} = \frac{V_0}{R + r} \sin \omega t$$

The power into the load resistor is then

$$P = I^2 R = V_0^2 \frac{R}{(R + r)^2} \sin^2 \omega t$$

The problem is to maximize this by varying r, for a fixed R.

$$\frac{dP}{dr} = V_0^2 R \sin^2 \omega t \frac{d}{dr} (R + r)^{-2} = V_0^2 R \sin^2 \omega t \frac{(-2)}{(R + r)^3}$$

This does not equal zero for any finite value of r, so the function P does not have a mathematical maximum, i.e., a point where the slope is zero. P, however, does have a physical maximum, a point where the function has its greatest value. This occurs when r is 0. There $P = (V_0^2/R) \sin^2 \omega t$. Averaged over one cycle, the power delivered to the load is $V_0^2/2R$. The steady undirectional voltage that would deliver the same power to this resistor is $(V_0/\sqrt{2})^2/R$. $V_0/\sqrt{2}$ is the root-mean-square voltage, V_{rms}, and $V_0^2/2R$ is the rms power.

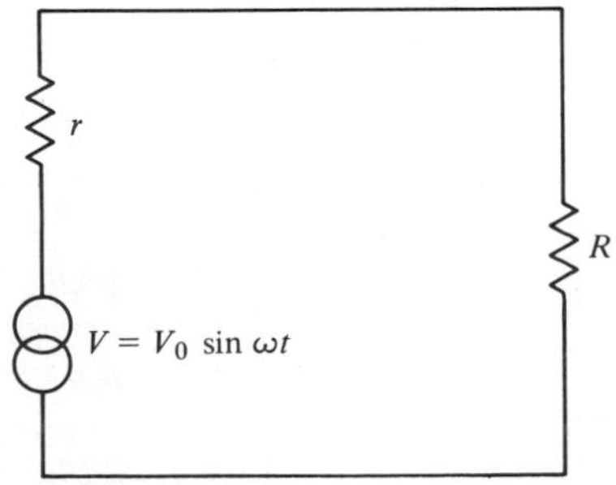

FIGURE 15-10

In practice is is not possible to obtain a generator with zero source impedance. The criterion then becomes that of obtaining a generator with as small an internal impedance as feasible, consistent with the economic factors. Note that the problem being considered is quite different from the one where it is desired to maximize the power delivered to R by varying R for a fixed r.

$$\frac{dP}{dR} = (V_0^2 \sin^2 \omega t)\left[\frac{(R+r)^2 - R2(R+r)}{(R+r)^4}\right] = 0$$

$$R + r = 2R \qquad R = r$$

Also $d^2P/dR^2 < 0$ at $R = r$. So, if it were possible to vary the load resistor R, then the maximum power delivered to R would be obtained for $R = r$. But, as we see, R is fixed at 377 Ω for free spaces, and this cannot be varied.

The fact that the product of ϵ_0 and μ_0 gives a velocity whereas their ratio gives a resistance squared depends on the fact that ϵ_0 and μ_0 were placed downstairs and upstairs, respectively, in Coulomb's law and the basic magnetostatic force equation. If one of these positions were altered then the ratio would give the speed of light.

3. The Poynting vector Care must be employed when sinusoidal quantities multiply each other if the quantities are represented by complex exponential functions, i.e., by either the real or the imaginary parts of these functions. Suppose, for example, that we wish to calculate the Poynting vector for a linearly polarized plane wave in free space:

$$\mathbf{E} = \hat{\mathbf{y}} E_0 \cos(\omega t - kx)$$

$$\mathbf{B} = \hat{\mathbf{z}} B_0 \cos(\omega t - kx)$$

Then

$$\mathbf{S} = \left(\frac{1}{\mu_0}\right) \mathbf{E} \times \mathbf{B} = \hat{\mathbf{x}} \frac{1}{\mu_0} E_0 B_0 \cos^2(\omega t - kx)$$

At $x = 0$ say, the instantaneous energy flux will be fluctuating but unidirectional:

$$\mathbf{S}(x = 0, t) = \hat{\mathbf{x}} \frac{1}{\mu_0} E_0 B_0 \cos^2 \omega t$$

while the energy flow averaged over any number of whole cycles will be

$$\langle \mathbf{S} \rangle = \hat{\mathbf{x}} \frac{1}{2\mu_0} E_0 B_0$$

This is the answer obtained when the calculation is made using the trigonometric functions.

Now let us employ the exponential functions. At $x = 0$ say:

$$\mathbf{E} = \hat{\mathbf{y}} E_0 e^{i\omega t}$$

$$\mathbf{B} = \hat{\mathbf{z}} B_0 e^{i\omega t}$$

It is no longer correct to write $\mathbf{S} = (1/\mu_0)\,\mathbf{E} \times \mathbf{B}$ for this implies

$$\mathbf{S} = \left(\frac{1}{\mu_0}\right) \mathcal{R}(\mathbf{E} \times \mathbf{B}) = \frac{1}{\mu_0} \mathcal{R}\,[(\hat{\mathbf{y}} E_0\, e^{i\omega t}) \times (\hat{\mathbf{z}} B_0 e^{i\omega t})]$$

$$= \frac{1}{\mu_0} \mathcal{R}\,(\hat{\mathbf{x}} E_0 B_0\, e^{2i\omega t}) = \hat{\mathbf{x}} \frac{1}{\mu_0} E_0 B_0 \cos 2\omega t$$

Not only is this instantaneous value wrong, but the average value is zero instead of $\hat{\mathbf{x}}(1/\mu_0)(E_0/\sqrt{2})(B_0/\sqrt{2})$.

Suppose we try $\mathbf{S} = (1/\mu_0)\,\mathbf{E} \times \mathbf{B}^*$, where the asterisk means complex conjugate. This really stands for

$$\mathbf{S} = \frac{1}{\mu_0} \mathcal{R}(\mathbf{E} \times \mathbf{B}^*) = \frac{1}{\mu_0} \mathcal{R}\,[(\hat{\mathbf{y}} E_0\, e^{i\omega t}) \times (\hat{\mathbf{z}} \mathbf{B}_0\, e^{-i\omega t})]$$

$$= \frac{1}{\mu_0} \mathcal{R}\,(\hat{\mathbf{x}} E_0 B_0)$$

This, also, does not yield the correct value for the *instantaneous* value of $\mathbf{S}$. (A similar result would be obtained with $(1/\mu_0)\,\mathbf{E}^* \times \mathbf{B}$.) But, except for a factor of $\frac{1}{2}$, it gives the correct value for the average power. It is common practice, therefore, to use the form

$$\langle \mathbf{S} \rangle = \mathcal{R}\left(\frac{1}{2\mu_0} \mathbf{E} \times \mathbf{B}^*\right)$$

for the time-averaged value of the Poynting vector.

PROBLEMS

1 A plane electromagnetic wave traveling toward $+x$ in vacuum has a frequency of 5×10^{15} Hz. What are the values of ω and k in $E = \hat{\mathbf{y}} E_0\, e^{i(\omega t - kx)}$? What is the wavelength in angstroms?

2 In the test $\psi_0\, e^{i(\omega t - kx)}$ was taken to represent a right-going sinusoidal wave. Would it be correct to take, instead,

(*a*) $\psi_0 e^{-i(\omega t - kx)}$ for this wave?

(*b*) How about $\psi_0 e^{i(\omega t + kx)}$?

(*c*) Or $\psi_0 e^{-i(\omega t + kx)}$?

3 (*a*) What is $\mathbf{B}$ for the wave of Prob. 1?

(*b*) What is $\nabla \times \mathbf{E}$ for this wave?

(*c*) Find $-\partial \mathbf{B}/\partial t$ for this wave.

(*d*) Compare the answers to (*b*) and (*c*).

4 Find $\nabla \times \mathbf{B}$ and $\partial \mathbf{E}/\partial t$ for the wave of Prob. 1 and compare.

5 Let $E_0 = 1\ \mu\text{V/m}$ in the wave of Prob. 1. Then, at $x = 0$ and $t = 0$, $E = \hat{\mathbf{y}} 10^{-6}$ V/m.

(a) What is E at $x = 0$ when $t = 1 \times 10^{-6}$ s?
(b) At $x = 0$ when $t = \frac{1}{2} \times 10^{-6}$ s?
(c) At $x = 0$ when $t = \frac{1}{4} \times 10^{-6}$ s?
(d) At $x = 0$ when $t = 1 \times 10^{-5}$ s?
(e) At $x = 0$ when $t = \frac{1}{2} \times 10^{-15}$ s?
(f) At $x = 0$ when $t = \frac{1}{4} \times 10^{-15}$ s?

6 Repeat parts (a) to (f) of Prob. 5 at exactly $x = 1$ m.

7 Repeat parts (a) to (f) of Prob. 5 at exactly $x = 3$ m.
(g) What would change by going to $x = 1.5$ m?

8 Calculate **S** and ⟨**S**⟩ for the wave of Prob. 1.

9 Consider a plane electromagnetic wave, traveling toward $+x$ in vacuum, which can be decomposed into the sum of two waves: $\mathbf{E} = \hat{\mathbf{y}}E_1 e^{i(\omega t - kx)} + \hat{\mathbf{z}}E_2 e^{i(\omega t - kx + \alpha)}$. Find **S** and ⟨**S**⟩. Are these the sums of the corresponding values for the component waves?

10 Consider a plane electromagnetic wave, traveling toward $+x$ in vacuum, which can be decomposed into the sum of two component waves: $\mathbf{E} = \hat{\mathbf{y}}E_1 e^{i(\omega t - kx)} + \hat{\mathbf{y}}E_2 i(\omega t - kx + \alpha)$. Find **S** and ⟨**S**⟩. Are these the sums of the corresponding values for the component waves?

11 Calculate (a) the electric energy density $u_{\text{elec}} = (\epsilon_0/2)E^2$ and
(b) the magnetic energy density $u_{\text{mag}} = (1/2\mu_0)B^2$ for the wave of Prob. 1.
(c) Show that when multiplied by c the sum of (a) and (b) equals the **S** for this wave. (See Prob. 8.)

12 To the wave of Prob. 1 add the wave $\hat{\mathbf{y}}E_0' e^{i(\omega t - kx)}$. Calculate **S** for the resultant wave (a) when $E_0' = E_0$, (b) when $E_0' \neq E_0$.

13 The only solutions to the wave equation, $\nabla^2 \psi - (1/c^2)(\partial^2 \psi/\partial t^2) = 0$, that we have dealt with in the text are sinusoidal plane waves moving, say, along the x direction. Their speed is c and the phase of ψ, i.e., $\omega t - kx - \alpha$, is a constant everywhere in a given yz plane, as is the amplitude ψ_0. To consider just one type of wave which does not fall into this category, take the wave front of constant phase to be, again, a yz plane; a ray, perpendicular to the wave front, is in the x direction, but let the amplitude of ψ vary in the yz plane. For example, suppose

$$\psi = \psi_0 e^{-\alpha y} e^{i(\omega t - kx)}$$

Find the velocity.

14 Take $\mathbf{E}_1 = \hat{\boldsymbol{\epsilon}}_+ E_1 e^{i(\omega t - kx)}$ and $E_2 = \hat{\boldsymbol{\epsilon}}_- E_2 e^{i(\omega t - kx)}$.
What is the resultant polarization in the following cases?
(a) $E_1 > E_2$
(b) $E_1 = E_2$
(c) $E_1 < E_2$

15 Given the sinusoidal plane wave

$$\mathbf{E} = [\hat{\mathbf{x}}(2 + i3) + \hat{\mathbf{y}}4 + \hat{\mathbf{z}}3]e^{i[\omega t + 1.8y - 2.4z]}$$

Find (*a*) the propagation constant k; (*b*) the direction cosines of **k**, that is n_x, n_y, n_z; (*c*) the wavelength λ; (*d*) the frequency f.

(*e*) Describe the direction in which the plane is moving.

(*f*) What sort of polarization does the wave have?

(*g*) Is the wave transverse?

16 A sinusoidal plane wave having $f = 10^6$ Hz propagates in a uniform, uncharged, dielectric with $\kappa = 2$. Find the wavelength and the velocity.

17 (*a*) In Prob. 15, if $\hat{\mathbf{z}}3$ became $\hat{\mathbf{z}}5$ in the amplitude factor would the wave be transverse?

(*b*) If $\hat{\mathbf{x}}(2 + i3)$ became $\hat{\mathbf{x}}(2 + i5)$ would it be transverse?

18 For a plane sinusoidal wave in vacuum, what are the magnitude and dimensions of E/H?

19 Take $\epsilon \approx \epsilon_0$ and $\mu \approx \mu_0$ for seawater while $g \approx 5$ mho/m. Find the velocity v and the ratio (E/cB) for $f = 10^3$ Hz, 10^7 Hz, 10^9 Hz, and 10^{15} Hz. These are typical audio, radio, microwave, and light frequencies, respectively. (1 kHz is called an audio frequency, but here we are considering an electromagnetic wave, not a sound wave of that frequency.) Also, find the attenuation distance δ at these frequencies.

20 Derive the "telegrapher's equation" (mentioned in Sec. 15-3, Uncharged, Conducting Medium) from Maxwell's equations when $\rho = 0$ and $\mathbf{J} = g\mathbf{E}$. Do this for both **E** and **B**.

21 In the case of good insulators ($Q \gg 1$) find an expression for the propagation constant in terms of the wavelength in vacuum and the dielectric constant.

22 Show that in good insulators the energy densities of the electrical and magnetic types are in the ratio

$$\frac{u_e}{u_m} \approx 1 - \frac{1}{2Q^2}$$

23 Find the ratio u_e/u_m in good conductors ($Q \ll 1$).

24 Starting from Maxwell's equations, derive the differential equations for **E** and **B** in the case when $\rho \neq 0$ but $\mathbf{J} = 0$, i.e., in the nonpractical case of a charged nonconducting medium.

25 Suppose, in the case of a plasma, that we follow the usual custom of setting the total current density $\mathbf{J}_t$ equal to the sum of the electron current density, $\mathbf{J}_e = gE$, and the "displacement current density," $\mathbf{J}_d = \partial \mathbf{D}/\partial t$. Show that one then obtains the paradoxical result, which seemingly meets no objections, that *the* current is less when there are electrons present than when there are none.

26 Derive the formula for the plasma frequency

$$f_p = \frac{1}{2\pi}\sqrt{\frac{ne^2}{\epsilon m}}$$

from (*a*) the equation of continuity, $\nabla \cdot \mathbf{J} = -\partial\rho/\partial t$, and (*b*) $\nabla \cdot \mathbf{J} = g\,\nabla \cdot \mathbf{E} = g(\rho/\epsilon)$.

27 What is the plasma frequency in the case of (*a*) a typical gas discharge where, say, $n = 10^{18}\ \text{m}^{-3}$ and (*b*) a portion of the earth's ionosphere where, say, $n = 10^{10}\ \text{m}^{-3}$?

28 In a plasma there are ions present as well as electrons. Why do the formulas—for conductivity, plasma frequency, etc.—contain reference only to the electrons?

29 Derive an equation for n, the index of refraction of a plasma, from the equation for the propagation constant. By *the* index of refraction is meant $n = c/v$, where v is the velocity of a sinusoidal wave in the plasma.

30 Let a sinusoidal plane wave be incident from below on a plasma whose density decreases with z. A ray of the original wave makes the angle θ with $\hat{\mathbf{z}}$. Does θ increase or decrease as the wave progresses upward?

16

TRANSMISSION LINES

16-1 CHARACTERISTIC IMPEDANCE

Throughout this book our primary interest has been in the fields in some region of space bounded by various conductors or dielectrics, not in the boundaries themselves. In Chap. 12, Sec. 4, however, when considering electric circuits, this approach became too cumbersome for even the simplest boundary conditions, and lumped circuit theory was used instead. This theory concentrates on the V's and I's in the circuit, quantities which are related to the **E** and **B** fields in the surrounding space by differential equations plus boundary conditions. In the present section still another method of analysis is employed, one which is, in a way, a combination of the other two. When the electric circuit is a transmission line the usual electric circuit parameters–resistance, capacitance, inductance–are considered as distributed rather than lumped, and an analysis then reveals the presence of waves of V and I, along the conductors, which are closely related to waves of **E** and **B** in the intervening space. This makes a logical extension of the field waves in space, considered in the previous section.

Figure 16-1a shows several common forms of transmission lines in corss section: an open or parallel wire line, a coaxial line, and a shielded pair. Figure 16-1b shows a longitudinal view of a transmission line. The symbolism of the diagram here is

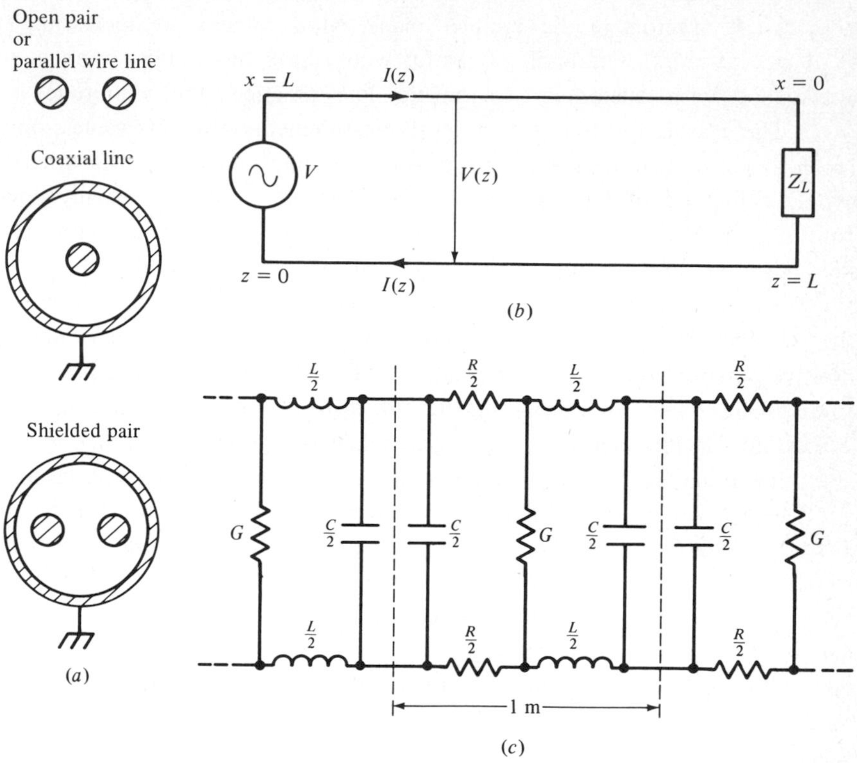

FIGURE 16-1

somewhat different from that in the ordinary circuit diagram, for there a drawn line between different circuit parameters or active elements is taken to mean an ideal connection having no resistance or inductance in itself, and possessing no capacitance relative to other circuit elements. In this section, however (and wherever transmission lines are dealt with), the two straight drawn lines represent the two nonideal conductors of the physical transmission line, regardless of type. As shown here, the line connects a sinusoidal voltage source V to a load impedance Z_L.

Figure 16-1*b* also indicates that the generator end of the line will be taken as $z = 0$, the load end as $z = L$. It will usually be convenient to measure distance from the load end; to facilitate this we employ the variable x, as shown. We will treat only steady-state, sinusoidal conditions; it then becomes possible to treat all quantities as functions only of z (or x); time is left out of the picture. Using the usual exponential notation we can take the current, e.g., to be given by $I(z)\, e^{(j\omega t)}$, it being understood that only the real part is being considered. Two explanations are necessary about conventions. (1) In this section I will represent the rms value of a phasor quantity, the current, and similarly for all other phasors (i.e., vectors in the complex plane, whose phases are increasing at the constant rate ω rad/s). Ordinarily, $I \cos \omega t$ would have the amplitude equal to I; but in accordance with customary transmission-line practice, the amplitude is given by $\sqrt{2}I$. The reason for this is that most measurements give rms values; only when an oscilloscope is used is it easier to measure the amplitude. (2) The other convention concerns the use of j rather than i to represent $\sqrt{-1}$. This is usually done to avoid confusion with the instantaneous current i. Since we are not dealing with $i(t)$ this danger does not exist here, but the notation is common and we will comply with it.

The V arrow in Fig. 16-1b shows the direction for the potential difference between the two lines at the plane $z = z$. In this figure the upper conductor would have a more positive potential than the lower conductor. The I arrows show such directions (opposite to each other in the two lines) that currents flowing in this sense are considered positive; if they flow in the opposite sense they are taken as negative. I is a function of z and, on a given line, it may be positive at one place and negative at another at any given instant.

In Fig. 16-1c the distributed parameters of the line are shown, per unit length of line. *Omitting the subscript* ℓ *(for per-unit length) in this section only,* R is the total series resistance per unit length, of both conductors, in ohms per meter; L is the total series inductance per unit length; C is the capacitance between the two conductors per unit length; and G is the conductance per unit length between them. (G represents the power loss in the dielectric medium rather than the flow of conventional current through this medium.) These parameters are constants at a given frequency, although they actually vary with frequency.

We now write the differential equations that determine the behavior of V and I. Take a section of the line (i.e., both conductors) that extends from z to $z + dz$. Then, by Kirchhoff's laws,

$$V(z + dz) - V(z) = -(R\,dz)I(z) - (j\omega L\,dz)I(z)$$

$$I(z + dz) - I(z) = -(G\,dz)V(z) - (j\omega C\,dz)V(z)$$

So

$$\frac{dV}{dz} = -(R + j\omega L)I$$

$$\frac{dI}{dz} = -(G + j\omega C)V$$

Treating the parameters as constants (we are considering only one given frequency), we now take the derivative of either equation and substitute into the value given by the other. This gives

$$\frac{d^2 V}{dz^2} - (R + j\omega L)(G + j\omega C)V = 0$$

$$\frac{d^2 I}{dz^2} - (R + j\omega L)(G + j\omega C)V = 0$$

Suppose we set

$$\gamma = \alpha + j\beta \equiv \sqrt{(R + j\omega L)(G + j\omega C)}$$

Then

$$\frac{d^2 V}{dz^2} - \gamma^2 V = 0$$

The general solution to this second-order differential equation involving two arbitrary constants is

$$V = V_1 e^{-\gamma z} + V_2 e^{\gamma z}$$

Similarly,

$$I = I_1 e^{-\gamma z} + I_2 e^{\gamma z}$$

The significance of these terms is best brought out by writing the expressions for the corresponding instantaneous quantities. Thus, from

$$v(z, t) = (\sqrt{2}\, V_1 e^{-\alpha z}) e^{j(\omega t - \beta z)} + (\sqrt{2}\, V_2 e^{\alpha z}) ej^{(\omega t + \beta z)}$$

it is seen that the first term represents a wave, traveling to the right with phase velocity ω/β, whose amplitude is attenuated exponentially as it moves along; the second term gives a similar wave, in general of different amplitude, attenuated as it travels to the left. The results for $i(z,t)$ are similar. In either case, the right-going wave is one initiated by the voltage source when the switch connecting the source to the line is closed at $t = 0$, while the left-going wave is produced (as we will soon see) by reflections of the right-going wave at the load.

In Fig. 16-2*a* we have shown the conditions near the source end just after the switch has first been closed. At the instant shown, the V and I waves, initiated at $z = 0$, have reached the section *a-a* down the line. To the right of *a-a* the voltage and current are still undisturbed from their original zero values, while to the left of *a-a* the voltage phasor will have the value $V = [V_1 e^{-\alpha z}]\, e^{-j\beta z}$ at the point $z = z$ (there is only one wave present); the current phasor will be $I = [I_1 e^{-\alpha z}]\, e^{-j\beta z}$ at that point. It is easy to understand that V_1 is determined by the signal generator at $z = 0$, but what determines I_1? In the steady state I_1 will be affected by the characteristics of the load impedance, but we are here not considering the steady state, since the wave has not yet had time to make the influence of the load impedance felt back at the source. Consequently, I_1 must be determined by the characteristics of the impedance line itself. Suppose we take an infinitely long line so that a wave, once initiated at the left end, goes on to the right indefinitely. It follows that the input terminals of the line present an impedance Z_0 to the generator such that I_1 is equal to V_1/Z_0. Z_0 is called the characteristic impedance of the line. See Fig. 16-2*b*.

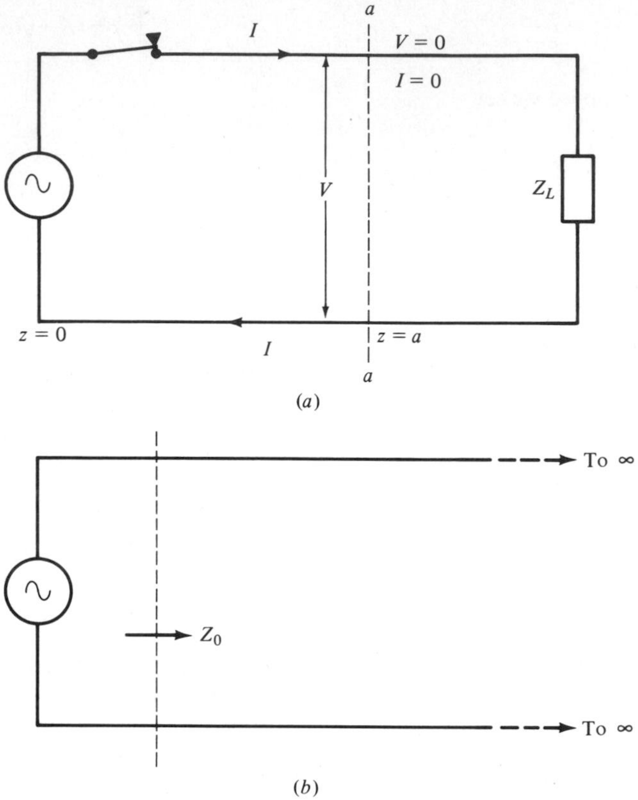

FIGURE 16-2

We now determine Z_0 in terms of the distributed parameters, per unit length, of the line. Note that for the infinite line $Z_0 = V_1/I_1$ is also given by $Z_0 = V/I$; i.e., Z_0 is the input impedance of the infinite line not only at $z = 0$ but also at any section $z = z$. For, removing the finite section between $z = 0$ and $z = z$ from the originally infinite line still leaves the remaining line infinitely long.

Repeating one of our original differential equations,

$$\frac{dV}{dz} = -(R + j\omega L)I$$

and setting $V = V_1 e^{-\gamma z}$, we have $dV/dz = -\gamma V$. So $-\gamma V = -(R + j\omega L)I$. Then

$$Z_0 = \frac{V}{I} = \sqrt{\frac{R + j\omega L}{G + j\omega C}}$$

Note that, though R is really R_ℓ and L, G, and C are also on a per-unit-length basis, this is not true of Z_0. Z_0 has ohms as its unit, and is *not* on a per-unit-length basis. It represents

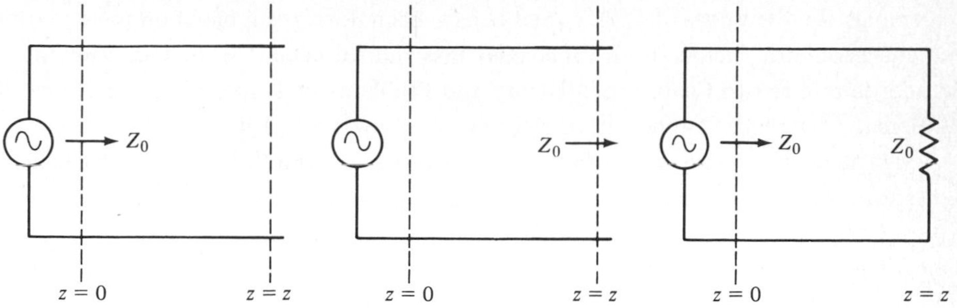

FIGURE 16-3

the input impedance of the entire line of infinite length as seen by the generator terminals. In Fig. 16-3*a* we see the input impedance Z_0 for the infinite line to the right of $z = 0$; in Fig. 16-3*b* the input impedance at $z = z$, i.e., looking to the right, is also Z_0, for the line is still infinitely long. Then if a lumped resistor, of value Z_0 ohms, is substituted in the latter figure for the infinite line to the right of $z = z$ we obtain the finite line of Fig. 16-3*c*. Thus, a line of finite length, having R,L,G,C as the per-unit-length parameters, will present in input impedance of Z_0 ohms to the generator terminals when the line itself is terminated in a lumped resistor of Z_0 ohms. The wave that starts down to the right has $Z_0 = V_1/I_1$, with I_1 determined by the line input impedance; and when the two waves, $V = V_1 e^{-\gamma z}$ and $I = I_1 e^{-\gamma z}$, reach the load resistor at $z = z$ their values are just the correct ones needed to satisfy Ohm's law, $Z_0 = V/I$. So steady-state conditions are established as soon as the wave has traversed the line once.

The complex quantity $\gamma = \alpha + j\beta$, called the complex propagation constant or the transmission constant of the line, has the dimensions $[\text{m}^{-1}]$. α is called the attenuation factor. Although it also has the dimensions $[\text{m}^{-1}]$, the unit for α is generally called the neper per meter (Np/m), where the neper (named after John Napier, the inventor of logarithms) is a dimensionless quantity. A line having an attenuation factor of α nepers per meter attenuates a traveling wave by a factor $e^{-\alpha L}$ in a distance L meters along the line. The attenuation is (αL) nepers when the amplitude is reduced by the factor, $e^{-\alpha L}$.

The quantity β, sometimes called the phase factor or the wavelength constant, is usually simply referred to as β. It, too, has the dimensions $[\text{m}^{-1}]$, but its unit is called the radian per meter (rad/m). The radian, of course, is the dimensionless quantity, representing angle, given by the ratio of arc length on a circle to its radius. From $e^{j(\omega t - \beta z)}$ it is seen that the wave phase velocity, the speed with which a given phase moves along the line, is ω/β; thus, $\beta = 2\pi/\lambda = 1/\bar{\lambda}$. Typical values for an open-wire transmission line are

R: 4×10^{-3} Ω/m	L: 2 μH/m	G: 0.2×10^{-9} mho/m
C: 6×10^{-12} pF/m	α: 3×10^{-6} Np/m	β: 2×10^{-5} rad/m

Formulas for the values of R, L, C, and G have been developed, based on the quantities ϵ, μ, and geometric factors, both for coaxial lines and for parallel wave lines. The interested reader is referred to Chap. 6 of "Theory and Problems of Transmission Lines," by R. A. Chipman (Schaum's Outline Series, McGraw-Hill, New York, 1968).

The shunt conductance of the line is such that $G \ll \omega C$, at almost all frequencies of interest, and G can generally be set equal to zero. Suppose we consider a frequency such that $\omega L/R \gg 1$. This common and important case gives, together with the restriction that $\omega C/G \gg 1$, a very great simplification. Thus,

$$Z_0 \equiv R_0 + jX_0 = \sqrt{\frac{R + j\omega L}{G + j\omega C}} \approx \sqrt{\frac{L}{C}}$$

$$\gamma \equiv \alpha + j\beta = \sqrt{(R + j\omega L)(G + j\omega C)} \approx \frac{1}{2}\left(\frac{R}{Z_0}\right) + j\omega\sqrt{LC}$$

$$\alpha \approx \frac{R}{2}\sqrt{\frac{C}{L}} \qquad \beta \approx \omega\sqrt{LC}$$

The value of α is very small compared to β and can usually be set equal to zero; i.e., most transmission lines are low-loss transmission lines. Only when we are specifically interested in the losses need we take α as nonzero. These results hold true for all frequencies above the kilohertz range for open pairs and above, say, 0.1 MHz for coaxial lines.

16-2 REFLECTION

Up to this point we have considered only a line of infinite length or its equivalent—a line of finite length terminated in its characteristic impedance Z_0. *Throughout this chapter we will restrict ourselves to the usual high-frequency case where Z_0 is a real quantity, $Z_0 = R_0$.* We now wish to consider the case where the load impedance, $Z_L = R_L + jX_L$, differs from R_0 (either because $X_L \neq 0$ or, even if $X_L = 0$, $R_L \neq R_0$). In this case, when the initial (right-going) V_1 and I_1 waves reach the load their ratio, $Z_0 = V_1/I_1$, does not have the correct value to satisfy Ohm's law for the load impedance: $Z_L \neq V_1/I_1$. In order to satisfy this law, every moment, at the load resistor an instantaneous change in the values of V and I at $z = \ell$ is necessary. This discontinuity in the values of V and I then propagates down the line to the left, back toward the source: two reflected waves are thus produced, one of V and one of I. Regardless of the conditions back at the source, additional reflections occur there, etc. When $Z_L \neq R_0$ an infinite sequence of reflected waves is produced. After a transient period (which theoretically takes an infinite time but, practically, is often essentially completed in microseconds) a steady-state condition is asymptotically approached in which the amplitudes of the voltage and current waves (each having right-going and left-going components) approach

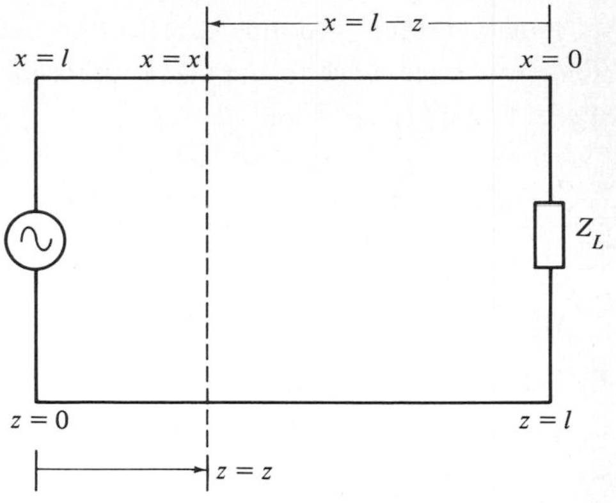

FIGURE 16-4

the values required by (the generalized, complex) Ohm's law at the load. The function served by the transient waves, in other words, is to bridge the gap between the initial values of V and I at the source to the final values of V and I at the load. Note that the final, steady-state, V_1 of the right-going voltage wave will differ from the V_1 of the initial right-going voltage wave.

In Fig. 16-4 we show this case, having the generalized load impedance Z_L. Going back to the general solution to represent the final steady-state conditions, we now have $V = V_1e^{-\gamma z} + V_2e^{\gamma z}$, representing right-going and left-going voltage waves at any point along the line. dV/dz must still be given at any point along the transmission line by $-(R + j\omega L)I$, as before; but now $dV/dz = -\gamma V_1e^{-\gamma z} + \gamma V_2e^{\gamma z}$. Equating the two expressions, and using the fact that $\gamma = \sqrt{(R + j\omega L)(G + j\omega C)}$, gives

$$I = \left(\frac{V_1}{Z_0}\right) e^{-\gamma z} + \left(-\frac{V_2}{Z_0}\right) e^{\gamma z} = I_1e^{-\gamma z} + I_2e^{\gamma z}$$

The amplitudes of the steady-state right-going and left-going current waves are here given in terms of the rms values of the voltage waves and Z_0, but not in terms of Z_L. However, the value of V_2 is affected by the load impedance. Note the minus sign in the relation between V_2 and I_2.

The input impedance of the line at any point, say $z = z$, is the input impedance of the section of line to the right of $z = z$ (i.e., looking to the right). It is given by $Z_{\text{in}} = V/I$. At the load end, in particular,

$$Z_L = \frac{V_1e^{-\gamma\ell} + V_2e^{\gamma\ell}}{(V_1/Z_0)e^{-\gamma\ell} - (V_2/Z_0)e^{\gamma\ell}}$$

It is now useful to define a reflection coefficient for the ratio of the steady-state reflected voltage wave to the steady-state incident voltage wave at the load end of the line: $\Gamma_L \equiv V_2 e^{(\gamma \ell)}/V_1 e^{-\gamma \ell}$ or

$$\boxed{\Gamma_L = \left(\frac{V_2}{V_1}\right) e^{2\gamma \ell}}$$

Γ_L, which is complex in general, is the reflection coefficient of the load, at the end of the line. In terms of Γ_L the previous Z_L equation then becomes

$$\boxed{Z_L = Z_0 \left(\frac{1 + \Gamma_L}{1 - \Gamma_L}\right)}$$

Solving this for Γ_L gives

$$\boxed{\Gamma_L = \frac{Z_L - Z_0}{Z_L + Z_0}}$$

(For the current wave, the reflection coefficient is the negative of this.)

We pause to consider some special cases. (*a*) When $Z_L = \infty$ (an open circuit at the load end), $\Gamma_L = 1$, and the reflected voltage wave has the same value and sense as the incident wave. (Both are steady-state waves, though this is also true of the initial incident and reflected waves.) The reflected current wave has the same value as the incident current wave, but is in the *opposite* sense. (*b*) When $Z_L = aZ_0$ with $1 < a < \infty$ then $0 < \Gamma_L < 1$. The reflected voltage wave has a smaller value than the incident voltage wave but is in the same sense. (*c*) When $Z_L = Z_0$ there is no reflection. (*d*) When $Z_L = aZ_0$ with $0 < a < 1$ then $-1 < \Gamma_L < 0$. The reflected voltage wave is smaller, and is in the opposite sense compared to the incident wave; the reflected current wave is in the same sense as the incident current wave. (*e*) When $Z_L = 0$ we have a short circuit at the load end. $\Gamma_L = -1$, so the reflected and incident voltage waves have the same value but are in the opposite senses relative to each other. The reflected current wave doubles the value of the incident wave alone.

All the special cases above assume a resistive load. But this need not be so. Z_L, for example, can be a pure inductance, or a pure capacitance, or any combination of these plus resistance. The effects in such cases will be brought out subsequently, but *for the moment we are considering Z_L real.*

We now switch from the variable z to the variable x, using $z = \ell - x$; x is, almost always, the more convenient distance to measure. At any point distant x from the load end, if we take $\gamma = j\beta$,

$$\begin{aligned}
V &= V_1 e^{-j\beta\ell} e^{j\beta x} + V_2 e^{j\beta\ell} e^{-j\beta x} \\
&= V_1 e^{-j\beta\ell}[e^{j\beta x} + \Gamma_L e^{-j\beta x}] \\
&= V_1 e^{-j\beta\ell}\left[e^{j\beta x} + \left(\frac{Z_L - Z_0}{Z_L + Z_0}\right) e^{-j\beta x}\right] \\
&= \frac{2V_1 e^{-j\beta\ell}}{Z_L + Z_0}(Z_L \cos \beta x + jZ_0 \sin \beta x)
\end{aligned}$$

$|V|$, the magnitude of V, is the quantity actually measured by a voltmeter, without regard to phase. Thus, taking both Z_0 and Z_L to be real,

$$|V| = \left(\frac{2V_1}{Z_L + Z_0}\right)\sqrt{Z_L^2 \cos^2 \beta x + Z_0^2 \sin^2 \beta x}$$

The variation of $|V|$ along the line is plotted in the left half of Fig. 16-5 for various values of Z_L. These graphs show voltage standing waves, with maxima at $x = 0, \lambda/2, \lambda, \ldots,$ when $Z_L > Z_0$; when $Z_L = Z_0$ there are no maxima; when $Z_L < Z_0$ the voltage maxima occur at $x = \lambda/4, 3\lambda/4, 5\lambda/4, \ldots$. These are all special cases, when both Z_0 and Z_L are real.

The right half of Fig. 16-5 shows the variation of $|I|$ along the line. This is obtained from $I = I_1 e^{-j\beta(\ell-x)} + I_2 e^{j\beta(\ell-x)}$ in a manner analogous to that employed for V. Note, again, that there Γ_L is the negative of the value given previously for V. The result of the calculation is

$$I = \left(\frac{2I_1 e^{-j\beta\ell}}{Z_L + Z_0}\right)(Z_0 \cos \beta x + jZ_L \sin \beta x)$$

$$|I| = \left(\frac{2I_1}{Z_L + Z_0}\right)\sqrt{Z_L^2 \sin^2 \beta x + Z_0^2 \cos^2 \beta x}$$

It is seen that the maxima of I occur where the minima of V are found, and vice versa.

We will continue the discussion of the peaks and troughs of the voltage standing wave, both with respect to values and position, for this is a most useful aspect of transmission line theory—both quantities are easy to measure and they yield considerable information. But first we wish to obtain a result that follows immediately. Dividing

$$V = \frac{2V_1 e^{-j\beta\ell}}{Z_L + Z_0}(Z_L \cos \beta x + jZ_0 \sin \beta x)$$

by

$$I = \frac{2I_1 e^{-j\beta\ell}}{Z_L + Z_0}(Z_0 \cos \beta x + jZ_L \sin \beta x)$$

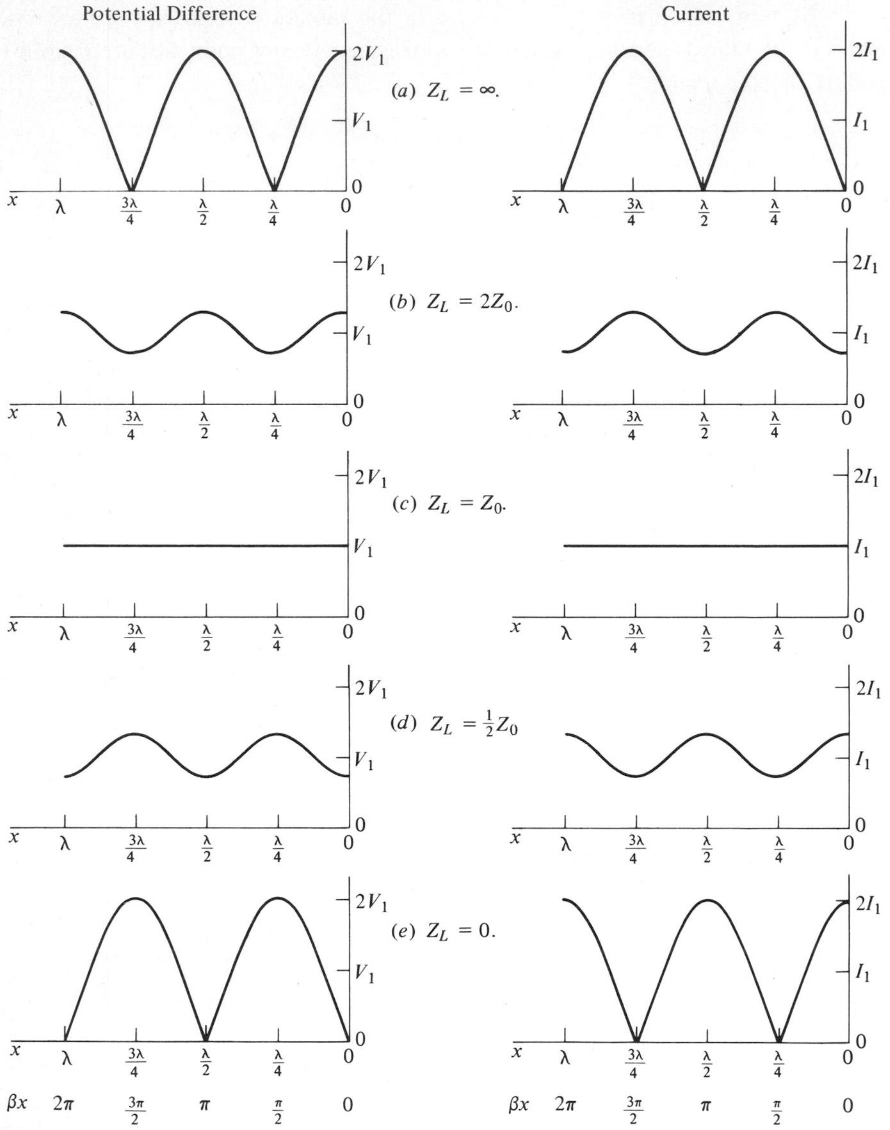

Potential Difference
Current
(a) $Z_L = \infty$.
(b) $Z_L = 2Z_0$.
(c) $Z_L = Z_0$.
(d) $Z_L = \frac{1}{2}Z_0$
(e) $Z_L = 0$.
$2V_1$
V_1
$2I_1$
I_1
0
x
λ
$\frac{3\lambda}{4}$
$\frac{\lambda}{2}$
$\frac{\lambda}{4}$
βx
2π
$\frac{3\pi}{2}$
π
$\frac{\pi}{2}$

FIGURE 16-5

gives

$$Z_{in} = Z_0 \left[\frac{Z_L \cos \beta x + jZ_0 \sin \beta x}{Z_0 \cos \beta x + jZ_L \sin \beta x} \right]$$

We will treat this further subsequently.

Returning to the standing wave expression for the magnitude of the potential difference between the conductors,

$$|V| = \left(\frac{2V_1}{Z_L + Z_0} \right) \sqrt{Z_L{}^2 \cos^2 \beta x + Z_0{}^2 \sin^2 \beta x}$$

suppose we consider the case when $Z_L < Z_0$, as in Fig. 16-5, part 1*d*. The minimum in $|V|$ that occurs at the load end $x = 0$ is given by $|V|_{min} = 2V_1 [Z_L/(Z_L + Z_0)]$; the maximum in $|V|$ (at $\beta x = \pi/2$) is $|V|_{max} = 2V_1 [Z_0/(Z_L + Z_0)]$. Thus, the ratio of the two, the so-called voltage standing wave ratio, VSWR, is

$$\text{VSWR} \equiv \frac{|V|_{max}}{|V|_{min}} = \frac{Z_0}{Z_L}$$

This ratio is, of course, greater than unity. For the case $Z_L = Z_0$ there are no maxima or minima and VSWR = 1. When $Z_L > Z_0$ the ratio is seen to be

$$\text{VSWR} \equiv \frac{|V|_{max}}{|V|_{min}} = \frac{Z_L}{Z_0}$$

again greater than unity. The measurement of the VSWR yields a quick method for determining the load resistance if the Z_0 of the line is known. The especially simple results for Z_L are only true, however, when Z_L is real.

We are still treating the case when Z_L is real. (Z_0 is assumed real throughout this section.) When $Z_L > Z_0$ then VSWR = Z_L/Z_0; but $Z_L = Z_0 \; [(1 + \Gamma_L)/(1 - \Gamma_L)]$, so VSWR = $(1 + \Gamma_L)/(1 - \Gamma_L)$ with $\Gamma_L > 0$. When $Z_L < Z_0$ then VSWR = Z_0/Z_L, so that VSWR = $(1 - \Gamma_L)/(1 + \Gamma_L)$ with $\Gamma_L < 0$. Both cases may be represented by

$$\boxed{\text{VSWR} = \frac{1 + |\Gamma_L|}{1 - |\Gamma_L|}}$$

We leave it to the problems to show that this equation holds true even when Z_L is taken as complex. In any case

$$\boxed{|\Gamma_L| = \frac{\text{VSWR} - 1}{\text{VSWR} + 1}}$$

A measurement of the VSWR therefore also determines the magnitude of the reflection coefficient at the load. It can be seen why this easy measurement is so very useful.

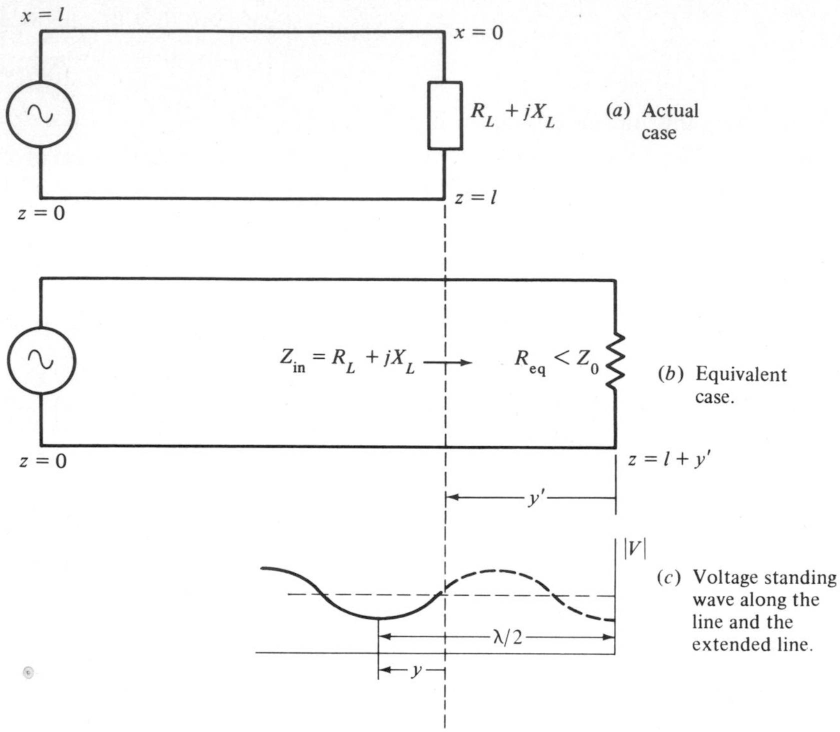

FIGURE 16-6

In all the cases treated above, the terminating impedance is a pure resistance. What of the cases when Z_L has a reactive component? Suppose, say, $Z_L = R_L + jX_L$. It is possible to replace this lumped impedance with a finite section of the line, continued to the right and terminated in a resistance R_{eq}. Thus, using the equation for Z_{in} above and letting y' be the length of a finite, equivalent, line to the right,

$$R_L + jX_L = Z_0 \frac{R_{eq} \cos \beta y' + jZ_0 \sin \beta y'}{Z_0 \cos \beta y' + jR_{eq} \sin \beta y'}$$

See Fig. 16-6 for the relation of y' to the other variables; y' increases going left. R_{eq} and y' will have unique values for a given Z_0, β, and $R_L + jX_L$. Consequently, whether for pure or partial reactive loads, the line acts as if it were effectively extended to the right of the load, and $x = 0$ no longer corresponds to a maximum or minimum point in the potential difference.

The standing wave to the left of $x = 0$ must be the same in cases (a) and (b) of Fig. 16-6 since the original line cannot distinguish between a lumped $R_L + jX_L$ at $x = 0$ and a $Z_{in} = R_L + jX_L$ there, obtained from the line extension to the right. Thus if $R_{eq} < Z_0$, there will be a minimum in $|V|$ at the equivalent load position, $z = \ell + |y'|$, as shown in the figure; if $R_{eq} > Z_0$, there is a maximum in $|V|$ at $z = \ell + |y'|$.

Let us solve the equation above for R_L and X_L. With VSWR representing $|V|_{max}/|V|_{min}$, and $R_{eq} < Z_0$, the VSWR $= Z_0/R_{eq}$. (We can limit ourselves to the case $R_{eq} < Z_0$ by taking y' either less than $\lambda/4$ or between $\lambda/4$ and $\lambda/2$.) Thus

$$R_L = \frac{(\text{VSWR})Z_0}{\sin^2 \beta y' + (\text{VSWR})^2 \cos^2 \beta y'} \qquad X_L = \frac{[(\text{VSWR})^2 - 1]Z_0 \sin \beta y' \cos \beta y'}{\sin^2 \beta y' + (\text{VSWR})^2 \cos^2 \beta y'}$$

But y' is not easily measured, whereas y, the distance from the actual load to the closest minimum (if $R_{eq} < Z_0$), is easy to determine. Since $y' = \lambda/2 - y$, the only effect of replacing y' by y is to change the sign of the cosine:

$$R_L = \frac{(\text{VSWR})Z_0}{\sin^2 \beta y + (\text{VSWR})^2 \cos^2 \beta y}$$

$$X_L = \frac{[1 - (\text{VSWR})^2]Z_0 \sin \beta y \cos \beta y}{\sin^2 \beta y + (\text{VSWR})^2 \cos^2 \beta y}$$

This gives the components of the load impedance in terms of easily measured quantities. It is a very useful set of equations because y and VSWR are much easier to measure than any impedance, whether R_L, X_L, or Z_0. When the load impedance is complex rather than real the only extra information we need to find Z_L (and Γ_ℓ) is the distance from the load to the first minimum. If there is a minimum closest to the load then the load is capacitive; if a maximum is closest it is inductive. The proof is left for the problems.

16-3 THE SMITH CHART

We now turn to the use of transmission lines as circuit elements. At a particular frequency the input impedance of an auxiliary transmission line may be adjusted over a very wide range by comparatively small adjustments of the length of the line and by choosing the proper load impedance. Such a transmission line, called a stub, may then be attached to the main transmission line—usually in parallel, rarely in series—in order to insert some desired impedance at a given point in the main line. So the question of the Z_{in} of the auxiliary line is an important one. Instead of dealing directly with the equation

$$Z_{in} = Z_0 \left[\frac{Z_L \cos \beta x + jZ_0 \sin \beta x}{Z_0 \cos \beta x + jZ_L \sin \beta x}\right]$$

however, it is convenient to employ a graphical method known as the Smith chart. [See "Transmission Line Calculator," by P. H. Smith, *Electronics*, 12:29 (1939).] The Smith chart is very widely used in connection with all sorts of transmission line problems; in fact, it is so widely used that only the boxed equations in this section, needed along with the chart, are widely known. The other equations in this section are largely ignored.

To see how this chart is constructed we go back to the basic equations, with Γ_L in general a complex quantity:

$$V = V_1 e^{-j\beta \ell} [e^{j\beta x} + \Gamma_L e^{-j\beta x}]$$

$$I = I_1 e^{-j\beta \ell} [e^{j\beta x} - \Gamma_L e^{-j\beta x}]$$

Then

$$Z_{in} = \frac{V}{I} = Z_0 \left(\frac{1 + \Gamma_L e^{-j2\beta x}}{1 + \Gamma_L e^{-j2\beta x}} \right)$$

Let us define

$$\boxed{\Gamma = \Gamma_L e^{-j2\beta x}}$$

Γ is the reflection coefficient of the line at any point x along its length. At that point the terminating impedance is the input impedance of the rest of the line to the right. Thus,

$$\boxed{Z_{in} = Z_0 \left(\frac{1 + \Gamma}{1 - \Gamma} \right)} \qquad \text{and} \qquad \boxed{\Gamma = \frac{Z_{in} - Z_0}{Z_{in} + Z_0}}$$

Setting $Z_{in}/Z_0 = r_{in} + jx_{in}$ (where x_{in} can be positive or negative), we have, in effect, normalized the input resistance and reactance at any point on the line in terms of the characteristic impedance; and letting $\Gamma = \Gamma_L e^{-j2\beta x} = p + jq$ gives

$$r_{in} + jx_{in} = \frac{(1 + p) + jq}{(1 - p) - jq}$$

This equation can be thrown into the following two forms, for its real and imaginary parts:

$$\left(p - \frac{r_{in}}{1 + r_{in}} \right)^2 + q^2 = \frac{1}{(1 + r_{in})^2}$$

$$(p - 1)^2 + \left(q - \frac{1}{x_{in}} \right)^2 = \frac{1}{x_{in}^2}$$

Note that the maximum value of $\sqrt{p^2 + q^2} = |\Gamma|$, the voltage reflection coefficient of the line at any point, is unity. For r_{in} = constant, the first equation yields a family of circles

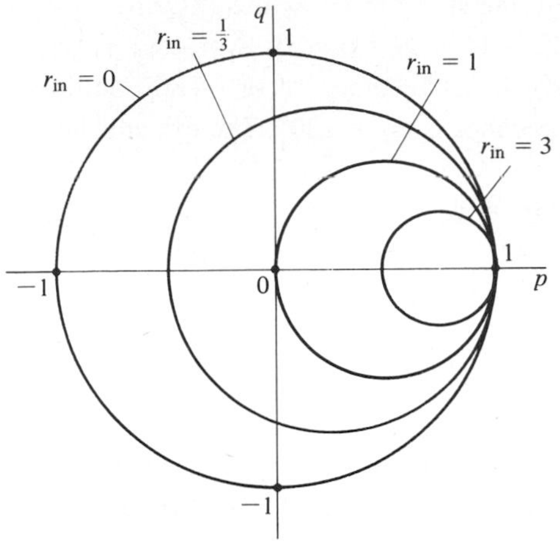

(a)

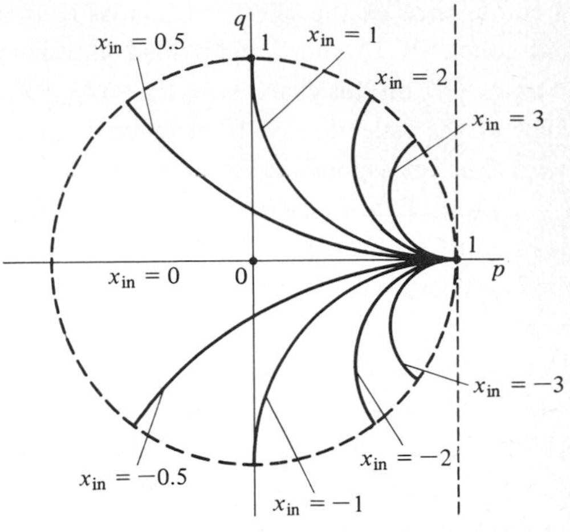

(b)

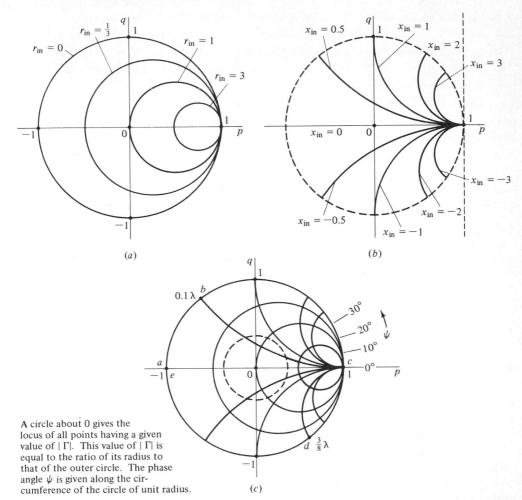

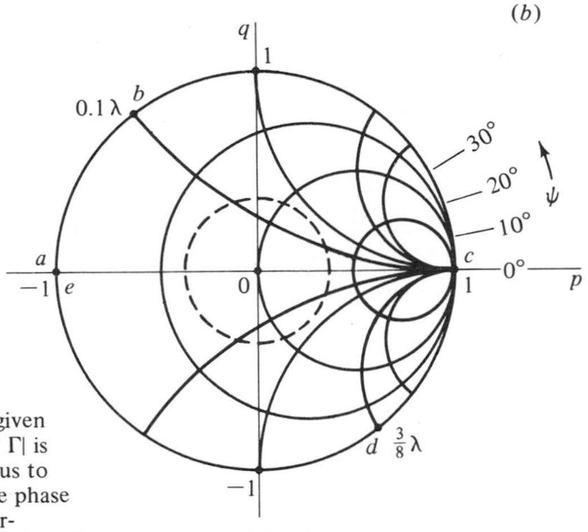

A circle about 0 gives the locus of all points having a given value of | Γ|. This value of | Γ| is equal to the ratio of its radius to that of the outer circle. The phase angle ψ is given along the circumference of the circle of unit radius.

(c)

FIGURE 16-7

in the (p,q) plane with the centers at $[p = r_{in}/(1 + r_{in}),\ q = 0]$, each having the radius $1/(1 + r_{in})$. Figure 16-7*a* shows several members of this family, all contained within a circle of unit radius, about the origin. This origin corresponds to zero reflection since with $p = q = 0$ we have $\Gamma = 0$. Similarly, the loci of points with x_{in} = constant are circles centered at $(p = 1,\ q = 1/x_{in})$ with radii $1/x_{in}$. These are shown in Fig. 16-7*b*. The Smith chart is obtained from the superposition of these two families, as shown in Fig. 16-7*c*.

In addition to the numbers labeling the values of r_{in} and x_{in}, the Smith chart also contains other pieces of useful information. Taking $\Gamma = |\Gamma|e^{j\psi}$, ψ is the angle of the reflection coefficient at any point βx along the line. ψ is given in degrees along the circumference of the chart starting, as is usual for angles, from 0° at the extreme right and going CW. A short-circuit load impedance, $Z_L = 0$ or $z_L = 0 + j0$, is plotted at the extreme left on this chart. Here $\Gamma_L = (Z_L - Z_0)/(Z_L + Z_0)$ becomes -1, or $1\underline{/180°}$, so the angle of the reflection coefficient for this load impedance is $\psi_L = 180°$. The extreme left point thus corresponds to $\beta x = 0$ for $Z_L = 0$.

From $\Gamma = \Gamma_L e^{-j2\beta x}$ we have $\psi = \psi_L - 2\beta x = \psi_L - 120°(x/\lambda)$. This gives the position of ψ, when $\psi_L = 180°$, for other values of (x/λ). A scale on the circumference of the chart gives (x/λ) starting with $x/\lambda = 0$ at the extreme left and, according to the above formula, going CW until it returns to the start at $x = 0.5\lambda$.

If the load impedance is open circuit then $\Gamma_L = 1\underline{/0°}$ and $\beta x = 0$ corresponds to the extreme right point, 180° from the previous position; but the calibration still increases CW as before. The βx scale has absolute significance for load impedance only when $Z_L = 0$; for other values of Z_L some constant must be added to the reading. The difference in readings between two different values of βx will be correct, however, regardless of the values of Z_L.

In Examples 2 and 3 we show how these βx scales are used when we are dealing with z_{in} and voltage minima, or with y_{in} and voltage maxima, rather than with Z_L.

The circle of unit radius corresponds to $\Gamma = 1$, since $\sqrt{p^2 + q^2} = 1$ at all points on it. $|\Gamma|$, the magnitude of the reflection coefficient at any given point within the chart, is obtained from the distance to the center of the chart, as a fraction of the radius to the edge of the chart. Then, since $|\Gamma| = |\Gamma_L|$, the VSWR at any point on the chart is given by

$$\boxed{\text{VSWR} = \frac{1 + |\Gamma|}{1 - |\Gamma|}}$$

At the center, O, $|\Gamma| = 0$ so VSWR = 1. Circles about O correspond to $0 < |\Gamma| < 1$, each giving some constant VSWR; at the edge of the diagram $|\Gamma| = 1$ and VSWR = ∞.

Suppose we consider a transmission line whose load impedance is $Z_L = 0$. The load reflection coefficient is

$$\Gamma_L = \frac{Z_L - Z_0}{Z_L + Z_0} = -1 = 1\underline{/180°}$$

On the Smith chart of Fig. 16-7*c*, the place where $\beta x = 0$ (i.e., at the load itself) is represented by the point a since here ψ, the angle of the reflection coefficient, is 180°; or, as read on the outer scale, we are zero wavelengths toward the generator. When $Z_L = 0$, Z_{in} is usually designated by Z_{sc}, standing for short-circuit input impedance. At a, $(Z_{\text{in}}/Z_0) = r_{\text{in}} + jx_{\text{in}} = 0 + j0$, as it should be: looking to the right, at the end of the line, we see a short circuit.

Next, let us consider a different point on the line, one where $0 < \beta x < \pi/2$ (that is, $0 < x < \lambda/4$). Point b on the Smith chart, for example, corresponds to a place on the line which is 0.1λ toward the generator from the load end with Γ_L still given by $1\underline{/180^\circ}$. On the outer circle we have $r_{\text{in}} = 0$; the other family of circles gives us $x_{\text{in}} = +j0.73$. So $Z_{\text{in}}/Z_0 = 0 + j0.73$—the line gives an inductive reactance at its input terminals. If $Z_0 = 50$ ohms, say, then $X_{\text{in}} = 36.5\ \Omega$. This checks with the results from

$$Z_{\text{in}} = Z_0 \left[\frac{Z_L \cos \beta x + jZ_0 \sin \beta x}{Z_0 \cos \beta x + jZ_L \sin \beta x} \right] = +jZ_0 \tan \beta x$$

This is inductive but, unlike the case with lumped parameters, the value of the inductance varies with the frequency.

Continuing down the line away from the short-circuited end, we reach a point where $\beta x = \pi/2$, or $x = \lambda/4$. This is a very important case. From the markings around the edge of the Smith chart, this point is represented by c. Here $r_{\text{in}} = 0$ but $jx_{\text{in}} = \infty$. (Indeed, $Z_{\text{in}} = +jZ_0 \tan \beta x \to \infty$ as $\beta x \to \pi/2$.) Although there is a short circuit at the load end of the stub, the input terminals represent an open circuit.

If the line is somewhat longer, $\pi/2 < \beta x < \pi$ (say $\beta x = 3\pi/4$, or $x = 3/8\lambda$), we obtain point d on the Smith chart. Here $Z_{\text{in}} = 0 - j\ell$: the line is a capacitor, with $|-jx_{\text{in}}| = 1$. For $Z_0 = 50\ \Omega$, $-j/\omega C = -j50\ \Omega$. $Z_{\text{in}} = +jZ_0 \tan \beta x$ verifies this result, for $\tan \beta x$ becomes negative when $\lambda/4 < x < \lambda/2$.

Finally, point e on the Smith chart corresponding to $x = \lambda/2\pi$ is the same as point a. The input terminals of a short-circuited line which is a half wavelength long have zero input impedance.

What if we make the line open-circuited? $Z_L = \infty$. (Z_{in} for this case is usually designated as Z_{oc}, meaning open-circuit load impedance.) Then

$$\Gamma_L = \frac{Z_L - Z_0}{Z_L + Z_0} = 1\underline{/0^\circ}$$

A point on the line right at the open-circuited load is represented by $1\underline{/0^\circ}$ on the Smith chart. This is at point c. The marking 0.25λ at this point does not apply; it is only a reference point. If, for example, we move to a point on the line λ/8 toward the generator, then adding 0.25λ to 0.125λ gives 0.375λ. Looking for this number on the rim of the chart gives point d again. Thus, with an open-circuited line less than λ/4 long we obtain a capacitive input impedance. Similarly, an open-circuited line of length λ/4 gives point a: a short circuit. The following table summarizes the results.

	Input impedance	
Length of line	$Z_L = 0$ ($Z_{\text{in}} = +jZ_0 \tan \beta x$)	$Z_L = \infty$ ($Z_{\text{in}} = -jZ_0 \cot \beta x$)
$0 < x < \lambda/4$	Inductive	Capacitive
$x = \lambda/4$	*Open circuit	Short circuit
$\lambda/4 < x < \lambda/2$	Capacitive	Inductive
$x = \lambda/2$	Short circuit	*Open circuit

*Asterisks explained in text below.

The asterisks in the table above are meant to indicate that the infinite impedance open circuit actually applies only to the case of a lossless line. In actual practice r_{in} becomes high, but finite, when the line has a nonvanishing R (per unit length). This is similar to the case of a lumped C in parallel with a lumped L, when the L has a small r in series with it. To deal with the two cases above we can no longer ignore the attenuation constant α. We will work one of these cases out, the one with $x = \lambda/4$ and $Z_L = 0$, leaving the other one for the problems.

We start with

$$Z_{\text{in}} = \frac{V}{I} = \frac{V_1 e^{-\gamma z} + V_2 e^{\gamma z}}{(V_1 e^{-\gamma z} - V_2 e^{\gamma z})/Z_0}$$

Putting $z = \ell - x$ and $\Gamma_L = (V_2/V_1)e^{2\gamma\ell}$, as before, gives

$$Z_{\text{in}} = Z_0 \frac{e^{\gamma x} + \Gamma_L e^{-\gamma x}}{e^{\gamma x} - \Gamma_L e^{-\gamma x}}$$

$$= Z_0 \frac{(1 - \Gamma_L)\sinh \gamma x + (1 + \Gamma_L)\cosh \gamma x}{(1 + \Gamma_L)\sinh \gamma x + (1 - \Gamma_L)\cosh \gamma x}$$

Here we have made use of the relations

$$e^{\gamma x} = \cosh \gamma x + \sinh \gamma x$$

$$e^{-\gamma x} = \cosh \gamma x - \sinh \gamma x$$

Substituting $Z_L = Z_0(1 + \Gamma_L)/(1 - \Gamma_L)$ into the above, we obtain

$$Z_{\text{in}} = Z_0 \frac{Z_0 \sinh \gamma x + Z_L \cosh \gamma x}{Z_L \sinh \gamma x + Z_0 \cosh \gamma x}$$

(Note that when $\alpha = 0$ then $\gamma = j\beta$ and, with $\sinh j\beta x = j \sin \beta x$ and $\cosh j\beta x = \cos \beta x$, we obtain our previous expression

$$Z_{\text{in}} = Z_0 \left[\frac{jZ_0 \sin \beta x + Z_L \cos \beta x}{jZ_L \sin \beta x + Z_0 \cos \beta x}\right]$$

But now, with $\alpha \neq 0$, this is no longer valid.)

When $Z_L = 0$, as it does for the short-circuited $\lambda/4$ line,

$$Z_{\text{in}} = Z_0 \tanh \gamma x = Z_0 \tanh(\alpha x + j\beta x)$$

$$= Z_0 \frac{\sinh \alpha x \cosh j\beta x + \cosh \alpha x \sinh j\beta x}{\cosh \alpha x \cosh j\beta x + \sinh \alpha x \sinh j\beta x}$$

But $\sinh j\beta x = j \sin \beta x$ and $\cosh j\beta x = \cos \beta x$, so

$$Z_{\text{in}} = Z_0 \frac{\sinh \alpha x \cos \beta x + j \cosh \alpha x \sin \beta x}{\cosh \alpha x \cos \beta x + j \sinh \alpha x \sin \beta x}$$

$\beta x = \pi/2$ when $x = \lambda/4$. Therefore $Z_{in} = Z_0$ (cosh αx/sinh αx). Now, we are considering low-loss lines, i.e., $\alpha \ll \beta$. Then $\alpha x \ll \pi/2$; so cosh $\alpha x = 1 + (\alpha x)^2/2! + \cdots \approx 1$, while sinh $(\alpha x) = (\alpha x) + (\alpha x)^3/3! + \cdots \approx \alpha x$. Thus $Z_{in} \approx Z_0\,(1/\alpha x)$. But $\alpha \approx 1/2\,(R/Z_0)$. So

$$Z_{in} \approx \frac{Z_0^2}{(R/2)x}$$

This is the relation that gives the finite input impedance for the $\lambda/4$ line with $Z_L = 0$ (instead of the infinite Z_{in} obtained in the table above if we set $R = 0$). Note that $R/2$ is the per-unit-length resistance of just one of the conductors, and $(R/2)x$ is the total resistance of one conductor.

Returning to the case where we can ignore the small loss in the line, the formula for Z_{in} reverts from the hyperbolic to the original trigonometric functions:

$$Z_{in} = Z_0 \left[\frac{Z_L \cos \beta x + jZ_0 \sin \beta x}{Z_0 \cos \beta x + jZ_L \sin \beta x}\right]$$

For a $\lambda/4$ line we have $\beta x = \pi/2$, so $Z_{in} = Z_0^2/Z_L$. A load having a high impedance

$$Z_L = |Z_L|e^{j\psi_L}$$

therefore, appears at the input terminals of the $\lambda/4$ line as an input impedance of low magnitude and reversed phase,

$$Z_{in} = \frac{Z_0^2}{|Z_L|} e^{-j\psi_L}$$

This property of inverting an impedance is very useful in matching two impedances together—say Z_1 is the impedance of a generator, and $Z_2 = Z_L$ is a load—because $|Z_{in}|$ and ψ_{in} may be readily varied over a considerable range by changing x between 0 and $\lambda/2$.

Examples

1. The radar TR circuit Fig. 16-8 shows a radar transmitter and receiver (TR) connected to a common antenna. The transmitter is operative for only a small fraction of the time (~0.001 of the time for a complete cycle of operation), but during that brief instant it is necessary to protect the receiver from the extremely high amplitude of the transmitter pulse. During that part of the time when the transmitter is inoperative, on the other hand, the weak signal reflected to the antenna (by an airplane, say) should all go to the receiver instead of having part of it going to the transmitter. Both these aims are accomplished by the TR system of Fig. 16-8.

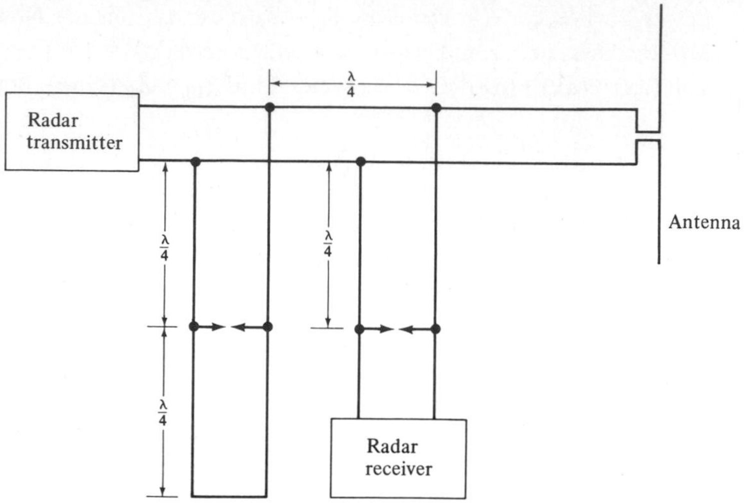

FIGURE 16-8

When the transmitter pulse is on, the pulse voltage is so high that the air insulation in an air gap between two small metal balls breaks down, a spark passes between them, and the gap is essentially a short circuit. There are two such spark gaps in this circuit, each indicated by $\rightarrow\leftarrow$, and each placed $\lambda/4$ of the way down individual stubs connected to the main transmission line and a distance $\lambda/4$ apart. The stub that is permanently short-circuited at $\lambda/2$ is now temporarily short-circuited at $\lambda/4$, so the input impedance to this shorted stub is very high. Effectively, then, it does not affect the transmitter when the large transmitter pulse (a sine wave inside a rectangular envelope) is going to the antenna.

Similarly, the spark gap in front of the receiver breaks down and the input impedance of the receiver stub is also very high. This stub does not affect the line very much either. Any energy from the main transmission line passing down this stub will go through the spark gap rather than through the higher-impedance receiver terminals. Thus, the receiver is protected from its transmitter.

When the transmitter is off and the sensitive receiver is attempting to pick up signals from the antenna, the two spark gaps are no longer conducting—they are open circuits. The stub short-circuited at $\lambda/2$ now presents a short circuit at its input terminals. The input impedance of this stub, in parallel with the left-going line to the transmitter, is also zero. Looking to the left at the junction of the receiver stub with the line ($\lambda/4$ from this stub) the line presents a very high impedance in parallel with the receiver stub. Consequently, the receiver stub is essentially connected by itself to the antenna.

2. The Smith Chart: Z_L from VSWR and V_{min} Figure 16-9 shows the configuration of a typical problem that may be met in practice. Say a generator feeds an antenna by means of a coaxial transmission line 1.72 m long; for measurement purposes a slotted section has been inserted between the generator and the transmission line, and is tied to the line by means of a connector. The line, the connector, and the slotted section all have a common characteristic impedance of 50 Ω. A minimum in the standing wave on the slotted section is observed 9 cm from the connector. The generator frequency is 750 MHz, and a voltage standing wave ratio of 3 is obtained along the slotted section. What impedance does the antenna present to the line at this frequency?

The magnitude of the reflection coefficient is

$$|\Gamma| = \frac{\text{VSWR} - 1}{\text{VSWR} + 1} = \frac{3 - 1}{3 + 1} = 0.5$$

The wavelength is

$$\lambda = \frac{c}{f} = \frac{3 \times 10^8}{750 \times 10^6} = 0.4\text{m}$$

The first voltage minimum, 0.09 m to the left of the connector, corresponds to $\frac{0.09}{0.4}$ of a wavelength, 0.225λ from the connector.

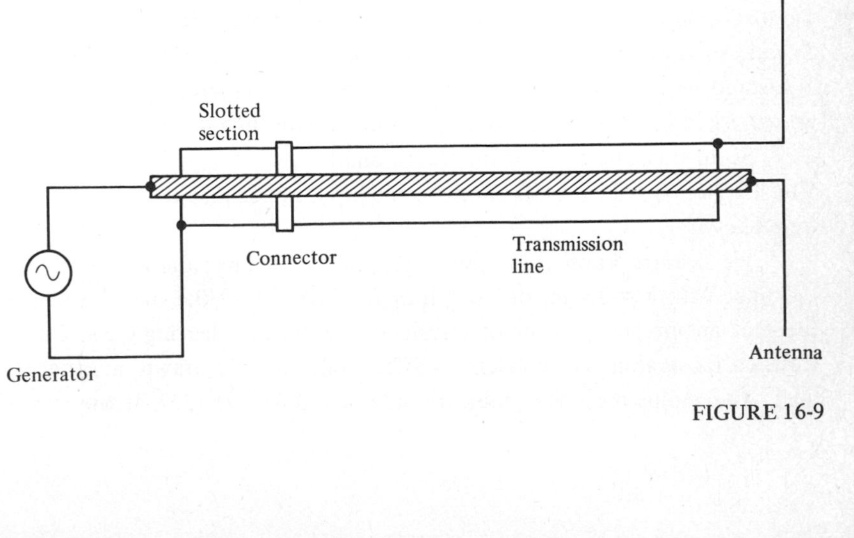

FIGURE 16-9

On the Smith chart of Fig. 16-10 the circle about the center with a radius equal to 0.5 that of the radius of the maximum circle gives the points having $|\Gamma| = 0.5$. On any such circle the point a, where it crosses the horizontal axis to the left of center, corresponds to a voltage minimum on the line, while the point b on the chart corresponds to a voltage maximum on the line. We now proceed to show this. We have

$$V = V_1 e^{-j\beta z} + V_2 e^{j\beta z}$$

$$= V_1 e^{-j\beta(\ell - x)} \left[1 + \left(\frac{V_2}{V_1} e^{j2\beta\ell} \right) e^{-j2\beta x} \right]$$

so

$$|V| = |V_1|(|1 + \Gamma_L e^{-j2\beta x}|) = |V_1|(|1 + \Gamma|)$$

The length of the line on the Smith chart from the extreme left point on the chart to any point on the circle is $|1 + \Gamma|$, and this has a minimum value for point a and a maximum value for point b.

Thus the value of ψ on *any* Smith chart for a point that lies at a voltage minimum on the line is $\psi = 180°$, i.e., at horizontal left. Similarly, a voltage maximum on the Smith chart lies along the $\psi = 0°$ direction. In the present case the voltage minimum on the slotted section of Fig. 16-9 lies at point a on the Smith chart of Fig. 16-10. The input to the transmission line proper in Fig. 16-9, at the connector, is 0.225λ toward the load, so on the Smith chart we go from point a, 0.225λ toward the load along the circle of constant Γ, arriving at point P on the chart. We may then read the $z_{in} = r_{in} + jx_{in}$ coordinates giving the normalized input impedance of the transmission line at the connector.

Since the point P on the chart may be read either as z_{in} or y_{in} it is necessary to remember one fact from the text: if the load is capacitive then a minimum lies closest to the load, while if the load is inductive then a maximum lies closest. Here a minimum lies closest, so the load is capacitive. The point P has a negative reactance or susceptance associated with it, so the former must be used. P, here, gives an impedance. In Example 3 we will see how to find the input admittance of this line, if that is desired.

So in this case at P on the Smith chart we read $r_{in} = 2.5 - jx_{in} = -j1.0$; thus Z_{in} is $50\ (2.5 - j1.0) = 125 - j50$ ohms for the input impedance to the transmission line at this frequency.

We need to know the input impedance at the antenna rather than at the input to the line. With $\lambda = 0.4$ m, the length of the line is 1.72/0.4 or 4.3 wavelengths long. We subtract an integral number of wavelengths from this, leaving 0.3λ. On the Smith chart we then move along the constant VSWR circle, already drawn, another 0.3λ toward the load. The radius must then indicate $0.225\lambda + 0.3\lambda = 0.525\lambda$ at the rim of the diagram.

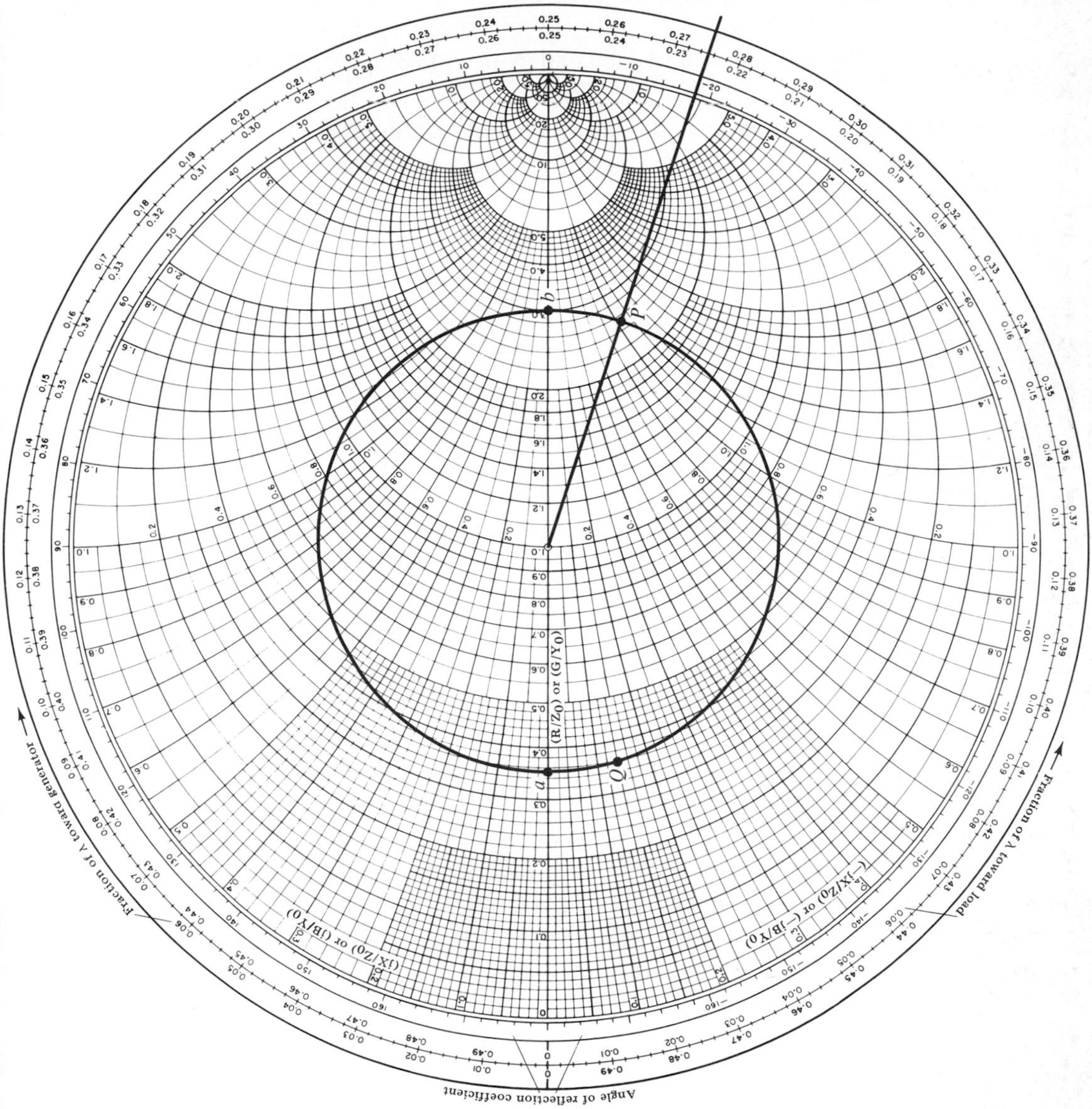

Fraction of λ toward generator
Fraction of λ toward load
Angle of reflection coefficient
(R/Z0) or (G/Y0)
(jX/Z0) or (jB/Y0)
(−jX/Z0) or (−jB/Y0)
a
b
P
Q

FIGURE 16-10

The calibration stops at 0.5λ, however, so we go to 0.025λ, giving the point Q. Here $r_{in} = 0.34$ and $-jx_{in} = -j0.14$. The antenna impedance is thus

$$50\,(0.34 - j0.14) = 17 - j7 \;\Omega$$

3. The Smith Chart: Admittance In dealing with impedances in parallel with each other it is convenient to change from an impedance Z to its inverse—the admittance $Y = 1/Z$. This is readily accomplished on a Smith chart. Suppose, for example, that $Z_{in} = 100 \;-j50$ ohms at a certain point of a line having a $Z_0 = 50$ ohms. Then $r_{in} = 2$ and $+jx_{in} = -j1$. Now $Y_{in} = Y_0\,(g_{in} + jb_{in})$, where $Y_0 = 1/Z_0 = \frac{1}{50}$ mho is the characteristic admittance; $g_{in} = r_{in}/(r_{in}^2 + x_{in}^2)$ is the normalized input conductance; and $b_{in} = -x_{in}/(r_{in}^2 + x_{in}^2)$ is the normalized input susceptance. So here

$$Y_{in} = 0.05\,(0.4 - j0.2)\text{ mho}$$

On the Smith chart of Fig. 16-11 the point P gives the normalized input impedance to the line as $r_{in} + jx_{in} = 2 - j1$. The point Q, equidistant from O but on the other side of the diameter of a circle from P, gives the normalized input admittance. The proof of this is left for one of the problems. Here $y_{in} = g_{in} + jb_{in} = 0.4 + j0.2$. Multiplying this by $Y_0 = \frac{1}{50}$ then gives Y_{in}.

This procedure is of great usefulness in stub matching, a procedure in which a stub (an auxiliary transmission line, usually either short-circuited or open-circuited at the far end) of appropriate length is placed in parallel with the line at a correct distance from an arbitrary load in ordeı to match the load to the line. Matching the load to the line means setting $|\Gamma_L| = 0$. We leave it to the problems to show that then the average power delivered to the load is a maximum. Figure 16-12 shows the geometry. To find Y_{stub} on a Smith chart we note that the short-circuited end of the stub corresponds to $Y = -j\infty$ since $Y = -jY_0 \cot \beta x$; this is represented on the Smith chart by the point at the right end of the horizontal diameter. (This is opposite the case using $Z_L = 0$ for a short-circuited load.) To find Y_{stub} we then go toward the generator, i.e., CW. (For an open-circuited stub, however, $Y = +jY_0 \tan \beta x$ gives $Y = 0$ at the open end. So the end of the stub is at the left end of the horizontal diameter of the Smith chart, and to find Y_{stub} in this case we go toward the generator, i.e., CW from here. The directions indicated on the Smith chart are always correct.)

Suppose, then, that we have a load impedance $Z_L = 100 - j50$ ohms, with $Z_0 = 50$ ohms. The point P of Fig. 16-11 (whose significance is now Z_L rather than Z_{in}, as above) then corresponds to $r_L = 2$ and $jx_L = -j1$; Q corresponds to $y_L = 0.4 + j2.0$, the load admittance. Q occurs at 0.037λ toward the generator on the chart. We wish to find the position, x, where a short-circuited stub, having the normalized $y_{stub} = 0 + jb_{stub}$, should be put in parallel with the line input admittance $y = g + jb$ such that

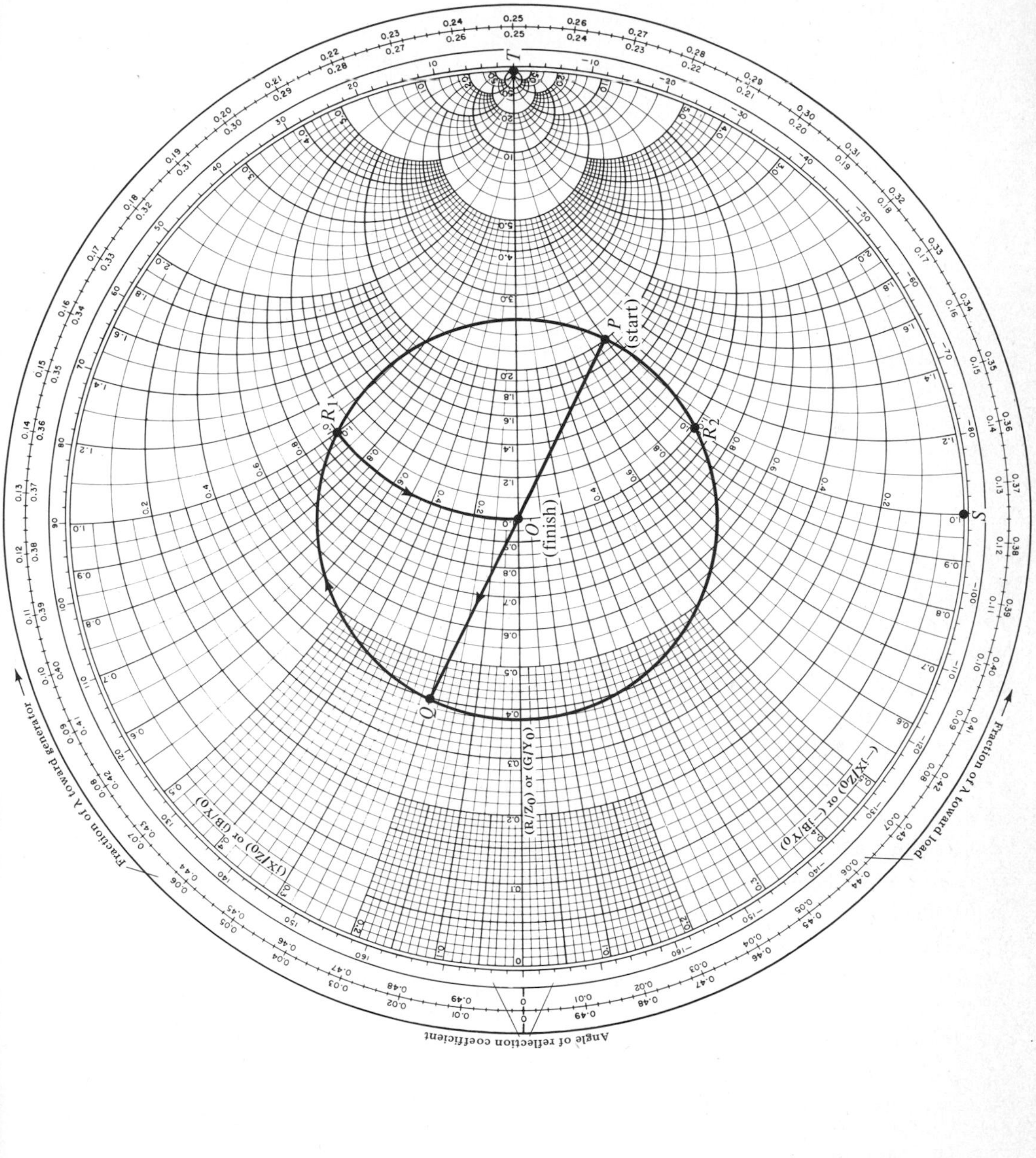

T
P
(start)
O
(finish)
Q
R_1
R_2
S
Fraction of λ toward generator
Fraction of λ toward load
Angle of reflection coefficient
(R/Z0) or (G/Y0)
(jX/Z0) or (jB/Y0)
(−jX/Z0) or (−jB/Y0)

FIGURE 16-11

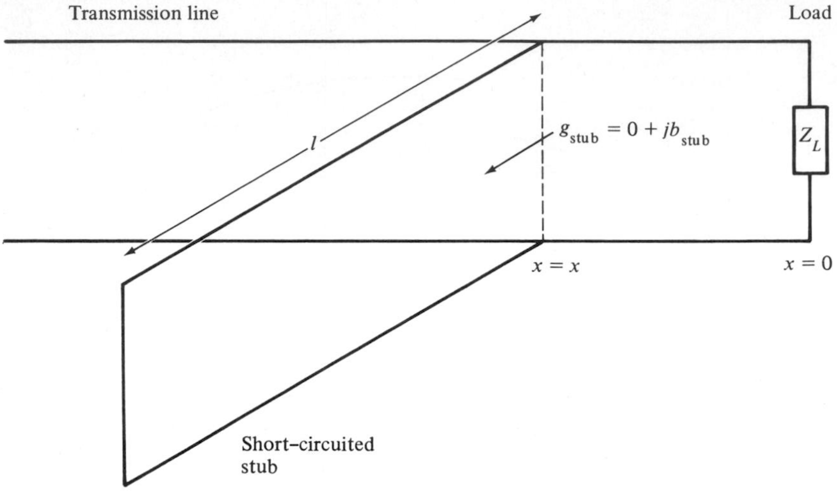

FIGURE 16-12

$$y_{\text{stub}} + y = g + j(b + b_{\text{stub}}) = 1 + j0$$

Further, we require the length of the stub.

In Fig. 16-11 the circle with O as center and OQ as radius gives the admittance at all the points along the line with this particular load. There are two points where the normalized conductance is unity. At R_1 normalized input admittance is $g + jb = 1 + j1$; this corresponds to a distance 0.162λ toward the generator. At R_2 we have $g + jb = 1 - j1$, at a distance 0.338λ toward the generator. R_1 gives a point on the line whose distance from the load is $0.162\lambda - 0.037\lambda = 0.125\lambda$. R_2 gives a point on the line distant $0.338\lambda - 0.037\lambda = 0.301\lambda$ from the load. Let us take R_1, the one closer to the load, as the place to put the stub. At $x = 0.125\lambda$ the normalized input admittance of the line and load to the right is $1 + j1$. So at $x = 0.125\lambda$ we want a short-circuited stub whose normalized input admittance is $y_{\text{stub}} = -j1$, for then the line, looking to the right toward the parallel combination, will see $(1 - j1) + (0 + j1) = 1 + j0$ for the total input admittance.

The point S on the chart having $y_{\text{stub}} = 0 - j1$ occurs at 0.375λ toward the generator; this gives the input admittance of the stub. The point T, on the chart with $y = 0 - j\infty$ at 0.25λ, corresponds to the short-circuited end of the stub. So the stub must be $0.375\lambda - 0.25\lambda = 0.125\lambda$ long.

PROBLEMS

1 The basic differential equations for the transmission line treated in this text were restricted to sinusoidal waves. To generalize this we could treat v and i as functions of z and t, giving

$$\frac{\partial v}{\partial z} = -Ri - L\frac{\partial i}{\partial t}$$

$$\frac{\partial i}{\partial z} = -Gv - C\frac{\partial v}{\partial t}$$

(*a*) Derive the partial differential equation obeyed by v.
(*b*) Repeat for i.
(*c*) Can you draw any conclusions by comparing (*a*) and (*b*)?

2 In Prob. 1, let $L = 0$ and $G = 0$. The resulting equation is known as the diffusion equation. If a steady voltage is applied at the transmitting end at $t = 0$ what will $v(t)$ be at the receiving end for $t > 0$?

3 A line has an attenuation of 2 Np. By what factor is the voltage reduced?

4 A voltage is attenuated by a factor of 10. How many nepers does this represent?

5 Suppose two voltage readings, V_1 and V_2, are made at places where the impedances are identical. Suppose $(V_2/V_1) = 3$ dB. How many nepers is this?

6 Using the typical values for R, L, G, C (per meter) for an open-wire transmission line mentioned in the text, find the values of $|Z_0|$ and the phase velocity at 1 kHz to go along with the α and β given there.

7 A transmission line has $Z_0 = 100$ ohms (resistive). If $Z_L = 50 + j150$ find Γ_L, using two methods: formula and Smith chart.

8 Show that the peripheral wavelength scales on the Smith chart give the electrical distance (in wavelengths) (*a*) from the voltage standing wave minimum, when impedance coordinates are used and (*b*) from the maximum, when admittance coordinates are used.

9 Derive an equation for the distance x_{min} from the first voltage standing wave minimum to the load in terms of the wavelength and the load reflection angle ψ_L.

10 A transmission line having $Z_0 = 50$ ohms is terminated in $Z_L = 25 - j75$ ohms. Find the reflection factor, the VSWR, and the position of the first voltage minimum.

11 Given $Z_0 = 50$ ohms and $Z_L = 100 + j200$ ohms. Find (*a*) $|\Gamma_L|$, (*b*) ψ, (*c*) the maximum $|Z_L|$ having the $|\Gamma_L|$ value found in (*a*) if any ψ is permissible, (*d*) the minimum $|Z_L|$ under these conditions.

12 Is the voltage standing wave on a transmission line having a resistive Z_0 and a complex Z_L sinusoidal? Suppose Z_L is also resistive, is it sinusoidal? Is the current standing wave sinusoidal?

13 (*a*) Find a relation between VSWR and $|z_{in}| \equiv |r_{in} + jx_{in}|_{max}$.
(*b*) Interpret this on a Smith chart.

14 Given that the VSWR = 2. On a Smith chart find the point corresponding to the voltage minimum in the standing wave.

15 Given VSWR = 3 and the first minimum in the voltage standing wave is $(1/8)\lambda$ from the load. Find: Z_L/Z_0.

16 The average power delivered to the right past a given point in a transmission line is $\langle P \rangle = \Re(\frac{1}{2} VI^*)$. For V substitute $(V_1 e^{j\beta\ell})e^{j\beta x}[1 + \Gamma_L e^{-j2\beta x}]$; then substitute a similar, but different, expression for I^* to obtain $\langle P\rangle = \frac{1}{2}\,(V_1^2/(Z_0)(1 - |\Gamma_L|^2)$. This shows that the average power is a maximum when $|\Gamma_L| = 0$, i.e., when VSWR = 1.

17 In Example 3 the short-circuited stub corresponding to point R_1 in Fig. 16-11 was located 0.125λ from the load and was 0.125λ long. Suppose, instead, that the point R_2 in Fig. 16-11 was taken.

(*a*) Where would the stub be then?

(*b*) If the stub were to be short-circuited how long should it be?

(*c*) If it were to be open-circuited how long should it be?

18 Show that for the general case of transmission line with losses, if the Heaviside condition $\omega L/R = \omega C/G$ is fulfilled, then $\alpha = \sqrt{RG}$, $\beta = \omega\sqrt{LC}$, and $v = 1/\sqrt{LC}$. The phase velocity is independent of frequency (with, in fact, the same value as that of a lossless line) so waves containing different frequency components are not distorted by dispersion. (This condition is difficult to fulfill in practice because L, R, C, G are functions of ω.)

19 A low-loss transmission line is half a wavelength long and is short-circuited at both ends.

(*a*) Suppose one of the conductors is cut at some point, leaving a small gap. A signal generator is connected to the two terminals created by the gap. What will the behavior of the impedance be between the two terminals at the gap as a function of the frequency of the generator?

(*b*) Suppose, instead, that no gap is cut in a conductor but that a generator is connected in parallel with the line, with one terminal going to one conductor at an arbitrary point and the other terminal of the generator going to the second conductor at the same point in the line. What will the behavior of Z_{in} of the line be now?

20 Prove that in a Smith chart the two points representing a z_{in} and its corresponding y_{in} are at the opposite ends of a diameter of a circle about the center of the chart.

21 Two parallel conductors, each of diameter d, have a separation s between their axes. Neglecting proximity effects, i.e., assuming $s \gg d$, (*a*) use the results of Chap. 4, Sec. 2, Prob. 9 to find the capacitance per unit length between them; (*b*) find the per-unit-length inductance from the previous result and transmission line theory; (*c*) find the corresponding quantities for a coaxial cable if a = smaller radius and b = larger radius.

22 Prove that when Z_L has a reactive component and R_{eq} is taken to be less than Z_0 then the load is capacitive if a minimum in the standing wave is closest to the load, and Z_L is inductive if a maximum is closest.

23 The distance from the load to the first minimum or maximum of the voltage standing wave is 0.15λ, and it is actually a minimum there. The VSWR is 3.

(*a*) Find Z_L.

(*b*) Is the load inductive or capacitive? Why?

(*c*) What is y_L? Use a Smith chart.

(*d*) How far from the load is the first maximum?

(*e*) Using a Smith chart, given that the first maximum is 0.4λ from the load, what is the load admittance?

24 The VSWR = 2. The first maximum comes before the first minimum: it is 0.2λ from the load. What is Z_L?

25 Use the expression

$$V = \left[\frac{2V_1 e^{-j\beta \ell}}{Z_L + Z_0}\right](Z_L \cos \beta x + jZ_0 \sin \beta x)$$

for the voltage at the input of a transmission line; and consider a $\lambda/4$ line as an impedance transformer between Z_1 and Z_L.

(*a*) From the potential difference at the load end (V_s, the secondary voltage) and the potential difference at the input (V_p, the primary voltage) find $|V_s/V_p|$ in terms of Z_L and Z_1.

(*b*) For an ordinary lumped circuit transformer $V_s/V_p = N_s/N_p$ while $I_s/I_p = N_p/N_s$. So $Z_s/Z_p = (N_s/N_p)^2$. How does the expression for V_s/V_p in terms of Z_s and Z_p for the line compare with the corresponding expression for the lumped circuits?

(*c*) Can the $\lambda/4$ line be considered a voltage transformer?

(*d*) What happens for an open-circuit load?

26 A transmitter generates a second harmonic as well as the fundamental frequency. How would one support, using metal, a transmission line to the load such that the second harmonic does not reach the load although the fundamental does?

17

REFLECTION AND REFRACTION

17-1 PLANE DISCONTINUITY BETWEEN TWO DIELECTRICS

In Fig. 17-1 a plane wave with propagation constant $\mathbf{k}_i$ is shown incident on the flat surface AA dividing a space into two regions, 1 and 2, with different dielectric constants. Suppose we take $\hat{\mathbf{n}}$ as *the* normal to AA, pointing in the same general sense as $\mathbf{k}_i$; then θ_i is the angle of incidence of the wave on the plane. On passing through the interface between the two dielectrics into the second medium the wave will be assumed to have the propagation constant $\mathbf{k}_t$, making the angle θ_t with $\hat{\mathbf{n}}$. If the incident wave is assumed plane polarized then, at any point P an infinitesimal distance from the boundary surface but inside medium 1, $\mathbf{E}_i = \mathbf{E}_{io}e^{i(\omega t - \mathbf{k}_i \cdot \mathbf{r})}$. Here $\mathbf{r}$ is the vector from the origin O to the point P and $\mathbf{E}_{io}$ is the amplitude vector of the incident wave. We will take O to lie somewhere in AA.

It was shown in Sec. 15-2 that the frequency of a wave remains unaltered in crossing a boundary. An infinitesimal distance from P, but inside medium 2, the wave must then have the form $\mathbf{E}_t = \mathbf{E}_{to}e^{i(\omega t - \mathbf{k}_t \cdot \mathbf{r})}$, with the same ω as in the $\mathbf{E}_i$ expression. Not only is the frequency the same in the two expressions but the argument of the exponent in the expression for $\mathbf{E}_t$ must be the same as that in the expression for $\mathbf{E}_i$; the

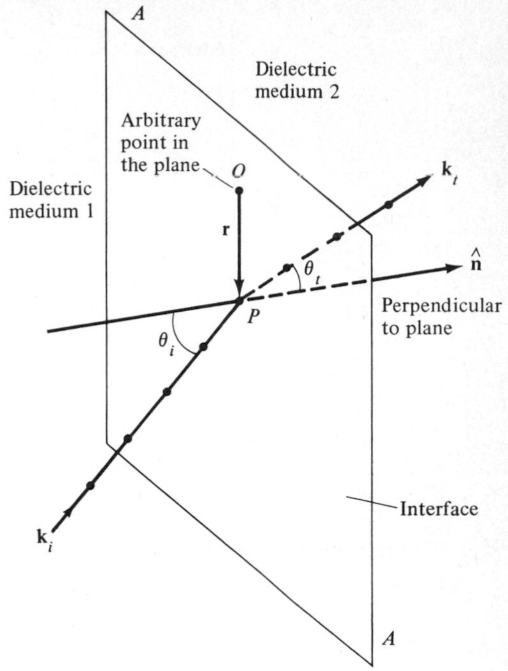

FIGURE 17-1

proof of this statement is left for one of the problems. Taking this for granted, it follows that at any instant of time $\mathbf{k}_i \cdot \mathbf{r} = \mathbf{k}_t \cdot \mathbf{r}$: the component of $\mathbf{k}_i$ along $\mathbf{r}$ must equal that of $\mathbf{k}_t$ along $\mathbf{r}$. But we may take O anywhere in AA; then the component of $\mathbf{k}_i$ parallel to any line in the interface must equal the component of $\mathbf{k}_t$ parallel to this line in the interface; in other words, the component of $\mathbf{k}_i$ parallel to the interface must equal the component of $\mathbf{k}_t$ parallel to this surface. So

$$k_i \sin \theta_i = k_t \sin \theta_t$$

$$\frac{\omega}{k_t} \sin \theta_i = \frac{\omega}{k_i} \sin \theta_t$$

$$v_t \sin \theta_i = v_i \sin \theta_t$$

Here v_i and v_t are the speeds of the wave in medium 1 and medium 2, respectively. The indices of refraction are $n_1 = c/v_i$ and $n_2 = c/v_t$; thus

$$\frac{c}{v_i} \sin \theta_i = \frac{c}{v_t} = \sin \theta_t$$

or

$$\boxed{n_1 \sin \theta_i = n_2 \sin \theta_t}$$

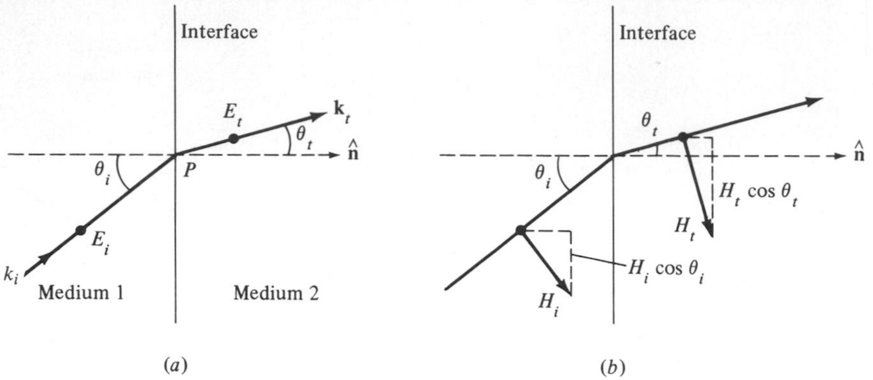

FIGURE 17-2

This is called Snell's law. If medium 1 is vacuum then $n_1 = 1$ and $n_2/n_1 = n$ is *the* index of refraction of medium 2; then

$$\frac{\sin \theta_i}{\sin \theta_t} = n$$

The vector $(\mathbf{k}_i - \mathbf{k}_t)$ is parallel to $\hat{\mathbf{n}}$, so all three vectors—$\mathbf{k}_i$, $\mathbf{k}$, and $\hat{\mathbf{n}}$—lie in one plane. This is called the plane of incidence. Suppose we now let the direction of polarization of the incident E wave be *normal* to the plane of incidence. In Fig. 17-2 this is indicated by the dots on the incident ray, representing a positive direction for the $\mathbf{E}$ vector outward from the page toward the reader. The incident ray here has $\mathbf{E}_i$ perpendicular to the plane of incidence but parallel to the surface dividing the two dielectrics.

In Chap. 6, Sec. 5 we derived the fact that the tangential component of $\mathbf{E}$ was continuous at a boundary surface. This was done, for static conditions, by utilizing one of Maxwell's equations:

$$\nabla \times \mathbf{E} = -\frac{\partial \mathbf{B}}{\partial t} = 0$$

But a slight modification of the procedure employed there yields the result that $E_{t1} = E_{t2}$ at any instant of time in all cases. From Fig. 17-2*a* we then have $E_i = E_t$ at P, where we can choose any instant of time. The dots are shown spaced apart, for convenience only; they are actually both at infinitesimal distances from P.

Figure 17-2 shows the conditions on the tangential components of $\mathbf{H}_i$ and $\mathbf{H}_t$ at P for the same instant. $\mathbf{H}_i$ is perpendicular both to $\mathbf{E}_i$ and to $\mathbf{k}_i$. It is polarized here *in* the plane of incidence (defined by $\mathbf{E}_i$ and $\hat{\mathbf{n}}$), as shown. So $H_i \cos \theta_i = H_t \cos \theta_t$. But

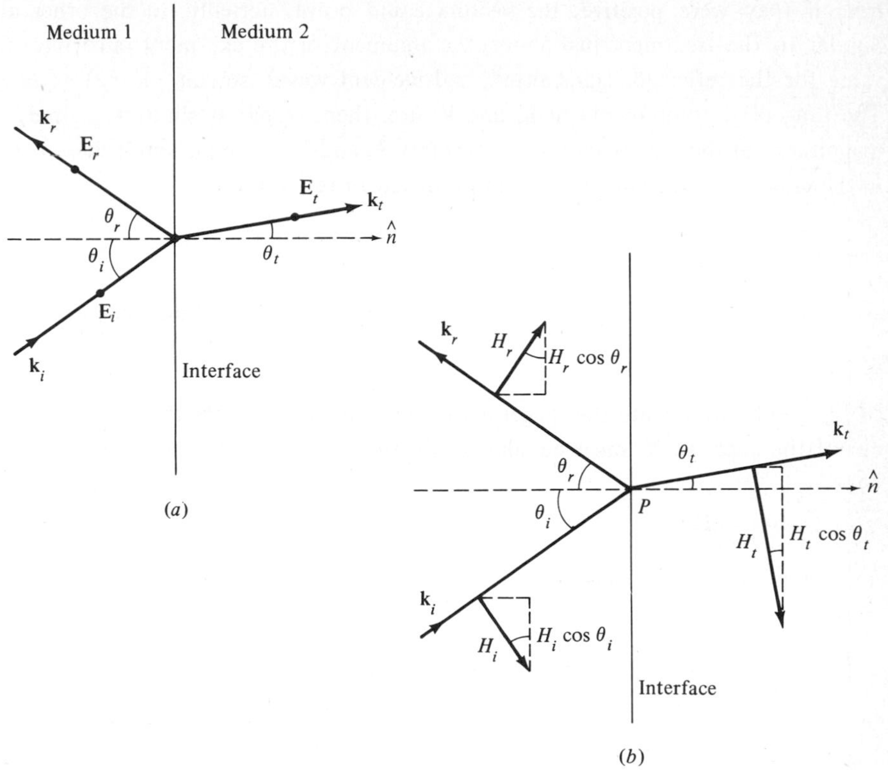

FIGURE 17-3

$$H_i = \frac{B_i}{\mu_1} = \frac{E_i/v_1}{\mu_1} = \frac{k_1}{\omega\mu_1}E_i$$

and, similarly,

$$H_t = \frac{k_2}{\omega\mu_2}E_t$$

Thus $(k_1/\omega\mu_1)E_i \cos\theta_i = (k_2/\omega\mu_2)E_t \cos\theta_t$. Taking $\mu_1 = \mu_2$ for dielectrics, and using the previous result $E_i = E_t$, we have $k_1 \cos\theta_i = k_2 \cos\theta_t$. This result disagrees with experiment; e.g., when $\theta_i = 0$ it gives $\theta_t = \cos^{-1}(k_1/k_2) \neq 0$. Something is wrong.

What is necessary to set things right is to change the very first thing—the picture! It is simply not possible, in general, to have a wave go from one medium to another by just changing the direction of the ray, with nothing else added or subtracted. In Fig. 17-3*a* we reproduce the previous results but add a reflected ray, $E_r = E_{r0}e^{i(\omega t - \mathbf{k}_r \cdot \mathbf{r})}$, with the same frequency. The dots represent the outward, positive, direction the vectors would have if they were positive; the vectors could point, actually, in the other direction. Similar to the treatment just above, the argument of the exponent must have the same value for the reflected, transmitted, and incident waves, so $(\omega t - \mathbf{k}_i \cdot \mathbf{r}) = (\omega t - \mathbf{k}_r \cdot \mathbf{r})$. The tangential components of $\mathbf{k}_i$ and $\mathbf{k}_r$ are, then, equal: $k_i \sin \theta_i = k_r \sin \theta_r$. But the magnitudes of the two propagation constants, k_i and k_r, are equal since these two rays are in the same medium. Therefore we have the law of reflection:

$$\boxed{\theta_i = \theta_r}$$

Equating tangential components of **E** at P we now have, at any instant of time,

$$E_i + E_r = E_t$$

In Fig. 17-1b we equate the tangential components of **H**. Note that E_r has been chosen out of the page, so H_r must be taken to the right, as shown. Substituting $\theta_r = \theta_i$ and also

$$H = \frac{B}{\mu} = \frac{E/v}{\mu} = \frac{k}{\omega\mu} E$$

for each of the three H fields; into

$$H_i \cos \theta_i - H_r \cos \theta_r = H_t \cos \theta_t$$

gives, together with the equation for the tangential E's,

$$\left(\frac{k_1}{\mu_1} \cos \theta_i\right) E_i = \left(\frac{k_1}{\mu_1} \cos \theta_i\right) E_r + \left(\frac{k_2}{\mu_2} \cos \theta_t\right) E_t$$

$$E_i = \qquad -E_r + \qquad E_t$$

First we solve for (E_r/E_i) and (E_t/E_i); then we set $\mu_1 = \mu_2$ (a usual, but not necessary, case) and put $k_1 = \omega n_1/c$ and $k_2 = \omega n_2/c$. This gives

$$\left(\frac{E_r}{E_i}\right) = \frac{(k_1/\mu_1) \cos \theta_i - (k_2/\mu_1) \cos \theta_t}{(k_1/\mu_1) \cos \theta_i + (k_2/\mu_2) \cos \theta_t} = \frac{n_1 \cos \theta_i - n_2 \cos \theta_t}{n_1 \cos \theta_i + n_2 \cos \theta_t}$$

$$\left(\frac{E_t}{E_i}\right) = \frac{2(k_1/\mu_1) \cos \theta_i}{(k_1/\mu_1) \cos \theta_i + k_2/\mu_2 \cos \theta_t} = \frac{2n_1 \cos \theta_i}{n_1 \cos \theta_i + n_2 \cos \theta_t}$$

But $n_1 = c/v_1 = (1/\sqrt{\epsilon_0 \mu_0})/(1/\sqrt{\epsilon_1 \mu_1}) \approx \sqrt{\epsilon_1/\epsilon_0} = \sqrt{\kappa_1}$; similarly $n_2 \simeq \sqrt{\kappa_2}$; and Snell's law now gives $\sqrt{\kappa_1} \sin \theta_i = \sqrt{\kappa_2} \sin \theta_t$, so $n_2 \cos \theta_t = \sqrt{\kappa_2}\sqrt{1 - \sin^2 \theta_t} = \sqrt{\kappa_2 - \kappa_1 \sin^2 \theta_i} = \sqrt{(\kappa_2 - \kappa_1) + \kappa_1 \cos^2 \theta_i}$. Thus

$$\left(\frac{E_r}{E_i}\right)_n = \frac{\cos \theta_i - \sqrt{\left(\dfrac{\kappa_2 - \kappa_1}{\kappa_1}\right) + \cos^2 \theta_i}}{\cos \theta_i + \sqrt{\left(\dfrac{\kappa_2 - \kappa_1}{\kappa_1}\right) + \cos^2 \theta_i}}$$

$$\left(\frac{E_t}{E_i}\right)_n = \frac{2\cos\theta_i}{\cos\theta_i + \sqrt{\left(\frac{\kappa_2 - \kappa_1}{\kappa_1}\right) + \cos^2\theta_i}}$$

This pair of equations is a variation of a set known as Fresnel's equations for plane polarized waves, here with the plane of polarization (of the **E** vector) normal to the plane of incidence. Fresnel's equations, themselves, are usually given in a form involving both θ_i and θ_t.

Figure 17-4 shows the conditions at the interface between two dielectrics when the incident wave is plane polarized *parallel* to the incident plane. We will assume a reflected wave from the start. If the **H** vectors are all taken into the plane of the page as in Fig. 17-4*b*, then the **E** vectors are as shown in Fig. 17-4*a*. So

$$E_i \cos\theta_i - E_r \cos\theta_r = E_t \cos\theta_t$$

$$H_i + H_r = H_t$$

Solving these gives

$$\left(\frac{E_r}{E_i}\right) = \frac{(k_2/\mu_2)\cos\theta_i - (k_1/\mu_1)\cos\theta_t}{(k_2/\mu_2)\cos\theta_i + (k_1/\mu_1)\cos\theta_t} = \frac{n_2\cos\theta_i - n_1\cos\theta_t}{n_2\cos\theta_i + n_1\cos\theta_t}$$

$$\left(\frac{E_t}{E_i}\right) = \frac{2(k_1/\mu_1)\cos\theta_i}{(k_2/\mu_2)\cos\theta_i + (k_1/\mu_1)\cos\theta_t} = \frac{2n_1\cos\theta_i}{n_2\cos\theta_i + n_1\cos\theta_t}$$

Putting $\mu_1 = \mu_2$, as before, and also $n_1 = \sqrt{\kappa_1}, n_2 = \sqrt{\kappa_2}$:

$$\left(\frac{E_r}{E_i}\right)_p = \frac{\frac{\kappa_2}{\kappa_1}\cos\theta_i - \sqrt{\frac{\kappa_2 - \kappa_1}{\kappa_1} + \cos^2\theta_i}}{\frac{\kappa_2}{\kappa_1}\cos\theta_i + \sqrt{\frac{\kappa_2 - \kappa_1}{\kappa_1} + \cos^2\theta_i}}$$

$$\left(\frac{E_t}{E_i}\right)_p = \frac{2\sqrt{\frac{\kappa_2}{\kappa_1}}\cos\theta_i}{\frac{\kappa_2}{\kappa_1}\cos\theta_i + \sqrt{\frac{\kappa_2 - \kappa_1}{\kappa_1} + \cos^2\theta_i}}$$

These are the modified Fresnel's equations for plane polarized waves with the plane of polarization (of **E**) parallel to the plane of incidence.

Figure 17-5*a* gives graphs of (E_{ro}/E_{io}) and (E_{to}/E_{io}) for the case of **E** plane polarized normal to the plane of incidence with $\kappa_2/\kappa_1 = 2$. Figure 17-5*b* refers to parallel polarization. Note that in both cases these are ratios of the amplitudes of the time-varying vectors.

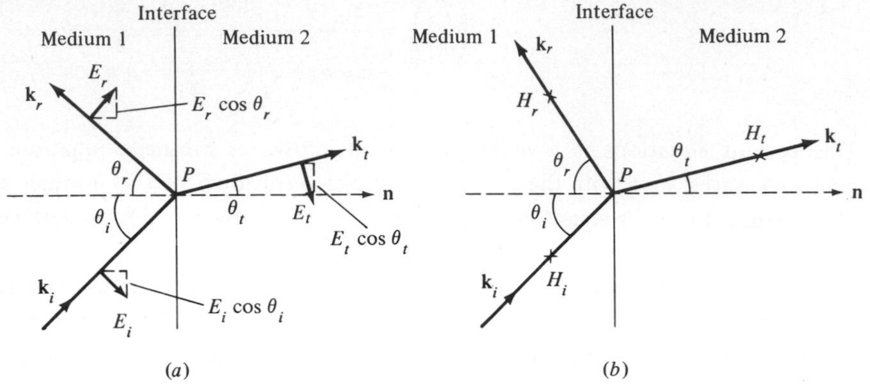

FIGURE 17-4

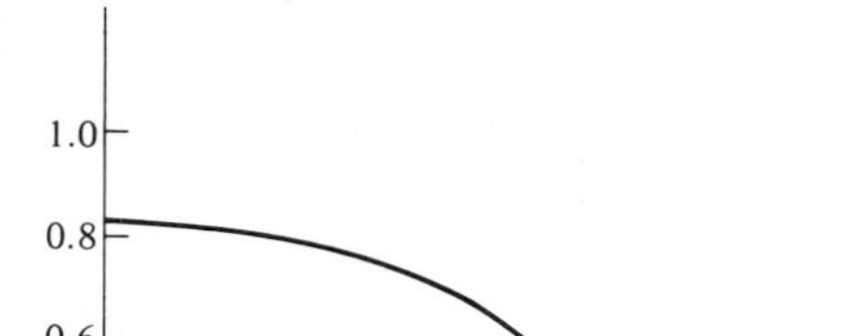

(a) Normal polarization.

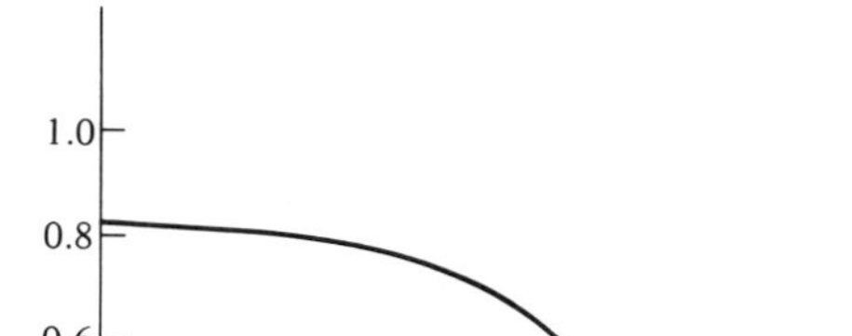

(b) Parallel polarization.

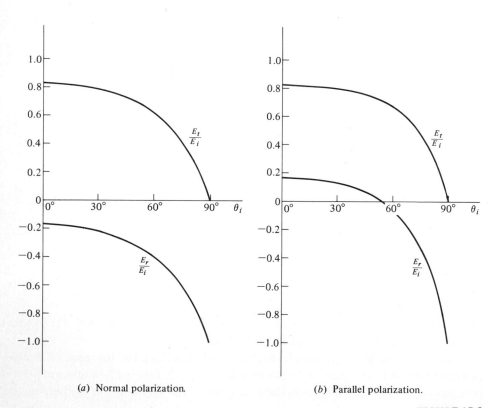

FIGURE 17-5

17-2 PLANE DISCONTINUITY BETWEEN A METAL AND A DIELECTRIC

We will only consider a plane polarized sinusoidal plane wave, traveling to the right along the x axis, incident normally on the surface of a good conductor.

To the left of the $x = 0$ plane the medium is a dielectric and we can take the incident wave as $\mathbf{E}_i = \hat{\mathbf{y}}E_{io}\, e^{i(\omega t-k_1x)}$, $\mathbf{B}_i = \hat{\mathbf{z}}B_{io}\, e^{i(\omega t-k_1x)}$ with $E_{io} = vB_{io}$. To the right of the $x = 0$ plane the medium is a metal and we have the transmitted wave

$$\mathbf{E}_t = \hat{\mathbf{y}}E_{to}e^{i[\omega t-(\beta-i\alpha)x]} = \hat{\mathbf{y}}(E_{to}e^{-\alpha x})e^{i(\omega t-\beta x)}$$

with $\beta = \alpha = \sqrt{\frac{1}{2}\mu g\omega}$. The complex propagation constant is $\mathbf{k} = |k|e^{-i\zeta}$, where $|k| = \sqrt{\mu g\omega}$ and $\zeta = 45°$. $\mathbf{B}_t$ is given by $\mathbf{B}_t = \hat{\mathbf{z}}[B_{to}e(-\alpha x)]e^{i[\omega t-\beta x-(\pi/4)]}$ with $|E_0/B_0| = \sqrt{Q}\,v'$. The procedure we must now follow is one, similar to that in the previous section, of equating the tangential components of $\mathbf{E}$ and $\mathbf{H}$ on the two sides of the plane interface. The results obtained here will have practical value when we come to consider waveguides.

Just as in the previous case, it is impossible here (even though we are exclusively considering normal incidence, $\theta_i = 0$) to have only an incident and a transmitted wave; a reflected wave is introduced by the boundary. Figure 17-6*a* shows the relationship between the vectors at two points an infinitesimal distance apart on either side of the boundary. We then have, at any instant, $E_i + E_r = E_t, H_i - H_r = H_t$. But $H_i = (k_1/\omega\mu_1)E_i$ and $H_r = (k_1/\omega\mu_1)E_r$; while

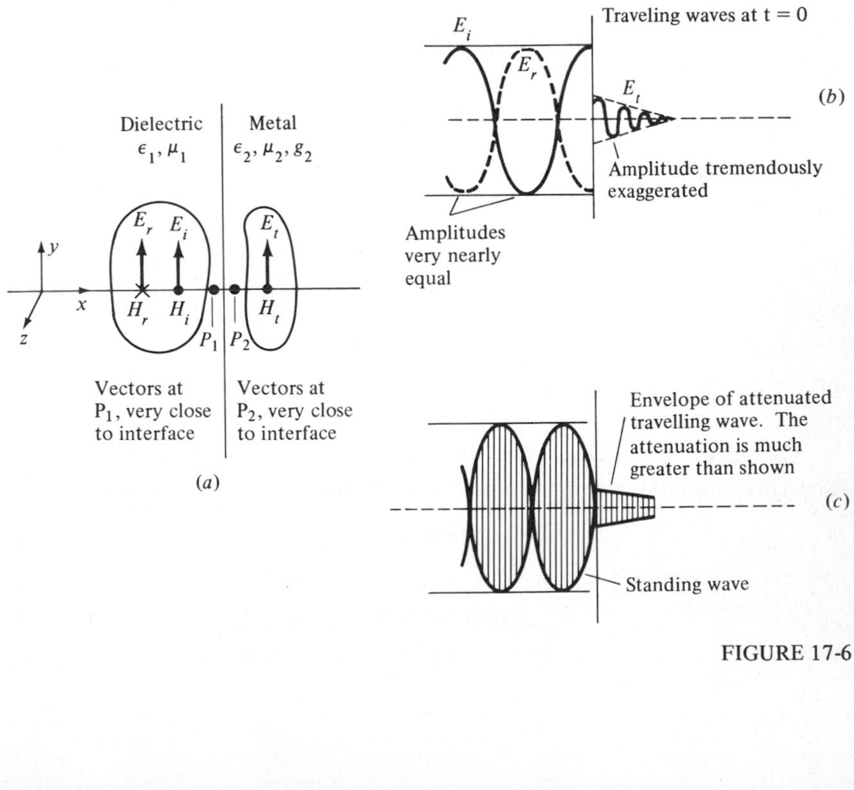

FIGURE 17-6

$$H_t = \left(\frac{1}{\mu_2}\right) B_t = \left(\frac{k}{\omega\mu_2}\right) E_t = \left[\frac{\sqrt{g\omega\mu_2} e^{i(\pi/4)}}{\omega\mu_2}\right] E_t$$

Then, with a nonferromagnetic metal, $\mu_1 \simeq \mu_2$ and

$$E_r \qquad -E_t = -E_i$$

$$kE_r + (\sqrt{g\omega\mu_2}\, e^{i\pi/4})\ E_t = kE_i$$

These give, if $a \equiv (1/k_1)\sqrt{g_2\, \omega\mu_2} e^{i\pi/4}$

$$\frac{E_r}{E_i} = \frac{-a+1}{a+1}$$

$$\frac{E_t}{E_i} = \frac{2}{a+1}$$

An insight into the practical significance of these equations is provided by taking a particular metal, say copper, with $g_2 = 5.8 \times 10^7$ mho-m^{-1}, at radio frequencies and at optical frequencies. When $f = 10^6$ Hz then $k = 2\pi/(3 \times 10^8/10^6) = 2\pi/300$; so

$$a = \frac{300}{2\pi} \sqrt{(5.8 \times 10^7)(2\pi \times 10^6)(4\pi \times 10^{-7})}\, e^{i\pi/4} = 10^6 e^{i\pi/4}$$

When $f = 5 \times 10^{14}$ Hz, similarly, $k = 3.3\pi \times 10^6$ and

$$a = \frac{10^{-6}}{3.3\pi} \sqrt{(5.8 \times 10^7)(\pi \times 10^{15})(4\pi \times 10^{-7})}\, e^{i\pi/4} = 50 e^{i\pi/4}$$

In either case we see that $|a| \gg 1$ and

$$\frac{E_r}{E_i} \simeq -1 \qquad \frac{E_t}{E_i} \simeq 0$$

The metal acts essentially like a perfect reflector at these frequencies. It is also clear that for ultraviolet frequencies, and higher, the magnitude of a becomes comparable with unity, and may even be less than unity; then, more and more, the wave is transmitted rather than reflected.

Because the reflected wave, at normal incidence, has a phase shift of 180° with an essentially undiminished amplitude, the total wave in the dielectric is a standing wave:

$$\begin{aligned}
\mathbf{E}_1 &= \mathbf{E}_i + \mathbf{E}_r = \hat{\mathbf{y}} E e^{i(\omega t - k_1 x)} - \hat{\mathbf{y}} E e^{i(\omega t + k_1 x)} \\
&= \hat{\mathbf{y}} E e^{i\omega t}(-2i \sin k_1 x) \\
\mathcal{R}(E_i) &= \mathcal{R}\,[-2\hat{\mathbf{y}} E \sin (k_1 x) e^{i(\omega t + \pi/2)}] \\
&= \hat{\mathbf{y}} 2E \sin (k_1 x) \sin (\omega t)
\end{aligned}$$

The traveling waves are shown in Fig. 17-6*b* at one particular instant; the standing waves are shown in Fig. 17-6*c*.

Within the metal the wave is

$$\mathbf{E}_t = \hat{\mathbf{y}} E e^{i[\omega t-(\beta-i\alpha)x]} = \hat{\mathbf{y}}[Ee^{-\alpha x}]e^{i[\omega t-\beta x]}$$

This traveling, attenuated, wave is also shown in Fig. 17-6*c*. Note that the wavelength is very much decreased with respect to the free space amplitude (by the factor $\sqrt{2Q} = \sqrt{2\epsilon\omega/g}$). The attenuation distance is $\delta = \sqrt{2/\mu g\omega}$, and δ/λ is very small, even at optical frequencies.

When the incident wave makes an angle other than 0° with the normal to the conductor, then the analysis is considerably more complicated and will not be dealt with here. It turns out that, regardless of the angle of incidence, the surfaces of constant amplitude are planes parallel to the surface of the conductor just as for normal incidence—the attenuation is along the normal to the surface. The surfaces of constant phase are planes which do not coincide, exactly, with the surface of constant amplitude; but for any good conductor the transmitted ray, whatever the angle of incidence, travels essentially along the normal to the surface and the two families of planes are almost precisely identical. The angular difference increases with frequency, but for copper, e.g., even at $f = 10^{11}$ Hz, the angle that the ray makes with the normal is only about $3\frac{1}{2}$ minutes of arc. For further details the interested reader is referred to "Basic Electromagnetic Theory" by D. T. Paris and F. K. Hurd (McGraw-Hill, New York, 1969, pp. 367-371).

Figure 17-7 shows the nature of the **E** and **B** fields that exist in the vicinity of a perfect conductor. Within the conductor, when $g \to \infty$, the skin depth $\delta \to 0$ and all the fields are zero. In the dielectric just outside the surface, from $\nabla \cdot D = \rho$ we obtain (as in Chap. 6, Sec. 5) $\mathbf{D} \cdot \hat{\mathbf{n}} = \sigma$. So a perpendicular component of **D**, and also of **E**, exists just as in the static case. The tangential component of **E** is continuous across the surface, so $E_{\text{tan}} = 0$. The currents in a perfect conductor all exist on the surface; otherwise **H** and **B** would be finite instead of zero. The surface current density, $\mathbf{j}_\ell$ amperes per meter of surface width, then determines **H**: $\nabla \times \mathbf{H} = \dot{\mathbf{D}} + \mathbf{J}$ here gives $H_{2,\text{tan}} - H_{1,\text{tan}} = j_c$, as in Chap. 7, Sec. 5. j_c is the component of $\mathbf{j}_\ell$ that is perpendicular to $H_{2,\text{tan}}$ and $H_{1,\text{tan}}$. Thus H_{tan} and B_{tan} exist in the dielectric near the surface. Continuity of B_{per} gives $B_{\text{per}} = H_{\text{per}} = 0$ just outside the surface.

When the conductor is good, though not perfect, the **E** and **B** fields are modified slightly from the above. E_{per} and B_{tan} now decay to some small value within the conductor in a finite distance from the surface and this is generally measured by δ. Within this skin depth, the region of transition **B** will be essentially tangential to the surface though there will be a small perpendicular component and **E** will be predominantly perpendicular though with a small tangential component. The energy within this layer will reside chiefly in the magnetic field; and there will be a 45° phase shift between **E** and

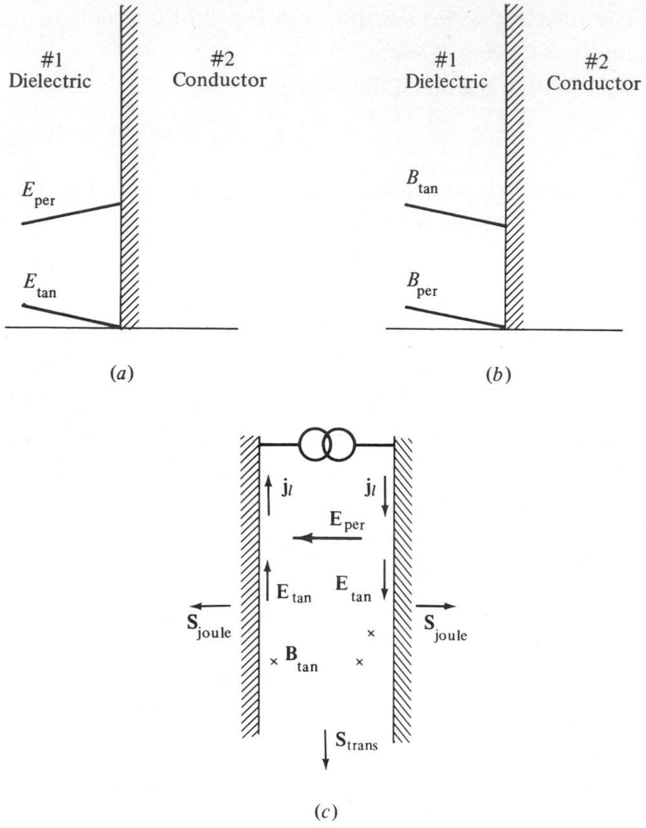

FIGURE 17-7

B. Figure 17-7*c* shows the geometric relation between the predominant vectors. These produce the large Poynting vector $\mathbf{S}_{trans}$ carrying power from the source to the load in the dielectric (and in the skin depth of the conductors). The small E_{tan} component together with B_{tan} provides the small $\mathbf{S}_{joule}$ supplying the power needed for the joule heating of the conductor.

17-3 PLANE DISCONTINUITY BETWEEN TWO METALS

Because of the tremendous attenuation of the transmitted wave, both within the metals and at their interface, this case is not of much interest. The great loss in energy of the transmitted wave occurs even when the metals are thin foils, since the attenuation at each boundary–air to metal, metal to metal, metal to air–is very large. For example, at the first one alone the transmitted power is only equal to about 0.2 of 1 percent of the incident power. See the case worked out in Sec. 17-2, above, where $|a| \approx 50$.

17-4 INHOMOGENEOUS MEDIUM

It is possible to study the reflection and refraction that occurs at the surface of discontinuity between a plasma and a dielectric; or between a plasma and a conductor; or even between two plasmas. In each case the plasma would, for simplicity, be taken uniform. One such case would be the reflection of low-frequency radio waves by the lowest Heaviside layer above the earth, beginning at an altitude of 50 km, say. For electromagnetic waves at an audio frequency of 15 kHz, e.g., the wavelength is large enough that there is a large change in the electron density in such a distance. The Heaviside layer then acts as if it had a sharp surface of discontinuity between it and the air dielectric beneath it. We may treat this case with the modified Fresnel equations developed in Sec. 17-1 above; here, however, $n_2/n_1 \neq \sqrt{\kappa_2/\kappa_1}$ but becomes a function of the wave frequency and the plasma frequency.

The lowest layer of the ionosphere has not been used extensively for this purpose because the skip distance which it makes available between transmitter and receiver is not very great. See Fig. 17-8*a*. Such a distance can generally be covered with more reliability by the direct surface wave, since the height of the *C* layer (and of the others also) varies with many factors. With the development of communication by reflection from an earth satellite, of course, the refracted sky waves, rather than the reflected ones, have become of increasing importance. These refracted waves continue upward until they meet the artificial satellite and are either passively reflected or, more commonly, actively retransmitted there, back toward the earth. This method of communication has now made all other methods of long distance transmission of signals obsolete.

The discontinuous change in n method of treatment is not the appropriate one for high-frequency sky waves. At a radio frequency of 3 MHz, for instance, the wavelength is only 100 m, an insufficient distance for the ionization density to change appreciably. The refraction of the incident wave that occurs here is due to the fact that the plasma is inhomogeneous. (Even if there were no ionization there would still be an inhomogeneous index of refraction because of the variation of molecule density with height. At ground level this gives $n \approx 1.0003$, while at 20,000 m we have $n \approx 1.00004$. This refraction is small compared to that in the plasma.) We will only treat the refraction and reflection of high-frequency waves by plasmas, as an illustration of an index of refraction which varies continuously over some distance instead of one whose value jumps discontinuously at one plane. Incidentally, for mid-frequency waves neither the continuous nor the discontinuous variation in n method is appropriate; what is employed there is one of the methods we are here bypassing: a series of strips of finite width, each having a constant n, with discontinuous change in n between adjacent strips.

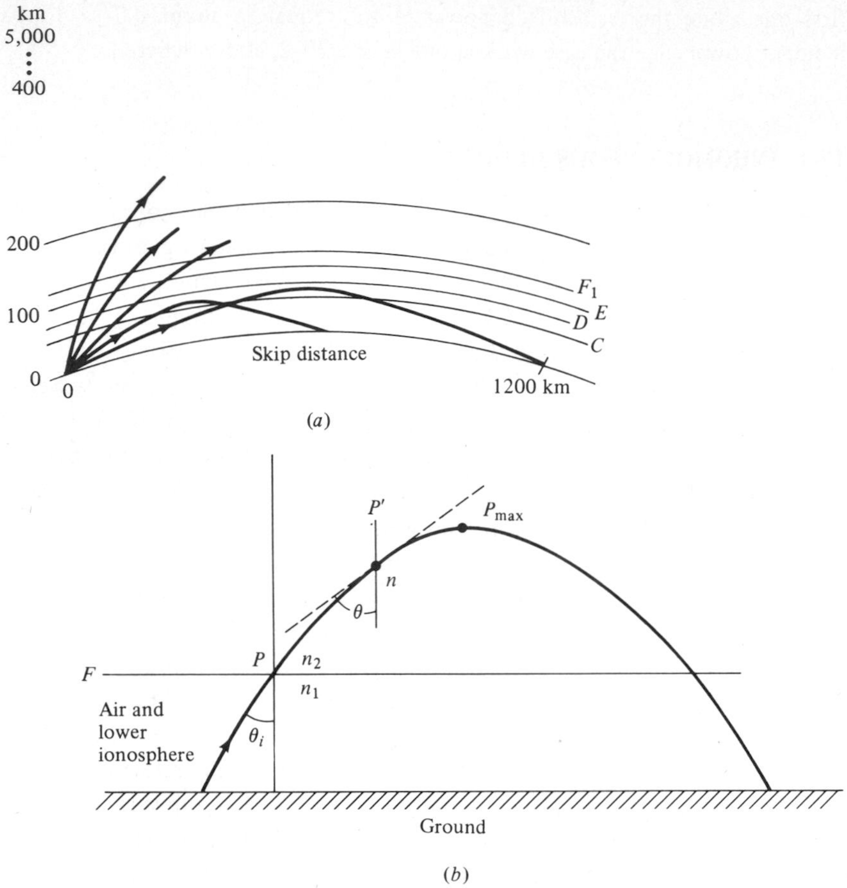

FIGURE 17-8

High-frequency refraction takes place in the upper, or F, layer of the ionosphere, where the electron density is a maximum. In the lower layers the changes in ionization are much smaller in absolute value, though they may be large relatively, and no appreciable refraction occurs for the high-frequency waves. We will consider a region where N, the number density of electrons, increases continuously with height. This effect is due to the intensity of the sun's ionizing radiation (a tenuous, fluctuating, stream of very high energy protons) which increases with altitude. In the higher regions of the F layer, N decreases with altitude since the density of molecules available to be ionized decreases with altitude. The regions of maximum densities, the layers, occur where these two effects buck each other the least.

In Fig. 17-8*b* we have a ray reaching the bottom of the F layer at P. It makes an angle θ_i with the normal, essentially the same as that it had at the transmitter. Taking a thin layer of plasma at P, we have $n_1 \sin\theta_i = n_2 \sin\theta_2$ where θ_2 is the exit angle. For the adjacent layer, farther along the trajectory $n_2 \sin\theta_2 = n_3 \sin\theta_3$. So $n_1 \sin\theta_i = n_3 \sin\theta_3$. This process may be continued, layer by layer, until we obtain $n_1 \sin\theta_i = n \sin\theta$ as the relation between P and any point P', a finite distance farther along the trajectory. Taking $n_1 = 1$ we have $n \sin\theta = \sin\theta_i$ with $n = n(\ell)$, where ℓ is the length of the curve from P to P'. The variation of θ with ℓ at P' may be found by taking derivatives with respect to ℓ for θ_i = constant.

The index of refraction is, from the wave velocity $n = c/v = \sqrt{1-(\omega_p/\omega)^2}$, less than unity. But $\omega_p = \sqrt{Ne^2/\epsilon m}$, so $n = \sqrt{1-(Ne^2/\epsilon m\omega^2)}$; since we are assuming that N increases with z we then have that n decreases with z. This is just the same as the case of the variation of n with height in normal air and just the opposite of the case of air density variation that leads to a mirage. A ray going generally upward from the earth is gradually bent away from the normal; just as in going discontinuously from a more dense to a less dense medium a ray is bent in this manner.

In order to obtain reflection back to earth it is necessary that the trajectory have a highest point, $P_{\max}$. Since there $\theta = 90°$, $\sin\theta_i = n \sin\theta$ becomes $n = \sin\theta_i$ at $P_{\max}$.

Examples

1. Total internal reflection Snell's law, $n_1 \sin\theta_1 = n_2 \sin\theta_2$, applies equally well whether the light ray is going from a less dense to a more dense medium ($n_1 < n_2$) or from a more dense to a less dense medium ($n_1 > n_2$). But only in the latter case can the interesting phenomenon of total internal reflection occur.

The modified Fresnel's equation derived above for the normalized reflected and transmitted ray amplitudes, (E_r/E_i) and (E_t/E_i), all contain the term $\sqrt{(\kappa_2/\kappa_1) - \sin^2\theta_i}$. This term is real only if $\sin^2\theta_i < (\kappa_2/\kappa_1)$, i.e., $\sin\theta_i < n_2/n_1$. When $\kappa_2 > \kappa_1$ (or $n_2 > n_1$) this condition is satisfied for all values of θ_i but when $\kappa_2 < \kappa_1$ (or $n_2 < n_1$) there is some maximum value of θ_i for which the condition is satisfied, and above this θ_i the condition is not satisfied. This is true for either type of plane polarized wave, so it is true in all cases. What happens when θ_i exceeds this maximum value, which is called the critical angle, θ_c?

For $\theta_i > \theta_c$ the radical becomes a pure imaginary number, $-iI$. (The reason for taking the minus will be discussed later in this example.) The only other terms that appear in the (E_r/E_i) equations are either $\cos\theta_i$ or $(\kappa_2/\kappa_1)\cos\theta_i$, and both are real. If we call this real number, indifferently, R then with either type of polarization

$$\left(\frac{E_r}{E_i}\right) = \frac{R+iI}{R-iI} = \frac{\sqrt{R^2+I^2}\,e^{+i\phi}}{\sqrt{R^2+I^2}\,e^{-i\phi}} = e^{+i\alpha}$$

So $|E_r| = |E_i|$ and the intensity (the average power transport) of the reflected wave will equal that of the incident wave for any $\theta \geqslant \theta_c$.

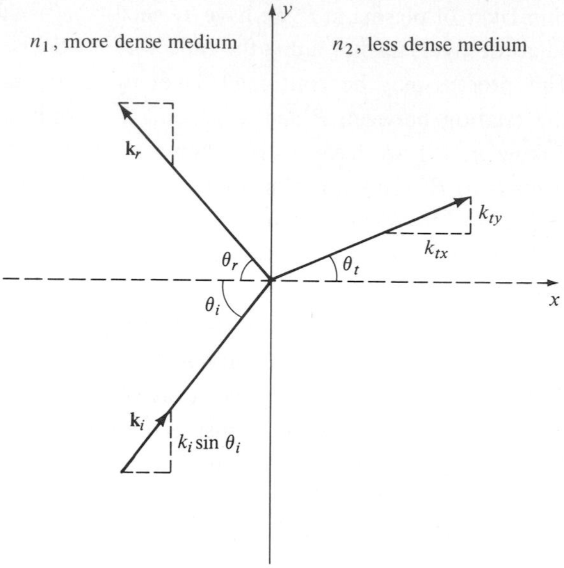

FIGURE 17-9

This fact means that the average power transport into the less-dense medium, $\langle \mathbf{S}_t \rangle$ joules per square meter per second (J m^{-2} s^{-1}), is zero; but it does *not* mean that $E_t = 0$, nor does it mean that there is no wave in the second medium. In fact, a wave does exist in the rare medium but it is a rather special type of wave: a surface wave.

Figure 17-9 shows the presumed situation with $\theta_i > \theta_c$. At present it is not necessary to concern oneself with the amplitudes of the waves, only with the phases. The phase factor of the incident wave of **E** is given by

$$\exp i(\omega t - \mathbf{k}_i \cdot \mathbf{r}) = \exp i[\omega t - (k_i \cos \theta_i)x - (k_i \sin \theta_i)y]$$

while that of the reflected wave is

$$\exp i[\omega t + (k_i \cos \theta_i)x - (k_1 \sin \theta_i)y]$$

[At the boundary both phase factors are the same: exp $i(\omega t - k_i y \sin \theta_i)$.] The corresponding expression for the transmitted wave of **E** into the less-dense medium is given by

$$\exp i(\omega t - \mathbf{k}_t \cdot \mathbf{r}) = \exp i(\omega t - k_{tx}x - k_{ty}y)$$

At all points on the surface of discontinuity the tangential component of **E** is continuous; this requires the phase factors on the two sides of the boundary to be equal for $x = 0$ and all values of y, i.e., $k_i \sin \theta_i = k_{ty}$ with $k_i = n_1/\lambda\!\!\!^{-}_0$. But $k_t^2 = k_{tx}^2 + k_{ty}^2 = 1/\lambda\!\!\!^{-}_2{}^2 = (n_2/\lambda\!\!\!^{-}_0)^2$, so

$$k_{tx} = \pm\sqrt{k_t^2 - k_{ty}^2} = \pm\sqrt{\left(\frac{n_2}{\lambdabar_0}\right)^2 - \left(\frac{n_1}{\lambdabar_0}\sin\theta_i\right)^2} = \pm\frac{n_2}{\lambdabar_0}\sqrt{1 - \left(\frac{n_1}{n_2}\right)^2 \sin^2\theta_i}$$

Since $\sin\theta_i > n_2/n_1$ here, the radical is imaginary—call it iA^2, i.e., $k_x = +i(n_2/\lambdabar_0)A^2$. Then the two possibilities for k_x give the following phase factors for the transmitted wave.

1. The plus gives

$$\exp i\left[\omega t - \left(+i\frac{n_2}{\lambdabar_0}A^2\right)x - \left(\frac{n_1}{\lambdabar_0}\sin\theta_i\right)y\right]$$
$$= \exp\left(\frac{n_2}{\lambdabar_0}A^2 x\right)\exp i\left(\omega t - \frac{n_1 y}{\lambdabar_0}\sin\theta_i\right)$$

This is a wave whose amplitude increases with penetration into the second medium so, while it is acceptable mathematically, it must be rejected on physical grounds.

2. The minus gives $\exp\,[(-n_2/\lambdabar_0)A^2x]\ \exp i[\omega t - (n_1 y/\lambdabar_0)\sin\theta_i]$. The amplitude of this wave decreases exponentially with penetration, which is acceptable. Inserting $A^2 \equiv \sqrt{(n_1/n_2)^2 \sin^2\theta_i - 1}$ we obtain

$$\exp\left[-\frac{n_2 x}{\lambdabar_0}\sqrt{\left(\frac{n_1}{n_2}\right)^2 \sin^2\theta_i - 1}\right]\exp i\left(\omega t - \frac{n_1 y}{\lambdabar_0}\sin\theta_i\right)$$

for the phase factor. The amplitude decreases exponentially in the x direction, as shown by the first factor. The second factor shows that the propagation is only along the y direction. This justifies our original statement that there is a wave in the less-dense medium. Figure 17-9 is, therefore, misleading as it stands: for $\theta_i > \theta_c$ it should be drawn with $\theta_t = 90°$.

2. Surface impedance Consider a block of metal with a surface at $x = 0$. A wave exists in the air, to the left. Suppose the electric field at the surface contains both longitudinal and tangential components. The latter, E_z, say, will be propagated within the surface as

$$E_z(x) \approx E_{z0}e^{-x/\delta}e^{i(\omega t - x/\delta)}$$

if we consider a traveling wave in the x direction. Here δ is the skin depth, $\delta = 1/\alpha = \sqrt{2/\mu g\omega}$. (Of course, if we take a very thin foil there would be a reflected wave also, traveling in the opposite direction; but if we go to a sufficiently high frequency, where the metal is more than $10\,\delta$ thick, say, there will be essentially only the one wave.) The planes of constant amplitude within the metal will be perpendicular to the x axis regardless of the angle θ_i.

From $\mathbf{J} = g\mathbf{E}$ we then have $J_z = J_0 e^{-x/\delta}e^{i(\omega t - x/\delta)}$. Then let $J_\ell = j$, for simplicity, represent the surface current density.

$$j = \int_0^\infty J_z \, dx = J_0 e^{i\omega t} \left[\frac{\delta}{1+i} \right]$$

is the current in the z direction that flows per meter width in the y direction. If the current were restricted only to the surface this would be a true surface current density; as it is, since there is a finite skin depth in the x direction, this is an equivalent surface current density.

We now define a surface impedance: $Z_\ell = E_{\tan}/j$ ohms, where $E_{\tan}$ is the amplitude of the tangential field intensity at the surface (here equal to E_{z0}) and j is the amplitude of the real or equivalent surface current density. Then

$$Z_\ell = \frac{E_{z0}}{J_0 [\delta/(1+i)]}$$

But $J_0 = gE_{z0}$, so $Z_\ell = (1 + i)/g\delta$. The real part, $1/g\delta$, is known as the surface resistance R_s, while the imaginary part is the surface reactance X_s.

$R_\ell = 1/g\delta$, the surface resistance, gives the total high-frequency resistance of a metal having unit length (from $E_{\tan}$ instead of V) and unit width (from j instead of I). If the unit for measuring length is changed from meters to anything else, say centimeters, the numerator and denominator of $E_{\tan}/j$ are affected in the same manner, so the value of Z_ℓ, and thus of R_ℓ, is invariant to such a scale factor. (But "ohm" would not be the proper unit for resistance in the CGS system.) Suppose, now, that we took a piece of this metal having unit width, unit length, and thickness δ and sent a steady, direct current through it, going from one small face to the opposite one. The dc resistance would be

$$R = \frac{\ell}{gA} = \frac{1}{g(\delta \times 1)} = \frac{1}{g\delta}$$

The dc resistance of the metal with thickness δ is, therefore, the same as its high-frequency ac resistance, the surface resistance. This has an important practical consequence. When the real or equivalent surface current density is j for the high-frequency metal then the current flowing through the dc analog, of thickness δ, is $I = j$ amperes. The power loss in the latter case is I^2R watts $= j^2 R_\ell$ watts for dc. For low-frequency ac, where the current distribution is still essentially uniform, the power loss with a current of amplitude j would be $j_{\text{eff}}^2 R_\ell$ watts. Consequently the power loss in the high-frequency case is also $j_{\text{eff}}^2 R_\ell$ W/m^2.

3. Intensities of reflected and refracted waves In Fig. 17-10 we have shown an incident beam, of rectangular cross section, $\ell \times w$, being reflected and refracted at an interface between two dielectrics. We wish to calculate the power reflection coefficient. This is defined as the ratio between (1) the reflected power passing normally back into medium

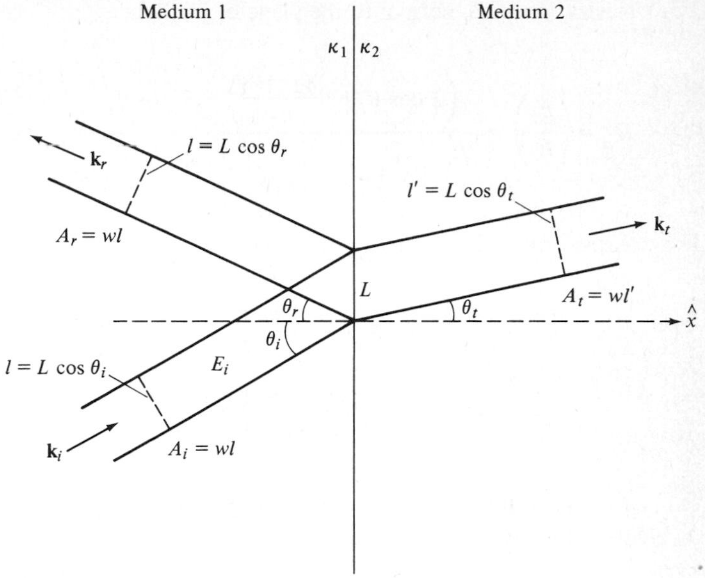

FIGURE 17-10

1, away from the interface area $L \times w$ and (2) the incident power being transported onto this interface area through medium 1. Similarly, the power transmission coefficient is the ratio of the power transmitted normally (through this area, into the second medium) to the incident power.

The Poynting vector for the incident wave is

$$\langle \mathbf{S}_i \rangle = \frac{1}{2\mu_0} \mathbf{E}_i \times \mathbf{B}_i^* = \hat{\mathbf{k}}_i \left(\frac{1}{2} \sqrt{\frac{\epsilon_1}{\mu_1}} E_i^2 \right)$$

and the component of this in the $\hat{\mathbf{x}}$ direction is

$$\langle \mathbf{S}_i \rangle_x = \frac{1}{2} \sqrt{\frac{\epsilon_1}{\mu_1}} E_i^2 \cos\theta_i$$

For the reflected wave, since $\theta_r = \theta_i$, the x component of the Poynting vector is

$$\langle \mathbf{S}_r \rangle_x = \frac{1}{2} \sqrt{\frac{\epsilon_1}{\mu_1}} E_r^2 \cos\theta_i$$

Then the power reflection coefficient is

$$R = \frac{\langle \mathbf{S}_r \rangle_x}{\langle \mathbf{S}_i \rangle_x} = \left(\frac{E_r}{E_i} \right)^2$$

For polarization of $\mathbf{E}_i$ normal to the plane of incidence

$$R_n = \left(\frac{E_r}{E_i}\right)^2 = \frac{\left(2\cos\theta_i + \frac{\kappa_2 - \kappa_1}{\kappa_1}\right) - 2\cos\theta_i\sqrt{\cos^2\theta_i + \frac{\kappa_2 - \kappa_1}{\kappa_1}}}{\left(2\cos^2\theta_i + \frac{\kappa_2 - \kappa_1}{\kappa_1}\right) + 2\cos\theta_i\sqrt{\cos^2\theta_i + \frac{\kappa_2 - \kappa_1}{\kappa_1}}}$$

For polarization of $\mathbf{E}_i$ parallel to the plane of incidence

$$R_p = \left(\frac{E_r}{E_i}\right)^2 = \frac{\left(\frac{\kappa_1^2 + \kappa_2^2}{\kappa_1^2}\cos^2\theta_i + \frac{\kappa_2 - \kappa_1}{\kappa_1}\right) - 2\frac{\kappa_2}{\kappa_1}\cos\theta_i\sqrt{\cos^2\theta_i + \frac{\kappa_2 - \kappa_1}{\kappa_1}}}{\left(\frac{\kappa_1^2 + \kappa_2^2}{\kappa_1^2}\cos^2\theta_i + \frac{\kappa_2 - \kappa_1}{\kappa_1}\right) + 2\frac{\kappa_2}{\kappa_1}\cos\theta_i\sqrt{\cos^2\theta_i + \frac{\kappa_2 - \kappa_1}{\kappa_1}}}$$

For the transmitted wave $\langle \mathbf{S}_t \rangle = \hat{\mathbf{k}}_t \frac{1}{2}\sqrt{\epsilon_2/\mu_2}E_t{}^2$ and the component of this in the $\hat{\mathbf{x}}$ direction is

$$\langle \mathbf{S}_t \rangle_x = \frac{1}{2}\sqrt{\frac{\epsilon_2}{\mu_2}}\,E_t{}^2\cos\theta_t$$

So the power transmission coefficient is

$$T = \frac{\langle \mathbf{S}_t \rangle_x}{\langle \mathbf{S}_i \rangle_x} = \frac{\frac{1}{2}\sqrt{\epsilon_2/\mu_2}E_t^2\cos\theta_t}{\frac{1}{2}\sqrt{\epsilon_1/\mu_1}E_i{}^2\cos\theta_i}$$

With $\mu_1 \approx \mu_2$ this becomes

$$T = \sqrt{\frac{\kappa_2}{\kappa_1}}\left(\frac{\cos\theta_t}{\cos\theta_i}\right)\left(\frac{E_t}{E_i}\right)^2$$

Using Snell's law $n_1 \sin\theta_i = n_2 \sin\theta_t$ we have

$$\cos\theta_t = \sqrt{1 - \sin^2\theta_t} = \sqrt{1 - \frac{\kappa_1}{\kappa_2}\sin^2\theta_i} = \sqrt{\left(\frac{\kappa_2 - \kappa_1}{\kappa_2}\right) + \left(\frac{\kappa_1}{\kappa_2}\right)\cos^2\theta_i}$$

Then, for polarization of $\mathbf{E}_i$ normal to the plane of incidence,

$$T_n = \sqrt{\frac{\kappa_2}{\kappa_1}}\,\frac{\sqrt{[(\kappa_2 - \kappa_1)/\kappa_2] + (\kappa_1/\kappa_2)\cos^2\theta_i}}{\cos\theta_i}$$

$$\frac{4\cos^2\theta_i}{[2\cos^2\theta_i + (\kappa_2 - \kappa_1)/\kappa_1] + 2\cos\theta_i\sqrt{\cos^2\theta_i + (\kappa_2 - \kappa_1)/\kappa_1}}$$

which becomes

$$T_n = \frac{4\cos\theta_i\sqrt{\cos^2\theta_i + \dfrac{\kappa_2-\kappa_1}{\kappa_1}}}{\left(2\cos^2\theta_i + \dfrac{\kappa_2-\kappa_1}{\kappa_1}\right) + 2\cos\theta_i\sqrt{\cos^2\theta_i + \dfrac{\kappa_2-\kappa_1}{\kappa_1}}}$$

For polarization of $\mathbf{E}_i$ parallel to the plane of incidence, similarly,

$$T_p = \frac{4\dfrac{\kappa_2}{\kappa_1}\cos\theta_i\sqrt{\cos^2\theta_i + \dfrac{\kappa_2-\kappa_1}{\kappa_1}}}{\left(\dfrac{\kappa_1^2+\kappa_2^2}{\kappa_1^2}\cos^2\theta_i + \dfrac{\kappa_2-\kappa_1}{\kappa_1}\right) + 2\dfrac{\kappa_2}{\kappa_1}\cos\theta_i\sqrt{\cos^2\theta_i + \dfrac{\kappa_2-\kappa_1}{\kappa_1}}}$$

For either type of polarization it is seen that $R + T = 1$.

PROBLEMS

1 (*a*) Under what conditions does $(E_r/E_i)_n = 0$ at a dielectric interface?
(*b*) Repeat (*a*) for $(E_r/E_i)_p$.
(*c*) The value of θ_i for which $E_r = 0$ is called the Brewster angle, θ_B. What is the phase shift of the reflected ray with respect to the incident ray for $\theta_i < \theta_B$ when **E** is polarized parallel to the plane of the incidence and $\kappa_2 > \kappa_1$?
(*d*) Repeat (*c*) for $\theta_i > \theta_B$.
(*e*) What is the phase shift of the reflected ray if **E** is polarized normal to the plane of incidence?

2 Form the dot product of two four-vectors, one the space-time four-vector, the other an extension of the **k** three-vector, to show that the phase factor of a traveling wave, $(\omega t - \mathbf{k}\cdot\mathbf{r})$, is an invariant.

3 (*a*) Find the speed of light in a liquid having an index of refraction equal to 1.4.
(*b*) Is it possible to have a beam of charged particles going through this liquid with a speed greater than the speed of light in the liquid?
(*c*) Greater than the speed of light in vacuum?

4 The index of refraction of a transparent dielectric may be measured by forming it into a prism. Light entering one face and leaving another, which makes the angle A with the first, will be deflected from its original direction by the angle D. Increasing θ_i, gradually, from an original small value causes D to increase to some maximum value D_{max}, then decrease. Prove that

$$n_2 = \frac{\sin\frac{1}{2}(A + D_{\text{max}})}{\sin\frac{1}{2}A}$$

5 (a) Compare the formulas for the Brewster angle θ_B and the critical angle θ_c, writing each in terms of κ_1 and $\kappa_2 > \kappa_1$. For the former the ray goes from medium 1 into medium 2; for the latter from 2 to 1.
(b) Let $\kappa = \kappa_2/\kappa_1$. Find θ_B and θ_c for $\kappa = 1, 2, 3, 4$.

6 Show that in Prob. 4, when θ_i gives D_{max}, the ray within the prism is symmetric with respect to the angle A; i.e., the bisector of A is perpendicular to the ray.

7 Using the same definition of κ in Prob. 5, find θ_B when the ray goes from medium 2 (more dense) to medium 1, again for $\kappa = 1, 2, 3, 4$.

8 Derive the Fresnel equations in the form in which they are often given:

$$\left(\frac{E_r}{E_i}\right)_n = \frac{\sin(\theta_t - \theta_i)}{\sin(\theta_t + \theta_i)}$$

$$\left(\frac{E_r}{E_i}\right)_p = \frac{\tan(\theta_t - \theta_i)}{\tan(\theta_t + \theta_i)}$$

Here one does not need to know κ_1 and κ_2; in the other form the knowledge of θ_t is unnecessary.

9 (a) Using the criterion $Q < 0.01$ for a good conductor, assume the ground (or earth) has average values of $g = 5 \times 10^{-3}$ mho-m^{-1} and $\kappa = 5$. What is the maximum frequency at which the ground is a good conductor?
(b) What is the penetration depth at this frequency?

10 Prove that when a sinusoidal wave is incident normally on a perfect conductor:
(a) **E** will form a standing wave with a node at the conductor;
(b) **B** will form a standing wave with an antinode at the conductor;
(c) the **E** standing wave will lag the **B** standing wave by 90° in time.
(d) What is the Poynting vector?

11 Suppose a wave, with **E** plane polarized normal to the plane of incidence, is incident on a perfect conductor at some nonzero angle with the normal. Then a standing wave for **E** occurs, as in normal incidence, with a node at the conductor.
(a) What is the distance between nodes along the normal to the conductor in terms of λ and θ?
(b) What is the wavelength along the normal, i.e., twice the distance between nodes in this direction? Call this λ_x.
(c) Where do the nodes occur?
(d) The antinodes?

12 (a) In Prob. 11, what is the velocity of the traveling wave in the direction tangential to the surface in terms of θ and v, the velocity of the wave along the beam?
(b) What is the wavelength of the tangential traveling wave in terms of λ and θ?
(c) Can this be larger than c?

13 Repeat the conditions of Prob. 11 but let **E** be plane-polarized parallel to the plane of incidence.
(a) Where do the nodes of **E** occur for the component of **E** perpendicular to the surface?

(*b*) Where are the antinodes for this component?

(*c*) Where do the nodes occur for the component of **E** tangential to the surface?

14 In discussing critical angle in Example 1 we derived the expression $E_r/E_i = e^{+i\alpha}$ for the case $\theta_i \geqslant \theta_c$.

(*a*) What is the physical significance of α?

(*b*) Since the value of α is different for the normal and parallel components of **E**, what does this imply for the polarization of the reflected wave?

15 In Prob. 14 obtain explicit expressions for α when **E** is polarized (*a*) normal and (*b*) parallel to the plane of incidence.

16 In Prob. 15 let the incident wave be plane polarized such that the components of **E** normal and parallel to the plane of incidence are equal. Find θ_i such that the reflected wave is circularly polarized, i.e., such that there is a difference of 90° between the two answers to Prob. 15.

17 Let s be the arc length along the trajectory in medium 2, an ionized gas, and let θ be the angle made by the tangent to the trajectory with the normal to the interface between media 1 and 2, where 1 is air. Starting with Snell's law, obtain a relation for $d\theta/ds$.

18 In determining the amplitudes of the transmitted and reflected waves at a discontinuity between dielectrics we made use of the conditions on the tangential (or parallel) components of **E** and **H** at the interface. We did this both for **E** polarized normal to the plane of incidence and for **E** polarized in the plane of incidence.

(*a*) Use the condition on the normal (or perpendicular) component of **B** (instead of the condition on **H**) together with the previous condition on **E** to determine the results for **E** polarized normal to the plane of incidence, if it is possible to do so.

(*b*) Repeat (*a*) for **E** polarized parallel to the plane of incidence, if it is possible to do so.

18
GUIDED WAVES

18-1 PARALLEL PLANES

Plane waves are the simplest ones to describe mathematically; and they are also of considerable interest because of their widespread use—in telephone communication over coaxial cables, for example. But there are other types of waves which are also important. One of these types, e.g., is that which exists within a waveguide used for conveying the microwaves between a radar transmitter-receiver and its antenna. The cross section of such a waveguide is generally rectangular, although for certain applications it could be circular. We will only consider the former case and we will limit ourselves to sinusoidal conditions.

Although the waves within a rectangular waveguide are not plane waves—when one considers only propagation along the axis of symmetry of the waveguide—it turns out that such waves may be decomposed into a sum of plane waves traveling in other directions, bouncing repeatedly off the walls of the waveguide. This is a viewpoint worth emphasizing, not only because it provides a connection with the previous work, but also because it aids in understanding the otherwise somewhat odd propagation properties of such waves. In undertaking a study of waveguides, therefore, it is a great simplification to

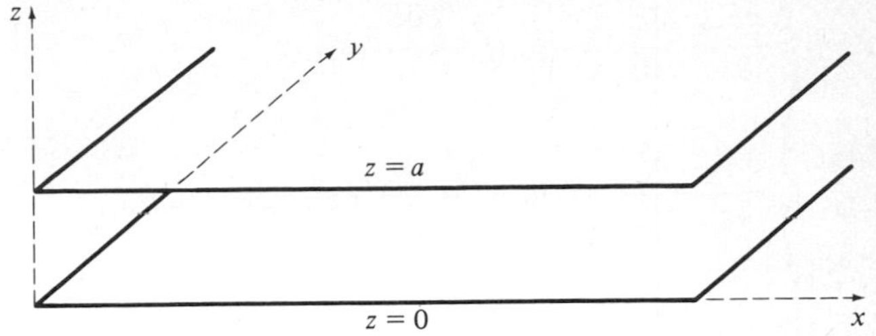

FIGURE 18-1

consider first the case of waves restricted to the slab of vacuum space between two infinite parallel, perfectly conducting planes spaced apart by the distance a, as in Fig. 18-1.

In the space between the planes $\rho = 0$ and $\mathbf{J} = 0$. We will seek only solutions to Maxwell's equations which are sinusoidal waves advancing along the $+x$ direction; i.e., all the components will be of the form $e^{j(\omega t - \beta_g x)}$, where the propagation constant β_g has been written with a subscript to distinguish it from the β that occurred previously in connection with plane waves. All the components of $\mathbf{E}$ and $\mathbf{B}$ will be taken to be independent of y. At the boundaries of this region (the inner surfaces of the perfectly conducting planes) the equation $\nabla \cdot \mathbf{B} = 0$ leads to $B_n = 0$ for the normal component of $\mathbf{B}$. This is so because $\mathbf{B} = 0$ within the conductor, where the conductivity g is very large, and an infinitesimal pillbox geometry, half in and half outside the conductor, then yields this result, just as in the case when there was no variation with time. Similarly, the Maxwell equation $\nabla \times \mathbf{E} = -\dot{\mathbf{B}}$ applied to an infinitesimal rectangle, half in and half outside the conductor, leads to $E_t = 0$, at the boundaries, for the tangential component of $\mathbf{E}$ even in the time-varying case. This is so because if $\mathbf{E} \neq 0$ within the perfectly conducting metal then infinite $\mathbf{J}$ would result.

It is now sufficient to consider the Maxwell equations $\nabla \times \mathbf{E} = -j\omega\mathbf{B}$ and $\nabla \times \mathbf{B} = +j\omega\epsilon_0\mu_0\mathbf{E}$. With our assumed behavior of $\mathbf{E}$ and $\mathbf{B}$, propagating in the x direction without attenuation (there is no loss of energy with a perfect conductor, nor does the energy density change here as the wave propagates), e.g.,

$$\frac{\partial E_z}{\partial y} = \frac{\partial}{\partial y}[E_z^{\,0}(z)\,e^{j(\omega t - \beta_g x)}] = 0$$

while

$$\frac{\partial E_z}{\partial x} = \frac{\partial}{\partial x}[E_z^{\,0}(z)\,e^{j(\omega t - \beta_g x)}] = -\,j\beta_g E_z$$

etc. Thus

$$\begin{cases} \dfrac{\partial E_z}{\partial y} - \dfrac{\partial E_y}{\partial z} = -j\omega B_x \rightarrow \dfrac{\partial E_y}{\partial z} = j\omega B_x \\[2ex] \dfrac{\partial E_x}{\partial z} - \dfrac{\partial E_z}{\partial x} = -j\omega B_y \rightarrow \dfrac{\partial E_x}{\partial z} + j\beta_g E_z = -j\omega B_y \\[2ex] \dfrac{\partial E_y}{\partial x} - \dfrac{\partial E_x}{\partial y} = -j\omega B_z \rightarrow \beta_g E_y = \omega B_z \end{cases}$$

$$\begin{cases} \dfrac{\partial B_z}{\partial y} - \dfrac{\partial B_y}{\partial z} = j\omega\epsilon_0\mu_0 E_x \rightarrow -\dfrac{\partial B_y}{\partial z} = j\omega\epsilon_0\mu_0 E_x \\[2ex] \dfrac{\partial B_x}{\partial z} - \dfrac{\partial B_z}{\partial x} = j\omega\epsilon_0\mu_0 E_y \rightarrow \dfrac{\partial B_x}{\partial z} + j\beta_g B_z = j\omega\epsilon_0\mu_0 E_y \\[2ex] \dfrac{\partial B_y}{\partial x} - \dfrac{\partial B_x}{\partial y} = j\omega\epsilon_0\mu_0 E_z \rightarrow -\beta_g B_y = \omega\epsilon_0\mu_0 E_z \end{cases}$$

These equations may be solved directly for the transverse components of **E** and **B** in terms of derivatives of the longitudinal components. Thus, putting the sixth into the second gives

$$B_y = \frac{+j\omega\epsilon_0\mu_0}{\omega^2\epsilon_0\mu_0 - \beta_g^2}\frac{\partial E_x}{\partial z}$$

and resubstituting this into the sixth gives

$$E_z = \frac{-j\beta_g}{\omega^2\epsilon_0\mu_0 - \beta_g^2}\frac{\partial E_x}{\partial z}$$

Similarly, putting the third into the fifth gives

$$E_y = \frac{-j\omega}{\omega^2\epsilon_0\mu_0 - \beta_g^2}\frac{\partial B_x}{\partial z}$$

and resubstituting this into the third gives

$$B_z = \frac{-j\beta_g}{\omega^2\epsilon_0\mu_0 - \beta_g^2}\frac{\partial B_x}{\partial z}$$

It is common practice to divide the solutions to these four equations into two sets. In the first set $E_x = 0$ but $B_x \neq 0$, so the solutions are called transverse electric, or TE, waves; in the second set $B_x = 0$ but $E_x \neq 0$ and the solutions are called transverse magnetic, or TM, waves. The general solution will be a sum of both types. Suppose we consider the first set first.

TE WAVES With $E_x = 0$ we obtain $B_y = 0$ and $E_z = 0$. For transverse electric waves between these two planes only the E_y component is present, and that one is not a function of y. But we know a differential equation that must be satisfied by each component of **E** and **B**: by taking the curls of $\nabla \times \mathbf{E} = -j\omega\mathbf{B}$ and $\nabla \times \mathbf{B} = +j\omega\epsilon_0\mu_0\mathbf{E}$ we obtain $\nabla^2\mathbf{E} = -\omega^2\epsilon_0\mu_0\mathbf{E}$ and $\nabla^2\mathbf{B} = -\omega^2\epsilon_0\mu_0\mathbf{B}$. So

$$\frac{\partial^2 E_y}{\partial x^2} + \frac{\partial^2 E_y}{\partial y^2} + \frac{\partial^2 E_y}{\partial z^2} = -\omega^2\epsilon_0\mu_0 E_y$$

The second term on the left equals zero here; the first term on the left is $(\partial/\partial x)[(\partial/\partial x)E_y{}^0\, e^{-j\beta_g x}]$ or $-\beta_g{}^2 E_y$, so

$$-\beta_g^2 E_y + \frac{\partial^2 E_y}{\partial z^2} = -\omega^2\epsilon_0\mu_0 E_y$$

$$\frac{\partial^2 E_y}{\partial z^2} = -(\omega^2\epsilon_0\mu_0 - \beta_g^2)E_y$$

E_y, the only component of **E** that exists in a TE wave, and one which is not a function of y, has its behavior in the z direction determined by the differential equation immediately above; it has been assumed to behave (like all the other components) sinusoidally in the longitudinal direction.

The solution to the $\partial^2 E_y/\partial z^2$ equation, as long as $\omega^2\epsilon_0\mu_0 - \beta_g{}^2 > 0$, is $E_y = [C_1 \sin(\sqrt{\omega^2\epsilon_0\mu_0 - \beta_g^2}\, z) + C_2 \cos(\sqrt{\omega^2\epsilon_0\mu_0 - \beta_g^2}\, z)]e^{-j\beta_g x}$, where the two constants must now be determined from the boundary conditions $E_{\text{par}} = 0$. Setting $E_y = 0$ for $z = 0$ requires that $C_2 = 0$. Setting $E_y = 0$ for $z = a$, at the top, then demands that $\sin(\sqrt{\omega^2\epsilon_0\mu_0 - \beta_g^2}\, a) = 0$, i.e., $\sqrt{\omega^2\epsilon_0\mu_0 - \beta_g^2}\, a = m\pi$ where m is an integer. Thus

$$\begin{cases} E_x = 0 \\ E_y = C_1 \sin\left(\dfrac{m\pi z}{a}\right) e^{-j\beta_g x} \\ E_z = 0 \end{cases}$$

We leave it to the problems to show that

$$\begin{cases} B_x = \left(\dfrac{-jm\pi}{\omega a}\right) C_1 \cos\left(\dfrac{m\pi z}{a}\right) e^{-j\beta_g x} \\ B_y = 0 \\ B_z = \left(\dfrac{\beta_g}{\omega}\right) C_1 \sin\left(\dfrac{m\pi z}{a}\right) e^{-j\beta_g x} \end{cases}$$

The different values of the integer m in this expression specify different modes. Actually, as we shall see later when discussing rectangular wave guides, the mode is specified by two integers, m and n. The first is connected with the distance a between the

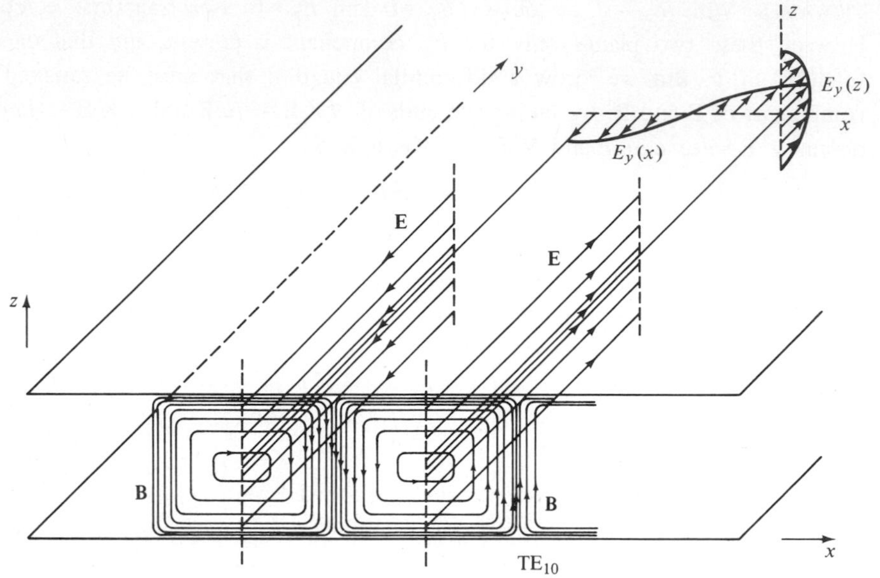

FIGURE 18-2

two conductors along the z direction; the second relates to the corresponding distance b between the two conductors along the y direction. In the present case, where b may be considered to be infinite, the value of n turns out to be zero. Therefore the modes for the two parallel planes are designated TE_{10}, TE_{20}, TE_{30}, ... Values of $m \leqslant 0$ need not be considered.

Figure 18-2 shows the configuration of the **E** and **B** fields at a given instant of time for the TE_{10} mode.

We would like to point out two facts explicitly. (1) Because of the sin $(m\pi z/a)$ and cos $(m\pi z/a)$ factors, **E** and **B** are *not* plane waves. In any yz plane, perpendicular to the x direction of propagation, the components of **E** and **B** do not all have given, fixed, values at one instant of time; the values depend on the position between the planes. (2) The waves have been assumed to be sinusoidally advancing along x, and the only way x appears in the components is in the factor $e^{-j\beta_g x}$.

TM WAVES Next we consider the second set of solutions, having $B_x = 0$ but $E_x \neq 0$. The general differential equations, applicable to either set, then show that $E_y = 0$ and $B_z = 0$. The three components of **E** and **B** that were nonvanishing for TE waves are all now equal to zero. We wish to determine the values of the other components (which equal zero for TE waves).

From $\nabla^2 \mathbf{B} = -\omega^2 \epsilon_0 \mu_0 \mathbf{B}$ we have

$$\frac{\partial^2 B_y}{\partial x^2} + \frac{\partial^2 B_y}{\partial y^2} + \frac{\partial^2 B_y}{\partial z^2} = -\omega^2 \epsilon_0 \mu_0 B_y$$

As before, this becomes

$$-\beta_g^2 B_y + \frac{\partial^2 B_y}{\partial z^2} = -\omega^2 \epsilon_0 \mu_0 B_y \qquad \text{or} \qquad \frac{\partial^2 B_y}{\partial z^2} = -(\omega^2 \epsilon_0 \mu_0 - \beta_g^2) B_y$$

If $\omega^2 \epsilon_0 \mu_0 - \beta_g{}^2 > 0$ the solution to this is

$$B_y = [C_3 \sin(\sqrt{\omega^2 \epsilon_0 \mu_0 - \beta_g^2}\, z + C_4 \cos(\sqrt{\omega^2 \epsilon_0 \mu_0 - \beta_g^2}\, z)] e^{-j\beta_g x}$$

Here, however, C_3 and C_4 cannot be determined directly from the boundary conditions since B_y is parallel to the bounding surfaces while the boundary condition on **B** applies to a perpendicular component. Instead, from $-\partial B_y/\partial_z = j\omega\epsilon_0\mu_0 E_x$ we obtain

$$E_x = \frac{j\sqrt{\omega^2 \epsilon_0 \mu_0 - \beta_g^2}}{\omega \epsilon_0 \mu_0} [C_3 \cos(\sqrt{\omega^2 \epsilon_0 \mu_0 - \beta_g^2}\, z) - C_4 \sin(\sqrt{\omega^2 \epsilon_0 \mu_0 - \beta_g^2}\, z)] e^{-j\beta_g x}$$

and E_x is a parallel component at the boundaries, to which the boundary condition $E_{\text{par}} = 0$ applies. Thus, $E_x = 0$ when $z = 0$, so C_3 must vanish. This makes

$$E_x = \left(\frac{-j\sqrt{\omega^2 \epsilon_0 \mu_0 - \beta_g^2}}{\omega \epsilon_0 \mu_0}\right) C_4 \sin(\sqrt{\omega^2 \epsilon_0 \mu_0 - \beta_g^2}\, z) e^{-j\beta_g x}$$

Setting $E_x = 0$ for $z = a$ then gives $\sqrt{\omega^2 \epsilon_0 \mu_0 - \beta_g^2}\, a = m\pi$ where m is an integer, as before. Thus

$$\begin{cases} B_x = 0 \\ B_y = C_4 \cos\left(\dfrac{m\pi z}{a}\right) e^{-j\beta_g x} \\ B_z = 0 \end{cases}$$

The equations for B_y in terms of $\partial E_x/\partial z$ and for E_z in terms of $\partial E_x/\partial z$ are then sufficient to determine the results for the **E** components:

$$\begin{cases} E_x = \left(\dfrac{-jm\pi/a}{\omega \epsilon_0 \mu_0}\right) C_4 \sin\left(\dfrac{m\pi z}{a}\right) e^{-j\beta_g x} \\ E_y = 0 \\ E_z = \left(\dfrac{-\beta_g}{\omega \epsilon_0 \mu_0}\right) C_4 \cos\left(\dfrac{m\pi z}{a}\right) e^{-j\beta_g x} \end{cases}$$

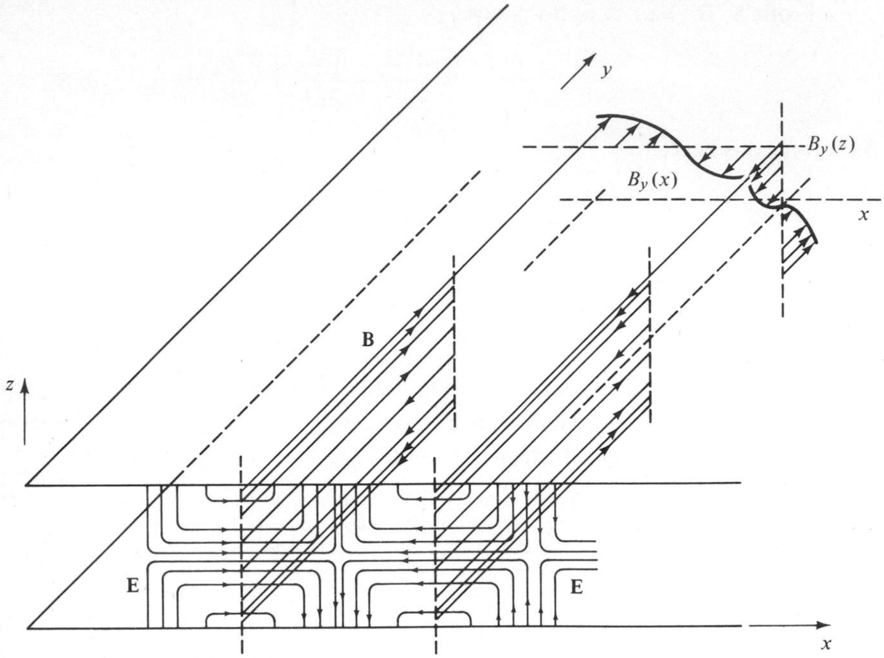

FIGURE 18-3

These equations are graphed in Fig. 18-3 for a TM_{10} wave. The reader may note a difference between the TE_{10} and TM_{10} waves in that the lines of **B** in the former form closed loops while the loops of **E** in the latter are not closed but terminate on the planes. This distinction is not significant, for it is indeed possible for the lines of **E** to be closed; and this is found in higher orders.

In either the TE or TM case a *yz* plane moves along the $+x$ direction with velocity $v_{\text{guide}} = \omega/\beta_g$. From the condition that holds true in either event, $\omega^2 \epsilon_0 \mu_0 - \beta_g{}^2 = m^2 \pi^2 / a^2$, we obtain

$$\beta_g = \sqrt{\omega^2 \epsilon_0 \mu_0 - m^2 \pi^2 / a^2}$$

Thus there is a critical angular frequency

$$\omega_c = \frac{1}{\sqrt{\epsilon_0 \mu_0}} \left(\frac{m\pi}{a} \right)$$

above this frequency β_g is real, as is v_g, and the wave progresses without attenuation (for perfectly conducting planes); below this frequency β_g becomes a pure imaginary and the common factor $e^{j(\omega t - \beta_g x)}$ in all the **E** and **B** components adopts the form $e^{-\alpha x} e^{j\omega t}$—no wave exists. Since $1/\sqrt{\epsilon_0 \mu_0}$ equals c, the velocity of an electromagnetic wave in vacuum, the critical frequency, itself, is given by

$$f_c = m\left(\frac{c}{2a}\right)$$

This is generally called the cutoff frequency. For a given mode (i.e., value of the integer m) it gives the minimum frequency that can propagate as a TE or TM wave between the planes. When such a wave does propagate, its velocity is given by

$$v_{\text{guide}} = \frac{\omega}{\sqrt{(\omega/c)^2 - (m\pi/a)^2}} = \left(\frac{f/f_c}{\sqrt{(f/f_c)^2 - 1}}\right)c$$

Figure 18-4 shows the variation of v_{guide} with f/f_c.

It is necessary to reconcile two facts: (1) the phase velocity of the wave along the guide v_{guide} is such that $v_{\text{guide}} > c$, and (2) relativity requires that any energy shall propagate with a group velocity v_{group} such that $v_{\text{group}} \leqslant c$. Consider then, as shown in Fig. 18-5, a *plane* wave propagating with velocity c in the $\hat{\mathbf{k}}_1$ direction, making an angle θ with the normal to the boundary planes. The parallel lines show the crests of several successive waves at a given instant of time; the ordinary wavelength λ, e.g., the one measured between crests along $\hat{\mathbf{k}}_1$, say, is given by $f\lambda = c$. The wavelength actually measured in all waveguides, however, is not this quantity but, rather λ_{guide}—a distance measured along the waveguide axis. Thus $\lambda_{\text{guide}} = \lambda/\sin\theta$. Now f, the frequency measured at a given point as the waves go by, is such that $f\lambda_{\text{guide}} = v_{\text{guide}}$ where v_{guide}, as above, is the velocity of propagation measured along the longitudinal waveguide axis. So $v_{\text{guide}} = (c/\sin\theta)$, and this quantity is never less than c.

In Fig. 18-5 there are shown waves moving not only in the $\hat{\mathbf{k}}_1$ direction but also waves moving in the $\hat{\mathbf{k}}_2$ direction, also making the angle θ with the normal to the surfaces. These are obtained by reflection of the incident waves on the perfectly conducting walls. Since the sum of the E_x components of the two waves must be zero at the boundaries ($E_x = 0$ just within the boundary and E_x is continuous across the boundary here) it follows that the E_x components of the two waves must be equal but opposite at the boundaries. The same holds true for the E_y components. So in Fig. 18-5 a point on the boundary that corresponds to a crest in one set corresponds to a trough in the other set.

Suppose we note the point of intersection P of the crest aa with the crest AA at one instant of time. At a later moment the first crest has moved to bb and the second has moved to BB: their point of intersection has become Q. The component of the velocity of each crest in the x direction is $c\sin\theta$, and this is also the velocity with which P moves to Q. These points correspond to a local maximization of energy density at two different instants. The velocity of the points of local minimization of energy density is also the same. In fact, this is the velocity of any given concentration of energy density; so it is also the group velocity with which the energy density moves. Then $v_{\text{group}} = c\sin\theta$. We see that $v_{\text{group}}v_{\text{guide}} = c^2$. At the cutoff frequency $v_{\text{group}} = v_{\text{guide}} = c$; at higher frequencies

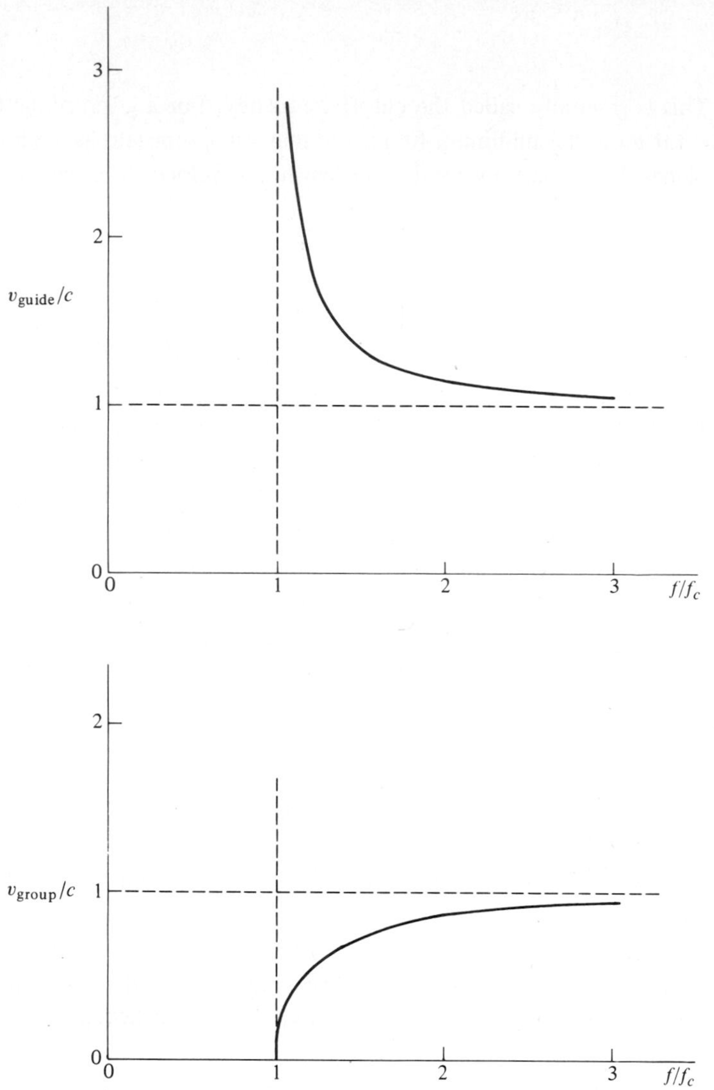

FIGURE 18-4

v_{group} is always less than c. The graph of Fig. 18-4*b* shows how (v_{group}/c) varies with (f/f_c). The requirement of relativity is seen to be satisfied.

It remains to show that the distribution of the **E** and **B** components in a *yz* plane is that actually found above in the case of the TE and TM waves. In Example 1 this is demonstrated for the case of TE waves. The two sets of plane traveling waves, in the $\hat{\mathbf{k}}_1$ and $\hat{\mathbf{k}}_2$ directions, produce a standing wave, in any *yz* plane, which moves in the $\hat{\mathbf{x}}$

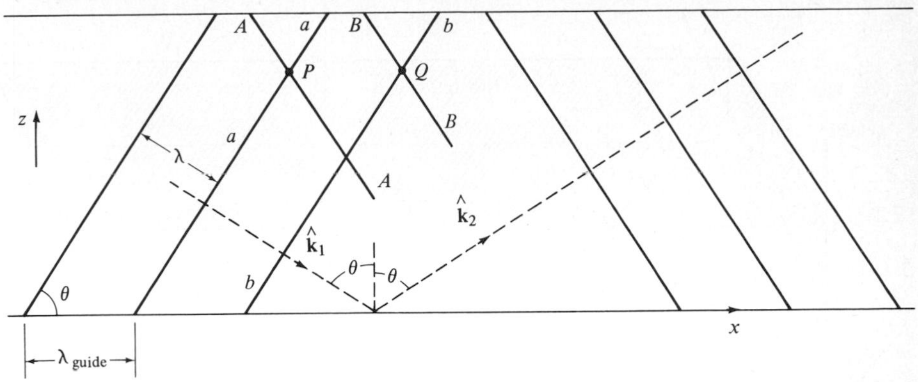

FIGURE 18-5

direction with the phase velocity magnitude v_{guide} and the group velocity magnitude v_{group}. This wave is, of course, *not* a plane wave since the values of E and B in the yz plane are not constant.

A distinction exists between the set of TE waves and the set of TM waves with respect to the $m = 0$ mode. For the TE waves all the components of **E** and **B** equal zero when $m = 0$; but for the TM waves we have, when $m = 0$, the transverse plane wave with

$$\begin{cases} E_x = 0 \\ E_y = 0 \\ E_z = \dfrac{-\beta_g}{\omega\epsilon_0\mu_0} C_4 e^{-j\beta_g x} \end{cases} \qquad \begin{cases} B_x = 0 \\ B_y = C_4 e^{-j\beta_g x} \\ B_z = 0 \end{cases}$$

Since $\beta_g{}^2 = \omega^2 \epsilon_0 \mu_0$ when $m = 0$, $\beta_g = \beta = 2\pi/\lambda_0$ then. Here λ_0 is the wavelength in vacuum of a plane wave with speed c and frequency f. (Note that $\beta_g \neq \beta$ when $m \neq 0$, which is why the subscript g was used.) So $E_z = -cC_4 e^{-j2\pi(x/\lambda)}$ and $B_y = C_4 e^{-j2\pi(x/\lambda)}$. This particular mode is more than just a transverse magnetic wave—both **E** and **B** are transverse, and the wave is a plane wave. The standing wave in the transverse plane has here degenerated to a constant. The wave is, therefore, called a TEM wave. It is, in fact, identical with the type of waves considered in Chap. 15 on infinite and semi-infinite media and in Chap. 16 along transmission lines. For the TEM wave $v_{\text{guide}} = v_{\text{group}} = c$, and the cutoff frequency is zero; i.e., all frequencies are permissible without attenuation.

The wave that propagates between the two planes considered here need not be restricted to one particular mode (i.e., value of m) or to one particular type (TE, TM, or TEM). Since the differential equations governing the propagation are all linear the sum of any solutions is also a solution. So mixtures of types and modes are also possible solutions.

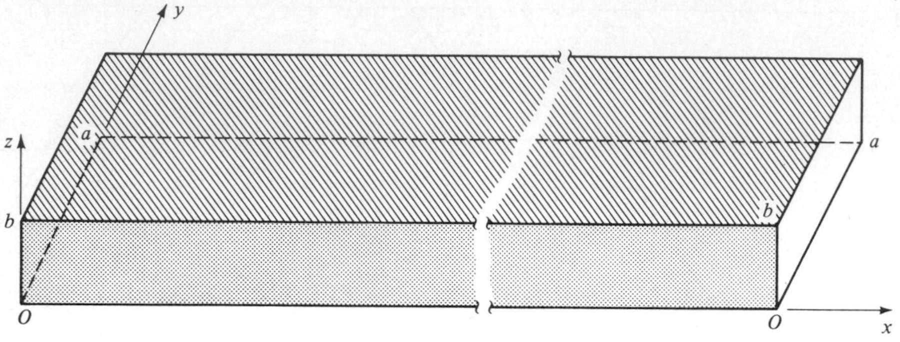

FIGURE 18-6

The fields which have been given above for the TE, TM, and TEM waves are those for **E** and **B**. These are the ones invariably used with guided waves. It is not difficult, however, to find those for ϕ and **A** in this case from $\mathbf{E} = -\nabla\phi - (\partial\mathbf{A}/\partial t)$ and $\mathbf{B} = \nabla \times \mathbf{A}$. Thus, if we set $\phi = 0$ for the TE wave we have $\mathbf{A} = \hat{\mathbf{y}}(-C/j\omega)\sin(m\pi z/a)e^{-j\beta x}$. (Like the previous waves this has the implicit $e^{j\omega t}$ factor, and only the real part is to be taken.) The appearance of **A** is similar to that of **E** in Fig. 17-2 but, because the j factor multiplying the amplitude is the same as $e^{j\pi/2}$, there is a displacement of 90° in time between the **E** and **A** patterns. For the TM wave we can also set $\phi = 0$ and easily derive an expression for **A**. In either case the Lorentz condition,

$$\nabla \cdot A + \frac{1}{c^2}\frac{\partial \phi}{\partial t} = 0$$

holds true: both terms vanish.

18-2 RECTANGULAR GUIDES

We now treat the case where, again, the propagation axis is along $\hat{\mathbf{x}}$ and the transmission region is limited between two xy planes; here, however, we also limit the transmission region to the finite range between two zx planes. See Fig. 18-6. Note that the range between 0 and a is now along the y axis, while that along the z axis is here between 0 and b. We have a rectangular waveguide, assumed to be a metal with zero resistivity so that there is no power loss and no wave attenuation.

Following, exactly, the pattern established in the previous case of two parallel planes we can write the six equations for the components of $\nabla \times \mathbf{E} = -j\omega\mathbf{B}$ and $\nabla \times \mathbf{B} = j\omega\epsilon_0\mu_0\mathbf{E}$ when each of the components of **E** and **B** depends on x only through the factor $e^{-j\beta_g x}$:

$$\begin{cases} \dfrac{\partial E_x}{\partial y} - \dfrac{\partial E_y}{\partial z} = -j\omega B_x \\ \dfrac{\partial E_x}{\partial z} + j\beta_g E_z = -j\omega B_y \\ -j\beta_g E_y - \dfrac{\partial E_x}{\partial y} = -j\omega B_z \end{cases} \qquad \begin{cases} \dfrac{\partial B_z}{\partial y} - \dfrac{\partial B_y}{\partial z} = j\omega\epsilon_0\mu_0 E_x \\ \dfrac{\partial B_x}{\partial z} + j\beta_g B_z = j\omega\epsilon_0\mu_0 E_y \\ -j\beta_g B_y - \dfrac{\partial B_x}{\partial y} = j\omega\epsilon_0\mu_0 E_z \end{cases}$$

We now obtain expressions for the transverse components in terms of transverse derivatives of the longitudinal components, proceeding exactly as before:

$$\begin{cases} B_y = \dfrac{j\omega\epsilon_0\mu_0}{\omega^2\epsilon_0\mu_0 - \beta_g^2}\dfrac{\partial E_x}{\partial z} + \dfrac{-j\beta_g}{\omega^2\epsilon_0\mu_0 - \beta_g^2}\dfrac{\partial B_x}{\partial y} \\ E_z = \dfrac{-j\beta_g}{\omega^2\epsilon_0\mu_0 - \beta_g^2}\dfrac{\partial E_x}{\partial z} + \dfrac{j\omega}{\omega^2\epsilon_0\mu_0 - \beta_g^2}\dfrac{\partial B_x}{\partial y} \\ E_y = \dfrac{-j\omega}{\omega^2\epsilon_0\mu_0 - \beta_g^2}\dfrac{\partial B_x}{\partial z} + \dfrac{-j\beta_g}{\omega^2\epsilon_0\mu_0 - \beta_g^2}\dfrac{\partial E_x}{\partial y} \\ B_z = \dfrac{-j\beta_g}{\omega^2\epsilon_0\mu_0 - \beta_g^2}\dfrac{\partial B_x}{\partial z} + \dfrac{-j\omega\epsilon_0\mu_0}{\omega^2\epsilon_0\mu_0 - \beta_g^2}\dfrac{\partial E_x}{\partial y} \end{cases}$$

We again divide the solutions to these four equations into two sets—TE waves with $E_x = 0$, and TM waves with $B_x = 0$.

TE WAVES We seek solutions to these equations of the form $B_x = B_x(y,z)$, i.e., the amplitude B_x is a function of y and z, but not of x. Using the method of separation of variables we set $B_x = Y(y)Z(z)$, where $Y(y)$ is a function only of y and $Z(z)$ is a function only of z. Next, we substitute this into the partial differential equation which B_x must satisfy (as in the case of the parallel planes):

$$\nabla^2 \mathbf{B} = -\omega^2\epsilon_0\mu_0 \mathbf{B}$$

so

$$\frac{\partial^2 B_x}{\partial x^2} + \frac{\partial^2 B_x}{\partial y^2} + \frac{\partial^2 B_x}{\partial z^2} = -\omega^2\epsilon_0\mu_0 B_x$$

and

$$-\beta_g^2 B_x + \frac{\partial^2 B_x}{\partial y^2} + \frac{\partial^2 B_x}{\partial z^2} = -\omega^2\epsilon_0\mu_0 B_x$$

or

$$(\omega^2\epsilon_0\mu_0 - \beta_g^2)B_x + \frac{\partial^2 B_x}{\partial y^2} + \frac{\partial^2 B_x}{\partial z^2} = 0$$

Thus,

$$(\omega^2\epsilon_0\mu_0 - \beta_g^2)YZ + Z\frac{d^2Y}{dy^2} + Y\frac{d^2Z}{dz^2} = 0$$

Dividing by YZ gives

$$(\omega^2 \epsilon_0 \mu_0 - \beta_g^2) + \frac{1}{Y}\frac{d^2 y}{dy^2} = -\frac{1}{Z}\frac{d^2 Z}{dz^2}$$

The left side of this equation is a function only of y and not z, while the right side is a function only of z and not y. For this to be true, at all values of y and z in the region, rather than just at one point, we must have each side equal to some given constant. We will take this constant $C_1^{\,2}$ to be positive, for this yields trigonometric functions which can be arranged to fit the boundary conditions easily:

$$(\omega^2 \epsilon_0 \mu_0 - \beta_g^2) + \frac{1}{Y}\frac{d^2 y}{dy^2} = C_1^2 \qquad \text{and} \qquad -\frac{1}{Z}\frac{d^2 Z}{dz^2} = C_1^2$$

Thus $\quad Y = C_2 \sin(\sqrt{\omega^2 \epsilon_0 \mu_0 - \beta_g^2 - C_1^2}\, y) + C_3 \cos(\sqrt{\omega^2 \epsilon_0 \mu_0 - \beta_g^2 - C_1^2}\, y)$

$$Z = C_4 \sin(C_1 z) + C_5 \cos(C_1 z)$$

and B_x is the product of these two factors.

We would now like to apply the boundary conditions, $E_{\text{par}} = 0$ and $B_{\text{per}} = 0$, to this expression in order to establish definite values for the parameters, but we cannot do this immediately, for B_x is parallel to the boundary surfaces rather than perpendicular to them. We, therefore, first obtain expressions for E_y and E_z from the bracketed equations above, knowing that $E_x = 0$ and $B_x = YZ$:

$$E_y = \left(\frac{-j\omega}{\omega^2 \epsilon_0 \mu_0 - \beta_g^2}\right) Y C_1 [C_4 \cos(C_1 z) - C_5 \sin(C_1 z)]$$

$$E_z = \left(\frac{j\omega}{\omega^2 \epsilon_0 \mu_0 - \beta_g^2}\right) Z\sqrt{\omega^2 \epsilon_0 \mu_0 - \beta_g^2 - C_1^2}\, [C_2 \cos(\sqrt{\omega^2 \epsilon_0 \mu_0 - \beta_g^2 - C_1^2}\, y)$$
$$- C_3 \sin(\sqrt{\omega^2 \epsilon_0 \mu_0 - \beta_g^2 - C_1^2}\, y)]$$

From the fact that $E_y = 0$ at $z = 0$ it follows that $C_4 = 0$ and from the fact that $E_z = 0$ at $y = 0$ it follows that $C_2 = 0$. Since $E_z = 0$ at $y = a$, $\sin(\sqrt{\omega^2 \epsilon_0 \mu_0 - \beta_g^2 - C_1^2}\, a) = 0$ and $\sqrt{\omega^2 \epsilon_0 \mu_0 - \beta_g^2 - C_1^2}\, a = m\pi$ (where m is an integer) or

$$\omega^2 \epsilon_0 \mu_0 - \beta_g^2 - C_1^2 = \left(\frac{m\pi}{a}\right)^2$$

Similarly, from the fact that $E_y = 0$ at $z = b$ it follows, since $\sin(C_1 b) = 0$, that $C_1 b = n\pi$ (n is also an integer); so $C_1 = (n\pi/b)$. Consequently

$$\omega^2 \epsilon_0 \mu_0 - \beta_g^2 = \left(\frac{m\pi}{a}\right)^2 + \left(\frac{n\pi}{b}\right)^2$$

Thus

$$Y = C_3 \cos\left(\frac{m\pi y}{a}\right) \qquad \text{and} \qquad Z = C_5 \cos\left(\frac{n\pi z}{b}\right)$$

Taking $C \equiv C_3 C_5$, this gives B_x in terms of one arbitrary constant C. Then B_y and B_z are determined from the four bracketed equations above; and so are E_y and E_z. The following results are obtained directly

For the TE case:

$$\begin{cases} E_x = 0 \\ E_y = \dfrac{j\omega(n\pi/b)}{(m\pi/a)^2 + (n\pi/b)^2} C \cos \dfrac{m\pi y}{a} \sin \dfrac{n\pi z}{b} \\ E_z = \dfrac{-j\omega(m\pi/a)}{(m\pi/a)^2 + (n\pi/b)^2} C \sin \dfrac{m\pi y}{a} \cos \dfrac{n\pi z}{b} \end{cases}$$

$$\begin{cases} B_x = C \cos \left(\dfrac{m\pi y}{a}\right) \cos \left(\dfrac{n\pi z}{b}\right) \\ B_y = \dfrac{j\beta_g(m\pi/a)}{(m\pi/a)^2 + (n\pi/b)^2} C \sin \left(\dfrac{m\pi y}{a}\right) \cos \left(\dfrac{n\pi z}{b}\right) \\ B_z = \dfrac{j\beta_g(n\pi/b)}{(m\pi/a)^2 + (n\pi/b)^2} C \cos \left(\dfrac{m\pi y}{a}\right) \sin \left(\dfrac{n\pi z}{b}\right) \end{cases}$$

Each of the components above must be multiplied by the factor $e^{j(\omega t - \beta_g x)}$, and the real parts of the resultant expressions are the final answers. Figure 18-7 shows several of these TE_{mn} modes where the m suffix is the integer in $(m\pi/a)$ along the y direction and n represents the integer in $(n\pi/b)$ along z. There is a universal convention that $a > b$, where a is the waveguide width to which the first integer m is applicable.

We leave it to Prob. 22 to show that there is no TE_{00} wave. For other values of m and n there is, as in the case of the parallel planes, a critical frequency below which β_g turns imaginary, with $e^{j\beta_g x} \to e^{-\alpha x}$. Thus, no waves can propagate except above the frequency

$$f_c = \frac{c}{2\pi} \sqrt{\left(\frac{m\pi}{a}\right)^2 + \left(\frac{n\pi}{b}\right)^2}$$

The cutoff wavelength is given by

$$\lambda_c = \frac{c}{f_c} = \frac{2}{\sqrt{(m/a)^2 + (n/b)^2}}$$

As in the case of the parallel planes, the phase velocity of propagation along the waveguide axis is c at the critical frequency; but at higher frequencies

$$\beta_g = \sqrt{\omega^2 \epsilon_0 \mu_0 - (m\pi/a)^2 - (n\pi/b)^2}$$

gives

$$\lambda_{\text{guide}} = \frac{2\pi}{\beta_g} = \frac{2\pi}{\sqrt{\omega^2 \epsilon_0 \mu_0 - (m\pi/a)^2 - (n\pi/b)^2}}$$

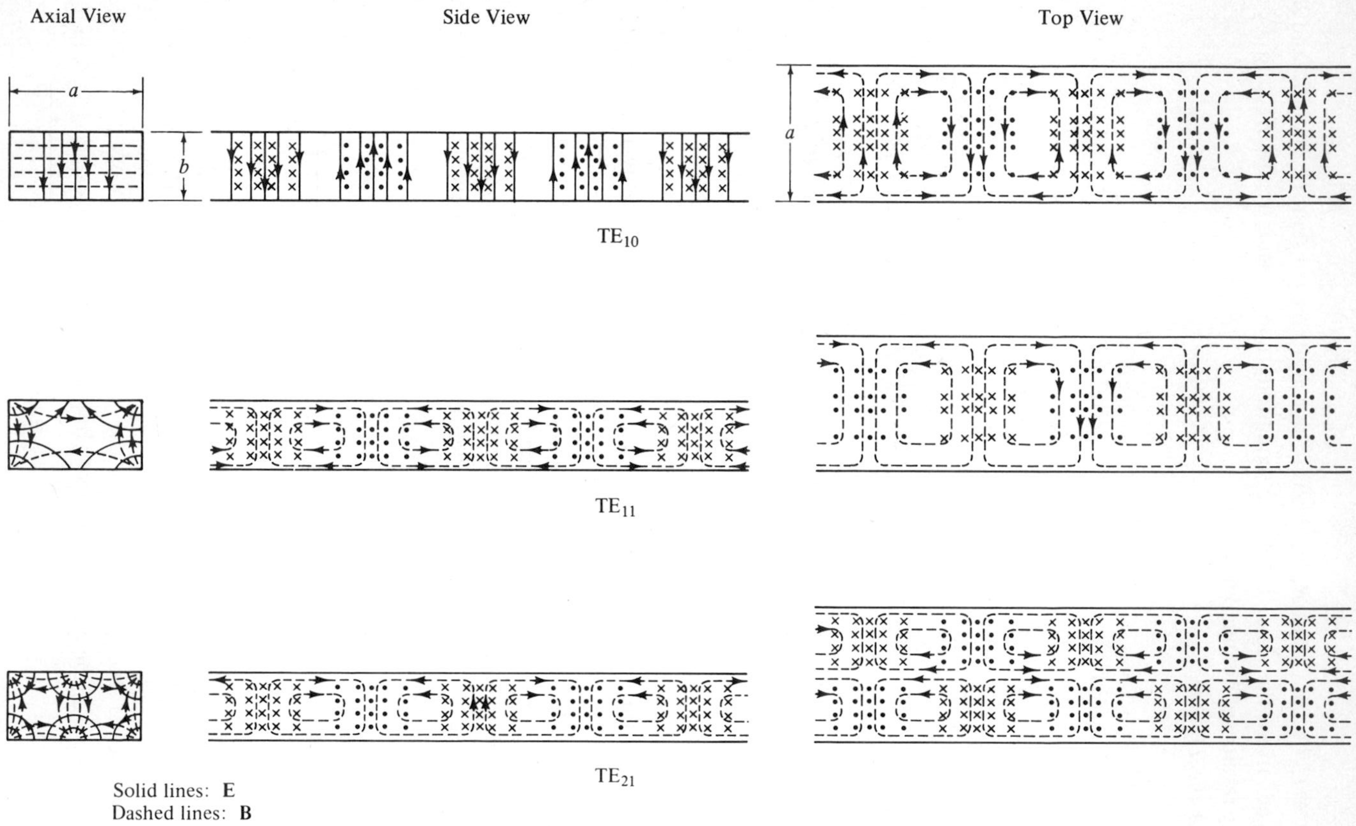
Axial View
Side View
Top View
a
b
a
TE$_{10}$
TE$_{11}$
TE$_{21}$
Solid lines: **E**
Dashed lines: **B**

FIGURE 18-7

and
$$v_{\text{guide}} = \frac{\omega}{\beta_g} = \frac{\omega}{\sqrt{\omega^2 \epsilon_0 \mu_0 - (m\pi/a)^2 - (n\pi/b)^2}}$$

Here again, $v_{\text{guide}} > c$.

TM WAVES The second set of waves of the type we seek (solutions to the bracketed four equations near the start of the discussion of rectangular guides) have $B_x = 0$ and $E_x = E_x(y,z)$. The method of separation of variables is again applied, here to $E_x(y,z) = Y(y)Z(z)$; but this case is simpler than the TE case because the boundary conditions can be applied directly to B_x.

Exactly as for TE waves we have

$$(\omega^2 \epsilon_0 \mu_0 - \beta_g^2) + \frac{1}{Y}\frac{d^2 y}{dy^2} = C_1^2 \qquad \text{and} \qquad -\frac{1}{Z}\frac{d^2 Z}{dz^2} = C_1^2$$

so
$$Y = C_2 \sin\left(\sqrt{\omega^2 \epsilon_0 \mu_0 - \beta_g^2 - C_1^2}\, y\right) + C_3 \cos\left(\sqrt{\omega^2 \epsilon_0 \mu_0 - \beta_g^2 - C_1^2}\, y\right)$$
$$Z = C_4 \sin(C_1 z) + C_5 \cos(C_1 z)$$

Since $Y = 0$ for $y = 0$ we have $C_3 = 0$; similarly, $C_5 = 0$. Since $Y = 0$ for $y = a$,

$$\sqrt{\omega^2 \epsilon_0 \mu_0 - \beta_g^2 - C_1^2}\, a = m\pi$$

Similarly $C_1 b = n\pi$. Then $\omega^2 \epsilon_0 \mu_0 - \beta_g^2 = (m\pi/a)^2 + (n\pi/b)^2$. If we set $C' \equiv C_2 C_4$, we have

$$E_x = C' \sin\left(\frac{m\pi y}{a}\right) \sin\left(\frac{n\pi z}{b}\right)$$

The bracketed four equations, mentioned previously, now yield the following results.

For the TM case:

$$\begin{cases} E_x = C' \sin\left(\frac{m\pi y}{a}\right) \sin\left(\frac{n\pi z}{b}\right) \\ E_y = \frac{-j\beta_g(m\pi/a)}{(m\pi/a)^2 + (n\pi/b)^2} C' \cos\left(\frac{m\pi y}{a}\right) \sin\left(\frac{n\pi z}{b}\right) \\ E_z = \frac{-j\beta_g(n\pi/b)}{(m\pi/a)^2 + (n\pi/b)^2} C' \sin\left(\frac{m\pi y}{a}\right) \cos\left(\frac{n\pi z}{b}\right) \end{cases}$$

$$\begin{cases} B_x = 0 \\ B_y = \frac{j\omega\epsilon_0\mu_0(n\pi/b)}{(m\pi/a)^2 + (n\pi/b)^2} C' \sin\left(\frac{m\pi y}{a}\right) \cos\left(\frac{n\pi z}{b}\right) \\ B_z = \frac{-j\omega\epsilon_0\mu_0(m\pi/a)}{(m\pi/a)^2 + (n\pi/b)^2} C' \cos\left(\frac{m\pi y}{a}\right) \sin\left(\frac{n\pi z}{b}\right) \end{cases}$$

These are the TM_{mn} components. Some are drawn in Fig. 18-8. For these waves there is again a cutoff frequency below which no propagation takes place. The formulas for f_c, and λ_c, and β_g, as well as λ_{guide} and v_{guide}, are identically the same as those obtained previously for the TE waves. Just as for TE waves, it is not possible to have TM waves with $m = n = 0$; see Prob. 22. One difference that does occur, however, is that now, with TM waves, it is not possible to have a wave with *either* $m = 0$ or $n = 0$. Thus, there is a TE_{10} wave but there is no TM_{10} wave for a rectangular waveguide. (Between infinite parallel planes, however, there is a TM_{10} mode possible.)

The fact that neither a TE_{00} wave nor a TM_{00} wave exists means that, unlike the case of propagation between conducting planes, no TEM wave can propagate through a rectangular waveguide. In fact, no TEM wave can propagate along a *hollow* waveguide of any singly connected cross-sectional shape—circular, elliptical, triangular, or any other shape—provided the fields are required to vanish in the bounding medium. The proof of this is left for Prob. 18. Note that a coaxial cable, which can transmit TEM waves, does not fall into this category: because of the central conductor the transmission cross section is not singly connected.

Like the case of transmission between parallel, perfectly conducting planes the waves transmitted along a waveguide are not plane waves; but here, too, these waves may be decomposed into pairs of plane waves bouncing back and forth between the boundary surfaces. The situation here is much more complicated, for we require two sets of such pairs—one with an xy cross section and one with an xz cross section—and it becomes impractical to pursue this viewpoint. Incidentally, instead of having the reflections occur at an air-metal interface it is possible to set up the system of waves in a dielectric and have internal reflection occur at the dielectric-air interface. Provided the angle of incidence exceeds the critical angle, total internal reflection prevents any attenuation of the wave at the boundaries. Such dielectric slab waveguides have, indeed, been made, and not only for microwave frequencies but also for optical frequencies. Of course, it is necessary to pick a dielectric in which there is not too much absorption of energy, within the dielectric itself, at the frequencies of interest.

Examples

1. Synthesis of plane waves into a guided wave In Fig. 18-9 there is shown a plane wave, with propagation vector $\mathbf{k}_1$, plane polarized with $\mathbf{E}_1$ normal to the plane of the page.

$$\mathbf{E}_1 = \mathbf{E}e^{j(\omega t - \mathbf{k}_1 \cdot \mathbf{r})} = \hat{\mathbf{y}}Ee^{j(\omega t - kx \sin\theta + kz\cos\theta)}$$

while $\mathbf{B}_1 = (\hat{\mathbf{k}}_1 \times \hat{\mathbf{y}})Be^{j(\omega t - \mathbf{k}_1 \cdot \mathbf{r})} = (\hat{\mathbf{x}}\cos\theta + \hat{\mathbf{z}}\sin\theta)Be^{j(\omega t - kx\sin\theta + kz\cos\theta)}$

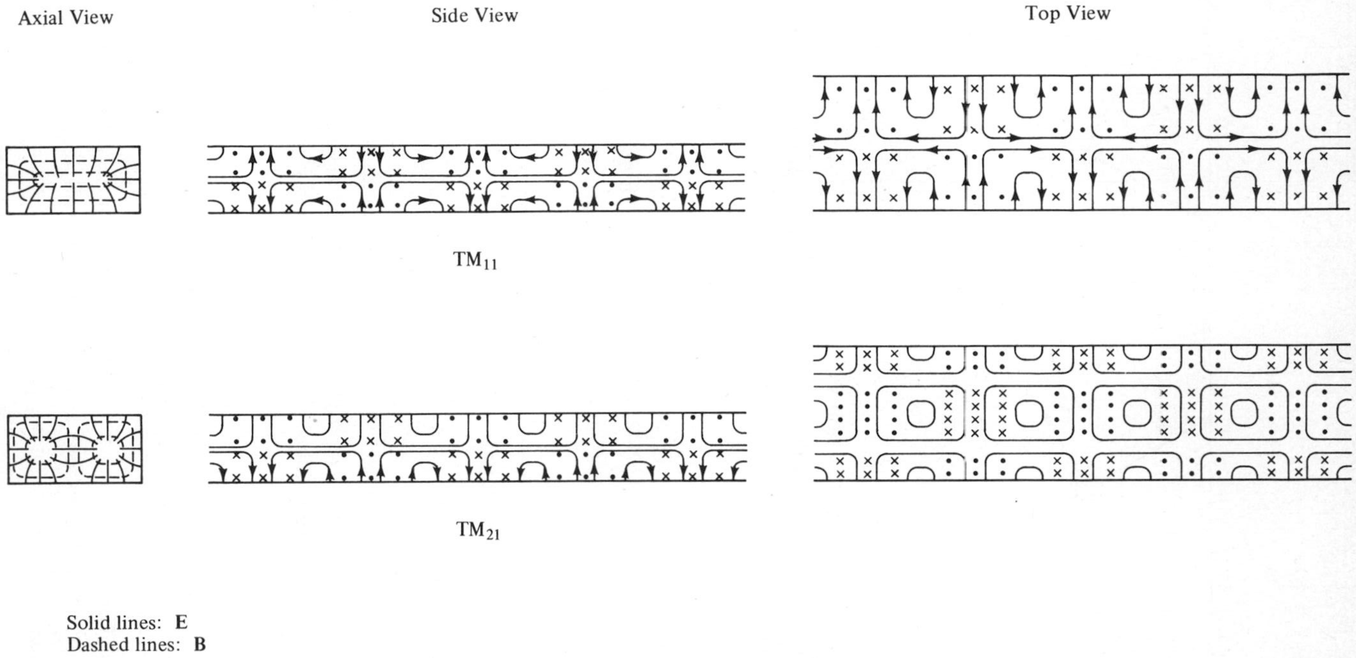
Axial View
Side View
Top View
TM$_{11}$
TM$_{21}$
Solid lines: **E**
Dashed lines: **B**

FIGURE 18-8

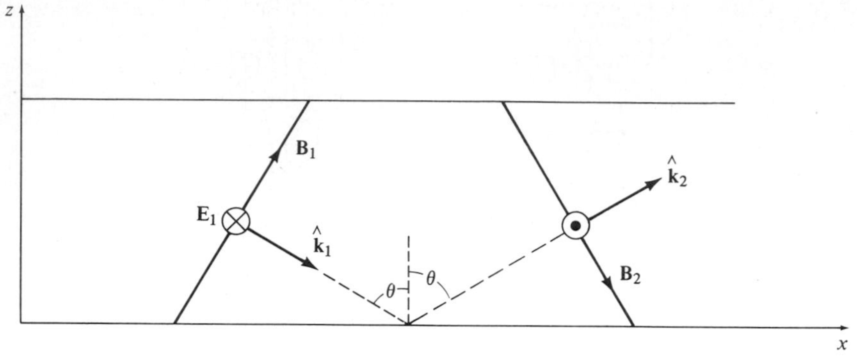

FIGURE 18-9

A second plane wave, with propagation vector $\mathbf{k}_2$, is also shown. Both $\mathbf{k}_1$ and $\mathbf{k}_2$ make the angle θ with the normal to the perfectly conducting xy plane. If we consider the first one to be an incident wave and the second one a reflected wave, there is a shift of 180° in the phase of $\mathbf{E}$ on reflection; the figure shows the positive direction for $\mathbf{E}_1$ as into the page and for $\mathbf{E}_2$ out of the page; and $\mathbf{B}_1$ and $\mathbf{B}_2$ have their positive directions determined by the $\mathbf{E}$ and $\mathbf{k}$. So $\mathbf{E}_2 = -\hat{\mathbf{y}}Ee^{j(\omega t - kx \sin\theta - kz\cos\theta)}$ and $\mathbf{B}_2 = (\hat{\mathbf{x}}\cos\theta - \hat{\mathbf{z}}\sin\theta)Be^{j(\omega t - kx\sin\theta + kz\cos\theta)}$. (Note that $\mathbf{E}$ and $\mathbf{B}$ are the same here as for the first wave, since we have assumed infinite conductivity in the reflecting planes, i.e., no losses.)

Suppose we take the real part of these wave expressions at a given instant, which we may take to be $t = 0$ for mathematical simplicity:

$$\mathbf{E}_1 = \hat{\mathbf{y}}E\cos(-kx\sin\theta + kz\cos\theta) \qquad E_2 = \hat{\mathbf{y}}E\cos(-kx\sin\theta - kz\cos\theta)$$

Now, at any given point in the region between the two conducting planes, both electric field vectors are actually present; the figure shows but one plane of each train of waves, while there are actually a large number of such planes in each train. So the electric field at any point is actually

$$\begin{aligned}\mathbf{E}_1 + \mathbf{E}_2 &= \hat{\mathbf{y}}E[\cos(-kx\sin\theta + kz\cos\theta) - \cos(-kx\sin\theta - kz\cos\theta)] \\ &= \hat{\mathbf{y}}E[2\sin(kx\sin\theta)\sin(kx\cos\theta)]\end{aligned}$$

The resultant $\mathbf{E}$ is tangential to the perfectly conducting surfaces, and it must, therefore, vanish at those surfaces. It does this at $z = 0$, but to vanish at $z = a$ it is necessary that $ka\cos\theta = m\pi$. Then

$$\mathbf{E} = \hat{\mathbf{y}}2E\sin\left(\sqrt{1 - \left(\frac{m\pi}{ka}\right)^2}\,kx\right)\sin\left(\frac{m\pi z}{a}\right)$$

If we compare this expression with the result in the text for a TE wave, viz. the real part of

$$\mathbf{E} = \hat{\mathbf{y}} C_1 \sin\left(\frac{m\pi z}{a}\right) e^{-j\beta_g x}$$

we see that this requires only that

$$C_1 \cos(\beta_g x) = 2E \sin\left(\sqrt{1 - \left(\frac{m\pi}{ka}\right)^2}\ kx\right)$$

for the two results to be identical. This must be true for all values of x and t. Then

$$\sqrt{1 - \left(\frac{m\pi}{ka}\right)^2}\ kx = \beta_g x + \frac{\pi}{2}$$

can be satisfied by taking the $(\pi/2)$ factor into the $e^{j\omega t}$: if we use $e^{j(\omega t - \beta_g x)}$ for the TE wave (i.e., a cosine space-wave at $t = 0$) we must use $e^{j(\omega t - \mathbf{k}_1 \cdot \mathbf{r} - \pi/2)}$ and $e^{j(\omega t - \mathbf{k}_2 \cdot \mathbf{r} - \pi/2)}$ for the plane waves (i.e., sine space-waves at $t = 0$). Then

$$\beta_g = \sqrt{k^2 - \frac{m^2\pi^2}{a^2}} = \sqrt{\omega^2 \epsilon_0 \mu_0 - \frac{m^2\pi^2}{2}}$$

as previously obtained in the text. Also, this gives C_1, for it is necessary that $C_1 = 2E$.

2. The charges and currents on the walls of the guides Instead of focusing attention on the fields within the waveguide, it is possible to consider guided waves from the viewpoint of the charges and currents on the boundary surfaces, as we did with transmission lines.

In general the charge density is obtained from $\nabla \cdot \mathbf{E} = (1/\epsilon_0)\rho$. But when the charge is restricted to a surface it is necessary to adapt this to σ instead of ρ; and it is advantageous to utilize the surface del operators we introduced when discussing dielectrics (Appendix 6). Thus, at a surface of discontinuity between two media

$$\nabla_{\text{surf}} \cdot \mathbf{E} = \frac{1}{\epsilon_0}\sigma$$

Now

$$\nabla_{\text{surf}} \cdot \mathbf{E} = -(\mathbf{E}_1 - \mathbf{E}_2) \cdot \hat{\mathbf{n}}_1$$

where $\hat{\mathbf{n}}_1$ points away from medium 1; if this is the metal then $\mathbf{E}_1 = 0$ and $\nabla_{\text{surf}} \cdot \mathbf{E} = \mathbf{E}_2 \cdot \hat{\mathbf{n}}_1 = \mathbf{E} \cdot \hat{\mathbf{n}}$, where $\mathbf{E}$ is in the air near the metal and $\hat{\mathbf{n}}$ points out of the metal. So

$$\nabla_{\text{surf}} \cdot \mathbf{E} = \frac{1}{\epsilon_0}\sigma$$

becomes

$$\mathbf{E} \cdot \hat{\mathbf{n}} = \frac{1}{\epsilon_0}\sigma$$

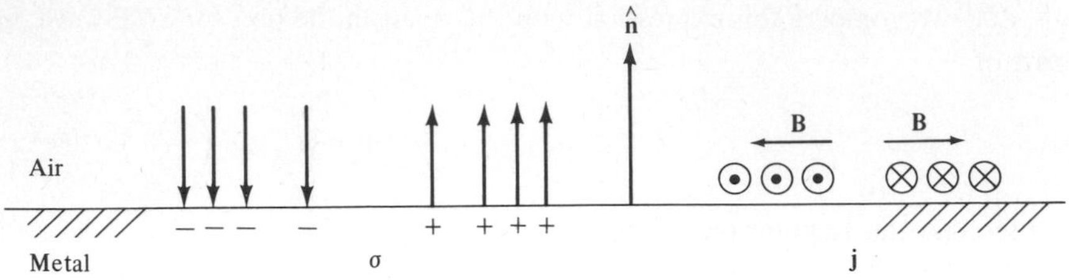

FIGURE 18-10

Similarly $\nabla \times \mathbf{B} = \mu_0 \mathbf{J} + \epsilon_0 \mu_0 (\partial \mathbf{E}/\partial t)$ becomes altered to $\nabla_{\text{surf}} \times \mathbf{B} = \mu_0 \mathbf{j}$ if $\mathbf{j}$ is a surface current density. (The $\partial \mathbf{E}/\partial t$ term contributes zero when the equation is integrated over the area of an infinitesimal rectangle.) But $\nabla_{\text{surf}} \times \mathbf{B} = (\mathbf{B}_1 - \mathbf{B}_2) \times \hat{\mathbf{n}}_1$; here this is $-\mathbf{B}_2 \times \hat{\mathbf{n}}_1 = -\mathbf{B} \times \hat{\mathbf{n}}$ where $\mathbf{B}$ is in the air near the metal. So $\nabla_{\text{surf}} \times \mathbf{B} = \mu_0 \mathbf{j}$ becomes $-\mathbf{B} \times \hat{\mathbf{n}} = \mu_0 \mathbf{j}$.

We start, then, with $\sigma = \epsilon_0 \mathbf{E} \cdot \hat{\mathbf{n}}$ and $\mathbf{j} = -(1/\mu_0)\mathbf{B} \times \hat{\mathbf{n}}$. See Fig. 18-10. Applying these to the case of guided waves between infinite, parallel, conducting planes we have the following.

For the TE case:

$$\sigma = 0$$

$$\mathbf{j} = \begin{cases} \hat{\mathbf{y}} \left[\dfrac{1}{\mu_0} \left(\dfrac{-jm\pi}{\omega a} \right) C_1 \right] e^{j(\omega t - \beta_g x)} & \text{on } z = 0 \\ \hat{\mathbf{y}} \left[\dfrac{1}{\mu_0} \left(\dfrac{-jm\pi}{\omega a} \right) C_1 (-1)^m \right] e^{j(\omega t - \beta_g x)} & \text{on } z = a \end{cases}$$

For the TM case:

$$\sigma = \begin{cases} (-\beta_g/\omega \epsilon_0 \mu_0) C_4 e^{j(\omega t - \beta_g x)} & \text{on } z = 0 \\ (-\beta_g/\omega \epsilon_0 \mu_0)(-1)^m C_4 e^{j(\omega t - \beta_g x)} & \text{on } z = a \end{cases}$$

$$\mathbf{j} = \begin{cases} \hat{\mathbf{x}} \left[-\dfrac{1}{\mu_0} C_4 \right] e^{j(\omega t - \beta_g x)} & \text{on } z = 0 \\ \hat{\mathbf{x}} \left[-\dfrac{1}{\mu_0} C_4 (-1)^m \right] e^{j(\omega t - \beta_g x)} & \text{on } z = a \end{cases}$$

Note that the $\mathbf{j}$ on the left is surface current density while j on the right is $\sqrt{-1}$.

3. Attenuation The entire discussion so far has been based on waveguides made of a material with $g = \infty$, i.e., ideal conductors having zero resistance. It is possible to achieve this condition using superconductors, but the necessity for lowering the temperature to the neighborhood of liquid helium, only a few degrees above absolute zero, makes this cumbersome and expensive. Thus it is far more usual to use either an ordinary metal like brass or to have plastic waveguides coated with extremely thin layers of gold (since at the microwave frequencies used for waveguides the penetration depth δ of the wave is very small). There is then some power lost in heat in the imperfectly conducting metal. This loss may be calculated by assuming that the fields are essentially the same as for an ideal conductor, the finite resistivity of the metal causing only a slight perturbation in their magnitudes and directions.

From Example 2 in the last section of the previous chapter we have the power lost per unit length, measured in the direction of current flow on the surface: $P_{\text{lost}} = j_{\text{eff}}^2 R_\ell$, where j_{eff} is the effective value of the equivalent surface current density and $R_\ell = 1/g\delta = \sqrt{\mu_0 \omega/2g}$ is the surface resistance. We will here calculate this for a TE wave between parallel, imperfectly conducting, planes. The results at the beginning of the present section give

$$j_{\text{eff}} = \frac{1}{\sqrt{2}}\left[\frac{1}{\mu_0}\left(\frac{m\pi}{\omega a}\right) C_1\right]$$

Thus

$$P_{\text{lost}} = \left(\frac{m\pi C_1}{\sqrt{2}\mu_0 \omega a}\right)^2 \sqrt{\frac{\mu_0 \omega}{2g}}$$

The factor C_1 depends on the transmitted power; from a practical viewpoint it is not so much the power lost that is of concern as the ratio: (power lost)/(power transmitted). This ratio is a very important factor in selecting just one type of wave and mode from the many possibilities. This is especially true in any case where either large amounts of power are being dealt with or where great sensitivity is desired. By taking the ratio we will also be able to eliminate the factor C_1.

We have left it to Prob. 3 to find the value of the Poynting vector, integrated over the cross-sectional area of the waveguide interior and averaged over one complete cycle for this particular case. The result obtained from this calculation, which is not very difficult, is

$$S_{\text{int}} = \frac{C_1^2}{4\mu_0}\left(\frac{\beta_g}{\omega}\right) a$$

Here S_{int} is

$$\int_\Sigma \langle \mathbf{S} \rangle \cdot \hat{\mathbf{x}}\, dy\, dz$$

Thus

$$\frac{P_{\text{lost}}}{S_{\text{int}}} = \left[\left(\frac{m\pi}{\sqrt{2}\mu_0 \omega a}\right)^2 \sqrt{\frac{\mu_0 \omega}{2g}}\right] \frac{4\mu_0 \omega}{\beta_g a} = \frac{2m^2\pi^2}{a^3} \frac{1}{\sqrt{2\mu_0 \omega g}} \frac{1}{\omega^2 \epsilon_0 \mu_0 - (m^2\pi^2)/a^2}$$

when we make use of the expression obtained for β_g at the beginning of this section. This formula tells us that the integer m should be as small as possible. When a similar calculation is made for waveguides, both for TE and TM modes, it is found that the ratio $P_{\text{lost}}/S_{\text{int}}$ is smaller for the TE_{10} mode than it is for any other mode. Consequently the TE_{10} is, by far, the mode most widely used.

Instead of utilizing the ratio $P_{\text{lost}}/S_{\text{int}}$ directly, it is possible to express the attenuation constant of a wave in terms of this ratio. If a voltage wave and a current wave each have the attenuation factor $e^{-\alpha x}$ then the power transferred by the wave has the attenuation factor $e^{-2\alpha x}$; similarly, if **E** and **B** in a waveguide each have the attenuation factor $e^{-\alpha x}$ then the averaged and integrated Poynting vector will have the attenuation factor $e^{-2\alpha x}$. Then the rate of decrease of this averaged and integrated Poynting vector with distance must equal the power loss along the waveguide per unit length. So

$$P_{\text{lost}} = -\frac{dS_{\text{int}}}{dx} = 2\alpha S_{\text{int}} \qquad \text{or} \qquad \alpha = \frac{P_{\text{lost}}}{2S_{\text{int}}}$$

The attenuation constant for the TE wave between parallel planes is, thus, when $m = 1$:

$$\alpha = \left(\frac{\pi^2}{a^3}\right)\frac{1}{\sqrt{2\mu_0 \omega g}}\left(\frac{1}{\omega^2 \epsilon_0 \mu_0 - \pi^2/a^2}\right)$$

The example of the parallel planes is one, primarily, of heuristic interest to us, for it is simpler to understand than the case of waveguides. Similar calculations can be made for α in the practical case of waveguides. We will not reproduce them here since they are considerably more complicated. Figure 18-11 shows the results of these calculations for a typical 2-by-1-in rectangular waveguide, in the form of a log-log graph of α, in decibels per foot vs. gigahertz frequency. Waveguide handbooks contain similar detailed information on α vs. f for the dimensions and other configurations. This is a topic of great practical importance.

PROBLEMS

1 In transmission between parallel conducting planes, (*a*) find a relative giving λ_{guide} in terms of λ and a; (*b*) what does λ_{guide} become at the cutoff frequency?

2 Prove, for TE waves between parallel conducting planes, that the results given in the text for B_x, B_y, B_z follow from the previous equations.

3 In the expressions for the components of **E** and **B** in the TE guided waves between parallel planes there is an arbitrary constant common to all the components.
(*a*) What determines the value of this constant?
(*b*) Obtain the Poynting vector as a function of z.
(*c*) Integrate (*b*) to give the average energy transported between the planes.

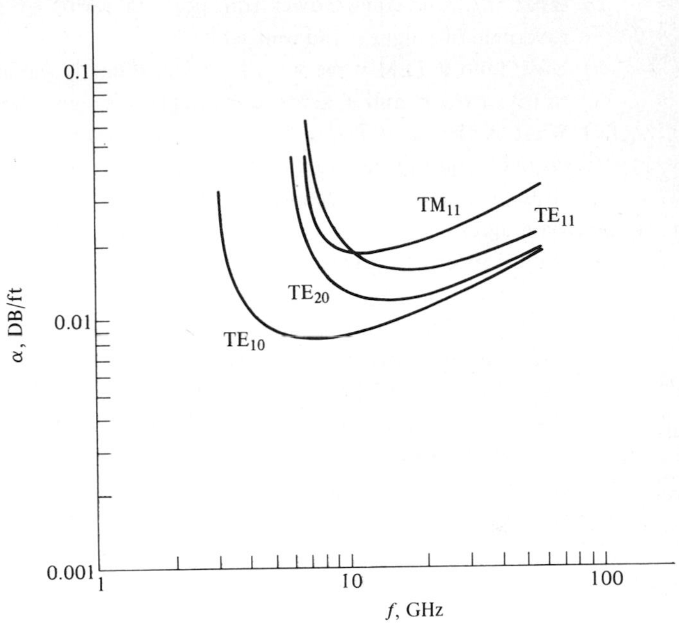

FIGURE 18-11
(Slightly modified from Fig. 8-11 of Edward C. Jordan and Keith G. Balmain, "Electromagnetic Waves and Radiating Systems," 2d ed., Prentice-Hall, Englewood Cliffs, N.J., 1968.)

(*d*) Does this result depend on m?

(*e*) Is there a component, in the answer to (*b*), along $\hat{\mathbf{z}}$?

4 Some of the expressions for the components of **E** and **B** contain the factor j in the amplitude factor (as distinct from the j in the exponent). What is its significance? How does it affect the sketches of the resulting waves?

5 Given the TE_{m0} and the TM_{m0} modes between parallel planes, (*a*) why is $m = 0$ ruled out?

(*b*) Why do we not need to consider negative values of m?

6 Draw a TE_{20} diagram, corresponding to the TE_{10} diagram of Fig. 18-2.

7 (*a*) What relation must exist between the C_1 of the TE_{m0} waves and the C_4 of the TM_{m0} waves (between parallel conducting planes) such that the energy transported across a plane will be the same for both cases?

(*b*) Will the average energy transport in the two cases have the same variation with z?

8 Show that the equations given in the text for the **E** components of TM waves between parallel planes follow from the equations for the **B** components.

9 Take a TE_{10} wave with $E_y = C_1$ halfway between the bounding planes.

(*a*) What is the average power transport, in terms of E_y, a, and v_{guide}, across a rectangle of height a and unit width?
(*b*) Now take a TEM wave with $E_y = C_1$. What is the average power transport, in terms of E_y, a, and c, across a rectangle of height a and unit width?
(*c*) What is the ratio $P_{\text{TEM}}/P_{\text{TE}}$, of (*b*) to (*a*), in terms of (f/f_c) where f_c is the cutoff frequency in case (*a*)?
(*d*) Evaluate this for $f = 2f_c$ and $f = 10f_c$.

10 Draw a diagram of the **E** and **B** distributions for a TM_{20} wave.

11 The equation $\cos\theta = (m\lambda/2a)$ for the angle between the normal to the parallel planes and the propagation vector of the plane waves equivalent to a TE_{m0} wave gives an upper limit to the number of modes which can exist for a given (λ/a).
(*a*) What number must (λ/a) be less than in order to allow at least one mode? At least two modes? Three? Four, five, six?
(*b*) How many modes could appear if $\lambda/a = 1.4$? If $\lambda/a = 0.8$?

12 Why are the lower-order modes, such as TE_{10} and TE_{20}, used in practice rather than higher-order modes, such as TE_{90}?

13 Derive the expression for **A**, the vector potential, for (*a*) a TE_{mn} wave and for (*b*) a TM_{mn} wave.

14 In deriving the equation for the TE waves by the method of separation of variables a constant was introduced, and it was chosen to be positive, $C_1{}^2$. Suppose it is taken to be negative: $-C_1{}^2$. Continue the derivation from that point.

15 Do the following waves exist: (*a*) TE_{10}, (*b*) TE_{01}, (*c*) TE_{00}, (*d*) TM_{10}, (*e*) TM_{01}, (*f*) TM_{00}? Give **A** in each case.

16 For either TE or TM waves the components have factors such as

$$\frac{(m\pi/a)}{(m\pi/a)^2 + (n\pi/b)^2}$$

multiplying the trigonometric and exponential factors. For given (m,n) these give relative values of the x,y,z components and they, therefore, have significance. Is there any significance to such factors when comparing a given field component, say E_y for a TE set of (m,n), with an E_y for a TM set of values for (m,n)?

17 For each (m,n) combination there is a cutoff frequency below which there is no transmission of energy in this mode.
(*a*) Show that, if f_c is the cutoff frequency for the (1,0) mode and $a = 2b$, then only one mode can exist between $f = f_c$ and $f = 2f_c$.
(*b*) Find the cutoff frequencies for the (2,0), (3,0), (0,2), (0,3), and (1,1) modes with $a = 2b$.

18 Prove that a TEM wave cannot propagate in a waveguide having a singly connected hollow cross section of arbitrary shape.

19 Given a rectangular waveguide with $a = 6$ cm and $b = 4$ cm. If a 3 GHz wave is to be transmitted, find the possible modes of propagation.

20 In Example 1 a TE wave was shown to be equivalent to two trains of plane waves, propagating at the proper angle with respect to the walls, if one has a phase shift of

180° relative to the other. Show that a TM wave may be synthesized, also, from two trains of plane waves.

21 Combine a TE_{mn} and TM_{mn} wave in a waveguide to give the following.

(*a*) $E_y = 0$, the other components of **E** and **B** being present. This wave has **E**, but not **B**, transverse to a nonaxial direction, viz., the $\hat{\mathbf{y}}$ axis. What relation must exist between the C of the equations giving the TE components and the C' of the equations giving the TM components?

(*b*) Repeat, to make $B_z = 0$. What is the relation between C and C'?

22 The components of **E** and **B** derived for the TE waves and the TM waves all contain the factor C. If we demand that the wave be TEM this gives $C = 0$. But the y and z components become (*a*) indeterminate if, also, $m = n = 0$ and (*b*) zero if either $m \neq 0$ or $n \neq 0$. Show that in case (*a*) the y and z components are also zero. *Hint:* any transverse component (say B_y) obeys an equation, in that case where $E_x = B_x = 0$, like

$$\frac{\partial^2 B_y}{\partial y^2} + \frac{\partial^2 B_y}{\partial z^2} = 0$$

Consider a transverse plane.

23 Find σ and **j** on the walls of a rectangular waveguide for the TE_{mn} and TM_{mn} modes.

19

RADIATION

Retarded Values

This chapter deals with the fields produced by sinusoidal currents whose frequencies are so high that the wavelengths associated with them are very small compared to the circuit dimensions. The sinusoidal currents are produced by charges undergoing sinusoidal displacement, sinusoidal velocity, and sinusoidal acceleration; and when the acceleration of the charges is high enough then, it turns out, they produce waves of ϕ, **A**, **E**, and **B** which spread out—spherically, cylindrically, or otherwise. It is simpler, as is usual, to deal with the potentials ϕ and **A** rather than with **E** and **B**; and here it is also customary. We would, therefore, like to spend some time discussing ϕ and **A**, for the previous emphasis has been the usual one on **E** and **B**.

In Chap. 11 we saw that when the Lorentz condition,

$$\nabla \cdot \mathbf{A} + \frac{1}{c^2}\frac{\partial \phi}{\partial t} = 0$$

is satisfied then the differential equations giving the potentials produced by the source charges and currents separate into one involving only ϕ and another involving only **A**:

$$\Box^2\phi = -\frac{1}{\epsilon_0}\rho \quad \text{and} \quad \Box^2\mathbf{A} = -\mu_0\mathbf{J}$$

In Chap. 14 it was shown that $(J^k) \equiv (\mathbf{J},c\rho)$ and $(A^k) \equiv (c\mathbf{A},\phi)$ are two four-vectors connected by the equation $\nabla^i\nabla_i A^k = \sqrt{\mu_0/\epsilon_0}\,J^k$ $(i,k = 1,2,3,4$; summation convention). Under static conditions this equation (really, four equations) gives the two $\Box^2$ equations above, with ρ and **J** independent of each other (as are ϕ and **A**). In the equation of continuity $\partial\rho/\partial t = 0$ and $\nabla \cdot \mathbf{J} = 0$: charge and current are independently conserved at all points of space. But under time-varying conditions the Lorentz condition $(\partial A^i/\partial x^i = 0)$ is a sufficient condition for the conservation of charge (the equation of continuity, $\partial J^i/\partial x^i = 0$). Then the $\Box^2\phi$ and $\Box^2\mathbf{A}$ equations are still separate of each other; this does not mean that ϕ and **A** are unrelated, for they must still be the components of a four-vector. Similarly, ρ and **J** are not independent of each other but are also the components of a four-vector which must satisfy the equation of continuity, this time when the individual terms do not vanish.

In the electrostatic case the solution to $\nabla^2\phi = -(1/\epsilon_0)\rho$ is

$$\phi(\mathbf{r}) = \frac{1}{4\pi\epsilon_0}\int_{\mathcal{V}} \frac{\rho(\mathbf{r}_s)}{R}\,d\tau_s$$

where $R = |\mathbf{R}| = |\mathbf{r} - \mathbf{r}_s|$; and in the magnetostatic case the solution to $\nabla^2\mathbf{A} = -\mu_0\mathbf{J}$ is

$$\mathbf{A}(\mathbf{r}) = \frac{\mu_0}{4\pi}\int_{\mathcal{V}} \frac{\mathbf{J}(\mathbf{r}_s)}{R}\,d\tau_s$$

These equations continue to hold true, approximately, in the quasistatic case where the variations with time are slow, as in ordinary circuit theory. How must these equations be generalized to give the solutions to $\Box^2\phi = -(1/\epsilon_0)\rho$ and $\Box^2\mathbf{A} = -\mu_0\mathbf{J}$ when arbitrary variations with time are permitted? A general, mathematical derivation of the answers to this question is given in "Classical Electricity and Magnetism" by W. K. H. Panofsky and M. Phillips (2d ed., pp. 242–245, Addison-Wesley, Reading, Mass.). However, this treatment requires a knowledge of Fourier integrals and Green's functions, and an exposition here would require too great a digression. The final answer, though, is one which is almost intuitively evident.

In Fig. 19-1 we wish to evaluate the potential ϕ at the point P at the time t. The charge that produces $\phi(\mathbf{r},t)$ is located in some region of space $\mathcal{V}$; we now let it take a finite time for a charge element in $\mathcal{V}$, $\rho\,d\tau_s$, to make its effect felt at P; and we will assume that this finite time is that required by an electromagnetic wave, traveling with

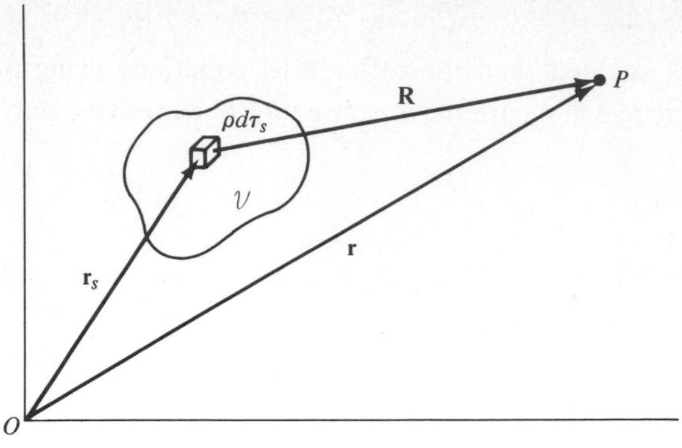

FIGURE 19-1

speed c in free space, to go from $\rho\, d\tau_s$ to P. The time of travel for the wave is then R/c. If we wish the potential at P, at time t, to be caused by a charge element $\rho\, d\tau_s$ we must use the ρ and the $d\tau_s$ at the previous, or retarded, time $t' = t - R/c$. Since R will be different for the various elements of $\mathcal{V}$, the retarded time t' will be different for the different elements also. If we indicate the retarded charge density $\rho(t')$ by $[\rho]$, then the solution to $\square^2\phi = -(1/\epsilon_0)\rho$ is

$$\phi(\mathbf{r}, t) = \frac{1}{4\pi\epsilon_0}\int_{\mathcal{V}} \frac{[\rho]}{R}\, d\tau_s$$

This is the answer obtained by a rigorous derivation, as in Panofsky and Phillips. Similarly, the solution to $\square^2\mathbf{A} = -\mu_0\mathbf{J}$ is

$$\mathbf{A}(r, t) = \frac{\mu_0}{4\pi}\int_{\mathcal{V}} \frac{[\mathbf{J}]}{R}\, d\tau_s$$

A different, simpler, derivation of this result can be given (following Feynman):

$$\square^2\phi = -\frac{1}{\epsilon_0}\rho \qquad \text{becomes} \qquad \nabla^2\phi - \frac{1}{c^2}\frac{\partial^2\phi}{\partial t^2} = 0$$

in free space. Suppose we solve this in spherical coordinates, limiting ourselves to the case of spherical symmetry so that ϕ is a function only of r. Then

$$\nabla^2\phi = \frac{1}{r}\frac{d^2}{dr^2}(r\phi)$$

and

$$\frac{1}{r}\frac{\partial^2}{\partial r^2}(r\phi) - \frac{1}{c^2}\frac{\partial^2\phi}{\partial t^2} = 0$$

can be written

$$\frac{\partial^2}{\partial r^2}(r\phi) - \frac{1}{c^2}\frac{\partial^2}{\partial t^2}(r\phi) = 0$$

But this is the one-dimensional wave equation

$$\frac{\partial^2 u}{\partial x^2} - \frac{1}{c^2}\frac{\partial^2 u}{\partial t^2} = 0$$

(where $u \to r\phi$ and $x \to r$), of which one solution is $u = f(x - ct)$. So here a solution is

$$r\phi = f(r - ct) \qquad \text{or} \qquad \phi = \frac{1}{r}f(r - ct) = \frac{1}{r}f\left(t - \frac{r}{c}\right)$$

The function f is arbitrary; if we take it to be sinusoidal then (the real part being implicit) we have

$$\phi = \frac{1}{r}e^{j\omega(t - r/c)} = \frac{1}{r}e^{j\omega[t]}$$

where $[t]$ is the retarded time $t - r/c$.

This result for ϕ represents a spherically symmetric wave traveling radially outward from the origin. Its amplitude decreases as $1/r$, so its intensity decreases as $1/r^2$; conversely, as $r \to 0$ the amplitude approaches an infinite value. Therefore, the radially outgoing wave can be *the* solution only when there is a source at the origin. When there is no source there, it is necessary to add the radially ingoing wave $(1/r)f(t + r/c)$ which corresponds to the advanced time $t'' = t + r/c$. This is the situation in true free space. If we insist on using the spherical coordinate system with its singular origin in free space, where there are no charges anywhere, then we must employ both the outgoing and incoming waves. At a given point these correspond to the retarded and advanced potential, respectively. But when there is a point charge at the origin (the center of the sphere) then we use only the outgoing wave (corresponding to the retarded time at any point in space). Now ϕ becomes infinite at the origin, but it should, since there is a point charge there.

Why do we use only the outgoing wave when there is a charge at the origin—why not only the incoming wave? Well, why did we use only the right-going wave on a transmission line when the generator was connected at the left end? There it was physically obvious that the left-going wave must be one obtained by reflection. The result here must be somewhat similar, but it is not at all obvious how the reflected wave arises.

The situation for $\square^2 \mathbf{A} = -\mu_0 \mathbf{J}$ is identical with that for ϕ. If we take, in either case, the waves to be spherically asymmetric then f becomes a function of (r,θ,ϕ) instead of only a function of r. The mathematics is then much more complicated, but the conclusions about ingoing and outgoing waves are identical.

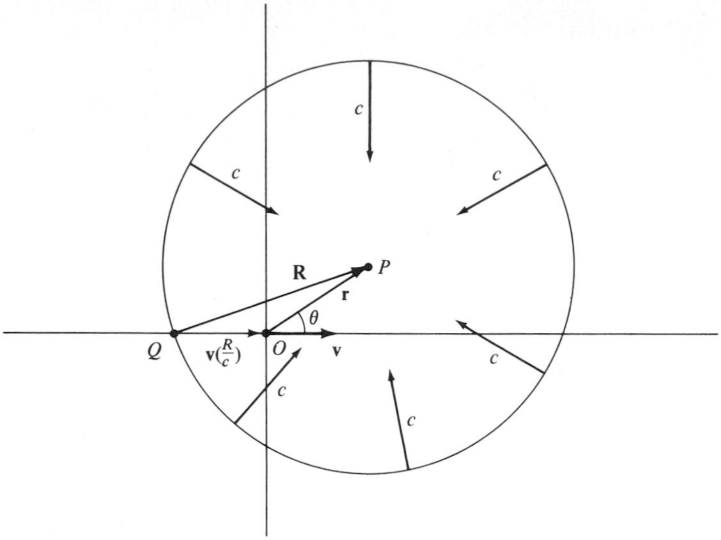

FIGURE 19-2

Liénard-Wiechert Potentials

We now apply these retarded potentials to the case of a charged particle moving with constant velocity $\mathbf{v} = \hat{\mathbf{x}}v$ along the $\hat{\mathbf{x}}$ axis. (This follows an argument given by Panofsky and Phillips, pp. 343–344.) Figure 19-2 shows an imagined, information-collecting sphere converging at the speed c on the point P such that, when the sphere reaches P at $t = t$, the charge is at the origin O. At some retarded time $t' = t - R/c$ the sphere crossed the charge at some point Q on the latter's trajectory before it reached O.

Consider that the charge is distributed over some finite volume; the extent of this volume is not germane to our argument, and nothing is changed if we decide to go over to an infinitesimal volume. Also, let dS be an infinitesimal area on the collection sphere. While the information-collecting sphere shrinks in radius by dR, during the time dt, the amount of charge crossed would be, if the charge were at rest relative to O (or P), $[\rho]\,dS\,dr$. But, since it is actually moving, it will be

$$dq = [\rho]\ ds(dR - \hat{\mathbf{R}} \cdot \mathbf{v}\,dt)$$

$$= [\rho]\,dS\,dR\left(1 - \frac{\hat{\mathbf{R}} \cdot \mathbf{v}}{c}\right) = [\rho]\,d\tau_s(1 - \boldsymbol{\beta} \cdot \hat{\mathbf{R}})$$

Then

$$[\rho]\,d\tau_s/R = dq/(R - \boldsymbol{\beta} \cdot \mathbf{R})$$

Inserting this into the retarded formula for $\phi(\mathbf{r},t)$ and $\mathbf{A}(\mathbf{r},t)$:

$$\phi = \frac{1}{4\pi\epsilon_0}\int \frac{dq}{R - \boldsymbol{\beta}\cdot\mathbf{R}} \qquad \mathbf{A} = \frac{\mu_0}{4\pi}\int \frac{v\,dq}{R - \boldsymbol{\beta}\cdot\mathbf{R}}$$

For a point charge the denominator assumes the value $[R - \boldsymbol{\beta}\cdot\mathbf{R}]$. Thus we finally obtain the formulas for the potentials at time t in terms of the retarded position and velocity of the moving charge:

$$\phi(\mathbf{r}, t) = \left(\frac{1}{4\pi\epsilon_0}\right)\frac{Q}{[R - \boldsymbol{\beta}\cdot\mathbf{R}]}$$

$$\mathbf{A}(\mathbf{r}, t) = \left(\frac{\mu_0}{4\pi}\right)\frac{Q\mathbf{v}}{[R - \boldsymbol{\beta}\cdot\mathbf{R}]}$$

These are known as the Liénard-Wiechert potentials.

When the velocity of a charge is not constant it is necessary to calculate the potentials and fields at $t = t$ in terms of its retarded position and velocity, but for the special case of constant velocity it is possible to find expressions in terms of the *present* position and velocity, at $t = t$. We leave it to the problems to show that $[R - \boldsymbol{\beta}\cdot\mathbf{R}] = r\sqrt{1-\beta^2 \sin^2\theta}$, where θ is shown in Fig. 19-2; then we can find ϕ and $\mathbf{A}$ directly. The results are identically the same as those found, relativistically, in Sec. 14-1, Chap. 14. (The primes of the result there refer to the O' observer and should be omitted here.)

The **E** and **B** equations have also been given in Sec. 14-1. They show that $\mathbf{E}(\mathbf{r},t)$ at P points from P to O (if $Q < 0$), where O is the position of Q at $t = t$; it does *not* point to the retarded position of the electron. For all values of β we have $\mathbf{B} = (1/c^2)\mathbf{v} \times \mathbf{E}$.

When $\beta \to 0$, the equations for **E** give the Coulomb field while those for **B** follow the law of Biot-Savart. But for $\beta \to 1$, the results are considerably different. In Probs. 1 and 5 of Sec. 14-3 we have given formulas for **E** in terms of r,β, and θ; a graph of E vs. θ for constant r gives a circle $\beta \ll 1$. When $\beta \approx 1$ the value of E is decreased greatly for θ near 0° and 180°, while it is increased greatly for θ near 90° and 270°. A high-speed electron passing near an observer gives only a brief pulse of any considerable magnitude, and **E** (and **B**, too) will be very nearly transverse. It is important to note that despite the highly transient nature of the pulse, in this case of large and constant $\boldsymbol{\beta}$, there is no energy lost by radiation: a calculation of $\mathbf{E} \times \mathbf{B}$ shows that the Poynting vector equals zero.

Acceleration

We now must consider the case of a charged particle which has, or had, acceleration. Suppose, for example, that we actually know the particle's trajectory and that it appears

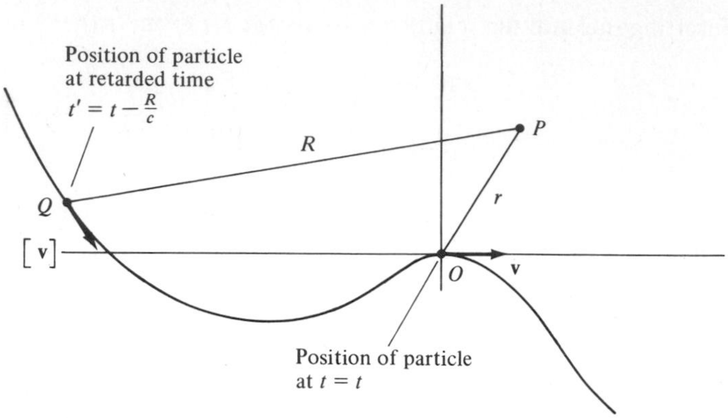

FIGURE 19-3

somewhat like that shown in Fig. 19-3. Here the particle's position at $t = t$ may be exactly the same as in Fig. 19-2 but now, because of the acceleration, the retarded position differs from the previous case and does not even lie on the x axis, in general. The formulas

$$\phi = \frac{1}{4\pi\epsilon_0} \frac{Q}{[R - \boldsymbol{\beta} \cdot \mathbf{R}]} \quad \text{and} \quad \mathbf{A} = \frac{\mu_0}{4\pi} \frac{Q\mathbf{v}}{R - \boldsymbol{\beta} \cdot \mathbf{R}}$$

continue to give ϕ and $\mathbf{A}$ at $P(\mathbf{r},t)$, *despite the particle's acceleration*, just as before, but the actual retarded value of R, of course, is different from the previous case. When it comes to **E** and **B** with acceleration, however, *even the formulas differ* from the previous case. For $\mathbf{E} = -\nabla\phi - \partial\mathbf{A}/\partial t$; and in taking $\partial\mathbf{A}/\partial t$ we can no longer treat $[\boldsymbol{\beta}]$ as constant. A detailed calculation of **E** and **B** follows, for this accelerated case.

A complication arises in taking the space and time derivatives of the Liénard-Wiechert potentials, for the operator ∇ taken at a given time $t = t$ compares the values of ϕ, say, at points in the neighborhood of P at that instant. For each of these points the retarded times, positions, and velocities of the charge are different. Similarly, the operator $\partial/\partial t$ taken at a given point P compares the values of **A** at different times in the infinitesimal time interval around $t = t$ and for each of these instants the retarded times, positions, and velocities of the charge are different. So the first task is to express ∇ and $\partial/\partial t$ in terms of the corresponding retarded quantities. It will be convenient, in this derivation, to let the retarded values of all quantities pertaining to the charge be designated by the subscript s. We will show that

$$\frac{\partial}{\partial t} = \frac{1}{[1 - \boldsymbol{\beta} \cdot \mathbf{R}]} \frac{\partial}{\partial t_s} \quad \text{and} \quad \nabla = \nabla_s - \frac{\hat{\mathbf{R}}}{c(1 - \boldsymbol{\beta} \cdot \hat{\mathbf{R}})} \frac{\partial}{\partial t_s}$$

We start with $\partial/\partial t$.

The quantity R may be expressed in two different ways:

$$R = \sqrt{(x - x_s)^2 + (y - y_s)^2 + (z - z_s)^2} \qquad \text{and} \qquad R = c(t - t_s)$$

We can then find two expressions for dR/dt and equate them. Thus

$$\frac{\partial R}{\partial t} = \left(\frac{\partial R}{\partial t_s}\right)_{r=\text{const}} \frac{\partial t_s}{\partial t}$$

and using the first expression for R, we find

$$\begin{aligned}\left(\frac{\partial R}{\partial t_s}\right)_{r=\text{const}} &= \frac{\partial R}{\partial x_s}\frac{\partial x_s}{\partial t_s} + \frac{\partial R}{\partial y_s}\frac{\partial y_s}{\partial t_s} + \frac{\partial R}{\partial z_s}\frac{\partial z_s}{\partial z_s} = \frac{-(x - x_s)}{R}\frac{\partial x_s}{\partial t_s} \\ &= -\frac{(x - x_s)}{R}\frac{\partial x_s}{\partial t_s} - \frac{(y - y_s)}{R}\frac{\partial y_s}{\partial t_s} - \frac{(z - z_s)}{R}\frac{\partial z_s}{\partial t_s} \\ &= -(\hat{\mathbf{R}} \cdot \mathbf{v})\end{aligned}$$

So

$$\frac{\partial R}{\partial t} = -(\hat{\mathbf{R}} \cdot \mathbf{v})\frac{\partial t_s}{\partial t}$$

Using the second expression for R:

$$\frac{\partial R}{\partial t} = c\left(1 - \frac{\partial t_s}{\partial t}\right)$$

Equating the two gives

$$\frac{\partial t_s}{\partial t} = \frac{1}{1 - \boldsymbol{\beta} \cdot \hat{\mathbf{R}}}$$

But $$\frac{\partial}{\partial t} = \left(\frac{\partial t_s}{\partial t}\right)\frac{\partial}{\partial t_s} \qquad \text{so} \qquad \frac{\partial}{\partial t} = \frac{1}{1 - \boldsymbol{\beta} \cdot \hat{\mathbf{R}}}\frac{\partial}{\partial t_s}$$

The distinction between $\partial/\partial t$ and $\partial/\partial t_s$ is appreciable only if $\beta \to 1$, and then only to the extent that $\boldsymbol{\beta}$ lies along $\hat{\mathbf{R}}$. We remind the reader that all quantities on the right are retarded.

To find the corresponding expression for ∇ we can compare two values for ∇R. First, $\nabla R = \nabla c(t - t_s) = -c\nabla t_s$. This unusual-looking expression is quite legitimate, for the delay time is a function of the position. Second, if we let ∇_s designate the gradient at constant t_s, then $\nabla R = \nabla_s R + (\partial R/\partial t_s)\nabla t_s$. But $\nabla_s R = \hat{\mathbf{R}}$; and using the result just obtained, $\partial R/\partial t_s = -(\hat{\mathbf{R}} \cdot \mathbf{v})$, we have $\nabla R = \hat{\mathbf{R}} - (\hat{\mathbf{R}} \cdot \mathbf{v})\nabla t_s$. Equating the two gives

$$\nabla t_s = -\frac{1}{c}\left(\frac{\hat{\mathbf{R}}}{1 - \boldsymbol{\beta} \cdot \hat{\mathbf{R}}}\right)$$

Putting this back into the expression for ∇R we obtain, if we set $s \equiv [R - \boldsymbol{\beta} \cdot \mathbf{R}]$,

$$\nabla = \nabla_s - \frac{\mathbf{R}}{cs} \frac{\partial}{\partial t_s}$$

Also, $\partial/\partial t = (R/s)(\partial/\partial t_s)$.

We may now proceed to evaluate

$$\mathbf{E} = -\nabla\phi - \frac{\partial \mathbf{A}}{\partial t} = -\nabla_s \phi + \left[\frac{\mathbf{R}}{cs}\right] \frac{\partial \phi}{\partial t_s} - \left[\frac{R}{s}\right] \frac{\partial \mathbf{A}}{\partial t_s}$$

since the formulas for ϕ and $\mathbf{A}$ refer to one given retarded moment (for a single point charge) and the operators ∇_s and $\partial/\partial t_s$ also refer to that constant retarded instant. We now have

$$\phi = \left(\frac{Q}{4\pi\epsilon_0}\right) \frac{1}{s} \qquad \text{and} \qquad \mathbf{A} = \left(\frac{Q}{4\pi\epsilon_0 c^2}\right) \frac{\mathbf{v}}{s}$$

It is convenient to continue to omit the brackets designating time retardation. So

$$\mathbf{E} = \frac{Q}{4\pi\epsilon_0} \left\{ \frac{1}{s^2} \nabla_s s - \frac{R}{cs^3} \dot{s} - \frac{R}{c^2 s^3} (s\dot{\mathbf{v}} - \mathbf{v}\dot{s}) \right\}$$

Now

$$\nabla_s s = \nabla_s (R - \boldsymbol{\beta} \cdot \mathbf{R}) = \hat{\mathbf{R}} - \boldsymbol{\beta}$$

It takes a bit more work to find

$$\dot{s} \cdot \frac{\partial s}{\partial t_s} = \frac{\partial R}{\partial t_s} - \boldsymbol{\beta} \cdot \frac{\partial \mathbf{R}}{\partial t_s} - \mathbf{R} \cdot \dot{\boldsymbol{\beta}}$$

The first term equals $-\hat{\mathbf{R}} \cdot (c\boldsymbol{\beta})$ while the second term is $-\boldsymbol{\beta} \cdot [\hat{\mathbf{R}}(\partial R/\partial t_s) + (\partial \hat{\mathbf{R}}/\partial t_s) R]$. Here we again substitute $-\hat{\mathbf{R}} \cdot (c\boldsymbol{\beta})$ for $\partial R/\partial t_s$, while for $\partial \hat{\mathbf{R}}/\partial t_s$ we use $-\hat{\mathbf{p}}(c\beta_{\text{per}})/R)$ where $\hat{\mathbf{p}}$ is a unit vector perpendicular to $\mathbf{R}$ and β_{per} is the component of $\boldsymbol{\beta}$ along this direction. The minus appears because $\mathbf{R}$ points from the charge to the observation point and we are discussing the motion of the tail end. The second term thus becomes $-\boldsymbol{\beta} \cdot \hat{\mathbf{R}}(-\hat{\mathbf{R}} \cdot c\boldsymbol{\beta}) - R\boldsymbol{\beta} \cdot (-c\beta_{\text{per}}\hat{\mathbf{p}}/R) = c\beta_{\text{par}}^2 + c\beta_{\text{per}}^2 = c\beta^2$ if β_{par} is the component of $\boldsymbol{\beta}$ parallel to $\mathbf{R}$. Then

$$\frac{\partial s}{\partial t_s} = -c(\boldsymbol{\beta} \cdot \hat{\mathbf{R}}) + c\beta^2 - \frac{1}{c}(\dot{\mathbf{v}} \cdot \mathbf{R})$$

Collecting the terms in $\mathbf{E}$ according to $\mathbf{R}$, $\boldsymbol{\beta}$, and $\dot{\mathbf{v}}$, we obtain the final result for $\mathbf{E}$, given below.

A similar calculation is made for **B** using

$$\nabla \times \mathbf{A} = \nabla_s \times \mathbf{A} - \frac{\mathbf{R}}{cs} \times \frac{\partial \mathbf{A}}{\partial t_s}$$

with

$$\mathbf{A} = \frac{\mu_0 Q}{4\pi} \frac{\mathbf{v}}{s}$$

Thus

$$\mathbf{B} = \frac{\mu_0 Q}{4\pi} \left\{ \nabla_s \times \frac{\mathbf{v}}{s} - \frac{\mathbf{R}}{cs} \times \frac{\partial}{\partial t_s} \left(\frac{\mathbf{v}}{s} \right) \right\}$$

$$= \frac{\mu_0 Q}{4\pi} \left\{ \nabla_s \left(\frac{1}{s} \right) \times \mathbf{v} + \frac{1}{s} (\nabla_s \times \mathbf{v}) - \frac{\mathbf{R}}{cs} \times \left(\frac{\dot{\mathbf{v}}}{s} - \frac{\mathbf{v}\dot{s}}{s^2} \right) \right\}$$

$$= \frac{\mu_0 Q}{4\pi} \left\{ -\frac{1}{s^2} (\nabla_s s) \times \mathbf{v} - \frac{\mathbf{R} \times \dot{\mathbf{v}}}{cs^2} + \frac{\mathbf{R} \times \mathbf{v}}{cs^3} (\dot{s}) \right\}$$

Using the previous expressions for $\nabla_s s$ and $\dot{s}$ we obtain the answer for B.

The final results of these two calculations are:

$$\mathbf{E} = \frac{Q}{4\pi\epsilon_0} \left[\left\{ \frac{(1-\beta^2)R}{s^3} \hat{\mathbf{R}} - \frac{(1-\beta^2)R}{s^3} \boldsymbol{\beta} \right\} \right.$$

$$\left. + \left\{ \frac{(\dot{\mathbf{v}} \cdot \mathbf{R})R}{c^2 s^3} \hat{\mathbf{R}} - \frac{(\dot{\mathbf{v}} \cdot \mathbf{R})R}{c^2 s^3} \boldsymbol{\beta} - \frac{R}{c^2 s^2} \dot{\mathbf{v}} \right\} \right]_{t'}$$

$$\mathbf{B} = \frac{\mu_0 Q}{4\pi} \left[\left\{ \frac{(1-\beta^2)Rc}{s^3} \boldsymbol{\beta} \times \hat{\mathbf{R}} \right\} + \left\{ \frac{(\dot{\mathbf{v}} \cdot \mathbf{R})R}{cs^3} \boldsymbol{\beta} \times \hat{\mathbf{R}} + \frac{\dot{\mathbf{v}} \times \mathbf{R}}{cs^2} \right\} \right]_{t'}$$

A number of conclusions may be drawn from these results for **E** and **B**, but first we will make a quick check. When $\dot{\boldsymbol{\beta}} = 0$ only the part of **E** in the first parenthesis remains and $\mathbf{E} = (Q/4\pi\epsilon_0 s^3)(1 - \beta^2)[\mathbf{R} - R\boldsymbol{\beta}]$. From Fig. 19-2 we see that $[\mathbf{R} - R\boldsymbol{\beta}] = \mathbf{r}$, the vector to P from O, the place where Q is at $t = t$. So $\mathbf{E} = (Q/4\pi\epsilon_0)[(1-\beta^2)\mathbf{r}/s^3]$. We leave it to the problems to show that

$$s = \sqrt{x^2 + (1-\beta^2)y^2 + (1-\beta^2)z^2} = r\sqrt{1 - \beta^2 \sin^2 \theta}$$

Then the present result becomes the same as that obtained previously for a particle moving with constant speed.

The **E** with $\dot{\boldsymbol{\beta}} = 0$, which we may call $\mathbf{E}_v$, varies as $1/r^2$. Even when $\dot{\boldsymbol{\beta}} \neq 0$ we have $E_v \propto r^{-2}$ if we interpret r as the distance from the place where the charge would be at $t = t$ if it had continued to move with constant velocity. The **B** when $\dot{\beta} = 0$ is also the same as the previous result and if we call it $\mathbf{B}_v$ then $c\mathbf{B}_v = \hat{\mathbf{R}} \times \mathbf{E}_v$. Note that though

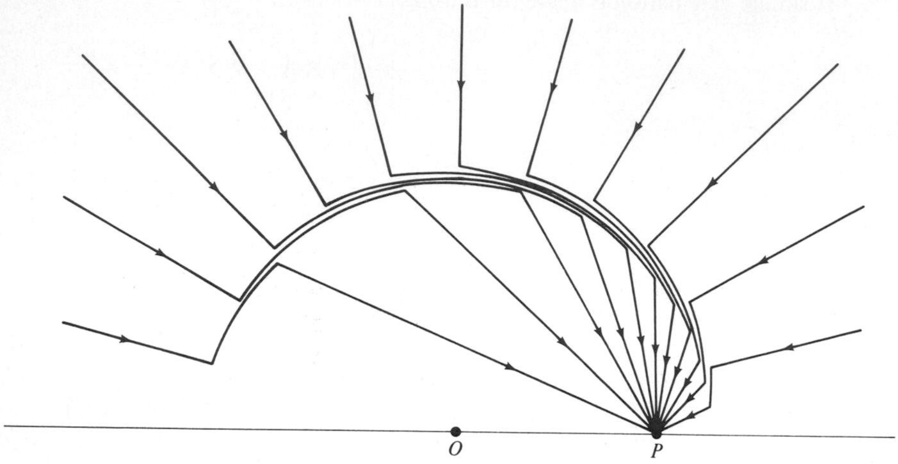

FIGURE 19-4
(This is Fig. 5.16 from Edward M. Purcell, "Electricity and Magnetism," McGraw-Hill, New York, 1965.)

$\mathbf{B}_v \propto r^{-2}$ where r is the present distance, $\mathbf{B}_v$ is orthogonal to $\hat{\mathbf{R}}_1$, the retarded unit vector. The Poynting vector for $\mathbf{E}_v$ and $\mathbf{B}_v$ is zero, as is g, the momentum density in the propagated wave.

When $\dot{\boldsymbol{\beta}} \neq 0$ we may write $\mathbf{E} = \mathbf{E}_v + \mathbf{E}_a$ and $\mathbf{B} = \mathbf{B}_v + \mathbf{B}_a$, where a denotes proportionality to $\dot{\mathbf{v}}$, the acceleration.

$$\mathbf{E}_a = \frac{Q}{4\pi\epsilon_0 c^2}\left[\frac{R(\dot{\mathbf{v}}\cdot\mathbf{R})(\hat{\mathbf{R}}-\boldsymbol{\beta})}{s^3} - \frac{R\dot{\mathbf{v}}}{s^2}\right]_{t'}$$

$$\mathbf{B}_a = \frac{\mu_0 Q}{4\pi c}\left[\frac{R(\dot{\mathbf{v}}\cdot\mathbf{R})(\boldsymbol{\beta}\times\hat{\mathbf{R}})}{s^3} + \frac{(\dot{\mathbf{v}}\times\mathbf{R})}{s^2}\right]$$

It is not difficult to show that $\mathbf{E}_a$ is perpendicular to $[\mathbf{R}]$; also that $c\mathbf{B}_a = \hat{\mathbf{R}} \times \mathbf{E}_a$. As with $\mathbf{B}_v$, $\mathbf{B}_a$ is orthogonal to the retarded $\hat{\mathbf{R}}$. Both $\mathbf{E}_a$ and $\mathbf{B}_a$ vary as $1/R$ for large R. When R is large the acceleration terms become large compared to the velocity terms. We leave it to the problems to show that the Poynting vector for $\mathbf{E}_a$ and $\mathbf{B}_a$ gives a finite radiated energy.

The fact that $\mathbf{E}_a$ is perpendicular to $[\hat{\mathbf{R}}]$ is brought out dramatically in some figures appearing on pages 164 and 165 of "Electricity and Magnetism," by Edward M. Purcell (McGraw-Hill, New York, 1965) and in an article, "Pictures of Dynamic Electric Fields," by R. Y. Tsien [in *Amer. J. Phys.*, **40**:46(1972)]. Figure 19-4 shows the lines of **E** of a negative charge, at rest at O prior to $t = 0$, that was then given an acceleration by a constant force for a time Δt to bring it up to a constant velocity $\beta = 0.2$. The particle is

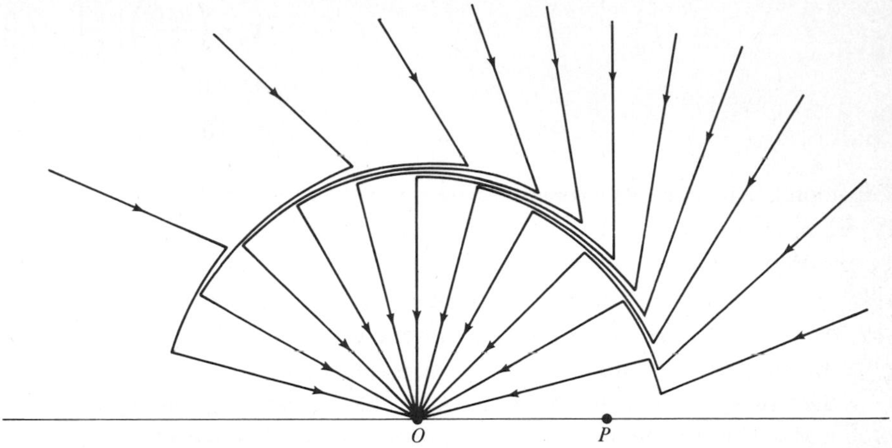

FIGURE 19-5
(This is Fig. 5.17 from Edward M. Purcell, "Electricity and Magnetism," McGraw-Hill, New York, 1965.)

now, at $t = t$, at P and moving to the right. At all points whose distance from O is greater than $R = t/c$, the news of the sudden acceleration cannot yet have arrived; at such points **E** must then point toward O. At all points whose distance from O is less than $R = t/c$ the information of the sudden acceleration did arrive prior to $t = t$, and **E** at such points is directed to P, the present position of the charge. The connections between these two different regions, as shown in the figure, are the arcs in a thin shell about the retarded position O. The lines of **E** here are not radial but are almost tangential. The situation is similar to a transverse shock wave, traveling radially outward with speed c from the point O.

Similarly, Fig. 19-5 illustrates the case of an electron moving with constant velocity $\beta = 0.95$ prior to $t = 0$, suffering a deceleration to rest produced by a constant force for a time Δt, and remaining at rest thereafter. Here the outer radial lines of **E** point to P, the place where the electron would be if it had continued to move with the same speed. The inner lines point to O, the actual present position. These are both $\mathbf{E}_v$ lines. The $\mathbf{E}_a$ lines are in the thin spherical shell about O and are almost transverse; they are *not* radial.

As we pointed out above, there is an interesting distinction between the behavior of ϕ and **A**, on the one hand, and **E** and **B**, on the other, when the particle accelerates. When acceleration occurs, $\mathbf{E}_a$ and $\mathbf{B}_a$ are added to the constant-velocity terms $\mathbf{E}_v$ and $\mathbf{B}_v$ and, in fact, the $\mathbf{E}_a$ and $\mathbf{B}_a$ terms are the only ones that contribute to **S** or g. Yet the ϕ and **A** expressions with acceleration are identically the same in the two cases:

$$\phi = \left(\frac{Q}{4\pi\epsilon_0}\right)\left[\frac{1}{s}\right]_{t'}$$

$$A_x = \left(\frac{\mu_0 Q}{4\pi}\right)\left[\frac{v}{s}\right]_{t'}$$

$$A_y = 0$$

$$A_z = 0$$

For nonrelativistic speeds s is replaced by r. **r** is the position of the field point (where ϕ, **A**, **E**, **B** are being found) at $t = t$; the particle is at the origin at $t = t$. If there is no acceleration, therefore, **A** points in the direction both of the *retarded and present velocity* vector; while, if there is acceleration, **A** points in the direction of the *retarded velocity* vector. But its magnitude varies inversely with the distance from the origin (the place where Q would be if there had been no acceleration).

Our view, all along, has been that **A**, in the magnetostatic and quasistatic cases, is produced exclusively by one term proportional to **J**, or **j**, or **I**; the additive term, $\nabla\chi$ from the gage transformation $\mathbf{A}' = \mathbf{A} + \nabla\chi$, has been set equal to zero, having no significance for any one particular case. The physical content of this $\nabla\chi$ term arises from the geometry of the currents; different **A**'s give different regions and boundaries for a given **B**, and the difference between any two possible **A**'s is a $\nabla\chi$ term.

In the radiative case, however, the situation is different. Here the values of ϕ, **A**, **E**, **B** at $t = t$ are effectively separated from the charges and currents that produced them at $t' = t - R/c$. It does not have any significance, now, to specify the particular current and charge distribution which produced a given **E** and **B** at some point in free space. Only the result is important, and one source configuration has no more validity than another. So here any χ, subject to one constraint, is permissible and any one χ is just as valid as any other. The one constraint arises from the Lorentz condition, which is sufficient to insure charge conservation; when

$$\nabla \cdot \mathbf{A}' + \frac{1}{c^2}\frac{\partial \phi'}{\partial t} = 0$$

is applied to $\mathbf{A}' = \mathbf{A} + \nabla\chi$ and $\phi' = \phi - \partial\chi/\partial t$, one obtains $\Box^2\chi = 0$. This wave equation is the identical one obeyed in free space by ϕ, **A**, **E**, **B**; so whatever the relation between χ and ϕ, **A**, **E**, **B** at one point in space, the same relation becomes propagated, unchanged, to all other points in free space. But the particular χ that is chosen, whatever it is, cannot be connected with $\mathbf{E}_a$ and $\mathbf{B}_a$; for, adding $-\partial\chi/\partial t$ to ϕ, and $\nabla\chi$ to **A**, adds nothing to **E** and **B**.

The fact is this—ϕ and **A** depend only on the retarded position and velocity, while **E** and **B** depend, also, on the retarded acceleration. But the energy flux **S** and the momentum density **g** only acquire nonvanishing values when there is acceleration. As seen in the previous chapter, **S** and **g** are not just three-vectors—they are two components of $\{T_{ik}\}$, the electromagnetic momentum-energy four-tensor. **E** and **B** are also the component of a four-tensor $\{F_{ik}\}$, while ϕ and **A** are only the components of a four-vector $\{A_i\}$. So it is not surprising that the connection between $\{T_{ik}\}$ and $\{F_{ik}\}$ is a more direct one than that between $\{T_{ik}\}$ and $\{A_i\}$.

Acceleration Parallel to the Velocity

Returning to the expression

$$\mathbf{E}_a = \frac{Q}{4\pi\epsilon_0 c^2}\left[\frac{R(\dot{\mathbf{v}}\cdot\mathbf{R})(\hat{\mathbf{R}}-\boldsymbol{\beta})}{s^3} - \frac{R\dot{\mathbf{v}}}{s^2}\right]_{t'=t-\frac{R}{c}}$$

there are two special cases of radiation that are fairly simple to develop and are also of great interest. These are the cases when the acceleration $\dot{\mathbf{v}}$ and the (normalized) velocity $\boldsymbol{\beta}$ are parallel and perpendicular, respectively. Taking the parallel case first, we throw $\mathbf{E}_a$ into the form

$$\mathbf{E}_a = \frac{Q}{4\pi\epsilon_0 c^2 s^3}[\mathbf{R}\times(\mathbf{R}\times\dot{\mathbf{v}}) - R\mathbf{R}\times(\boldsymbol{\beta}\times\dot{\mathbf{v}})]$$

$$= \frac{QR^2}{4\pi\epsilon_0 c^2 s^3}[\hat{\mathbf{R}}\times\{(\hat{\mathbf{R}}-\boldsymbol{\beta})\times\dot{\mathbf{v}}\}]$$

When the (retarded) velocity and acceleration are parallel this simplifies to

$$\mathbf{E}_a = \frac{Q}{4\pi\epsilon_0 c^2 s^3}[\mathbf{R}\times(\mathbf{R}\times\dot{\mathbf{v}})]$$

$$= \frac{Q}{4\pi\epsilon_0 c^2 s^3}[(\mathbf{R}\cdot\dot{\mathbf{v}})\mathbf{R} - R^2\dot{\mathbf{v}}]$$

Then

$$\mathbf{B}_a = \frac{1}{c}\hat{\mathbf{R}}\times\mathbf{E}_a = \left(\frac{Q}{4\pi\epsilon_0 c^3 s^3}\right)R(\dot{\mathbf{v}}\times\mathbf{R})$$

So the Poynting vector is

$$\mathbf{S} = \frac{1}{\mu_0}\mathbf{E}_a\times\mathbf{B}_a = \frac{1}{\mu_0 c}E_a^2\hat{\mathbf{R}}$$

From the last expression for $\mathbf{E}_a$,

$$E_a^2 = \frac{Q^2R^2}{(4\pi\epsilon_0)^2c^4s^6}[R^2\dot{v}^2 - (\mathbf{R}\cdot\dot{\mathbf{v}})^2]$$

Now $\cos\phi = (\mathbf{R}\cdot\dot{\mathbf{v}})/R\dot{v}$, so $\sin^2\phi = [R^2\dot{v}^2 - (\mathbf{R}\cdot\dot{\mathbf{v}})^2]/R^2\dot{v}^2$

then

$$E_a^2 = \frac{Q^2}{(4\pi\epsilon_0)^2c^4}\frac{R^4}{s^6}\dot{v}^2\sin^2\theta$$

and

$$\mathbf{S} = \hat{\mathbf{R}} \left(\frac{Q^2}{16\pi^2 \epsilon_0 c^3} \right) \frac{R^4}{s^6} \dot{\mathbf{v}}^2 \sin^2 \theta$$

We are interested in the power radiated per steradian in a given direction at the time t. Care is needed here: the power that crosses a surface in the dt equals the power lost by the charge in the time dt_s. If U is the particle's energy, $-(\partial U/\partial t_s)\, d\Omega\, dt_s$ is the energy lost by the particle into the solid angle $d\Omega$ at the time t_s, during the time interval dt_s; and if $d\Sigma$ is the area bounding $d\Omega$ at distance R from the particle, then the energy radiated through $d\Sigma$ at the time t during the time interval dt is $S\, d\Sigma\, dt$. Equating the two:

$$-\frac{\partial U}{\partial t_s} d\Omega\, dt_s = S\, d\Sigma\, dt$$

Then

$$-\frac{\partial U}{\partial t_s} d\Omega = S \frac{\partial t}{\partial t_s} d\Sigma = \left(SR^2 \frac{\partial t}{\partial t_s} \right) d\Omega \equiv \frac{dP}{d\Omega} d\Omega$$

Here $dP/d\Omega$ is the radiated power per steradian crossing the surface $d\Sigma$ at distance R. (If $\Delta\Omega$ is some particular, small, finite solid angle then $(dP/d\Omega)\Delta\Omega$ is the power ΔP radiated into that particular solid angle $\Delta\Omega$.) Thus, $dP/d\Omega = SR^2(\partial t/\partial t_s)$. From our previous equation,

$$\frac{\partial}{\partial t} = \frac{1}{1 - \boldsymbol{\beta} \cdot \hat{\mathbf{R}}} \frac{\partial}{\partial t_s}$$

we have

$$\frac{\partial t}{\partial t_s} = 1 - \beta \cos \theta$$

Also $s \equiv R - \boldsymbol{\beta} \cdot \mathbf{R} = R(1 - \beta \cos \theta)$. So

$$\frac{dP}{d\Omega} = \left(\frac{Q^2 \dot{\mathbf{v}}^2}{16\pi^2 \epsilon_0 c^3} \right) \frac{\sin^2 \theta}{(1 - \beta \cos \theta)^5}$$

When $\beta \to 0$, then

$$\frac{dP}{d\Omega} = \left(\frac{Q^2 \dot{\mathbf{v}}^2}{16\pi^2 \epsilon_0 c^3} \right) \sin^2 \theta$$

Integrating this over all solid angles then gives

$$P = \left(\frac{1}{4\pi\epsilon_0} \right) \frac{2Q^2 \dot{\mathbf{v}}^2}{3c^3}$$

This is known as the Larmor formula for the radiated power. It can be shown that this formula gives the power for low velocities even when the velocity and acceleration are not

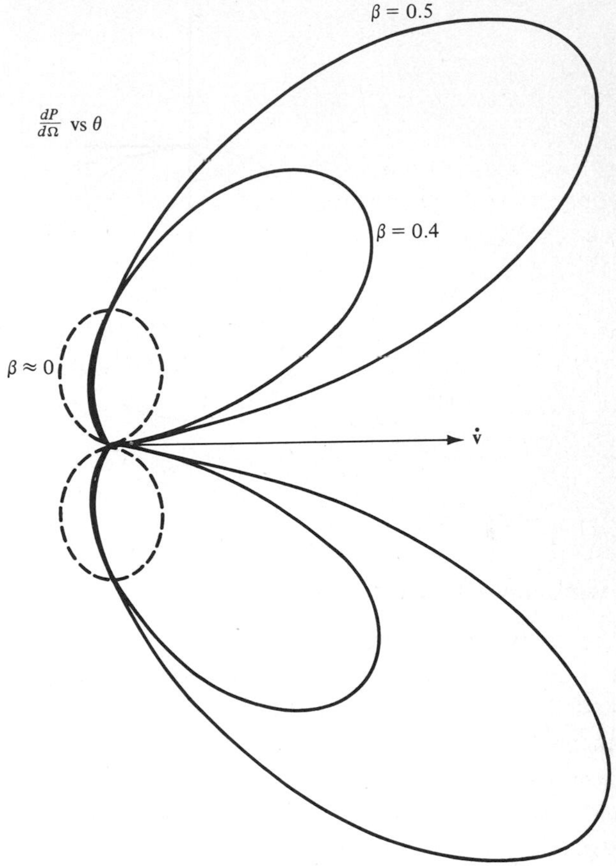

FIGURE 19-6

parallel. Figure 19-6 shows how $dP/d\Omega$, the power radiated per unit steradian, varies with θ, the angle between any given direction and the direction of acceleration. P, the total radiated power, is then proportional to the area enclosed by the curve. The curves of Fig. 19-6 show $dP/d\Omega$ for two cases when β is not very small. The radiation then peaks at the edge of a cone in the forward direction; again, there is nothing radiated directly forward. P increases markedly with β.

The radiation when there is deceleration, i.e., negative acceleration, is known as bremsstrahlung–braking radiation. For the determination of the amplitudes of the various frequency components it is necessary to know the details of the variation of acceleration with time. This is very important in practice, but we will not pursue it further. Figures 19-4 and 19-5 apply to this important case of a charge both accelerating and moving along the same direction.

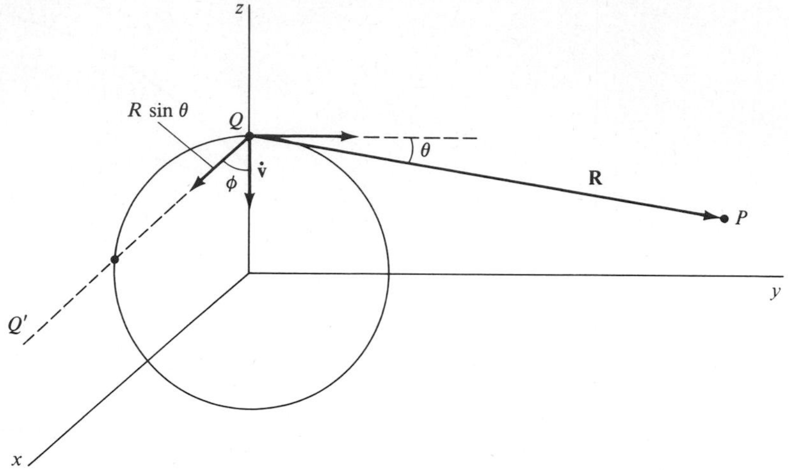

FIGURE 19-7

Acceleration Perpendicular to the Velocity

The second case of great practical importance is that of radiation by charged particles moving in circular orbits. (Strictly speaking this is only approximately so, for when radiation occurs the trajectories become spirals, since the particles are left with less energy. We assume the radiated energy is very small compared to the initial particle energy.)

Suppose the circular orbit is in the yz plane, as shown in Fig. 19-7; let the field point P be such that the vector from the source point Q on the orbit to P makes the angle θ with the $\hat{\mathbf{y}}$ direction. θ is in a skew plane, not parallel to any coordinate plane, in general; QQ' is the intercept of this plane with the zx plane, and ϕ is the angle between this intercept and $-\hat{\mathbf{z}}$. This geometry is that given by Marion in his book, "Classical Electromagnetic Radiation" (Academic Press, New York, 1965), and the following derivation is based on this. We have, as above,

$$\mathbf{E}_a = \frac{QR^2}{4\pi\epsilon_0 c^2 s^3}\left[\hat{\mathbf{R}} \times \{(\hat{\mathbf{R}} - \boldsymbol{\beta}) \times \dot{\mathbf{v}}\}\right]$$

Set $\mathbf{b} \equiv \hat{\mathbf{R}} - \boldsymbol{\beta}$, so

$$\mathbf{E}_a = \frac{QR^2}{4\pi\epsilon_0 c^2 s^3}\,[\hat{\mathbf{R}} \times (\mathbf{b} \times \dot{\mathbf{v}})] \qquad \mathbf{B}_a = \frac{1}{c}\,\hat{\mathbf{R}} \times \mathbf{E}_a$$

Then, as before,

$$\mathbf{S} = \frac{1}{\mu_0 c}\,\mathbf{E}_a \times (\hat{\mathbf{R}} \times \mathbf{E}_a) = \hat{\mathbf{R}}\,\frac{E_a^2}{\mu_0 c}$$

Setting $k \equiv QR^2/4\pi\epsilon_0 c^2 s^3$ we have

$$E_a^2 = k^2 [\hat{\mathbf{R}} \times (\mathbf{b} \times \dot{\mathbf{v}})]^2 = k^2 [\mathbf{b}(\hat{\mathbf{R}} \cdot \dot{\mathbf{v}}) - \dot{\mathbf{v}}(\hat{\mathbf{R}} \cdot \mathbf{b})]^2$$
$$= k^2 [b^2(\hat{\mathbf{R}} \cdot \dot{\mathbf{v}})^2 + \dot{v}^2(\hat{\mathbf{R}} \cdot \mathbf{b})^2 - 2(\mathbf{b} \cdot \dot{\mathbf{v}})(\hat{\mathbf{R}} \cdot \mathbf{b})(\hat{\mathbf{R}} \cdot \dot{\mathbf{v}})]$$
$$= k^2 [(1 - 2\hat{\mathbf{R}} \cdot \boldsymbol{\beta} + \beta^2)(\dot{v} \sin\theta \cos\phi)^2 + \dot{v}^2(1 - \beta\cos\theta)^2$$
$$- 2(\dot{v} \sin\theta \cos\phi)(\dot{v} \sin\theta \cos\phi)(1 - \beta\cos\theta)]$$
$$= k^2 \dot{v}^2 [(1 - \beta\cos\theta)^2 - (1 - \beta^2)\sin^2\theta \cos^2\phi]$$

Thus
$$\mathbf{S} = \hat{\mathbf{R}} \frac{Q^2 R^4 \dot{v}^2}{16\pi^2 \epsilon_0 c^3 s^6} [(1 - \beta\cos\theta)^2 - (1 - \beta^2)\sin^2\theta \cos^2\phi]$$

On comparing this expression with the corresponding one obtained for the previous case of $\mathbf{v}$ parallel to $\dot{\mathbf{v}}$, one sees that the only change is the replacement of $\sin^2\theta$ by $[(1 - \beta\cos\theta)^2 - (1 - \beta^2)\sin^2\theta\cos^2\phi]$. So the expression for $(dP/d\Omega)$ is obtained immediately from the previous result by making this substitution:

$$\frac{dP}{d\Omega} = \left(\frac{Q^2 \dot{v}^2}{16\pi^2 \epsilon_0 c^3}\right) \frac{(1 - \beta\cos\theta)^2 - (1 - \beta^2)\sin^2\theta\cos^2\phi}{(1 - \beta\cos\theta)^5}$$

Here, however, θ and ϕ are cyclic and, generally, $dP/d\Omega$ is a complicated function of the time. Figure 19-8, from Marion's book, shows the radiation pattern in the plane of the orbit, $\phi = 0$, for several values of β. The intensity in the backward lobes has been multiplied by 10β in these graphs to make them more noticeable.

Tsien's article in the *American Journal of Physics*, quoted above, also has several drawings of the lines of **E** for this case. Figure 19-9 shows these results for $\beta = 0.2$, $\beta = 0.5$, and $\beta = 0.95$. In all three cases the circular path is centered on the x in the diagram. Note how much more pronounced the kinks become as $\beta \to 1$; these kinks give the pronounced portions where **E** is orthogonal to $[\hat{\mathbf{R}}]$. To see how the field changes with time it is only necessary to rotate each diagram clockwise. This may be done in a most effective manner by putting copies of these graphs on a phonograph turntable. Note, also, how in the near field the radial $\mathbf{E}_v$ component predominates, while in the far field the tangential $\mathbf{E}_a$ field takes over. This is seen most clearly in the $\beta = 0.5$ case, but it is true for all three.

The case of circular motion is important because it is applicable to the circular machines–cyclotrons, betatrons, synchrotrons, etc.–used for accelerating elementary particles. This so-called synchrotron radiation puts an upper effective limit on the energies which may be attained in such machines.

Electric Dipole Radiation

Consider a very short element of a conductor along the z axis carrying a sinusoidal current. This small section of length Δz can then be taken to have the same current

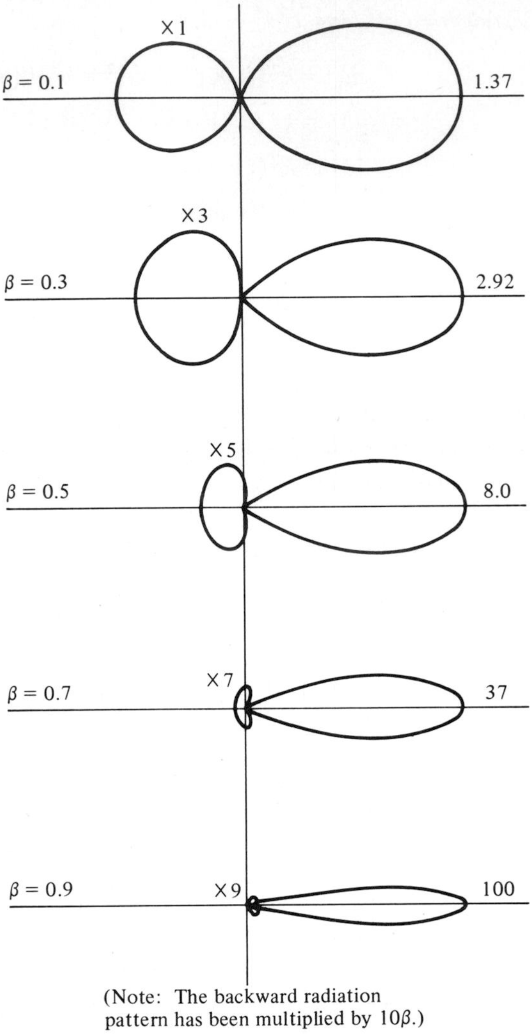

FIGURE 19-8
(From Jerry B. Marion, "Classical Electromagnetic Radiation," Academic Press, New York, 1965.)

throughout its length: $i = Ie^{j\omega t}$. This element is called a differential antenna. See Fig. 19-10. We will calculate the **E** and **B** produced by this Hertz antenna at an arbitrary point in space, first finding **A**, then determining ϕ from the Lorentz condition $\nabla \cdot \mathbf{A} + (1/c^2)(\partial\phi/\partial t) = 0$, and finally using $\mathbf{E} = -\nabla\phi - \partial\mathbf{A}/\partial t$, $\mathbf{B} = \nabla \times \mathbf{A}$.

We have, in general, $A = (\mu_0/4\pi)\int_v [J/r]\, d\tau_s$. For the filamentary conductor here this becomes, if the distance from the center of the small section of wire to the field point r is large relative to Δz,

$$\mathbf{A} = \frac{\mu_0}{4\pi}\frac{[i]\,\Delta z}{r}$$

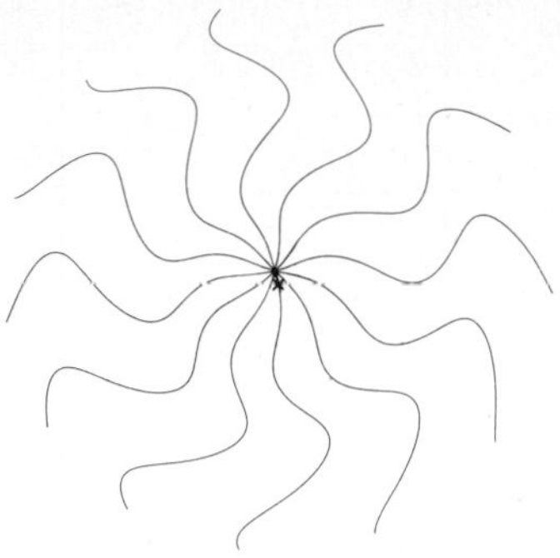

Electric field lines of a charge moving at tangential speed $\beta = 0.20$ on a circular path centered on the x

(a)

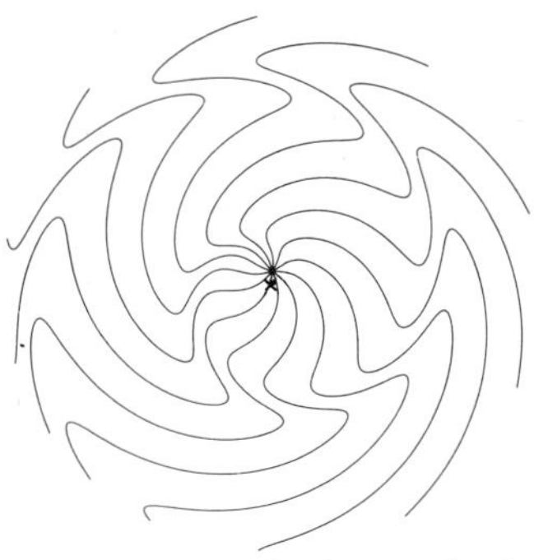

Electric field lines of a charge moving at tangential speed $\beta = 0.95$ on a circular path centered on the x

(b)

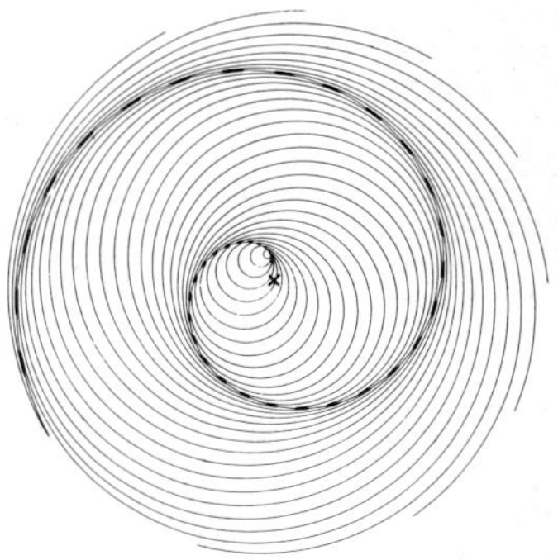

Electric field lines of a charge moving at tangential speed $\beta = 0.50$ on a circular path centered on the x

(c)

FIGURE 19-9
[*From R. Y. Tsien, "Pictures of Dynamic Electric Fields," Amer. J. Phys.*, **40:**46 (*1972*).]

Thus,

$$\mathbf{A} = \hat{\mathbf{z}}\left(\frac{\mu_0 I \Delta z}{4\pi}\right)\frac{1}{r}\, e^{j\omega(t-r/c)}$$

This is a spherical outward-going wave. For a small, finite, section of conductor this expression is only valid for $r \gg \Delta z$; but if the length of conductor is a differential this formula for **A** is good at all points. Setting $A_0 \equiv \mu_0 I\, \Delta z/4\pi$ and $k \equiv \omega/c = 2\pi/\lambda$ we may write this

$$\mathbf{A} = \hat{\mathbf{z}} A_0 \,\frac{e^{j(\omega t - kr)}}{r}$$

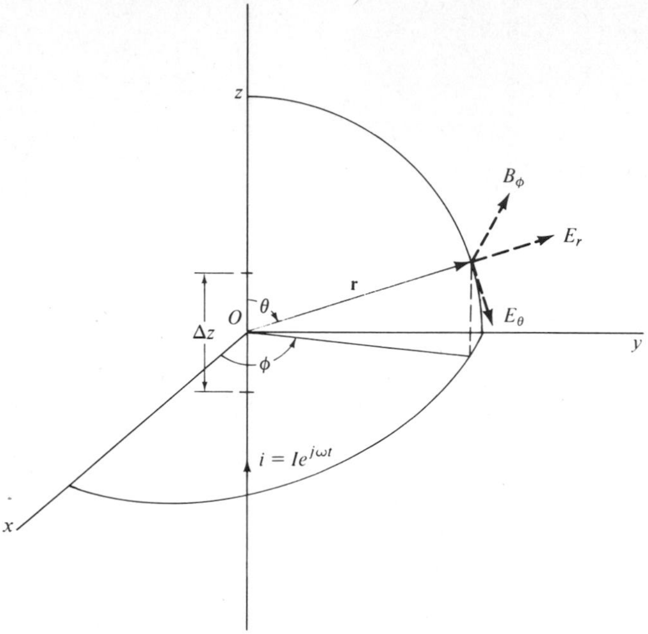

FIGURE 19-10

A has only a $\hat{\mathbf{z}}$ component, like the current that produces it, but **E** and **B** have other components, so we now write **A** in spherical coordinates:

$$\mathbf{A} = \hat{\mathbf{r}}A_0 \frac{e^{j(\omega t - kr)}}{r} \cos\theta - \hat{\boldsymbol{\theta}}A_0 \frac{e^{j(\omega t - kr)}}{r} \sin\theta = \hat{\mathbf{r}}A_r + \hat{\boldsymbol{\theta}}A_\theta$$

Next, from the Lorentz condition,

$$\frac{\partial\phi}{\partial t} = -c^2\,\nabla\cdot\mathbf{A} = -\frac{\omega^2}{k^2}\left\{-A_0\cos\theta\left(\frac{1}{r^2} + j\frac{k}{r}\right)e^{j(\omega t - kr)}\right\}$$

so

$$\phi = -\frac{j\omega A_0\cos\theta}{k^2}\left(\frac{1}{r^2} + j\frac{k}{r}\right)e^{j(\omega t - kr)}$$

Then

$$-\nabla\phi = -j\frac{\omega}{k^2}A_0 e^{j(\omega t - kr)}\left\{\hat{\mathbf{r}}\left(\frac{2}{r^3} + j\frac{2k}{r^2} - \frac{k^2}{r}\right)\cos\theta + \hat{\boldsymbol{\theta}}\left(\frac{1}{r^3} + j\frac{k}{r^2}\right)\sin\theta\right\}$$

and with

$$-\frac{\partial\mathbf{A}}{\partial t} = -j\omega A_0 e^{j(\omega t - kr)}\left\{\hat{\mathbf{r}}\left(\frac{1}{r}\right)\cos\theta - \hat{\boldsymbol{\theta}}\left(\frac{1}{r}\right)\sin\theta\right\}$$

we have

$$\mathbf{E} = -\nabla\phi - \frac{\partial \mathbf{A}}{\partial t} = A_0 e^{j(\omega t - kr)} \left\{ \hat{\mathbf{r}} \cos\theta \left(\frac{-j2\omega}{k^2 r^3} + \frac{2\omega}{kr^2} \right) + \hat{\boldsymbol{\theta}} \sin\theta \left(\frac{-j\omega}{k^2 r^2} + \frac{\omega}{kr^2} + \frac{j\omega}{r} \right) \right\}$$

Similarly,

$$\mathbf{B} = \nabla \times \mathbf{A} = \hat{\boldsymbol{\phi}} \frac{1}{r} \left\{ \frac{\partial}{\partial r}(rA_\theta) - \frac{\partial A_r}{\partial \theta} \right\} = \hat{\boldsymbol{\phi}} A_0 \sin\theta\, e^{j(\omega t - kr)} \left(\frac{1}{r^2} + \frac{jk}{r} \right)$$

These equations may be simplified considerably by restricting the range of r such that one term or another predominates. In the expression for E_r, for example, there are two terms: $(2\omega/kr^2)[-j/(kr) + 1]$. If $kr \gg 1$ this becomes approximately $2\omega/kr^2$. This condition, which is also $r \gg \lambda$, gives the far field (or radiation zone) approximation for E_r; the E_θ formula simplifies likewise to ω/r, so that

$$\mathbf{E} = A_0 e^{j(\omega t - kr)} \left\{ \hat{\mathbf{r}} \frac{2\omega}{kr^2} \cos\theta + \hat{\boldsymbol{\theta}} \frac{j\omega}{r} \sin\theta \right\}$$

But then the $\hat{\mathbf{r}}$ term is negligible compared to the $\hat{\boldsymbol{\theta}}$ term, and similarly for $\mathbf{B}$. So

Electric dipole
$$\begin{cases} \mathbf{E}_{\text{far}} = \hat{\boldsymbol{\theta}} \dfrac{j\omega A_0}{r} \sin\theta\, e^{j(\omega t - kr)} \\[2ex] \mathbf{B}_{\text{far}} = \hat{\boldsymbol{\phi}} \dfrac{jkA_0}{r} \sin\theta\, e^{j(\omega t - kr)} \end{cases}$$

We see that under this condition $\mathbf{E}$, $\mathbf{B}$, and $\hat{\mathbf{r}}$ form an orthogonal set of vectors. Because $\mathbf{E}_{\text{far}}$ and $\mathbf{B}_{\text{far}}$ both vary inversely with r rather than inversely as the square of r, there will be a finite amount of radiation energy propagated radially outward. The $\mathbf{E}_{\text{far}}$ and $\mathbf{B}_{\text{far}}$ expressions are those for spherical waves having the same phase velocity, $c = \omega/k$. Like the case of the plane electromagnetic wave, $E_{\text{far}} = cB_{\text{far}}$; also the two fields are in phase with each other (though advanced 90° in time phase with respect to the current that produced them).

The time-averaged Poynting vector for this so-called electric dipole radiation is

$$\langle \mathbf{S} \rangle = \frac{1}{2\mu_0} \mathcal{R}\, [\mathbf{E} \times \mathbf{B}^*] = \frac{1}{2\mu_0} (\hat{\mathbf{r}} E_\theta B_\phi^* - \hat{\boldsymbol{\theta}} E_r B_\phi^*)$$

We will here employ the exact expressions for E_r, E_θ, and B_ϕ, valid for all values of r; thus

$$\langle \mathbf{S} \rangle = \frac{A_0^2}{2\mu_0} \mathcal{R} \left\{ \hat{\mathbf{r}} \left(\frac{-j\omega}{k^2 r^5} + \frac{\omega k}{r^2} \right) \sin^2\theta - \hat{\boldsymbol{\theta}} j2\omega \left(\frac{1}{k^2 r^5} + \frac{1}{r^3} \right) \sin\theta \cos\theta \right\}$$

(If we had used the radiation zone expressions for **E** and **B** then the $\hat{\boldsymbol{\theta}}$ component of $\langle \mathbf{S} \rangle$ would be missing, the $\hat{\mathbf{r}}$ component being the same. But this would yield the same ultimate result since the real part of the $\hat{\boldsymbol{\theta}}$ component of the $\langle \mathbf{S} \rangle$ is zero.) Thus

$$\langle \mathbf{S} \rangle = \hat{\mathbf{r}} \left(\frac{A_0^2}{2\mu_0} \right) \frac{\omega k}{r^2} \sin^2 \theta$$

Note that the angular variation here is the same as that found previously for the case of acceleration parallel to velocity; this must, of course, be so for charges whose velocity and acceleration vary sinusoidally along z. From $\langle \mathbf{S} \rangle$ we can find $dP/d\Omega$, the power radiated per unit solid angle. Integrating this over the surface of a sphere gives the total time-averaged radiated power as

$$\langle P \rangle = \frac{4\pi \omega k A_0^2}{3\mu_0}$$

The radiation zone values of **E** and **B** are, predominantly, the ones of interest: the mutually transverse nature of **r**, **E**, and **B** yields radiated energy. We will now investigate the nature of these fields for the conditions at the other extreme, i.e., when $kr \ll 1$ (or $r \ll \lambda$). Then the near-field approximation holds:

$$\begin{cases} E_{\text{near}} = \dfrac{-j\omega A_0}{k^2} \left(\dfrac{\hat{\mathbf{r}} 2 \cos \theta + \hat{\boldsymbol{\theta}} \sin \theta}{r^3} \right) e^{j(\omega t - kr)} \\ B_{\text{near}} = A_0 \left(\dfrac{\hat{\boldsymbol{\phi}} \sin \theta}{r^2} \right) e^{j(\omega t - kr)} \end{cases}$$

Because the variation of $\mathbf{E}_{\text{near}}$ with r and θ for the differential antenna is identically the same, here, as in the case of a static dipole this near field is often called the static zone. Actually, the entire original expression is that which would be obtained from an oscillating, rather than a static, electric dipole. This oscillating electric dipole was first studied by Hertz and it is for this reason that it is called a Hertz dipole. We consider this further in Example 1 below. We note that while $\mathbf{B}_{\text{near}}$ is again transverse, this is not true of $\mathbf{E}_{\text{near}}$: it has a radial, or longitudinal, component. Thus, a differential antenna, or Hertz dipole, emits a TM wave in the near field and a TEM wave in the far field.

In the intermediate zone, where $r \approx \lambda$, the values of **E** and **B** cannot be simplified further and we must employ the general expressions, each containing several terms. This is sometimes called the induction zone and **E**, **B** are then called the induction fields. In calculating the radiated energy with these formulas it is found that five of the terms make no contribution, leaving only two that do so: the $1/r$ terms.

Figure 19-11 is taken from "Electromagnetic Fields and Waves," by P. Lorrain and D. Carson (2d ed., W. H. Freeman and Co., San Francisco, 1970). It shows lines of **E** for the oscillating electric dipole at four instants of time. The dipole, central to each diagram,

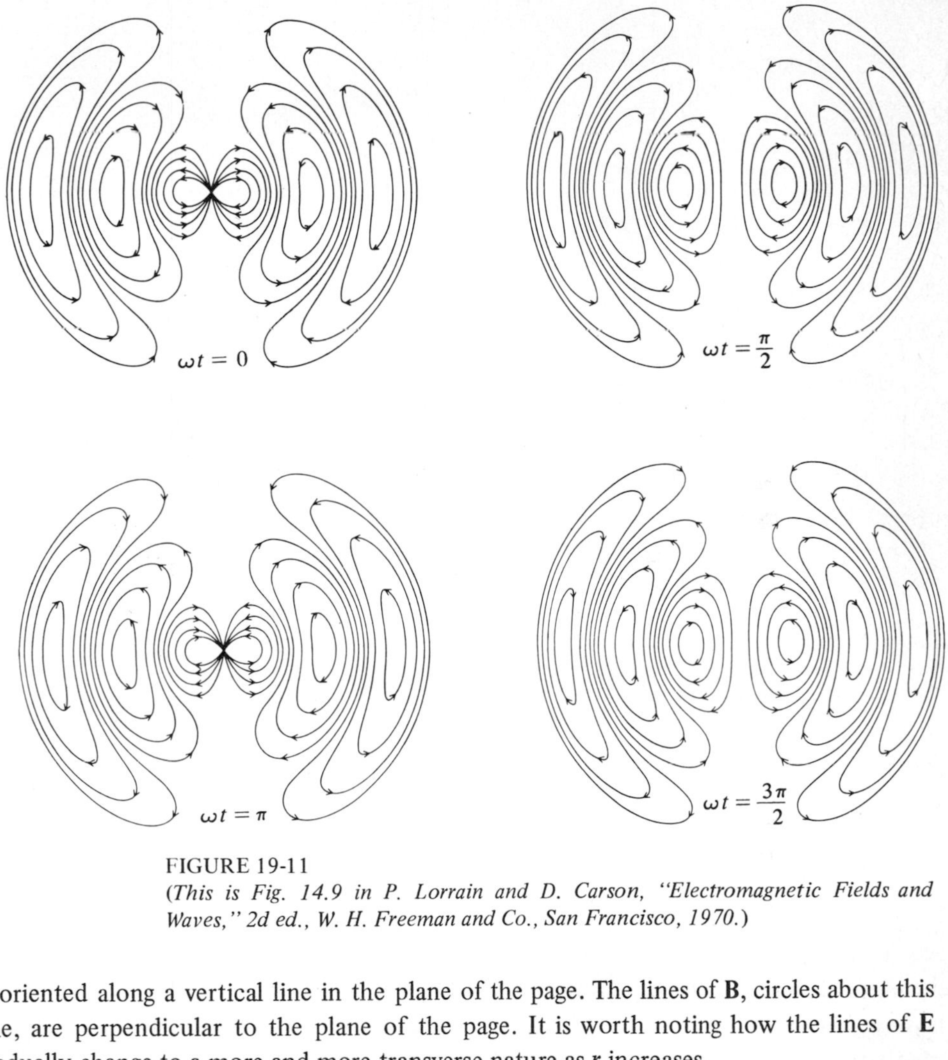

FIGURE 19-11
(This is Fig. 14.9 in P. Lorrain and D. Carson, "Electromagnetic Fields and Waves," 2d ed., W. H. Freeman and Co., San Francisco, 1970.)

is oriented along a vertical line in the plane of the page. The lines of **B**, circles about this line, are perpendicular to the plane of the page. It is worth noting how the lines of **E** gradually change to a more and more transverse nature as r increases.

The diagrams for the oscillating electric dipole are all made with the implicit assumption that the charge velocity is nonrelativistic. We do not have any diagrams for the case of the relativistic oscillating electric dipole, but they will be altered radically. This can be deduced from Fig. 19-12. These three diagrams are taken from the same R. Y. Tsien article in the January 1972 issue of the *American Journal of Physics*, mentioned above. Incidentally, these diagrams are all based on similar ones, made earlier, by E. M. Purcell in his book, "Electricity and Magnetism," Berkeley Physics Course (vol. 2, McGraw-Hill, New York, 1965). They show the lines of **E** for a single oscillating electric

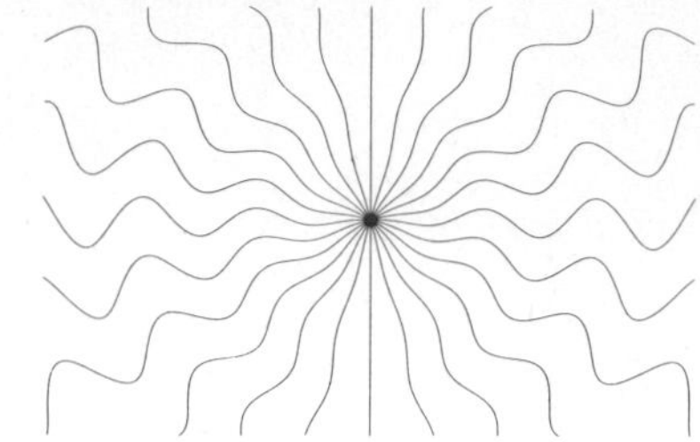

Electric field lines of a charge undergoing one-dimensional simple harmonic motion with $\beta_{max} = 0.10$, $t_0 = 0$

(a)

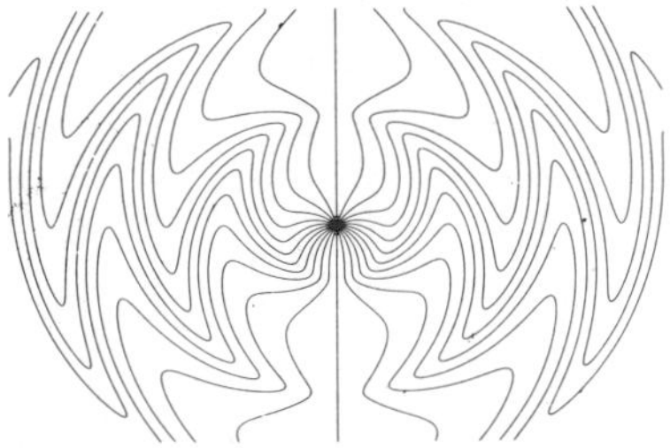

Electric field lines of a charge undergoing one-dimensional simple harmonic motion with $\beta_{max} = 0.50$, $t_0 = 0$

(b)

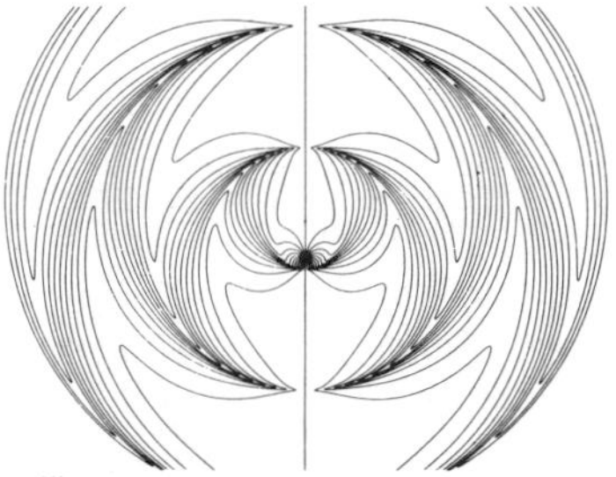

Electric field lines of a charge undergoing one-dimensional simple harmonic motion with $\beta_{max} = 0.90$, $t_0 = \frac{1}{2}\pi/\omega$ so that $y(t_0) = a$

(c)

FIGURE 19-12
[*From R. Y. Tsien, "Pictures of Dynamic Electric Fields," Amer. J. Phys.*, **40**:*46* *(1972)*.]

charge. Figure 19-12*a* has $\beta_{max} = 0.1$, very mildly relativistic, with the charge shown at the instant when it is at the center of its oscillation. In Fig. 19-12*b* we are definitely relativistic with $\beta_{max} = 0.5$ and the charge again at the center, while in Fig. 19-12*c* $\beta_{max} = 0.9$—very relativistic. Here the charge is shown at its maximum excursion.

The requirement that the velocity be nonrelativistic can be shown to be equivalent to the condition that the system dimensions be small relative to the minimum wavelength of the emitted radiation. This assumes the classical case, where the frequency of the radiation is the same as that of the oscillator. To see this, let v be the maximum velocity of the charge. Then $T = d/v$ is the shortest time required to cover the distance d, while the time for an electromagnetic wave to cover this distance is $\tau = d/c$. So $\tau \ll T$ means that the time for a signal to cover the system dimensions is negligible relative to T, the time it

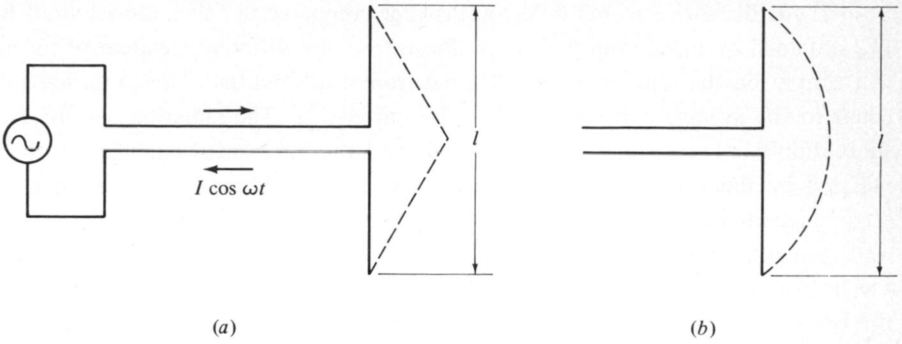

FIGURE 19-13

would take the particle to do so while going at its fastest. But the maximum frequency with which the charge could be moved between its extreme positions is $f_{max} = 1/T$. Then $\tau \ll T$ gives $(d/c) \ll (1/f_{max})$ or $d \ll \lambda_{min}$. Q.E.D. Thus, in Fig. 19-11 the amplitude is small relative to the wavelength.

The (Electric) Dipole Antenna

The differential antenna is important theoretically, for by combining various combinations of such antennas one can determine the potentials and fields for any configuration, at least in principle. When a differential antenna is considered by itself there is an accumulation of charge at the tip of each end of the differential length; this is, in fact, the connection with the Hertz oscillating dipole. When such a differential antenna is only one part of an antenna, the question of whether or not there is a charge accumulation at any point depends on the currents of adjacent differential antenna units; usually there is a varying current over the antenna so that the effects of adjacent units do not exactly cancel each other and there is a charge at various points. In the Hertz oscillating electric dipole viewpoint, the charges of adjacent dipoles would not be exactly the same so they would not quite cancel each other at any point.

Suppose we consider the antenna configuration shown in Fig. 19-13a: a center-fed linear antenna whose length is very short compared to the wavelength: $l \ll \lambda$. This, rather than the differential antenna, is the practical basic dipole antenna. The dotted lines in the figure show an assumed distribution of current along the dipole antenna, the current amplitude at the central feed points being I; the amplitude is taken to vary linearly down to zero at the two ends. The exact theoretical distribution of the current along the antenna, obtained from Maxwell's equations plus the boundary conditions, has not been obtained in most cases. The radiation itself alters the values from the generally assumed values—whether constant, linear, or sinusoidal—but the effect is usually small and here we neglect it.

If the distance r to the field point is large compared to l then the values of $\mathbf{E}_{\text{far}}$ and $\mathbf{B}_{\text{far}}$ obtained by integrating the contributions from the different elements of the antenna will simply be the same as those obtained from a differential antenna having a current equal to the average current here, i.e., the current $\frac{1}{2}I$. The radiation pattern of such a short dipole antenna will be the same as that of the differential antenna but the power radiated by the antenna will be a fourth of what it would be for a differential antenna (i.e., constant-current amplitude) with the same current at the central feed points.

Sometimes only the upper half of the dipole antenna is employed, the equivalent of the bottom half being obtained by reflection in a horizontal conducting plane through the center. The power radiated by such a short monopole is restricted to half the region available to the dipole and, since the radiation patterns are the same in the two cases, the total power emitted will be one-half the dipole case. The surface of the earth approximates a conducting plane and may be used for this purpose.

The Half-wave Dipole Antenna

Figure 19-13b shows a dipole antenna also, but here the length of the antenna equals half a wavelength. Each side of the vertical dipole equals $\lambda/4$ and the current distribution is assumed to be a sinusoidal standing wave having a maximum amplitude at the center and zero amplitude at the ends: $i = I \cos(2\pi z/\lambda)$. The current $I \cos \omega t$ is established at the center of the antenna by the generator and feed-in wires; at any point along the antenna the current is then $I \cos(2\pi z/\lambda) \cos \omega t$. In general, if the antenna is not an integral number of half wavelengths long, the standing wave of the current will still be sinusoidal and will be zero at each end of the antenna. The amplitude of the standing wave at the center feed-in leads will not be a maximum, however, nor will it be zero. The maximum will occur at two symmetric points–one on each arm of the antenna–part way between the center and the ends.

We are interested in the far fields. Thus, changing Δz to dz in the far-field results for the differential antenna, we integrate z from $-l/2$ to $+l/2$ while keeping the r fixed at $r \cong R$ in the denominator, where the actual small variation of r has little effect. In the phase factor, however, the variation of r must be permitted: $r \cong R - z \cos\theta$. Therefore

$$d\mathbf{E}_{\text{far}} \cong \hat{\boldsymbol{\theta}} \frac{j\omega}{R} \left(\frac{\mu_0 I \, dz \cos \dfrac{2\pi z}{\lambda}}{4\pi} \right) \sin\theta \, e^{j(\omega t - kR + kz\cos\theta)}$$

$$\mathbf{E}_{\text{far}} \cong \hat{\boldsymbol{\theta}} \frac{j\omega\mu_0 I \sin\theta \, e^{j(\omega t - kr)}}{4\pi R} \int_{-\lambda/4}^{\lambda/4} \cos\left(\frac{2\pi z}{\lambda}\right) e^{j(kz\cos\theta)} dz$$

Putting in the value for the integral, obtainable from tables,

$$\mathbf{E}_{\text{far}} \cong \hat{\boldsymbol{\theta}} \frac{jI}{2\pi R} \sqrt{\frac{\mu_0}{\epsilon_0}}\, e^{j(\omega t - kR)} \frac{\cos\,[(\pi/2)\cos\theta]}{\sin\theta}$$

Similarly

$$\mathbf{B}_{\text{far}} = \hat{\boldsymbol{\phi}} \frac{jI}{2\pi R} \mu_0 e^{j(\omega t - kR)} \frac{\cos\,[(\pi/2)\cos\theta]}{\sin\theta}$$

These give the average Poynting vector as

$$\langle \mathbf{S} \rangle = \frac{I^2}{8\pi^2 R^2} \sqrt{\frac{\mu_0}{\epsilon_0}} \frac{\cos^2\,[(\pi/2)\cos\theta]}{\sin^2\theta}$$

Figure 19-14 gives a graph of E_{far} and $\langle S \rangle$ vs. θ. The power is radiated predominantly near the horizontal plane.

Magnetic Dipole Radiation

When the current in a magnetic dipole is made to oscillate sinusoidally with time, instead of being constant, we obtain the so-called magnetic dipole radiation. This radiation is to be distinguished from the synchrotron radiation, discussed earlier in this section, which is obtained here even if the charged particles are going at constant speed. Consider the current of Fig. 19-15; actually, since we are primarily interested in the radiation from this source, this should be called an antenna rather than a circuit. The field point P is taken in the $y = 0$ plane for simplicity. Since we will use the spherical coordinate system this involves no loss in generality. The antenna is energized by the distant oscillator and the twisted pair (or very close, parallel, lead-in wires). These are necessary physically, but do not enter the calculations. We will first find the vector potential and then, from this, determine **E** and **B**. This is similar to the procedure for the differential antenna (or the Hertz electric dipole—see Example 1 below), except that here we may set the electric potential ϕ equal to zero in the wave zone, $a \ll \lambda$. The justification for this is left to Prob. 20.

We have

$$\mathbf{A} = \frac{\mu_0}{4\pi} \oint \frac{[i]\, d\mathbf{r}_s}{R} = \hat{\boldsymbol{\phi}} \frac{\mu_0 I a}{4\pi} \int_0^{2\pi} \frac{\cos\phi\, d\phi}{R} e^{j\omega(t - R/c)}$$

when the current is sinusoidal, $i = Ie^{j\omega t}$, and we take the retardation effect into account. With $\omega R/c = kR$ we have

$$e^{j\omega(t-R/c)} = e^{j\omega(t-r/c)} e^{jk(r-R)}$$

so

$$\mathbf{A} = \hat{\boldsymbol{\phi}} \frac{\mu_0 I a}{4\pi} e^{j\omega(t-r/c)} \int_0^{2\pi} \frac{e^{jk(r-R)}}{R} \cos\phi\, d\phi$$

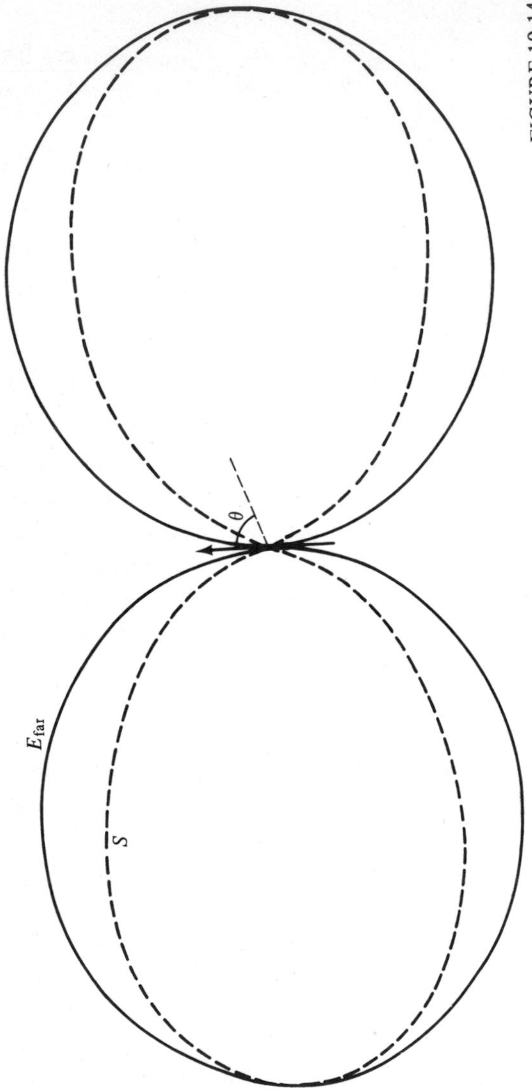

FIGURE 19-14

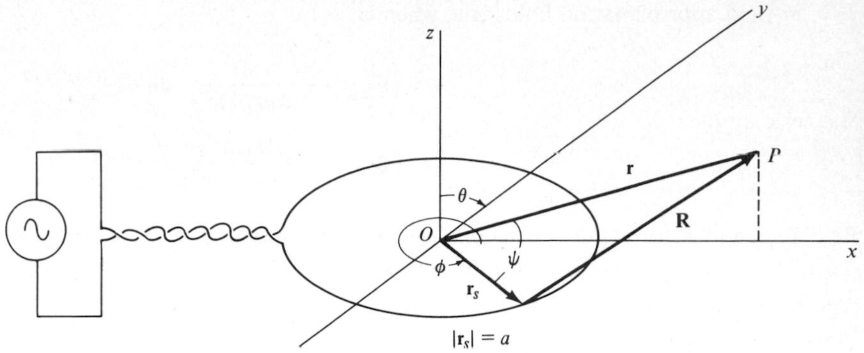

FIGURE 19-15

Then, if $a \ll \lambda$, we can expand the exponential in a power series:

$$e^{jk(r-R)} = 1 + jk(r-R) + \cdots$$

Dividing and multiplying by r, this gives

$$\mathbf{A} = \hat{\boldsymbol{\phi}} \frac{\mu_0 Ia}{4\pi r} e^{j\omega(t-r/c)} \int_0^{2\pi} \left\{ \frac{r}{R} + jkr\left(\frac{r}{R} - 1\right) + \cdots \right\} \cos\phi \, d\phi$$

We leave it to one of the problems to show that $r/R \approx 1 + (a/r)\sin\theta\cos\phi$. Substituting into the integral then gives

$$\mathbf{A} = \hat{\boldsymbol{\phi}} \left(\frac{\mu_0}{4\pi}\right) \frac{I\pi a^2 e^{j(\omega t - kr)}}{r} \left(\frac{1}{kr} + j1\right) k \sin\theta$$

This is the vector potential in the wave zone ($a \ll \lambda$) of a real, oscillating, magnetic dipole. For an ideal, point, oscillating magnetic dipole, however, this expression is good everywhere. Designating the ordinary, static, magnetic dipole moment by $\mathbf{M}$ we have

$$\mathbf{M} = \hat{\mathbf{z}} I\pi a^2$$

Then

$$\mathbf{m} = \mathbf{M}e^{j\omega t} \quad \text{and} \quad [\mathbf{m}] = \mathbf{M}e^{j(\omega t - kr)}$$

so

$$\mathbf{A} = \frac{\mu_0}{4\pi} \frac{[\mathbf{m}] \times \hat{\mathbf{r}}}{r} \left(\frac{1}{kr} + j1\right) k$$

The values of $\mathbf{E}$ and $\mathbf{B}$ are now given by $\mathbf{E} = -\partial\mathbf{A}/\partial t$ and $\mathbf{B} = \nabla \times \mathbf{A}$. Thus,

$$\mathbf{E} = -\hat{\boldsymbol{\phi}} j\omega \left(\frac{\mu_0}{4\pi}\right) \frac{I\pi a^2 e^{j(\omega t - kr)}}{r} \left(\frac{1}{kr} + j1\right) k \sin\theta$$

$$\mathbf{B} = \left(\frac{\mu_0}{4\pi}\right) I\pi a^2 e^{j(\omega t - kr)} \left\{ \hat{\mathbf{r}} \frac{2k\cos\theta}{r^2} \left(\frac{1}{kr} + j1\right) + \hat{\boldsymbol{\theta}} \sin\theta \left(\frac{1}{r^3} + \frac{jk}{r^2} - \frac{k^2}{r}\right) \right\}$$

The far-field approximation holds true when $kr \gg 1$:

Magnetic dipole
$$\begin{cases} \mathbf{E}_{\text{far}} = \hat{\boldsymbol{\phi}} \dfrac{M}{4\pi\epsilon_0 c\lambda\!\!\!^{-2} r} \sin\theta e^{j(\omega t - kr)} \\ \mathbf{B}_{\text{far}} = -\hat{\boldsymbol{\theta}} \dfrac{\mu_0}{4\pi} \dfrac{M \sin\theta}{\lambda\!\!\!^{-2} r} e^{j(\omega t - kr)} \end{cases}$$

$\mathbf{E}_{\text{far}}$, $\mathbf{B}_{\text{far}}$, and $\hat{\mathbf{r}}$ form the orthogonal set of a TEM wave with $E_{\text{far}} = cB_{\text{far}}$. Further, the results here, for the magnetic dipole, can be converted to the previous ones for the Hertz dipole by letting $M/c \to P$: then

$$(cB_{\text{far}})_{\text{mag}} \leftrightarrow (E_{\text{far}})_{\text{elec}} \qquad (E_{\text{far}})_{\text{mag}} \leftrightarrow -(cB_{\text{far}})_{\text{elec}}$$

The value of A in the far field (for a point magnetic dipole: everywhere) can now be written, since $[\dot{\mathbf{m}}] = j\omega[\mathbf{m}] = jck[\mathbf{m}]$, as

$$\mathbf{A} = \frac{\mu_0}{4\pi} \frac{[\dot{\mathbf{m}}] \times \hat{\mathbf{r}}}{cr}$$

To obtain this from the $\mathbf{A}$ for the electric dipole one must not only set $\mathbf{p} \to \mathbf{m}/c$, but one must also take only the transverse component.

Similarly, the near-field values are obtained when $kr \ll 1$:

$$\begin{cases} \mathbf{E}_{\text{near}} = \dfrac{-1}{4\pi\epsilon_0 c^2} \omega M \left(\dfrac{\hat{\boldsymbol{\phi}} \sin\theta}{r^2} \right) e^{j(\omega t - kr + \pi/2)} \\ \mathbf{E}_{\text{near}} = \dfrac{\mu_0 M}{4\pi} \left(\dfrac{\hat{\mathbf{r}} 2 \cos\Theta + \hat{\boldsymbol{\theta}} \sin\theta}{r^3} \right) e^{j(\omega t - kr)} \end{cases}$$

Here $\mathbf{B}_{\text{near}}$ is the same as the $\mathbf{B}$ of the static magnetic dipole multiplied by the sinusoidal factor. The electromagnetic wave in the near field is a TE wave. These results may be obtained from those of the near field of the Hertz dipole by letting $P \to M/c$: then

$$(c\mathbf{B}_{\text{near}})_{\text{mag}} \leftrightarrow (\mathbf{E}_{\text{near}})_{\text{elec}} \quad \text{while} \quad (\mathbf{E}_{\text{near}})_{\text{mag}} \leftrightarrow -(c\mathbf{B}_{\text{near}})_{\text{elec}}$$

Thus, in both the far field and the near field the magnetic dipole radiation is mathematically the same as the electric dipole radiation; the only physical distinction between the two cases, because $\mathbf{E}$ and $c\mathbf{B}$ are interchanged, comes in the geometry of the two types of dipole antennas. (The term "dipole antenna" is also widely used for the configuration of Fig. 19-12 regardless of the antenna length. This is unfortunate since the radiation pattern of an $m(\lambda/2)$ antenna, with $m = 2,3,4, \ldots$, is very different from that of a true dipole antenna; even for $m = 1$ the two are actually different, although the patterns are then somewhat similar in appearance. The terminology, however, is solidly entrenched.) Figure 19-10, which shows the lines of $\mathbf{E}$ for an oscillating electric dipole therefore also gives the lines of $\mathbf{B}$ for an oscillating magnetic dipole. The lines of $\mathbf{B}$ for the case of the electric dipole are circles in planes perpendicular to the paper. The lines of **E** for the magnetic dipole are also circles in planes perpendicular to the paper, but they are opposite in direction to the **B** of the electric dipole.

Electric Quadrupole Radiation

We have used the simplest possible source configurations to produce either the electric dipole or the magnetic dipole radiation. As in the corresponding static cases, however, the electric and magnetic potentials of an *arbitrary* configuration of charges and currents may be written in the form of a multipole expansion. Then, if the electric dipole moment is $\mathbf{p} = \int_{\mathcal{V}} \rho \mathbf{r}_s \, d\tau_s$, we obtain an electric dipole term in the radiation with the **p** of the simple Hertz dipole (see Example 1 below) replaced by the **p** of the particular distribution. Similarly, we obtain a magnetic dipole term in the radiation from an arbitrary configuration, with the magnetic dipole moment defined either by

$$\mathbf{m} = \int_{\mathcal{V}} \tfrac{1}{2}(\mathbf{r}_s \times \mathbf{J})\, d\tau_s$$

for a volume distribution or by

$$\frac{I}{2} \oint \mathbf{r}_s \times d\mathbf{r}_s$$

for a filamentary collection of currents. The derivations are more complicated (see Example 2 below), but the final result of the formulas for the *far-field* radiation in all cases (dipole, quadrupole, octupole, etc.) is the same as for the simplest configurations—provided only that the proper values of the various moments are employed. For electric quadrupole radiation the pertinent quantities would be the second moments, $\int \rho x_s^2 \, d\tau_s$, $\int \rho x_s y_s \, d\tau_s$, $\int \rho x_s z_s \, d\tau_s$, ... Suppose we take a simple case of this type.

Figure 19-16*a* shows two electric dipoles, having equal but opposite moments $\mathbf{p}_1 = \hat{\mathbf{z}} qs$ and $\mathbf{p}_2 = -\hat{\mathbf{z}} qs$, such that their centers are slightly displaced—they do not quite cancel each other. Here the displacement is along the line of their dipole moments, so the quadrupole is designated as linear. For each of the four constituent charges there is a common sinusoidal variation of the magnitude with time: $q = Qe^{j\omega t}$. Two small linear current elements, center-fed and carrying oppositely directed currents, would approximate this condition. The only nonvanishing second moment in this case would be

$$P_{zz} = Qs^2 e^{j\omega t} + (-2Q)\, 0^2 e^{j\omega t} + Qs^2 e^{j\omega t} = 2Qs^2 e^{j\omega t}$$

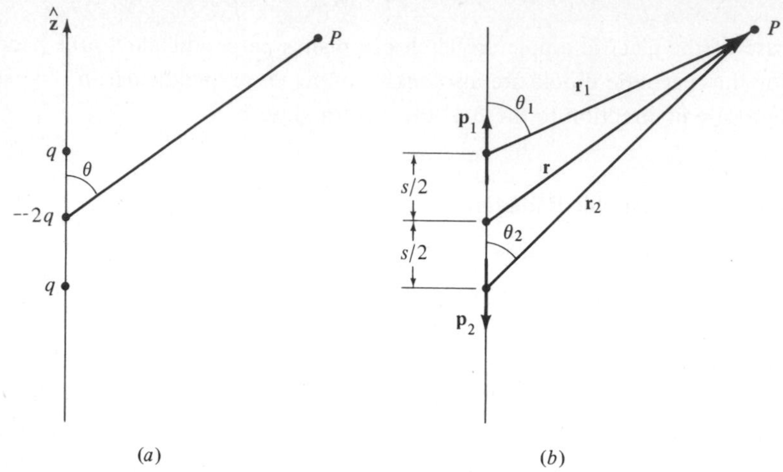

FIGURE 19-16

For simplicity we will only calculate the fields in the wave zone. These are the only pertinent fields required when the formulas are taken over to the multipole expansion of an arbitrary distribution. Then, instead of finding **A** for this combination of four charges and calculating **E** and **B** from it, we may use the previous results for $\mathbf{E}_{far}$ and $\mathbf{B}_{far}$ for the electric dipole radiation:

$$\mathbf{E}_{far} = -\hat{\boldsymbol{\theta}}\frac{P}{(4\pi\epsilon_0)\lambda\!\!\!^{-2} r}\sin\theta e^{j(\omega t - kr)}$$

$$\mathbf{B}_{far} = -\hat{\boldsymbol{\phi}}\left(\frac{\mu_0}{4\pi}\right)\frac{cP}{\lambda\!\!\!^{-2} r}\sin\theta e^{j(\omega t - kr)}$$

In Fig. 19-16*b* we have two such dipoles, with their centers displaced by $\pm s/2$ from the origin, respectively. To add the two **E**'s produced at some field point by these two dipoles we may let $\theta_1 = \theta_2 = \theta$ in the two numerators and we may let $r_1 = r_2 = r$ in the two denominators, but the phase factors of the two expressions may be appreciably different. Thus,

$$\mathbf{E}_{far} = -\hat{\boldsymbol{\theta}}\frac{P}{(4\pi\epsilon_0)\lambda\!\!\!^{-2} r}\sin\theta e^{j\omega t}\left(e^{-jkr_1} - e^{-jkr_2}\right)$$

But $\qquad r_1 \approx r - (s/2)\cos\theta \qquad$ and $\qquad r_2 \approx r + (s/2)\cos\theta$

$$\mathbf{E}_{far} = -\hat{\boldsymbol{\theta}}\frac{P}{(4\pi\epsilon_0)\lambda\!\!\!^{-2} r}\sin\theta e^{j(\omega t - kr)}[e^{jk(s/2)\cos\theta} - e^{-jk(s/2)\cos\theta}]$$

Taking $e^{jk(s/2)\cos\theta} \approx 1 + jk(s/2)\cos\theta + \cdots$ and $e^{-jk(s/2)\cos\theta} \approx 1 - jk(s/2)\cos\theta + \cdots$, the difference, when $ks \ll 1$, becomes $jks\cos\theta$. Then, setting $P_{zz} = 2Qs^2$,

Electric quadrupole

$$\begin{cases} \mathbf{E}_{\text{far}} = -\hat{\boldsymbol{\theta}} \dfrac{jP_{zz}}{(4\pi\epsilon_0)2\lambda\!\!\!^{-3}r} \sin\theta\cos\theta e^{j(\omega t-kr)} \\ \mathbf{B}_{\text{far}} = -\hat{\boldsymbol{\phi}} \left(\dfrac{\mu_0}{4\pi}\right) \dfrac{jcP_{zz}}{2\lambda\!\!\!^{-3}r} \sin\theta\cos\theta e^{j(\omega t-kr)} \end{cases}$$

The result for $\mathbf{B}_{\text{far}}$ was obtained in identical fashion.

A very important distinction becomes apparent here, between the static and radiative cases. (1) In the former the quadrupole **E** and **B** vary as r^{-4} while the dipole fields vary as r^{-3}, so the farther one goes from a static source the smaller the quadrupole field becomes relative to the dipole field. In the radiative case, however, both the dipole and the quadrupole field fall off as r^{-1} so there is no diminution of one, relative to the other with distance. In fact, this is true also of all the higher orders of multipole radiation.

(2) In the case of radiation, however, we see that at a given point the quadrupole fields become smaller and smaller relative to the dipole fields as the wavelength increases. As $\lambda \to \infty$ we approach the static case.

The angular pattern for electric quadrupole radiation, $dP/d\Omega$, is a figure of revolution which has four symmetric lobes; there are maxima at 45° and 135° for θ and zeros at 0°, 90°, and 180°. Dipole radiation has two symmetric lobes, with a maximum at $\theta = 90°$ and minima at $\theta = 0°$ and 180°. The names monopole, dipole, quadrupole, etc., coming from 2^n with $n = 0,1,2,\ldots$, designate the various multipole terms of ϕ or **A** for a static distribution. These fall off with distance according to $r^{-(n+1)}$. We now see that in the radiative case the names are also appropriate, since there are 2^n lobes in the xz plane, say, of the angular pattern. However, if one considers the whole three-dimensional instead of a cross section including the axis then the number of lobes is given by 2^{n-1}: the dipole pattern has but one lobe, the quadrupole two, etc.

Magnetic Quadrupole Radiation

Suppose we take a one-turn coil that acts as a source for magnetic dipole radiation, and place it axially above, and close to, another such coil in which the current is 180° out of phase with that in the first. The two dipoles–almost, but not quite, canceling each other–produce magnetic quadrupole radiation. See Fig. 19-17. (This is only one type of quadrupole that can be formed from the dipoles–the displacement could have been lateral instead of axial, or a combination of the two could also serve.) The use of the superposition principal to obtain the far field of the electric quadrupole from those of the component dipoles can be taken over here–lock, stock, and barrel–for the magnetic case since the **E** field of the electric dipole is the **B** field for the magnetic dipole. In the electric case the quadrupole result may be obtained by multiplying the dipole field by $(js\cos\theta)/\lambda\!\!\!^{-}$. The same factor is applicable in the magnetic case. Thus for magnetic quadrupole radiation we have

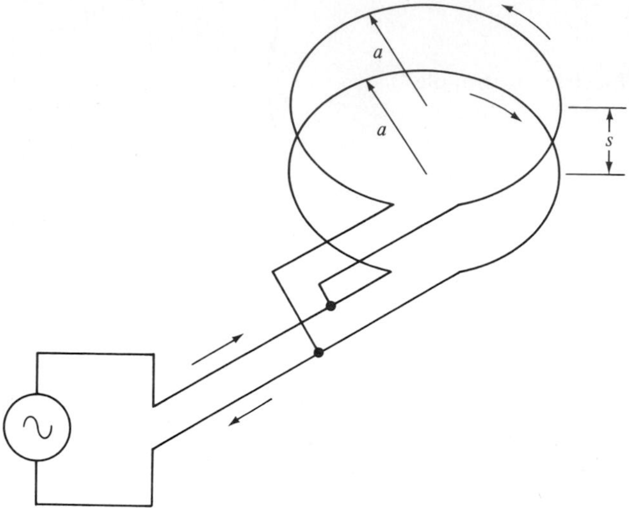

FIGURE 19-17

Magnetic quadrupole

$$\begin{cases} \mathbf{E}_{\text{far}} = \hat{\boldsymbol{\phi}} \dfrac{jMs}{(4\pi\epsilon_0)c\lambda\!\!\!^{-3} r} \sin\theta \cos\theta e^{j(\omega t - kr)} \\ \mathbf{B}_{\text{far}} = -\hat{\boldsymbol{\theta}} \left(\dfrac{\mu_0}{4\pi}\right) \dfrac{jMs}{\lambda\!\!\!^{-3} r} \sin\theta \cos\theta e^{j(\omega t - kr)} \end{cases}$$

The result for $(c\mathbf{B})$ here is obtainable from the $\mathbf{E}$ for electric quadrupole radiation by letting $P \longleftrightarrow M/c$; and the result for $\mathbf{E}$ here is obtainable from the $(-c\mathbf{B})$ for electric quadrupole radiation by using the same switch: $M/c \longleftrightarrow P$. (Another way of doing this is to let $M \longleftrightarrow P$ and $1/\epsilon_0 \longleftrightarrow \mu_0$; then $\mathbf{B}_{\text{mag}} \longleftrightarrow \mathbf{E}_{\text{elec}}$ while $\mathbf{E}_{\text{mag}} \longleftrightarrow -\mathbf{B}_{\text{elec}}$.

The angular distribution here is identical with that for electric quadrupole radiation. This similarly is true, except for the switch between **E** and **B**, for all the higher multipole terms of the radiation also. It is customary, therefore, to restrict one's consideration to the electric multipole radiation.

Examples

1. The relation between the Hertz dipole and the differential antenna The differential antenna treated in the text can be shown to be equivalent to the Hertz oscillating dipole shown in Fig. 19-18. Here we take two spheres, small compared to the distance between them, and connect them with a filamentary wire of zero resistance. Take the charge on a sphere to vary sinusoidally, $q = Qe^{j\omega t}$. Then the dipole moment is also sinusoidal, $\mathbf{p} = (Qe^{j\omega t})\mathbf{s} = (Q\mathbf{s})e^{j\omega t} = \mathbf{P}e^{j\omega t}$, while the current is $i = dq/dt = (j\omega Q)e^{j\omega t} = Ie^{j\omega t}$. So

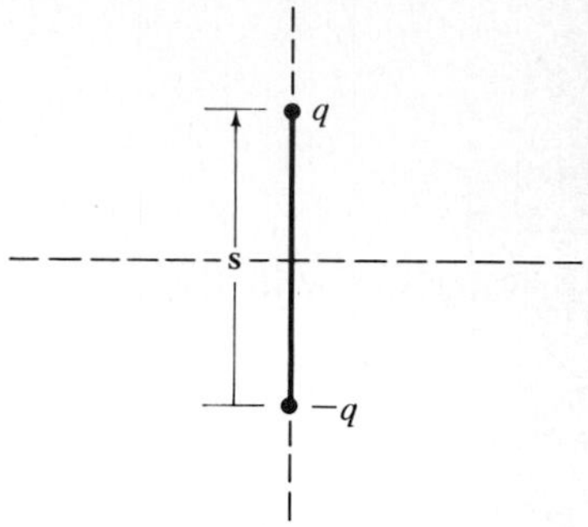

FIGURE 19-18

the expression for the current for the oscillating dipole is the same as for the differential antenna, but here $I = j\omega Q$. If $s \to 0$ while $q \to \infty$ such that p remains fixed we obtain the idealized point Hertz dipole.

The amplitude factor A_0 for the vector potential **A** of the differential antenna was found to be $A_0 = \mu_0 I \, \Delta z/4\pi$; setting $I = j\omega Q$ and $\Delta z = s$ then gives

$$A_0 = \frac{\mu_0 j\omega Qs}{4\pi} = \frac{j(\mu_0 \omega P)}{4\pi}$$

The previous amplitude expression, $-j\omega A_0/k^2$ for the near field of **E** then becomes

$$\frac{-j\omega}{k^2}\left(\frac{j\mu_0 \omega P}{4\pi}\right) = \frac{\mu_0}{4\pi} c^2 P = \frac{1}{4\pi\epsilon_0} P$$

so that

$$\begin{cases} \mathbf{E}_{\text{near}} = \dfrac{P}{4\pi\epsilon_0}\left(\dfrac{\hat{\mathbf{r}} 2 \cos\theta + \hat{\boldsymbol{\theta}} \sin\theta}{r^3}\right) e^{j(\omega t - kr)} \\ \mathbf{E}_{\text{near}} = \dfrac{\mu_0}{4\pi} \omega P \left(\dfrac{\hat{\boldsymbol{\phi}} \sin\theta}{r^2}\right) e^{j(\omega t - kr + \pi/2)} \end{cases}$$

We see that $\mathbf{E}_{\text{near}}$ is the same as the **E** of the static dipole multiplied by the sinusoidal factor.

The corresponding expressions for the far field are then

$$\begin{cases} \mathbf{E}_{\text{far}} = -\hat{\boldsymbol{\theta}} \dfrac{P}{(4\pi\epsilon_0)\lambda\!\!\!^{-2} r} \sin\theta e^{j(\omega t - kr)} \\ \mathbf{B}_{\text{far}} = -\hat{\boldsymbol{\phi}} \left(\dfrac{\mu_0}{4\pi}\right)\dfrac{c\mathrm{P}}{\lambda\!\!\!^{-2} r} \sin\theta e^{j(\omega t - kr)} \end{cases}$$

These can also be written more succinctly in terms of the retarded dipole moment $[\mathbf{p}] = \mathbf{P} e^{j(\omega t - kr)}$.

$$\begin{cases} \mathbf{E}_{\text{far}} = \left(\dfrac{1}{4\pi\epsilon_0 c^2}\right) \dfrac{([\ddot{\mathbf{p}}] \times \hat{\mathbf{r}}) \times \hat{\mathbf{r}}}{r} \\ \mathbf{B}_{\text{far}} = \left(\dfrac{\mu_0}{4\pi c}\right) \dfrac{[\ddot{\mathbf{p}}] \times \hat{\mathbf{r}}}{r} \end{cases}$$

We can also rewrite the expression for **A** in terms of [**p**]. Thus

$$\mathbf{A} = \hat{\mathbf{z}} j\omega \left(\frac{\mu_0}{4\pi}\right) P e^{j(\omega t - kr)} = j\omega \frac{\mu_0}{4\pi r} [\mathbf{p}]$$

so

$$\mathbf{A} = \left(\frac{\mu_0}{4\pi}\right) \frac{[\dot{\mathbf{p}}]}{r}$$

2. The distinction between the static and radiative multipole expansions of A We start with

$$\mathbf{A}(\mathbf{r}, t) = \frac{\mu_0}{4\pi} \int_{\mathcal{V}} \frac{\mathbf{J}(\mathbf{r}_s, t - R/c)}{R} \, d\tau_s$$

where **J** is sinusoidal. See Fig. 19-19. Then, if $\mathbf{J}_0$ is the amplitude of the current density oscillation,

$$\mathbf{A}(\mathbf{r}, t) = \frac{\mu_0}{4\pi} \int_{\mathcal{V}} \frac{\mathbf{J}_0(\mathbf{r}_s)}{R} e^{j\omega(t - R/c)} d\tau_s$$

$$= \frac{\mu_0}{4\pi} e^{j\omega t} \int_{\mathcal{V}} \frac{\mathbf{J}_0}{R} e^{-jkR} \, d\tau_s$$

In the near zone $kR \ll 1$ and $e^{-jkR} \approx 1$, so

$$\mathbf{A}_{\text{near}}(\mathbf{r}, t) = \frac{\mu_0}{4\pi} e^{j\omega t} \int_{\mathcal{V}} \frac{\mathbf{J}(\mathbf{r}_s)}{R} \, d\tau_s$$

This is the quasistatic approximation: $e^{j\omega t}$ times the result for the static case. We merely multiply the static result by the sinuosidal time factor common to all the current elements in order to obtain the instantaneous time-varying potential at the field point P.

In the far zone we have $r_s \ll \lambda \ll r$. Because $r_s \ll r$ we have

$$\frac{1}{R} = \frac{1}{r} \left\{ 1 + \frac{(\hat{\mathbf{r}} \cdot \mathbf{r}_s)}{r} - \frac{r_s^2 - 3(\hat{\mathbf{r}} \cdot \mathbf{r}_s)^2}{2r^2} + \cdots \right\} \approx \frac{1}{r}$$

and

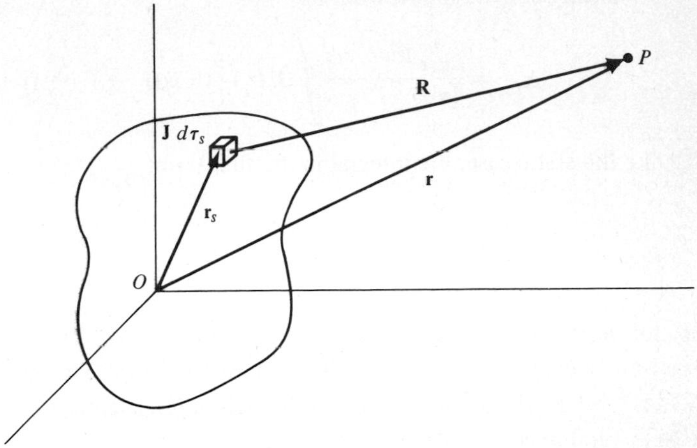

FIGURE 19-19

$$e^{-jkR} = \exp\left\{-jk\left[r - (\hat{\mathbf{r}}\cdot\mathbf{r}_s) + \frac{r_s^2 - (\hat{\mathbf{r}}\cdot\mathbf{r}_s)^2}{2r} + \cdots\right]\right\}$$

$$\approx e^{-jkr} e^{jk(\hat{\mathbf{r}}\cdot\mathbf{r}_s)}$$

so

$$\mathbf{A}_{\text{far}}(\mathbf{r}, t) \approx \frac{\mu_0}{4\pi r} e^{j(\omega t - kr)} \int_{\mathcal{V}} \mathbf{J}_0(\mathbf{r}_s) e^{jk(\hat{\mathbf{r}}\cdot\mathbf{r}_s)}\, d\tau_s$$

Further, because $r_s \ll \lambda$ we have $kr_s \ll 1$ so the exponential factor becomes

$$\sum_{n=0}^{\infty} \frac{(jk\hat{\mathbf{r}}\cdot\mathbf{r}_s)^n}{n!}$$

Then

$$\mathbf{A}_{\text{far}} = \frac{\mu_0}{4\pi} \frac{e^{j(\omega t - kr)}}{r} \int_{\mathcal{V}} \mathbf{J}_0(\mathbf{r}_s) \sum_0^{\infty} \frac{(jk\hat{\mathbf{r}}\cdot\mathbf{r}_s)^n}{n!}\, d\tau_s$$

Note that here the multipole terms do not contain any factors r^{-n}, unlike the case in the static development; here we have an expansion in inverse powers of λ. Only the terms which vary as $1/r$ contribute a finite radiated power.

Writing out the first few terms,

$$\mathbf{A}_{\text{far}} = \frac{\mu_0}{4\pi} \frac{e^{j(\omega t - kr)}}{r} \int_{\mathcal{V}} \mathbf{J}_0(\mathbf{r}_s) \left\{ 1 + jk(\hat{\mathbf{r}} \cdot \mathbf{r}_s) - \frac{k^2}{2}(\hat{\mathbf{r}} \cdot \mathbf{r}_s)^2 + \cdots \right\} d\tau_s$$

Unlike the static case, the integral in the first term,

$$\int_{\mathcal{V}} \mathbf{J}_0(\mathbf{r}_s) d\tau_s$$

is not necessarily equal to zero, though in certain cases this may be true. Using the continuity equation it can be shown [see "Classical Theory of Electric and Magnetic Fields" by R. H. Good, Jr., and T. J. Nelson (Academic Press, New York, 1971)] that this integral may be written

$$j\omega \int_{\mathcal{V}} \rho_0(\mathbf{r}_s) r_s d\tau_s$$

Here the integral is readily recognized as the first moment of the charge density, i.e., the electric dipole $\mathbf{p}_0$. Thus, the first term of the expansion here is

$$\mathbf{A}_{\text{far}} = \frac{\mu_0}{4\pi} \frac{e^{j(\omega t - kr)}}{r} \dot{\mathbf{p}} = \frac{\mu_0}{4\pi} \frac{[\dot{\mathbf{p}}]}{r}$$

Then $\mathbf{E}_{\text{far}}$ and $\mathbf{B}_{\text{far}}$ are obtained from this expression by taking $\mathbf{E} = -\nabla\phi - \partial\mathbf{A}/\partial t$ [with $\phi = -(c^2/j\omega)\,\nabla \cdot \mathbf{A}$] and $B = \nabla \times \mathbf{A}$. The results are found to be identical with the $[\ddot{\mathbf{p}}]$ expression in the previous example if all the $1/r^2$ terms are ignored (because they do not contribute to the radiation). Thus, this term of the multipole expansion of **A** is called the electric dipole term.

The second term of the multipole expansion contains both the electric quadrupole and the magnetic dipole radiation. The former comes from the symmetric portion of the term, the latter from the antisymmetric. The magnetic quadrupole radiation comes from part of the third term, etc., but we shall not go into the details.

3. Two simple radiation configurations We have previously considered two elementary examples of radiation: the Hertz oscillating dipole and a single charge in a circular orbit with constant angular velocity. Each of these cases may be modified in a simple way to bring out some points of interest. We will consider (*a*) a single charge oscillating sinusoidally along a straight line and (*b*) two equal charges, at the ends of a diameter, in a circular orbit with equal and constant angular velocity.

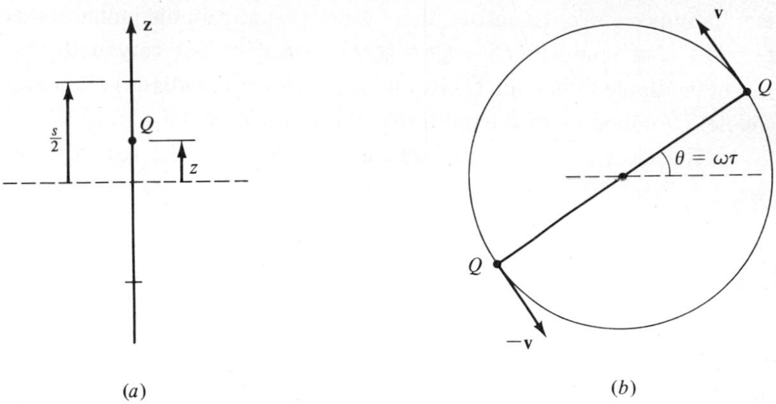

FIGURE 19-20

a. The single oscillating charge. Figure 19-20*a* differs from Fig. 19-19: the system now has a small charge Q, whose position varies sinusoidally along the z axis. But there is no monopole radiation term in the monopole expansion of **A**. (Even if there were, no radiation would be produced by it since the charge is constant.) The lowest order multipole (i.e., order in λ^{-n}) that needs to be considered is the dipole. The formula for the dipole term of **A** is

$$\mathbf{A} = \frac{\mu_0}{4\pi}\frac{[\dot{\mathbf{p}}]}{r}$$

Previously, for Fig. 19-17, we had $\mathbf{p} = \mathbf{P}e^{j\omega t}$, $\dot{\mathbf{p}} = j\omega\mathbf{p}$, $[\mathbf{p}] = \mathbf{P}e^{j(\omega t - kz)}$, and $[\dot{\mathbf{p}}] = j\omega[\mathbf{p}]$; where $\mathbf{P} = \hat{\mathbf{z}}Qs$. So there $A = (\mu_0/4\pi r)\hat{\mathbf{z}}j\omega P e^{j(\omega t - kz)}$.

For Fig. 19-20*a* we now have $\mathbf{p} = \hat{\mathbf{z}}qz = \hat{\mathbf{z}}Q(s/2)e^{j\omega t}$. So now $\mathbf{P} = \hat{\mathbf{z}}\frac{1}{2}Qs$. Though the method for obtaining the variable dipole moment is different from the previous case, the net result is that the moment now is half its previous value. All the results that obtained before will then hold true here if half the previous **p** is used.

b. The two charges in a common circular orbit. This is indicated in Fig. 19-20*b*. Again, the dipole radiation is the lowest order to be considered. We treat the electric dipole radiation first: $\mu_0/4\pi r[\dot{\mathbf{p}}]$. From

$$\mathbf{p} = \sum_{n=1}^{2} q_n(\mathbf{r}_s)_n$$

we have $\mathbf{p} = \hat{\mathbf{x}}[Qa\cos\theta + Qa\cos(\pi+\theta)] + \hat{\mathbf{y}}[Qa\sin\theta + Qa\sin(\pi+\theta)] = 0$, independent of the time. So $[\dot{\mathbf{p}}] = 0$ and there is no electric dipole radiation.

Next we consider the magnetic dipole and electric quadrupole radiation. But the magnetic dipole radiation may be obtained directly from the electric dipole radiation by employing $P \to M/c$. Therefore, there is no magnetic dipole radiation here either. (We have $\mathbf{m} = \mathbf{M} = \hat{\mathbf{z}} I\pi a^2$, and $I = Qv + Qv = 2Qv$; so $\mathbf{m}$ does not vary with time and there is no magnetic dipole radiation. There will be synchrotron radiation, however. This comes from the next component of the multipole expansion.)

The electric quadrupole radiation comes from the second moments. In Chap. 6, Sec. 6-1, Prob. 20, we had

$$\frac{1}{4\pi\epsilon_0 r^3} \sum_{s=1}^{2} q_s r_s^2 \left(\frac{3\cos^2\theta_s - 1}{2}\right)$$

for the quadrupole term of the static expansion of ϕ. That term could also be written

$$\frac{1}{4\pi\epsilon_0 r^5} \sum_{i,j} Q_{ij} a_{ji}$$

where

$$Q_{ij} = \sum_{s=1}^{N} Q_s [3(\mathbf{r}_s)_i (\mathbf{r}_s)_j - (\mathbf{r}_s)^2 \delta_{ij}]$$

is the (i,j) component of the second moment of the distribution, and $a_{ij} = \frac{1}{6}[3r_i r_j - r^2 \delta_{ij}]$ is a function only of the test point.

In the present case we have

$$\begin{cases} q_1 = Q \\ r_1 = a \\ x_1 = a\cos\omega t \\ y_1 = a\sin\omega t \\ z_1 = 0 \end{cases} \qquad \begin{cases} q_2 = Q \\ r_2 = a \\ x_2 = -a\cos\omega t \\ y_2 = -a\sin\omega t \\ z_2 = 0 \end{cases}$$

Thus

$$\begin{aligned} Q_{11} &= q_1(3x_1^2 - r_1^2) + q_2(3x_2^2 - r_2^2) \\ &= Q(3a^2\cos^2\omega t - a^2) + Q(3a^2\cos^2\omega t - a^2) \\ &= 2Qa^2(3\cos^2\omega t - 1) = Qa^2(3\cos 2\omega t + 1) \end{aligned}$$

Also,

$$\begin{aligned} Q_{12} &= q_1(3x_1 y_1) + q_2(3x_2 y_2) \\ &= Q(3a\cos\omega t)(a\sin\omega t) + Q(-a\cos\omega t)(-a\sin\omega t) \\ &= 6Qa^2 \sin\omega t\cos\omega t = 3Qa^2 \sin 2\omega t \end{aligned}$$

In this manner we obtain

$$\{Q_{ij}\} = Qa^2 \begin{bmatrix} 3\cos 2\omega t + 1 & 3\sin 2\omega t & 0 \\ 3\sin 2\omega t & -3\cos 2\omega t + 1 & 0 \\ 0 & 0 & -2 \end{bmatrix}$$

$$= 3Qa^2 \begin{bmatrix} \cos 2\omega t & \sin 2\omega t & 0 \\ \sin 2\omega t & -\cos 2\omega t & 0 \\ 0 & 0 & 0 \end{bmatrix} + \begin{bmatrix} 1 & 0 & 0 \\ 0 & 1 & 0 \\ 0 & 0 & -2 \end{bmatrix}$$

The physical reason for the appearance of the double-frequency component is clear from the diagram: one cycle of radiation will be completed in half a period of particle revolution. To find the quadrupole radiation the static component matrix may be ignored; the complex matrix of which the first matrix is the real part is, then,

$$Q_{ij} = 3Qa^2 \begin{bmatrix} 1 & j & 0 \\ j & -1 & 0 \\ 0 & 0 & 0 \end{bmatrix} e^{j2\omega t}$$

Since in the text we have not actually carried out the derivation of the electric quadrupole radiation to the very end, to obtain a specific formula for **A**, we will leave the development in this case incomplete at this point. There are actually only a few more steps required. The interested reader may consult the text by Good and Nelson mentioned in the previous example.

PROBLEMS

1 If the Lorentz condition $\nabla \cdot \mathbf{A} + (1/c^2)(\partial\phi/\partial t) = 0$ is imposed, then the potentials satisfy the equations $\Box^2 \phi = -\rho/\epsilon_0$ and $\Box^2 \mathbf{A} = -\mu_0 \mathbf{J}$ in free space. The **A** and ϕ are said to give the Lorentz gage. Suppose that, instead of the Lorentz condition, one imposes the equation $\nabla \cdot \mathbf{A} = 0$. The ϕ and **A** that now result are called the Coulomb gage.

Find the differential equations that ϕ and **A** must satisfy in the Coulomb gage in free space.

2 (*a*) Show that the Lorentz condition, $\partial A^i/\partial x^i = 0$, is a sufficient condition for the conservation of charge, $\partial J^i/\partial x^i = 0$.

(*b*) Is it a necessary condition?

3 A spherical shell of charge, of differential thickness dr, radiates sinusoidally along the radial direction. Find the Poynting vector.

4 (*a*) Prove that $[R - \boldsymbol{\beta} \cdot \mathbf{R}] = r\sqrt{1 - \beta^2 \sin^2\theta}$, where θ is shown in Fig. 19-2.

(*b*) Prove that if $s = \sqrt{x^2 + (1 - \beta^2)y^2 + (1 \quad \beta^2)z^2}$ then $s = r\sqrt{1 - \beta^2 \sin^2\theta}$.

5 (*a*) Express the Coulomb potential, $(q/4\pi\epsilon_0)1/r_0$, in a form-invariant fashion that is valid in any frame moving with the four-velocity u^i relative to the proper frame, $u^i = (\gamma\mathbf{u},\gamma c)$. Let $R^i = \{\mathbf{r},r\}$ be the four-distance.

(*b*) Let $s = r - \mathbf{r}\cdot\boldsymbol{\beta}$. Find the components of the relativistic $\{\phi\}$ of (*a*) in terms of s. These are the Liénard-Wiechert potentials, obtained here relativistically instead of, classically, by retardation.

6 Let $\mathbf{p} = \mathbf{p}(t - R/c)$. Show that for $\partial\mathbf{p}/\partial r$ in any operation we may substitute $-(1/c)(\partial\mathbf{p}/\partial t)$. (This offers a great simplification in calculating **E** and **B** from **A**.)

7 (*a*) Graph the Liénard-Wiechert potentials at $R = 1$ (if $\mathbf{v}$ is along the $+\hat{\mathbf{x}}$ direction) for all values of α, the angle between $\boldsymbol{\beta}$ and $\mathbf{R}$, the retarded position vector (see Fig. 19-2). Take $\beta = 0$ and $\beta = 0.8$.

(*b*) Using the results of Prob. 4, graph these potentials at $r = 1$ for all values of θ when $\beta = 0, 0.8$.

8 Show that the Poynting vector equals zero for a charged particle moving with constant velocity.

9 (*a*) Find a relation between ϕ and **A** in the wave zone of an oscillating system.

(*b*) Is this true in the static zone also?

(*c*) Is it possible to have $\phi = 0$, $\mathbf{A} \neq 0$?

(*d*) Is it possible to have $\phi \neq 0$, $\mathbf{A} = 0$?

10 Show that $\mathbf{B} = (1/c^2)\mathbf{v} \times \mathbf{E}$ for a uniformly moving charge.

11 For the case of a charged particle moving with constant speed around a circle the angular dependence of the radiation was derived in the text:

$$\frac{dP}{d\Omega} = \left(\frac{Q^2\dot{v}^2}{16\pi^2\epsilon_0 c^3}\right)\frac{(1-\beta\cos\theta)^2 - (1-\beta^2)\sin^2\theta\cos^2\phi}{(1-\beta\cos\theta)^5}$$

If the radius is a and the angular velocity is ω, find the radiated power

$$P = \int_{\text{sphere}} \left(\frac{dP}{d\Omega}\right) d\Omega$$

12 For a charge moving with arbitrary velocity we set $\mathbf{E} = \mathbf{E}_v + \mathbf{E}_a$ and $\mathbf{B} = \mathbf{B}_v + \mathbf{B}_a$. Find the dependence on the distance r that is characteristic of the Poynting vector caused by (*a*) $\mathbf{E}_a$ and $\mathbf{B}_a$, (*b*) $\mathbf{E}_a$ and $\mathbf{B}_v$, (*c*) $\mathbf{E}_v$ and $\mathbf{B}_a$, (*d*) $\mathbf{E}_v$ and $\mathbf{B}_v$.

(*e*) Show that $\mathbf{E}_a$ is perpendicular to $[\mathbf{R}]$.

13 Show that the phase velocity of the scalar potential produced by an electric dipole in the induction field can be greater than c.

14 In the text the Larmor formula for the radiated power was obtained for $\beta \to 0$ when the particle's velocity and acceleration were parallel. Show that this result is obtained when $\beta \to 0$ regardless of the relation between the directions of $\mathbf{v}$ and $\dot{\mathbf{v}}$.

15 Let the circulating particle of Prob. 11 have a mass m.

(*a*) Find a formula for the ratio of the radiated power P to the kinetic energy of the particle K. (This is $-\dot{K}/K$; the formula should be multiplied by one-half if we wish to consider the total energy, $W = P + K = 2K$, of a harmonic oscillator.)

(*b*) What is the minimum frequency of rotation for which $-\dot{K}/K$ becomes 1 percent for an electron when $\beta = 0.1$? $\beta = 0.95$?

(*c*) For a proton?

16 From the formula for the average Poynting vector $\langle \mathbf{S} \rangle$ for electric dipole radiation find the formula for $dP/d\Omega$, the average angular (per steradian) radiated power.

17 (*a*) Find the radiation resistance of an oscillating electric dipole ($P = I_{\text{rms}}^2 R$).

(*b*) Repeat for a short dipole antenna. (For a half-wave antenna the computation is much more complicated; the result is 73 Ω.)

18 (*a*) What is the dependence of the radiated power of an oscillating electric dipole on the wavelength?

(*b*) Of an oscillating magnetic dipole?

(*c*) Of an oscillating electric quadrupole?

(*d*) For oscillating dipoles and quadrupoles one can go from the electric to the magnetic case by switching the lines of **E** and **B**. Will that hold true, also, for higher-ordered multipoles?

19 (*a*) Calculate the time-averaged Poynting vector for a Hertz dipole using the *near-field* expressions for **E** and **B**.

(*b*) Does (*a*) give the correct value for the radiation?

(*c*) What is the ratio $|\mathbf{E}/\mathbf{B}|$ in the near field?

(*d*) In the far field?

20 (*a*) Show that $\nabla \cdot \mathbf{A} = 0$ in the wave zone of the oscillating magnetic dipole.

(*b*) Justify the statement in the text that the electric potential ϕ for a magnetic dipole may be set equal to zero in the wave zone, $a \ll \lambda$.

(*c*) Is $\phi = 0$ in the near field as well?

(*d*) The prototype of the oscillating magnetic dipole is a one-turn loop of wire fed by an oscillator. This is a conduction current and in the wave zone the potential is then a constant, taken to be zero for convenience. Suppose the current is a convection current: a betatron ring with sinusoidal particle velocity. Will $\phi = 0$ in the wave zone? In the near field?

21 (See Prob. 15) Let the energy lost per cycle by radiation of a nonrelativistic harmonic oscillator of mass m and charge Q be very small compared to the oscillator's total energy. Assume a damping force F that causes energy loss, and assume no radiation. Find F such that the power loss by damping equals that actually lost by radiation.

22 Fill in the following table to give the dependence of **E** and **B** on frequency for the electric and magnetic oscillating dipoles. Consider the near fields and the far fields separately.

	Electric dipole		Magnetic dipole	
	Near field	Far field	Near field	Far field
E				
B				

23 The monopole antenna is a widely used variation of the dipole antenna. Only the upper half of the center-fed dipole antenna is used, of height $h = l/2$. It is end-fed, near the ground. The earth is used as a reflecting plane (assuming it to be a good conductor) to duplicate the other half of the antenna as an image.

Find the radiation resistance of a short monopole antenna of height h.

24 Show that $r/R \approx 1 + (a/r) \sin\theta \cos\phi$ for $r \gg a$. (See Fig. 19-14.)

25 If W is the radiated power of an electric dipole, find the rms electric field at a point (r,θ,ϕ) in the wave zone. (The ϕ is immaterial because of symmetry.)

26 Can the vector potential of a magnetic multipole be obtained from that of the electric multipole by switching $\mathbf{m}/c$ for $\mathbf{p}$? By switching $\mathbf{m}$ for $\mathbf{p}$ and μ_0 for $1/\epsilon_0$?

REFERENCE TABLE I

FUNDAMENTAL CONSTANTS

Planck's constant	h	6.63×10^{-34} J s
	$\hbar$	6.6×10^{-16} eV s
Charge quantum	e	1.602×10^{-19} C
Electron mass	m	9.109×10^{-31} kg
Proton mass	m_{pro}	1.6725×10^{-27} kg
Boltzmann's constant	k	1.3805×10^{-23} J K^{-1}
		8.62×10^{-5} eV K^{-1}
Permittivity of free space	ϵ_0	8.854×10^{-12} F m^{-1}
		$\left(\frac{1}{4\pi\epsilon_0} \approx 9 \times 10^9 \text{ m F}^{-1}\right)$
Permeability of free space	μ_0	$4\pi \times 10^{-7}$ H m^{-1}
Avogadro's constant	N_A	6.0225×10^{26} kmol^{-1}
Gravitational constant	G	6.67×10^{-11} N m^2 kg^{-2}
Velocity of light in vacuum	c	2.99×10^8 m s^{-1}
Bohr magneton	m_B	9.27×10^{-24} A m^2
Electron magnetic moment	m_e	9.28×10^{-24} A m^2
Proton magnetic moment	m_p	1.41×10^{-26} A m^2
Faraday's constant	F	96,490 C/chemical equivalent*
Rydberg constant	R_∞	1.097×10^7 m^{-1}
Gas constant	R	8310 J K^{-1} kmol^{-1}
Stefan-Boltzmann constant	σ	5.67×10^{-8} W m^{-2} K^{-4}
Fine structure constant	α	7.297×10^{-3}

*The chemical equivalent = M/z = (molecular weight)/valence.

REFERENCE TABLE II

II-A UNITS: DEFINED VALUES

1 meter	1,650,763.73 wavelengths of the Kr^{86} transition $2p_{10} - 5d_5$.
1 kilogram	The mass of a standard stored in Paris.
1 second	9,192,631,770 periods of the Cs frequency.
1 ampere	$F/l = 2/D \times 10^{-7}$ N for 1 A, D meters.
1 kelvin	273.16 K = the temperature of the triple point of water. (The freezing point is 273.15 K = 0°C.)
1 atomic mass unit	$\frac{1}{12}$ the mass of the C^{12} nuclide (= 1.66 × 10^{-27} kg = 1.49 × 10^{-10} J = 931 MeV).
1 normal atmosphere	101,325 N m^{-2} (= 1.01325 bars)
Standard acceleration of gravity	9.80665 m s^{-2}.

1 calorie	4.1840 J (thermochemical); 4.1868 J (International Steam Table).
1 liter	10^{-3} m^3 (= 0.2642 U.S. gal = 0.220 Imperial gallons).
1 angstrom	10^{-10} m.
1 inch	0.0254 m (from this it follows that 1 ft = 0.305 cm and 1 km = 0.621 mi).
1 pound (mass)	0.453,592,37 kg (it follows that 1 oz (mass) = 28.35 g (mass), and 1 oz (force) = 28.35 g (force) in the sense that 1 kg (force) = 9.8 N).
1 light-year	9.5×10^{15} m.
1 parsec	3×10^{16} m = 3.2 light-years.

II-B MKSA-CGS CONVERSION FACTORS FOR UNITS

The table below gives the number of cgs units that are equivalent to 1 mksa unit. This number of cgs units would measure the same quantity as would be measured by 1 mksa unit. The equivalence is an equality only when the dimensions in the two systems are identical. The inverse of the values would give the number of units required to measure a given quantity measured by 1 mksa unit.

Quantity	Mksa unit	Cgs units		
		Esu	Emu	Gaussian
Capacitance, C	1 F (farad, μF = microfarad = 10^{-6} F, pF = picofarad = 10^{-12} F)	$(10^{-9}\ c^2)$ cm = 9×10^{11} statfarad	10^{-9} abfarad	9×10^{11} esu
Charge, q	1 C (coulomb)	$(0.1\ c)$ statcoulomb	0.1 abcoulomb	3×10^9 esu
Conductivity, g	1 mho-m^{-1}	$(10^{-11}\ c^2)$ statmho cm^{-1}	10^{-11} abmho cm^{-1}	10^{-11} emu
Current, I	1 A (ampere)	$(10^{-1}\ c)$ statampere	0.1 abampere	0.1 emu
D field, D	1 C m^{-2}	$(4\pi \times 10^{-5}\ c)$ dyne-stat-coulomb^{-1}	$(4\pi \times 10^{-5})$ dyne abcoulomb^{-1}	3.77×10^{-6} esu
Electric field, E	1 V m^{-1}	$(10^6/c)$ dyne-statcoulomb^{-1}	10^6 dyne-abcoulomb^{-1}	$\left(\frac{1}{30{,}000}\right)$ esu
Energy, U	1 J (joule)	10^7 erg	10^7 erg	10^7 erg
Energy density, u	1 J m^{-3}	10 erg cm^{-3}	10 erg cm^{-3}	10 erg cm^{-3}
Energy transport density, S	1 J m^{-2} s^{-1}	10^3 erg cm^{-2} sec^{-1}	10^3 erg cm^{-2} s^{-1}	10^3 erg cm^{-2} s^{-1}
Force, F	1 N (newton)	10^5 dyne	10^5 dyne	10^5 dyne
Frequency, f	1 Hz (hertz, kHz = 10^3 Hz, Mhz = 10^6 Hz)	1 Hz	1 Hz	1 Hz
H field, H	1 A m^{-1} or A-turn m^{-1}	$(4\pi \times 10^{-3}\ c)$ statampere cm^{-1}	$(4\pi \times 10^{-3})$ Oe	0.0126 Oe
Inductance, L	1 H (henry, μH = microhenry = 10^{-6} H, pH = picohenry = 10^{-12} H)	$(10^9/c^2)$ statohm s	10^9 abohm s	10^9 emu
Length, ℓ	1 m (meter)	10^2 cm	10^2 cm	10^2 cm
Magnetic field, B	1 T (tesla)	$(10^4/c)$ statvolt s cm^{-2}	10^4 G (gauss)	10^4 G
Magnetic flux, Φ	1 Wb (weber)	$(10^8/c)$ statvolt s	10^8 Mx (maxwell)	10^8 Mx
Magnetization, M	1 A m^{-1}	$(10^{-3}\ c)$ statamp cm^{-1}	10^{-3} Oe (oersted)	10^{-3} Oe
Magnetomotance, $\mathfrak{M}$	1 A-turn	$0.4\pi c$ statampere-turn	0.4π Gb (gilbert)	1.26 Gb
Mass, m	1 kg (kilogram)	10^3 g	10^3 g	10^3 g
Mole, mol	1 kg mol	10^3 g mol	10^3 g mol	10^3 g mol
Permeability, μ	1 H m^{-1}	κ_m used, but not μ_0	κ_m used, but not μ_0	κ_m, same as in mksa

(Continued)

Quantity	Mksa unit	Cgs units		
		Esu	Emu	Gaussian
Permittivity, ϵ	1 F m^{-1}	κ used, but not ϵ_0	κ used, but not ϵ_0	κ, same as in mksa
Polarization, P	1 C m^{-2}	$(10^{-5}\ c)$ statcoulomb cm^{-2}	10^{-5} abcoulomb cm^{-2}	3×10^5 esu
Potential, ϕ	1 V (volt)	$(10^8/c)$ statvolts	10^8 abvolts	$(\frac{1}{300})$ esu
Power, P	1 W (watt)	10^7 erg s^{-1}	10^7 erg sec^{-1}	10^7 erg sec^{-1}
Reluctance, $\mathcal{R}$	1 A-turn Wb^{-1}	$36\pi \times 10^{11}$ $statohm^{-1}$ s^{-1}	$4\pi \times 10^{-9}$ Gb Mx^{-1}	1.26×10^{-8} emu
Resistance, R	1 Ω (ohm)	$(10^9/c^2)$ statohm	10^9 abohms	1.11×10^{-12} esu
Temperature, T	1 K (kelvin, also °K or deg K)	1 K	1 K	1 K
Time, t	1 s (second)	1 s	1 s	1 s
Vector potential, A	1 W m^{-1}	$(10^6/c)$ statvolt s cm^{-1}	10^6 G cm	10^6 emu
Voltage, V	1 V	$(10^8/c)$ statvolts	10^8 abvolts	$(\frac{1}{300})$ esu
Volume, $\mathcal{V}$	1 m^3	10^6 cm^3	10^6 cm^3	10^6 cm^3

REFERENCE TABLE III

MKSA DIMENSIONS

This table lists the exponents α, β, γ, δ to which the four basic dimensions—[m], [kg], [s], [A], respectively—must be raised in order to obtain the dimensions of various electromagnetic quantities. For example, the exponents for the dimensions of permittivity are listed as −3, −1, 4, 2; so $[\epsilon_0] = [m^{-3}\ kg^{-1}\ s^4\ A^2]$. Here, as elsewhere in the book, we distinguish between dimensions (e.g., length) and units (e.g., meter) by using a square bracket for dimensions. Thus, [m] means the dimension which has the meter for its unit.

$[m^{\alpha}\ kg^{\beta}\ s^{\gamma}\ A^{\delta}]$

Capacitance	Symbol	Unit	α	β	γ	δ
Capacitance	C	F	−2	−1	4	2
Charge	q	C	0	0	1	1
Charge density	ρ	$C\ m^{-3}$	−3	0	1	1
Conductance	G	mho	−2	−1	3	2

Capacitance	Symbol	Unit	α	β	γ	δ
Conductivity	g	mho-m^{-1}	−3	−1	3	2
Current	I	A	0	0	0	1
Current density	$\mathbf{J}$	A m^{-2}	−2	0	0	1
D field	D	C m^{-2}	−2	0	1	1
Electric field	E	V-m^{-1}	1	1	−3	−1
Energy	U	J	2	1	−2	0
Energy density	u	J m^{-3}	−1	1	−2	0
Energy flux	S	J m^{-2} s^{-1}	0	1	−3	0
Force	F	N	1	1	−2	0
H field	H	A-turns m^{-1}	−1	0	0	1
Hertz vector	Π	V-m	3	1	−3	−1
Impedance	Z	Ω	2	1	−3	−2
Inductance	L	H	2	1	−2	−2
Length	ℓ	m	1	0	0	0
Linear charge density	λ	C m^{-1}	−1	0	1	1
Magnetic field	B	T	0	1	−2	−1
Magnetic flux	Φ	W	2	1	−2	−1
Magnetization	M	A m^{-1}	−1	0	0	1
Magnetomotance	$\mathfrak{M}$	A-turns	0	0	0	1
Mass	m	kg	1	0	0	0
Momentum	p	m kg s^{-1}	1	1	−1	0
Momentum density	g	m^{-2} kg s^{-1}	−2	1	−1	0
Momentum flux	f	m^{-1} kg s^{-2}	−1	1	−2	0
Permeability	μ	H m^{-1}	1	1	−2	−2
Permittivity	ϵ	F m^{-1}	−3	−1	4	2
Polarizability	α	F m^{-2}	0	−1	4	2
Polarization	P	C m^{-2}	−2	0	1	1
Potential	ϕ	V	2	1	−3	−1
Potential energy	U	J	2	1	−2	0
Power	P	W	2	1	−3	0
Poynting vector	S	W m^{-2}	0	1	−3	0
Reluctance	$\mathcal{R}$	A-turns Wb^{-1}	−2	−1	2	2
Resistance	R	Ω	2	1	−3	−2
Resistivity	ρ	Ω m	3	1	2	1
Surface charge density	σ	C m^{-2}	−2	0	1	1
Surface current density	j, J_ℓ	A m^{-1}	−1	0	0	1
Time	t	s	0	0	1	0
Vector potential	A	T	1	1	−2	−1
Voltage	V	V	2	1	−3	−1
Work	W	J	2	1	−2	0

REFERENCE TABLE IV

MKSA-CGS CONVERSION FACTORS IN EQUATIONS

To convert an equation written in rationalized mksa units into gaussian cgs form, multiply each symbol appearing in the equation by the conversion factor listed below. If a quantity appearing in the equation has no listed conversion factor then that factor is unity. To go in the reverse direction, from a cgs to an mksa equation, multiply each quantity by the inverse of the factor listed. The table lists the combination of two factors which enter: (1) a dimensional adjustment and (2) a rationalization conversion. The table here is then sufficient for converting an equation from one system to another. But for comparing a quantity measured in the two systems it is necessary to use a third factor, a size-of-units comparison. This is given in Reference Table II-B.

Example: the mksa equation

$$\mathbf{F} = \frac{1}{4\pi\epsilon_0}\frac{q_1 q_2}{R^2}\hat{\mathbf{R}}_{12}$$

becomes

$$\mathbf{F} = \frac{1}{4\pi\epsilon_0}\frac{(\sqrt{4\pi\epsilon_0}q_1)(\sqrt{4\pi\epsilon_0}q_2)}{R^2}\hat{\mathbf{R}}_{12}$$

or

$$\mathbf{F} = \frac{q_1 q_2}{R^2}\hat{\mathbf{R}}_{12}$$

Quantity	Symbol	Conversion factor
Capacitance	C	$4\pi\epsilon_0$
Charge	q	$\sqrt{4\pi\epsilon_0}$
Charge density	ρ, σ, λ	$\sqrt{4\pi\epsilon_0}$
Conductivity	g	$4\pi\epsilon_0$
Current	I	emu, $\sqrt{4\pi/\mu_0}$; esu, $\sqrt{4\pi\epsilon_0}$
Current density	J, j	emu, $\sqrt{4\pi/\mu_0}$; esu, $\sqrt{4\pi\epsilon_0}$
D field	D	$\sqrt{\epsilon_0/4\pi}$
Electric dipole moment	p	$\sqrt{4\pi\epsilon_0}$
Electric field	E	$1/\sqrt{4\pi\epsilon_0}$
H field	H	$1/\sqrt{4\pi\mu_0}$
Inductance	L	$\mu_0 c^2/4\pi$
Magnetic dipole moment	B	$\sqrt{4\pi/\mu_0}$
Magnetic field	B	$\sqrt{\mu_0/4\pi}$
Magnetic flux	Φ	$\sqrt{\mu_0/4\pi}$
Magnetization	M	$\sqrt{4\pi/\mu_0}$
Permeability	μ	$\kappa_m \mu_0/\mu$
Permittivity	ϵ	$\kappa\epsilon_0/\epsilon$
Polarization	P	$\sqrt{4\pi\epsilon_0}$
Potential	ϕ	$1/\sqrt{4\pi\epsilon_0}$
Reluctance	$\mathcal{R}$	$\mu/\kappa_m \mu_0$
Resistance	R	$1/4\pi\epsilon_0$
Resistivity	ρ	$1/4\pi\epsilon_0$
Vector potential	A	$\sqrt{\mu_0/4\pi}$
Voltage	V	$1/\sqrt{4\pi\epsilon_0}$

ANSWERS TO ODD-NUMBERED PROBLEMS

CHAP. 1

Sec. 1-1

1. $\hat{\mathbf{x}}(1/y) - \hat{\mathbf{y}}(x/y^2)$ 3. $\hat{\mathbf{x}}y + \hat{\mathbf{y}}x$ 5. $\hat{\mathbf{x}}e^{-x} \sin y + \hat{\mathbf{y}}e^{-x} \cos y$

7. $\cos^{-1} (0.1\sqrt{10})$ 9. 1 11. $2\sqrt{2}$

13. $\hat{\mathbf{X}}(\partial\phi/\partial X) + \hat{\mathbf{Y}}(\partial\phi/\partial Y) + \hat{\mathbf{Z}}(\partial\phi/\partial Z)$; yes.

15. $\hat{\mathbf{X}}[(3 \cos^2 \theta + \sin^2 \theta)(\partial\phi/\partial X) + (-3 \sin \theta \cos \theta + \sin \theta \cos \theta)(\partial\phi/\partial Y)]$
$+ \hat{\mathbf{Y}}[(-3 \sin \theta \cos \theta + \sin \theta \cos \theta)(\partial\phi/\partial X) + (3 \sin^2 \theta + \cos^2 \theta)(\partial\phi/\partial Y)] + \hat{\mathbf{Z}}(\partial\phi/\partial Z)$.

Not form-invariant under rotation. Yes, form-invariant under translation. Yes, magnitude invariant after rotation. Very little significance.

17. $\dfrac{\partial F_x}{\partial y} = \dfrac{\partial F_y}{\partial x}, \dfrac{\partial F_y}{\partial z} = \dfrac{\partial F_z}{\partial y}, \dfrac{\partial F_z}{\partial x} = \dfrac{\partial F_x}{\partial z}$

Sec. 1-2

1. $\frac{3}{2}$ 3. 44 5. ±44 7. 3 9. 0

Sec. 1-3

1. 0 3. 10 5. 0 7. 0 9. $yz + zx + xy$

Sec. 1-4

1. (*a*) $\hat{\mathbf{r}} = \hat{\mathbf{x}} \cos\phi + \hat{\mathbf{y}} \sin\phi$, $\hat{\boldsymbol{\phi}} = -\hat{\mathbf{x}} \sin\phi + \hat{\mathbf{y}} \cos\phi$, $\hat{\mathbf{z}} = \hat{\mathbf{z}}$

   (*b*) $\hat{\mathbf{x}} = \hat{\mathbf{r}} \cos\phi - \hat{\boldsymbol{\phi}} \sin\phi$, $\hat{\mathbf{y}} = \hat{\mathbf{r}} \sin\phi + \hat{\boldsymbol{\phi}} \cos\phi$, $\hat{\mathbf{z}} = \hat{\mathbf{z}}$

   (*c*) $\hat{\mathbf{r}} = \hat{\mathbf{x}} \sin\theta \cos\phi + \hat{\mathbf{y}} \sin\theta \sin\phi + \hat{\mathbf{z}} \cos\theta$

   $\hat{\boldsymbol{\theta}} = \hat{\mathbf{x}} \cos\theta \cos\phi + \hat{\mathbf{y}} \cos\theta \sin\phi - \hat{\mathbf{z}} \sin\theta$

   $\hat{\boldsymbol{\phi}} = -\hat{\mathbf{x}} \sin\phi + \hat{\mathbf{y}} \cos\phi$

   (*d*) $\hat{\mathbf{x}} = \hat{\mathbf{r}} \sin\theta \cos\phi + \hat{\boldsymbol{\theta}} \cos\theta \cos\phi - \hat{\boldsymbol{\phi}} \sin\phi$

   $\hat{\mathbf{y}} = \hat{\mathbf{r}} \sin\theta \sin\phi + \hat{\boldsymbol{\theta}} \cos\theta \sin\phi + \hat{\boldsymbol{\phi}} \cos\phi$

   $\hat{\mathbf{z}} = \hat{\mathbf{r}} \cos\theta - \hat{\boldsymbol{\theta}} \sin\theta$

3. $3\pi R^2 La$

5. (*a*) 1 (*b*) 0

7. 3

9. 0

11. 0, $\pi R^2 La$, $2\pi R^2 La$

Sec. 1-5

1. $\frac{135}{4}\sqrt{10}$

3. 25.3

5. $-\frac{1}{12}$

Sec. 1-6

1. 0

3. $\hat{\mathbf{y}}(v_m/h)$

5. 2ω

7. $\nabla \times \mathbf{F} = \hat{\mathbf{z}}$. A circular velocity whose speed is proportional to the radius.

9. $\nabla \times \mathbf{F} = \hat{\mathbf{z}}$, as in Prob. 7, but $\mathbf{F}_7 \neq \mathbf{F}_9$.

11. 0 when $r \neq 0$; ∞ when $r = 0$.

Sec. 1-7

1. 1 3. 0 5. 0

7. Use Stokes' theorem with $\mathbf{F} = -\hat{\mathbf{x}}(y/2) + \hat{\mathbf{y}}(x/2)$.

9. 2; equal in magnitude but opposite in sign.

11. $$\int_{\Sigma} (\nabla f) \times d\mathbf{S} = -\oint_{c} f\, d\mathbf{r}$$

Sec. 1-8

1. (*a*) 0 (*b*) 0 (*c*) $2(a+c)$ (*d*) $\sum_{n=2}^{N} n(n-1)a_n x^{n-2}$

3. $-\dfrac{1}{4\sqrt{x^3}}$; $\dfrac{3}{4\sqrt{x}}$; $\dfrac{15\sqrt{x}}{4}$

5. $-k^2\phi$

7. Yes. No. Yes.

9. $\dfrac{\partial^2 f}{\partial r^2} + \dfrac{1}{r}\dfrac{\partial f}{\partial r} + \dfrac{1}{r^2}\dfrac{\partial^2 f}{\partial \phi^2}$

11. 10

13. 24

CHAP. 2

Sec. 2-2

1. 10^{36}

3. $-\dfrac{1}{4\pi\epsilon_0}\dfrac{eQ}{r^2}\hat{\mathbf{r}}$ newton

5. $4 \times 10^{-22}\,\hat{\mathbf{x}}$ newton

7. $7.2 \times 10^{-22}\,\hat{\mathbf{x}}$ newtons; i.e., larger. Even though the four nearer charges have a larger effect than the four more distant charges, the change in the effect of the distant charges is much larger.

9. $\dfrac{1}{4\pi\epsilon_0}\dfrac{qQ}{c^2}\hat{\mathbf{x}}$ newton

11. 0

13. $\dfrac{qR^2\sigma}{\epsilon_0 r^2}\hat{\mathbf{x}}$ newton

15. (*a*) (10,0,0) (*b*) No. This would be true only for a single source charge.

Sec. 2-3

1. $\dfrac{\lambda}{4\pi\epsilon_0 D}[\hat{\mathbf{x}}(\sin\theta_1 + \sin\theta_2) + \hat{\mathbf{y}}(\cos\theta_1 - \cos\theta_2)]$

3. $\left(\dfrac{\pm\lambda}{4\pi\epsilon_0}\right)\dfrac{1}{d(\ell + d)}\hat{\mathbf{y}}$

5. 1.44×10^{21} N/C; 230 N; 51.7 lb

7. $\frac{\rho r}{3\pi_0}\hat{\mathbf{r}}$

9. $-2.10 \times 10^{-9}\,\hat{\mathbf{x}}$ N/C

11. (a) 9.2×10^{-8} N (b) 10^{22}

13. $\left(\frac{\lambda a}{2\epsilon_0}\right)\frac{z}{(a^2+z^2)^{3/2}}\,\hat{\mathbf{z}}$

15. $\frac{q}{4\pi\epsilon_0}\left\{\frac{(x-1)}{[(x-1)^2+y^2]^{3/2}} - \frac{(x+1)}{[(x+1)^2+y^2]^{3/2}}\right\}$

17. $-\hat{\mathbf{y}}\,\frac{q}{2\pi\epsilon_0}\,\frac{1}{(y^2+1)^{3/2}}$

Sec. 2-4

1. $\begin{cases} \mathbf{E} = 0 & r < r_i \\ \mathbf{E} = \left(\frac{\sigma r_i}{\epsilon_0 r}\right)\hat{\mathbf{r}} = \left(\frac{\lambda}{2\pi\epsilon_0 r}\right)\hat{\mathbf{r}} & r_i \leqslant r \leqslant r_0 \\ \mathbf{E} = 0 & r > r_0 \end{cases}$

3. $\frac{2a}{5\epsilon_0}\sqrt{r}\,\hat{\mathbf{r}},\ r \leqslant R;\ \left(\frac{2aR^{5/2}}{5\epsilon_0}\right)\frac{\hat{\mathbf{r}}}{r^2},\qquad r \geqslant R$

5. $\left(\frac{2a}{3\epsilon_0}\right)\sqrt{r}\,\hat{\mathbf{r}},\ r \leqslant R;\ \left(\frac{2a\sqrt{R^3}}{3\epsilon_0}\right)\frac{1}{r}\hat{\mathbf{r}}\qquad r \geqslant R$

7. No. The surface integrals require further information for evaluation:

$$\int_{-a/2}^{a/2}\int_{-a/2}^{a/2} E_x(\text{front})\,dy\,dz$$

with $E_x \neq$ constant.

9. Consider a gaussian surface inside the metal and infinitesimally distant from the cavity wall.

11. Consider a gaussian surface within the cavity.

13. Direct integration, with $E = kr^{-n}$ and $a \equiv \frac{R+r}{R-r}$ gives $a^{2-n} = \frac{[f(a)+ng(a)]}{[h(a)+nj(a)]}$. Expand the right: $F(n) + G(n)\left(\frac{1}{a}\right) + H(n)\left(\frac{1}{a^2}\right) + \cdots$ and equate. Rule out nonintegral n by graphing, to show no other solutions.

15. $\frac{-Ze}{4\pi\epsilon_0}\left[\frac{1 - e^{-\alpha r} - \alpha r\left(1+\frac{\alpha r}{2}\right)e^{-\alpha r}}{r^2}\right]\hat{\mathbf{r}}$

17. Take a cone from the point to the surface and find the **E** caused by the charge within it. Then take the extended cone on the other side and find the **E** of its charge. Do this for all the possible cones.

CHAP. 3

Sec. 3-2

1. $\hat{\mathbf{x}} \dfrac{\mu_0 I_1 I_2}{4\pi} \displaystyle\int_0^{2\pi}\int_0^{2\pi} \dfrac{aR^2 \cos(\alpha_1 - \alpha_2)\, d\alpha_1\, d\alpha_2}{\{2R^2[1 - \cos(\alpha_1 - \alpha_2)] + a^2\}^{3/2}}$

3. $0.2\ \mathrm{N\ m^{-1}}$

5. 1 abamp ↔ 10 A

7. $d\mathbf{f} = -\hat{\mathbf{x}} \dfrac{\mu_0 I_1\, dI_1}{4\pi a}$

9. 10^{-4} N, attractive

Sec. 3-3

1. (*a*) $\hat{\boldsymbol{\phi}} \dfrac{\mu_0 I}{4\pi D}(\sin\alpha_1 - \sin\alpha_2)$. Here $\alpha_1 \geqslant 0,\ \alpha_2 \leqslant 0$.

   (*b*) The same, but with $\alpha_1, \alpha_2 < 0$.

3. $\hat{\mathbf{z}} \dfrac{\mu_0 I a^2}{2(a^2 + D^2)^{3/2}}$

5. $\dfrac{\mu_0 I a^2}{4r^5} [\hat{\mathbf{x}} 3xz + \hat{\mathbf{y}} 3yz + \hat{\mathbf{z}}(3z^2 - r^2)]$

7. $\dfrac{\mu_0 I a^2}{4r^3} [\hat{\mathbf{r}} 2\cos\theta + \hat{\boldsymbol{\theta}} \sin\theta]$

9. $\nabla^2 B_i + \alpha^2 B_i = 0$

11. $\hat{\mathbf{z}} \dfrac{\mu_0 n I}{2}(\cos\alpha_1 + \cos\alpha_2)$

13. $\dfrac{\mu_0}{2} J_\ell^2 \quad \mathrm{N\ m^{-2}}$

15. vBw; $-\hat{\mathbf{y}} vB$

Sec. 3-4

1. ∞; ∞; ∞

3. (*a*) $\hat{\mathbf{z}}(1 - n)B_0^2/r^{n+1}$; (*b*) not for $n = 1$. Must use the more basic integral definition of curl. (*c*) $0 < n \leqslant 1$ (*d*) none (*e*) $n > 1$.

5. $\dfrac{\mu_0 I}{2\pi r}$

7. $\left(\dfrac{\mu_0 I}{2\pi r_0^2}\right) r$ $\qquad r < r_0$

$\left(\dfrac{\mu_0 I}{2\pi}\right)\dfrac{1}{r}$ $\qquad r_0 \leqslant r \leqslant r_1$

$\left[\dfrac{\mu_0 I r_2^2}{2\pi(r_2^2 - r_1^2)}\right]\dfrac{1}{r} - \left[\dfrac{\mu_0 I}{2\pi(r_2^2 - r_1^2)}\right] r$ $\qquad r_1 \leqslant r \leqslant r_2$

$0, r_2 \leqslant r$

9. $\frac{1}{2}\mu_0 \mathbf{J}_\ell \times \hat{\mathbf{n}}$

11. 16

13. Zero

15. $\dfrac{\mu_0 I r^2}{R^2}$

17. $\hat{\mathbf{z}}\,\frac{1}{32}\mu_0 \rho\omega R^2$

CHAP. 4

Sec. 4-1

1. $\nabla \times \mathbf{G} = 0$; yes

3. $-\frac{1}{2}(x^2 + y^2 + z^2) + k$

5. $\phi = \dfrac{\rho}{2\epsilon_0}\left(R^2 - \dfrac{1}{3}r^2\right)$ $\qquad r \leqslant R$

$\phi = \dfrac{\rho R^3}{3\epsilon_0}\left(\dfrac{1}{r}\right)$ $\qquad r \geqslant R$

7. $\phi = \left(\dfrac{\sigma R}{\epsilon_0}\right)$ $\qquad r \leqslant R$

$\phi = \left(\dfrac{\sigma R^2}{\epsilon_0}\right)\dfrac{1}{r}$ $\qquad r \geqslant R$

9. 1.44×10^{-9} m

11. 1.6×10^{-19} J

13. 1.6×10^{-19} J; 1 eV

15. Very close to 1.44×10^{-9} m

17. $\left(\dfrac{1}{4\pi\epsilon_0}\right)\dfrac{q}{\sqrt{(x - x_s)^2 + (y - y_s)^2 + (z - z_s)^2}}$

19. (*a*) $\left(\dfrac{1}{4\pi\epsilon_0}\right)\dfrac{3.2 \times 10^{-19}}{\sqrt{(x-2)^2 + (y-1)^2}}$ V

(*b*) 2×10^{-9} V

(*c*) $(2.88 \times 10^{-9})\dfrac{\hat{\mathbf{x}}(x-2) + \hat{\mathbf{y}}(y-1)}{[(x-2)^2 + (y-1)^2]^{3/2}}$ V/m

(*d*) $1.03 \times 10^{-9}(-\hat{\mathbf{x}} + \hat{\mathbf{y}})$ V/m

21. (*a*) $\left(\dfrac{\lambda}{4\pi\epsilon_0}\right) \ln \left[\dfrac{\dfrac{s^2}{4} + R^2 + sR\cos\phi}{\dfrac{s^2}{4} + R^2 - sR\cos\phi}\right]$ (*b*) $\left(\dfrac{\lambda}{4\pi\epsilon_0}\right) \ln \left[\dfrac{\left(x + \dfrac{s}{2}\right)^2 + y^2}{\left(x - \dfrac{s}{2}\right)^2 + y^2}\right]$

23. $\lambda \ln \left[\dfrac{z + s/2 + r_1}{z - s/2 + r_2}\right]$

25. Start with the fact that for a positive charge to be in stable equilibrium at P the value of ϕ must be greater everywhere near P than it is at ϕ. Find **E**. Then obtain a contradiction. At a saddle point, rather than a minimum, there is unstable equilibrium. At a maximum point the positive particle would also be unstable. (This ignores the question of whether ϕ can have maxima or minima, discussed elsewhere.)

27. 3×10^6 V; 3×10^3 V

Sec. 4-2

1. 0.89 pF
3. 0.018 μF
5. $C_1 + C_2$
7. $\dfrac{2\pi\epsilon_0 L}{\ln(R_2/R_1)}$
9. $\dfrac{2\pi\epsilon_0}{\cosh^{-1}(s/d)}$
11. $\dfrac{s}{\ln[(2a + s)/(2a - s)]}$
13. 200 V across the 1-μF capacitor, 100 V across the 2-μF capacitor
15. 715 μF
17. 6.44×10^8 cm; the two are equal
19. $C = K\theta$

Sec. 4-3

1. 3 3. 0.5 J 5. 40 J/m^3 7. 5.2×10^{-17} J

9. (*a*) 45,000 J in^{-3}; (*b*) 22,500 J in^{-3}; (*c*) The potential difference drops about 5 percent, so (*a*) is approximately true.

11. 2.48×10^4 m (approximately 16 mi)

13. ∞

15. 10^{12} J

Sec. 4-4

1. Tension; pressure; tension; pressure
3. $(\epsilon_0/2)E^2 A$, tensile force
5. $-\dfrac{q^2}{4\pi\epsilon_0 a^2}\hat{\mathbf{x}}$ newton
7. $\dfrac{1}{4\pi\epsilon_0}\dfrac{q^2}{(2a)^2}\hat{\mathbf{x}}$ newton

CHAP. 5

Sec. 5-1

1. $\mathbf{A} \approx \hat{\mathbf{z}}\dfrac{\mu_0 I}{4\pi}\left[\ln\left(\dfrac{4L^2}{R^2}+1\right)+1-\dfrac{r^2}{R^2}\right]$
3. $\dfrac{\mu_0 I}{2\pi}\ln\left(\dfrac{r_2}{r_1}\right)$
5. $\mathbf{A} = 0 \qquad \mathbf{B} = \hat{\mathbf{z}}\dfrac{\mu_0 I}{2(r_1+r_2)}$
7. $\mathbf{B} = \dfrac{\mu_0 I}{4\pi\sqrt{ar^3}}\left\{\hat{\mathbf{r}}\left[-kz\left(K-\dfrac{2-k^2}{2-2k^2}E\right)\right]+\hat{\mathbf{z}}\left[kr\left(K+\dfrac{(a+r)k^2-2r}{2r(1-k^2)}E\right)\right]\right.$
9. $\hat{\mathbf{z}}\dfrac{\mu_0 I}{2\pi(a+r)}\left[K+\left(\dfrac{a+r}{a-r}\right)E\right]$
11. $\dfrac{8\mu_0 I}{\sqrt{125}\,a}$
13. $\dfrac{\mu_0 I}{4\pi}\ln\left(\dfrac{\ell_2+z_2}{\ell_1-z_1}\right)$
15. 0
17. $A_{\phi 1} = B_0 r_0\left[-\dfrac{2}{3}(8+\sqrt{2})\left(\dfrac{r}{r_0}\right)^2+(4+\sqrt{2})\left(\dfrac{r}{r_0}\right)\right]$

    $A_{\phi 2} = B_0 r_0\left[\dfrac{2}{3}\sqrt{\dfrac{r}{r_0}}+\dfrac{r_0}{3r}\right]$
19. 0

Sec. 5-2

1. $\mu_0 n_1 A_1 n_2 \ell_2$
3. (*a*) $\dfrac{\mu_0 N^2 d}{2\pi}\ln\left[\dfrac{2R+d}{2R-d}\right]$; (*b*) $\dfrac{\mu_0 N^2 d^2}{2\pi R}$
5. $M = \pm\sqrt{k_1 k_2}\sqrt{L_1 L_2}$
7. $\dfrac{\mu_0 a}{2\pi}\ln\dfrac{b+c}{c}$
9. (*a*) $L_1 + L_2$; (*b*) $\dfrac{L_1 L_2}{L_1 + L_2}$
11. $\dfrac{\mu_0 N^2 r^2}{2R}$

13. $\dfrac{\mu_0 \pi r_1{}^2 r_2{}^2}{2z^3}$

15. $\dfrac{\mu_0}{\pi}\left(\dfrac{1}{4} + \ln \dfrac{D}{r}\right)$

17. $\mu_0 r_1 r_2 \left[\left(\dfrac{2}{k} - k\right) K(k) - \dfrac{2}{k} E(k)\right]$

19. (a) $\dfrac{\mu_0}{4\pi} \displaystyle\int_0^{2\pi}\int_0^{2\pi} \frac{r_1 r_2 d\alpha_1 d\alpha_2 (\cos \alpha_1 \cos \alpha_2 + \sin \alpha_1 \sin \alpha_2)}{\sqrt{D^2 + r_1{}^2 + r_2{}^2 - 2r_1 r_2 (\cos \alpha_1 \cos \alpha_2 + \sin \alpha_1 \sin \alpha_2)}}$

(b) $\dfrac{\mu_0}{2} \displaystyle\int_0^{2\pi} \frac{r_1 r_2 \cos \theta \, d\theta}{\sqrt{D^2 + r_1{}^2 + r_2{}^2 - 2r_1 r_2 \cos \theta}}$

21. (a) $-\mu_0 k \sqrt{r_1 r_2} \displaystyle\int_0^{\pi 2} \frac{\cos 2\phi \, d\phi}{\sqrt{1 - k^2 \sin^2 \phi}}$

(b) $\dfrac{\pi \mu_0}{2D^3} r_1{}^2 r_2{}^2$

Sec. 5-3

1. $\frac{1}{2}\pi \mu_0 a^2 n^2 I^2$

3. $\dfrac{\mu_0 I^2}{4\pi} \ln\left(\dfrac{b}{a}\right)$

5. (a) $\mathbf{J}_\ell = \hat{\boldsymbol{\Phi}} \sigma \omega r \sin \theta$

(b) $\mathbf{A} = -\hat{\boldsymbol{\phi}} \dfrac{\mu_0 \sigma \omega r^2}{4\sqrt{2\pi}} \displaystyle\int_{\theta=0}^{\pi}\int_{\phi=0}^{\pi} \frac{\sin^2 \theta \sin \phi d\theta d\phi}{\sqrt{1 - \sin \theta \sin \theta_t \sin \phi - \cos \theta \cos \theta_t}}$

7. $\dfrac{\mu_0 N^2 I^2}{4}\left(\dfrac{b^2}{a}\right)$

9. Magnetic: 40×10^{-16} J/m^3;
Kinetic: 8×10^{-16} J/m^3

11. 0.35J, 0.15J

13. (a) $\frac{1}{2}\pi \mu_0 a^2 n^2 I^2$; (b) $\pi \mu_0 a^2 n^2$

15. $L_1 L_2 \geqslant M^2$

17. $V_m \displaystyle\sum_{k=1}^{n} I_k \Phi_k$

19. (a) $-I_T I_S \delta M$; (b) $-I_T I_S \delta M - M I_S \delta I_T$; (c) $M I_S \delta I_T$;

(d) $\delta V_m = M I_S \delta I_T$ and $\delta V_m = M T_S \delta I_T + L I_T \delta I_T + I_T I_S \delta M$

so $L_T \delta I_T + I_S \delta M = 0$;

$$(e)\ \delta V_b = -I_S \delta M\left[I_T - \left(\frac{M}{L_T}\right) I_S\right], \delta V_m = -\left(\frac{M}{L_T}\right) I_S^2 \delta M$$

21. $(a)(L_1 L_2 - M^2)\dfrac{d^2 i_1}{dM^2} - 4M\dfrac{di_1}{dM} - 2i_1 = 0$

$(b)(1 - k^2)\dfrac{d^2 y}{dk^2} - 4k\dfrac{dy}{dk} - 2y = 0$

$(c) y = \dfrac{c_0 + c_1 k}{1 - k^2}$

$(d) i_1 = i_{10}\left[\dfrac{1 - (1/f)k}{1 - k^2}\right], i_2 = i_{20}\left[\dfrac{1 - (f)k}{1 - k^2}\right]$

The former is graphed in Fig. 5-17*a*; the latter is obtained by selecting the curve for f instead of $(1/f)$.

23. $(a)\ i_1^2 + \left(\dfrac{2Mi_2}{L_1}\right) i_1 - i_{10}^2 = 0 \qquad (b) -\dfrac{k}{f} + \sqrt{1 + (k/f)^2}$

25. (*a*) No. (*b*) No. (*c*) Yes, if $L_T I_T = MI_S$, where the test circuit is the isolated one.

Sec. 5-4

1. (*a*) 1,450 N/cm^2 (*b*) 2,100 lb/in^2

   (*c*) 24,000 lb/in^2
   Stainless steel has a very high resistance.

   (*d*) No. $B \neq 0$ on one side. This is required to make it applicable.

3. (*a*) A pull toward the neighbor turn. (*b*) Outward. (*c*) No.

   (*d*) The radial force is much greater.

5. $\dfrac{dx}{2}\left[\left(2T_{yx} + \dfrac{\partial T_{yx}}{\partial x} dx\right) dy\ dz\right] = \dfrac{dy}{2}\left[\left(2T_{xy} + \dfrac{\partial T_{xy}}{\partial y} dy\right)\right] dz\ dx$

   so $T_{xy} = T_{yx}$; etc.

7. $\dfrac{1}{2\mu_0} B^2, \dfrac{1}{2\mu_0}(-B^2), \dfrac{1}{2\mu_0}(-B^2)$

CHAP. 6

Sec. 6-1

1. $\dfrac{p}{4\pi\epsilon_0}\left[\hat{\mathbf{x}}\dfrac{3xz}{r^5} + \hat{\mathbf{y}}\dfrac{3yz}{r^5} + \hat{\mathbf{z}}\dfrac{2z^2 - x^2 - y^2}{r^5}\right]$

3. $\dfrac{1}{4\pi\epsilon_0}\left[\dfrac{3(\mathbf{p}\cdot\hat{\mathbf{r}})\hat{\mathbf{r}}-\mathbf{p}}{r^3}\right]$

5. $\mathbf{T}=\mathbf{p}\times\mathbf{E}$

7. $(\mathbf{p}\times\mathbf{E})+[\mathbf{r}\times(\mathbf{p}\cdot\nabla)\mathbf{E}]$

9. $(q/2)(r_+{}^2-r_-{}^2)(-1+3\cos^2\theta_+)$

11. Zero

13. $\mathbf{F}=p_{\text{eff}}\nabla E$, with $p_{\text{eff}}=p\cos\theta$

15. $\dfrac{1}{4\pi\epsilon_0 R^3}[(\mathbf{p}_1\cdot\mathbf{p}_2)-3(\mathbf{p}_1\cdot\hat{\mathbf{R}})(\mathbf{p}_2\cdot\hat{\mathbf{R}})]$

where $\hat{\mathbf{R}}$ is, arbitrarily, either $\hat{\mathbf{R}}=\hat{\mathbf{r}}_{12}$ or $\hat{\mathbf{R}}=\hat{\mathbf{r}}_{21}$

17. $\mathbf{F}_{12}=\dfrac{3}{4\pi\epsilon_0 R^4}\{[(\mathbf{p}_1\cdot\mathbf{p}_2)-5(\mathbf{p}_1\cdot\hat{\mathbf{R}})(\mathbf{p}_2\cdot\hat{\mathbf{R}})]\hat{\mathbf{R}}+(\mathbf{p}_1\cdot\hat{\mathbf{R}})\mathbf{p}_2+(\mathbf{p}_2\cdot\hat{\mathbf{R}})\mathbf{p}_1\}$

19. Figure 6-7*b*: $\hat{\mathbf{R}}\dfrac{3p_1p_2}{4\pi\epsilon_0 R^4}$ (repulsive)

Figure 6-7*c*: $-\hat{\mathbf{R}}\dfrac{6p_1p_2}{4\pi\epsilon_0 R^4}$ (attractive)

21. 5; 2 (Three on the diagonal minus one condition, $\Sigma_i Q_{ii}=0$. Three parameters are required to specify the axis.); this becomes 1 for a figure with rotational symmetry.

Sec. 6-2

1. (*a*) 0 (*b*) $-\left(\dfrac{Qa}{4\pi\kappa_0}\right)\dfrac{1}{r^4}$ (*c*) $+\left(\dfrac{Q}{4\pi\kappa_0 a}\right)\dfrac{1}{r^2}$

3. $\mathbf{E}_{\text{out}}=\dfrac{PR^3}{3\epsilon_0}\left[\hat{\mathbf{r}}\dfrac{2\cos\theta}{r^3}+\hat{\boldsymbol{\theta}}\dfrac{\sin\theta}{r^3}\right]$; $\phi=\dfrac{PR^3\cos\theta}{3\epsilon_0 r^2}$

5. (*a*) $\phi_h=+Ez-\dfrac{1}{3\epsilon_0}(\mathbf{P}\cdot\mathbf{r})$; (*b*) $\mathbf{E}_h=\mathbf{E}+\dfrac{1}{3\epsilon_0}\mathbf{P}$; (*c*) $\mathbf{P}'=\mathbf{P}$ since $\mathbf{P}$ is fixed, but $\mathbf{E}'$ is altered from $\mathbf{E}$ by the external field of the sphere. See Prob. 3.

7. $\left(\dfrac{0.06}{Z}\right)$

9. $\mathbf{E}=0$; $\phi=0$; no, because the integral formula for ϕ is invalid if charges exist at infinity.

11. πP; 0

13. $\phi = \begin{cases} -P/(2\epsilon_0), z < 0 \\ P/(2\epsilon_0), z > 0 \end{cases}$ $\quad \mathbf{E} = \begin{cases} 0, Z \neq 0 \\ -\infty, Z = 0 \end{cases}$

15. $\nabla \cdot \mathbf{P} = 0 \qquad \nabla \times \mathbf{P} = -\hat{\mathbf{y}} \dfrac{w\alpha E_0 N_0}{kT} \exp\left(-\dfrac{wz}{kT}\right)$

$\nabla \cdot \mathbf{P} = -\dfrac{w\alpha E_0 N_0}{kT} \exp\left(-\dfrac{wZ}{kT}\right) \qquad \nabla \times \mathbf{P} = 0$

17. $\phi_{in} = \phi_{out} = \dfrac{PR}{3\epsilon_0}$; $\mathbf{E}_{in} = -\dfrac{\mathbf{P}}{3\epsilon_0}$; $\mathbf{E}_{out} = +\dfrac{2\mathbf{P}}{3\epsilon_0}$

Sec. 6-3

1. $\left(\dfrac{\kappa - 1}{\kappa}\right) Q$

3. $-\left(\dfrac{\kappa - 1}{\kappa}\right) Q$

5. $\left(\dfrac{\kappa + 1}{2}\right) \dfrac{\epsilon_0 A}{\alpha}$

7. $\left(\dfrac{2\kappa}{1 + \kappa}\right) \dfrac{\epsilon_0 A}{\alpha}$

9. $(a)\, \mathbf{E}_{in} = 0,\ \mathbf{E}_{out} = \hat{\mathbf{r}} \left[E_{ext}\left(1 + \dfrac{2R^3}{r^3}\right) \cos\theta\right] + \hat{\boldsymbol{\theta}} \left[E_{ext}\left(-1 + \dfrac{R^3}{r^3}\right) \sin\theta\right]$;

$(b)\, \phi_{in} = 0,\ \phi_{out} = E_{ext} \cos\theta \left(-r + \dfrac{R^3}{r^2}\right)$

11. $\left(1 - \dfrac{2a}{d}\right)\phi$

13. (*a*) $\kappa = \infty$, because $E_{tot} = 0$ inside; (*b*) $\mathbf{E} = \mathbf{E}_{tot} = 0$ for a metal in electrostatics, $\mathbf{P} = 0$ for a metal (no dipoles), $\mathbf{D} = 0$ for a metal (there is a σ_f giving a discontinuity in $\mathbf{D}$); (*c*) $\mathbf{E}_{in} = 0$, $\mathbf{P} = 3[(\kappa - 1)/(\kappa + 2)]\epsilon_0 \mathbf{E}_{ext} \rightarrow 3\epsilon_0 \mathbf{E}_{ext}$, $\mathbf{D} = \epsilon_0 \mathbf{E} + \mathbf{P} = 3\epsilon_0 \mathbf{E}_{ext}$. Outside, the metal sphere gives the same result as the dielectric sphere with $\kappa = \infty$; inside, they differ in the values of **P** and **D**.

15. $(a)\, \mathbf{E}_h = \mathbf{E} = \mathbf{E}_{ext}$; $(b)\, \mathbf{E} + \dfrac{1}{3\epsilon_0}\mathbf{P}$; $(c)\, \mathbf{E} + \dfrac{1}{\epsilon_0}\mathbf{P}$; $(d)\, \mathbf{E}$; $(e)\, \mathbf{E} + \dfrac{1}{3\epsilon_0}\mathbf{P}$; $(f)\, \mathbf{E} + \dfrac{1}{\epsilon_0}\mathbf{P}$

17. $(a)\left(\dfrac{R + 2}{3}\right)\mathbf{E}$; $(b)\, \mathbf{E}_{ext}$

19. $\sigma_p = 0.256 \times 10^{-6}\ \text{C m}^{-2}$; $\mathbf{E}_{slab} = -\hat{\mathbf{n}}\, 0.15 \times 10^5\ \text{V m}^{-1}$

Sec. 6-4

1. $-\dfrac{Z^2 e^2 d}{4\pi\epsilon_0 r^3}$; 0

3. $4\pi\epsilon_0 r^3 \mathbf{E}$; $4\pi\epsilon_0 r^3$

5. $2.69 \times 10^{19}\,\text{cm}^{-3}$; $2.43 \times 10^{19}\,\text{cm}^{-3}$

7. $\left(\frac{M}{\rho}\right)\frac{\kappa - 1}{\kappa + 1}$

9. $\frac{N_A}{3}\left(\alpha + \frac{p^2}{3kT}\right)$

11. 0

13. Stronger; no

Sec. 6-5

1. $\tan\theta_2 = \left(\frac{\kappa_2}{\kappa_1}\right)\tan\theta_1$

3. $\hat{\mathbf{x}}2\epsilon_0 + \hat{\mathbf{y}}4\epsilon_0 + \hat{\mathbf{z}}3\epsilon_0$

5. $\hat{\mathbf{x}}2 + \hat{\mathbf{y}}2 + \hat{\mathbf{z}}2; -\hat{\mathbf{x}} + \hat{\mathbf{z}}$

7. $(\hat{\mathbf{x}} + \hat{\mathbf{y}} + \hat{\mathbf{z}}); (-\hat{\mathbf{x}} + \hat{\mathbf{z}}); (\hat{\mathbf{y}} + 2\hat{\mathbf{z}})$

9. Away

11. $\mathbf{D}_1 = \epsilon_0(\hat{\mathbf{x}} - \hat{\mathbf{y}}2 + \hat{\mathbf{z}}3)$; $\mathbf{E}_2 = \hat{\mathbf{x}} - \hat{\mathbf{y}}2 + \hat{\mathbf{z}}1.5$; $\mathbf{D}_2 - \epsilon_0(\hat{\mathbf{x}}2 - \hat{\mathbf{y}}4 + \hat{\mathbf{z}}3)$

13. $\frac{Qa}{4\pi\epsilon_0\kappa_0 r^3}$; $\frac{Q}{4\pi}\frac{\kappa_0 r - a}{\kappa_0 r^3}$; $\frac{Q}{4\pi r^2}$; no

15. $\left(\frac{\epsilon_2 - \epsilon_1}{\epsilon_2}\right)E_{1n} = \left(\frac{\epsilon_2 - \epsilon_1}{\epsilon_1}\right)E_{2n}$

17. No. These are completely different cases.

CHAP. 7

Sec. 7-1

1. $-\hat{\mathbf{x}}\left(\frac{\mu_0 IS}{4\pi}\right)\frac{y}{r^3} + \hat{\mathbf{y}}\left(\frac{\mu_0 IS}{4\pi}\right)\frac{x}{r^3}$

3. $\frac{\mu_0 m}{4\pi}\left[\hat{\mathbf{x}}\frac{3xz}{r^5} + \hat{\mathbf{y}}\frac{3yz}{r^5} + \hat{\mathbf{z}}\frac{(2z^2 - x^2 - y^2)}{r^5}\right]$

5. $\mathbf{T}_m = m \times \mathbf{B}$

7. (*a*) $+\frac{\mu_0}{4\pi}\left[\frac{(\mathbf{m}_1 \cdot \mathbf{m}_2) - 3(\mathbf{m}_1 \cdot \hat{\mathbf{R}})(\mathbf{m}_2 \cdot \hat{\mathbf{R}})}{R^3}\right]$

   (*b*) The negative of (*a*); (*c*) For finding the force on the isolated dipole $\mathbf{F} = -\nabla V_m = -\nabla V_{tot}$ where V_m is that given in (*a*). For finding the force on the nonisolated dipole $\mathbf{F} = +\nabla V_m = -\nabla V_{tot}$ where V_m is that in (*b*); (*d*) Same as (*a*).

9. 0

11. $\hat{\boldsymbol{\theta}}\,\frac{\mu_0 m_1 m_2}{4\pi r^3}$ N m in the *xz* plane

13. Attracted; repelled

15. Zero; no

17. $m\cos\theta\,\nabla|\mathbf{B}| = m_{eff}\nabla B$

19. $r = \sqrt{\frac{3}{2}}\, a \qquad r = 10\sqrt{\frac{3}{2}}\, a \qquad \mathbf{A} = 0$ on the axis.

Sec. 7-2

1. $r = 4\pi\epsilon_0 \dfrac{\hbar^2}{m_e e^2} = 5.34 \times 10^{-11}\,\text{m}$

$v = 2.2 \times 10^6\ \text{m s}^{-1}$

$f = 6.55 \times 10^{15}\ \text{s}^{-1}$

$I = 1.05 \times 10^{-3}\ \text{A}$

3. $\nabla \cdot (\mathbf{m} \times \mathbf{A})$

5. $U_{\text{elec}} = 900\, U_{\text{mag}}$

7. $0;\ \hat{\boldsymbol{\phi}} M \sin\theta$

9. $\mathbf{J}_m = 0;$

$$\mathbf{j}_m = \begin{cases} \text{top:} & -\hat{\mathbf{y}}M \\ \text{side:} & \hat{\mathbf{z}}M \sin\phi \\ \text{bottom:} & \hat{\mathbf{y}}M \end{cases}$$

11. $0;\ \dfrac{-\hat{\mathbf{y}}Mz + \hat{\mathbf{z}}My}{\sqrt{\left(\dfrac{b^4}{a^4}\right) x^2 + y^2 + z^2}}$

13. (*a*) $\nabla \cdot \mathbf{M} = 0,\ \nabla \times \mathbf{M} = -\hat{\mathbf{y}} \dfrac{w\alpha_M B_0 N_0}{kT} \exp\left(-\dfrac{wz}{kT}\right)$

(*b*) $\nabla \cdot \mathbf{M} = -\dfrac{w\alpha_M B_0 N_0}{kT} \exp\left(-\dfrac{wz}{kT}\right),\ \nabla \times \mathbf{M} = 0$

(*c*) Here $\mathbf{J}_m$ exists, and decreases exponentially with height, for a horizontal **B**. In the electrostatic case ρ_p exists, and decreases exponentially with height, for a vertical **E**.

15. (*a*) Top: $-\hat{\mathbf{y}}\alpha_M B_0 N_0 \exp(-wh/kT)$

Side: $\hat{\mathbf{z}}\alpha_M B_0 N_0 (\sin\phi) \exp(-wz/kT)$

Bottom: $\hat{\mathbf{y}}\alpha_M B_0 N_0$

(*b*) Top and bottom: 0

Side: $\hat{\boldsymbol{\phi}}\alpha_M B_0 N_0 \exp(-wz/kT)$

Sec. 7-3

1. (*a*) $\mathbf{B} - \mu_0 \mathbf{M}$ (*b*) $\mathbf{B}$ (*c*) $\mathbf{B} - \frac{2}{3}\mu_0 \mathbf{M}$

3. (*a*) ∞ (*b*) $L_{\ell(\text{int})} = \dfrac{\mu}{8\pi} \qquad L_{\ell(\text{ext})} = \dfrac{\mu}{2\pi} \ln\left(\dfrac{b}{a}\right)$ (*c*) $\dfrac{\mu_0}{8\pi}\left[1 + 4\ln\left(\dfrac{b}{a}\right)\right]$

(*d*) $\mu\pi a^2 n$ (*e*) $\mu_0 \pi a^2 n$ (*f*) $\mu_0 \to \mu$, to the extent that the flux goes through the medium instead of through vacuum.

5. (*a*) $\dfrac{\mu_0}{2}\mathbf{M}$; (*b*) $\mu_0 \mathbf{M}$

7. $\mathbf{B} = \mu_0 M_0 \left(\dfrac{r}{a}\right);\ \mathbf{H} = 0$

9. $\dfrac{nIa^2}{2}\left[\dfrac{r}{\left(r^2 - \dfrac{\ell^2}{4}\right)^2}\right]$

11. $\mathbf{B}_{\text{in}} = \hat{\mathbf{z}}\frac{2}{3}\mu_0 M_0$; $\mathbf{B}_{\text{out}} = \hat{\mathbf{r}}\left(\dfrac{\mu_0 M_0 R^3}{3}\right)\dfrac{2\cos\theta}{r^3} + \hat{\boldsymbol{\theta}}\left(\dfrac{\mu_0 M_0 R^3}{3}\right)\dfrac{\sin\theta}{r^3}$

13. $m = NI\pi a^2$; $r = 2.3\ell$

15. 12.8 A-turns/m

Sec. 7-4

1. (*a*) 12 (*b*) 0.08 (*c*) Yes

3. (*a*) $Nm_0 = \left(\dfrac{\sqrt{3}}{3}\right) Nm$; (*b*) 0.0025; (*c*) 0.12 Nm_0

5. $\langle r^2 \rangle/6$

7. $-\dfrac{\mu_0 NZe^2\langle R^2\rangle}{6m_e}$

9. 1.4×10^9; 2.8×10^9; the microwave region

11. 1.02; 1.04; no; small

13. 9.4×10^5 G

(*a*) $y = x/3$; (*b*) $y = x/3$; (*c*) 0.55; (*d*) 0.95; (*e*) 1

Sec. 7-5

1. $H_{1n} = H_{2n} + (M_{2n} - M_{1n})$

3. $B_{1n} = \dfrac{0.1}{\sqrt{2}}$, $B_{1t} = \dfrac{0.1}{\sqrt{2}}$; $B_{2n} = \dfrac{0.1}{\sqrt{2}}$, $B_{2t} = \dfrac{0.100001}{\sqrt{2}}$ T

$H_{1n} = \dfrac{10^6}{4\sqrt{2}\pi}$, $H_{1t} = \dfrac{10^6}{4\sqrt{2}\pi}$; $H_{2n} = \dfrac{999{,}999}{4\sqrt{2}\pi}$, $H_{2t} = \dfrac{10^6}{4\sqrt{2}\pi}$ A-turns/m

5. Up; 0.000285°; 0.0171′; 1.026″

7. 1′42.6″ up

9. $\chi = -0.9$; $B_{2n} = \dfrac{0.1}{\sqrt{2}}$, $B_{2t} = \dfrac{0.01}{\sqrt{2}}$ T; $\mu = 0.1\mu_0$

$H_{2n} = \dfrac{10^7}{4\sqrt{2}\pi}$, $H_{2t} = \dfrac{10^6}{4\sqrt{2}\pi}$ A-turns/m

11. 0.99; $B_{2n} = \dfrac{0.1}{\sqrt{2}}$, $B_{2t} = \dfrac{10}{\sqrt{2}}$ T;

$H_{2n} = \dfrac{10^4}{4\sqrt{2}\pi}$, $H_{2t} = \dfrac{10^6}{4\sqrt{2}\pi}$ A-turns/m

13. Yes; no; only if $\chi_m = +\infty$ or -1 ($\chi_m' = 1$ or $-\infty$) can the shielding be perfect. The mechanism in a metal is different: equal and opposite charges set up a field which just obliterates the external field.

15. $\mathbf{B} = \hat{\boldsymbol{\phi}} \dfrac{\mu NI}{2\pi R + (\kappa_m - 1)L}$; $\mathbf{H}_m = \hat{\boldsymbol{\phi}} \dfrac{NI}{2\pi R + (\kappa_m - 1)L}$

$\mathbf{H} = \hat{\boldsymbol{\phi}} \dfrac{\kappa_m NI}{2\pi R + (\kappa_m - 1)L}$; same

17. 13.7 percent, 86.3 percent

CHAP. 8

Sec. 8-1

1. Add a second image charge, in addition to q' of Example 2; $q'' = (a/d)q$ at O.

3. $q' = -\left(\dfrac{a}{d}\right) q$ at $d_1 = \dfrac{a^2}{d}$

$q'' = +\left(\dfrac{a}{d}\right) q + Q$ at $r = 0$

5. $[(qd)s]/[(s^2 + d^2)^{3/2}]ds$; $R = \sqrt{2}/2$

7. $-q$ at $(-a,b)$, q at $(-a,-b)$, $-q$ at $(a,-b)$.

9. (a) $q' = -q$ at $x = -d$ (b) $q'' = \left(\dfrac{a}{2d}\right) q$ at $x = d - \dfrac{a^2}{2d}$

(c) $q''' = -\left(\dfrac{a}{2d}\right) q$ at $x = -d + \dfrac{a^2}{2d}$

11. $(x - a)^2 + y^2 = r^2$

13. $d = 2r \sqrt{\dfrac{D^2}{4r^2} - 1}$ $\dfrac{r_-}{r_+} = \dfrac{D}{2r} + \sqrt{\dfrac{D^2}{4r^2} - 1}$

15. $\dfrac{\pi\epsilon_0}{\ln\left(\dfrac{D}{2r} + \sqrt{\dfrac{D^2}{4r^2} - 1}\right)}$

17. $\dfrac{q^2}{16\pi\epsilon_0 d}$

Sec. 8-2

1. $\left[\dfrac{-2ab\lambda}{\pi(a^2 + b^2)^2}\right] r$

3. $\hat{\mathbf{y}}$ 1000 V/m

5. (a) $r = \dfrac{2k_1^2}{1 \quad \cos\phi}$ (b) $r = \dfrac{2k_2^2}{1 + \cos\phi}$

7. $\lambda \propto x^{-1/2}$

9. $0 \leqslant \theta \leqslant 3\pi/2$, with equipotentials $r^{2/3} = K_1/\sin(2\theta/3)$ and field lines given by $r^{2/3} = K_2/\cos(2\theta/3)$.

11. $u = \dfrac{+2ay}{(x+a)^2 + y^2}$, $v = \dfrac{a^2 - (x^2 + y^2)}{(x+a)^2 + y^2}$

13. $\lambda \propto R^{-1/3}$

15. (a) $\dfrac{u^2}{\sin^2 x} - \dfrac{v^2}{\cos^2 x} = -1$, hyperbolas

(b) $\dfrac{u^2}{\cosh^2 y} + \dfrac{v^2}{\sinh^2 y} = 1$, ellipses

Sec. 8-3

1. 43.1, 53.6, 18.4, 25.0, 7.0, 9.7

3. 51.2, 23.4, 9.0

5.

	1	2	3	4	5	6
A	90.0	83.4	79.8	78.4	78.2	78.2
B	76.8	63.6	57.5	55.6	56.2	56.6
C	53.4	37.0	31.2	30.2	34.2	35.8
D					14.8	18.6
E					6.7	9.0
F					4.2	3.0
G					1.6	1.2

(The right half is symmetric with this.)

7.

	1	2	3	4	5	6
A	90.0	83.4	79.8	78.4	78.2	78.2
B	76.8	63.6	57.5	55.6	56.2	56.6
C	53.4	37.0	31.2	30.2	34.2	35.8
D					14.4	18.0
E					5.4	7.2

9. $h = \left(\dfrac{q}{mg}\right) \dfrac{\lambda}{2\pi\epsilon_0} \ln\left(\dfrac{r_0}{r}\right)$

11. (a) $mg \sin\theta$ (b) $|-\nabla h| = \tan\theta$ (c) $\theta \approx 0^\circ$

13. $RC = \dfrac{\epsilon}{g}$

15. (a) $10\epsilon_0$ (b) $10\epsilon_0$ (c) same.

17. The values listed below are obtained from the equation $\phi = \frac{100}{\sinh \pi} \sin\left(\frac{\pi x}{a}\right) \sinh\left(\frac{\pi y}{a}\right)$

y					
$5a/6$	29.51	51.12	59.03	51.12	29.51
$4a/6$	17.81	29.98	34.62	29.98	17.81
$3a/6$	9.96	17.25	19.92	17.25	9.96
$2a/6$	5.40	9.36	10.81	9.36	5.40
$a/6$	2.37	4.10	4.74	4.10	2.37
0					
0	$\frac{a}{6}$	$\frac{2a}{6}$	$\frac{3a}{6}$	$\frac{4a}{6}$	$\frac{5a}{6}$ x

CHAP. 9

Sec. 9-1

1. g/ne

3. (a) $R_1 + R_2$ (b) $R_1R_2/(R_1 + R_2)$

5. $R_1 = R_{01} - \sqrt{R_{02}(R_{01} - R_{s1})}$ $R_2 = R_{02} - \sqrt{R_{02}(R_{01} - R_{s1})}$

$R_3 = \sqrt{R_{02}(R_{01} - R_{s1})}$

7. 3 Ω

9. $\frac{3}{4}$ A

11. $R_A = \frac{R_1R_2 + R_2R_3 + R_3R_1}{R_2}$ $R_B = \frac{R_1R_2 + R_2R_3 + R_3R_1}{R_3}$

$R_C = \frac{R_1R_2 + R_2R_3 + R_3R_1}{R_1}$

13. $\frac{b - a}{4\pi abg}$

15. 26.8 Ω

17. (a) 8.43×10^{28} (b) 7.4×10^{-6} cm s^{-1}

19. $\ln\left(\frac{R}{r}\right)/2\pi g\ell$

CHAP. 10

Sec. 10-1

1. $H = 8 \times 10^5 B$

3. (a) Not possible (b) 731 A-turns

5. 3.9×10^{-4} Wb

7. Opposite

9\. (a) $\int_1^2 \mathbf{H} \cdot d\mathbf{B}$; (b) $\oint \mathbf{H} \cdot d\mathbf{B}$

11\. $\mu = (1 + \chi_m)\mu_0$ if $\mathbf{M} = \chi_m \mathbf{H}$. Either one of two viewpoints is possible. (a) The last relation is not valid for a permanent magnet, so μ is not defined. (b) We simply set $\mu = K_m \mu_0$. Then K_m is negative here.

13\.

	(a)	(b)
Φ	$\kappa_m \Phi_0$	Φ_0
L	$\kappa_m L_0$	$\kappa_m L_0$
I	I_0	$(1/\kappa_m) I_0$
B	$\kappa_m B_0$	B_0
H	H_0	$(1/\kappa_m) H_0$

15\. No;

$$H_a = \left[\frac{2\pi R}{2\pi R + (\kappa_m - 1)\ell_a}\right] \kappa_m H_0\,;$$

$$H_i = \left[\frac{2\pi R}{2\pi R + (\kappa_m - 1)\ell_a}\right] H_0$$

17\. (a) 0

(b) M on the top, $-M$ on the bottom.

(c) $\pm M\pi r^2$

(d) $\dfrac{M\pi r^2}{4\pi}\left(\dfrac{1}{R_+} - \dfrac{1}{R_-}\right) \approx \dfrac{M[\ell(\pi r^2)] \cos\theta}{4\pi R^2}$

19\. (a) 5 T

(b) 6383 A-turns/m

(c) 3000 A-turns

(d) 3×10^6 A-turns/Wb

CHAP. 11

Sec. 11-1

1\. $-\omega B w \ell \cos\theta \cos\omega t$

3\. $\omega B w \ell \sin\omega t$

5\. $-\omega_1 B w \ell \cos\omega_1 t \cos\omega_2 t + \omega_2 B w \ell \sin\omega_1 t \sin\omega_2 t$

7\. 1.34 V; $-\hat{\mathbf{y}}$

9\. 4.71 V

11\. (a) $\omega_L = \dfrac{e}{2m}\mathbf{B}$ (b) parallel for electron, antiparallel for proton

(c) $\omega_c = \dfrac{e}{m}\mathbf{B}$ so $\omega_L = \frac{1}{2}\omega_c$ and in same sense (d) 5×10^9 G

13\. (a) $v\ell B/R$ (b) $v^2 \ell^2 B^2/R$ (c) $v^2 \ell^2 B^2/R$

15\. (a) [illegible]$\pi^2 f e r_0^2 B_0 \cos 2\pi f t$ (b) $4\pi^2 f r_0^2 B_0$

17. 5.8×10^4

19. $P = (\frac{1}{4}gB^2)r^2$; no.

21. See *American Journal of Physics*, **31**:428 (1963) and **29**:635 (1961).

Sec. 11-2

1. $g/\epsilon\omega$

3. 22.3 μA; no

5. $1/\omega CR$ ($=X_C/R = 1/Q$, where X_C is capacitive reactance and Q = quality factor)

7. $(a)\ \frac{\dot{q}}{2}(1 - \cos\theta),\ -\frac{\dot{q}}{2}[1 - \cos(\theta + d\theta)]$ $\quad (b)\ \frac{\dot{q}}{2r}\sin^2\theta\ ds$

$(c)\ \hat{\mathbf{\Phi}}\ \frac{\mu_0}{4\pi r^2}\dot{q}\sin^2\theta$ $\quad (d)$ same

9. $\mathbf{B} \approx \hat{\mathbf{\Phi}}\ \frac{\mu_0}{2\pi R}\left(\frac{d}{2R}\right)$

11. $\int \dot{\mathbf{D}} \cdot d\mathbf{S} \neq 0 \rightarrow B_\phi = B_\phi(t) \neq 0 \qquad \nabla \times \mathbf{E} = 0 \rightarrow B_\phi \neq B_\phi(t)$

13. $(a)\ C(\phi_1 - \phi_2)\exp\left(-\frac{g}{\epsilon}t\right) \quad (b)\ -\left(\frac{g}{\epsilon}\right)\frac{C}{A}(\phi_1 - \phi_2)\exp\left(-\frac{g}{\epsilon}t\right) \quad (c)\ 0$

15. $\mathbf{E} = \nabla \times \nabla \times \boldsymbol{\pi}_e + \Box^2\boldsymbol{\pi}_e;\ \mathbf{B} = \epsilon_0\mu_0\ \nabla \times \dot{\boldsymbol{\pi}}_e$

17. (a) 1.6×10^3 Hz (an audio frequency)

(b) 1.4×10^{14} Hz (almost an optical frequency)

19. $(a)\ -4\pi r^2 J \quad (b)\ 0$

Sec. 11-3

1. $(a)\ u_1 = \frac{1}{\mu_0}B_1^2\sin^2(\omega t - kz) \qquad \mathbf{S}_1 = \hat{\mathbf{z}}\frac{1}{\mu_0}cB_1^2\sin^2(\omega t - kz)$ $\qquad$ Yes

$(b)\ u_2 = \frac{1}{\mu_0}B_1^2\sin^2(\omega t - kz) + \epsilon_0 E_1 E_2\sin(\omega t - kz) + \frac{\epsilon_0}{2}E_2^2$

$\mathbf{S}_2 = \hat{\mathbf{z}}\frac{1}{\mu_0}[cB_1^2\sin^2(\omega t - kz) + E_2 B_1\sin(\omega t - kz)]$ $\qquad$ No.

(c) $\mathbf{s} = \hat{\mathbf{z}} c \left(\frac{\epsilon_0}{2} E_2{}^2 \right)$

3. $\hat{\mathbf{z}} \frac{2}{9} Q \mu_0 M R^2$

5. (a) $(\pi R^2 h)(\epsilon_0/2)E^2$ (b) $(\pi R^2 h)(\epsilon_0/16c^2)R^2 \dot{E}^2$ (c) $(\dot{E}/E)^2(R^2/8c^2)$

(d) $t \gg \sqrt{2}R/(4c)$, i.e., $ct \gg R$ (e) 10^{-9} s

7. (a) $(\pi R^2)\epsilon_0 E\dot{E}$ (b) $-\hat{\mathbf{r}}(\frac{1}{2}\epsilon_0 E\dot{E})R$ (c) $\frac{1}{2}\epsilon_0 E\dot{E}R(2\pi Rh)$ (d) Same

(e) Radial inward (rather than along the direction of the charging current!)

9. (a) $\hat{\mathbf{z}} c\epsilon_0 E^2 \sin^2 \frac{2\pi}{\lambda}(z - ct)$ (b) $\hat{\mathbf{z}} \frac{1}{2} c\epsilon_0 E^2$

11. 10^3 V/m; 1/30 G

13. $\mathbf{S} = \hat{\boldsymbol{\phi}} \left(\frac{Qm}{16\pi^2 \epsilon_0} \right) \frac{\sin\theta}{r^5}$ $\quad \mathbf{g} = \hat{\boldsymbol{\phi}} \left(\frac{Q\mu_0 m}{16\pi^2} \right) \frac{\sin\theta}{r^5}$

15. (a) $\partial_x E_x + \partial_y E_y + \partial_z E_z = 0$; the first two terms = 0 by definition, so $\partial_z E_z = 0$; but $\partial_x E_z = \partial_y E_z = 0$ by definition, so E_z = constant and variations can only occur in x and y components. Similarly for **B**.
(b) If $\nabla \cdot \mathbf{A} = 0$, the same argument follows. But we are also free to take $\nabla \cdot \mathbf{A} \neq 0$; then it is not true.

17. (a) $\partial_z E_z = 0$, $\partial_z B_z = 0$, $\dot{B}_x = 0$, $\dot{B}_y = -\partial_z E_x$, $\dot{B}_z = 0$, $\partial_z B_y = -\frac{1}{c^2}\dot{E}_x$, $\partial_z B_x = 0$ (seven equations).

(b) From the second and fifth equations in (a), B_z = constant. Take $B_z = 0$. From the third and seventh, $B_x = 0$. So $\mathbf{B} = \hat{\mathbf{y}}B$, $\mathbf{E} = \hat{\mathbf{x}}E$, e.g., if wave direction is $\pm\hat{\mathbf{z}}$.

19. (a) $\mathbf{B} = \hat{\mathbf{y}} E_0 \sqrt{\epsilon_0 \mu_0} \sin(2\pi/\lambda)(z - ct)$ (b) c

21. $Z_0 = \sqrt{\mu_0/\epsilon_0}$; 377 Ω

23. (a) $\hat{\mathbf{y}} E_0 \sin(kz - \omega t)$

(b) $\hat{\mathbf{x}} \frac{\sqrt{3}}{2} E_0 \sin(kz - \omega t) + \hat{\mathbf{y}} \frac{1}{2} E_0 \sin(kz - \omega t)$

(c) $\hat{\mathbf{x}} E_0 \sin(kz - \omega t) + \hat{\mathbf{y}} E_0 \sin(kz - \omega t)$

25. (a) 300 m (b) 0.0021 m^{-1} (c) linear, along x (d) 3.33×10^{-8} T

27. (a) $\mathbf{E} = -j\left[\omega \mathbf{A} + \frac{c^2}{\omega} \nabla\nabla \cdot \mathbf{A}\right]$, $\mathbf{B} = \nabla \times \mathbf{A}$ (b) yes (c) the equation of continuity gives $\phi = -(c^2/j\omega)\nabla \cdot \mathbf{A}$.

CHAP. 12

Sec. 12-1

1. $A_1 = \dfrac{V_0}{2L\sqrt{\dfrac{R^2}{4L^2} - \dfrac{1}{LC}}}$ $\quad I = \dfrac{(V_0/L)}{\sqrt{\dfrac{R^2}{4L^2} - \dfrac{1}{LC}}} e^{-(R/2L)t} \sinh \sqrt{\dfrac{R^2}{4L^2} - \dfrac{1}{LC}}\, t$

3. (a) $I = \dfrac{V_0}{\sqrt{R^2 + X_C^2}} e^{j[\omega t + \tan^{-1}(X_C/R)]}$

(b) $I = \dfrac{V_0}{R} e^{-t/RC}$

5. $V_0(1 - e^{-t/RC})$

7. $I = \dfrac{(R_2R_3 - R_1R_4)V}{R_1R_2R_3 + R_2R_3R_4 + R_3R_4R_1 + R_4R_1R_2 + R_A(R_1 + R_2)(R_3 + R_4)}$

9. (a) $I = \dfrac{V_0}{\sqrt{R^2 + X_L^2}} e^{j[\omega t - \tan^{-1}(X_L/R)]}$ $\quad$ (b) $I = \dfrac{V_0}{R}(1 - e^{-t/(L/R)})$

11. (a) $1\Big/\sqrt{1 + Q^2\left(\gamma - \dfrac{1}{\gamma}\right)^2}$ $\quad$ (b) $1/\sqrt{1 + 4Q^2\delta^2}$

(c) $\pm 1/(2Q)$ $\quad$ (d) $Q = f_0/(f_+ - f_-)$

13. (a) $\frac{1}{2}V_0I_0 \cos\theta$ (b) The same. (c) Change one to its complex conjugate.

15. $I_{\text{rms}} = \dfrac{\sqrt{2}}{2} I_0$

17. (a) $\dfrac{Z}{R} = \dfrac{1 + jQ\left(\dfrac{1}{\gamma} - \gamma\right)}{1 + Q^2\left(\dfrac{1}{\gamma} - \gamma\right)^2}$ $\quad$ (b) $\dfrac{|Z|}{R} = \dfrac{1}{\sqrt{1 + 4Q^2\delta^2}}$

(c) $\dfrac{|I|}{\left(\dfrac{V_0}{R}\right)} = \sqrt{1 + 4Q^2\delta^2}$ $\quad$ (d) Minimum $\quad$ (e) Yes

19. $|V| = V_0\left(\dfrac{R}{r + R}\right)\dfrac{1}{\sqrt{1 + 4Q^2\delta^2}}$

21. (a) 30 A (b) 3 A (c) A 30-A ideal current generator in parallel.

CHAP. 13

Sec. 13-1

1. (*a*) $+0.5 \times 10^{-16}$ s (*b*) $+0.5 \times 10^{-12}$ s (*c*) $+0.5 \times 10^{-8}$ s (*d*) $+0.5 \times 10^{-4}$ s

3. (*a*) $\sqrt{2}/2$ s (*b*) 10^{-4} s (*c*) 10^{-10} s

5. 594 m; 356 m

7. $x - x_1 = \dfrac{(X - X_2) + \beta(cT - cT_2)}{\sqrt{1 - \beta^2}}$ $\quad t - t_1 = \dfrac{\beta(X - X_2) + (cT - cT_2)}{\sqrt{1 - \beta^2}}$

9. No. A' reads $\dfrac{-(\ell_0 v/c^2)}{\sqrt{1 - \beta^2}}$

   No. C' reads $\dfrac{+(\ell_0 v/c^2)}{\sqrt{1 - \beta^2}}$

11. No. *D* will say *A* was not synchronized properly.

13. $\beta = 0.0837$; $\beta = 0.999{,}999{,}875$

15. $u_x = \dfrac{U_X + v}{1 + \dfrac{vU_X}{c^2}}$ $\quad U_X = \dfrac{u_x - v}{1 - \dfrac{vu_x}{c^2}}$

17. $u_y = \dfrac{\sqrt{1 - \beta^2}\, U_Y}{1 + \dfrac{vU_X}{c^2}}$ $\quad U_Y = \dfrac{\sqrt{1 - \beta^2}\, u_y}{1 - \dfrac{vu_x}{c^2}}$

    $u_z = \dfrac{\sqrt{1 - \beta^2}\, U_Z}{1 + \dfrac{vU_X}{c^2}}$ $\quad U_Z = \dfrac{\sqrt{1 - \beta^2}\, u_z}{1 - \dfrac{vu_x}{c^2}}$

19. $U = \dfrac{c}{n} + v\left(1 - \dfrac{1}{n^2}\right)$

21. $\tan\left(\dfrac{\theta_t}{2}\right) = \sqrt{\dfrac{1 - \beta}{1 + \beta}}\tan\left(\dfrac{\theta_r}{2}\right)$

23. (*a*) $\Delta\,(cT_{\text{approaching}}) = \Delta\,(ct_0)\sqrt{\dfrac{1 - \beta}{1 + \beta}}$ (*b*) $f_{\text{approaching}} = f_0\sqrt{\dfrac{1 + \beta}{1 - \beta}}$

    (*c*) $\Delta(cT_{\text{receding}}) = \Delta(ct_0)\sqrt{\dfrac{1 + \beta}{1 - \beta}}$ (*d*) $f_{\text{receding}} = f_0\sqrt{\dfrac{1 - \beta}{1 + \beta}}$

25. (*a*) $s = ct_0$ (*b*) 10 light-years (*c*) 10 years (*d*) 8 years (*e*) 8 light-years

Sec. 13-2

1. 52.2 MeV

3. $$\beta = \frac{1}{1 + \left(\frac{q\phi}{m_0 c^2}\right)} \sqrt{\left(\frac{q\phi}{m_0 c^2}\right)\left[\left(\frac{q\phi}{m_0 c^2}\right) + 2\right]}$$

5. $$\frac{2m_0}{\sqrt{1-\beta^2}}$$

7. 2.45×10^{20}; 6×10^{14}

9. $(h/m_0 c)(1 - \cos\theta)$

11. (*a*) $$\sinh^{-1}\left(\frac{p_X}{m_0 c}\right) = \sinh^{-1}\left(\frac{p_x}{m_0 c}\right) + \sinh^{-1}\left(\frac{p_v}{m_0 c}\right)$$

(*b*) $$\cosh^{-1}\left(\frac{E_X}{m_0 c^2}\right) = \cosh^{-1}\left(\frac{E_x}{m_0 c^2}\right) + \cosh^{-1}\left(\frac{E_v}{m_0 c^2}\right)$$

(*c*) $$\sinh^{-1}\sqrt{\frac{K_X}{2m_0 c^2}} = \sinh^{-1}\sqrt{\frac{K_x}{2m_0 c^2}} + \sinh^{-1}\sqrt{\frac{K_v}{2m_0 c^2}}$$

13. $$\left[\frac{m_0}{\sqrt{1-\beta^2}}\right]\mathbf{a} + \left\{\left[\frac{m_0}{(1-\beta^2)^{3/2}}\right]\frac{v a_v}{c^2}\right\}\mathbf{v}$$

15. $$A_X = \left[\frac{(1-\beta^2)^{3/2}}{(1+\beta\delta_x)^3}\right] a_x$$

$$A_Y = \left[\frac{1-\beta^2}{(1+\beta\delta_x)^2}\right] a_y - \left[\frac{\beta\delta_y(1-\beta^2)}{(1+\beta\delta_x)^3}\right] a_x$$

$$A_Z = \left[\frac{1-\beta^2}{(1+\beta\delta_x)^2}\right] a_z - \left[\frac{\beta\delta_z(1-\beta^2)}{(1+\beta\delta_x)^3}\right] a_x$$

17. $(1-\beta^2)F_{0y}L_{0x} - F_{0x}L_{0y}$

Yes, $T = dL/dt$. If the forces were applied simultaneously for (x, ct), giving no net angular displacement of the lever, they were not simultaneous for (X, cT). The first torque applied would produce an angular displacement; the second would no longer be equal and opposite, and afterward the figure does not correspond to the facts.

19. $$\frac{E_m}{c^2} = \frac{1}{2M_0}(M_0{}^2 + m_0{}^2 - \mu_0{}^2) \qquad \frac{E_\mu}{c^2} = \frac{1}{2M_0}(M_0{}^2 + \mu_0{}^2 - m_0{}^2)$$

21. $W = \sqrt{1-\beta^2}\, W_0$

23. 4.24×10^9 kg s^{-1}

CHAP. 14

Sec. 14-1

1. $\dfrac{-e}{4\pi\epsilon_0 (r')^2}\left[\dfrac{1-\beta^2}{(1-\beta^2 \sin^2 \theta)^{3/2}}\right]\hat{\mathbf{r}}'$

3. $(a)\ \dfrac{\lambda_0}{2\pi\epsilon_0 D}$ $\quad (b)\ \dfrac{\lambda_0}{\sqrt{1-\beta^2}2\pi\epsilon_0 D}$

5. $\left(\dfrac{-e}{4\pi\epsilon_0}\right)\dfrac{(1-\beta^2)\mathbf{r}'}{[(r')^2-(\mathbf{r}' \times \boldsymbol{\beta})^2]^{3/2}}$

7. $(a)\ \dfrac{\lambda_0{}^2}{2\pi\epsilon_0 D}$ $\quad (b)\ \dfrac{\lambda_0{}^2}{(1-\beta^2)2\pi\epsilon_0 D}$ $\quad (c)\ \dfrac{\mu_0{}^2\lambda^2 v^2}{2\pi D}$ $\quad (d)\ \dfrac{\lambda_0{}^2}{2\pi\epsilon_0 D}$ $\quad (e)$ The same

9. $(a)\ \lambda_{0p}{}^2/2\pi\epsilon_0 D$ (repulsive) $\quad (b)\ \lambda_n\lambda_{0p}/2\pi\epsilon_0 D$ (attractive)

   $(c)\ \dfrac{\lambda_n\lambda_{0p}}{2\pi\epsilon_0 D}$ (attractive) $\quad (d)\ \dfrac{\lambda_n{}^2}{(1-\beta^2)2\pi\epsilon_0 D}$ (repulsive)

11. $\dfrac{\lambda_n{}^2\beta^2}{(1-\beta^2)2\pi\epsilon_0 D}$

13. 8 V/m

15. Yes, m_0 becomes m $\quad R=\dfrac{p}{eB}=\dfrac{m_0 v}{\sqrt{1-\beta^2}eB}$ $\quad \omega=\dfrac{eB}{m}=\dfrac{\sqrt{1-\beta^2}\,eB}{m_0}$

17. $(a)\ \beta=\dfrac{-cB}{E}$ (**v** must be in the $\hat{\mathbf{x}}$ direction and we must have $cB<E$)

    $(b)\ \sqrt{E^2-c^2B^2}$

    $(c)\ \beta=-\dfrac{E}{cB}$ (**v** must be in the $\hat{\mathbf{x}}$ direction and we must have $E<cB$)

    $(d)\ \dfrac{1}{c}\sqrt{c^2B^2-E^2}$

19. Along the direction perpendicular to **E** and **B**,

$$\beta=-\frac{1}{2|\mathbf{E}\times c\mathbf{B}|}\left[(E^2+c^2B^2)\pm\sqrt{(E^2+c^2B^2)^2-4|\mathbf{E}\times c\mathbf{B}|}\right]$$

$$=-\frac{1}{2|\mathbf{E}\times c\mathbf{B}|}\left[(E^2+cB^2)\pm\sqrt{(E^2-c^2B^2)^2+4(\mathbf{E}\cdot c\mathbf{B})}\right]$$

The two solutions come from $E_z'/E_y'=B_z'/B_y'$, but only the one which gives $\beta<1$ is to be used. There are an infinite number of other solutions. If the direction parallel to $\mathbf{E}'$ and $\mathbf{B}'$ is, say, 45° from $\hat{\mathbf{y}}'$ toward $\hat{\mathbf{z}}'$, then any velocity along this direction may be added to the above.

21. (*a*) $\mathbf{v}' = \hat{\mathbf{x}}\dfrac{I}{\lambda}$ (*b*) $\mathbf{E}' = \hat{\mathbf{r}}'\left(\dfrac{\lambda}{2\pi\epsilon_0 r'}\right)\sqrt{1-\dfrac{I^2}{\lambda^2 c^2}}$ (cylindrical r')

(*c*) $\mathbf{v}'' = \hat{\mathbf{x}}\dfrac{\lambda c^2}{I}$ (*d*) $\mathbf{B}'' = \hat{\boldsymbol{\phi}}'\left(\dfrac{\mu_0 I}{2\pi r'}\right)\sqrt{1-\dfrac{\lambda^2 c^2}{I^2}}$ (cylindrical r')

(*e*) We can make $\mathbf{B}' = 0$ when $I \leqslant \lambda c$, i.e., for a convection current. We can make $\mathbf{E}' = 0$ when $I \geqslant \lambda c$, i.e., for a conduction current.

23. $q\phi + \sqrt{(c\mathbf{p} - q\mathbf{A})^2 + m_0{}^2 c^4}$

25. $\mathbf{E}' = \gamma\mathbf{E} - (\gamma - 1)(\hat{\mathbf{v}} \cdot \mathbf{E})\hat{\mathbf{v}} + \gamma(\boldsymbol{\beta} \times \mathbf{B})$, $\mathbf{B}' = \gamma\mathbf{B} - (\gamma - 1)(\hat{\mathbf{v}} \cdot \mathbf{B})\hat{\mathbf{v}} - \gamma(\hat{\boldsymbol{\beta}} \times \mathbf{E})$

Sec. 14-2

1. Yes

3. (*a*) $\{\gamma\mathbf{v}, \gamma c\}$ (*b*) From $\{c\mathbf{p}, E\}$ (*c*) No, e.g., $\{\gamma\boldsymbol{\beta}, \gamma\}$

(*d*) c^2 for the first above, one for the second

5. (*a*) $\{\gamma\mathbf{F}, \gamma(\mathbf{F} \cdot \boldsymbol{\beta})\}$

(*b*) This is arbitrary. For example, if the above is taken as contravariant then $\{-\gamma\mathbf{F}, \gamma(\mathbf{F} \cdot \boldsymbol{\beta})\}$ could be taken as the covector. But it could also be vice versa.

7. No; no.

9. *Three-space:* Three (ϵ^{123}, ϵ^{231}, ϵ^{312}) are positive; three (ϵ^{321}, ϵ^{132}, ϵ^{213}) are negative; 21 are 0; $3^3 = 27$ elements.
Four-space: 12 (ϵ^{1234}, ϵ^{1342}, ϵ^{1423}, ϵ^{2341}, ϵ^{2413}, ϵ^{2134}, ϵ^{3412}, ϵ^{3124}, ϵ^{3241}, ϵ^{4123}, ϵ^{4231}, ϵ^{4312}) are positive; 12 are negative; 232 are 0; $4^4 = 256$ elements.

11. $W^I = \dfrac{\partial X^I}{\partial x^j} w^j$

13. (*a*) +1: rotation. −1: rotation plus reflection of one or three coordinate axes.

(*b*) Same as (*a*). (*c*) Same as (*a*).

15. $\left\{\dfrac{(1-\beta^2)\mathbf{a} + (\boldsymbol{\beta} \cdot \mathbf{a})\boldsymbol{\beta}}{(1-\beta^2)^2}, \dfrac{\boldsymbol{\beta} \cdot \mathbf{a}}{(1-\beta^2)^2}\right\}$

17. $v = \dfrac{v_1 + v_2}{1 + (v_1 v_2/c^2)}$

19. $\dfrac{\partial x^i}{\partial X^J}\dfrac{\partial X^K}{\partial x^i} = \delta_J{}^K$

21. (*a*) Yes. (*b*) Yes. (*c*) Not in general, but possible.

Sec. 14-3

1. $f_4 = \dfrac{1}{c}\,\mathbf{J}\cdot\mathbf{E}$

3. $\dfrac{\partial F^{ij}}{\partial x^j} = c\mu_0 J^i \qquad \dfrac{\partial F^{ij}}{\partial x_k} + \dfrac{\partial F^{jk}}{\partial x_i} + \dfrac{\partial F^{ki}}{\partial x_j} = 0$

5. $F_{ik} = \dfrac{\partial A_i}{\partial x^k} - \dfrac{\partial A_k}{\partial x^i}$

7. $$\begin{pmatrix} 0 & E_z & -E_y & cB_x \\ -E_z & 0 & E_x & cB_y \\ E_y & -E_x & 0 & cB_z \\ -cB_x & -cB_y & -cB_z & 0 \end{pmatrix}$$

9. $\dfrac{\partial D^{ij}}{\partial x^j} = 0$

11. $\mathbf{J}^{(b)} = \left\{[(\nabla\times\mathbf{M}) + \dot{\mathbf{P}}],\ c(-\nabla\cdot\mathbf{P})\right\} \qquad \dfrac{\partial P_{ij}}{\partial x_j} = -\dfrac{1}{c}J_i{}^{(b)}$

13. $\dfrac{\partial \mathbf{P}}{\partial t}$

15. (*a*) $q[\mathbf{E} + (\mathbf{v}\times\mathbf{B})] = \dfrac{d}{dt}\left[\dfrac{m_0\mathbf{v}}{(1-\beta^2)^{1/2}}\right]$ (*b*) $qF^{ij}v_j = m_0c^2\dfrac{dv^i}{ds}$

17. $P'_x = P_x \qquad P'_y = \dfrac{P_y + \frac{v}{c^2}M_z}{\sqrt{1-\beta^2}} \qquad P'_z = \dfrac{P_z - \frac{v}{c^2}M_y}{\sqrt{1-\beta^2}}$

$M'_x = M_x \qquad M'_y = \dfrac{M_y - \frac{v}{c^2}P_z}{\sqrt{1-\beta^2}} \qquad M'_z = \dfrac{M_z + \frac{v}{c^2}P_y}{\sqrt{1-\beta^2}}$

19. vm/c^2

CHAP. 15

Sec. 15-1 through Sec. 15-5

1. $\omega = \pi \times 10^{16} \mathrm{s}^{-1}$ $\quad k = \dfrac{\pi}{3} \times 10^8 \mathrm{m}^{-1}$ $\quad \lambda = 600\text{Å}$

3. (*a*) $\mathbf{B} = \hat{\mathbf{z}} \left(\dfrac{E_0}{c}\right) e^{(\omega t - kx)}$ $\quad$ (*b*) $\nabla \times \mathbf{E} = \hat{\mathbf{z}}(-ik)E_0 e^{i(\omega t - kx)}$

   (*c*) $-\dfrac{\partial B}{\partial t} = \hat{\mathbf{z}} \left(-i\dfrac{\omega}{c}\right) E_0 e^{i(\omega t - kx)}$ $\quad$ (*d*) Equal

5. (*a*) The same (*b*) The same (*c*) The same (*d*) The same (*e*) $-\hat{\mathbf{y}}10^{-6}$ V/m

   (*f*) Zero

7. (*a*) $\hat{\mathbf{y}}10^{-6}$ V/m (*b*) The same (*c*) The same (*d*) The same

   (*e*) $-\hat{\mathbf{y}}10^{-6}$ V/m (*f*) Zero (*g*) Nothing

9. $\mathbf{S} = \hat{\mathbf{x}} \sqrt{\dfrac{\epsilon_0}{\mu_0}} [E_1{}^2 \cos^2(\omega t - kx) + E_2{}^2 \cos^2(\omega t - kx + \alpha)]$

   $\langle S \rangle = \hat{\mathbf{x}} \sqrt{\dfrac{\epsilon_0}{\mu_0}} \dfrac{(E_1{}^2 + E_2{}^2)}{2} = \hat{\mathbf{x}} \sqrt{\dfrac{\epsilon_0}{\mu_0}} [(E_1)^2_{\text{rms}} + (E_2)^2_{\text{rms}}]$

   Yes

11. (*a*) $u_{\text{elec}} = 4.44 \times 10^{-24} \cos^2(\omega t - kx)$ $\quad$ W/m^3

    (*b*) $u_{\text{mag}} = 4.44 \times 10^{-24} \cos(\omega t - kx)$ $\quad$ W/m^3

    (*c*) $S = 2.67 \times 10^{-15} \cos^2(\omega t - kx)$ $\quad$ W m^{-2} s^{-1}

13. $\dfrac{c}{\sqrt{1 + \dfrac{\alpha^2 c^2}{\omega^2}}}$

15. (*a*) $k = 3$ $\quad$ (*b*) $n_x = 0, n_y = -0.6, n_z = 0.8$

    (*c*) $\lambda = 2.09$ m $\quad$ (*d*) $f = 143$ MHz

    (*e*) A ray is perpendicular to $\hat{\mathbf{x}}$, makes 53.1° with $-\hat{\mathbf{y}}$ and 36.9° with $+\hat{\mathbf{z}}$

    (*f*) Elliptical polarization (*g*) Yes

17. (*a*) No (*b*) Yes

19.

f, Hz	v, m/s	E/cB	δ, m
10^3	4.7×10^4	$0.0001e^{i\pi/4}$	7
10^7	4.7×10^6	$0.01e^{i\pi/4}$	0.07
10^9	4.7×10^7	$0.1e^{i\pi/4}$	0.007
10^{15}	3×10^8	$e^{i0.00005}$	0.08

21. $(1/\lambda\!\!\!^{-}_0)\sqrt{\kappa}$

23. $u_e/u_m = Q$

25. $\mathbf{J}_t = -i\dfrac{ne^2}{m\omega}E + i\omega\epsilon_0 E$

27. (*a*) 10^{10} Hz (*b*) 10^6 Hz

29. $n = \sqrt{1 - \left(\dfrac{f_p}{f}\right)^2}$

CHAP. 16

Sec. 16-1 through Sec. 16-3

1. (*a*) $\dfrac{\partial^2 v}{\partial z^2} = LC\dfrac{\partial^2 v}{\partial t^2} + (LG + RC)\dfrac{\partial v}{\partial t} + RGv$

   (*b*) $\dfrac{\partial^2 i}{\partial z^2} = LC\dfrac{\partial^2 i}{\partial t^2} + (LG + RC)\dfrac{\partial i}{\partial t} + RGi$

   (*c*) No. The boundary conditions for v and i are not necessarily the same.

3. $e^{-2} = 0.1355$

5. 26.06 Np

7. $0.745\,\underline{/64.5^\circ}$

9. $(x_{min}/\lambda) = \frac{1}{4} + \dfrac{\psi_L}{4\pi}$

11. (*a*) 0.83 (*b*) 23° (*c*) 500 Ω (*d*) 5 Ω

13. (*a*) VSWR = $|z_{in}|_{max}$

    (*b*) The point where the circle $|\Gamma_L|$ = constant intersects the right half of the horizontal diameter.

15. $1\underline{/-53^\circ}$

17. (*a*) At 0.301 λ from the load toward the generator (*b*) 0.375 λ (*c*) 0.125 λ

19. (*a*) Very low impedance at resonance and behavior like a series resonant circuit at neighboring frequencies.
(*b*) Very high impedance at resonance and behavior like a parallel resonant circuit at neighboring frequencies.

21. (*a*) $\dfrac{\pi\epsilon_0}{\cosh^{-1}(s/d)} \to \dfrac{\pi\epsilon_0}{\ln(2s/d)}$

(*b*) $\dfrac{\mu_0}{\pi}\cosh^{-1}(s/d) \to \dfrac{\mu_0}{\pi}\ln(2s/d)$

(*c*) $C = \dfrac{2\pi\epsilon_0}{\ln(b/a)} \qquad L = \dfrac{\mu_0}{2\pi}\ln(b/a)$

23. (*a*) $0.8 - j1.0$

(*b*) Capacitive if a minimum is closest to the load.

(*c*) $0.49 + j0.61$

(*d*) 0.4λ

(*e*) $0.49 + j0.61$

25. (*a*) $\left|\dfrac{V_s}{V_p}\right| = \dfrac{Z_L}{Z_0} = \dfrac{Z_L}{\sqrt{Z_1 Z_L}} = \sqrt{\dfrac{Z_L}{Z_1}}$

(*b*) Lumped: $\dfrac{V_s}{V_p} = \sqrt{\dfrac{Z_s}{Z_p}}$ Line: $\dfrac{V_s}{V_p} = -j\sqrt{\dfrac{Z_L}{Z_1}}$

(*c*) Yes

(*d*) $\dfrac{V_s}{V_p} = \dfrac{Z_0}{\left(\dfrac{R}{2}\right)x}$

CHAP. 17

Sec. 17-1 through Sec. 17-4

1. (*a*) $\kappa_1 = \kappa_2$ (*b*) $\cos\theta_i = \sqrt{\kappa_1/(\kappa_1 + \kappa_2)}$ (*c*) $0°$ (*d*) $180°$ (*e*) $180°$

3. (*a*) 2.14×10^8 m s^{-1} (*b*) Yes, this is how one obtains Cerenkov radiation.
(*c*) No, according to special relativity this is not possible.

5. (*a*) $\cos\theta_B = \sqrt{\dfrac{\kappa_1}{\kappa_1 + \kappa_2}} \qquad \cos\theta_c = \sqrt{\dfrac{\kappa_2 - \kappa_1}{\kappa_2}}$

(*b*)

κ	θ_B	θ_c
1	45°	90°
2	50° 50′	45°
3	60°	35° 30′
4	63° 25′	30°

7.

κ	θ_B
1	45°
2	35° 20′
3	30°
4	27°

9. (*a*) 2×10^5 Hz (*b*) 16 m

11. (*a*) $\dfrac{\lambda}{2\cos\theta}$ (*b*) $\lambda_x = \dfrac{\lambda}{\cos\theta}$ (*c*) At $0, \dfrac{\lambda_x}{2}, \lambda_{x_1}, \dfrac{3\lambda_x}{2}, \ldots$

(*d*) $\dfrac{\lambda_x}{4}, \dfrac{3\lambda_x}{4}, \dfrac{5\lambda_x}{4}, \ldots$

13. (*a*) $\dfrac{\lambda_x}{4}, \dfrac{3\lambda_x}{4}, \dfrac{5\lambda_x}{4}, \ldots$ (*b*) $0, \dfrac{\lambda_x}{2}, \lambda_x, \dfrac{3\lambda_x}{2}, \ldots$ (*c*) $0, \dfrac{\lambda_x}{2}, \lambda_x, \dfrac{3\lambda_x}{2}, \ldots$

(*d*) $\dfrac{\lambda_x}{4}, \dfrac{3\lambda_x}{4}, \dfrac{5\lambda_x}{4}, \ldots$

15. (*a*) $2\tan^{-1}\left(\dfrac{\sqrt{\sin^2\theta_i - \kappa_2/\kappa_1}}{\cos\theta_i}\right)$ (*b*) $2\tan^{-1}\left(\dfrac{\sqrt{\sin^2\theta_i - \dfrac{\kappa_2}{\kappa_1}}}{\dfrac{\kappa_2}{\kappa_1}\cos\theta_i}\right)$

17. $\dfrac{d\theta}{ds} = -\dfrac{\tan\theta}{n}\left(\dfrac{dn}{ds}\right)$

CHAP. 18

Sec. 18-1 through Sec. 18-2

1. (*a*) $\lambda_{\text{guide}} = (\lambda/a)/\sqrt{1-(m/2)^2(\lambda/a)^2}$

(*b*) $(\lambda/a) = (2/m)$ there, so $\lambda_{\text{guide}} = \infty$

3. (*a*) The power (*b*) $(C_1{}^2/2\mu_0)(\beta_g/\omega)\sin^2(m\pi z/a)$ (*c*) $(C_1{}^2/4\mu_0)(\beta_g/\omega)a$ (*d*) no

(*e*) there is a $\hat{\mathbf{z}}$ component that is purely imaginary, so it does not contribute to $\langle \mathbf{S} \rangle$.

5. (*a*) All components of **E** and **B** vanish.

(*b*) All components of **E** and **B** are reversed in sign, i.e., the wave is merely displaced by a half cycle.

7. (*a*) $C_1 = cC_4$ (*b*) for TE $\langle \mathbf{S} \rangle$ varies as $\sin^2 (m\pi z/a)$, for TM as $\cos^2 (m\pi z/a)$

9. (*a*) $E_y{}^2 a/(4v_{\text{guide}})$ (*b*) $E_y{}^2 a/(2c)$ (*c*) $2(v_{\text{guide}}/c) = \dfrac{2(f/f_c)}{\sqrt{\left(\dfrac{f}{f_c}\right)^2 - 1}}$

(*d*) 2.31, 2.01

11. (*a*) 2.00, 1.00, 0.67, 0.50, 0.40, 0.33 (*b*) 1, 2

13. (*a*) $A_x = 0$

$$A_y = -\frac{\dfrac{n\pi}{b}}{\left(\dfrac{m\pi}{a}\right)^2 + \left(\dfrac{n\pi}{b}\right)^2} C \cos\left(\frac{m\pi y}{a}\right) \sin\left(\frac{n\pi z}{b}\right) e^{j(\omega t - \beta_g x)}$$

$$A_z = \frac{\dfrac{m\pi}{a}}{\left(\dfrac{m\pi}{a}\right)^2 + \left(\dfrac{n\pi}{b}\right)^2} C \sin\left(\frac{m\pi y}{a}\right) \cos\left(\frac{n\pi z}{b}\right) e^{j(\omega t - \beta_g x)}$$

(*b*) $A_x = -\dfrac{1}{j\omega} C \sin\left(\dfrac{m\pi y}{a}\right) \sin\left(\dfrac{n\pi z}{b}\right) e^{j(\omega t - \beta_g x)}$

$$A_y = \frac{\left(\dfrac{\beta_g}{\omega}\right)\dfrac{m\pi}{a}}{\left(\dfrac{m\pi}{a}\right)^2 + \left(\dfrac{n\pi}{b}\right)^2} C \cos\left(\frac{m\pi y}{a}\right) \sin\left(\frac{n\pi z}{b}\right) e^{j(\omega t - \beta_g x)}$$

$$A_z = \frac{\left(\dfrac{\beta_g}{\omega}\right)\dfrac{n\pi}{b}}{\left(\dfrac{m\pi}{a}\right)^2 + \left(\dfrac{n\pi}{b}\right)^2} C \sin\left(\frac{m\pi y}{a}\right) \cos\left(\frac{n\pi z}{b}\right) e^{j(\omega t - \beta_g x)}$$

15. (*a*) Yes. $A_z = [C/(\pi/a)] \sin (\pi y/a) e^{j(\omega t - \beta_g x)}$

(*b*) Yes. $A_y = - C/(\pi/b) \sin (\pi z/b) e^{j(\omega t - \beta_g x)}$

(*c*) No. $A_x = A_y = A_z = 0$ (*d*) No. $A_x = A_y = A_z = 0$

(*e*) No. $A_x = A_y = A_z = 0$ (*f*) No. $A_x = A_y = A_z = 0$

17. (*a*) $\frac{c}{2}\sqrt{\left(\frac{m}{a}\right)^2+\left(\frac{n}{a/2}\right)^2}$ has values 1, 2, 2.24, in order of increasing magnitude.

(*b*) $2f_c, 3f_c, 4f_c, 6f_c, 2.24f_c$

19. $TE_{10}(f_c = 2.5 \text{ GHz})$ $\quad TE_{01}(f_c = 3.75 \text{ GHz})$

$TE_{11}(f_c = 4.5 \text{ GHz})$ $\quad TM_{11}(f_c = 4.5 \text{ GHz})$.

21. (*a*) $C' = \left[\left(\frac{n}{m}\right)\left(\frac{a}{b}\right)v_{\text{guide}}\right]C$ (*b*) $C' = \left[\left(\frac{n}{m}\right)\left(\frac{a}{b}\right)v_{\text{group}}\right]C$

23. Let $D \equiv (m\pi/a)^2 + (n\pi/b)^2$, $E \equiv e^{j(\omega t-\beta_g x)}$.

For TE_{mn}, $y = 0$: $\sigma = \epsilon_0 j\omega\left(\frac{n\pi}{b}\right)\frac{C}{D}\sin\left(\frac{n\pi z}{b}\right)E$

$$\mathbf{j} = \frac{-C}{\mu_0}\left[\hat{\mathbf{z}}\cos\left(\frac{n\pi z}{b}\right) - \hat{\mathbf{x}}j\beta_g\left(\frac{n\pi}{b}\right)\frac{1}{D}\sin\left(\frac{n\pi z}{b}\right)\right]E$$

$$z = 0:\ \sigma = -\epsilon_0 j\omega\left(\frac{m\pi}{a}\right)\frac{C}{D}\sin\left(\frac{m\pi y}{a}\right)E$$

$$\mathbf{j} = -\frac{C}{\mu_0}\left[-\hat{\mathbf{y}}\cos\left(\frac{m\pi y}{a}\right) + \hat{\mathbf{x}}j\beta_g\left(\frac{m\pi}{a}\right)\frac{1}{D}\sin\left(\frac{m\pi y}{a}\right)\right]E$$

For TM_{mn}, $y = 0$: $\sigma = -\epsilon_0 j\beta_g\left(\frac{m\pi}{a}\right)\frac{C'}{D}\sin\left(\frac{n\pi z}{b}\right)E$

$$\mathbf{j} = \frac{1}{\mu_0}\hat{\mathbf{x}}(-j\omega\epsilon_0\mu_0)\left(\frac{m\pi}{a}\right)\frac{C'}{D}\sin\left(\frac{n\pi z}{b}\right)E$$

$$z = 0:\ \sigma = -\epsilon_0 j\beta_g\left(\frac{n\pi}{b}\right)\frac{C'}{D}\sin\left(\frac{m\pi}{a}\right)E$$

$$\mathbf{j} = -\frac{1}{\mu_0}\hat{\mathbf{x}}(j\omega\epsilon_0\mu_0)\left(\frac{n\pi}{b}\right)\frac{C'}{D}\sin\left(\frac{m\pi y}{a}\right)E.$$

The results for $y = a$ equal $(-1)^m$ times those for $y = 0$; similarly, the results for $z = b$ equal $(-1)^m$ times those for $z = 0$; i.e., they are either the same or the negative of those given, depending on whether m is even or odd.

CHAP. 19

1. $\nabla^2\phi = -\frac{\rho}{\epsilon_0}$ $\quad \Box^2\mathbf{A} - \frac{1}{c^2}\nabla\dot{\phi} = -\mu_0\mathbf{J}$

3. Zero

5. (a) $\phi^i = \left(\dfrac{q}{4\pi\epsilon_0}\right)\dfrac{u^i}{u^i R_j}$ (b) $\{\phi^i\} \equiv \{c\mathbf{A}, \phi\} = \left\{c\left(\dfrac{\mu_0}{4\pi}\right)\dfrac{q\mathbf{u}}{s}, \left(\dfrac{q}{4\pi\epsilon_0}\right)\dfrac{1}{s}\right\}$

7. (a) $\dfrac{\phi}{\left(\dfrac{Q}{4\pi\epsilon_0}\right)} = \dfrac{1}{1-\beta\cos\alpha}$ $\dfrac{\mathbf{A}}{\left(\dfrac{\mu_0 Qc}{4\pi}\right)} = \dfrac{\beta}{1-\beta\cos\alpha}$

(b) $\dfrac{\phi}{\left(\dfrac{Q}{4\pi\epsilon_0}\right)} = \dfrac{1}{\sqrt{1-\beta^2\sin^2\theta}}$ $\dfrac{\mathbf{A}}{\left(\dfrac{\mu_0 Qc}{4\pi}\right)} = \dfrac{\boldsymbol{\beta}}{\sqrt{1-\beta^2\sin^2\theta}}$

9. (a) $\phi = c(\hat{\mathbf{r}} \cdot \mathbf{A})$ (b) no (c) yes (d) no

11. $\dfrac{2Q^2 a^2 \omega^4}{(4\pi\epsilon_0)3c^3(1-\beta^2)^2}$

13. $\phi = \dfrac{P}{4\pi\epsilon_0}\cos\theta\left(\dfrac{1}{r^2} + j\dfrac{k}{r}\right)e^{j(\omega t - kr)}$

$= \dfrac{P\cos\theta}{4\pi\epsilon_0 r^2}\sqrt{1+k^2r^2}\, e^{j(\omega t - kr + \tan^{-1}kr)}$

In the far field, $kr \gg 1$, $\phi \to (jPk\cos\theta/4\pi\epsilon_0 r)\, e^{j(\omega t - kr)}$ with $v = \omega/k = c$. In the near field, $kr \ll 1$, $\phi \to (P\cos\theta/4\pi\epsilon_0 r^2)\, e^{j\omega t}$ and there is no v. In the induction field, $kr \approx 1$, $\phi \to (\sqrt{2}P\cos\theta/4\pi\epsilon_0 r^2)\, e^{j\omega[t-(r/c)+(1/\omega)(\pi/4)]}$. The third term in the exponent, for a given ω, has the same effect as it would if, instead, the denominator of the second term there were increased.

15. (a) $\dfrac{4Q^2\omega^2}{(4\pi\epsilon_0)3mc^2(1-\beta^2)^2}$ (b) 4.5×10^9 Hz, 4.5×10^8 Hz

(c) 2.4×10^6 Hz, 2.4×10^5 Hz

17. (a) $80(s/\lambda\!\!\!^{-})^2$ ohms (b) $20(\ell/\lambda\!\!\!^{-})^2$ ohms

19. (a) Zero (b) No, one must use the exact expressions

(c) $1/kr$ (d) c

21. $4Q^2\dddot{z}/(4\pi\epsilon_0 6c^3)$

23. $40(h/\lambda\!\!\!^{-})^2 = 10(\ell/\lambda\!\!\!^{-})^2$

25. $\dfrac{6.7\sqrt{W}\sin\theta}{r}$ V/m

INDEX

INDEX